The Nature of Life: Classical and Contemporary Perspectives from Philosophy and Science

Bringing together the latest scientific advances and some of the most enduring subtle philosophical puzzles and problems, this book collects original historical and contemporary sources to explore the wide range of issues surrounding the nature of life.

Selections ranging from Aristotle and Descartes to Sagan and Dawkins are organized around four broad themes covering classical discussions of life, the origins and extent of natural life, contemporary artificial life creations, and the definition and meaning of "life" in its most general form. Each section is preceded by an extensive introduction connecting the various ideas discussed in individual chapters and providing helpful background material for understanding them. With its interdisciplinary perspective, this fascinating collection is essential reading for scientists and philosophers interested in astrobiology, the origin of life, synthetic biology, and the philosophy of life.

MARK A. BEDAU is Professor of Philosophy and Humanities at Reed College, Oregon, Director of the Initiative for Science, Society, and Policy in Denmark, and a regular Visiting Professor at the European School of Molecular Medicine in Milan, Italy. He is an internationally recognized leader in the philosophical and scientific study of living systems and has published and lectured extensively on issues concerning emergence, evolution, life, mind, and the social and ethical implications of creating life from nonliving materials.

CAROL E. CLELAND is Professor of Philosophy at the University of Colorado (Boulder). She is a central figure in the emerging science of astrobiology and has published in major science as well as philosophy journals. Her research focuses on issues in scientific methodology, historical science (especially earth and planetary science), biology (especially microbiology, origins of life, the nature of life, and astrobiology), and the theory of computation.

The Nature of Life: Classical and Contemporary Perspectives from Philosophy and Science

Edited by

Mark A. Bedau
Reed College, Oregon, USA

Carol E. Cleland
University of Colorado, Boulder, USA

CAMBRIDGE
UNIVERSITY PRESS

CAMBRIDGE
UNIVERSITY PRESS

University Printing House, Cambridge CB2 8BS, United Kingdom

One Liberty Plaza, 20th Floor, New York, NY 10006, USA

477 Williamstown Road, Port Melbourne, VIC 3207, Australia

314–321, 3rd Floor, Plot 3, Splendor Forum, Jasola District Centre, New Delhi - 110025, India

79 Anson Road, #06-04/06, Singapore 079906

Cambridge University Press is part of the University of Cambridge.

It furthers the University's mission by disseminating knowledge in the pursuit of education, learning and research at the highest international levels of excellence.

www.cambridge.org
Information on this title: www.cambridge.org/9781108722063

First published 2010
First paperback edition 2018

A catalogue record for this publication is available from the British Library

Library of Congress Cataloging in Publication data
The nature of life : classical and contemporary perspectives from philosophy and science / edited by Mark A. Bedau, Carol Cleland.
　p.　cm.
Includes bibliographical references and index.
ISBN 978-0-521-51775-1 (Hardback)
1. Science–Philosophy.　2. Philosophy of nature.　I. Bedau, Mark.　II. Cleland, Carol.　III. Title.
Q175.N333 2010
501–dc22

2010004399

ISBN　978-0-521-51775-1　Hardback
ISBN　978-1-108-72206-3　Paperback

Contents

Preface

This book is a collection of readings about the nature of life. The idea for it was born when we first met and discovered our mutual interest in the nature of life, a shared background in philosophy of science and philosophy of biology, and a complementary scientific expertise in the origin of life and astrobiology (Cleland) and artificial life and synthetic biology (Bedau). We both wanted to have a book like this, so we decided to compile it together. Our interests and orientation led us to include material in four general areas: (i) classical philosophical and scientific discussions about the nature of life, (ii) contemporary scientific and philosophical discussions of the origins of life, and of chemical possibilities for unfamiliar forms of natural life, (iii) discussions of contemporary artificial life creations, including not just computer simulations but also self-reproducing robots, protocells, and other synthetic biology constructions created in the wet lab, and (iv) attempts by contemporary scientists and philosophers to describe and explain the nature of life in its most general, non-Earthcentric, form. Producing this book confirmed for us the value of combining multiple perspectives on life.

We hope that this book will inform philosophers about the latest scientific advances and introduce scientists to subtle philosophical puzzles and problems, and thereby foster new, well-informed and thoughtful philosophical and scientific reflection about the nature of life. In our opinion, genuine progress in understanding life crucially depends upon combining both scientific and philosophical perspectives on life.

Our book is aimed at a broad audience. Some of the chapters are quite accessible and others are fairly technical. Each section is preceded by an introduction connecting the various ideas discussed in individual chapters and providing helpful background material for understanding them. We hope that philosophers interested in life (including those working in philosophy of biology, philosophy of science, and philosophy of mind) will pay especially close attention to the scientific chapters, and that scientists interested in the nature of life (including biologists, chemists, physicists, astronomers, geologists, and astrobiologists) will pay especially close attention to the philosophical chapters. Graduate and undergraduate students in any of these areas will also find the book useful as a text in interdisciplinary courses on life. Anyone from the general public who is curious about up-to-date scientific and philosophical perspectives on life should enjoy the book as well.

Mark Bedau, Portland, Oregon
Carol Cleland, Boulder, Colorado

Acknowledgments

The editors would like to thank the authors and publishers who gave us permission to collect their work in this anthology. We owe a special thanks to Chris Shields, for permitting us to publish part of his new translation of *De Anima*, forthcoming from Oxford University Press.

We would also like to thank many friends and colleagues who have shared and fostered our growing fascination with life and our attempts to understand it, and who helped us while developing this book. Thanks to Bob Hanna for help with the Kant selections, and to Meg Scharle for help on Aristotle. Thanks to Kellyn Bardeen for collecting permissions, copy-editing, and producing the final manuscript, and to Emily Parke for myriad crucial assistance. Thanks to Ben Pagelar and Michael Zerella at the Boulder campus of the University of Colorado for help with assembling the bibliography. We owe an important debt to the up and coming philosophers and scientists in Carol Cleland's 2006 graduate seminar on life for enthusiastically and industriously working through the articles in the original Table of Contents, helping us to decide which should be included as chapters in the final version. Thanks to the NASA Astrobiology Institute for providing partial funding for this project through a grant to the University of Colorado's Astrobiology Center, and to Reed College for a Stillman Drake grant that supported this project. Thanks also to the Center for Advanced Computation at Reed College, for hospitality during a crucial final stage in the preparation of the book.

Sources

Grateful acknowledgment is made to the following publishers and authors:

1. Selections from *De Anima*, written by Aristotle with translation and commentary by Christopher Shields. Copyright © 2009, Oxford University Press (forthcoming). Reprinted by permission of Oxford University Press and Christopher Shields.

2. "Treatise on Man" by René Descartes, from *The Philosophical Writing of Descartes*, vol. 1, translated by John Cottingham, Robert Stoothoff, and Dugald Murdoch. Copyright © 1985, Cambridge University Press. Reprinted by permission of Cambridge University Press, John Cottingham, and Dugald Murdoch.

3. Selections from *Critique of the Teleological Power of Judgment*, written by Immanuel Kant and translated by Paul Guyer and Eric Matthews. Copyright © 2001, Cambridge University Press. Reprinted by permission of Cambridge University Press, Paul Guyer, and Eric Matthews.

4. Selections from *What is Life? The Physical Aspect of the Living Cell* by Erwin Schrödinger. Copyright © 1967, Cambridge University Press. Reprinted by permission of Cambridge University Press.

5. Excerpt from *Life: Its Nature, Origin, and Development*, written by A. I. Oparin and translated by Ann Synge. Copyright © 1964, Ann Synge. Reprinted by permission of Charlotte Synge and Liz Smith, Academic Press, and Oliver and Boyd Ltd., a subsidiary of Longman Publishing.

6. Chapter 1 from *This is Biology: The Science of the Living World* by Ernst Mayr. Copyright © 1997, Harvard University Press. Reprinted by permission of Harvard University Press.

7. Selections from *The Principles of Life*, written by Tibor Gánti, with commentary by James Griesemer and Eörs Szathmáry. Copyright

© 2003, Oxford University Press. Reprinted by permission of Oxford University Press.

8. "The Origin of Life: A Review of Facts and Speculation" by Leslie E. Orgel, from *Trends in Biochemical Sciences 23 (1998)*. Copyright © 1998, Elsevier Science. Reprinted by permission of Elsevier Science and Alice Orgel.

9. Section "A Simpler Origin for Life" by Robert Shapiro, from *Scientific American (June 2007)*. Text copyright © 2007, Scientific American; figures copyright © 2007 Jennifer C. Christiansen. Reprinted by permission of Scientific American, Jennifer C. Christiansen, and Robert Shapiro.

10. "Are the Different Hypotheses on the Emergence of Life as Different as They Seem?" by Iris Fry, from *Biology and Philosophy 10(4) (1995)*. Copyright © 2004, Springer Science and Business Media. Reprinted by permission of Springer Science and Business Media, and by permission of Iris Fry.

11. "The Universal Nature of Biochemistry" by Norman Pace, from *Proceedings of the National Academy of Sciences 98(3) (2001)*. Text copyright © 2001, National Academy of Sciences; figure copyright © 1997, American Association for the Advancement of Science. Reprinted by permission of the National Academy of Sciences, the American Association for the Advancement of Science, and Norman Pace.

12. "Is There a Common Chemical Model for Life in the Universe?" by Steven A. Benner, Alonso Ricardo, and Matthew A. Carrigan, from *Current Opinion in Chemical Biology 8 (2004)*. Copyright © 2004, Elsevier Ltd. Reprinted by permission of Elsevier Ltd., Steven A. Benner, Alonso Ricardo, and Matthew A. Carrigan.

13. "Searching for Life in the Universe: Lessons from Earth" by Kenneth H. Nealson, from *Annals of the New York Academy of Sciences 950(1) (2001)*.

14. "The Possibility of Alternative Microbial Life on Earth" by Carol E. Cleland and Shelley D. Copley, from *International Journal of Astrobiology 4 (2005)*. Copyright © 2005, Cambridge University Press. Reprinted by permission of Cambridge University Press, Carol E. Cleland, and Shelley D. Copley.

15. Introduction to *The Limits of Organic Life in Planetary Systems* by the National Research Council of the National Academies. Copyright © 2007, National Academy of Sciences. Reprinted by permission of the National Academy of Sciences.

16. "Learning from Functionalism: Prospects for Strong Artificial Life" by Elliott Sober, from *Artificial Life II*, edited by C. Langton, C. Taylor, J. D. Farmer, and S. Rasmussen. Copyright © 2003, Westview Press, A Member of the Perseus Books Group. Reprinted by permission of Westview Press and Elliott Sober.

17. "Life, 'Artificial Life,' and Scientific Explanation" by Marc Lange, from *Philosophy of Science 63(2) (1996)*. Copyright © 1996, Philosophy of Science Association. Reprinted by permission of the University of Chicago Press on behalf of the Philosophy of Science Association, and by permission of Marc Lange.

18. "Alien Life: How Would We Know?" by Margaret A. Boden, from *International Journal of Astrobiology 2(2) (2003)*. Copyright © 2003, Cambridge University Press. Reprinted by permission of Cambridge University Press and Margaret A. Boden.

19. "Automatic Design and Manufacture of Robotic Life Forms" by Hod Lipson and Jordan P. Pollack, from *Nature 406 (2000)*. Copyright © 2000, Nature Publishing Group. Reprinted by permission of Nature Publishing Group and Hod Lipson.

20. "A Giant Step Towards Artificial Life?" by David Deamer, from *Trends in Biotechnology 23 (2005)*. Copyright © 2005, Elsevier Ltd. Reprinted by permission of Elsevier Ltd. and David Deamer.

21. "Approaches to Semi-Synthetic Minimal Cells: A Review" by Pier Luigi Luisi, Francesca Ferri, and Pasquale Stano, from *Naturwissenschaften 93 (1) (2006)*. Copyright © 2005, Springer-Verlag. Reprinted by permission of Springer-Verlag, Pier Luigi Luisi, Francesca Ferri, and Pasquale Stano. Table reprinted by permission of Professor A. Moya, Institut Cavanilles de Biodiversitat i Biologia Evolutiva, Universitat de Valencia.

22. "Creating 'Real Life'" by Evelyn Fox Keller, from *Making Sense of Life: Explaining Biological Development with Models, Metaphors and Machines* by Evelyn Fox Keller. Copyright © 2002, President and Fellows of Harvard College. Reprinted by permission of Harvard University Press and Evelyn Fox Keller.

23. "The Definition of Life" by Carl Sagan, from *Encyclopædia Britannica, 14th Edition*. Copyright © 1970, Encyclopædia Britannica Incorporated. Reprinted by permission of Encyclopædia Britannica Incorporated.

24. "The Seven Pillars of Life" by Daniel E. Koshland, Jr., from *Science 295 (2002)*. Copyright © 2002, American Association for the Advancement of Science. Reprinted by permission of the American Association for the Advancement of Science.

25. "A Universal Definition of Life: Autonomy and Open-Ended Evolution" by Kepa Ruiz-Mirazo, Juli Peretó, and Alvaro Moreno, from *Origins of Life and Evolution of the Biosphere 34(3) (2004)*. Copyright © 2004, Kluwer Academic Publishers. Reprinted by permission of Kluwer Academic Publishers, subsidiary of Springer Science and Business Media, and by permission of Kepa Ruiz-Mirazo, Juli Peretó, and Alvaro Moreno.

26. "Does 'Life' Have a Definition?" by Carol Cleland and Christopher Chyba, from *Planets and Life: The Emerging Science of Astrobiology*, edited by W. T. Sullivan III and J. A. Baross. Text copyright © 2007, Cambridge University Press; image copyright © British Library Board. Reprinted by permission of Cambridge University Press, The British Library Board, Carol Cleland, and Christopher Chyba.

27. "Sentient Symphony," from *What is Life?* by Lynn Margulis and Dorion Sagan. Copyright © 1995, Lynn Margulis and Dorion Sagan. Reprinted by permission of the authors.

About the authors

ARISTOTLE (384 BC–322 BC), a student of Plato and teacher of Alexander the Great, was a prominent Greek philosopher and one of the most important founding figures in Western philosophy. With the scope of his work encompassing morality, science, logic, aesthetics, metaphysics and politics, he was the first to create a comprehensive system of Western philosophy. In addition, Aristotle is sometimes regarded as the originator of the scientific study of life, as he was among the first to articulate a systematic and comprehensive method for biological investigation. His own studies of animals were extensive, and several of his biological observations waited more than two millennia to be confirmed by modern science.

MARK A. BEDAU is a Professor of Philosophy and Humanities at Reed College, editor-in-chief of the MIT Press journal *Artificial Life*, co-founder of the European Center for Living Technology (Venice), Adjunct Professor of Systems Science at Portland State University, and Visiting Professor at the European School of Molecular Medicine (Milan). His areas of research include emergence, biological evolution and adaptation, the nature of life, artificial life and protocells, the evolution of technology, autonomous intelligent discovery and optimization processes, and the social and ethical implications of creating life in the laboratory.

STEVEN A. BENNER is a Distinguished Fellow at the Foundation for Applied Molecular Evolution and the Westheimer Institute for Science and Technology. He received his bachelor's degree and M.S. in Molecular Biophysics and Biochemistry from Yale University, and his Ph.D. in Chemistry from Harvard University. He co-founded several fields, including synthetic biology, paleogenetics, evolutionary bioinformatics, and dynamic combinatorial chemistry. His research combines the traditions of natural history with those of the physical sciences in order to gain a better understanding of life.

MARGARET A. BODEN is Research Professor of Cognitive Science at the University of Sussex. She is a Fellow of the British Academy, and of the American Association for Artificial Intelligence, along with its British and European equivalents. In 2002, she was awarded an Order of the British Empire "for services to cognitive science," and, in addition to her Cambridge Sc.D. and Harvard Ph.D., she has honorary doctorates from the Universities of Bristol and Sussex and the Open University. Her work is highly interdisciplinary, and has been translated into 20 languages. Her latest book is *Mind as Machine: A History of Cognitive Science* (2006).

MATTHEW A. CARRIGAN is a Research Fellow at the Foundation for Applied Molecular Evolution in Gainesville, Florida. His research centers on the origin of life and synthetic biology, with a particular focus on the synthesis and stability of the sugar backbone of RNA and the polymerization of ribonucleotides. Current efforts are directed toward isolating an RNA ribozyme capable of catalyzing RNA polymerization.

CHRISTOPHER CHYBA is a Professor of Astrophysical Sciences and International Affairs at Princeton University. He served on the White House staff from 1993 to 1995, working in the National Security Council and the National Security Division of the Office of Science and Technology Policy. His security-related research focuses on nuclear proliferation, nuclear weapons policy, and biological terrorism, while his planetary science and astrobiology research focuses on the search for life in the solar system. He received the Presidential Early Career Award in 1996 and was named a MacArthur Fellow in 2001. In addition to his Ph.D. in Astronomy, he holds an M.Phil. in History and Philosophy of Science from Cambridge University.

CAROL E. CLELAND is a Professor of Philosophy at the University of Colorado in Boulder. She is a member of the NASA Astrobiology Institute (NAI) and the Center for Astrobiology at the University of Colorado. Her research is in the areas of philosophy of science, logic, and metaphysics. She has published articles on space and time, supervenience, events and causation, hypercomputation, the Church-Turing thesis, standards of evidence in science, the methodology of historical and experimental science, the problem of defining "life," and the possibility of alternative microbial life on Earth. She is currently finishing a book, *The Quest for a Universal Theory of Life: Searching for Life as We Don't Know it.*

SHELLEY D. COPLEY is a Professor of Molecular, Cellular and Developmental Biology at the University of Colorado in Boulder. Her research centers on the molecular evolution of catalysts and metabolic pathways throughout the history of life on Earth, from the emergence of proto-metabolic networks that supported the RNA world to the evolution of novel pathways for the degradation of toxic pollutants introduced into the environment in the last century.

JOHN COTTINGHAM is a Professor of Philosophy at the University of Reading in England. He is also the president of the British Society for the Philosophy of Religion, the director of the three-year research program Impartiality and Partiality in Ethics, and an editor for *Ratio*, an international journal of analytic philosophy. His main research interests include early modern philosophy (particularly Descartes), ethics, and philosophy of religion.

RICHARD DAWKINS is a British ethologist, evolutionary biologist, and popular science author well known for his criticism of religion, creationism, and intelligent design. He currently holds the position of Charles Simonyi Professor of the Public Understanding of Science at Oxford University. His influential first book, *The Selfish Gene* (1976), popularized the gene-centered view of evolution and introduced the term *meme* to the discourse on cultural evolution. He is a fellow of the Royal Society and has received the Faraday Award and the Kistler Prize. His most recent book is *The God Delusion* (2006).

DAVID DEAMER is a Research Professor of Biomolecular Engineering at the University of California, Santa Cruz. His research interests include nanopore analysis of nucleic acids and the biophysics behind processes of self-assembly, particularly those related to the structure and function of biological membranes. He is currently writing a book on the origin of cellular life to be published in 2009 by the University of California Press.

RENÉ DESCARTES (1596–1650) was a highly influential French philosopher who is generally regarded as the father of modern philosophy. In his pivotal work, *Meditations on First Philosophy*, he famously put forth the *cogito* argument as a step in his investigation into the rational foundations of philosophical inquiry and natural science. Descartes made significant contributions to mathematics and science during his lifetime, inventing the Cartesian coordinate system, founding analytic geometry, and discovering the law of refraction. As a strong proponent of the theory of mechanism, he was also a major player in the Scientific Revolution.

FRANCESCA FERRI is currently working at the Department of Neuroscience at the University of Parma. Her present research interests include self-identity, embodiment, coherence, and disorders involving the embodied self.

IRIS FRY teaches the history and philosophy of biology in the Department of Humanities and Arts at Technion at the Israel Institute of Technology. She was trained in chemistry and biochemistry at the Hebrew University in Jerusalem and studied the history and philosophy of science at the Universities of Haifa and Tel Aviv. Her publications discuss the concept of teleology in the philosophy of Kant, the history of evolutionary ideas, and the origin of life. Her first book, *The Origin of Life— Mystery or Scientific Problem?*, was published in Israel in 1997. Her second book, *The Emergence of Life on Earth: A Historical and Scientific Overview*, was published in 2000 by Rutgers University Press.

TIBOR GÁNTI is a chemical engineer, theoretical biologist and D.Sc. in biology. His main work is the creation of the chemoton theory, which contains the theory of fluid (chemical) automata and its application to living systems. This theory makes possible the precise separation of living entities from nonliving entities and the construction of an exact, quantitative model of the minimal living system (the chemoton model), while supplying the theoretical foundations for investigating the origin of life, biogenesis, and the artificial synthesis of living systems.

PAUL GRIFFITHS is a Philosopher of Science with a focus on biology and psychology. He received his education from Cambridge University and the Australian National University and is currently a University Professorial Research Fellow in Philosophy at the University of Sydney, Australia, and a Professor of Philosophy of Science at the ESRC Centre for Genomics in Society at the University of Exeter, UK. He is also a Fellow of the Australian Academy of the Humanities, an adjunct member of the Pittsburgh HPS faculty, and a member of the Australian Health Ethics Committee of NHMRC.

PAUL GUYER is a Professor of Philosophy and a Florence R. C. Murray Professor in Humanities at the University of Pennsylvania. He is the author and editor of numerous books on the philosophy of Kant, and the general co-editor of the Cambridge Edition of the Works of Kant, in which he co-translated the *Critique of Pure Reason*, the *Critique of the Power of Judgment*, and Kant's *Notes and Fragments*. Guyer is a fellow of the American Academy of Arts and Sciences and has been awarded the Alexander von Humboldt-Stiftung Research Prize. He is currently writing a history of modern aesthetics.

IMMANUEL KANT (1724–1804) was a German philosopher who is widely regarded as one of the most influential thinkers of modern Europe and the late Enlightenment. Grappling with problems in philosophy of mind, Kant developed the doctrine of Transcendental Idealism, which he regarded as a compromise between the conflicting traditions of Rationalism and Empiricism. This doctrine was first set out in his epochal work, *The Critique of Pure Reason* (1781), which is now uniformly regarded as one of the greatest works in the history of philosophy. Kant also exerted a profound influence on moral philosophy with his contention that morality is required by reason itself.

STUART A. KAUFFMAN is a Professor of Biological Sciences, Physics, Astronomy, and Philosophy at the University of Calgary, in addition to being Director of the Institute for Biocomplexity and Informatics (IBI) and Chair of Informatics Circle of Research Excellence (iCORE). His interests include developmental genetics, theoretical biology, evolution, and the origin of life. He is a MacArthur Fellow and received the Accademia Nazionale dei Lincei Gold Medal in 1997, and the Herbert A. Simon Award in 2000.

EVELYN FOX KELLER received her Ph.D. in theoretical physics from Harvard University and worked for a number of years at the interface between physics and biology. She currently holds the position of Professor Emerita of History and Philosophy of Science in the MIT Program in Science, Technology and Society. She has authored many works, including: *Refiguring Life: Metaphors of Twentieth Century Biology* (1995), *The Century of the Gene* (2000), and *Making Sense of Life:*

Explaining Biological Development with Models, Metaphors, and Machines (2002). Keller has received a number of awards and honorary doctorates, and is a MacArthur Fellow, a member of the American Philosophical Society, a member of the American Academy of Arts and Sciences, and, most recently, the recipient of the Blaise Pascal Research Chair in Paris.

DANIEL E. KOSHLAND, JR. (1920–2007) was a member of the National Academy of Sciences and the editor-in-chief of the leading science journal *Science* from 1985–1995. During World War II, he was one of the scientists working on the Manhattan Project purifying plutonium. In the early part of his career, his work with enzymes led him to propose the "induced fit" model of enzyme catalysis, overturning a century-old consensus on enzyme function. Later in his career, he spearheaded a major reorganization of the study of biology at the University of California at Berkeley, bringing departmental divisions into alignment with advances in genetics and other life sciences. In 1990, he won the National Medal of Science, and his more recent work focused on methane production in blue-green algae.

MARC LANGE is Bowman and Gordon Gray Distinguished Term Professor of Philosophy at the University of North Carolina at Chapel Hill. His interests include scientific explanation, inductive confirmation, natural kinds, disease, necessity and possibility. His most recent books are *Laws and Lawmakers* (2009), which concerns the concept of a law of nature, and *Locality, Fields, Energy, and Mass: An Introduction to the Philosophy of Physics* (2002). He is a member of the American Philosophical Association and the Philosophy of Science Society.

HOD LIPSON is an Associate Professor of Mechanical & Aerospace Engineering and Computing & Information Science at Cornell University in Ithaca, NY. He received his Ph.D. in 1998 from Technion at the Israel Institute of Technology, and continued on to a postdoctoral fellowship at Brandeis University and MIT. At present, Lipson directs the Computational Synthesis group, which focuses on novel ways of generating the automatic design, fabrication and adaptation of virtual and physical machines. He has led work in areas such as evolutionary robotics, multi-material functional rapid prototyping, machine self-replication, and programmable self-assembly.

PIER LUIGI LUISI is a Professor of Biophysics at the University of Rome (*Roma Tre*), and Emeritus Professor of Macromolecular Chemistry at the Swiss Federal Institute of Technology (ETH) in Zurich. His scientific interests mainly concern the origins of life, and studying the macromolecular chemistry, enzymology, and self-assembly of structures such as micelles, reverse micelles, liposomes and giant vesicles. In the last few years, his research has included work on the chemical implementation of autopoiesis, the construction of minimal cells, and the origin of cellular complexity and self-reproduction in supramolecular structures.

LYNN MARGULIS is a Distinguished University Professor in the Department of Geosciences at the University of Massachusetts-Amherst. A member of the National Academy of Science, she investigates and publishes on evolution through symbiogenesis and the Gaia hypothesis (of J. E. Lovelock). She is best known for her endosymbiotic theory of the origin of certain eukaryotic organelles, a theory that is now generally accepted by the scientific community. In 1999, Lynn received the Presidential Medal of Science from William J. Clinton for her contributions to the understanding of living organisms.

ERIC MATTHEWS is Emeritus Professor of Philosophy and Honorary Research Professor of Medical and Psychiatric Ethics in the University of Aberdeen, Scotland. He has pursued an interest in the French philosophers Henri Bergson, Jean-Paul Sartre, and Maurice Merleau-Ponty and has translated a number of works of German philosophy for Cambridge University Press. His recent research concerns the ethics of communicating genetic information and the ethical and policy issues in the upcoming crisis of an aging population. His most recent book is *Body Subjects and Disordered Minds: Treating the Whole Person in Psychiatry* (2007).

ERNST MAYR (1904–2005) is widely regarded as one of the foremost evolutionary biologists of the twentieth century. His work contributed to the conceptual revolution that led to the modern evolutionary synthesis of Mendelian genetics, systematics, and Darwinian evolution. He also contributed to the development of the biological species concept that is currently in use. Among other honors, Mayr received the National Medal of Science in 1970, the Balzan Prize in 1983, the International Prize for Biology in 1994, and the Crafoord Prize in 1999.

ALVARO MORENO is a Professor of Philosophy of Science at the University of the Basque Country, in Spain, where he founded a research group specializing in complex systems, philosophy of biology, and artificial life. He has organized several international workshops and authored many publications on the relationship between artificial life and artificial intelligence.

DUGALD MUDOCH received his higher education at the Universities of Glasgow, Uppsala, and Oxford, and is currently a Professor of Philosophy at the University of Stockholm. His research interests lie in the areas of the philosophy of science, epistemology, and the history of modern philosophy, with a special interest in the philosophy of Descartes. At present, he is working on a project concerning the nature of reasons for belief.

KENNETH H. NEALSON is the Wrigley Professor of Environmental Sciences at the University of Southern California, and a Distinguished Investigator at the J. Craig Venter Institute of Environmental Genomics. His research focuses on the role of microbes in the biogeochemical cycle(s) of the elements, with special focus on iron and manganese. He has also pursued the definition of life and its detection, considering extreme environments on Earth and probing how to effectively search for life in extraterrestrial environments.

ALEKSANDR IVANOVICH OPARIN (1894–1980) pioneered the idea that life arose from inorganic matter through the spontaneous generation of simple organic compounds from inorganic molecules in the prebiotic environment. In his work, he formulated a number of pathways through which biologically important compounds such as amino acids, volatile fatty acids and carbohydrates could have formed in the chemically reductive atmosphere he hypothesized for early Earth. In 1979, he received the Lomonosov Gold Medal from the USSR Academy of Sciences for his work in biochemistry.

LESLIE E. ORGEL (1937–2007) was a distinguished biologist and one of the fathers of the RNA-world theory and the concept of panspermia. Born in London, he received his undergraduate degree and doctorate from Oxford University before pursuing research at Cambridge, the University of Chicago, and the California Institute of Technology. He assisted NASA in analyzing the findings of the Viking missions to Mars and was one of five principal investigators in the NASA-sponsored NSCORT program in exobiology. He was a fellow of the Royal Society and a member of both the National Academy of Sciences and the American Academy of Arts and Sciences.

NORMAN R. PACE is a Distinguished Professor at the University of Colorado in Boulder and an investigator for NASA's Astrobiology Institute. His research interests focus on RNA biochemistry, the application of molecular tools to problems in microbial ecology, and the development of methods for analyzing phylogenetic and quantitative aspects of natural microbial populations without the necessity for laboratory cultivation. He is a member of the National Academy of Sciences and a fellow of the American Association for the Advancement of Science. In 2008, he received Lifetime Achievement Awards from both the RNA Society and the International Society for Microbial Ecology.

JULI PERETÓ is a Professor of Biochemistry and Molecular Biology at the University of València, and researcher at the Cavanilles Institute for Biodiversity and Evolutionary Biology. From 2005 to 2008, Peretó served as the secretary of the International Society for the Study of the Origin of Life (ISSOL – The International Astrobiology Society). His research interests include the evolution of metabolic networks, the minimal genome concept, and the history of ideas concerning the artificial synthesis of life and its natural origin.

JORDAN P. POLLACK is a Professor at Brandeis University in the Computer Science Department as well as the Volen Center for Complex Systems. He directs a research laboratory seeking a fundamental and replicable understanding of how nature can "design" biologically complex machinery. In conjunction with Hod Lipson, Pollack led the Golem project (Genetically Organized Lifelike Electro Mechanics) to evolve simple electro-mechanical systems from scratch to yield physical locomoting machines.

ALONSO RICARDO is a postdoctoral fellow working at Szostak Lab, a research group aimed at creating the functional components of a protocell. He earned his Ph.D. from the University of Florida, working under Steven Benner for his doctorate. His research interests center on bioorganic chemistry, the origins of life, and using mRNA display as a tool for discovering new antibiotics.

KEPA RUIZ-MIRAZO is currently a Ramón y Cajal Research Fellow at the University of the Basque Country, Spain, working in the Department of Logic and Philosophy of Science in addition to the Biophysics Research Unit (CSIC-UPV/EHU). With a degree in Physics and a Ph.D. in Complex Systems, his present research activity cuts across several disciplines, centering on the definition of life, protocell models (*in vitro* and *in silico*), and major transitions in the origin of living systems.

CARL SAGAN (1934–1996) was an astronomer, astrochemist, and author who pioneered exobiology, promoted the search for extraterrestrial intelligence, and successfully popularized astronomy, astrophysics and other natural sciences. He played a leading role in the American space program from its inception, acting as an advisor to NASA and being directly involved in the *Apollo*, *Mariner*, *Viking*, and *Galileo* missions. A best-selling author and Pulitzer Prize winner, Sagan received twenty-two honorary degrees from American colleges and universities for his contributions to science, literature, education, and the preservation of the environment. As one of the first scientists to suggest the possibility of a nuclear winter, he also received multiple awards for his work on the long-term consequences of nuclear war and for his attempts to reverse the nuclear arms race.

DORION SAGAN is an American science writer and General Partner of Sciencewriters, an organization whose goal is to advance science through "enchantment in the form of the finest possible books, videos, and other media". He has written or co-authored many essays and more than fourteen books on evolution and other topics in biology, and is a recipient of an Ed-Press Award from the Educational Press Association of America for excellence in Educational Journalism. A recent book, *Death and Sex*, won first place at the 2010 New York Book Show in the general trade nonfiction category.

ERWIN SCHRÖDINGER (1887–1961) was an Austrian physicist best known for his contributions to the understanding of quantum mechanics, particularly the Schrödinger Equation, for which he received the Nobel Prize in 1933. He also devised the famous "Schrödinger's cat" thought experiment to illustrate what he saw as a problem with the Copenhagen interpretation of quantum physics. In 1944, he wrote *What is Life?* in which he derived the concept of a complex molecule as a carrier for genetic information. This book directly influenced James Watson and Francis Crick whose work eventually led to the discovery of DNA in 1953.

ROBERT SHAPIRO is Professor Emeritus and Senior Research Scientist in the Department of Chemistry at New York University. He is author or co-author of over 125 publications, primarily in the areas of DNA chemistry and the origin of life. In his research, he and his co-workers have studied the ways in which environmental chemicals can damage hereditary material, causing changes that can lead to mutations and cancer. His research has been supported by numerous grants from the National Institute of Health, the Department of Energy, the National Science Foundation, and other organizations.

CHRISTOPHER SHIELDS is a Professor of Classical Philosophy at the University of Oxford and Tutorial Fellow at Lady Margaret Hall. His interests lie in classical philosophy, metaphysics, and philosophy of mind. He is the author of *Order in Multiplicity* (1999), *Classical Philosophy: A Contemporary Introduction* (2003), *Aristotle* (2007), and coauthor of *The Philosophy of Thomas Aquinas* (2004) along with Robert Pasnau. He also has a forthcoming translation (with commentary) of Aristotle's *De Anima* in the Clarendon Aristotle Series of Oxford University Press.

ELLIOTT SOBER is Hans Reichenbach Professor of Philosophy and William F. Vilas Research Professor at University of Wisconsin-Madison. His research is based in the philosophy of science, with a particular focus on the philosophy of evolutionary biology. Sober's books include *The Nature of Selection—Evolutionary Theory in Philosophical Focus* (1984), *Reconstructing the Past—Parsimony, Evolution, and Inference* (1988), *Unto Others—The Evolution and Psychology of Unselfish Behavior* (1998, coauthored with David Sloan Wilson), and *Evidence and Evolution—the Logic Behind the Science* (2008).

PASQUALE STANO is a research assistant to Professor Pier Luigi Luisi at the University of Rome (*Roma Tre*). His research interests concern the properties and reactivity of vesicles and other supramolecular assemblies, and the use of liposomes as cellular models. He is currently involved in the EU project "Synthcells."

KIM STERELNY is an Australian philosopher and Professor of Philosophy. He is affiliated with both the Philosophy Program in the Research School of Social Sciences (RSSS) at Australian National University and the Philosophy Program at the Victoria University of Wellington. He is the editor of *Biology and Philosophy*, the main specialist journal of the philosophy of biology, co-editor of the MIT Press series, *Life and Mind*, and the winner of several international prizes, including the Lakatos Award and the Jean-Nicod Prize, in the philosophy of science. In addition, he is a Fellow of the Australian Academy of the Humanities. His main research interests are philosophy of biology, philosophy of psychology, and philosophy of mind.

ROBERT STOOTHOFF is an Emeritus Professor in the School of Philosophy and Religious Studies at the University of Canterbury, New Zealand. His work, *The Philosophical Writings of Descartes*, co-edited with John Cottingham and Dugald Murdoch and published in three volumes by Cambridge University Press, has become the standard edition of Descartes' philosophical work. Stoothoff also coauthored the entry on theories of meaning in the *Encyclopedia of Philosophy* (1988) along with Jack Copeland.

ANN SYNGE (1916–1998) studied medicine in Cambridge and Dublin before marrying Nobel Prize winner Dr. Richard L. M. Synge, with whom she had eight children. She started translating from Russian in 1954 when she worked on A. I. Oparin's book, *Life: Its Nature, Origin, and Development*. She also translated *Protein Biosynthesis and Problems of Heredity Development and Aging* by Zhores A. Medvedev (1966). Later on, she taught biology and worked actively for nuclear disarmament and world peace.

Introduction

This book is a collection of classic and contemporary readings by philosophers and scientists on the nature of life. Philosophers have pondered the question "what is life?" since at least the time of the ancient Greek philosopher Aristotle. In recent years the question has taken on increasing scientific importance. Molecular biologists and biochemists investigating the origin of life or trying to synthesize chemical life in the laboratory from basic molecular building blocks want to know at what stage an ensemble of nonliving molecules turns into a primitive living thing. Astrobiologists charged with designing instrument packages for spacecraft to detect extraterrestrial life struggle with the question of which characteristics of familiar Earth life (metabolism? reproduction? Darwinian evolution? carbon-based chemistry?) are universal indicators of life. Even computer scientists find themselves mired in questions about the nature of life when they speculate whether lifelike systems constructed of software (purely informational or digital structures) or hardware (metal, plastic, and silicon) could ever literally be alive. Many of these pressing questions are notable for their lack of obvious scientific solutions; one cannot answer them merely by performing more experiments or constructing additional lifelike systems. Although they arise out of science, these questions are deeply philosophical.

The twentieth-century American philosopher W. V. O. Quine proposed that epistemology could be viewed as an "enterprise within natural science" (Quine, 1975, p. 68). In a similar spirit, the editors of this volume believe that collaboration between scientists and philosophers provides the best hope for achieving a compelling answer to the question of the nature of life. Contemporary science has greatly expanded our understanding of the complexities of natural life, and it has provided many intriguing new examples of lifelike artificial systems. But an intellectually satisfying answer to the question "what is life?" requires more than this. It requires analyzing, evaluating, and systematizing disparate information emerging from a multiplicity of scientific disciplines. It is natural for philosophy to play a central role in this process. Philosophers are trained in logical and conceptual analysis. They have expertise in sorting out foundational issues, rooting out subtle inconsistencies, transforming vague generalizations into logically precise principles, tracing logical relations among concepts, principles, hypotheses, and empirical evidence, and evaluating the strength of arguments and theories. Philosophy brings conceptual clarity and logical rigor to scientific theorizing. At the same time, science grounds philosophical reflection on empirical evidence gleaned from careful observation and experiment. Together, contemporary philosophy and science hold forth the promise of finding a satisfactory answer to the age-old question of the nature of life.

The contents of this anthology are the outcome of many lively discussions among the editors and their scientific and philosophical colleagues concerning which issues about life matter most. The chapters selected represent only a small portion of a vast philosophical and scientific literature on life; readers are urged to consult the extensive Supplementary Bibliography at the end for many excellent and important additional writings on life. We tried to be representative, that is, to include classical writings that covered the historically most influential scientific and philosophical ideas, and the leading contemporary scientific and philosophical positions. Chapters were selected also on the basis of their promise for shedding light on central contemporary issues and controversies about life, such as how recent discoveries in microbiology have challenged traditional conceptions of the nature of life. We also made a point to include different approaches to methodology, such as whether the best answer to the question "what is life?" is a definition or a theory, or something else entirely.

We divided the anthology into four sections, corresponding to four central areas of contemporary philosophical and scientific research: Classical Discussions of Life (Section I), The Origin and Extent of Natural Life (Section II), Artificial Life and Synthetic Biology (Section III), and Defining and Explaining Life (Section IV). Each section is preceded by an introduction that conveys the current scientific and philosophical debate about the nature of life, underscoring lines of agreements and disagreements, and raising questions that, in our view, need greater attention. We hope that, taken together, the readings in this book will help inform and guide further scientific and philosophical investigations into the nature of life.

The father of classical physics, Issac Newton, famously observed in a letter to fellow English scientist Robert Hooke, "if I have seen further, it is by standing on the shoulders of giants." Newton's remark reflects the truism that most cutting-edge intellectual work builds upon earlier work. So, it is no surprise that many contemporary debates about life have deep historical roots. The chapters in Section I provide an historical context for the discussions in the rest of the book. We begin with the birth of philosophy in ancient Greece and the rise of modern philosophy and science in the sixteenth and seventeenth centuries, and continue with subsequent seminal scientific and philosophical developments up through the early twentieth century. These readings put contemporary thought about life into historical context and remind us of some great lessons from this history. They also reveal the extent to which contemporary perspectives and problems reflect their historical origins. Here is one illustration: A central theme running through the writings of the ancient Greek philosopher Aristotle up to the present is the idea that living things have functional characteristics, such as metabolism and development, which distinguish them from non-living things. As the chapters in Section I underscore, one of the great controversies about the nature of life bequeathed to us by Aristotle is whether these *prima facie* "teleological" (purposive, goal-directed or self-causing) characteristics are primitive or somehow "reducible to" nonteleological (structural or compositional) characteristics. Many different approaches (essentialist, mechanistic, vitalist, organismic, Darwinian, information-theoretic, thermodynamic, and chemical) to understanding the *prima facie* teleological characteristics of life are

explored in the classic works included in Section I. Although sometimes disguised in innocuous-sounding contemporary scientific terminology, this debate reoccurs throughout this whole anthology.

Section II provides an overview of our current scientific understanding of the origin and extent of natural forms of life. Most of the authors are scientists who share the view that an understanding of life lies at the level of molecules and biochemical processes, as opposed to higher-level organizational and functional properties. This is reflected in the ways in which they interpret and articulate traditional questions about the nature of life. For example, what was "nutrition" for Aristotle and "metabolism" for early modern biologists becomes "chemical self-organization," and what was "reproduction" for Aristotle and "Darwinian evolution by natural selection" for nineteenth-century biologists becomes "replication" by means of "genetic [informational] structures." The reader is encouraged to consider whether this is the proper level of analysis for theorizing about life in general.

The authors in Section II discuss a variety of important issues about life within the context of recent scientific developments in molecular biology, biochemistry, microbiology, and astrobiology: Is the transition from a nonliving ensemble of molecules to a primitive living thing gradual or abrupt? Can an account of the nature of life be given independently of an account of the origin of life? Which came first, chemical self-organization or genetic structures? Was the original genetic molecule RNA or something else, e.g., a mineral surface? What are the alternative molecular and biochemical possibilities for life? How can we design methods and instruments for searching for truly "weird" forms of life elsewhere in the solar system, given that all known life on Earth is so similar at the molecular and biochemical level? Could the contemporary Earth contain undetected alternative forms of microbial life descended from a separate origin? A couple of chapters in this section are authored by philosophers who ferret out philosophical assumptions underlying scientific investigations into the origin and extent of life. They also evaluate the bearing of the results of these scientific investigations on some traditional and contemporary philosophical questions about life. Philosophers will want to know the material in this section, so that their discussions about life are scientifically informed.

Any contemporary attempt to understand the nature of life should be informed by important recent scientific developments. Artificial life and synthetic biology are motivating significant philosophical and scientific reflection about the nature of life. Section III concerns the implications of attempts in artificial life and synthetic biology to recreate life in the laboratory and to make lifelike systems out of hardware or software. The scientific state of the art in artificial life and synthetic biology is presented in some chapters. In other chapters philosophers and historians reflect on the larger implications of those scientific developments. The strong thesis of artificial life and synthetic biology is that we can construct new kinds of systems (using software, hardware, or wetware) that literally are alive—as alive as any other form of life we know. It contrasts with a weaker thesis that construes the lifelike systems constructed by artificial life researchers as useful theoretical tools for exploring properties of living systems, but not literally alive. Because research in artificial life aims to create novel kinds of lifelike systems, scientists must ask what really makes something alive. Is it the material stuff out of which it is composed (cytoplasm, flesh and bones, etc.)? Or is it the kinds of metabolic processes in which those materials are participating, in a transitory and fleeting way? Attempts to create lifelike systems or even life itself from nonliving materials also focus attention on a variety of foundational issues, such as whether there is a sharp distinction between life and non-life, or whether there is an open-ended array of alternative kinds of systems that are more or less alive. Artificial life and synthetic biology also raise a fundamental epistemological issue: What kind of evidence should we use when explaining why a given kind of chemical system is literally alive? For example, what weight should we give to generalizations derived only from all known forms of life on Earth?

Section IV illustrates the main philosophical and scientific approaches to answering the general question "what is life?" Achieving a satisfactory understanding of the nature of life involves explaining a range of familiar and striking phenomena. As these chapters illustrate, there is not a consensus about which characteristics of familiar life constitute the "signs," "hallmarks," or "puzzles" most in need of explanation. We need to figure out what evidential status to give to our current preconceptions about life. It is not obvious

what epistemological authority they should have. Part of the problem is that our experience with natural life is limited to familiar Earth life. As the discussions in Section II reveal, we have compelling scientific reasons for believing that all known life on Earth descends from a common ancestor. We also have good scientific reasons for believing that life could have been at least modestly different, but we have very little idea how different it could have been. This makes it difficult to identify which characteristics of familiar Earth life are found universally in all actual examples of life. For all we know, familiar Earth life could constitute an unrepresentative example of all possible forms of life. However, not everyone views this problem as insurmountable. The development of artificial ("hard" and "soft") life systems and the creation of novel microorganisms in the laboratory by synthetic biologists give us powerful new tools for exploring and expanding our concept of life, and perhaps even increasing the sample size of kinds of life available to us for empirical investigation. The discussions in Section IV also address foundational theoretical issues about life such as the role of reduction and emergence in explanations of life and the question of whether life is a compositional, structural, or functional kind, or some combination thereof. The chapters in this section explore different perspectives on these and related issues within the context of recent scientific and philosophical advances.

The question of the proper methodology for constructing and evaluating views about the nature of life is a central concern of many of the chapters in Section IV written by philosophers. One issue is the evidential status of signs, hallmarks, and puzzles about life. What role should these play when investigating the nature of life? Are they test cases for evaluating accounts of life? A striking feature of investigations into the nature of life is the ambiguous and uncertain application of various methodologies. Another important question concerns the widespread use of definitions in explaining life. Definitions can be distinguished from criteria for life, and from life's signs and hallmarks. They can take different forms, including pragmatic and operational, although many scientists and philosophers prefer logically complete definitions providing necessary and sufficient conditions for life. Some authors in Section IV reject the project of defining life. They contend that definitions cannot provide satisfactory answers to "what is" questions about natural kinds, such as water and life. On their view an empirically well-grounded, scientific theory of life is needed. But how does a theory of life differ from a highly complex definition of life? There is also the question of the status of computer simulations of complex living systems. Some authors argue that simulations can be fruitfully construed as distinctively *constructive* definitions or theories of life. Many of the chapters in Section IV address these issues. Scientists engaged in theorizing about life might find the material in this section especially useful.

The four sections of this anthology are knit together by three central open philosophical and scientific questions about the nature of life:

1. What are the central characteristic phenomena exhibited by all forms of life?
2. What are the best descriptions and explanations of the nature of life?
3. What is the proper way to construct and evaluate views about the nature of life?

In our view, attempting to answer any of these questions in isolation generates a tangle of thorny philosophical and scientific issues that can be resolved only by addressing the two other questions. In fact, we find it difficult to draw a bright red line separating scientific and philosophical questions about life. Fundamental progress in either discipline requires knowing and appreciating the other discipline. This book aims to foster mutual understanding and appreciation among philosophical and scientific perspectives on life. In the long run, a good measure of its success will be the advances in understanding of the nature of life that are eventually made by its readers.

REFERENCE

Quine, W. V. O. (1975). The nature of empirical knowledge. In S. Guttenplan, S. (Ed.), *Mind and language* (pp. 57–81). Oxford: Oxford University Press.

Section I
Classical discussions of life

Humans have long been puzzled about the nature of life—how living things are similar to and different from nonliving things, both natural and artificial, and whether the characteristics that are universal in familiar Earth life are genuinely essential to all possible forms of life. The chapters in this section provide classical historical perspectives on present day philosophical and scientific debates about life. These perspectives have an often underappreciated and sometimes even unrecognized influence on current philosophical and scientific thought. Cutting-edge contemporary ideas are sometimes not so novel after all! More importantly, however, sometimes the older debates, which typically focus on more general, and hence more fundamental, conceptual issues, can provide unexpected insights into present-day controversies.

This section begins with the writings of three intellectual giants: Aristotle, René Descartes, and Immanuel Kant. Best known today for their philosophical work, each also made important contributions to the development of modern science. Each holds a different view about the nature of life. As the remaining chapters in this book illustrate, the differences between them are still relevant today.

One central theme running from the writings of the ancient Greek philosopher Aristotle right up to the present is the idea that living things have distinctive *functional* characteristics. Aristotle also thought that living things are distinguished from inanimate objects by the ability to self-organize (develop from fertilized eggs) and maintain this self-organization against both internal and external perturbations. Unlike scientists and contemporary philosophers of science, Aristotle distinguished four different kinds of "cause:" material, efficient, formal, and final (discussed below). Aristotle thought that life forms had material causes (material

composition) and formal causes (organization or structure), but he also thought that all life forms had final causes (teleological and functional explanations). Aristotle's concept of life is fundamentally different from the contemporary scientific concept of chemical substances such as water, because water is distinguished from other chemical substances, e.g., nitric acid, by a unique molecular composition and structure (H_2O); water is not viewed as having a teleological or functional explanation.

As the chapters in this section underscore, one of the great controversies about the nature of life is whether its *prima facie* teleological characteristics are primitive or analyzable in terms of (i.e., "reducible to") nonteleological (e.g., compositional or structural) characteristics. Aristotle thought not. Some like-minded people today agree with Aristotle that the striking goal-directed characteristics of life (nutrition, development, growth, maintenance, repair, sensation, and reproduction) cannot be explained without an appeal to natural ends or intrinsic purposiveness. The question of how to provide a naturalistic account of the *prima facie* teleological characteristics of organisms has been at the heart of philosophical and scientific debates about life since the time of Aristotle.

Writing at the dawn of modern physics, just before the birth of Isaac Newton, Descartes (1596–1650) attacked the long-standing Aristotelian tenet that life is intrinsically teleological. Comparing organisms to the intricate artificial mechanisms (clocks, church organs, and elaborate fountains) popular during his day, Descartes argued that organisms are just exceedingly complex machines. He believed that the teleological aspects of life were fully analyzable using only the principles and concepts of the newly emerging physics. Descartes's view of living systems is reflected today in

The Nature of Life, ed. M. A. Bedau and C. E. Cleland. Published by Cambridge University Press. © M. A. Bedau and C. E. Cleland 2010.

versions of "hard" artificial life that take seriously the possibility of robotic life forms composed of mechanical and electronic parts (see, e.g., Ch. 19). At the same time, it is worth noting that some have argued that hard ALife provides a strong argument *against* the dominant Cartesian orientation in classical and connectionist cognitive science. Michael Wheeler (2007), for example, gives a Heideggerian interpretation of the emphasis on embodiment and dynamical systems in contemporary hard artificial life.

Less than 200 years later, Kant (1723–1804) concluded that Descartes was wrong about the capacity of classical physics to explain the teleological characteristics of life. According to Kant, the purposiveness exhibited by organisms is fundamentally different from that exhibited by the most elaborate artificial mechanisms. Unlike a mechanical device, an organism is both "cause and effect of itself." When Kant famously declared that there would never be a Newton of biology, he was calling attention to his belief that the teleological characteristics of life could never be explained mechanistically. At the time he was writing, in the heyday of classical physics, this amounted to the claim that these characteristics would never have a physical explanation.

Kant thought that teleology is the central characteristic of life forms that distinguishes them from the nonliving. Even those who disagree with Kant's explanation of teleology might still agree that teleology is one of life's deepest hallmarks. Kant's writings on life brought to the forefront the previously underappreciated difficulty of reconciling the *prima facie* teleological aspects of life (the appearance of design) with the nonteleological concepts of classical physics. Many different approaches to resolving this conflict were explored in the ensuing years. Some tried to circumvent it by developing distinctively biological concepts or principles, whereas others appealed to non–classical (twentieth century) physics or chemistry.

As Ernst Mayr discusses in this section (Ch. 6), vitalism is a historically important view that proposed a distinctively biological concept, or, depending upon the version, principle of life. Vitalism holds that life is conferred upon nonliving matter by a special kind of animate substance ("protoplasm") or organizing energy or force ("vital spark" or "élan vital"). Vitalists thus agree with Aristotle and Kant that the teleological properties of life cannot be explained in terms of

classical physics. They depart from Kant, however, in contending that they can be explained scientifically in terms of distinctively biological (namely, vitalist) concepts or principles. With the advent of biochemistry and molecular biology in the twentieth century, vitalism lost its appeal to biologists (Oparin, Ch. 5), and is now scorned as unscientific. But as philosopher Marc Lange discusses in Section III, an unanalyzed concept of vitality could still end up playing a fundamental theoretical role (analogous to that of mass in Newtonian physics or proton in contemporary physics) in explaining the hallmarks of life. (Lange does not, however, endorse this view.)

With the advent of Charles Darwin's theory of evolution by natural selection in the mid-nineteenth century, some philosophers and scientists concluded that the teleological aspects of life could be fully explained within the framework of classical physics after all, without importing suspect concepts such as vitality. As Richard Dawkins's contemporary defense of universal Darwinism in Section IV illustrates, the Darwinian view that evolution by natural selection centrally explains life remains popular. More general evolutionary views, inspired by Darwin but not limited to natural selection, have also been developed; see, for example, Ruiz-Mirazo and colleagues and Bedau in Section IV. Not everyone, however, finds evolutionary approaches to the nature of life convincing. Contemporary Darwinian biology takes populations of organisms, classified by common ancestry (species, genus, families, etc., or, in recent years, lineages), as the basic unit of analysis. Hence, some Darwinian definitions of "life" do not classify sterile hybrids such as mules as cases of life, and some even reject the idea that a fertile individual organism counts as a living thing. The same is true of some of the more general evolutionary accounts. For example, Mark Bedau (Ch. 31) asserts that individual organisms, including both horses and mules, are all "secondary" forms of life, and the "primary" form of life is the whole biosphere undergoing "open-ended evolution," because the evolving biosphere can explain what he identifies as the familiar hallmarks and puzzles about life. The point is that there is a tendency for most contemporary evolutionary accounts of life to focus on the evolutionary histories of evolving populations of well-adapted organisms, and consider of secondary importance each individual organism and its individual teleological properties (self-organization, self-maintenance, self-repair, etc.). For some philosophers and scientists, this makes evolutionary accounts unsatisfactory.

The chapters by Erwin Schrödinger, Alexander Oparin, Ernst Mayr, and Tibor Gánti represent a different approach to making good scientific sense of the teleological characteristics of life. None of these authors believe that the purposive properties of individual organisms can be fully understood in terms of classical physics and Darwin's theory of evolution by natural selection. While concurring that Darwin's theory is crucial to understanding important aspects of life on Earth, they differ on how central it is to explaining the *nature* of life. Gánti takes the most extreme position, contending that Darwinian evolution is not essential to life, considered generally, even though it is important for understanding the history of life on Earth. Schrödinger, Mayr, and Oparin, in contrast, take Darwinian evolution to be necessary but not sufficient for life. They believe that something more is required.

In Chapter 4 Schrödinger, a theoretical physicist, appeals to concepts and principles from the "new" physics of his day (specifically, statistical thermodynamics and quantum mechanics), developed during the early part of the twentieth century. He attributes the teleological features of life to its ability to maintain itself in a state of disequilibrium by extracting energy from its environment. In fact, he views this open chemical metabolism as perhaps the most essential hallmark of individual life forms. This aspect of Schrödinger's work anticipates more contemporary accounts of life in terms of dissipative structures and far from equilibrium systems; as an example, see Stuart Kauffman's chapter in Section IV.

Oparin, Mayr, and Gánti, in contrast, appeal to twentieth century chemistry. Departing the furthest from the machine metaphor, Oparin (Ch. 5) and Mayr (Ch. 6) contend that life is a product of the gradual chemical evolution of a "primordial soup" into a highly complex, tightly integrated chemical system able to exert a degree of control over its parts not found in any mechanical or electrical device. Gánti (Ch. 7) also analyzes life as a chemical system but, unlike Oparin and Mayr, portrays it as machine-like in a novel way. He describes life forms as "fluid" chemical automata; on his view, living systems are chemical automata

("chemotons"). Gánti's chemoton model is a forerunner of the more abstract autopoietic model of life proposed by Maturana and Varela (1980), which is not restricted to chemical compounds. The chemoton and autopoietic models of life focus on explicating minimal life as an autocatalytic network separated from its surroundings by a boundary, and both admit the possibility of life forms that cannot evolve. Autopoesis is discussed by Luisi and colleagues (Ch. 21) in Section III and Ruiz-Mirazo and colleagues (Ch. 25) in Section IV.

Oparin, Mayr, and Gánti disagree about the chemical possibilities for life. Mayr and Gánti are open to inorganic (e.g., silicon-based) forms of chemical life, whereas Oparin specifically restricts life to organic (carbon-containing) compounds. In addition, Oparin and Gánti tie the nature of life to the origin of life, contending that the former cannot be understood independently of the latter. This approach is not uncommon today. It is reflected in discussions of the nature of life that appeal to theories of the origin of life such as the RNA world; see, for example, Pace (Ch. 11) and the NRC report (Ch. 15) in Section II. Nevertheless, it is not obvious that a theory of the nature of life presupposes an understanding of the origin of life (see the introduction to Section II). The origin and extent of life are covered in detail in Section II of this book.

Each of the classical discussions of life in this section attempts to place the phenomena of life within the framework of the empirical and theoretical understanding of nature during its day. Taken together, the chapters exhibit a wide range of explanatory principles and frameworks. Aristotle's explanatory principles are his four "causes" (see below). Descartes proposed to explain all the phenomena of life within a purely physical and mechanistic framework, often by postulating invisible micro-mechanical entities and processes. Kant concluded that Descartes's purely mechanistic framework could never explain life's autonomous purposiveness. As Mayr recounts in Chapter 6, vitalists posited non-physical vital substances or forces, to explain the hallmarks of life. Schrödinger focused on how life's metabolism sustains a complex and robust organization in the face of the second law of thermodynamics. He also presaged how DNA and RNA govern metabolic processes, and how this control is inherited when life reproduces. Oparin and Gánti both attempted to understand how the simplest and earliest forms of life emerged from nonliving materials.

Unlike some contemporary literature, the classical discussions in this section typically ignore methodological questions about the proper way to evaluate explanations of the nature of life. Of course, even though they do not *discuss* methodologies, the discussions still *display* them. The methodologies displayed include cultural preconceptions and armchair conceptual analysis, as well as empirical investigations of fundamental biological, chemical, and physical mechanisms. The examples set by these authors raise concrete questions about the proper methodology for investigating the nature of life. This issue is revisited in later sections, especially Section IV.

This overview of Section I ends with brief biographies and background information about the authors of each chapter in this section.

ARISTOTLE

Best known as a philosopher, Aristotle (384–322 BC) was also one of the first scientists. Unlike his teacher Plato, Aristotle emphasized the importance of observation in theorizing about the world. He was especially fascinated by life, and dedicated considerable time to studying it. His devoted student Alexander the Great reputedly sent him exotic animals that he encountered on his conquests. Aristotle's writings on ethics and metaphysics draw extensively from his work in biology, and vice versa. The selection from Aristotle included in this section is taken from his work *De Anima*, sometimes translated as *"On the Soul."*

Aristotle sets the stage for subsequent debates about the nature of life in Chapter 1 by distinguishing the "mineral [inanimate] kingdom" from the "animal and vegetable kingdom" and "defining" life *functionally* in terms of its capacities or powers, and what he called "souls" (a capacity for a set of activities). One cannot understand Aristotle's notion of a soul independently of his theory of explanation. Aristotle distinguishes four different kinds of factors that could be cited in an explanation: material, efficient, formal, and final. These explanatory factors are traditionally called "causes," but this can be very misleading because they differ from how we think of causes today. Efficient causation has survived to the present day; this is the causal trigger (e.g., flipping a switch) that brings about an effect (the lighting of a room). We also still recognize material causes, e.g., the disposition of a wine glass to shatter if struck. For Aristotle, however, the capacities that distinguish life from inanimate matter critically involve formal causation (which constitutes the *essence* of something) and final causation (which constitutes its natural *purpose*), neither of which has an exact analog in modern science. For more on Aristotle's view of causation and explanation, see his *Physics*, Book II, Section 3. (For the philosophically unsophisticated, Falcon (2008) provides an accessible discussion of Aristotle's highly complex ideas about causation and explanation.)

For Aristotle, something is alive if and only if it has a certain kind of "soul." It is important to realize that what Aristotle means by "soul" is very different from the Christian theological notion of the soul. Aristotle's soul is not a disembodied spirit but a set of animating capacities produced by certain natural internal faculties (Matthews, 1996). On Aristotle's account, plants as well as animals have souls. Aristotle called special attention to the capacities of nutrition, perception, locomotion, and thought. Two of life's fundamental capacities involve the mind: sensation and rational thought. So, for Aristotle, anything with a mind is necessarily alive. There is a minor renaissance today in more ambitious connections between life and mind (e.g., Thompson, 2007).

The capacities that Aristotle associated with life are multifaceted. For example, nutrition involves the capacity to grow and reproduce; perception involves the capacity for pleasure, pain, desire, and appetites; thought involves understanding, reason, and imagination. There are various kinds of dependencies among these capacities. Aristotle thought that the most fundamental life-conferring capacities are (self-)nutrition and reproduction; all life forms (plants as well as animals) possess them, and some (e.g., human beings) possess additional capacities (the set that make up the "rational" soul). Aristotle further argued that reproduction is more basic than nutrition. His writings anticipate contemporary debates over whether metabolism or replication is more fundamental to life; see, for example, Orgel (Ch. 8), Shapiro (Ch. 9), Pace (Ch. 11), Boden (Ch. 18), Dawkins (Ch. 29), Kauffman (Ch. 30), and Bedau (Ch. 31).

RENÉ DESCARTES

The great seventeenth-century French philosopher René Descartes (1596–1650) is foundational not only to modern philosophy but also to mathematics. He developed the Cartesian coordinate system of analytic geometry, which allows geometrical shapes to be represented algebraically and was crucial to the invention of calculus. Descartes's writings played a pivotal role in the downfall of Aristotelian-oriented scholastic philosophy and science. Indeed, he is commonly called the "father of modern philosophy." In addition to his work in philosophy and mathematics, Descartes made important contributions to a number of fields of science. His views about the nature of life were strongly influenced by his detailed studies of animal physiology. Chapter 2 includes his most extensive discussion of the nature of life in *Treatise on Man*.

Breaking with Aristotle, Descartes sharply distinguished mind ("rational soul") from life ("body"). His arguments for the distinction are powerful and do not rest upon the limited science of his day. They gave rise to the infamous "mind-body problem" that still exercises many philosophers.

Descartes is easier to read than Aristotle in large part because Descartes laid the philosophical foundations for modern science. Modeling science on what would become classical (Newtonian) physics, he rejected any appeal to purposes—natural or divine—in explaining natural phenomena. For Descartes, living systems ("bodies") do not differ in nature from mechanical devices such as clocks and fountains; like the latter, the former can be fully understood in terms of efficient causation, namely, the pushing and pulling of different physical parts (organs, muscles, tendons) on each other. In short, living systems are just machines. Variants of this view of life are still popular today. Contemporary molecular biologists have a tendency to characterize organisms as complicated molecular "machines" (see the chapters in Section II). In addition, some ("hard") versions of artificial life view certain machines made with mechanical and electronic parts as alive.

IMMANUEL KANT

Broadly educated in science and philosophy, the great German scholar Immanuel Kant (1724–1804) made important contributions to both. His work in astronomy, which included formulating the nebular hypothesis and deducing that the Milky Way and possibly other nebula are huge disks of stars, is foundational to modern astronomy. He is best remembered, however, for his philosophical writings. Kant's writings on life are found in the *Critique of the Power of Teleological Judgment*, from which the excerpts in Chapter 3 of this section are taken.

By the eighteenth century, Newtonian physics was considered the foundation for all of science. Kant became convinced, however, that it could not explain life. Breaking with Descartes, he argued that an organism ("organized being") is not a mere machine ("artifact"). While organisms share with artifacts the appearance of design, organisms nonetheless differ from artifacts in a fundamental way: They are "natural ends" (self-causing rather than externally caused). Kant's work represents a return to the Aristotelian idea that life involves a special kind of causation—intrinsic natural purposiveness.

Kant's writings are notoriously difficult to understand, and there is disagreement among Kant scholars over how to interpret them. He sometimes says that organisms *are* natural ends and sometimes says only that they must be "regarded" as such by us. Moreover, his claims about natural purposes sometimes seem to conflict with each other, e.g., his assertion that we must endeavor to explain *everything* in nature in mechanical terms. According to the most widely accepted interpretation, Kant is claiming only that organisms cannot be *comprehended* by the human mind (and hence scientifically *explained*) as mechanical devices. But this does not mean that they are not *in fact* such devices. On this interpretation the teleological aspects of life are a product of the limitations of the human intellect, as opposed to an objective feature of a mind-independent world of nature.

It seems clear that Kant was trying to reconcile the teleological properties of life with the science of his day. His successors took up the challenge. They concurred that the teleological aspects of life cannot be explained within the framework of classical physics, but they were reluctant to ascribe them to what amounts to a defect of the human intellect. So, they sought a solution in new scientific theories and developments, such as Darwin's theory of evolution by natural selection, twentieth-century physics and chemistry, and even mathematics (complexity theory and chaos theory). For a survey of contemporary perspectives on natural teleology, see Allen *et al.* (1997).

ERWIN SCHRÖDINGER

An Austrian physicist, Erwin Schrödinger (1887–1961) received the Nobel Prize in 1933 for the Schrödinger equation, one of the central principles of quantum mechanics. Schrödinger was also fascinated by life. In 1944 he wrote *What is Life?*, a now classic book that arose from a series of public lectures on life given in Ireland during World War II. Key excerpts from this book are included here as Chapter 4. Schrödinger believed that the teleological aspects of life could be explained through the concepts and principles of non-classical physics, especially those of thermodynamics and quantum mechanics. In addition to famously anticipating the structure of the "hereditary substance" (the DNA molecule) as a "large aperiodic crystal," he provocatively suggests that new principles of thermodynamics might be required for understanding the purposive aspects of life, which he identifies with the ability of individual organisms to maintain order by extracting energy from their environments (i.e., metabolism). Schrödinger's speculations about the importance of metabolism in life have been revisited in recent years (e.g., Boden, Ch. 18), and are the inspiration behind attempts to explicate life in terms of far from equilibrium systems and dissipative structures; a good example is Kauffman (Ch. 30), who proposes a fourth law of thermodynamics as central to understanding life.

ALEXANDR OPARIN

Oparin (1894–1980), a Russian biochemist, is the father of "protein-first" metabolic accounts of the origin of life. Oparin believed that a genetic system capable of evolution by natural selection could not arise unless a primitive metabolic system was already in place. He thus opposed "genes-first" accounts of the origin of life; see Shapiro (Ch. 9) for a contemporary discussion of this position in the context of the currently popular (genes-first) RNA world theory. Oparin's ideas evolved over the years. Chapter 5 in this section is taken from a book, entitled *Life*, written in his later years.

Like Kant, Oparin rejects the machine analogy for life. He is also adamantly opposed to vitalism. Oparin identifies organizational complexity and purposiveness as the most essential properties of life, and he stresses that they can be fully explained in terms of twentieth-

century chemistry. Oparin believes that the molecular possibilities of life are limited to carbon compounds. (The chapters in Section II contain the latest scientific developments along these lines.) According to Oparin, life developed on Earth through the gradual chemical evolution of organic (carbon-based) compounds from a "primordial soup" consisting of simpler organic molecules. Because the British geneticist and evolutionary biologist J. B. S. Haldane (1937) proposed a similar hypothesis at around the same time as Oparin first developed these ideas, this theory has become known as the "Oparin–Haldane hypothesis." Oparin was one of the first to claim that a theory of the nature of life cannot be separated from a theory of the origin of life. On his view, understanding the nature of life presupposes understanding the transition from a nonliving ensemble of organic molecules to a living chemical system, which amounts to understanding how life originates from nonliving chemicals.

ERNST MAYR

Born in Germany, the evolutionary biologist Ernst Mayr (1904–2005) immigrated to the United States in his twenties. Mayr played a key role in the development of the modern evolutionary synthesis of Darwinian natural selection with Mendelian genetics. One of his important contributions was redefining the concept of a species, which was previously based upon morphological similarities, as a group of individuals that can breed among themselves and are reproductively isolated.

Mayr was also interested in the more general question of what distinguishes life from nonlife. Like Kant and Oparin, he rejected the machine analogy, and like Oparin, he was opposed to vitalism. Chapter 6 in this section is taken from a book, *This is Biology*, written for the general public in his later years. It includes an extended discussion of the history of vitalism and mechanism as well as a discussion of the works of Kant, Haldane (whose views are very similar to Oparin's), and Schrödinger, indicating where they went wrong and what they got right. As a consequence, Mayr's chapter provides an interesting survey from a biologist's point of view of nineteenth-century and early-to-mid-twentieth-century thought about life.

According to Mayr, the "characteristics of living organisms that distinguish them categorically from inanimate systems" are the capacities for (1) evolution, (2) self-reproduction, (3) growth and differentiation through a genetic program, (4) metabolism, (5) self-regulation, (6) response to stimuli from the environment, and (7) change at both phenotypic and genotypic levels. Unlike Oparin, Mayr believes that certain distinguishing characteristics of life (such as hierarchical organization and purposiveness) cannot be fully explained in terms of physico-chemical mechanisms at the molecular level. He contends that these unique characteristics of life literally "emerge" *de novo* at higher levels of organization and integration; they cannot be fully "reduced to" or "predicted from" a knowledge of their lower level parts. Thus Mayr believes that he can make good scientific sense of the intuitive idea that an organism, as a whole, is more than the sum of its parts.

Known as organicism, Mayr's account distinguishes life from "inert matter" in terms of organizational characteristics. Unlike Oparin, he does not restrict life to organic compounds; the key to life is chemical organization, as opposed to chemical composition. Organicism faces the problem of making good scientific sense of the mysterious concept of "emergence" without falling into something closely resembling vitalism. Some scholars have sought a solution to this problem in complexity theory, a fairly new, interdisciplinary field of research growing out of work in theoretical computer science and mathematics. For a collection of recent philosophical and scientific perspectives on emergence, including those based on the contemporary study of complex systems, see Bedau and Humphreys (2008).

TIBOR GÁNTI

The Hungarian chemical engineer Tibor Gánti (b. 1933) has made many underappreciated contributions to our understanding of the nature of life. A recently published collection (Gánti, 2003) of English translations of his most important papers is the source of the seventh and final chapter in this section. Gánti proposes criteria of life and a model of minimal chemical life. While his hallmarks of life are similar to those proposed by others, there are some significant differences. It is instructive to compare them with, for example, the hallmarks listed by Mayr. One novelty is that Gánti divides the hallmarks of life

into two categories: *absolute* life criteria, which are necessary and sufficient for individual organisms to be alive, and *potential* life criteria, which are necessary only for life to populate and be sustained indefinitely on a planet. As the reader will discover, this distinction gives Gánti the conceptual resources to claim that something could be alive and yet not part of an evolving system, while still agreeing with Oparin and Mayr that the capacity to evolve is essential for life to adapt to the environment and diversify.

Gánti's views about the nature of life are informed by his experience as a chemist, and he emphasizes life's chemical requirements. But since he wants to formulate a conception of life that applies as broadly as possible, his chemical analysis is abstract and functional. The result is Gánti's celebrated chemoton model. The chemoton model depicts minimal chemical cellular life necessarily as an autocatalytic (self-sustaining) chemical network that integrates three kinds of chemical subsystems: a metabolism, a container, and a heritable chemical program. Because Gánti does not restrict life to organic compounds, the chemoton model represents a metabolism-first but not a protein-first perspective (see the introduction to Section II). The three subsystems cannot function separately, and they cooperate to create a unified whole that can exhibit the hallmarks of life. More or less similar chemical triads are presupposed in many contemporary experimental and theoretical investigations into creating new forms of life in the laboratory (Rasmussen *et al.*, 2008), although Gánti's emphasis on stoichiometric coupling of the functional triad is usually dropped and the internal program usually does more work.

REFERENCES

Allen, C., Bekoff, M., & Lauder, G. (Eds.) (1997). *Nature's purposes: Analyses of function and design in biology.* Cambridge: MIT Press.

Bedau, M. A. & Paul Humphreys, P. (Eds.) (2008). *Emergence: Contemporary readings in philosophy and science.* Cambridge: MIT Press.

Falcon, A. (2008). Aristotle on causality. In E. N. Zalta (Ed.), *The Stanford encyclopedia of philosophy* (Fall 2008 edition). Available online at http://plato.stanford.edu/archives/fall2008/entries/aristotle-causality/(accessed February 2009).

Gánti, T. (2003). *The principles of life*. New York: Oxford University Press. Commentary by James Grisemer and Eörs Szathmáry.

Haldane, J. B. S. (1937). What is life? In J. B. S. Haldane, *Adventures of a biologist* (pp. 49–64). New York: Macmillan.

Matthews, G. B. (1996). Aristotle on life. In M. C. Nussbaum & A. O. Rorty (Eds.), *Essays on Aristotle's* De anima (pp. 185–193). Oxford: Clarendon Press.

Maturana, H. & Varela, F. (1980). *Autopoiesis and cognition: The realization of the living*. Boston: D. Reidel.

Rasmussen, S., Bedau, M. A., McCaskill, J. S., & Packard, N. H. (2008). A roadmap to protocells. In S. Rasmussen, M. A. Bedau, L. Chen, *et al.* (Eds.), *Protocells: Bridging nonliving and living matter* (pp. 71–100). Cambridge: MIT Press.

Thompson, E. (2007). *Mind in life: Phenomenology, and the sciences of the mind*. Cambridge: Harvard University Press.

Wheeler, M. (2007). *Reconstructing the cognitive world: The next step*. Cambridge: MIT Press.

1 · De Anima (selections)

ARISTOTLE

DA II 1 T

[412a1] Let this much be said about what has been handed down concerning the soul by our predecessors. Let us start anew, as if from the beginning, endeavoring to determine what the soul is and what its most common account would be.

Among the things which are, we call one kind substance. Belonging to this is, first, matter, which in itself is not some this; another is shape and form, in terms of which something is already called some this; and the third is what comes from these. Matter is potentiality, while form is actuality; and actuality is spoken of in two ways, first as knowledge, and second as contemplating.

Substances seem most of all to be bodies, and among these, natural bodies, since these are the sources of the others. Among natural bodies, some have life and some do not. We mean by 'life' that which has through itself nourishment, growth, and decay.

It would follow that every natural body having life is a substance, and a substance as a composite. But since every such body would also be a body of this sort, that is, one having life, the soul could not be a body; for the body is not among those things said of a subject, but rather is spoken of as a subject and as matter. It is necessary, then, that the soul is a substance as the form of a natural body which has life in potentiality. But substance is actuality; hence, the soul will be an actuality of a certain sort of body.

Actuality is spoken of in two ways, first as knowledge, and second as contemplating. Evidently, then, the soul is actuality as knowledge is. For both sleeping and waking depend upon the soul's being present; and as waking is analogous to contemplating, sleeping is analogous to the having of knowledge without exercising it. And in the same individual, having knowledge occurs prior to contemplating. Hence, the soul is the first actuality of a natural body which has life in potentiality.

This sort of body would be one which is organic. [412b] And even the parts of plants are organs, although altogether simple ones. For example, the leaf is a shelter of the outer covering, and the outer covering of the fruit; even the roots are analogous to the mouth, since both draw in nutrition. Hence, if it is necessary to say something which is common to every soul, it would be that the soul is the first actuality of an organic natural body.

Consequently, it is not necessary to ask whether the soul and body are one, just as it is not necessary to ask this concerning the wax and the seal, nor generally concerning the matter of each thing and that of which it is the matter. For while one and being are spoken of in several ways, what is properly so spoken of is the actuality.

It has now been said in general what the soul is: the soul is a substance corresponding to the account; and this is the essence of such and such a body. It is as if some tool were a natural body, e.g., an axe; in that case being an axe would be its substance, and this would also be its soul. If this were separated, it would no longer be an axe, aside from homonymously. But as things are, it is an axe. For the soul is not the essence and structure of this sort of body, but rather of a certain sort of natural body, one having a source of motion and rest in itself.

What has been said must also be considered when applied to parts. For if an eye were an animal, its soul would be sight, since this would be the substance of the eye corresponding to the account. The eye is the matter of sight; if sight is lost, it is no longer an eye,

The Nature of Life, ed. M. A. Bedau and C. E. Cleland. Published by Cambridge University Press. © M. A. Bedau and C. E. Cleland 2010.

except homonymously, in the way that a stone eye or painted eye is.

What has been said in the case of parts must of course be understood as applying to the whole living body. For there is an analogy: as one part is to one part, so the whole perceptive faculty is to the whole of the body which is capable of perception, insofar as it is capable of perception. The body which has lost its soul is not the one which is potentially alive; this is rather the one which has a soul. The seed, however, and the fruit, is such a body in potentiality.

Hence, as cutting and seeing are actualities, so too is waking an actuality; [413a] and as sight and the potentiality of a tool are, so too is the soul. The body is a being in potentiality. But just as an eye is a pupil plus sight, so an animal is the soul plus the body.

Therefore, that the soul is not separable from the body, or some parts of it if it naturally has parts, is not unclear. For the actuality of some parts belong to the parts of the body themselves. Even so, nothing hinders some parts from being separable, because of their not being the actualities of any body.

It is still unclear, however, whether the soul is the actuality of the body in the way that a sailor is of a ship.

So let the soul be defined in outline and sketched out.

DA II 2 T

[413a11] Because what is sure and better known as conforming to reason comes to be from what is unsure but more obvious, one must proceed anew in this way concerning the soul. For it is not only necessary that a defining account make clear *that something is*, which is what most definitions state, but it must also contain and make manifest the cause. As things are, statements of definitions are like conclusions. For example: "what is squaring? It is an equilateral rectangle being equal to an oblong figure." But this sort of definition is a statement of the conclusion. The one who states that squaring is the discovery of a mean states the cause of the matter.

We say, then, taking up the beginning of the inquiry, that what is ensouled is distinguished from what is not ensouled by living. But living is spoken of in several ways. And should even one of these belong to something, we say that it is alive: reason, perception, motion and rest with respect to place, and further the motion attendant upon nourishment, decay and growth.

For this reason, even plants, all of them, seem to be alive, since they seem to have in themselves a potentiality and the sort of principle through which they grow and decay, in opposite directions. For it is not the case that they grow upward but not downward; rather they grow in both directions and in all ways, those, that is, which are always nourished and continue to live as long as they are able to receive nourishment.

This capacity can be separated from the others, but among mortals the others cannot be separated from this. This is clear in the case of plants. For no other capacity of soul belongs to them.

[413b] Living, then, belongs to what lives because of this principle, but something is an animal primarily because of perception. For even those things which do not move or change place, but which have perception, we call animals and not merely living things. The primary form of perception which belongs to all animals is touch. But just as the nutritive capacity can be separated from touch and from the whole of perception, so touch can be separated from these other senses. By nutritive we mean the sort of part belonging to the soul of which even plants partake. But all animals are seen to have the sense of touch. The reason why each of these two things turns out to be the case we shall state later.

For now let just this much be said: the soul is the principle of the capacities mentioned and is delimited by them, namely, nutrition, perception, thought, and motion. It is not difficult to see whether each of these is a soul or a part of a soul, and if a part, whether in such a way as to be separate in account alone or also in place. But in some cases there is a difficulty. For just as in the case of plants, some, when divided, evidently go on living even when separated from one another, there being one soul in actuality in each plant, but many in potentiality, so we see this occurring in other characteristics of the soul in the case of insects cut into two. For each of the parts has perception and motion with respect to place, but if perception, then also imagination and desire. For wherever there is perception, there is also both pain and pleasure; and wherever these are, of necessity is appetite.

But concerning reason and the capacity for contemplation nothing is yet clear. Still, reason seems to be a different kind of soul, and it alone admits of being separated, in the way the everlasting is from the perishable.

It is clear from these things, though, that the remaining parts of the soul are not separable, just as some assert. That they differ in account is, however, clear; for what it is to be the perceptive faculty is different from what it is to be the faculty of belief, since perceiving differs from believing, and so on for each of the other faculties mentioned.

Further, all of these faculties belong to some animals, and some of them to others, and only one to still others. And this will provide a differentiation among animals. **[414a]** It is necessary to investigate the reason why later. Nearly the same thing holds for the senses: some animals have them all, others have some of them, and still others have one, the most necessary, touch.

That by which we live and perceive is spoken of in two ways, just as is that by which we know. We speak in one case of knowledge and in the other of the soul, because we maintain that we know by means of each of these. Likewise we are healthy in one way by health and in another way by some part of, or the whole of, the body.[1] On one of these ways of speaking, knowledge and health is each a shape and a sort of form, an account, and so as to be an actuality of what is capable of receiving them, in the one case of what is capable of knowledge and in the other of what is capable of health. For the actuality of productive things seems to reside in what is affected and is disposed to receive it.

Consequently, the soul is in a primary way that by which we live and perceive and think, so that it will be a sort of account and a form, rather than matter and a substrate. For substance is spoken of in three ways, just as we said, one of which is the form, another the matter, and another what comes from both; and of these the matter is potentiality and the form actuality. Since what comes from both is an ensouled thing, the body is not the actuality of the soul, but the soul is the actuality of some body.

For this reason, those to whom it seems that the soul is neither without the body nor a kind of body understand the situation rightly. For it is not a body, but is something belonging to a body; and because of this it is present in a body, and in a body of a certain sort. The situation is not as our predecessors supposed when they fitted the soul into the body without additionally specifying in which body or in which sort, even though whatever just happens to show up does not receive whatever it happens upon.

In conformity with reason, it occurs this way: the actuality of each thing comes about naturally in what has it in potentiality, that is, in its appropriate matter. It is clear from these considerations, then, that the soul is a kind of actuality and an account of what has a potentiality to be of this sort.

DA II 3 T

[414a29] Among the capacities of the soul, some things have all of them, just as we have said, others some of them, and still others only one. The capacities we mentioned were: nutrition, perception, desire, motion with respect to place, and understanding. Plants have only nutrition; other things have both this and perception. **[414b]** But if perception, then also desire: desire is appetite, spirit, and wish. And all animals have at least one kind of perception, touch. And that which has perception also has both pleasure and pain, as well as both the pleasurable and the painful; and what has these also has appetite, since appetite is a desire for what is pleasurable. They have, additionally, perception of nourishment; for touch is perception of nourishment, since all living things are nourished by dry, wet, hot, and cold things, and touch is perception of these. Touch is perception of other sensibles co-incidentally. For neither sound nor color nor smell contributes anything to nourishment, whereas flavor is among the objects of touch.

It will be necessary to be completely clear about these matters later. For now let this much be said: those living things which have touch also have desire.

But regarding imagination things are not clear. One must inquire into that later.

In addition to these things, some things have a capacity to move with respect to place; and others have reasoning and understanding, for example humans, and anything there may be of a similar sort or more elevated.

It is clear, then, that there could in the same way be one account belonging to both soul and figure. For in the one case a figure is nothing beyond a triangle and the other figures following in a series, and in the other a soul is nothing beyond the things mentioned. There could, however, be a common account covering figures which fits them all, though it will be unique to none; and the same holds for the souls mentioned. For this reason, it is ludicrous to seek a common account covering these cases, or other cases, an account which is not distinctive to anything which exists, which does not correspond to any proper and indivisible species, all the while neglecting what is distinctive. Consequently, one must ask individually what the soul of each is, for example, what the soul of a plant is, and what the soul of a man or a beast is.

What holds in the case of the soul is similar to what holds concerning figures: for both figures and the ensouled, what is prior is present in potentiality in what follows in the series, for example, the triangle in the square, and the nutritive in the perceptive. We must investigate the reason why they are thus in a series. [415a] For the perceptive faculty is not without the nutritive, though the nutritive faculty is separated from the perceptive in plants. Again, without touch, none of the other senses is present, though touch is present without the others; for many animals have neither sight nor hearing nor a sense of smell. Also, among things capable of perceiving, some have motion in respect of place, while others do not. Lastly, and most rarely, some have reasoning and understanding. Among perishable things, those with reasoning also have all the remaining capacities, though it is not the case that those with each of the remaining capacities also have reasoning. Rather, some do not have imagination, while others live by this alone. A different account will deal with the theoretical mind.

It is clear, therefore, that the account of each of the capacities will also be the most appropriate account concerning the soul.

[415a14] It is necessary for anyone who is going to conduct an inquiry into these things to grasp what each of them is, and then to investigate in the same way things closest to them as well as other features. And if one ought to say what each of these is, for example, what the intellective or perceptive or nutritive faculty is, then one should say beforehand what thinking is and what perceiving is, since actualities and actions are prior in account to potentialities. But if this is so, and their corresponding objects are prior to them,[2] it would for the same reason be necessary to offer determinations concerning, for instance, the objects of nutrition, perception, and thinking.

The result is that one must speak first of nutrition and generation; for the nutritive soul also belongs to the others as well. This is both the first and most common capacity of the soul, in virtue of which living belongs to all things alive, a capacity whose functions are generating and making use of nutrition. For the most natural among the functions belonging to living things, at least those which are complete and neither deformed nor spontaneously generated, is this: to make another such as itself, an animal an animal and a plant a plant, so that it may, insofar as it is able, partake of the everlasting and the divine. [415b] For that is what everything desires, and for the sake of that everything does whatever it does in accordance with nature. (That for the sake of which is spoken of in two ways: that on account of which and that for which.) Since, then, these things are incapable of sharing in the everlasting and the divine by existing continuously (because among perishable things nothing can remain the same and one in number), each has a share insofar as it is able to partake in this, some more and some less, and remains not as itself but such as it is, not one in number but one in form.

The soul is the cause and principle of the living body. As these things are spoken of in many ways, so the soul is spoken of as a cause in three of the ways delineated: the soul is a cause as the source of motion, as that for the sake of which, and as the substance of ensouled bodies.

That it is a cause as substance is clear: substance is the cause of being for all things, and for living things, living is being, while the cause and principle of living is the soul. Further, actuality is the formula of that which is potentially.

It is evident that the soul is a cause as that for the sake of which: just as the mind acts for the sake of something, in the same way nature does so as well; and this is its end. And in living beings, the soul is naturally such a thing.[3] For all ensouled natural bodies are organs of the soul—as it is for the bodies of animals so is it for the bodies of plants—since they are for the sake of the soul.[4] That for the sake of which is spoken of in two ways: that on account of which and that for which.

Moreover, the soul is also that from which motion in respect of place first arises, though not all living things have this capacity. There are also alteration and growth in virtue of the soul; for perception seems to be a sort of alteration, and nothing lacking a soul perceives. The same holds for both growth and decay; for nothing which is not nourished decays or grows, and nothing is nourished which lacks a share of life.

Empedocles was not right when he added that downward growth occurs for plants, when they take root, **[416a]** because earth is naturally borne in this direction, and upward growth occurs because fire, in like manner, naturally moves upward. Nor even does he understand up and down rightly. For up and down are not the same for individuals as for the universe. Rather, as the head is in animals, so the roots are in plants, if, that is, it is in terms of their functions that one ought to say that organs are the same or different. Moreover, what is it that holds fire and earth together, even though they are borne in opposite directions? For they will be torn apart if there is nothing which hinders them. If there is something, however, that will be the soul, i.e. the cause of growing and being nourished.

Fire's nature seems to some to be without qualification the cause of nourishment and growth, since among bodies fire alone is evidently something which is nourished and grows. On this basis, one might suppose fire to be what accomplishes this in plants and animals. Fire, however, is a sort of co-cause, and most surely not a cause without qualification; the cause is, rather, the soul. For fire's growth carries on without limit, so long as there is something combustible. By contrast, for all things naturally constituted, there is a limit and a formula of both size and growth. These things belong to the soul, and not to fire, i.e., to the formula rather than to the matter.

Since the same capacity of soul is both nutritive and generative, it is necessary to determine what concerns nutrition first; for it is in virtue of this function that it is marked off from the other capacities. Nutrition seems to proceed contrary to contrary, though not in every single case, but in only those contraries which have not only generation from one another, but also growth. For many things are generated from one another, but not all of them are quantities, as, for example, the healthy from the sick. Nor even among growing contraries does it appear that nourishment is reciprocally one from the other: whereas water is nourishment for fire, fire does not nourish water.

Now then, in the case of simple bodies it seems most true that the one is nourishment and the other nourished. Yet there is a difficulty. Some say that like is nourished by like just as like grows by like. By contrast, as we said earlier, it seems to others that contrary is nourished by contrary, since like is unaffected by like,

while nourishment requires change, i.e., to be digested, and every change is into its opposite or an intermediary. Further, nourishment is somehow affected by what is nourished, but what is nourished is not affected by nourishment, **[416b]** just as a carpenter is not affected by the matter, but it is affected by him. The carpenter changes only from idleness into activity.

It makes a difference whether nourishment is what is added last or first. If it is both, in one instance undigested and in the other digested, it would be possible to call either nutrition. For insofar as it is undigested, contrary is nourished by contrary; and insofar as it is digested, like is nourished by like. As a consequence, evidently each side will be in one way correct and in another way incorrect.

Since nothing which does not partake of life is nourished, what is nourished would be the ensouled body, insofar as it is ensouled, with the result that nourishment is relative—and not coincidentally—to what is ensouled.

Still, there is a difference between being nourishment and being able to augment something. For insofar as an ensouled thing is a particular quantity, something is capable of augmenting it, while insofar as it is some this and a substance, something is nourishment for it. For what is ensouled preserves its substance and exists as long as it is nourished; and it is capable of generating not the very thing which is nourished, but rather something nourished as it is, since its substance already exists and nothing generates itself, but rather preserves itself.

Consequently, this principle of soul is a capacity of the sort which preserves the thing which has it, as the sort of thing it is, while nutrition equips it to become active. Hence, whatever has been deprived of nutrition cannot exist.[5]

Since it is right to call each thing by the name derived from its end, and here the end is to generate another such as itself, it would be right to call this primary soul *generative* of another such as itself.

Since these are three things—what is nourished, that by which it is nourished, and what nourishes—so too are these three: the primary soul, as that which nourishes; the body which has the primary soul, as that which is nourished; and that by which it is nourished, the nourishment. But that by which something is nourished is two-fold, just as that by which one steers is both the hand and the rudder, the one both initiating movement and itself moving, and the other merely moving. It is necessary that all nourishment be able to be digested; and what is hot effects digestion. For this reason, everything ensouled contains heat.

So, it has been said in outline what nourishment is. It will be necessary to make it completely clear in the appropriate treatise.

NOTES

This chapter originally appeared as Book II 412ᵃ 13–416ᵇ in Aristotle, *De anima*, with translation and commentary by Christopher Shields, Oxford: Oxford University Press, forthcoming.

ENDNOTES

1 Omitting hô(i) w/Bywater at 414a7.

2 Omitting dei tetheôrêkenai w/W.

3 Reading zôsin in 415b17 w/V.

4 Reading empsucha in 415b18 w/Torstrik, relying ultimately on Fᴾ.

5 This translation, following Torstrick, transposes 416b23–25 to 416b20.

2 · Treatise on Man

RENÉ DESCARTES

These men will be composed, as we are, of a soul and a body.[1] First I must describe the body on its own; then the soul, again on its own; and finally I must show how these two natures would have to be joined and united in order to constitute men who resemble us.

I suppose the body to be nothing but a statue or machine made of earth,[2] which God forms with the explicit intention of making it as much is possible like us. Thus God not only gives it externally the colours and shapes of all the parts of our bodies, but also places inside it all the parts required to make it walk, eat, breathe, and indeed to imitate all those of our functions which can be imagined to proceed from matter and to depend solely on the disposition of our organs.

We see clocks, artificial fountains, mills, and other such machines which, although only man-made, have the power to move of their own accord in many different ways. But I am supposing this machine to be made by the hands of God, and so I think you may reasonably think it capable of a greater variety of movements than I could possibly imagine in it, and of exhibiting more artistry than I could possibly ascribe to it.

Now I shall not pause to describe the bones, nerves, muscles, veins, arteries, stomach, liver, spleen, heart, brain, or any of the various other parts from which this machine must be composed. For I am supposing that they are entirely like the parts of our own bodies which have the same names, and I assume that if you do not already have sufficient first-hand knowledge of them, you can get a learned anatomist to show them to you—at any rate, those which are large enough to be seen with the naked eye. As for the parts which are too small to be seen, I can inform you about them more easily and clearly by speaking of the movements which depend on them. Thus I need only give

an orderly account of these movements in order to tell you which of our functions they represent . . .[3]

The parts of the blood which penetrate as far as the brain, serve not only to nourish and sustain its substance, but also and primarily to produce in it a certain very fine[4] wind, or rather a very lively and pure flame, which is called the *animal spirits*. For it must be noted that the arteries which carry blood to the brain from the heart, after dividing into countless tiny branches which make up the minute tissues that are stretched like tapestries at the bottom of the cavities of the brain, come together again around a certain little *gland*[5] situated near the middle of the substance of the brain, right at the entrance to its cavities. The arteries in this region have a great many little holes through which the finer parts of the blood can flow into this gland . . . These parts of the blood, without any preparation or alteration except for their separation from the coarser parts and their retention of the extreme rapidity which the heat of the heart has given them, cease to have the form of blood, and are called the "animal spirits".

Now in the same proportion as the animal spirits enter the cavities of the brain, they pass from there into the pores of its substance, and from these pores into the nerves. And depending on the varying amounts which enter (or merely tend to enter) some nerves more than others, the spirits have the power to change the shape of the muscles in which the nerves are embedded, and by this means to move all the limbs. Similarly, you may have observed in the grottos and fountains in the royal gardens that the mere force with which the water is driven as it emerges from its source is sufficient to move various machines, and even to make them play certain instruments or utter certain

The Nature of Life, ed. M. A. Bedau and C. E. Cleland. Published by Cambridge University Press. © M. A. Bedau and C. E. Cleland 2010.

words depending on the various arrangements of the pipes through which the water is conducted.

Indeed, one may compare the nerves of the machine I am describing with the pipes in the works of these fountains, its muscles and tendons with the various devices and springs which serve to set them in motion, its animal spirits with the water which drives them, the heart with the source of the water, and the cavities of the brain with the storage tanks. Moreover, breathing and other such activities which are normal and natural to this machine, and which depend on the flow of the spirits, are like the movements of a clock or mill, which the normal flow of water can render continuous. External objects, which by their mere presence stimulate its sense organs and thereby cause them to move in many different ways depending on how the parts of its brain are disposed, are like visitors who enter the grottos of these fountains and unwittingly cause the movements which take place before their eyes. For they cannot enter without stepping on certain tiles which are so arranged that if, for example, they approach a Diana who is bathing they will cause her to hide in the reeds, and if they move forward to pursue her they will cause a Neptune to advance and threaten them with his trident; or if they go in another direction they will cause a sea-monster to emerge and spew water onto their faces; or other such things according to the whim of the engineers who made the fountains. And finally, when a *rational soul* is present in this machine it will have its principal seat in the brain, and reside there like the fountain-keeper who must be stationed at the tanks to which the fountain's pipes return if he wants to produce, or prevent, or change their movements in some way . . .[6]

Next, to understand how the external objects which strike the sense organs can prompt this machine to move its limbs in numerous different ways, you should consider that the tiny fibres (which, as I have already told you, come from the innermost region of its brain and compose the marrow of the nerves) are so arranged in each part of the machine that serves as the organ of some sense that they can easily be moved by the objects of that sense. And when they are moved, with however little force, they simultaneously pull the parts of the brain from which they come, and thereby open the entrances to certain pores in the internal surface of the brain. Through these pores the animal spirits in the cavities of the brain immediately begin to

Fig. 2.1.

make their way into the nerves and so to the muscles which serve to cause movements in the machine quite similar to those we are naturally prompted to make when our senses are affected in the same way.

Thus, for example (in Fig. 2.1), if fire A is close to foot B, the tiny parts of this fire (which, as you know, move about very rapidly) have the power also to move the area of skin which they touch. In this way they pull the tiny fibre *cc* which you see attached to it, and simultaneously open the entrance to the pore *de*, located opposite the point where this fibre terminates—just as when you pull one end of a string, you cause a bell hanging at the other end to ring at the same time.

When the entrance to the pore or small tube *de* is opened in this way, the animal spirits from cavity F enter and are carried through it—some to muscles which serve to pull the foot away from the fire, some to muscles which turn the eyes and head to look at it, and some to muscles which make the hands move and the whole body turn in order to protect it . . .

Now I maintain that when God unites a rational soul to this machine (in a way that I intend to explain later) he will place its principal seat in the brain, and will make its nature such that the soul will have different sensations corresponding to the different ways in which the entrances to the pores in the internal surface of the brain are opened by means of the nerves.

Suppose, firstly, that the tiny fibres which make up the marrow of the nerves are pulled with such force that they are broken and separated from the part of the body to which they are joined, with the result that the structure of the whole machine becomes somehow less perfect. Being pulled in this way, the fibres cause a movement in the brain which gives occasion for the soul (whose place of residence must remain constant) to have the sensation of *pain*.

Now suppose the fibres are pulled with a force almost as great as the one just mentioned, but without their being broken or separated from the parts to which they are attached. Then they will cause a movement in the brain which, testifying to the good condition of the other parts of the body, will give the soul occasion to feel a certain bodily pleasure which we call "*titillation*". This, as you see, is very close to pain in respect of its cause but quite opposite in its effect.

Again, if many of these tiny fibres are pulled equally and all together, they will make the soul perceive that the surface of the body touching the limb where they terminate is *smooth*; and if the fibres are pulled unequally they will make the soul feel the surface to be uneven and *rough*.

And if the fibres are disturbed only slightly and separately from one another, as they constantly are by the heat which the heart transmits to the other parts of the body, the soul will have no more sensation of this than of any other normal function of the body. But if this stimulation is increased or decreased by some unusual cause, its increase will make the soul have a sensation of *heat*, and its decrease a sensation of *cold*. Finally, according to the various other ways in which they are stimulated, the fibres will cause the soul to perceive all the other qualities belonging to touch in general, such as *moisture, dryness, weight* and the like.

It must be observed, however, that despite the extreme thinness and mobility of these fibres, they are not thin and mobile enough to transmit to the brain all the more subtle motions that take place in nature. In fact the slightest motions they transmit are ones involving the coarser parts of terrestrial bodies. And even among these bodies there may be some

whose parts, although rather coarse, can slide against the fibres so gently that they compress them or cut right through them without their action passing to the brain. In just the same way there are certain drugs which have the power to numb or even destroy the parts of the body to which they are applied without causing us to have any sensation of them at all ...[7]

It is time for me to begin to explain how the animal spirits make their way through the cavities and pores of the brain of this machine, and which of the machine's functions depend on these spirits.

If you have ever had the curiosity to examine the organs in our churches, you know how the bellows push the air into certain receptacles (which are called, presumably for this reason, wind-chests). And you know how the air passes from there into one or other of the pipes, depending on the different ways in which the organist moves his fingers on the keyboard. You can think of our machine's heart and arteries, which push the animal spirits into the cavities of its brain, as being like the bellows of an organ, which push air into the wind-chests; and you can think of external objects, which stimulate certain nerves and cause spirits contained in the cavities to pass into some of the pores, as being like the fingers of the organist, which press certain keys and cause the air to pass from the wind-chests into certain pipes. Now the harmony of an organ does not depend on the externally visible arrangement of the pipes or on the shape of the wind-chests or other parts. It depends solely on three factors: the air which comes from the bellows, the pipes which make the sound, and the distribution of the air in the pipes. In just the same way, I would point out, the functions we are concerned with here do not depend at all on the external shape of the visible parts which anatomists distinguish in the substance of the brain, or on the shape of the brain's cavities, but solely on three factors: the spirits which come from the heart, the pores of the brain through which they pass, and the way in which the spirits are distributed in these pores. Thus my sole task here is to give an orderly account of the most important features of these three factors ...[8]

Now, the substance of the brain being soft and pliant, its cavities would be very narrow and almost all closed (as they appear in the brain of a corpse) if no spirits entered them. But the source which produces these spirits is usually so abundant that they enter these cavities in sufficient quantity to have the force

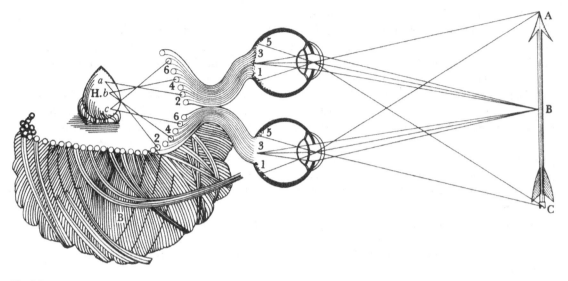

Fig. 2.2.

to push out against the surrounding matter and make it expand, thus tightening all the tiny nerve-fibres which come from it (in the way that a moderate wind can inflate the sails of a ship and tighten all the ropes to which the sails are attached). It follows that at such times the machine is disposed to respond to all the actions of the spirits, and hence it represents the body of a man who is *awake*. Or at least the spirits have enough force to push against some parts of the surrounding matter in this way, and so make it tight, while the other parts remain free and relaxed (as happens in parts of a sail when the wind is a little too weak to fill it). At such times the machine represents the body of a man who is *asleep* and who has *various dreams* as he sleeps . . .

But before I speak in greater detail about *sleep* and *dreams*, I must have you consider the most noteworthy events that take place in the brain during the time of waking: namely, how ideas of objects are formed in the place assigned to the *imagination* and to the "*common*" *sense*, how the ideas are retained in the *memory*, and how they cause *movement in all the parts of the body* . . .

In order . . . to see clearly how ideas are formed of the objects which strike the senses, observe in this diagram (Fig. 2.2) the tiny fibres 12, 34, 56 and the like, which make up the optic nerve and stretch from the back of the eye at 1, 3, 5 to the internal surface of

the brain at 2, 4, 6. Now assume that these fibres are so arranged that if the rays coming, for example, from point A of the object happen to press upon the back of the eye at point 1, they pull the whole of fibre 12 and enlarge the opening of the tiny tube marked 2. In the same way, the rays which come from point B enlarge the opening of the tiny tube 4, and likewise for the others. We have already described how, depending on the different ways in which the points 1, 3, 5 are pressed by these rays, a figure is traced on the back of the eye corresponding to that of the object ABC. Similarly, it is obvious that, depending on the different ways in which the tiny tubes 2, 4, 6 are opened by the fibres 12, 34, 56, etc., a corresponding figure must also be traced on the internal surface of the brain.

Suppose next that the spirits which tend to enter each of the tiny tubes 2, 4, 6, and the like, do not come indifferently from all points on the surface of gland H, but only from certain of these points: those coming from point *a* on this surface, for example, tend to enter tube 2, those from points *b* and *c* tend to enter tubes 4 and 6, and likewise for the others. As a result, at the same instant that the openings to these tubes expand, the spirits begin to leave the corresponding points on the gland more freely and more rapidly than they did previously. Thus, just as figure corresponding to that of the object ABC is traced on the internal surface of the brain according to the different ways in which tubes 2, 4, 6 are opened, likewise that figure is traced on the surface of the gland according to the ways in which the spirits leave from points *a, b, c*.

And note that by "figures" I mean not only things which somehow represent the position of the edges and surfaces of objects, but also anything which, as I said above, can give the soul occasion to perceive movement, size, distance, colours, sounds, smells and other such qualities. And I also include anything that can make the soul feel pleasure, pain, hunger, thirst, joy, sadness and other such passions. For it is easy to understand that tube 2, for example, may be opened in different ways—in one way by the action which I said causes sensory perception of the colour red, or of tickling, and in another way by the action which I said causes sensory perception of the colour white, or of pain; and the spirits which leave from point *a* will tend to move towards this tube in a different manner according to differences in its manner of opening, and likewise for the others.

Now among these figures, it is not those imprinted on the external sense organs, or on the internal surface of the brain, which should be taken to be ideas—but only those which are traced in the spirits on the surface of the gland H (*where the seat of the imagination and the "common" sense is located*).[9] That is to say, it is only the latter figures which should be taken to be the forms or images which the rational soul united to this machine will consider directly when it imagines some object or perceives it by the senses.

And note that I say "imagines or perceives, by the senses". For I wish to apply the term "idea" generally to all the impressions which the spirits can receive as they leave gland H. These are to be attributed to the "common" sense when they depend on the presence of objects; but they may also proceed from many other causes (as I shall explain later), and they should then be attributed to the imagination.

Here I could add something about how the traces of these ideas pass through the arteries to the heart, and thus radiate through all the blood; and about how certain actions of a mother may sometimes even cause such traces to be imprinted on the limbs of the child being formed in her womb. But I shall content myself with telling you more about how the traces are imprinted on the internal part of the brain (marked B on Fig. 2.2) which is the seat of the *memory*.

To this end, suppose that after the spirits leaving gland H have received the impression of some idea, they pass through tubes 2, 4, 6 and the like, into the pores or gaps lying between the tiny fibres which make up part B of the brain. And suppose that the spirits are strong enough to enlarge these gaps somewhat, and to bend and arrange in various ways any fibres they encounter, according to the various ways in which the spirits are moving and the different openings of the tubes into which they pass. Thus they also trace figures in these gaps, which correspond to those of the objects. At first they do this less easily and perfectly than they do on gland H, but gradually they do it better and better, as their action becomes stronger and lasts longer, or is repeated more often. That is why these figures are no longer so easily erased, and why they are preserved in such a way that the ideas which were previously on the gland can be formed again long afterwards without requiring the presence of the objects to which they correspond. And this is what *memory* consists in . . .[10]

But before going on to describe the rational soul, I should like you once again to give a little thought to everything I have said about this machine. Consider, in the first place, that I have supposed in it only organs and mechanisms of such a type that you may well believe very similar ones to be present both in us and in many animals which lack reason as well. Regarding those which can be seen clearly with the naked eye, the anatomists have already observed them all. And as for what I have said about the way in which the arteries carry the spirits into the head, and about the difference between the internal surface of the brain and its central substance, the anatomists will, if they simply make closer observations, be able to see sufficient indications of this to allay any doubts about these matters too. Nor will they be able to have doubts about the tiny doors or valves which I have placed in the nerves where they enter each muscle, if they take care to note that nature generally has formed such valves at all the places in our bodies where some matter regularly goes in and may tend to come out, as at the entrances to the heart, gall-bladder, throat, and large intestine, and at the main divisions of all the veins. Again, regarding the brain, they will not be able to imagine anything more plausible than that it is composed of many tiny fibres variously interlaced; for, in view of the fact that every type of skin and flesh appears to be similarly composed of many fibres or threads, and that the same thing is observed in all plants, such fibrous composition is apparently a common property of all bodies that can grow and be nourished by the union and joining together of the minute parts of other bodies. Finally, as for the rest of the things I have assumed which cannot be perceived by any sense, they are all so simple and commonplace, and also so few in number, that if you compare them with the diverse composition and marvellous artistry which is evident in the structure of the visible organs, you will have more reason to think I have omitted many that are in us than to think I have introduced any that are not. And, knowing that nature always acts by the simplest and easiest means, you will perhaps conclude that it is possible to find some which are more similar to the ones she in fact uses than to those proposed here.

I should like you to consider, after this, all the functions I have ascribed to this machine—such as the digestion of food, the beating of the heart and arteries, the nourishment and growth of the limbs, respiration, waking and sleeping, the reception by the external sense organs of light, sounds, smells, tastes, heat and other such qualities, the imprinting of the ideas of these qualities in the organ of the "common" sense and the imagination, the retention or stamping of these ideas in the memory, the internal movements of the appetites and passions, and finally the external movements of all the limbs (movements which are so appropriate not only to the actions of objects presented to the senses, but also to the passions and the impressions found in the memory, that they imitate perfectly the movements of a real man). I should like you to consider that these functions follow from the mere arrangement of the machine's organs every bit as naturally as the movements of a clock or other automaton follow from the arrangement of its counter-weights and wheels. In order to explain these functions, then, it is not necessary to conceive of this machine as having any vegetative or sensitive soul or other principle of movement and life, apart from its blood and its spirits, which are agitated by the heat of the fire burning continuously in its heart—a fire which has the same nature as all the fires that occur in inanimate bodies.

NOTES

This chapter originally appeared in John Cottingham, Robert Stoothoff and Dugald Murdoch (trans.), *The philosophical writing of Descartes*, vol. I, pp. 99–108, Cambridge, UK: Cambridge University Press, 1985.

ENDNOTES

1 By 'these men', Descartes means the fictional men he introduced in an earlier (lost) part of the work. Their description is intended to cast light on the nature of real men in the same way that the description of a 'new world' in *The World*, ch. 6, is intended to cast light on the real world; See also *Discourse*, part 5.

2 By 'earth' Descartes means the third 'element', which he had discussed in *The World*, ch. 5.

3 There follows a description of digestion, the formation and circulation of the blood, the action of the heart and respiration. Cf. *Discourse*, part 5, and *Passions*, Part 1, art. 3–10, and *Description of the Human Body*. For an English version of material omitted here arid below, see *Descartes: Treatise on Man*, tr. T. S. Hall (Cambridge: Harvard U.P., 1972).

4 Fr. *subtil*, by which Descartes means 'composed of very small, fast-moving particles'.

5 The pineal gland, which Descartes later identifies as the seat of the imagination and the 'common' sense, p. 106 below). See also *Passions* where the gland is identified as the seat of the soul.

6 There follows a description of the way in which the animal spirits bring about muscular movements, breathing, swallowing, etc. See *Passions*, Part I.

7 There follows an account of the other external senses (taste, smell, hearing and sight) and of internal sensations (hunger, thirst, joy and sadness). For Descartes' theory of vision, see *Optics*, for the other external senses, see *Principles*, Part 4, art. 192–4, and for the internal sensations, see *Passions, passim*.

8 There follows a description of the animal spirits and how their state is affected by digestion, respiration, and other bodily functions; of the pores of the brain; and of the movement of the spirits through the pores.

9 See note 3, p. 100 above.

10 There follows an account of the way in which the animal spirits form ideas on the surface of the pineal gland, and produce bodily movements like those of real men, despite the absence of any soul. See *Passions*, Part 1, art. 13–16, 21–4.

3 · Critique of the teleological power of judgment (selections)

IMMANUEL KANT

ON THE OBJECTIVE PURPOSIVENESS WHICH IS MERELY FORMAL, IN DISTINCTION TO THAT WHICH IS MATERIAL

All geometrical figures that are drawn in accordance with a principle display a manifold and often admired objective purposiveness, namely that of serviceability for the solution of many problems in accordance with a single principle, and indeed of each of them in infinitely many different ways. The purposiveness here is evidently objective and intellectual, not, however, merely subjective and aesthetic. For it expresses the suitability of the figure for the generation of many shapes aimed at purposes, and is cognized through reason. But the purposiveness still does not make the concept of the object itself possible, i.e., it is not regarded as possible merely with respect to this use.

In such a simple figure as the circle there lies the basis for the solution of a host of problems, for each of which by itself much preparation would be required, and which as it were arises from this figure itself as one of its many splendid properties. If, e.g., the problem is to construct a triangle from a given baseline and the angle opposite to it, then it is indeterminate, i.e., it can be solved in infinitely many ways. But the circle comprehends them all, as the geometrical locus for all triangles that satisfy this condition. Or two lines are supposed to intersect in such a way that the rectangle constructed from the two parts of the one is equal to the rectangle from the two parts of the other: the solution of this problem looks as if it will be very difficult. But all the lines that intersect within the circle the circumference of which bounds each of them are of themselves divided into this proportion. The other curves yield in turn other purposive solutions that were not thought of at all in the rule that constitutes their construction. All conic sections, by themselves and in comparison with one another, are fruitful in principles for the solution of a host of possible problems, as simple as the definition is which determines their concept.—It is a true joy to see the eagerness with which the ancient geometers investigated the properties of such lines without being distracted by the question of limited minds: for what is this knowledge useful?, e.g., that of the parabola, without knowing the law of terrestrial gravitation, which would have given them its application to the trajectory of heavy bodies (whose gravitational direction in their motion can be seen as parallel); or of the ellipse, without suspecting that there is also gravity in heavenly bodies, and without knowing its law at different distances from points of attraction, which makes them describe these lines in free movement. While these geometers, unbeknownst to themselves, were working for posterity, they delighted in a purposiveness in the essence of things, which they could yet exhibit fully *a priori* in its necessity. Plato, himself a master of this science, was led by such an original constitution of things, in the discovery of which we can dispense with all experience, and by the mental capacity for drawing the harmony of things out of their supersensible principle (to which pertain the properties of numbers, with which the mind plays in music), to the enthusiasm that elevated him beyond the concepts of experience to ideas, which seemed to him explicable only by means of an intellectual communion with the origin of all things.[1] No wonder that he

---

The Nature of Life, ed. M. A. Bedau and C. E. Cleland. Published by Cambridge University Press. © M. A. Bedau and C. E. Cleland 2010.

banned from his school those who were ignorant of geometry, for he thought he could derive that which Anaxagoras inferred from objects of experience from the pure intuition internal to the human mind.[2] For in the necessity of that which is purposive and so constituted as if it were intentionally arranged for our use, but which nevertheless seems to pertain originally to the essence of things, without any regard to our use, lies the ground for the great admiration of nature, not outside of us so much as in our own reason; in which case it is surely excusable that through misunderstanding this admiration gradually rose to enthusiasm.

This intellectual purposiveness, however, although it is objective (not, like the aesthetic, subjective), can nevertheless be conceived, as far as its possibility is concerned, as merely formal (not real), i.e., as purposiveness that is not grounded in a purpose, for which teleology would be necessary, but only in general. The figure of a circle is an intuition that can be determined by the understanding in accordance with a principle; the unity of this principle, which I assume arbitrarily and, as a concept, make into a ground, when applied to a form of intuition (to space) which is to be found in me merely as representation and indeed *a priori*, makes comprehensible the unity of many rules resulting from the construction of that concept, which are purposive in many respects, without an *end* or any other ground having to be the basis of this purposiveness. This is different from finding order and regularity in a sum of *things* outside of me enclosed in certain bounds, e.g., among the trees, flower beds and paths in a garden, which I cannot hope to deduce *a priori* from my demarcation of a space in accordance with an arbitrary rule: for these are existing things, which must be given empirically in order to be cognized, and not a mere representation in me determined in accordance with an *a priori* principle. Hence the latter (empirical) purposiveness, as real, is dependent on the concept of an end.

But the reason for the admiration of a purposiveness perceived in the essence of things (insofar as their concept can be constructed) can be quite easily and indeed quite rightly understood. The many rules, the unity of which (from a principle) arouses this admiration, are one and all synthetic, and do not follow from a concept of the object, e.g., from that of a circle, but need this object to be given in intuition. But it thereby comes to seem as if this unity empirically possesses an external ground, distinct from our power

of representation, for its rules, and thus as if the correspondence of the object with the need for rules, which is characteristic of the understanding, is in itself contingent, hence possible only by means of an end expressly aimed at it. Now, of course, this very harmony, since it is, in spite of all this purposiveness, cognized not empirically but *a priori*, should bring it home to us that space, by the determination of which (by means of the imagination, in accordance with a concept) the object is alone possible, is not a property of the object outside of me, but merely a kind of representation in me, and thus that I introduce the purposiveness into the figure that I draw in accord with a concept, i.e., into my own way of representing that which is given to me externally, whatever it may be in itself, thus that I am not instructed empirically about this purposiveness by the object, and consequently do not need for this purposiveness any particular end outside of me in the object. But since this reflection already requires a critical use of reason, and hence cannot be immediately contained in the judging of the object in accordance with its properties, the latter gives me immediately nothing other than the unification of heterogeneous rules (united even in that which is diverse in them) in one principle, which, without needing a particular ground lying *a priori* beyond my concept and, in general, my representation, can nevertheless be cognized by me *a priori* as truthful. Now astonishment is a mental shock at the incompatibility of a representation and the rule that is given through it with the principles already grounded in the mind, which thus produces a doubt as to whether one has seen or judged correctly; but admiration is an astonishment that continually recurs despite the disappearance of this doubt. The latter is consequently an entirely natural effect of that purposiveness observed in the essence of things (as appearances), which also cannot be criticized insofar as the compatibility of that form of sensible intuition (which is called space) with the faculty of concepts (the understanding) is not only inexplicable for us insofar as it is precisely thus and not otherwise, but also enlarges the mind, allowing it, as it were, to suspect something lying beyond those sensible representations, in which, although unknown to us, the ultimate ground of that accord could be found. Indeed, it is not necessary for us to know this if it is merely a matter of the formal purposiveness of our *a priori* representations; but even just being compelled to look in that direction fills us with admiration for the object that forces us to do so.

It is customary to call the properties of geometrical shapes as well as of numbers that have been mentioned beauty, on account of a certain *a priori* purposiveness, not expected from the simplicity of their construction, for all sorts of cognitive use, and to speak of this or that beautiful property of, e.g., a circle, which is discovered in this way or that. But it is not an aesthetic judging by means of which we find it purposive, not a judging without a concept, which makes noticeable a merely subjective purposiveness in the free play of our cognitive faculties, but an intellectual judging in accordance with concepts, which gives us distinct cognition of an objective purposiveness, i.e., serviceability for all sorts of (infinitely manifold) purposes. One would have to call it a relative perfection rather than a beauty of the mathematical figure. The designation of an intellectual beauty can also not be allowed at all, for otherwise the word "beauty" would have to lose all determinate meaning, or intellectual satisfaction would have to lose all preeminence over sensible satisfaction. It would be better to be able to call a demonstration of such properties beautiful, since by means of this the understanding, as the faculty of concepts, and the imagination, as the faculty for exhibiting them, feel themselves strengthened *a priori* (which, together with the precision which is introduced by reason, is called its elegance): for here at least the satisfaction, although its ground lies in concepts, is subjective, whereas perfection is accompanied with an objective satisfaction.

ON THE RELATIVE PURPOSIVENESS OF NATURE IN DISTINCTION FROM INTERNAL PURPOSIVENESS

Experience leads our power of judgment to the concept of an objective and material purposiveness, i.e., to the concept of an end of nature, only if there is a relation of the cause to the effect to be judged[a] which we can understand as lawful only insofar as we find ourselves capable of subsuming the idea of the effect under the causality of its cause as the underlying condition of the possibility of the former. But this can happen in two ways: either if we regard the effect immediately as a product of art or if we regard it only as material for the art of other possible natural beings,

thus if we regard it either as an end or as a means for the purposive use of other causes. The latter purposiveness is called usefulness (for human beings) or advantageousness (for every other creature), and is merely relative; while the former is an internal purposiveness of the natural being.

Rivers, e.g., carry with them all sorts of soil helpful for the growth of plants, which they sometimes deposit in the middle of the land, sometimes in their deltas. On many coasts, the tide spreads this silt on the land, or deposits it on the bank, and, particularly if human beings help prevent the ebb from carrying it away again, the fruitful land increases, and the vegetable kingdom wins a place where previously fish and shellfish dwelt. Most of these sorts of extension of the land have been produced by nature, and it continues to do so, although slowly.—Now the question arises, is this to be judged as an end of nature, because it is useful for human beings?—for its usefulness for the vegetable kingdom cannot be brought into the balance, because just as much is taken away from the creatures of the sea as is added to the land.[3]

Or, to give an example of the advantageousness of certain natural things as means for other creatures (if one presupposes them as ends): no soil is more favorable to pine trees than a sandy soil. Now the ancient sea, before it withdrew from the land, left so many sandy tracts behind in our northern regions that on this soil, otherwise so useless for any cultivation, extensive pine forests grew up, for the irrational eradication of which we frequently blame our ancestors; and here one can ask whether this ancient deposit of sandy strata was an end of nature for the sake of the pine forests that were possible there. This much is clear: that if one assumes this to be an end of nature, then one would also have to admit that the sand is an end, though only a relative one, for which in turn the ancient beach and the withdrawal of the sea were the means; for in the series of subordinated members of a connection of ends every intermediate member must be considered as an end (although not as the final end), for which its proximate cause is the means. In the same way, if cattle, sheep, horses, etc. were even to exist in the world, then there had to be grass on the earth, and saltwort had to grow in the desert if camels were to thrive, and these and other herbivorous animals had to be found if there were to be wolves, tigers and lions. Hence the objective purposiveness which is grounded on advantageousness is not an objective purposiveness of the things in themselves, as if the sand in itself, as an effect of its cause, the sea, could not be comprehended without ascribing a purpose to the latter and without considering the effect, namely the sand, as a work of art. It is a merely relative purposiveness, contingent in the thing itself to which it is ascribed; and although in the examples we have given the species of grasses themselves are to be judged as organized products of nature, hence as rich in art, nevertheless in relation to the animals which they nourish they are to be regarded as mere raw materials.

If, however, the human being, through the freedom of his causality, finds things in nature completely advantageous for his often foolish aims (colorful bird feathers for the decoration of his clothing, colored soils or juices of plants for painting himself), but sometimes also to his rational ends, as the horse for riding or the ox and in Minorca even the ass and the swine for plowing, one cannot assume here even a relative end of nature (for this use). For the human's reason knows how to bring things into correspondence with his own arbitrary inspirations, to which he was by no means predestined by nature. Only if one assumes that human beings have to live on the earth would there also have to be at least no lack of the means without which they could not subsist as animals and even as rational animals (in however low a degree); but in that case those things in nature which are indispensable for this purpose would also have to be regarded as natural ends.

From this it can readily be seen that external purposiveness (advantageousness of one thing for another) can be regarded as an external natural end only under the condition that the existence of that for which it is advantageous, whether in a proximate or a distant way, is in itself an end of nature. This, however, can never be made out by mere contemplation of nature; thus it follows that relative purposiveness, although it gives hypothetical indications of natural ends, nevertheless justifies no absolute teleological judgments.

In cold lands the snow protects the seeds from frost; it facilitates communication among humans (by means of sleds); the Laplanders find animals there that bring about this communication (reindeer), which find adequate nourishment in a sparse moss, which they must even scrape out from under the snow, and yet are easily tamed and readily deprived of the freedom in which they could otherwise maintain themselves quite well.

For other peoples in the same icy regions the sea contains a rich supply of animals which, even beyond the nourishment and clothing that they provide and the wood which the sea as it were washes up for them for houses, also supplies them with fuel for warming their huts. Now here is an admirable confluence of so many relations of nature for one end: and this is the Greenlander, the Lapp, the Samoyed, the Yakut, etc. But one does not see why human beings have to live there at all. Thus to say that moisture falls from the air in the form of snow, that the sea has its currents which float the wood that has grown in warmer lands there, and that great sea animals filled with oil exist because the cause that produces all these natural products is grounded in the idea of an advantage for certain miserable creatures would be a very bold and arbitrary judgment. For even if all of this natural usefulness did not exist, we would find nothing lacking in this state of things for the adequacy of natural causes; rather, even merely to demand such a predisposition and to expect such an end of nature would seem to us presumptuous and ill-considered (for only the greatest incompatibility among human beings could have forced them into such inhospitable regions).[4]

ON THE SPECIAL CHARACTER OF THINGS AS NATURAL ENDS

In order to see that a thing is possible only as an end, i.e., that the causality of its origin must be sought not in the mechanism of nature, but in a cause whose productive capacity is determined by concepts, it is necessary that its form not be possible in accordance with mere natural laws, i.e., ones that can be cognized by us through the understanding, applied to objects of the senses, alone; rather even empirical cognition of their cause and effect presupposes concepts of reason.[5] Since reason must be able to cognize the necessity in every form of a natural product if it would understand the conditions connected with its generation, the contingency of their form with respect to all empirical laws of nature in relation to reason is itself a ground for regarding their causality as if it were possible only through reason; but this is then the capacity for acting in accordance with ends (a will); and the object which is represented as possible only on this basis is represented as possible only as an end.

If someone were to perceive a geometrical figure, for instance a regular hexagon, drawn in the sand in an apparently uninhabited land, his reflection, working with a concept of it, would become aware of the unity of the principle of its generation by means of reason, even if only obscurely, and thus, in accordance with this, would not be able to judge as a ground of the possibility of such a shape the sand, the nearby sea, the wind, the footprints of any known animals, or any other non-rational cause, because the contingency of coinciding with such a concept, which is possible only in reason, would seem to him so infinitely great that it would be just as good as if there were no natural law of nature, consequently no cause in nature acting merely mechanically, and as if the concept of such an object could be regarded as a concept that can be given only by reason and only by reason compared with the object, thus as if only reason can contain the causality for such an effect, consequently that this object must be thoroughly regarded as an end, but not a natural end, i.e., as a product of art (*vestigium hominis video*[b]).[6]

But in order to judge something that one cognizes as a product of nature as being at the same time an end, hence a natural end, something more is required if there is not simply to be a contradiction here. I would say provisionally that a thing exists as a natural end if it is cause and effect of itself (although in a twofold sense); for in this there lies a causality the likes of which cannot be connected with the mere concept of a nature without ascribing an end to it, but which in that case also can be conceived without contradiction but cannot be comprehended. We will first elucidate the determination of this idea of a natural end by means of an example before we fully analyze it.[7]

First, a tree generates another tree in accordance with a known natural law. However, the tree that it generates is of the same species; and so it generates itself as far as the species is concerned, in which it, on one side as effect, on the other as cause, unceasingly produces itself, and likewise, often producing itself, continuously preserves itself, as species.

Second, a tree also generates itself as an individual. This sort of effect we call, of course, growth; but this is to be taken in such a way that it is entirely distinct from any other increase in magnitude in accordance with mechanical laws, and is to be regarded as equivalent, although under another name, with generation. This plant first prepares the matter that it adds to itself with a quality peculiar to its species, which could not be provided by the mechanism of

nature outside of it, and develops itself further by means of material which, as far as its composition is concerned, is its own product. For although as far as the components that it receives from nature outside of itself are concerned, it must be regarded only as an educt, nevertheless in the separation and new composition of this raw material there is to be found an originality of the capacity for separation and formation in this sort of natural being that remains infinitely remote from all art when it attempts to reconstitute such a product of the vegetable kingdom from the elements that it obtains by its decomposition or from the material that nature provides for its nourishment.

Third, one part of this creature also generates itself in such a way that the preservation of the one is reciprocally dependent on the preservation of the others. An eye from the leaf of one tree grafted into the twig of another brings forth a growth of its own kind in an alien stock, and similarly a scion attached to another trunk. Hence one can regard every twig or leaf of one tree as merely grafted or inoculated into it, hence as a tree existing in itself, which only depends on the other and nourishes itself parasitically. At the same time, the leaves are certainly products of the tree, yet they preserve it in turn, for repeated defoliation would kill it, and its growth depends upon their effect on the stem. The self-help of nature in the case of injury in these creatures, where the lack of a part that is necessary for the preservation of the neighboring parts can be made good by the others; the miscarriages or malformations in growth, where certain parts form themselves in an entirely new way because of chance defects or obstacles, in order to preserve that which exists and bring forth an anomalous creature: these I mention only in passing, although they belong among the most wonderful properties of organized creatures.

THINGS, AS NATURAL ENDS, ARE ORGANIZED BEINGS

According to the characterization of the previous section, a thing that is to be cognized as a natural product but yet at the same time as possible only as a natural end must be related to itself reciprocally as both cause and effect, which is a somewhat improper and indeterminate expression, in need of a derivation from a determinate concept.

The causal nexus, insofar as it is conceived merely by the understanding, is a connection that constitutes a series (of causes and effects) that is always descending; and the things themselves, which as effects presuppose others as their causes, cannot conversely be the causes of these at the same time. This causal nexus is called that of efficient causes (*nexus effectivus*). In contrast, however, a causal nexus can also be conceived in accordance with a concept of reason (of ends), which, if considered as a series, would carry with it descending as well as ascending dependency, in which the thing which is on the one hand designated as an effect nevertheless deserves, in ascent, the name of a cause of the same thing of which it is the effect. In the practical sphere (namely, of art) such a connection can readily be found, e.g., the house is certainly the cause of the sums that are taken in as rent, while conversely the representation of this possible income was the cause of the construction of the house.[8] Such a causal connection is called that of final causes (*nexus finalis*). The first could perhaps more aptly be called the connection of real causes, and the second that of ideal ones, since with this terminology it would immediately be grasped that there cannot be more than these two kinds of causality.

Now for a thing as a natural end it is requisite, first, that its parts (as far as their existence and their form are concerned) are possible only through their relation to the whole. For the thing itself is an end, and is thus comprehended under a concept or an idea that must determine *a priori* everything that is to be contained in it. But insofar as a thing is conceived of as possible only in this way it is merely a work of art, i.e., the product of a rational cause distinct from the matter (the parts), the causality of which (in the production and combination of the parts) is determined through its idea of a whole that is thereby possible (thus not through nature outside of it).[9]

But if a thing, as a natural product, is nevertheless to contain in itself and its internal possibility a relation to ends, i.e., is to be possible only as a natural end and without the causality of the concepts of a rational being outside of it, then it is required, second, that its parts be combined into a whole by being reciprocally the cause and effect of their form. For in this way alone is it possible in turn for the idea of the whole conversely (reciprocally) to determine the form and combination of all the parts: not as a cause—for then it would be a

product of art—but as a ground for the cognition of the systematic unity of the form and the combination of all of the manifold that is contained in the given material for someone who judges it.

For a body, therefore, which is to be judged as a natural end in itself and in accordance with its internal possibility, it is required that its parts reciprocally produce each other, as far as both their form and their combination is concerned, and thus produce a whole out of their own causality, the concept of which, conversely, is in turn the cause (in a being that would possess the causality according to concepts appropriate for such a product) of it in accordance with a principle; consequently the connection of efficient causes could at the same time be judged as an effect through final causes.

In such a product of nature each part is conceived as if it exists only through all the others, thus as if existing for the sake of the others and on account of the whole, i.e., as an instrument (organ), which is, however, not sufficient (for it could also be an instrument of art, and thus represented as possible at all only as an end); rather it must be thought of as an organ that produces the other parts (consequently each produces the others reciprocally), which cannot be the case in any instrument of art, but only of nature, which provides all the matter for instruments (even those of art): only then and on that account can such a product, as an organized and self-organizing being, be called a natural end.

In a watch one part is the instrument for the motion of another, but one wheel is not the efficient cause for the production of the other: one part is certainly present for the sake of the other but not because of it. Hence the producing cause of the watch and its form is not contained in the nature (of this matter), but outside of it, in a being that can act in accordance with an idea of a whole that is possible through its causality. Thus one wheel in the watch does not produce the other, and even less does one watch produce another, using for that purpose other matter (organizing it); hence it also cannot by itself replace parts that have been taken from it, or make good defects in its original construction by the addition of other parts, or somehow repair itself when it has fallen into disorder: all of which, by contrast, we can expect from organized nature.—An organized being is thus not a mere machine, for that has only a motive power, while the organized being possesses in itself a formative power, and indeed one that it communicates to the matter, which does not have it (it organizes the latter): thus it has a self-propagating formative power, which cannot be explained through the capacity for movement alone (that is, mechanism).[10]

One says far too little about nature and its capacity in organized products if one calls this an analog of art: for in that case one conceives of the artist (a rational being) outside of it. Rather, it organizes itself, and in every species of its organized products, of course in accordance with some example in the whole, but also with appropriate deviations, which are required in the circumstances for self-preservation. Perhaps one comes closer to this inscrutable property if one calls it an analog of life: but then one must either endow matter as mere matter with a property (hylozoism) that contradicts its essence, or else associate with it an alien principle standing in communion with it (a soul), in which case, however, if such a product is to be a product of nature, organized matter as an instrument of that soul is already presupposed, and thus makes that product not the least more comprehensible, or else the soul is made into an artificer of this structure, and the product must be withdrawn from (corporeal) nature. Strictly speaking, the organization of nature is therefore not analogous with any causality that we know.[c] Beauty in nature, since it is ascribed to objects only in relation to reflection on their outer intuition, thus only to the form of their surfaces, can rightly be called an analog of art. But inner natural perfection, as is possessed by those things that are possible only as natural ends and hence as organized beings, is not thinkable and explicable in accordance with any analogy to any physical, i.e., natural capacity that is known to us; indeed, since we ourselves belong to nature in the widest sense, it is not thinkable and explicable even through an exact analogy with human art.

The concept of a thing as in itself a natural end is therefore not a constitutive concept of the understanding or of reason, but it can still be a regulative concept for the reflecting power of judgment, for guiding research into objects of this kind and thinking over their highest ground in accordance with a remote analogy with our own causality in accordance with ends; not, of course, for the sake of knowledge of nature or of its original ground, but rather for the sake of the very same practical faculty of reason in us in

analogy with which we consider the cause of that purposiveness.

Organized beings are thus the only ones in nature which, even if considered in themselves and without a relation to other things, must nevertheless be thought of as possible only as its ends, and which thus first provide objective reality for the concept of an end that is not a practical end but an end of nature, and thereby provide natural science with the basis for a teleology, i.e., a way of judging its objects in accordance with a particular principle the likes of which one would otherwise be absolutely unjustified in introducing at all (since one cannot at all understand the possibility of such a kind of causality *a priori*).

ON THE PRINCIPLE FOR THE JUDGING OF THE INTERNAL PURPOSIVENESS IN ORGANIZED BEINGS

This principle, or its definition, states: An organized product of nature is that in which everything is an end and reciprocally a means as well.[12] Nothing in it is in vain, purposeless, or to be ascribed to a blind mechanism of nature.

As for what occasions it, this principle is of course to be derived from experience, that is, experience of the kind that is methodically undertaken and is called observation; but the universality and necessity that it asserts of such a purposiveness cannot rest merely on grounds in experience, but must have as its ground some sort of *a priori* principle, even if it is merely regulative and even if that end lies only in the idea of the one who judges and never in any efficient cause. One can thus call this principle a maxim for the judging of the inner purposiveness of organized beings.

It is well known that the anatomists of plants and animals, in order to investigate their structure and to understand for what reason and to what end they have been given such a disposition and combination of parts and precisely this internal form, assume as indispensably necessary the maxim that nothing in such a creature is in vain, and likewise adopt it as the fundamental principle of the general doctrine of nature that nothing happens by chance. In fact, they could just as little dispense with this teleological principle as they could do without the universal physical principle, since, just as in the case of the abandonment of the latter there would remain no experience at all, so in the case of the abandonment of the former principle there would remain no guideline for the observation of a kind of natural thing that we have conceived of teleologically under the concept of a natural end.

For this concept leads reason into an order of things entirely different from that of a mere mechanism of nature, which will here no longer satisfy us. An idea has to ground the possibility of the product of nature. However, since this is an absolute unity of the representation, while the matter is a multitude of things, which by itself can provide no determinate unity of composition, if that unity of the idea is even to serve as the determining ground *a priori* of a natural law of the causality of such a form of the composite, then the end of nature must extend to everything that lies in its product. For once we have related such an effect in the whole to a supersensible determining ground beyond the blind mechanism of nature, we must also judge it entirely in accordance with this principle; and there is no ground for assuming that the form of such a thing is only partially dependent on the latter, for in such a case, in which two heterogeneous principles are jumbled together, no secure rule for judging would remain at all.

It might always be possible that in, e.g., an animal body, many parts could be conceived as consequences of merely mechanical laws (such as skin, hair, and bones); Yet the cause that provides the appropriate material, modifies it, forms it, and deposits it in its appropriate place must always be judged teleologically, so that everything in it must be considered as organized, and everything is also, in a certain relation to the thing itself, an organ in turn.

ON THE PRINCIPLE OF THE TELEOLOGICAL JUDGING OF NATURE IN GENERAL AS A SYSTEM OF ENDS

We have said above that the external purposiveness of natural things offers no sufficient justification for using them at the same time as ends of nature, as grounds for the explanation of their existence, and using their contingently purposive effects, in the idea, as grounds for their existence in accordance with the principle of final causes. Thus because rivers promote communication among peoples in inland countries, and mountains contain the sources of rivers and stores of snow for their maintenance in times of drought, while the slope of the land carries these waters down and allows the land to drain, one cannot immediately take these to be natural ends: for even though this configuration of the surface of the earth was quite necessary for the origination and preservation of the vegetable and animal kingdoms, yet there is nothing in it the possibility of which would require the assumption of a causality in accordance with ends. The same is true of plants that humans use for their needs or diversion, and of animals, such as camels, cattle, horses, dogs, etc., which are so widely used, partly for nourishment and partly for service, and are in great part indispensable. In things that one has no cause to regard as ends for themselves, an external relationship can be judged to be purposive only hypothetically.

To judge a thing to be purposive on account of its internal form is entirely different from holding the existence of such a thing to be an end of nature. For the latter assertion we need not only the concept of a possible end, but also cognition of the final end (*scopus*) of nature, which requires the relation of nature to something supersensible, which far exceeds all of our teleological cognition of nature; for the end of the existence of nature itself must be sought beyond nature. The internal form of a mere blade of grass can demonstrate its merely possible origin in accordance with the rule of ends in a way that is sufficient for our human faculty for judging. But if one leaves this aside and looks only to the use that other natural beings make of it, then one abandons the contemplation of its internal organization and looks only at its external purposive relations, where the grass is necessary to the livestock, just as the latter is necessary to the human being as the means for his existence; yet one does not see why it is necessary that human beings exist (a question which, if one thinks about the New Hollanders or the Fuegians,[13] might not be so easy to answer); thus one does not arrive at any categorical end, but all of this purposive relation rests on a condition that is always to be found further on, and which, as unconditioned, (the existence of a thing as a final end) lies entirely outside of the physical-teleological way of considering the world. But then such a thing is also not a natural end; for it (or its entire species) is not to be regarded as a natural product.

It is therefore only matter insofar as it is organized that necessarily carries with it the concept of itself as a natural end, since its specific form is at the same time a product of nature. However, this concept necessarily leads to the idea of the whole of nature as a system in accordance with the rule of ends, to which idea all of the mechanism of nature in accordance with principles of reason must now be subordinated (at least in order to test natural appearance by this idea).[14] The principle of reason is appropriate for it only subjectively, i.e., as the maxims that everything in the world is good for something, that nothing in it is in vain; and by means of the example that nature gives in its organic products, one is justified, indeed called upon to expect nothing in nature and its laws but what is purposive in the whole.

It is self-evident that this is not a principle for the determining but only for the reflecting power of judgment, that it is regulative and not constitutive, and that by its means we acquire only a guideline for considering things in nature, in relation to a determining ground that is already given, in accordance with a new, lawful order, and for extending natural science in accordance with another principle, namely that of final causes, yet without harm to the mechanism of nature. Moreover, it is by no means determined by this whether something that we judge in accordance with this principle is an intentional end of nature—whether grass exists for cattle or sheep, and these and the other things in nature for human beings. It is even good for us to consider in this light things that are unpleasant and in certain relations contrapurposive for us. Thus one could say, e.g., that the vermin that plague humans in their clothes, hair, or bedding are, in accordance with a wise dispensation of nature, an incentive for cleanliness, which is in itself already an important means for the preservation of health. Or the mosquitoes and other stinging insects that make the wilds of America so trying for the savages are so many goads to spur these primitive people to drain the swamps and let light into the thick, airless forests and thereby as well as by the cultivation of the soil to make their abode more salubrious. If it is treated in this way, then even what seems to the human being to be contrary to nature in his internal organization provides an entertaining and sometimes also instructive prospect on a teleological order of things, to which merely physical consideration alone, without such a principle, would

not lead us. Just as some judge that a tapeworm is given to the human or the animal in which it resides as if it were to make good a certain defect in its organs, so I would ask whether dreams (from which our sleep is never free, although we rarely remember them) might not be a purposive arrangement in nature, since, when all the motive forces in the body have relaxed, they serve to move the vital organs internally by means of the imagination and its great activity (which in this condition often amount to an affect);[15] and in the case of an overfilled stomach, where this movement during nocturnal sleep is all the more necessary, they commonly play themselves out with all the more liveliness; consequently, without this internal motive force and exhausting unrest, on account of which we often complain about dreams (which nevertheless are in fact perhaps a remedy), sleep, even in a healthy condition, might well amount to a complete extinction of life.[16] Even beauty in nature, i.e., its agreement with the free play of our cognitive faculties in the apprehension and judging of its appearance, can be considered in this way as an objective purposiveness of nature in its entirety, as a system of which the human being is a member, once the teleological judging of nature by means of natural ends, which have been made evident to us by organized beings, has justified us in the idea of a great system of the ends of nature. We may consider it as a favor[d] that nature has done for us that in addition to usefulness it has so richly distributed beauty and charms, and we can love it on that account, just as we regard it with respect because of its immeasurability, and we can feel ourselves to be ennobled in this contemplation—just as if nature had erected and decorated its magnificent stage precisely with this intention.

In this section we have meant to say nothing except that once we have discovered in nature a capacity for bringing forth products that can only be conceived by us in accordance with the concept of final causes, we may go further and also judge to belong to a system of ends even those things (or their relation, however purposive) which do not make it necessary to seek another principle of their possibility beyond the mechanism of blindly acting causes; because the former idea already, as far as its ground is concerned, leads us beyond the sensible world, and the unity of the supersensible principle must then be considered as valid in the same way not merely for certain species of natural beings but for the whole of nature as a system.

ON THE PRINCIPLE OF TELEOLOGY
AS AN INTERNAL PRINCIPLE OF
NATURAL SCIENCE

The principles of a science are either internal to it, and are then called indigenous (*principia domestica*), or they are based on principles that can find their place only outside of it, and are *foreign* principles (*peregrina*). Sciences that contain the latter base their doctrines on auxiliary propositions (*lemmata*), i.e., they borrow some concept, and along with it a basis for order, from another science.[17]

Every science is of itself a system; and it is not enough that in it we build in accordance with principles and thus proceed technically; rather, in it, as a freestanding building, we must also work architectonically, and treat it not like an addition and as a part of another building, but as a whole by itself, although afterwards we can construct a transition from this building to the other or vice versa.

Thus if one brings the concept of God into natural science and its context in order to make purposiveness in nature explicable, and subsequently uses this purposiveness in turn to prove that there is a God, then there is nothing of substance in either of the sciences, and a deceptive fallacy casts each into uncertainty by letting them cross each other's borders.

The expression "an end of nature" is already enough to preclude this confusion so that there is no mix-up between natural science and the occasion that it provides for the teleological judging of its objects and the consideration of God, and thus a theological derivation; and one must not regard it as unimportant whether one exchanges the former expression for that of a divine purpose in the order of nature or even passes off the latter as more fitting and more suitable for a pious soul because in the end it must come down to deriving every purposive form in nature from a wise author of the world; rather, we must carefully and modestly restrict ourselves to the expression that says only exactly as much as we know, namely that of an end of nature. For even before we ask after the cause of nature itself, we find within nature and the course of its generation products generated in accordance with the known laws of experience within it, in accordance with which natural science must judge its objects and thus seek within itself for their causality in accordance with

the rule of ends. Hence natural science must not jump over its boundaries in order to bring within itself as an indigenous principle that to whose concept no experience at all can ever be adequate and upon which we are authorized to venture only after the completion of natural science.

Natural properties that can be demonstrated *a priori* and whose possibility can thus be understood from general principles without any assistance from experience, even though they are accompanied with a technical purposiveness, can nevertheless, because they are absolutely necessary, not be counted at all as part of the teleology of nature as a method of solving its problems that belongs within physics. Arithmetical and geometrical analogies as well as universal mechanical laws, no matter how strange and admirable the unification of different and apparently entirely independent rules in a single principle in them may seem, can make no claim on that account to be teleological grounds of explanation within physics; and even if they deserve to be taken into consideration within the general theory of the purposiveness of things in nature, this would still belong elsewhere, namely in metaphysics, and would not constitute any internal principle of natural science: whereas in the case of the empirical laws of natural ends in organized beings it is not merely permissible but is even unavoidable to use the teleological way of judging as the principle of the theory of nature with regard to a special class of its objects.

Now in order to remain strictly within its own boundaries, physics abstracts entirely from the question of whether the ends of nature are intentional or unintentional; for that would be meddling in someone else's business (namely, in that of metaphysics). It is enough that there are objects that are explicable only in accordance with natural laws that we can think only under the idea of ends as a principle, and which are even internally cognizable, as far as their internal form is concerned, only in this way. In order to avoid even the least suspicion of wanting to mix into our cognitive grounds something that does not belong in physics at all, namely a supernatural cause, in teleology we certainly talk about nature as if the purposiveness in it were intentional, but at the same time ascribe this intention to nature, i.e., to matter, by which we would indicate (since there can be no misunderstanding here, because no intention in the strict sense of the term can be attributed to any lifeless matter) that this term signifies here only a principle of the reflecting, not of the determining power of judgment, and is thus not meant to introduce any special ground for causality, but is only meant to add to the use of reason another kind of research besides that in accordance with mechanical laws, in order to supplement the inadequacy of the latter even in the empirical search for all the particular laws of nature. Hence in teleology, insofar as it is connected to physics, we speak quite rightly of the wisdom, the economy, the forethought, and the beneficence of nature, without thereby making it into an intelligent being (since that would be absurd); but also without daring to set over it, as its architect, another, intelligent being, because this would be presumptuous;^{e} rather, such talk is only meant to designate a kind of causality in nature, in accordance with an analogy with our own causality in the technical use of reason, in order to keep before us the rule in accordance with which research into certain products of nature must be conducted.

Why, then, does teleology usually not constitute a proper part of theoretical natural science, but is instead drawn into theology as a propaedeutic or transition? This is done in order to keep the study of the mechanism of nature restricted to what we can subject to our observation or experiments, so that we could produce it ourselves, like nature, at least as far as the similarity of its laws is concerned; for we understand completely only that which we ourselves can make and bring about in accordance with concepts. Organization, however, as the internal end of nature, infinitely surpasses all capacity for a similar presentation by art; and as far as natural arrangements that are held to be externally purposive are concerned (e.g., wind, rain, etc.), physics can very well consider their mechanism, but it cannot present their relation to ends, insofar as this is supposed to be a condition necessarily belonging to their cause, at all, because this necessity in the connection pertains entirely to the combination of our concepts and not to the constitution of things.

WHAT IS AN ANTINOMY OF THE POWER OF JUDGMENT?

The determining power of judgment by itself has no principles that ground concepts of objects. It is no autonomy, for it merely subsumes under given laws or concepts as principles. For that very reason it is not

exposed to any danger from its own antinomy and from a conflict of its principles. Thus the transcendental power of judgment, which contains the conditions for subsuming under categories, was not by itself nomothetic, but merely named the conditions of sensible intuition under which a given concept, as a law of the understanding, could be given reality (application)—about which it could never fall into disunity with itself (at least in the matter of principles).^{18}

But the reflecting power of judgment is supposed to subsume under a law that is not yet given and which is in fact only a principle for reflection on objects for which we are objectively entirely lacking a law or a concept of the object that would be adequate as a principle for the cases that come before us. Now since no use of the cognitive faculties can be permitted without principles, in such cases the reflecting power of judgment must serve as a principle itself, which, since it is not objective, and cannot be presupposed as a sufficient ground for cognition of the intention of the object, can serve as a merely subjective principle for the purposive use of the cognitive faculties, namely for reflecting on one kind of objects. In relation to such cases, the reflecting power of judgment therefore has its maxims, indeed necessary ones, for the sake of the cognition of natural laws in experience, in order to arrive by their means at concepts, even if these are concepts of reason, if it needs these merely in order to come to know nature as far as its empirical laws are concerned.—Now between these necessary maxims of the reflecting power of judgment there can be a conflict, hence an antinomy, on which is based a dialectic which, if each of the two conflicting maxims has its ground in the nature of the cognitive faculties, can be called a natural dialectic and an unavoidable illusion which we must expose and resolve in the critique so that it will not deceive us.

REPRESENTATION OF THIS ANTINOMY

Insofar as reason has to do with nature, as the sum of the objects of the outer senses, it can be grounded on laws which are in part prescribed *a priori* to nature by the understanding itself, and which can in part be extended beyond what can be foreseen by empirical determinations encountered in experience. For the application of the first sort of laws, namely the universal laws of material nature in general, the power of judgment needs no special principle of reflection: for in that case it is determining, since an objective principle is given to it by the understanding. But as far as the particular laws that can only be made known to us by experience are concerned, there can be such great diversity and dissimilarity among them that the power of judgment itself must serve as a principle even in order merely to investigate the appearances of nature in accordance with a law and spy one out, because it requires one for a guideline if it is to have any hope of an interconnected experiential cognition in accordance with a thoroughgoing lawfulness of nature or of its unity in accordance with empirical laws. Now in the case of this contingent unity of particular laws the power of judgment can set out from two maxims in its reflection, one of which is provided to it by the mere understanding a priori, the other of which, however, is suggested by particular experiences that bring reason into play in order to conduct the judging of corporeal nature and its laws in accordance with a special principle. It may then seem that these two sorts of maxims are not consistent with each other, thus that a dialectic will result that will make the power of judgment go astray in the principle of its reflection.

The first maxim of the power of judgment is the thesis: All generation of material things and their forms must be judged as possible in accordance with merely mechanical laws.

The second maxim is the antithesis: Some products of material nature cannot be judged as possible according to merely mechanical laws (judging them requires an entirely different law of causality, namely that of final causes).

Now if one were to transform these regulative principles for research into constitutive principles of the possibility of the objects themselves, they would run:

Thesis: All generation of material things is possible in accordance with merely mechanical laws.

Antithesis: Some generation of such things is not possible in accordance with merely mechanical laws.

In this latter quality, as objective principles for the determining power of judgment, they would contradict one another, and hence one of the two propositions would necessarily be false; but that would then be an antinomy, though not of the power of judgment, but rather a conflict in the legislation of reason. However, reason can prove neither the one nor the other of these fundamental principles, because we can have no determining principle a priori of the possibility of things in accordance with merely empirical laws of nature.

By contrast, the maxims of a reflecting power of judgment that were initially expounded do not in fact contain any contradiction. For if I say that I must judge the possibility of all events in material nature and hence all forms, as their products, in accordance with merely mechanical laws, I do not thereby say that they are possible only in accordance with such laws (to the exclusion of any other kind of causality); rather, that only indicates that I should always reflect on them in accordance with the principle of the mere mechanism of nature, and hence research the latter, so far as I can, because if it is not made the basis for research then there can be no proper cognition of nature. Now this is not an obstacle to the second maxim for searching after a principle and reflecting upon it which is quite different from explanation in accordance with the mechanism of nature, namely the principle of final causes, on the proper occasion, namely in the case of some forms of nature (and, at their instance, even the whole of nature). For reflection in accordance with the first maxim is not thereby suspended, rather one is required to pursue it as far as one can; it is also not thereby said that those forms would not be possible in accordance with the mechanism of nature. It is only asserted that human reason, in the pursuit of this reflection and in this manner, can never discover the least basis for what is specific in a natural end, although it may well be able to discover other cognitions of natural laws; in which case it will remain undetermined whether in the inner ground of nature itself, which is unknown to us, physical-mechanical connection and connection to ends may not cohere in the same things, in a single principle: only our reason is not in a position to unify them in such a principle, and thus the power of judgment, as a reflecting (on a subjective ground) rather than as a determining (according to an objective principle of the possibility of things in themselves) power of judgment, is forced to think of another principle than that of the mechanism of nature as the ground of the possibility of certain forms in nature.

PREPARATION FOR THE RESOLUTION OF THE ABOVE ANTINOMY

We can by no means prove the impossibility of the generation of organized products of nature through the mere mechanism of nature, because since the infinite manifold of particular laws of nature that are contingent for us are only cognized empirically, we have no insight into their primary internal ground, and thus we cannot reach the internal and completely sufficient principle of the possibility of a nature (which lies in the supersensible) at all. Whether, therefore, the productive capacity of nature may not be as adequate for that which we judge as formed or combined in accordance with the idea of ends as well as for that which we believe to need merely the machinery of nature, and whether in fact things as genuine natural ends (as we must necessarily judge them) must be based in an entirely different kind of original causality, which cannot be contained at all in material nature or in its intelligible substratum, namely, an architectonic understanding: about this our reason, which is extremely limited with regard to the concept of causality if the latter is supposed to be specified *a priori*, can give us no information whatever.—However, with respect to our cognitive faculty, it is just as indubitably certain that the mere mechanism of nature is also incapable of providing an explanatory ground for the generation of organized beings. It is therefore an entirely correct fundamental principle for the reflecting power of judgment that for the evident connection of things in accordance with final causes we must conceive of a causality different from mechanism, namely that of an (intelligent) world-cause acting in accordance with ends, no matter how rash and indemonstrable that would be for the determining power of judgment. In the first case, the principle is a mere maxim of the power of judgment, in which the concept of that causality is a mere idea, to which one by no means undertakes to concede reality, but uses only as a guideline for reflection, which thereby always remains open for any mechanical explanatory grounds, and never strays from the sensible world; in the second case, the fundamental principle would be an objective principle, which would be prescribed by reason and to which the power of judgment would be subjected as determining, in which case, however, it would stray beyond the sensible world into that which transcends it, and would perhaps be led astray.

All appearance of an antinomy between the maxims of that kind of explanation which is genuinely physical (mechanical) and that which is teleological (technical) therefore rests on confusing a fundamental principle of the reflecting with that of the determining power of judgment, and on confusing the autonomy of the former (which is valid merely subjectively for the use of our reason in regard to the particular laws of experience) with the heteronomy of the latter, which has to conform to the laws given by the understanding (whether general or particular).

ON THE VARIOUS SYSTEMS CONCERNING THE PURPOSIVENESS OF NATURE

No one has doubted the correctness of the fundamental principle that certain things in nature (organized beings) and their possibility must be judged in accordance with the concept of final causes, even if one requires only a guideline for coming to know their constitution through observation without rising to the level of an investigation into their ultimate origin. The question can thus be only whether this fundamental principle is merely subjectively valid, i.e., merely a maxim of our power of judgment, or is an objective principle of nature, according to which there would pertain to it, in addition to its mechanism (in accordance with mere laws of motion) yet another kind of causality, namely that of final causes, under which the first kind (that of moving forces) would stand only as intermediate causes.

Now one could leave this question or problem for speculation entirely untouched and unsolved, for if we are satisfied with speculation within the boundaries of the mere cognition of nature, the above maxims are sufficient for studying nature as far as human powers reach and for probing its most hidden secrets. It must therefore be a certain presentiment of our reason, or a hint as it were given to us by nature, that we could by means of that concept of final causes step beyond nature and even connect it to the highest point in the series of causes if we were to abandon research into nature (even though we have not gotten very far in that), or at least set it aside for a while, and attempt to discover first where that stranger in natural science, namely the concept of natural ends, leads.

Now here, to be sure, the maxim that was not disputed above must lead to a wide array of controversial problems: whether the connection of ends in nature proves a special kind of causality in it; or whether, considered in itself and in accordance with objective principles, it is not instead identical with the mechanism of nature or dependent on one and the same ground, where, however, since in many products of nature this ground is often too deeply hidden for our research, we attempt to ascribe it to nature by analogy with a subjective principle, namely that of art, i.e., causality in accordance with ideas—an expedient that also succeeds in many cases, although it certainly seems to fail in some, but in any case never justifies us in introducing into natural science a special kind of agency distinct from causality in accordance with merely mechanical laws of nature. Insofar as we would call the procedure (the causality) of nature a technique, on account of the similarity to ends that we find in its products, we would divide this into intentional technique (*technica intentionalis*) and unintentional technique (*technica naturalis*). The former would mean that the productive capacity of nature in accordance with final causes must be held to be a special kind of causality; the latter that it is at bottom entirely identical with the mechanism of nature, and that the contingent coincidence with our concepts of art and their rules, as a merely subjective condition for judging nature, is falsely interpreted as a special kind of natural generation.

If we now speak of the systems for the explanation of nature with regard to final causes, one must note that they all controvert one another dogmatically, i.e., concerning objective principles of the possibility of things, whether through intentionally or even entirely unintentionally acting causes, but not concerning the subjective maxims for merely judging about the causes of such purposive products—in which case disparate principles could well be united with each other, unlike the former case, where contradictorily opposed principles cancel each other out and cannot subsist together.

The systems with regard to the technique of nature, i.e., of its productive force in accordance with the rule of ends, are twofold: those of the idealism or of the realism of natural ends. The former is the assertion that all purposiveness in nature is unintentional, the latter that some purposiveness in nature (in organized beings) is intentional, from which there can also be inferred as a hypothesis the consequence that the technique of nature is also intentional, i.e., an end, as far as concerns all its other products in relation to the whole of nature.

1. The idealism of purposiveness (I always mean objective purposiveness here) is now either that of the accidentality or of the fatality of the determination of nature in the purposive form of its products. The first principle concerns the relation of matter to the physical ground of its form, namely the laws of motion; the second concerns the hyperphysical ground of matter and the whole of nature. The system of accidentality, which is ascribed to Epicurus[19] or Democritus,[20] is, if taken literally, so obviously absurd that it need not detain us; by contrast, the system of fatality (of which Spinoza[21] is made the author, although it is to all appearance much older), which appeals to something supersensible, to which our insight therefore does not reach, is not so easy to refute, since its concept of the original being is not intelligible at all. But this much is clear: that on this system the connection of ends in the world must be assumed to be unintentional (because it is derived from an original being, but not from its understanding, hence not from any intention on its part, but from the necessity of its nature and the unity of the world flowing from that), hence the fatalism of purposiveness is at the same time an idealism of it.

2. The realism of the purposiveness of nature is also either physical or hyperphysical. The first bases ends in nature on the analog of a faculty acting in accordance with an intention, the life of matter (in it, or also through an animating inner principle, a world-soul); and is called hylozoism.[22] The second derives them from the original ground of the world-whole, as an intentionally productive (originally living) intelligent being; and it is theism.[f]

NONE OF THE ABOVE SYSTEMS ACCOMPLISHES WHAT IT PRETENDS TO DO

What do all these systems want? They want to explain our teleological judgments about nature, but go to work in such a way that some of them deny the truth of these judgments, thus declaring them to be an idealism of nature (represented as an art), while the others acknowledge them to be true, and promise to demonstrate the possibility of a nature in accordance with the idea of final causes.

1. On the one hand, the systems that contend for the idealism of final causes in nature concede to its principle a causality according to laws of motion (through which natural things purposively exist), but they deny intentionality to it, i.e., they deny that nature is intentionally determined to its purposive production, or, in other words, that an end is the cause. This is Epicurus's kind of explanation, on which the difference between a technique of nature and mere mechanism is completely denied, and blind chance is assumed to be the explanation not only of the correspondence of generated products with our concepts of ends, hence of technique, but even of the determination of the causes of this generation in accordance with laws of motion, hence of their mechanism, and thus nothing is explained, not even the illusion in our teleological judgments, and hence the putative idealism in them is not demonstrated at all.

On the other hand, Spinoza would suspend all inquiry into the ground of the possibility of the ends of nature and deprive this idea of all reality by allowing them to count not as products of an original being but as accidents inhering in it, and to this being, as the substratum of those natural things, he ascribes not causality with regard to them but merely subsistence, and (on account of the unconditional necessity of this being, together with all natural things as accidents inhering in it), he secures for the natural forms the unity of the ground that is, to be sure, requisite for all purposiveness, but at the same time he removes their contingency, without which no unity of purpose can be thought, and with that removes everything intentional, just as he removes all understanding from the original ground of natural things.

However, Spinozism does not accomplish what it wants. It wants to provide a basis for the explanation of the connection of ends (which it does not deny) in the things of nature, and names merely the unity of the subject in which they all inhere. But even if one concedes to it this sort of existence for the beings of the world, still that ontological unity is not immediately

a unity of end, and in no way makes the latter comprehensible. The latter is a quite special mode of the former, which does not follow at all from the connection of the things (the beings of the world) in one subject (the original being), but which throughout implies relation to a cause that has understanding; and even if all these things were united in a simple subject, still no relation to an end would be exhibited unless one conceives of them, first, as internal effects of the substance, as a cause, and, second, of the latter as a cause through its understanding. Without these formal conditions all unity is mere natural necessity, and, if it is nevertheless ascribed to things that we represent as external to one another, blind necessity. If, however, one would call purposiveness in nature that which the academy called the transcendental perfection of things (in relation to their own proper essence), in accordance with which everything must have in itself everything that is necessary in order to be that kind of thing and not any other, then that is merely a childish game played with words instead of concepts. For if all things must be conceived as ends, thus if to be a thing and to be an end are identical, then there is at bottom nothing that particularly deserves to be represented as an end.

From this it is readily seen that by tracing our concept of the purposiveness in nature back to the consciousness of ourselves in one all-comprehending (yet at the same time simple) being, and seeking that form merely in the unity of the latter, Spinoza must have intended to assert not the realism but merely the idealism of nature; but he could not accomplish even this, for the mere representation of the unity of the substratum can never produce the idea of even an unintentional purposiveness.

2. Those who intend not merely to assert but also to explain the realism of natural ends believe themselves able to understand a special kind of causality, namely that of intentionally acting causes, at least as far as its possibility is concerned; otherwise they could not undertake to try to explain it. For even the most daring hypothesis can be authorized only if at least the possibility of that which is assumed to be its ground is certain, and one must be able to insure the objective reality of its concept.

However, the possibility of a living matter (the concept of which contains a contradiction, because lifelessness, *inertia*, constitutes its essential characteristic), cannot even be conceived;[23] the possibility of an animated matter and of the whole of nature as an animal can be used at all only insofar as it is revealed to us (for the sake of an hypothesis of purposiveness in nature at large), in experience, in the organization of nature in the small, but its possibility can by no means be understood *a priori*. There must therefore be a circle in the explanation if one would derive the purposiveness of nature in organized beings from the life of matter and in turn is not acquainted with this life otherwise than in organized beings, and thus cannot form any concept of its possibility without experience of them. Hylozoism thus does not accomplish what it promises.

Theism, finally, is just as incapable of dogmatically establishing the possibility of natural ends as a key to teleology, although among all the grounds for explaining this it has the advantage that by means of the understanding that it ascribes to the original being it can best rid the purposiveness of nature of idealism and introduce an intentional causality for its generation.

For in order to be justified in placing the ground of the unity of purpose in matter beyond nature in any determinate way, the impossibility of placing this in matter through its mere mechanism would first have to be demonstrated in a way sufficient for the determining power of judgment. But we cannot say more than that given the constitution and the limits of our cognitive capacities (by means of which we cannot understand the primary internal ground of even this mechanism) we must by no means seek for a principle of determinate purposive relations in matter; rather, for us there remains no other way of judging the generation of its products as natural ends than through a supreme understanding as the cause of the world. But that is only a ground for the reflecting, not for the determining power of judgment, and absolutely cannot justify any objective assertion.

THE CAUSE OF THE IMPOSSIBILITY OF A DOGMATIC TREATMENT OF THE CONCEPT OF A TECHNIQUE OF NATURE IS THE INEXPLICABILITY OF A NATURAL END

We deal with a concept dogmatically (even if it is supposed to be empirically conditioned) if we consider it as contained under another concept of the object, which constitutes a principle of reason, and determine it in accordance with the latter. But we deal with it merely critically if we consider it only in relation to our cognitive faculties, hence in relation to the subjective conditions for thinking it, without undertaking to decide anything about its object. The dogmatic treatment of a concept is thus that which is lawful for the determining, the critical that which is lawful merely for the reflecting power of judgment.

Now the concept of a thing as a natural end is a concept that subsumes nature under a causality that is conceivable only by means of reason, in order to judge, in accordance with this principle, about that which is given by the object in experience. But in order to use it dogmatically for the determining power of judgment, we would first have to be assured of the objective reality of this concept, for otherwise we would not be able to subsume any natural thing under it. The concept of a thing as a natural end, however, is certainly an empirically conditioned concept, i.e., one that is possible only under certain conditions given in experience, but it is still not a concept that can be abstracted from experience, but one that is possible only in accordance with a principle of reason in the judging of the object. It thus cannot be understood and dogmatically established at all as in accordance with such a principle of its objective reality (i.e., that an object is possible in accordance with such a principle); and we do not know whether it is merely a rationalistic and objectively empty concept (*conceptus ratiocinans*) or a concept of reason that grounds cognition and is confirmed by reason (*conceptus ratiocinatus*). Thus it cannot be treated dogmatically for the determining power of judgment, i.e., not merely can it not be determined whether or not things of nature, considered as natural ends, require for their generation a causality of an entirely special kind (that in accordance with intentions), but this question cannot even be raised, because the objective reality of the concept of a natural end is not demonstrable by means of reason at all (i.e., it is not constitutive for the determining, but is merely regulative for the reflecting power of judgment).

That this concept is not demonstrable is clear from the fact that as a concept of a natural product it includes natural necessity and yet at the same time a contingency of the form of the object (in relation to mere laws of nature) in one and the same thing as an end; consequently, if there is not to be a contradiction here, it must contain a basis for the possibility of this thing in nature and yet at the same time a basis of the possibility of this nature itself and its relation to something that is not empirically cognizable nature (supersensible) and thus is not cognizable at all for us, in order to be judged in accordance with another kind of causality than that of the mechanism of nature, if its possibility is to be determined. Thus, since the concept of a thing as a natural end is excessive for the determining power of judgment if one considers the object by means of reason (although it may be immanent for the reflecting power of judgment with regard to objects of experience), and thus it cannot be provided with objective reality for determining judgments, it is thereby comprehensible how all the systems that can even be sketched for the dogmatic treatment of the concept of natural ends and of nature as a whole connected by final causes cannot decide anything about it, whether objectively affirmative or objectively negative; because if things are subsumed under a concept that is merely problematic, the synthetic predicates of such a concept (here, e.g., whether the end of nature which we conceive for the generation of things is intentional or unintentional) must yield the same sort of (problematic) judgments of the object, whether they are affirmative or negative, since one does not know whether one is judging about something or nothing. The concept of a causality through ends (of art) certainly has objective reality, as does that of a causality in accordance with the mechanism of nature. But the concept of a causality of nature in accordance with the rule of ends, even more the concept of a being the likes of which is not given to us in experience at all, namely that of a original ground of nature, can of course be thought without contradiction, but is not good for any dogmatic determinations, because since it cannot be drawn from experience and is not requisite for the possibility of experience its objective reality cannot be guaranteed by anything. But even if it could be, how could I count things that are definitely supposed to be products of divine art among the products of nature, whose incapacity for producing such things in accordance with its laws is precisely that which has made necessary the appeal to a cause that is distinct from it?

THE CONCEPT OF AN OBJECTIVE PURPOSIVENESS OF NATURE IS A CRITICAL PRINCIPLE OF REASON FOR THE REFLECTING POWER OF JUDGMENT

To say that the generation of certain things in nature or even of nature as a whole is possible only through a cause that is determined to act in accordance with intentions is quite different from saying that because of the peculiar constitution of my cognitive faculties I cannot judge about the possibility of those things and their generation except by thinking of a cause for these that acts in accordance with intentions, and thus by thinking of a being that is productive in accordance with the analogy with the causality of an understanding. In the first case I would determine something about the object, and I am obliged to demonstrate the objective reality of a concept that has been assumed; in the second case, reason merely determines the use of my cognitive faculties in accordance with their special character and with the essential conditions as well as the limits of their domain. The first principle is thus an objective fundamental principle for the determining, the second a subjective fundamental principle merely for the reflecting power of judgment, hence a maxim that reason prescribes to it.

It is in fact indispensable for us to subject nature to the concept of an intention if we would even merely conduct research among its organized products by means of continued observation; and this concept is thus already an absolutely necessary maxim for the use of our reason in experience. It is obvious that once we have adopted such a guideline for studying nature and found it to be reliable we must also at least attempt to apply this maxim of the power of judgment to the whole of nature, since by means of it we have been able to discover many laws of nature which, given the limitation of our insights into the inner mechanism of nature, would otherwise remain hidden from us. But with regard to the latter use this maxim of the power of judgment is certainly useful, but not indispensable, because nature as a whole is not given to us as organized (in the strictest sense of the term adduced above).[24] By contrast, this maxim of the reflecting power of judgment is essential for those products of nature which must be judged only as intentionally formed thus and not otherwise, in order to obtain even an experiential cognition of their internal constitution; because even the thought of them as organized things is impossible without associating the thought of a generation with an intention.

Now the concept of a thing whose existence or form we represent as possible under the condition of an end is inseparable from the concept of its contingency (according to natural laws). Hence natural things which we find possible only as ends constitute the best proof of the contingency of the world-whole, and are the only basis for proof valid for both common understanding as well as for philosophers of the dependence of these things on and their origin in a being that exists outside of the world and is (on account of that purposive form) intelligent; thus teleology cannot find a complete answer for its inquiries except in a theology.

But what does even the most complete teleology prove in the end? Does it prove anything like that such an intelligent being exists? No; it proves nothing more than that because of the constitution of our cognitive faculties, and thus in the combination of experience with the supreme principles of reason, we cannot form any concept at all of the possibility of such a world except by conceiving of such an intentionally acting supreme cause. Objectively, therefore, we cannot establish the proposition that there is an intelligent original being; we can establish it only subjectively for the use of our power of judgment in its reflection upon the ends in nature, which cannot be conceived in accordance with any other principle than that of an intentional causality of a highest cause.

If we would establish the supreme proposition dogmatically, from teleological grounds, then we would be trapped by difficulties from which we could not extricate ourselves. For then these inferences would have to be based on the proposition that the organized beings in the world are not possible except through an intentionally acting cause. But then we could not avoid asserting that because the causal connection of these things can be pursued and their lawfulness cognized only under the idea of ends we would also be justified in presupposing that this is a necessary condition for every thinking and cognizing being, thus that it is a condition that depends on the object and not just on our own subject. But we cannot get away with such an assertion. For since we do not actually observe ends in nature as intentional, but

merely add this concept as a guideline for the power of judgment in reflection on the products of nature, they are not given to us through the object. It is even impossible for us to justify *a priori* the assumption of the objective reality of such a concept. There is thus left nothing but a proposition resting only on subjective conditions, namely those of a reflecting power of judgment appropriate to our cognitive faculties, which, if one were to express it as objectively and dogmatically valid, would say: There is a God; but all that is allowed to us humans is the restricted formula: We cannot conceive of the purposiveness which must be made the basis even of our cognition of the internal possibility of many things in nature and make it comprehensible except by representing them and the world in general as a product of an intelligent cause (a God).

Now if this proposition, grounded on an indispensably necessary maxim of our power of judgment, is completely sufficient for every speculative as well as practical use of our reason in every human respect, I would like to know what we lose by being unable to prove it valid for higher beings, on purely objective grounds (which unfortunately exceed our capacity)? For it is quite certain that we can never adequately come to know the organized beings and their internal possibility in accordance with merely mechanical principles of nature, let alone explain them; and indeed this is so certain that we can boldly say that it would be absurd for humans even to make such an attempt or to hope that there may yet arise a Newton who could make comprehensible even the generation of a blade of grass according to natural laws that no intention has ordered; rather, we must absolutely deny this insight to human beings.[25] But for us to judge in turn that even if we could penetrate to the principle of nature in the specification of its universal laws known to us there could lie hidden no ground sufficient for the possibility of organized beings without the assumption of an intention underlying their generation would be presumptuous: for how could we know that? Probabilities count for nothing here, where judgments of pure reason are at stake.—Thus we cannot make any objective judgment at all, whether affirmative or negative, about the proposition that there is an intentionally acting being as a world-cause (hence as an author) at the basis of what we rightly call natural ends; only this much is certain, namely, that if we are to judge at least in accordance with what it is granted to us to understand through our own nature (in accordance with the conditions and limits of our reason), we absolutely cannot base the possibility of those natural ends on anything except an intelligent being—which is what alone is in accord with the maxims of our reflecting power of judgment and is thus a ground which is subjective but ineradicably attached to the human race.

REMARK

This consideration, which would certainly deserve to be elaborated in detail in transcendental philosophy, can come in here only as a digression, for elucidation (not for the proof of what has here been expounded).

Reason is a faculty of principles, and in its most extreme demand it reaches to the unconditioned, while understanding, in contrast, is always at its service only under a certain condition, which must be given. Without concepts of the understanding, however, which must be given objective reality, reason cannot judge at all objectively (synthetically), and by itself it contains, as theoretical reason, absolutely no constitutive principles, but only regulative ones. One soon learns that where the understanding cannot follow, reason becomes excessive, displaying itself in well-grounded ideas (as regulative principles) but not in objectively valid concepts; the understanding, however, which cannot keep up with it, but which would yet be necessary for validity for objects, restricts the validity of those ideas of reason solely to the subject, although still universally for all members of this species, i.e., understanding restricts the validity of those ideas to the condition which, given the nature of our (human) cognitive faculty or even the concept that we can form of the capacity of a finite rational being in general, we cannot and must not conceive otherwise, but without asserting that the basis for such a judgment lies in the object. We will adduce examples, which are certainly too important as well as too difficult for them to be immediately pressed upon the reader as proven propositions, but which will still provide material to think over and can serve to elucidate what is our proper concern here.

It is absolutely necessary for the human understanding to distinguish between the possibility and the actuality of things. The reason for this lies in the subject and the nature of its cognitive faculties. For if two entirely heterogeneous elements were not required

for the exercise of these faculties, understanding for concepts and sensible intuition for objects corresponding to them, then there would be no such distinction (between the possible and the actual). That is, if our understanding were intuitive, it would have no objects except what is actual. Concepts (which pertain merely to the possibility of an object) and sensible intuitions (which merely give us something, without thereby allowing us to cognize it as an object) would both disappear. Now, however, all of our distinction between the merely possible and the actual rests on the fact that the former signifies only the position of the representation of a thing with respect to our concept and, in general, our faculty for thinking, while the latter signifies the positing of the thing in itself (apart from this concept).[26] Thus the distinction of possible from actual things is one that is merely subjectively valid for the human understanding, since we can always have something in our thoughts although it does not exist, or represent something as given even though we do not have any concept of it. The propositions, therefore, that things can be possible without being actual, and thus that there can be no inference at all from mere possibility to actuality, quite rightly hold for the human understanding without that proving that this distinction lies in the things themselves. For that the latter cannot be inferred from the former, hence that those propositions are certainly valid of objects insofar as our cognitive faculty, as sensibly conditioned, is concerned with objects of these senses, but are not valid of objects in general, is evident from the unremitting demand of reason to assume some sort of thing (the original ground) as existing absolutely necessarily, in which possibility and actuality can no longer be distinguished at all, and for which idea our understanding has absolutely no concept, i.e., can find no way in which to represent such a thing and its way of existing. For if understanding thinks it (it can think it as it will), then it is represented as merely possible. If understanding is conscious of it as given in intuition, then it is actual without understanding being able to conceive of its possibility. Hence the concept of an absolutely necessary being is an indispensable idea of reason but an unattainable problematic concept for the human understanding. It is still valid, however, for the use of our cognitive faculties in accordance with their special constitution, thus not for objects and thereby for every cognitive being: because I cannot presuppose that in every such being thinking and intuiting, hence the possibility and actuality of things, are two different conditions for the exercise of its cognitive faculties. For an understanding to which this distinction did not apply, all objects that I cognize would be (exist), and the possibility of some that did not exist, i.e., their contingency if they did exist, as well as the necessity that is to be distinguished from that, would not enter into the representation of such a being at all. What makes it so difficult for our understanding with its concepts to be the equal of reason is simply that for the former, as human understanding, that is excessive (i.e., impossible for the subjective conditions of its cognition) which reason nevertheless makes into a principle belonging to the object.—Now here this maxim is always valid, that even where the cognition of them outstrips the understanding, we should conceive all objects in accordance with the subjective conditions for the exercise of our faculties necessarily pertaining to our (i.e., human) nature; and, if the judgments made in this way cannot be constitutive principles determining how the object is constituted (as cannot fail to be the case with regard to transcendent concepts), there can still be regulative principles, immanent and secure in their use and appropriate for the human point of view.

Just as in the theoretical consideration of nature reason must assume the idea of an unconditioned necessity of its primordial ground, so, in the case of the practical, it also presupposes its own unconditioned (in regard to nature) causality, i.e., freedom, because it is aware of its moral command. Now since here, however, the objective necessity of the action, as duty, is opposed to that which it, as an occurrence, would have if its ground lay in nature and not in freedom (i.e., in the causality of reason), and the action which is morally absolutely necessary can be regarded physically as entirely contingent (i.e., what necessarily should happen often does not), it is clear that it depends only on the subjective constitution of our practical faculty that the moral laws must be represented as commands (and the actions which are in accord with them as duties), and that reason expresses this necessity not through a be (happening) but through a should-be: which would not be the case if reason without sensibility (as the subjective condition of its application to objects of nature) were considered, as far as its causality is concerned, as a cause in an intelligible world, corresponding completely with the moral law, where there would be no distinction between what should be done and what is done, between a practical law concerning that which is possible through us and the theoretical law concerning that which is actual through us. Now, however, although an intelligible world, in which everything would be actual merely because it is (as something good) possible, and even freedom, as its formal condition, is a transcendent concept for us, which is not serviceable for any constitutive principle for determining an object and its objective reality, still, in accordance with the constitution of our (partly sensible) nature, it can serve as a universal regulative principle for ourselves and for every being standing in connection with the sensible world, so far as we can represent that in accordance with the constitution of our own reason and capacity, which does not determine the constitution of freedom, as a form of causality, objectively, but rather makes the rules of actions in accordance with that idea into commands for everyone and indeed does so with no less validity than if it did determine freedom objectively.

Likewise, as far as the case before us is concerned, it may be conceded that we would find no distinction between a natural mechanism and a technique of nature, i.e., a connection to ends in it, if our understanding were not of the sort that must go from the universal to the particular, and the power of judgment can thus cognize no purposiveness in the particular, and hence make no determining judgments, without having a universal law under which it can subsume the particular. But now since the particular, as such, contains something contingent with regard to the universal, but reason nevertheless still requires unity, hence lawfulness, in the connection of particular laws of nature (which lawfulness of the contingent is called purposiveness), and the *a priori* derivation of the particular laws from the universal, as far as what is contingent in the former is concerned, is impossible through the determination of the concept of the object, thus the concept of the purposiveness of nature in its products is a concept that is necessary for the human power of judgment in regard to nature but does not pertain to the determination of the objects themselves, thus a subjective principle of reason for the power of judgment which, as regulative (not constitutive), is just as necessarily valid for our human power of judgment as if it were an objective principle.

ON THE SPECIAL CHARACTER OF THE HUMAN UNDERSTANDING, BY MEANS OF WHICH THE CONCEPT OF A NATURAL END IS POSSIBLE FOR US

In the remark, we have adduced special characteristics of our cognitive faculty (even the higher one) which we may easily be misled into carrying over to the things themselves as objective predicates; but they concern ideas for which no appropriate objects can be given in experience, and which could therefore serve only as regulative principles in the pursuit of experience. It is the same with the concept of a natural end, as far as the cause of the possibility of such a predicate is concerned, which can only lie in the idea; but the consequence that answers to it (the product) is still given in nature, and the concept of a causality of the latter, as a being acting in accordance with ends, seems to make the idea of a natural end into a constitutive principle of nature; and in this it differs from all other ideas.

This difference, however, consists in the fact that the idea at issue is not a principle of reason for the understanding, but for the power of judgment, and is thus merely the application of an understanding in general to possible objects of experience, where, indeed, the judgment cannot be determining, but merely reflecting, hence where the object is, to be sure, given in experience, but where it cannot even be determinately (let alone completely appropriately) judged in accordance with the idea, but can only be reflected upon.

What is at issue is therefore a special character of our (human) understanding with regard to the power of judgment in its reflection upon things in nature. But if that is the case, then it must be based on the idea of a possible understanding other than the human one (as in the *Critique of Pure Reason* we had to have in mind another possible intuition if we were to hold our own to be a special kind, namely one that is valid of objects merely as appearances),[27] so that one could say that certain products of nature, as far as their possibility is concerned, must, given the particular constitution of our understanding, be considered by us as intentional and generated as ends, yet without thereby demanding that there actually is a particular cause that has the representation of an end as its determining ground, and thus without denying that another (higher) understanding than the human one might be able to find the ground of the possibility of such

products of nature even in the mechanism of nature, i.e., in a causal connection for which an understanding does not have to be exclusively assumed as a cause.

What is at issue here is thus the relation of our understanding to the power of judgment, the fact, namely, that we have to seek a certain contingency in the constitution of our understanding in order to notice this as a special character of our understanding in distinction from other possible ones.

This contingency is quite naturally found in the particular, which the power of judgment is to subsume under the universal of the concepts of the understanding; for through the universal of our (human) understanding the particular is not determined, and it is contingent in how many different ways distinct things that nevertheless coincide in a common characteristic can be presented to our perception. Our understanding is a faculty of concepts, i.e., a discursive understanding, for which it must of course be contingent on what and how different might be the particular that can be given to it in nature and brought under its concepts. But since intuition also belongs to cognition, and a faculty of a complete spontaneity of intuition would be a cognitive faculty distinct and completely independent from sensibility, and thus an understanding in the most general sense of the term, one can thus also conceive of an intuitive understanding (negatively, namely merely as not discursive), which does not go from the universal to the particular and thus to the individual (through concepts), and for which that contingency of the agreement of nature in its products in accordance with particular laws for the understanding, which makes it so difficult for ours to bring the manifold of these to the unity of cognition, is not encountered—a job that our understanding can accomplish only through the correspondence of natural characteristics with our faculty of concepts, which is quite contingent, but which an intuitive understanding would not need.

Our understanding thus has this peculiarity for the power of judgment, that in cognition by means of it the particular is not determined by the universal, and the former therefore cannot be derived from the latter alone; but nevertheless this particular in the manifold of nature should agree with the universal (through concepts and laws), which agreement under such circumstances must be quite contingent and without a determinate principle for the power of judgment.

Nevertheless, in order for us to be able at least to conceive of the possibility of such an agreement of the things of nature with the power of judgment (which we represent as contingent, hence as possible only through an end aimed at it), we must at the same time conceive of another understanding, in relation to which, and indeed prior to any end attributed to it, we can represent that agreement of natural laws with our power of judgment, which for our understanding is conceivable only through ends as the means of connection, as necessary.

Our understanding, namely, has the property that in its cognition, e.g., of the cause of a product, it must go from the analytical universal (of concepts) to the particular (of the given empirical intuition), in which it determines nothing with regard to the manifoldness of the latter, but must expect this determination for the power of judgment from the subsumption of the empirical intuition (when the object is a product of nature) under the concept. Now, however, we can also conceive of an understanding which, since it is not discursive like ours but is intuitive, goes from the synthetically universal (of the intuition of a whole as such) to the particular, i.e., from the whole to the parts, in which, therefore, and in whose representation of the whole, there is no contingency in the combination of the parts, in order to make possible a determinate form of the whole, which is needed by our understanding, which must progress from the parts, as universally conceived grounds, to the different possible forms, as consequences, that can be subsumed under it. In accordance with the constitution of our understanding, by contrast, a real whole of nature is to be regarded only as the effect of the concurrent moving forces of the parts. Thus if we would not represent the possibility of the whole as depending upon the parts, as is appropriate for our discursive understanding, but would rather, after the model of the intuitive (archetypical) understanding, represent the possibility of the parts (as far as both their constitution and their combination is concerned) as depending upon the whole, then, given the very same special characteristic of our understanding, this cannot come about by the whole being the ground of the possibility of the connection of the parts (which would be a contradiction in the discursive kind of cognition), but only by the representation of a whole containing the ground of the possibility of its form and of the

connection of parts that belongs to that. But now since the whole would in that case be an effect (product) the representation of which would be regarded as the cause of its possibility, but the product of a cause whose determining ground is merely the representation of its effect is called an end, it follows that it is merely a consequence of the particular constitution of our understanding that we represent products of nature as possible only in accordance with another kind of causality than that of the natural laws of matter, namely only in accordance with that of ends and final causes, and that this principle does not pertain to the possibility of such things themselves (even considered as phenomena) in accordance with this sort of generation, but pertains only to the judging of them that is possible for our understanding. From this we at the same time understand why in natural science we are far from being satisfied with an explanation of the products of nature by means of causality in accordance with ends, since in the latter we are required to judge the generation of nature as is appropriate for our faculty for judging them, i.e., the power of reflecting judgment, and not according to the things themselves as is appropriate for the determining power of judgment. And further, it is not at all necessary here to prove that such an *intellectus archetypus* is possible, but only that in the contrast of it with our discursive, image-dependent understanding (*intellectus ectypus*) and the contingency of such a constitution we are led to that idea (of an *intellectus archetypus*), and that this does not contain any contradiction.

Now if we consider a material whole, as far as its form is concerned, as a product of the parts and of their forces and their capacity to combine by themselves (including as parts other materials that they add to themselves), we represent a mechanical kind of generation. But from this there arises no concept of a whole as an end, whose internal possibility presupposes throughout the idea of a whole on which even the constitution and mode of action of the parts depends, which is just how we must represent an organized body. But from this, as has just been shown, it does not follow that the mechanical generation of such a body is impossible; for that would be to say the same as that it is impossible (i.e., self-contradictory) to represent such a unity in the connection of the manifold for every understanding without the idea of that connection being at the same time its generating cause, i.e., without intentional production. Nevertheless, this would in fact follow if we were justified in regarding material beings as things in themselves. For then the unity that constitutes the ground of the possibility of natural formations would be merely the unity of space, which is however no real ground of generatings but only their formal condition; although it has some similarity to the real ground that we seek in that in it no part can be determined except in relation to the whole (the representation of which is thus the basis of the possibility of the parts).[28] But since it is still at least possible to consider the material world as a mere appearance, and to conceive of something as a thing in itself (which is not an appearance) as substratum, and to correlate with this a corresponding intellectual intuition (even if it is not ours), there would then be a supersensible real ground for nature, although it is unknowable for us, to which we ourselves belong, and in which that which is necessary in it as object of the senses can be considered in accordance with mechanical laws, while the agreement and unity of the particular laws and corresponding forms, which in regard to the mechanical laws we must judge as contingent, can at the same time be considered in it, as object of reason (indeed the whole of nature as a system) in accordance with teleological laws, and the material world would thus be judged in accordance with two kinds of principles, without the mechanical mode of explanation being excluded by the teleological mode, as if they contradicted each other.

From this we may also understand what we could otherwise easily suspect but only with difficulty assert as certain and prove, namely, that the principle of a mechanical derivation of purposive products of nature could of course subsist alongside the teleological principle, but could by no means make the latter dispensable; i.e., one could investigate all the thus far known and yet to be discovered laws of mechanical generation in a thing that we must judge as an end of nature, and even hope to make good progress in this, without the appeal to a quite distinct generating ground for the possibility of such a product, namely that of causality through ends, ever being canceled out; and absolutely no human reason (or even any finite reason that is similar to ours in quality, no matter how much it exceeds it in degree) can ever hope to understand the generation of even a little blade of grass from merely mechanical causes. For if the teleological

connection of causes and effects is entirely indispensable for the possibility of such an object for the power of judgment, even merely for studying it with the guidance of experience; if for outer objects, as appearances, a sufficient ground related to causes cannot even be found, but this, which also lies in nature, must still be sought only in its supersensible substratum, from all possible insight into which we are cut off: then it is absolutely impossible for us to draw from nature itself any explanatory grounds for purposive connections, and in accordance with the constitution of the human cognitive faculty it is necessary to seek the highest ground of such connections in an original understanding as cause of the world.

ON THE UNIFICATION OF THE PRINCIPLE OF THE UNIVERSAL MECHANISM OF MATTER WITH THE TELEOLOGICAL PRINCIPLE IN THE TECHNIQUE OF NATURE

It is of infinite importance to reason that it not allow the mechanism of nature in its productions to drop out of sight and be bypassed in its explanations; for without this no insight into the nature of things can be attained. As soon as it is granted to us that a highest architect immediately created the forms of nature as they have always existed or has predetermined those which in their course are continuously formed in accordance with one and the same model, our cognition of nature is not thereby in the least advanced, because we do not know the mode of action of such a being and the ideas which should contain the principles of the possibility of natural beings at all, and we cannot explain nature from that being as if from above (*a priori*). But if, in order to explain the forms of the objects of experience from below (*a posteriori*), we appeal from them to a cause acting in accordance with ends because we believe that we find purposiveness in these forms, then our explanation would be entirely tautological, and reason would be deceived with words, not to mention that where we stray into excess with this sort of explanation, where knowledge of nature cannot follow us, reason is seduced into poetic enthusiasm, although the avoidance of this is precisely reason's highest calling.

On the other hand, it is an equally necessary maxim of reason not to bypass the principle of ends in the products of nature, because even though this principle does not make the way in which these products have originated more comprehensible, it is still a heuristic principle for researching the particular laws of nature, even granted that we would want to make no use of it for explaining nature itself, since although nature obviously displays an intentional unity of purpose we still always call that a merely natural end, i.e., we do not seek the ground of its possibility beyond nature. But since the question of the latter must ultimately still arise, it is just as necessary to conceive of a particular kind of causality for it that is not, unlike the mechanism of natural causes, found in nature, since to the receptivity to various and different forms than those of which matter is capable in accordance with that mechanism there must still be added the spontaneity of a cause (which thus cannot be matter) without which no ground of those forms could be given. Of course, before reason takes this step, it must proceed carefully, and not attempt to explain every technique of nature, i.e., a productive capacity in it which displays purposiveness of form for our mere apprehension in itself (as in the case of regular bodies), as teleological, but must instead always regard these as possible merely mechanically; but to exclude the teleological principle entirely, and always to stick with mere mechanism even where purposiveness, for the rational investigation of the possibility of natural forms by means of their causes, undeniably manifests itself as a relation to another kind of causality, must make reason fantastic and send it wandering about among figments of natural capacities that cannot even be conceived, just as a merely teleological mode of explanation which takes no regard of the mechanism of nature makes it into mere enthusiasm.

The two principles cannot be united in one and the same thing in nature as fundamental principles for the explanation (deduction) of one from the other, i.e., as dogmatic and constitutive principles of insight into nature for the determining power of judgment. If, e.g., I assume that a maggot can be regarded as a product of the mere mechanism of matter (a new formation that it produces for itself when its elements are set free by putrefaction), I cannot derive the very same product from the very same matter as a causality acting according to ends. Conversely, if I assume that the same product is a natural end, I cannot count on a mechanical mode of generation for it and take that as a

constitutive principle for the judging of its possibility, thus uniting both principles. For one kind of explanation excludes the other, even on the supposition that objectively both grounds of the possibility rest on a single one, but one of which we take no account. The principle which is to make possible the unifiability of both in the judging of nature in accordance with them must be placed in what lies outside of both (hence outside of the possible empirical representation of nature) but which still contains the ground of both, i.e., in the supersensible, and each of these two kinds of explanation must be related to that. Now since we can have no concept of this except the undetermined concept of a ground that makes the judging of nature in accordance with empirical laws possible, but cannot determine this more precisely by any predicate, it follows that the unification of the two principles cannot rest on a ground for the explanation (explication) of the possibility of a product in accordance with given laws for the determining power of judgment, but only on a ground for the elucidation (exposition) of this for the reflecting power of judgment.[29]—For to explain means to derive from a principle, which one must therefore cognize distinctly and be able to provide. Now of course the principle of the mechanism of nature and that of its causality according to ends in one and the same product of nature must cohere in a single higher principle and flow from it in common, because otherwise they could not subsist alongside one another in the consideration of nature. But if this objectively common principle, which also justifies the commonality of the maxims of natural research that depend upon it, is such that it can be indicated but can never be determinately cognized and distinctly provided for use in actual cases, then from such a principle there can be drawn no explanation, i.e., a distinct and determinate derivation of the possibility of a natural product that is possible in accordance with those two heterogeneous principles. Now, however, the common principle of the mechanical derivation on the one side and the teleological on the other is the supersensible, on which we must base nature as phenomenon. But from a theoretical point of view, we cannot form the least affirmative determinate concept of this. Thus how in accordance with this, as a principle, nature (in accordance with its particular laws) constitutes a principle for us, which could be cognized as possible in accordance with the principle of generation from physical as well as from final causes, can by no means be explained; rather, if it happens that we are presented with objects of nature the possibility of which we cannot conceive in accordance with the principle of mechanism (which always has a claim on any natural being) without appeal to teleological principles, then we can only presuppose that we may confidently research the laws of nature (as far as the possibility of their product is cognizable from one or the other principle of our understanding) in accordance with both of these principles, without being troubled by the apparent conflict between the two principles for judging this product; for at least the possibility that both may be objectively unifiable in one principle (since they concern appearances that presuppose a supersensible ground) is secured.

Thus even though the mechanism as well as the teleological (intentional) technicism of nature can stand, with regard to one and the same product and its possibility, under a common higher principle for the particular laws of nature, still, since this principle is transcendent, we cannot, given the limitation of our understanding, unite both principles in the explanation of one and the same natural generation, even if the inner possibility of this product is only intelligible through a causality according to ends (as is the case with organized matter). The above fundamental principle of teleology thus stands, namely, that given the constitution of the human understanding, only intentionally acting causes for the possibility of organic beings in nature can be assumed, and the mere mechanism of nature cannot be adequate at all for the explanation of these products of it—even though nothing is to be decided with regard to the possibility of such things themselves by means of this fundamental principle.

That is, since this is only a maxim of the reflecting, not of the determining power of judgment, and hence is valid only subjectively for us, not objectively for the possibility of this sort of thing itself (where both sorts of generation could well cohere in one and the same ground); since, further, without the concept of a mechanism of nature that is also to be found together with any teleologically conceived kind of generation such a generation could not be judged as a product of nature at all, the above maxim leads to the necessity of a unification of both principles in the judging of things as natural ends, but not in order to put one wholly or partly in place of the other. For in

the place of that which (at least for us) can only be conceived of as possible in accordance with an intention no mechanism can be assumed; and in the place of that which can be cognized as necessary in accordance with the latter, no contingency, which would require an end as its determining ground, can be assumed; rather, the one (mechanism) can only be subordinated to the other (intentional technicism), which, in accordance with the transcendental principle of the purposiveness of nature, can readily be done.

For where ends are conceived as grounds of the possibility of certain things, there one must also assume means the laws of the operation of which do not of themselves need anything that presupposes an end, which can thus be mechanical yet still be a cause subordinated to intentional effects. Hence even in organic products of nature, but even more if, prodded to do so by their infinite multitude, we assume that intentionality in the connection of natural causes in accordance with particular laws is also (at least as a permissible hypothesis) the universal principle of the reflecting power of judgment for the whole of nature (the world), we can conceive a great and even universal connection of the mechanical laws with the teleological ones in the productions of nature, without confusing the principles for judging it with one another and putting one in the place of the other, because in a teleological judging of matter, even if the form which it assumes is judged as possible only in accord with an intention, still its nature, in accordance with mechanical laws, can also be subordinated as a means to that represented end; likewise, since the ground of this unifiability lies in that which is neither the one nor the other (neither mechanism nor connection to an end) but is the supersensible substratum of nature, of which we can cognize nothing, the two ways of representing the possibility of such objects are not to be fused into one for our (human) reason, but rather we cannot judge them other than as a connection of final causes grounded in a supreme understanding, by which nothing is taken away from the teleological kind of explanation.

But now since how much the mechanism of nature as a means contributes to each final end in it is entirely undetermined and for our reason also forever undeterminable, and, on account of the above mentioned intelligible principle of the possibility of a nature in general, it can be assumed that nature is completely possible in accordance with both of the universally consonant laws (the physical laws and those of final causes), although we can have no insight at all into the way in which this happens, we also do not know how far the mechanical mode of explanation that is possible for us will extend, but are only certain of this much, namely, that no matter how far we ever get with that, it will still always be inadequate for things that we once acknowledge as natural ends, and, given the constitution of our understanding, we must always subordinate all such mechanical grounds to a teleological principle.

Now on this is grounded the authorization and, on account of the importance that the study of nature in accordance with the principle of mechanism has for our theoretical use of reason, also the obligation to give a mechanical explanation of all products and events in nature, even the most purposive, as far as it is in our capacity to do so (the limits of which within this sort of investigation we cannot determine), but at the same time never to lose sight of the fact that those which, given the essential constitution of our reason, we can, in spite of those mechanical causes, subject to investigation only under the concept of an end of reason, must in the end be subordinated to causality in accordance with ends.

NOTES

This chapter originally appeared as sections §62–68 and §69–78 in Paul Guyer (Ed./trans.) and Eric Matthews (trans.), *Critique of the teleological power of judgment*, pp. 235–284 (notes pp. 388–391), Cambridge, UK: Cambridge University Press, 2001.

REFERENCES

For early materials concerning Kant's work, including original editions of Kant's works and works by innumerable early expositors and opponents, consult: Erich Adickes, *German Kantian bibliography: Bibliography of writings by and on Kant which have appeared in Germany up to the end of 1887*, New York: Burt Franklin, 1970.

1. Barnes, J. (1984). *The complete works of Aristotle*, trans. R. P. Hardie and R. K. Gaye. Princeton: Princeton University Press.

2. Cooper, J. M. (Ed.) (1997). *Plato, complete works*, trans. G. M. A. Grube. Indianapolis: Hackett.

3. Förster, E. & Rosen, M. (Eds.) (1995). *Opus postumum*, au. I. Kant. Cambridge, UK: Cambridge University Press.

4. Freeman, K. (1966). *Ancilla to the pre-Socratic philosophers*. Cambridge, MA: Harvard University Press.

5. Hastie, W. (1969). *The universal natural history and theory of heaven*. Ann Arbor, MI: University of Michigan Press.

6. Kant, I. (1998). *Critique of pure reason,* trans. P. Guyer and A. W. Wood. Cambridge, UK: Cambridge University Press.

7. Keyes, C. W. (trans.) (1928). *Volume XVI: De re publica, de legibus*, au. M. T. Cicero. Cambridge, MA: Harvard University Press.

8. Oates, W. J. (1957). *The Stoic and Epicurean philosophers: The complete extant writings of Epicurus, Epictetus, Lucretius, Marcus Aurelius*. New York: Modern Library.

ENDNOTES

Lettered notes are from Kant's original text.

1. For Kant's view of Plato's theory of ideas, see *Critique of Pure Reason* (1998), pp. 369–675, A313–19/B.

2. See Socrates's criticism of Anaxagoras: "This man made no use of Mind, nor gave it any responsibility for the management of things, but mentioned as causes air and ether and water and many other strange things" (*Paedo* 98b–c in Cooper 1997, p. 85).

a. Since in pure mathematics there can never be an issue of the existence of things, but only of their possibility, namely the possibility of an intuition corresponding to their concept, and hence there can never be an issue of cause and effect, all of the purposiveness that has been noted there must therefore be considered merely as formal, never as a natural end.

3. For Kant's account of the geological effects of rivers and seas, see *Only Possible Basis for a Demonstration of God*, 2:128–9, and *Physical Geography* (edited from Kant's notes by F. T. Rink in 1802), 9:296–9.

4. To this paragraph, compare *Toward Perpetual Peace*, 8:363–5. Kant had argued as early as *Only Possible Basis*, 2:131, that it is a mistake to infer immediately from the fact that certain natural conditions seem advantageous to human beings to the conclusion that they have been purposively designed to be so.

5. In the *Only Possible Basis*, Kant had distinguished between the case in which a harmonious diversity of effects arises from a single law, which he held to be characteristic of inanimate nature, from the case in which a harmony of effects arises from distinct types of causation, which he took to be characteristic

of animate objects in nature (2:107–8); however, he then went on to argue that our failure to explain everything in nature according to mechanical laws should not be taken as a valid ground for a belief in a supersensible ground of nature, because such a supersensible ground—for which, of course, the work does argue—could clearly achieve all of its intended effects in nature through the mechanical operations of the laws that it institutes as the very conditions of the possibility of nature (see 2:114–5).

b I see it as a trace of a human being.

6. In a discussion of the value of learning over material goods, Cicero writes: "Who in truth would consider anyone... happier than one who is set free from all perturbations of mind, or more secure in his wealth than one who possesses only what, as the saying goes, he can carry away with him out of a shipwreck.... In this connection the remark made by Plato, or perhaps by someone else, seems to me particularly apt. For when a storm at sea had driven him to an unknown land and stranded him on a deserted shore, and his companions were frightened on account of their ignorance of the country, he, according to the story, noticed certain geometrical figures traced in the sand, and immediately cried out, 'Be of courage; I see the tracks of men.' He drew his inference, evidently, not from the cultivation of the soil, which he also observed, but from the indications of learning" (*De Re Publica*, I.XVII.28–30 in Keyes (trans.) 1928, pp. 51–53). According to this edition, the saying is also found in Vitruvius, *De Architectura*, VI, where it is attributed to Aristippus (435–366 B.C.) rather than Plato.

7. Kant first used the example of the tree to illustrate problems in explanation at *Only Possible Basis*, 2:114–5.

8. Kant argues that it is only from the case of our own intentional actions that we can originally form the concept of purposiveness in the 1788 essay *On the Use of Teleological Principles in Philosophy*, 8:181. But Kant's example is ancient: Aristotle used the example of building a house in his illustration of the four causes at *Physics*, book II, chapter 3, 195b3–5, 20–21.

9. Kant stresses both the analogy between a living thing and a work of art and the limits of this analogy, about to be explicated, at *Religion within the Boundaries of Mere Reason*, 6:64–5n, a note which ends with an allusion back to the present argument.

10. For Kant's concept of the nature and limits of mechanical forces, see *Metaphysical Foundations of Natural Science*, "General Observation on Dynamics," especially 4:525, 530,532–3.

c One can, conversely, illuminate a certain association, though one that is encountered more in the idea than in reality, by means of an analogy with the immediate ends of nature that have been mentioned. Thus, in the case of a recently undertaken fundamental transformation of a great people into a state, the word organization has frequently been quite appropriately used for the institution of the magistracies, etc., and even of the entire body politic. For in such a whole each member should certainly be not merely a means, but at the same time also an end, and, insofar as it contributes to the possibility of the whole, its position and function should also be determined by the idea of the whole.[11]

11. This note could refer to either of the two recent revolutions that Kant had followed with great interest, the American Revolution of 1776–83, or the French Revolution, which was in its early months when Kant wrote the present work. But it would seem somewhat strange for Kant to describe the American Revolution as the transformation of a "great nation" (*Volk*) into a state, since the American population, not all of whom in any case supported the revolution, would not be likely to be thought of as a single "people" or "nation." Several years later, when Kant spoke of the danger of anarchy as destroying the organization of a people before it could be replaced with another, he clearly had the later stages of the French Revolution in mind (see *Theory and Practice*, 8:302n.).

12. For a precursor to this definition, see *Teleological Principles*, 8:181. For later versions of it, see passages in the *Opus Postumum*, such as "The definition of an organic body is that it is a body, every part of which is *there for the sake of the other* (reciprocally as end and, at the same time, means). It is easily seen that this is a mere idea, which is not assured of reality *a priori* (i.e., that such a thing could exist)" (21:210; Förster and Rosen 1995, p. 64; see also pp. 100 and 146).

13. "New Holland" was an eighteenth-century designation for Australia prior to its colonization by the British, so Kant's term "New Hollanders" refers to the aboriginal people of Australia. By "Fuegians" he refers to the aboriginal inhabitants of Tierra del Fuego, Argentina.

14. Here Kant introduces his central argument that once we have been led to introduce the idea of purposive systematicity by our experience of organisms, it is then natural for us to attempt to see whether we might not also see the whole of nature as a purposive system. This idea was anticipated in *Pure Reason*, A 691/B 719. In *Teleological Principles*, Kant argues that the idea of a purpose for the whole of nature, introduced by the analogy between organized beings and human artistic production, has to be

"empirically conditioned" by what we actually observe in nature (8:182). This restriction is part of what is expressed in the present work by calling the principle of the purposiveness of nature a regulative rather than constitutive principle.

15. In *Anthropology from a Pragmatic Point of View*, Kant defines an affect as "a feeling of pleasure or displeasure in his present state that does not let [a person] rise to reflection (to rational consideration of whether he should give himself up to it or refuse it)" (§ 73, 7:251); in other words, an affect is an emotional state that threatens the ability of reason to control our conduct.

16. Kant frequently discussed the nature of dreams; see *Anthropologic Collins*, 25:101–2; *Anthropologie Friedländner*, 25:1283–8; *Menschenkunde*, 25:995–7, *Anthropologie Mrongovius*, 25:1283–8, and *Anthropology from a Pragmatic Point of View*, § 37, 7:189–90.

^d In the aesthetic part it was said that we would look on nature with favor insofar as we have an entirely free (disinterested) satisfaction in its form. For in this mere judgment of taste there is no regard for the end for which these natural beauties exist, whether to arouse pleasure in us or without any relation to us as ends. In a teleological judgment, however, we do attend to this relation, and then we can regard it as a favor of nature that by means of the exhibition of so many beautiful shapes it would promote culture.

17. To this entire section, compare Kant's discussion of the "Rules of a Revised Method of Physico-Theology" in *Only Possible Basis*, 2:126–7.

^e The German word *vermessen* ("presumptuous") is a good, meaningful word. A judgment in which we forget to take the proper measure of our powers (of understanding) can sound very modest and yet make great claims and be very presumptuous. Most of the judgments by means of which we purport to exalt the divine wisdom are like this, since in them we ascribe intentions to the works of creation and preservation that are really intended to do honor to our own wisdom as subtle thinkers.

18. Kant alludes here to the account of judgment that he gave in *Pure Reason*, A 131–235/B 164–194.

19. See *Letter to Herodotus:* "The motions of the heavenly bodies and their turnings and eclipses and risings and settings, and kindred phenomena to these, must not be thought to be due to any being who controls and ordains or has ordained them... Nor again, must we believe that they, which are but fire agglomerated in a mass, possess blessedness, and voluntarily take upon themselves these movements" (Oates 1957, p. 13).

20. Democritus of Abdera flourished ca. 460 BC. Large numbers of titles of his works on nature and causes are reported by ancient sources, but none of his surviving fragments actually expound what Kant here calls the system of causality (see Freeman 1966, pp. 91–120). Aristotle says, "if we look at the ancients, natural science would seem to be concerned with the *matter*. (It was only very slightly that Empedocles and Democritus touched on form and essence)" (*Physics*, book II, chapter 2, 194a18–20 in Barnes (Ed.), p. 331).

21. In his reference to Spinoza, Kant could have in mind statements like the following from Spinoza's *Ethics:* "I shall show... that neither intellect nor will pertain to God's nature. Of course I know there are many who think they can demonstrate that a supreme intellect and a free will pertain to God's nature. For they say they know nothing they can ascribe to God more perfect than what is the highest perfection in us" (*Ethics*, part one, proposition 17, scholium, in Curley (Ed.) 1985, p. 426); or *"Things could have been produced by God in no other way, and in no other order than they have been produced. For all things have necessarily followed from God's given nature, and have been determined from the necessity of God's nature to exist and produce an effect in a certain way"* (*Ethics*, part one, proposition 33, in Curley (Ed.) 1985, p. 436).

22. For further comments on hylozoism, see *Dreams of a Spirit-Seer* (1766), 2:330, and *Metaphysical Foundations of Natural Science*, 4:544.

^f One sees from this that in most speculative matters of pure reason the philosophical schools have usually tried all of the solutions that are possible for a certain question concerning dogmatic assertions. Thus for the sake of the purposiveness of nature either lifeless matter or a lifeless God as well as living matter or a living God have been tried. Nothing is left for us except, if need be, to give up all these objective assertions and to weigh our judgment critically, merely in relation to our cognitive faculty, in order to provide its principle with the non-dogmatic but adequate validity of a maxim for the reliable use of reason.

23. See *Metaphysik L₁*, 28:275.

24. Kant presumably refers to the strict sense of organization as reciprocal causation described in §§ 63 and 64 above.*

* Ed. note: §§ 2 and 3 in this volume.

25. See *Universal Natural History*, 1:229–30 (Hastie 1969, pp. 28–9).

26. See Kant's account of the empirical conditions for the application of the concepts of possibility and actuality in *Pure Reason*, "The postulates of empirical thinking in general," A 218–35/B 265–87.

27. In *Pure Reason*, Kant distinguishes the human intellect, which requires both intuitions and concepts for knowledge, from an intellectual intuition in which concepts would also give their own objects, at B 145, B 150, A 252/B 308–9, and A 256/B 311–2. For an early statement of the discursive nature of human understanding, see R 1832 (1772–75?), 16:131.

28. Here Kant refers to the doctrine of the *Critique of Pure Reason* that space is not an aggregate constituted out of regions of space that exist independently of it, as parts, but is rather a whole into which particular regions are introduced only by introducing limits into it; see A 25/B 39.

29. Kant's comment on these terms in *Pure Reason*, A 720/B 758, suggests that it is not clear what the precise differences between them are.

4 · What is Life? (selections)

ERWIN SCHRÖDINGER

THE GENERAL CHARACTER AND THE PURPOSE OF THE INVESTIGATION

This little book arose from a course of public lectures, delivered by a theoretical physicist to an audience of about 400 which did not substantially dwindle, though warned at the outset that the subject-matter was a difficult one and that the lectures could not be termed popular, even though the physicist's most dreaded weapon, mathematical deduction, would hardly be utilized. The reason for this was not that the subject was simple enough to be explained without mathematics, but rather that it was much too involved to be fully accessible to mathematics. Another feature which at least induced a semblance of popularity was the lecturer's intention to make clear the fundamental idea, which hovers between biology and physics, to both the physicist and the biologist.

For actually, in spite of the variety of topics involved, the whole enterprise is intended to convey one idea only—one small comment on a large and important question. In order not to lose our way, it may be useful to outline the plan very briefly in advance.

The large and important and very much discussed question is:

How can the events *in space and time* which take place within the spatial boundary of a living organism be accounted for by physics and chemistry?

The preliminary answer which this little book will endeavor to expound and establish can be summarized as follows:

The obvious inability of present-day physics and chemistry to account for such events is no reason at all for doubting that they can be accounted for by those sciences.

STATISTICAL PHYSICS. THE FUNDAMENTAL DIFFERENCE IN STRUCTURE

That would be a very trivial remark if it were meant only to stimulate the hope of achieving in the future what has not been achieved in the past. But the meaning is very much more positive, viz. that the inability, up to the present moment, is amply accounted for.

To-day, thanks to the ingenious work of biologists, mainly of geneticists, during the last 30 or 40 years, enough is known about the actual material structure of organisms and about their functioning to state that, and to tell precisely why, present-day physics and chemistry could not possibly account for what happens in space and time within a living organism.

The arrangements of the atoms in the most vital parts of an organism and the interplay of these arrangements differ in a fundamental way from all those arrangements of atoms which physicists and chemists have hitherto made the object of their experimental and theoretical research. Yet the difference which I have just termed fundamental is of such a kind that it might easily appear slight to anyone except a physicist who is thoroughly imbued with the knowledge that the laws of physics and chemistry are statistical throughout.[1] For it is in relation to the statistical point of view that the structure of the vital parts of living organisms differs so entirely from that of any piece of matter that we physicists and chemists have ever handled physically in our laboratories or mentally at our writing desks.[2] It is well-nigh unthinkable that the laws and regularities thus discovered should happen to apply immediately to the behaviour of

The Nature of Life, ed. M. A. Bedau and C. E. Cleland. Published by Cambridge University Press. © M. A. Bedau and C. E. Cleland 2010.

systems which do not exhibit the structure on which those laws and regularities are based.

The non-physicist cannot be expected even to grasp—let alone to appreciate the relevance of—the difference in "statistical structure" stated in terms so abstract as I have just used. To give the statement life and colour, let me anticipate, what will be explained in much more detail later, namely, that the most essential part of a living cell—the chromosome fibre—may suitably be called *an aperiodic crystal*. In physics we have dealt hitherto only with *periodic crystals*. To a humble physicist's mind, these are very interesting and complicated objects; they constitute one of the most fascinating and complex material structures by which inanimate nature puzzles his wits. Yet, compared with the aperiodic crystal, they are rather plain and dull. The difference in structure is of the same kind as that between an ordinary wallpaper in which the same pattern is repeated again and again in regular periodicity and a masterpiece of embroidery, say a Raphael tapestry, which shows no dull repetition, but an elaborate, coherent, meaningful design traced by the great master.

In calling the periodic crystal one of the most complex objects of his research, I had in mind the physicist proper. Organic chemistry, indeed, in investigating more and more complicated molecules, has come very much nearer to that "aperiodic crystal" which, in my opinion, is the material carrier of life. And therefore it is small wonder that the organic chemist has already made large and important contributions to the problem of life, whereas the physicist has made next to none.

THE NAÏVE PHYSICIST'S APPROACH TO THE SUBJECT

After having thus indicated very briefly the general idea—or rather the ultimate scope—of our investigation, let me describe the line of attack.

I propose to develop first what you might call "a naïve physicist's ideas about organisms", that is, the ideas which might arise in the mind of a physicist who, after having learnt his physics and, more especially, the statistical foundation of his science, begins to think about organisms and about the way they behave and function and who comes to ask himself conscientiously whether he, from what he has learnt, from the point of view of his comparatively simple and clear and humble science, can make any relevant contributions to the question.

It will turn out that he can. The next step must be to compare his theoretical anticipations with the biological facts. It will then turn out that—though on the whole his ideas seem quite sensible—they need to be appreciably amended. In this way we shall gradually approach the correct view—or, to put it more modestly, the one that I propose as the correct one.

Even if I should be right in this, I do not know whether my way of approach is really the best and simplest. But, in short, it was mine. The "naïve physicist" was myself. And I could not find any better or clearer way towards the goal than my own crooked one.

WHY ARE THE ATOMS SO SMALL?

A good method of developing "the naïve physicist's ideas" is to start from the odd, almost ludicrous, question: Why are atoms so small? To begin with, they are very small indeed. Every little piece of matter handled in everyday life contains an enormous number of them. Many examples have been devised to bring this fact home to an audience, none of them more impressive than the one used by Lord Kelvin: Suppose that you could mark the molecules in a glass of water; then pour the contents of the glass into the ocean and stir the latter thoroughly so as to distribute the marked molecules uniformly throughout the seven seas; if then you took a glass of water anywhere out of the ocean, you would find in it about a hundred of your marked molecules.[3]

The actual sizes of atoms[4] lie between about $\frac{1}{5000}$ and $\frac{1}{2000}$ of the wavelength of yellow light. The comparison is significant, because the wavelength roughly indicates the dimensions of the smallest grain still recognizable in the microscope. Thus it will be seen that such a grain still contains thousands of millions of atoms.

Now, why are atoms so small?

Clearly, the question is an evasion. For it is not really aimed at the size of the atoms. It is concerned with the size of organisms, more particularly with the size of our own corporeal selves. Indeed, the atom is small, when referred to our civic unit of length, say the yard or the meter. In atomic physics one is accustomed to use the so-called Ångström (abbr. A.), which is the 10^{10}th part of a meter, or in decimal notation 0.0000000001 metre. Atomic diameters range between 1 and 2 A. Now those civic units (in relation to which the atoms are so small) are closely related to the size of our bodies. There is a story tracing the yard back to the humor of an English king whom his councillors asked what unit to adopt—and he stretched out his arm sideways and said: "Take the distance from the middle of my chest to my fingertips, that will do all right." True or not, the story is significant for our purpose. The king would naturally indicate a length comparable with that of his own body, knowing that anything else would be very inconvenient. With all his predilection for the Ångström unit, the physicist prefers to be told that his new suit will require six and a half yards of tweed—rather than sixty-five thousand millions of Ångströms of tweed.

It thus being settled that our question really aims at the ratio of two lengths—that of our body and that of the atom—with an incontestable priority of independent existence on the side of the atom, the question truly reads: Why must our bodies be so large compared with the atom?

I can imagine that many a keen student of physics or chemistry may have deplored the fact that every one of our sense organs, forming a more or less substantial part of our body and hence (in view of the magnitude of the said ratio) being itself composed of innumerable atoms, is much too coarse to be affected by the impact of a single atom. We cannot see or feel or hear the single atoms. Our hypotheses with regard to them differ widely from the immediate findings of our gross sense organs and cannot be put to the test of direct inspection.

Must that be so? Is there an intrinsic reason for it? Can we trace back this state of affairs to some kind of first principle, in order to ascertain and to understand why nothing else is compatible with the very laws of Nature?

Now this, for once, is a problem which the physicist is able to clear up completely. The answer to all the queries is in the affirmative.

THE WORKING OF AN ORGANISM REQUIRES EXACT PHYSICAL LAWS

If it were not so, if we were organisms so sensitive that a single atom, or even a few atoms, could make a perceptible impression on our senses—Heavens, what would life be like! To stress one point: an organism of that kind would most certainly not be capable of developing the kind of orderly thought which, after passing through a long sequence of earlier stages, ultimately results in forming, among many other ideas, the idea of an atom.

Even though we select this one point, the following considerations would essentially apply also to the functioning of organs other than the brain and the sensorial system. Nevertheless, the one and only thing of paramount interest to us in ourselves is that we feel and think and perceive. To the physiological process which is responsible for thought and sense all the others play an auxiliary part, at least from the human point of view, if not from that of purely objective biology. Moreover, it will greatly facilitate our task to choose for investigation the process which is closely accompanied by subjective events, even though we are ignorant of the true nature of this close parallelism. Indeed, in my view, it lies outside the range of natural science and very probably of human understanding altogether.

We are thus faced with the following question: Why should an organ like our brain, with the sensorial system attached to it, of necessity consist of an enormous number of atoms, in order that its physically changing state should be in close and intimate correspondence with a highly developed thought? On what grounds is the latter task of the said organ incompatible with being, as a whole or in some of its peripheral parts which interact directly with the environment, a mechanism sufficiently refined and sensitive to respond to and register the impact of a single atom from outside?

The reason for this is, that what we call thought (1) is itself an orderly thing, and (2) can only be applied to material, i.e. to perceptions or experiences, which have a certain degree of orderliness. This has two consequences. First, a physical organization, to be in close correspondence with thought (as my brain is with my thought) must be a very well-ordered organization, and that means that the events that happen within it must obey strict physical laws, at least to a very high degree of accuracy. Secondly, the physical impressions made upon that physically well-organized system by other bodies from outside, obviously correspond to the perception and experience of the corresponding thought, forming its material, as I have called it. Therefore, the physical interactions between our system and others must, as a rule, themselves possess a certain degree of physical orderliness, that is to say, they too must obey strict physical laws to a certain degree of accuracy.

PHYSICAL LAWS REST ON ATOMIC STATISTICS AND ARE THEREFORE ONLY APPROXIMATE

And why could all this not be fulfilled in the case of an organism composed of a moderate number of atoms only and sensitive already to the impact of one or a few atoms only?

Because we know all atoms to perform all the time a completely disorderly heat motion, which, so to speak, opposes itself to their orderly behaviour and does not allow the events that happen between a small number of atoms to enroll themselves according to any recognizable laws. Only in the co-operation of an enormously large number of atoms do statistical laws begin to operate and control the behaviour of these *assemblées* with an accuracy increasing as the number of atoms involved increases. It is in that way that the events acquire truly orderly features. All the physical and chemical laws that are known to play an important part in the life of organisms are of this statistical kind; any other kind of lawfulness and orderliness that one might think of is being perpetually disturbed and made inoperative by the unceasing heat motion of the atoms.

THEIR PRECISION IS BASED ON THE LARGE NUMBER OF ATOMS INTERVENING. FIRST EXAMPLE (PARAMAGNETISM)

Let me try to illustrate this by a few examples, picked somewhat at random out of thousands, and possibly not just the best ones to appeal to a reader who is learning for the first time about this condition of things—a condition which in modern physics and chemistry is as fundamental as, say, the fact that

organisms are composed of cells is in biology, or as Newton's Law in astronomy, or even as the series of integers, 1, 2, 3, 4, 5, ... in mathematics. An entire newcomer should not expect to obtain from the following few pages a full understanding and appreciation of the subject, which is associated with the illustrious names of Ludwig Boltzmann and Willard Gibbs and treated in text-books under the name of "statistical thermodynamics."

If you fill an oblong quartz tube with oxygen gas and put it into a magnetic field, you find that the gas is magnetized. The magnetization is due to the fact that the oxygen molecules are little magnets and tend to orientate themselves parallel to the field, like a compass needle. But you must not think that they actually all turn parallel. For if you double the field, you get double the magnetization in your oxygen body, and that proportionality goes on to extremely high field strengths, the magnetization increasing at the rate of the field you apply.

This is a particularly clear example of a purely statistical law. The orientation the field tends to produce is continually counteracted by the heat motion, which works for random orientation. The effect of this striving is, actually, only a small preference for acute over obtuse angles between the dipole axes and the field. Though the single atoms change their orientation incessantly, they produce on the average (owing to their enormous number) a constant small preponderance of orientation in the direction of the field and proportional to it (Fig. 4.1). This ingenious explanation is due to the French physicist P. Langevin. It can be checked in the following way. If the observed weak magnetization is really the outcome of rival tendencies, namely, the magnetic field, which aims at combing all the molecules parallel, and the heat motion, which makes for random orientation, then it ought to be possible to increase the magnetization by weakening the heat motion, that is to say, by lowering the temperature, instead of reinforcing the field. That is confirmed by experiment, which gives the magnetization inversely proportional to the absolute temperature, in quantitative agreement with theory (Curie's law). Modern equipment even enables us, by lowering the temperature, to reduce the heat motion to such insignificance that the orientating tendency of the magnetic field can assert itself, if not completely, at least sufficiently to produce a substantial fraction of "complete

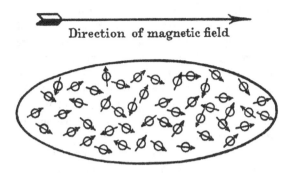

Fig. 4.1. Paramagnetism.

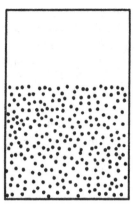

Fig. 4.2. Sinking fog.

magnetization". In this case we no longer expect that double the field strength will double the magnetization, but that the latter will increase less and less with increasing field, approaching what is called "saturation". This expectation too is quantitatively confirmed by experiment.

Notice that this behavior entirely depends on the large numbers of molecules which co-operate in producing the observable magnetization. Otherwise, the latter would not be constant at all, but would, by fluctuating quite irregularly from one second to the next, bear witness to the vicissitudes of the contest between heat motion and field.

SECOND EXAMPLE (BROWNIAN MOVEMENT, DIFFUSION)

If you fill the lower part of a closed glass vessel with fog, consisting of minute droplets, you will find that the upper boundary of the fog gradually sinks, with a well-defined velocity, determined by the viscosity of the air and the size and the specific gravity of the droplets. But if you look at one of the droplets under the microscope you find that it does not permanently sink with constant velocity, but performs a very irregular movement, the so-called Brownian movement, which corresponds to a regular sinking only on the average.

Now these droplets are not atoms, but they are sufficiently small and light to be not entirely insusceptible to the impact of one single molecule of those which hammer their surface in perpetual impacts. They are thus knocked about and can only on the average follow the influence of gravity (Figs. 4.2, 4.3).

Fig. 4.3. Brownian movement of a sinking droplet.

This example shows what funny and disorderly experience we should have if our senses were susceptible to the impact of a few molecules only. There are bacteria and other organisms so small that they are strongly affected by this phenomenon. Their

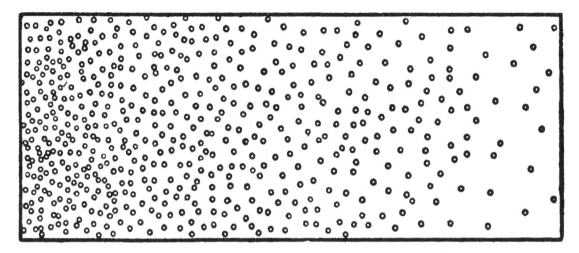

Fig. 4.4. Diffusion from left to right in a solution of varying concentration.

movements are determined by the thermic whims of the surrounding medium; they have no choice. If they had some locomotion of their own they might nevertheless succeed in getting from one place to another—but with some difficulty, since the heat motion tosses them like a small boat in a rough sea.

A phenomenon very much akin to Brownian movement is that of *diffusion*. Imagine a vessel filled with a fluid, say water, with a small amount of some colored substance dissolved in it, say potassium permanganate, not in uniform concentration, but rather as in Fig. 4.4, where the dots indicate the molecules of the dissolved substance (permanganate) and the concentration diminishes from left to right. If you leave this system alone a very slow process of "diffusion" sets in, the permanganate spreading in the direction from left to right, that is, from the places of higher concentration towards the places of lower concentration, until it is equally distributed through the water.

The remarkable thing about this rather simple and apparently not particularly interesting process is that it is in no way due, as one might think, to any tendency or force driving the permanganate molecules away from the crowded region to the less crowded one—like the population of a country spreading to those parts where there is more elbow-room. Nothing of the sort happens with our permanganate molecules. Every one of them behaves quite independently of all the others, which it very seldom meets. Every one of them,

whether in a crowded region or in an empty one, suffers the same fate of being continually knocked about by the impacts of the water molecules and thereby gradually moving on in an unpredictable direction—sometimes towards the higher, sometimes towards the lower, concentrations, sometimes obliquely. The kind of motion it performs has often been compared with that of a blindfolded person on a large surface, imbued with a certain desire of "walking", but without any preference for any particular direction, and so changing his line continuously.

That this random walk of the permanganate molecules, the same for all of them, should yet produce a regular flow towards the smaller concentration and ultimately make for uniformity of distribution, is at first sight perplexing—but only at first sight. If you contemplate in Fig. 4.4 thin slices of approximately constant concentration, the permanganate molecules which in a given moment are contained in a particular slice will, by their random walk, it is true, be carried with equal probability to the right or to the left. But precisely in consequence of this, a plane separating two neighboring slices will be crossed by more molecules coming from the left than in the opposite direction, simply because to the left there are more molecules engaged in random walk than there are to the right. And as long as that is so the balance will show up as a regular flow from left to right, until a uniform distribution is reached.

When these considerations are translated into mathematical language the exact law of diffusion is reached in the form of a partial differential equation

$$\frac{\partial \rho}{\partial t} = D\nabla^2 \rho,$$

which I shall not trouble the reader by explaining, though its meaning in ordinary language is again simple enough.[6] The reason for mentioning the stern "mathematically exact" law here, is to emphasize that its physical exactitude must nevertheless be challenged in every particular application. Being based on pure chance, its validity is only approximate. If it is, as a rule, a very good approximation, that is only due to the enormous number of molecules that co-operate in the phenomenon. The smaller their number, the larger the quite haphazard deviations we must expect—and they can be observed under favourable circumstances.

THIRD EXAMPLE (LIMITS OF ACCURACY OF MEASURING)

The last example we shall give is closely akin to the second one, but has a particular interest. A light body, suspended by a long thin fibre in equilibrium orientation, is often used by physicists to measure weak forces which deflect it from that position of equilibrium, electric, magnetic or gravitational forces being applied so as to twist it around the vertical axis. (The light body must, of course, be chosen appropriately for the particular purpose.) The continued effort to improve the accuracy of this very commonly used device of a "torsional balance", has encountered a curious limit, most interesting in itself. In choosing lighter and lighter bodies and thinner and longer fibres—to make the balance susceptible to weaker and weaker forces—the limit was reached when the suspended body became noticeably susceptible to the impacts of the heat motion of the surrounding molecules and began to perform an incessant, irregular "dance" about its equilibrium position, much like the trembling of the droplet in the second example. Though this behaviour sets no absolute limit to the accuracy of measurements obtained with the balance, it sets a practical one. The uncontrollable effect of the heat motion competes with the effect of the force to be measured and makes the single deflection observed insignificant. You have to multiply observations, in order to eliminate the effect of the Brownian movement of your instrument. This example is, I think, particularly illuminating in our present investigation. For our organs of sense, after all, are a kind of instrument. We can see how useless they would be if they became too sensitive.

THE $\sqrt{n}$ RULE

So much for examples, for the present. I will merely add that there is not one law of physics or chemistry, of those that are relevant within an organism or in its interactions with its environment, that I might not choose as an example. The detailed explanation might be more complicated, but the salient point would always be the same and thus the description would become monotonous.

But I should like to add one very important quantitative statement concerning the degree of inaccuracy to be expected in any physical law, the so-called $\sqrt{n}$ law. I will first illustrate it by a simple example and then generalize it.

If I tell you that a certain gas under certain conditions of pressure and temperature has a certain density, and if I expressed this by saying that within a certain volume (of a size relevant for some experiment) there are under these conditions just n molecules of the gas, then you might be sure that if you could test my statement in a particular moment of time, you would find it inaccurate, the departure being of the order of $\sqrt{n}$. Hence if the number $n = 100$, you would find a departure of about 10, thus relative error = 10%. But if $n = 1$ million, you would be likely to find a departure of about 1000, thus relative error $= \frac{1}{10}\%$. Now, roughly speaking, this statistical law is quite general. The laws of physics and physical chemistry are inaccurate within a probable relative error of the order of $1/\sqrt{n}$, where n is the number of molecules that co-operate to bring about that law—to produce its validity within such regions of space or time (or both) that matter, for some considerations or for some particular experiment.

You see from this again that an organism must have a comparatively gross structure in order to enjoy the benefit of fairly accurate laws, both for its internal life and for its interplay with the external world. For otherwise the number of co-operating particles would be too small, the "law" too inaccurate. The particularly exigent demand is the square root. For though a million is a reasonably large number, an accuracy of just 1 in 1000 is not overwhelmingly good, if a thing claims the dignity of being a "Law of Nature".

THE GENERAL PICTURE OF THE HEREDITARY SUBSTANCE

From these facts emerges a very simple answer to our question, namely: Are these structures, composed of comparatively few atoms, capable of withstanding for long periods the disturbing influence of heat motion to which the hereditary substance is continually exposed? We shall assume the structure of a gene to be that of a huge molecule, capable only of discontinuous change, which consists in a rearrangement of the atoms and leads to an isomeric[7] molecule. The rearrangement may affect only a small region of the gene, and a vast number of different rearrangements may be possible. The energy thresholds, separating the actual configuration from any possible isomeric ones, have to be high enough (compared with the average heat energy of an atom) to make the change-over a rare event. These rare events we shall identify with spontaneous mutations.

The later parts of this chapter will be devoted to putting this general picture of a gene and of mutation (due mainly to the German physicist M. Delbrück) to the test, by comparing it in detail with genetical facts. Before doing so, we may fittingly make some comment on the foundation and general nature of the theory.

THE UNIQUENESS OF THE PICTURE

Was it absolutely essential for the biological question to dig up the deepest roots and found the picture on quantum mechanics? The conjecture that a gene is a molecule is to-day, I dare say, a commonplace. Few biologists, whether familiar with quantum theory or not, would disagree with it. In § 32[8] we ventured to put it into the mouth of a pre-quantum physicist, as the only reasonable explanation of the observed permanence. The subsequent considerations about isomerism, threshold energy, the paramount role of the ratio $W: kT$ in determining the probability of an isomeric transition—all that could very well be introduced on a purely empirical basis, at any rate without drawing explicitly on quantum theory. Why did I so strongly insist on the quantum-mechanical point of view,

though I could not really make it clear in this little book and may well have bored many a reader?

Quantum mechanics is the first theoretical aspect which accounts from first principles for all kinds of aggregates of atoms actually encountered in Nature. The Heitler–London bondage is a unique, singular feature of the theory, not invented for the purpose of explaining the chemical bond. It comes in quite by itself, in a highly interesting and puzzling manner, being forced upon us by entirely different considerations. It proves to correspond exactly with the observed chemical facts, and, as I said, it is a unique feature, well enough understood to tell with reasonable certainty that "such a thing could not happen again" in the further development of quantum theory.

Consequently, we may safely assert that there is no alternative to the molecular explanation of the hereditary substance. The physical aspect leaves no other possibility to account for its permanence. If the Delbrück picture should fail, we would have to give up further attempts. That is the first point I wish to make.

SOME TRADITIONAL MISCONCEPTIONS

But it may be asked: Are there really no other endurable structures composed of atoms except molecules? Does not a gold coin, for example, buried in a tomb for a couple of thousand years, preserve the traits of the portrait stamped on it? It is true that the coin consists of an enormous number of atoms, but surely we are in this case not inclined to attribute the mere preservation of shape to the statistics of large numbers. The same remark applies to a neatly developed batch of crystals we find embedded in a rock, where it must have been for geological periods without changing.

That leads us to the second point I want to elucidate. The cases of a molecule, a solid, a crystal are not really different. In the light of present knowledge they are virtually the same. Unfortunately, school teaching keeps up certain traditional views, which have been out of date for many years and which obscure the understanding of the actual state of affairs.

Indeed, what we have learnt at school about molecules does not give the idea that they are more closely akin to the solid state than to the liquid or gaseous state. On the contrary, we have been taught to distinguish carefully between a physical change, such as melting or evaporation in which the molecules are preserved (so that, for example, alcohol, whether solid, liquid, or a gas, always consists of the same molecules, C_2H_6O), and a chemical change, as, for example, the burning of alcohol,

$$C_2H_6O + 3O_2 = 2CO_2 + 3H_2O,$$

where an alcohol molecule and three oxygen molecules undergo a rearrangement to form two molecules of carbon dioxide and three molecules of water.

About crystals, we have been taught that they form threefold periodic lattices, in which the structure of the single molecule is sometimes recognizable, as in the case of alcohol and most organic compounds, while in other crystals, e.g., rock-salt (NaCl), NaCl molecules cannot be unequivocally delimited, because every Na atom is symmetrically surrounded by six Cl atoms, and vice versa, so that it is largely arbitrary what pairs, if any, are regarded as molecular partners.

Finally, we have been told that a solid can be crystalline or not, and in the latter case we call it amorphous.

DIFFERENT "STATES" OF MATTER

Now I would not go so far as to say that all these statements and distinctions are quite wrong. For practical purposes they are sometimes useful. But in the true aspect of the structure of matter the limits must be drawn in an entirely different way. The fundamental distinction is between the two lines of the following scheme of "equations":

$$\text{molecule} = \text{solid} = \text{crystal.}$$
$$\text{gas} = \text{liquid} = \text{amorphous.}$$

We must explain these statements briefly. The so-called amorphous solids are either not really amorphous or not really solid. In "amorphous" charcoal fibre the rudimentary structure of the graphite crystal has been disclosed by X-rays. So charcoal is a solid, but also crystalline. Where we find no crystalline structure we have to regard the thing as a liquid with very high "viscosity" (internal friction). Such a substance discloses by the absence of a well-defined melting temperature and of a latent heat of melting that it is not a true solid. When heated it softens gradually and eventually liquefies without discontinuity. (I remember that at the end of the first Great War we were given in Vienna an asphalt-like substance as a

substitute for coffee. It was so hard that one had to use a chisel or a hatchet to break the little brick into pieces, when it would show a smooth, shell-like cleavage. Yet, given time, it would behave as a liquid, closely packing the lower part of a vessel in which you were unwise enough to leave it for a couple of days.)

The continuity of the gaseous and liquid state is a well-known story. You can liquefy any gas without discontinuity by taking your way "around" the so-called critical point. But we shall not enter on this here.

THE DISTINCTION THAT REALLY MATTERS

We have thus justified everything in the above scheme, except the main point, namely, that we wish a molecule to be regarded as a solid = crystal.

The reason for this is that the atoms forming a molecule, whether there be few or many of them, are united by forces of exactly the same nature as the numerous atoms which build up a true solid, a crystal. The molecule presents the same solidity of structure as a crystal. Remember that it is precisely this solidity on which we draw to account for the permanence of the gene!

The distinction that is really important in the structure of matter is whether atoms are bound together by those "solidifying" Heitler–London forces or whether they are not. In a solid and in a molecule they all are. In a gas of single atoms (as e.g., mercury vapor) they are not. In a gas composed of molecules, only the atoms within every molecule are linked in this way.

THE APERIODIC SOLID

A small molecule might be called "the germ of a solid". Starting from such a small solid germ, there seem to be two different ways of building up larger and larger associations. One is the comparatively dull way of repeating the same structure in three directions again and again. That is the way followed in a growing crystal. Once the periodicity is established, there is no definite limit to the size of the aggregate. The other way is that of building up a more and more extended aggregate without the dull device of repetition. That is the case of the more and more complicated organic molecule in which every atom, and every group of atoms, plays an individual role, not entirely equivalent to that of many others (as is the case in a periodic structure). We might quite properly call that an aperiodic crystal or solid and express our hypothesis by saying: We believe a gene—or perhaps the whole chromosome fibre[9]—to be an aperiodic solid.

THE VARIETY OF CONTENTS COMPRESSED IN THE MINIATURE CODE

It has often been asked how this tiny speck of material, the nucleus of the fertilized egg, could contain an elaborate code-script involving all the future development of the organism? A well-ordered association of atoms, endowed with sufficient resistivity to keep its order permanently, appears to be the only conceivable material structure, that offers a variety of possible ("isomeric") arrangements, sufficiently large to embody a complicated system of "determinations" within a small spatial boundary. Indeed, the number of atoms in such a structure need not be very large to produce an almost unlimited number of possible arrangements. For illustration, think of the Morse code. The two different signs of dot and dash in well-ordered groups of not more than four allow of thirty different specifications. Now, if you allowed yourself the use of a third sign, in addition to dot and dash, and used groups of not more than ten, you could form 29,524 different "letters"; with five signs and groups up to 25, the number is 372,529,029,846,191,405.

It may be objected that the simile is deficient, because our Morse signs may have different composition (e.g., • – – and • • –) and thus they are a bad analog for isomerism. To remedy this defect, let us pick, from the third example, only the combinations of exactly 25 symbols and only those containing exactly 5 out of each of the supposed 5 types (5 dots, 5 dashes, etc.). A rough count gives you the number of combinations as 62,330,000,000,000, where the zeros on the right stand for figures which I have not taken the trouble to compute.

Of course, in the actual case, by no means "every" arrangement of the group of atoms will represent a possible molecule; moreover, it is not a question of a code to be adopted arbitrarily, for the code-script must itself be the operative factor bringing about the development. But, on the other hand, the number chosen in the example (25) is still very small, and we have envisaged only the simple arrangements in one line. What we wish to illustrate is simply that with the molecular picture of the gene it is no longer inconceivable that the miniature code should precisely correspond with a highly complicated and specified plan of development and should somehow contain the means to put it into operation.

COMPARISON WITH FACTS: DEGREE OF STABILITY; DISCONTINUITY OF MUTATIONS

Now let us at last proceed to compare the theoretical picture with the biological facts. The first question obviously is, whether it can really account for the high degree of permanence we observe. Are threshold values of the required amount—high multiples of the average heat energy kT—reasonable, are they within the range known from ordinary chemistry? That question is trivial; it can be answered in the affirmative without inspecting tables. The molecules of any substance which the chemist is able to isolate at a given temperature must at that temperature have a lifetime of at least minutes. (That is putting it mildly; as a rule they have much more.) Thus the threshold values the chemist encounters are of necessity precisely of the order of magnitude required to account for practically any degree of permanence the biologist may encounter; for we recall from § 36[10] that thresholds varying within a range of about 1:2 will account for lifetimes ranging from a fraction of a second to tens of thousands of years.

But let me mention figures, for future reference. The ratios W/kT mentioned by way of example in § 36,[11] viz.

$$\frac{W}{kT} = 30, 50, 60,$$

producing lifetimes of

$$\frac{1}{10} \text{ second, 16 months, 30,000 years,}$$

respectively, correspond at room temperature with threshold values of

$$0.9, 1.5, 1.8 \text{ electron-volts.}$$

We must explain the unit "electron-volt", which is rather convenient for the physicist, because it can be visualized. For example, the third number (1.8) means that an electron, accelerated by a voltage of about 2 volts, would have acquired just sufficient energy to effect the transition by impact. (For comparison, the battery of an ordinary pocket flash-light has 3 volts.)

These considerations make it conceivable that an isomeric change of configuration in some part of our molecule, produced by a chance fluctuation of the vibrational energy can actually be a sufficiently rare event to be interpreted as a spontaneous mutation. Thus we account, by the very principles of quantum mechanics, for the most amazing fact about mutations, the fact by which they first attracted de Vries's attention, namely, that they are "jumping" variations, no intermediate forms occurring.

STABILITY OF NATURALLY SELECTED GENES

Having discovered the increase of the natural mutation rate by any kind of ionizing rays, one might think of attributing the natural rate to the radio-activity of the soil and air and to cosmic radiation. But a quantitative comparison with the X-ray results shows that the "natural radiation" is much too weak and could account only for a small fraction of the natural rate.

Granted that we have to account for the rare natural mutations by chance fluctuations of the heat motion, we must not be very much astonished that Nature has succeeded in making such a subtle choice of threshold values as is necessary to make mutation rare. For we have, earlier in these lectures, arrived at the conclusion that frequent mutations are detrimental to evolution. Individuals which, by mutation, acquire a gene configuration of insufficient stability, will have little chance of seeing their "ultra-radical", rapidly mutating descendancy survive long. The species will be freed of them and will thus collect stable genes by natural selection.

THE SOMETIMES LOWER STABILITY OF MUTANTS

But, of course, as regards the mutants which occur in our breeding experiments and which we select, *qua* mutants, for studying their offspring, there is no reason to expect that they should all show that very high stability. For they have not yet been "tried out"— or, if they have, they have been "rejected" in the wild breeds—possibly for too high mutability. At any rate, we are not at all astonished to learn that actually some of these mutants do show a much higher mutability than the normal "wild" genes.

TEMPERATURE INFLUENCES UNSTABLE GENES LESS THAN STABLE ONES

This enables us to test our mutability formula, which was

$$t = \tau e^{W/kT}.$$

(It will be remembered that t is the time of expectation for a mutation with threshold energy W.) We ask: How does t change with the temperature? We easily find from the preceding formula in good approximation the ratio of the value of t at temperature $T + 10$, to that at temperature T

$$\frac{t_T + 10}{t_T} = e^{-10 W/kT^2}$$

The exponent being now negative, the ratio is, naturally, smaller than 1. The time of expectation is diminished by raising the temperature, the mutability is increased. Now that can be tested and has been tested with the fly *Drosophila* in the range of temperature which the insects will stand. The result was, at first sight, surprising. The *low* mutability of wild genes was distinctly increased, but the comparatively *high* mutability occurring with some of the already mutated genes was not, or at any rate was much less, increased. That is just what we expect on comparing our two formulae. A large value of W/kT, which according to the first formula is required to make t large (stable gene), will, according to the second one, make for a small value of the ratio computed there, that is to say for a considerable increase of mutability with temperature. (The actual values of the ratio seem to lie between about $\frac{1}{2}$ and $\frac{1}{5}$. The reciprocal, 2.5, is what in an ordinary chemical reaction we call the van't Hoff factor.)

HOW X-RAYS PRODUCE MUTATION

Turning now to the X-ray-induced mutation rate, we have already inferred from the breeding experiments, first (from the proportionality of mutation rate, and dosage), that some single event produces the mutation; secondly (from quantitative results and from the fact that the mutation rate is determined by the integrated ionization density and independent of the wavelength), this single event must be an ionization, or similar process, which has to take place inside a certain volume of only about 10 atomic-distances-cubed, in order to produce a specified mutation. According to our picture, the energy for overcoming the threshold must obviously be furnished by that explosion-like process, ionization or excitation. I call it explosion-like, because the energy spent in one ionization (spent, incidentally, not by the X-ray itself, but by a secondary electron it produces) is well known and has the comparatively enormous amount of 30 electron-volts. It is bound to be turned into enormously increased heat motion around the point where it is discharged and to spread from there in the form of a "heat wave", a wave of intense oscillations of the atoms. That this heat wave should still be able to furnish the required threshold energy of 1 or 2 electron-volts at an average "range of action" of about ten atomic distances, is not inconceivable, though it may well be that an unprejudiced physicist might have anticipated a slightly lower range of action. That in many cases the effect of the explosion will not be an orderly isomeric transition but a lesion of the chromosome, a lesion that becomes lethal when, by ingenious crossings, the uninjured partner (the corresponding chromosome of the second set) is removed and replaced by a partner whose corresponding gene is known to be itself morbid—all that is absolutely to be expected and it is exactly what is observed.

THEIR EFFICIENCY DOES NOT DEPEND ON SPONTANEOUS MUTABILITY

Quite a few other features are, if not predictable from the picture, easily understood from it. For example, an unstable mutant does not on the average show a much higher X-ray mutation rate than a stable one. Now, with an explosion furnishing an energy of 30 electron-volts you would certainly not expect that it makes a lot of difference whether the required threshold energy is a little larger or a little smaller, say 1 or 1.3 volts.

REVERSIBLE MUTATIONS

In some cases a transition was studied in both directions, say from a certain "wild" gene to a specified mutant and back from that mutant to the wild gene. In such cases the natural mutation rate is sometimes nearly the same, sometimes very different. At first sight one is puzzled, because the threshold to be overcome seems to be the same in both cases. But, of course, it need not be, because it has to be measured from the energy level of the starting configuration, and that may be different for the wild and the mutated gene. (See Fig. 12 on p. 55[12], where "1" might refer to the wild allele, "2" to the mutant, whose lower stability would be indicated by the shorter arrow.)

On the whole, I think, Delbrück's "model" stands the tests fairly well and we are justified in using it in further considerations.

A REMARKABLE GENERAL CONCLUSION FROM THE MODEL

Let me refer to the last phrase in § 46,[13] in which I tried to explain that the molecular picture of the gene made it at least conceivable "that the miniature code should be in one-to-one correspondence with a highly complicated and specified plan of development and should somehow contain the means of putting it into operation". Very well then, but how does it do this? How are we going to turn "conceivability" into true understanding?

Delbrück's molecular model, in its complete generality, seems to contain no hint as to how the hereditary substance works. Indeed, I do not expect that any detailed information on this question is likely to come from physics in the near future. The advance is proceeding and will, I am sure, continue to do so, from biochemistry under the guidance of physiology and genetics.

No detailed information about the functioning of the genetical mechanism can emerge from a description of its structure so general as has been given above. That is obvious. But, strangely enough, there is just one general conclusion to be obtained from it, and that, I confess, was my only motive for writing this book.

From Delbrück's general picture of the hereditary substance it emerges that living matter, while not eluding the "laws of physics" as established up to date, is likely to involve "other laws of physics" hitherto unknown, which, however, once they have been revealed, will form just as integral a part of this science as the former.

ORDER BASED ON ORDER

This is a rather subtle line of thought, open to misconception in more than one respect. All the remaining pages are concerned with making it clear. A preliminary insight, rough but not altogether erroneous, may be found in the following considerations:

It has been explained in Chapter 1 that the laws of physics, as we know them, are statistical laws.[14] They have a lot to do with the natural tendency of things to go over into disorder.

But, to reconcile the high durability of the hereditary substance with its minute size, we had to evade the tendency to disorder by "inventing the molecule," in fact, an unusually large molecule which has to be a masterpiece of highly differentiated order, safeguarded by the conjuring rod of quantum theory. The laws of chance are not invalidated by this "invention", but their outcome is modified. The physicist is familiar with the fact that the classical laws of physics are modified by quantum theory, especially at low temperature. There are many instances of this. Life seems to be one of them, a particularly striking one. Life seems to be orderly and lawful behavior of matter, not based exclusively on its tendency to go over from order to disorder, but based partly on existing order that is kept up.

To the physicist—but only to him—I could hope to make my view clearer by saying: The living organism seems to be a macroscopic system which in part of its behaviour approaches to that purely mechanical (as contrasted with thermodynamical) conduct to which all systems tend, as the temperature approaches the absolute zero and the molecular disorder is removed.

The non-physicist finds it hard to believe that really the ordinary laws of physics, which he regards as the prototype of inviolable precision, should be based on the statistical tendency of matter to go over into disorder. I have given examples in Chapter 1.[15] The general principle involved is the famous Second Law of Thermodynamics (entropy principle) and its equally famous statistical foundation. In §§ 56–60[16] I will try to sketch the bearing of the entropy principle

on the large-scale behaviour of a living organism—forgetting at the moment all that is known about chromosomes, inheritance, and so on.

LIVING MATTER EVADES THE DECAY TO EQUILIBRIUM

What is the characteristic feature of life? When is a piece of matter said to be alive? When it goes on "doing something", moving, exchanging material with its environment, and so forth, and that for a much longer period than we would expect an inanimate piece of matter to "keep going" under similar circumstances. When a system that is not alive is isolated or placed in a uniform environment, all motion usually comes to a standstill very soon as a result of various kinds of friction; differences of electric or chemical potential are equalized, substances which tend to form a chemical compound do so, temperature becomes uniform by heat conduction. After that the whole system fades away into a dead, inert lump of matter. A permanent state is reached, in which no observable events occur. The physicist calls this the state of thermodynamical equilibrium, or of "maximum entropy."

Practically, a state of this kind is usually reached very rapidly. Theoretically, it is very often not yet an absolute equilibrium, not yet the true maximum of entropy. But then the final approach to equilibrium is very slow. It could take anything between hours, years, centuries, ... To give an example—one in which the approach is still fairly rapid: If a glass filled with pure water and a second one filled with sugared water are placed together in a hermetically closed case at constant temperature, it appears at first that nothing happens, and the impression of complete equilibrium is created. But after a day or so it is noticed that the pure water, owing to its higher vapor pressure, slowly evaporates and condenses on the solution. The latter overflows. Only after the pure water has totally evaporated has the sugar reached its aim of being equally distributed among all the liquid water available.

These ultimate slow approaches to equilibrium could never be mistaken for life, and we may disregard them here. I have referred to them in order to clear myself of a charge of inaccuracy.

IT FEEDS ON "NEGATIVE ENTROPY"

It is by avoiding the rapid decay into the inert state of 'equilibrium', that an organism appears so enigmatic; so much so, that from the earliest times of human thought some special non-physical or supernatural force (*vis viva*, entelechy) was claimed to be operative in the organism, and in some quarters is still claimed.

How does the living organism avoid decay? The obvious answer is: By eating, drinking, breathing and (in the case of plants) assimilating. The technical term is *metabolism*. The Greek word (μεταβάλλειυ) means change or exchange. Exchange of what? Originally the underlying idea is, no doubt, exchange of material (e.g., the German for metabolism is Stoffwechsel). That the exchange of material should be the essential thing is absurd. Any atom of nitrogen, oxygen, sulphur, etc., is as good as any other of its kind; what could be gained by exchanging them? For a while in the past our curiosity was silenced by being told that we feed upon energy. In some very advanced country (I don't remember whether it was Germany or the USA or both) you could find menu cards in restaurants indicating, in addition to the price, the energy content of every dish. Needless to say, taken literally, this is just as absurd. For an adult organism the energy content is as stationary as the material content. Since, surely, any calorie is worth as much as any other calorie, one cannot see how a mere exchange could help.

What then is that precious something contained in our food which keeps us from death? That is easily answered. Every process, event, happening—call it what you will; in a word, everything that is going on in Nature means an increase of the entropy of the part of the world where it is going on. Thus a living organism continually increases its entropy—or, as you may say, produces positive entropy—and thus tends to approach the dangerous state of maximum entropy, which is death. It can only keep aloof from it, i.e., alive, by continually drawing from its environment negative entropy—which is something very positive as we shall immediately see. What an organism feeds upon is negative entropy. Or, to put it less paradoxically, the essential thing in metabolism is that the organism succeeds in freeing itself from all the entropy it cannot help producing while alive.

WHAT IS ENTROPY?

What is entropy? Let me first emphasize that it is not a hazy concept or idea, but a measurable physical quantity just like the length of a rod, the temperature at any point of a body, the heat of fusion of a given crystal or the specific heat of any given substance. At the absolute zero point of temperature (roughly –273 °C) the entropy of any substance is zero. When you bring the substance into any other state by slow, reversible little steps (even if thereby the substance changes its physical or chemical nature or splits up into two or more parts of different physical or chemical nature) the entropy increases by an amount which is computed by dividing every little portion of heat you had to supply in that procedure by the absolute temperature at which it was supplied—and by summing up all these small contributions. To give an example, when you melt a solid, its entropy increases by the amount of the heat of fusion divided by the temperature at the melting-point. You see from this, that the unit in which entropy is measured is cal./°C. (just as the calorie is the unit of heat or the centimeter the unit of length).

THE STATISTICAL MEANING OF ENTROPY

I have mentioned this technical definition simply in order to remove entropy from the atmosphere of hazy mystery that frequently veils it. Much more important for us here is the bearing on the statistical concept of order and disorder, a connection that was revealed by the investigations of Boltzmann and Gibbs in statistical physics. This too is an exact quantitative connection, and is expressed by

$$\text{entropy} = k \, \log \, D,$$

where k is the so-called Boltzmann constant $(= 3.2983.10^{-24} \text{ cal./°C})$, and D a quantitative measure of the atomistic disorder of the body in question. To give an exact explanation of this quantity D in brief non-technical terms is well-nigh impossible. The disorder it indicates is partly that of heat motion, partly that which consists in different kinds of atoms or molecules being mixed at random, instead of being neatly separated, e.g., the sugar and water molecules in the example quoted above. Boltzmann's equation is well illustrated by that example. The gradual "spreading out" of the sugar over all the water available increases the disorder D, and hence (since the logarithm of D increases with D) the entropy. It is also pretty clear that any supply of heat increases the turmoil of heat motion, that is to say increases D and thus increases the entropy; it is particularly clear that this should be so when you melt a crystal, since you thereby destroy the neat and permanent arrangement of the atoms or molecules and turn the crystal lattice into a continually changing random distribution.

An isolated system or a system in a uniform environment (which for the present consideration we do best to include as a part of the system we contemplate) increases its entropy and more or less rapidly approaches the inert state of maximum entropy. We now recognize this fundamental law of physics to be just the natural tendency of things to approach the chaotic state (the same tendency that the books of a library or the piles of papers and manuscripts on a writing desk display) unless we obviate it. (The analog of irregular heat motion, in this case, is our handling those objects now and again without troubling to put them back in their proper places.)

ORGANIZATION MAINTAINED BY EXTRACTING "ORDER" FROM THE ENVIRONMENT

How would we express in terms of the statistical theory the marvellous faculty of a living organism, by which it delays the decay into thermodynamical equilibrium (death)? We said before: "It feeds upon negative entropy", attracting, as it were, a stream of negative entropy upon itself, to compensate the entropy increase it produces by living and thus to maintain itself on a stationary and fairly low entropy level.

If D is a measure of disorder, its reciprocal, $1/D$, can be regarded as a direct measure of order. Since the logarithm of $1/D$ is just minus the logarithm of D, we can write Boltzmann's equation thus:

$$-(\text{entropy}) = k \log(1/D).$$

Hence the awkward expression "negative entropy" can be replaced by a better one: Entropy, taken with the negative sign, is itself a measure of order. Thus the device by which an organism maintains itself stationary at a fairly high level of orderliness (= fairly low level of

entropy) really consists in continually sucking orderliness from its environment. This conclusion is less paradoxical than it appears at first sight. Rather could it be blamed for triviality. Indeed, in the case of higher animals we know the kind of orderliness they feed upon well enough, viz. the extremely well-ordered state of matter in more or less complicated organic compounds, which serve them as foodstuffs. After utilizing it they return it in a very much degraded form—not entirely degraded, however, for plants can still make use of it. (These, of course, have their most powerful supply of "negative entropy" in the sunlight.)

NEW LAWS TO BE EXPECTED IN THE ORGANISM

What I wish to make clear in this last chapter is, in short, that from all we have learnt about the structure of living matter, we must be prepared to find it working in a manner that cannot be reduced to the ordinary laws of physics. And that not on the ground that there is any "new force" or what not, directing the behavior of the single atoms within a living organism, but because the construction is different from anything we have yet tested in the physical laboratory. To put it crudely, an engineer, familiar with heat engines only, will, after inspecting the construction of a dynamo, be prepared to find it working along principles which he does not yet understand. He finds the copper familiar to him in kettles used here in the form of long, long wires wound in coils; the iron familiar to him in levers and bars and steam cylinders is here filling the interior of those coils of copper wire. He will be convinced that it is the same copper and the same iron, subject to the same laws of Nature, and he is right in that. The difference in construction is enough to prepare him for an entirely different way of functioning. He will not suspect that the dynamo is driven by a ghost because it is set spinning by the turn of a switch, without furnace and steam.

REVIEWING THE BIOLOGICAL SITUATION

The unfolding of events in the life cycle of an organism exhibits an admirable regularity and orderliness, unrivalled by anything we meet with in inanimate matter. We find it controlled by a supremely well-ordered group of atoms, which represent only a very small fraction of the sum total in every cell. Moreover, from the view we have formed of the mechanism of mutation we conclude that the dislocation of just a few atoms within the group of "governing atoms" of the germ cell suffices to bring about a well-defined change in the large-scale hereditary characteristics of the organism.

These facts are easily the most interesting that science has revealed in our day. We may be inclined to find them, after all, not wholly unacceptable. An organism's astonishing gift of concentrating a "stream of order" on itself and thus escaping the decay into atomic chaos—of "drinking orderliness" from a suitable environment—seems to be connected with the presence of the "aperiodic solids", the chromosome molecules, which doubtless represent the highest degree of well-ordered atomic association we know of—much higher than the ordinary periodic crystal—in virtue of the individual role every atom and every radical is playing here.

To put it briefly, we witness the event that existing order displays the power of maintaining itself and of producing orderly events. That sounds plausible enough, though in finding it plausible we, no doubt, draw on experience concerning social organization and other events which involve the activity of organisms. And so it might seem that something like a vicious circle is implied.

SUMMARIZING THE PHYSICAL SITUATION

However that may be, the point to emphasize again and again is that to the physicist the state of affairs is not only not plausible but most exciting, because it is unprecedented. Contrary to the common belief, the regular course of events, governed by the laws of physics, is never the consequence of one well-ordered configuration of atoms—not unless that configuration of atoms repeats itself a great number of times, either as in the periodic crystal or as in a liquid or in a gas composed of a great number of identical molecules.

Even when the chemist handles a very complicated molecule *in vitro* he is always faced with an enormous number of like molecules. To them his laws apply. He might tell you, for example, that one minute after he has started some particular reaction half of the molecules will have reacted, and after a second minute three-quarters of them will have done so. But whether any particular molecule, supposing you could follow its

course, will be among those which have reacted or among those which are still untouched, he could not predict. That is a matter of pure chance.

This is not a purely theoretical conjecture. It is not that we can never observe the fate of a single small group of atoms or even of a single atom. We can, occasionally. But whenever we do, we find complete irregularity, co-operating to produce regularity only on the average. We have dealt with an example in Chapter 1.[17] The Brownian movement of a small particle suspended in a liquid is completely irregular. But if there are many similar particles, they will by their irregular movement give rise to the regular phenomenon of diffusion.

The disintegration of a single radioactive atom is observable (it emits a projectile which causes a visible scintillation on a fluorescent screen). But if you are given a single radioactive atom, its probable lifetime is much less certain than that of a healthy sparrow. Indeed, nothing more can be said about it than this: As long as it lives (and that may be for thousands of years) the chance of its blowing up within the next second, whether large or small, remains the same. This patent lack of individual determination nevertheless results in the exact exponential law of decay of a large number of radioactive atoms of the same kind.

THE STRIKING CONTRAST

In biology we are faced with an entirely different situation. A single group of atoms existing only in one copy produces orderly events, marvellously tuned in with each other and with the environment according to most subtle laws. I said, existing only in one copy, for after all we have the example of the egg and of the unicellular organism. In the following stages of a higher organism the copies are multiplied, that is true. But to what extent? Something like 10^{14} in a grown mammal, I understand. What is that! Only a millionth of the number of molecules in one cubic inch of air. Though comparatively bulky, by coalescing they would form but a tiny drop of liquid. And look at the way they are actually distributed. Every cell harbours just one of them (or two, if we bear in mind diploidy). Since we know the power this tiny central office has in the isolated cell, do they not resemble stations of local government dispersed through the body, communicating with each other with great ease, thanks to the code that is common to all of them?

Well, this is a fantastic description, perhaps less becoming a scientist than a poet. However, it needs no poetical imagination but only clear and sober scientific reflection to recognize that we are here obviously faced with events whose regular and lawful unfolding is guided by a "mechanism" entirely different from the "probability mechanism" of physics. For it is simply a fact of observation that the guiding principle in every cell is embodied in a single atomic association existing only in one (or sometimes two) copy—and a fact of observation that it results in producing events which are a paragon of orderliness. Whether we find it astonishing or whether we find it quite plausible, that a small but highly organized group of atoms be capable of acting in this manner, the situation is unprecedented, it is unknown anywhere else except in living matter. The physicist and the chemist, investigating inanimate matter, have never witnessed phenomena which they had to interpret in this way. The case did not arise and so our theory does not cover it—our beautiful statistical theory of which we were so justly proud because it allowed us to look behind the curtain, to watch the magnificent order of exact physical law coming forth from atomic and molecular disorder; because it revealed that the most important, the most general, the all-embracing law of entropy increase could be understood without a special assumption *ad hoc*, for it is nothing but molecular disorder itself.

TWO WAYS OF PRODUCING ORDERLINESS

The orderliness encountered in the unfolding of life springs from a different source. It appears that there are two different "mechanisms" by which orderly events can be produced: the "statistical mechanism" which produces "order from disorder" and the new one, producing "order from order". To the unprejudiced mind the second principle appears to be much simpler, much more plausible. No doubt it is. That is why physicists were so proud to have fallen in with the other one, the "order-from-disorder" principle, which is actually followed in Nature and which alone conveys an understanding of the great line of natural events, in

the first place of their irreversibility. But we cannot expect that the "laws of physics" derived from it suffice straightaway to explain the behaviour of living matter, whose most striking features are visibly based to a large extent on the "order-from-order" principle. You would not expect two entirely different mechanisms to bring about the same type of law—you would not expect your latch-key to open your neighbour's door as well.

We must therefore not be discouraged by the difficulty of interpreting life by the ordinary laws of physics. For that is just what is to be expected from the knowledge we have gained of the structure of living matter. We must be prepared to find a new type of physical law prevailing in it. Or are we to term it a non-physical, not to say a super-physical, law?

THE NEW PRINCIPLE IS NOT ALIEN TO PHYSICS

No. I do not think that. For the new principle that is involved is a genuinely physical one: It is, in my opinion, nothing else than the principle of quantum theory over again. To explain this, we have to go to some length, including a refinement, not to say an amendment, of the assertion previously made, namely, that all physical laws are based on statistics.

This assertion, made again and again, could not fail to arouse contradiction. For, indeed, there are phenomena whose conspicuous features are visibly based directly on the "order-from-order" principle and appear to have nothing to do with statistics or molecular disorder.

The order of the solar system, the motion of the planets, is maintained for an almost indefinite time. The constellation of this moment is directly connected with the constellation at any particular moment in the times of the Pyramids; It can be traced back to it, or vice versa. Historical eclipses have been calculated and have been found in close agreement with historical records or have even in some cases served to correct the accepted chronology. These calculations do not imply any statistics, they are based solely on Newton's law of universal attraction.

Nor does the regular motion of a good clock or of any similar mechanism appear to have anything to do with statistics. In short, all purely mechanical events seem to follow distinctly and directly the "order-from-order" principle. And if we say "mechanical", the term must be taken in a wide sense. A very useful kind of clock is, as you know, based on the regular transmission of electric pulses from the power station.

I remember an interesting little paper by Max Planck on the topic "The Dynamical and the Statistical Type of Law" ("Dynamische und Statistische Gesetzmässigkeit"). The distinction is precisely the one we have here labelled as "order from order" and "order from disorder". The object of that paper was to show how the interesting statistical type of law, controlling large-scale events, is constituted from the "dynamical" laws supposed to govern the small-scale events, the interaction of the single atoms and molecules. The latter type is illustrated by large-scale mechanical phenomena, as the motion of the planets or of a clock, etc.

Thus it would appear that the "new principle", the order-from-order principle, to which we have pointed with great solemnity as being the real clue to the understanding of life, is not at all new to physics. Planck's attitude even vindicates priority for it. We seem to arrive at the ridiculous conclusion that the clue to the understanding of life is that it is based on a pure mechanism, a "clock work" in the sense of Planck's paper. The conclusion is not ridiculous and is, in my opinion, not entirely wrong, but it has to be taken "with a very big grain of salt".

THE MOTION OF A CLOCK

Let us analyse the motion of a real clock accurately. It is not at all a purely mechanical phenomenon. A purely mechanical clock would need no spring, no winding. Once set in motion, it would go on for ever. A real clock without a spring stops after a few beats of the pendulum, its mechanical energy is turned into heat. This is an infinitely complicated atomistic process. The general picture the physicist forms of it compels him to admit that the inverse process is not entirely impossible: A springless clock might suddenly begin to move, at the expense of the heat energy of its own cog wheels and of the environment. The physicist would have to say: The clock experiences an exceptionally intense fit of Brownian movement. We have seen in Chapter 1 (§ 9)[18] that with a very sensitive torsional balance (electrometer or galvanometer) that sort of

thing happens all the time. In the case of a clock it is, of course, infinitely unlikely.

Whether the motion of a clock is to be assigned to the dynamical or to the statistical type of lawful events (to use Planck's expressions) depends on our attitude. In calling it a dynamical phenomenon we fix attention on the regular going that can be secured by a comparatively weak spring, which overcomes the small disturbances by heat motion, so that we may disregard them. But if we remember that without a spring the clock is gradually slowed down by friction, we find that this process can only be understood as a statistical phenomenon.

However insignificant the frictional and heating effects in a clock may be from the practical point of view, there can be no doubt that the second attitude, which does not neglect them, is the more fundamental one, even when we are faced with the regular motion of a clock that is driven by a spring. For it must not be believed that the driving mechanism really does away with the statistical nature of the process. The true physical picture includes the possibility that even a regularly going clock should all at once invert its motion and, working backward, re-wind its own spring—at the expense of the heat of the environment. The event is just "still a little less likely" than a "Brownian fit" of a clock without driving mechanism.

CLOCKWORK AFTER ALL STATISTICAL

Let us now review the situation. The "simple" case we have analyzed is representative of many others—in fact of all such as appear to evade the all-embracing principle of molecular statistics. Clockworks made of real physical matter (in contrast to imagination) are not true "clock-works". The element of chance may be more or less reduced, the likelihood of the clock suddenly going altogether wrong may be infinitesimal, but it always remains in the background. Even in the motion of the celestial bodies irreversible frictional and thermal influences are not wanting. Thus the rotation of the Earth is slowly diminished by tidal friction, and along with this reduction the moon gradually recedes from the Earth, which would not happen if the Earth were a completely rigid rotating sphere.

Nevertheless the fact remains that "physical clock-works" visibly display very prominent "order-from-order" features—the type that aroused the physicist's excitement when he encountered them in the organism. It seems likely that the two cases have after all something in common. It remains to be seen what this is and what is the striking difference which makes the case of the organism after all novel and unprecedented.

NERNST'S THEOREM

When does a physical system—any kind of association of atoms—display "dynamical law" (in Planck's meaning) or "clock-work features"? Quantum theory has a very short answer to this question, viz. at the absolute zero of temperature. As zero temperature is approached the molecular disorder ceases to have any bearing on physical events. This fact was, by the way, not discovered by theory, but by carefully investigating chemical reactions over a wide range of temperatures and extrapolating the results to zero temperature—which cannot actually be reached. This is Walther Nernst's famous "Heat-Theorem", which is sometimes, and not unduly, given the proud name of the "Third Law of Thermodynamics" (the first being the energy principle, the second the entropy principle).

Quantum theory provides the rational foundation of Nernst's empirical law, and also enables us to estimate how closely a system must approach to the absolute zero in order to display an approximately "dynamical" behavior. What temperature is in any particular case already practically equivalent to zero?

Now you must not believe that this always has to be a very low temperature. Indeed, Nernst's discovery was induced by the fact that even at room temperature entropy plays an astonishingly insignificant role in many chemical reactions. (Let me recall that entropy is a direct measure of molecular disorder, viz. its logarithm.)

THE PENDULUM CLOCK IS VIRTUALLY AT ZERO TEMPERATURE

What about a pendulum clock? For a pendulum clock room temperature is practically equivalent to zero. That is the reason why it works "dynamically". It will continue to work as it does if you cool it (provided that you have removed all traces of oil!). But it does not continue to work, if you heat it above room temperature, for it will eventually melt.

THE RELATION BETWEEN CLOCKWORK AND ORGANISM

That seems very trivial but it does, I think, hit the cardinal point. Clockworks are capable of functioning "dynamically", because they are built of solids, which are kept in shape by London–Heitler forces, strong enough to elude the disorderly tendency of heat motion at ordinary temperature.

Now, I think, few words more are needed to disclose the point of resemblance between a clockwork and an organism. It is simply and solely that the latter also hinges upon a solid—the aperiodic crystal forming the hereditary substance, largely withdrawn from the disorder of heat motion. But please do not accuse me of calling the chromosome fibres just the "cogs of the organic machine"—at least not without a reference to the profound physical theories on which the simile is based.

For, indeed, it needs still less rhetoric to recall the fundamental difference between the two and to justify the epithets novel and unprecedented in the biological case.

The most striking features are: first, the curious distribution of the cogs in a many-celled organism, for which I may refer to the somewhat poetical description in § 64;[19] and secondly, the fact that the single cog is not of coarse human make, but is the finest masterpiece ever achieved along the lines of the Lord's quantum mechanics.

NOTES

This chapter originally appeared as chapters 1, 5, 6 and 7 in Erwin Schrödinger, *What is life? The physical aspect of the living cell*, pp. 3–18 and 56–85, Cambridge, UK: Cambridge University Press, 1967.

ENDNOTES

1 This contention may appear a little too general. The discussion must be deferred to the end of this book, §§ 67 and 68. (Ed. note: See pages 67–68 of this volume.)

2 This point of view has been emphasized in two most inspiring papers by F. G. Donnan, *Scientia,* vol. 24, no. 78, p. 10, 1918 ('La science physico-chimique décrit-elle d'une façon adéquate les phénomènes bio-logiques?'); *Smithsonian Report for 1929,* p. 309 ('The mystery of life').

3 You would not, of course, find exactly 100 (even if that were the exact result of the computation). You might find 88 or 95 or 107 or 112, but very improbably as few as 50 or as many as 150. A "deviation" or "fluctuation" is to be expected of the order of the square root of 100, i.e. 10. The statistician expresses this by stating that you would find 100 ± 10. This remark can be ignored for the moment, but will be referred to later, affording an example of the statistical $\sqrt{n}$ law.

4 According to present-day views an atom has no sharp boundary, so that "size" of an atom is not a very well-defined conception. But we may identify it (or, if you please, replace it) by the distance between their centres in a solid or in a liquid—not, of course, in the gaseous state, where that distance is, under normal pressure and temperature, roughly ten times as great.

5 A gas is chosen, because it is simpler than a solid or a liquid; the fact that the magnetization is in this case extremely weak, will not impair the theoretical considerations.

6 To wit: The concentration at any given point increases (or decreases) at a time rate proportional to the comparative surplus (or deficiency) of concentration in its infinitesimal environment. The law of heat conduction is, by the way, of exactly the same form, "concentration" having to be replaced by "temperature".

7 For convenience I shall continue to call it an isomeric transition, though it would be absurd to exclude the possibility of any exchange with the environment.

8 Ed. Note. Not reprinted in this volume.

9 That it is highly flexible is no objection; so is a thin copper wire.

10 Ed. note: Section 36 of *The Meaning of Life* has not been reprinted in this volume.

11 Ed. note: Section 36 of *The Meaning of Life* has not been reprinted in this volume.

12 Ed. note: Page 55 and figure 12 of *The Meaning of Life* have not been reprinted in this volume.

13 Ed. note: See pages 59 of this volume.

14 To state this in complete generality about 'the laws of physics' is perhaps challengeable. The point will be discussed in Chapter VII.

15 Ed. note: See pages 50–57 of this volume.

16 Ed. note: See pages 62–64 of this volume.

17 Ed. note: See pages 50–57 of this volume.

18 Ed. note: See pages 50–57 of this volume.

19 Ed. note: See pages 65–66 of this volume.

5 · The nature of life

ALEXANDER OPARIN

THE EXTENT OF LIFE

LIFE—the word is so easy to understand yet so enigmatic for any thoughtful person. One would have thought that the meaning of this word would have been clear and the same for all ages and all peoples. Nevertheless, we know that, during the many centuries of human cultural history, there have been irreconcilable conflicts as to how it should properly be understood.

Even the question of what is alive, which of the objects in the world around us are imbued or endowed with life, the extent of the realm of life or its scope, have been defined and are still defined in various totally different ways. We have, as it were, a whole multicoloured spectrum of different opinions. At the one end of this spectrum we find the views of those philosophers and scientists who believe that life is a general property, inalienable from all matter, and who therefore extend the realm of life to cover all objects in the universe.

On the other hand, the philosophers of the opposite end of the spectrum arbitrarily restrict the scope of life to the compass of human existence, or may even maintain that life is the prerogative of one single thinking subject.

The first of these opinions owes its origin to the ancient Greek hylozoists. According to Aristotle, even Thales the founder of the Miletian school of philosophy (who was alive about 600 BC) considered magnets to be animate on account of their ability to attract iron. More than 2000 years later, in the seventeenth century, the Dutch philosopher and materialist, Spinoza, maintained that stones think and that all natural bodies are animate, while even 100 years later still (also in Holland), the French philosopher Robinet published a book called *On Nature* in which he acknowledged that all matter was living and even considered the stars in heaven as living organic bodies.

Even in our own times, many engineers and physicists are ready to consider the most complicated modern mechanisms and automata as being alive, just as Descartes compared organisms to water clocks or mills, or La Mettrie referred to man as an extremely well-educated machine. Some present-day chemists and geneticists follow Diderot in an attempt to assign life even to individual molecules of organic substances.

On the other hand, it is understood by anyone that, if some writer or philosopher speaks, in one of his works, about the significance or value of life, or about its aim, then he is referring only to human life, to that "striving towards good" which, according to Tolstoï, constitutes the main aspiration of life and is understood as such by all men.

This last expression of opinion is taken from Tolstoï's treatise *On Life* (1886/1887). In it Tolstoï censures the experimental scientists or, as he calls them, the scribes[1] for using the actual word "life" in that by cunning sophistry they have invented a conventional scientific Volapük in which words do not correspond with what all ordinary laymen understand by them. Tolstoï justifiably recommends that "a man is bound, by every word, to mean that which all indisputably understand alike."

It seems to me that if we follow this wise advice we shall be able to find a way out of the present confusing labyrinth of contradictory opinions on the question of the delimitation of the realm of life, although the way out which we shall find will be far from the same as that of which Tolstoï approved. Any ordinary person looking at the world around him will infallibly sort it into the kingdom of the lifeless or inorganic, and that of living things. In all places and at all times he will see that life is not simply scattered about all over the place, but only exists in individual organisms which are separate from their environment; so that the sum of these organisms constitutes the realm of life, the world of living things. This world exhibits a colossal variety, including plants, animals and microbes, which are very diverse and which, at first glance, would hardly seem to have anything in common. Nevertheless, anybody, even without any scientific experience, can easily observe what they do have in common and what enables one to include in the one category of living being a man and a tree, a whale and a tiny beetle or blade of grass, a bird and a shapeless mollusc.

When the simple glass polisher from Amsterdam, Leeuwenhoek, first saw microbes of various kinds through his magnifying glass, he unhesitatingly designated them as living things (*viva animalcula*), although some of them, such as the cocci, which Leeuwenhoek drew with his own hand, could not move and had none of the other external features of living things.

On perceiving that living things have something in common, which relates them to one another, one distinguishes them from the objects of the inorganic world which lack that "something" i.e., which lack life. Thus, even by his unaided observation of the world around him, any ordinary person can establish the most elementary, but also the most general definition of the extent of life or the area embraced by its natural realm. Life is a property of any organism, from the highest to the lowest, but it does not exist in inorganic natural objects, no matter how complicated their structure may be. It is very possible that, in the unbounded extent of the universe, there exists a multitude of extremely complicated and highly evolved forms of movement and organisation of matter, of which we, as yet, have no suspicion, but it would be quite unjustifiable to call any of these forms "life" if they differed in essential principles from that life which is represented on our planet by a whole multitude of organisms of different forms. It would be better to think up a special new word to denote these forms of organisation when it is required.

We have thus marked out the region of nature, the category of objects which are pertinent to our enquiries about life. This means that, in what follows, we can avoid many of the mistakes which are rather common in scientific literature, by keeping strictly to the terms of reference laid down above. Of course, what has been given is not, by any means, a definition of life. To give that one would have to solve the problem of the nature of that "something" which is characteristic only of the world of living things and which is absent from the objects of inorganic nature.

THE CONFLICT BETWEEN IDEALISM AND MATERIALISM AS TO THE ESSENTIAL NATURE OF LIFE

From the earliest times, even until the present day, this problem of the essential nature of life has always been a battlefield in the embittered war which has been waged between the two irreconcilable philosophic camps of idealism and materialism.

The representatives of the idealist camp see, as the essence of life, some sort of eternal supramaterial origin which is inaccessible to experiment. This is the "psyche" of Plato, the "entelechy" of Aristotle, the "immortal soul" or "particle of divinity" of various religious doctrines and faiths, Kant's *"inneres Prinzip der Kausalität"*, the *"Weltgeist"* of the Hegelians, the "life force" of the vitalists and the "dominant" of the neovitalists, and other such concepts.

From this point of view, matter, in the sense of that objective reality which we observe directly and study experimentally, is in itself and as such, lifeless and inert. It serves only as the material from which the spirit or soul creates a living being, gives it form, adapts its structure to functional needs, endows it with the power of breathing and moving and, in general, makes it alive. And when the soul leaves the organism and death takes place, there remains but the lifeless material envelope, a rotten, decomposing corpse.

This concept of death as the departure from the body of the soul, which constitutes the essence of life, is, in fact, the basis of a definition of life which is widely held and even appears in a number of encyclopedias, namely, that life is the contrary of death. This, however, pushes out of sight the fact that the living can only properly be contrasted with the lifeless, not with the dead. It is obvious that the dead body is a product of life for, in the absence of life, that is, in an inorganic world, a corpse could never occur on its own.

Even if one starts from idealistic premises one can, of course, make an objective study of particular organisms and their organs, but it is inherently impossible to reach an understanding of the essence of life itself by experimental, materialistic means, as this essence is of a supramaterial or spiritual nature. Only by means of speculative introspection can one approach an understanding of that divine principle which we bear within ourselves. We can only passively contemplate all the rest of the world of living things and marvel at the wisdom of the Creator Who made them. And naturally there can be no question of man making any change or transformation in living nature if one adopts this position.

Materialists approach the problem of the essence of life from a diametrically opposite viewpoint. Basing their arguments on the facts obtained by science they assert that life, like all the rest of the world, is material and does not require for its understanding the acceptance of a spiritual origin which is not amenable to experimental study. On the contrary, objective study of the world around us is, for the materialist, not only a hopeful way of leading us to an understanding of the very essence of life, but it also enables us to alter living nature purposefully in a way favorable to mankind.

Wide circles of biological scientists, either consciously or intuitively, base their investigations on a materialistic concept of living nature and, in following this line, they are always enriching the science of life by their work and bringing us closer to an understanding of the essence of life.

MECHANICAL AND DIALECTICAL CONCEPTS OF LIFE

However, even within the limits of the materialistic concept of life, its essence may be interpreted in various ways.

According to the mechanistic view, which prevailed in the scientific world of last century, and which is partly retained even now, the understanding of life in general comprises simply a complete explanation in terms of physics and chemistry, a complete account of all living phenomena as physical and chemical processes. If one adopts this position there is no place for any specifically biological laws of nature. In reality there is only one law which governs both the inorganic world and all the phenomena occurring in living organisms. This is, in fact, to deny that there is any qualitative difference between organisms and inorganic objects. We thus reach a position where we must say either that inorganic objects are alive or that life does

not really exist. Thus, by a logical development of the mechanistic outlook which has been explained, we are forced to a conclusion which is fundamentally opposed to the view which we adopted earlier. However, in contradistinction to this, one must be quite clear that acceptance of the material nature of life does not mean that one must deny that it has specific characteristics and that living things show qualitative differences from inorganic objects. One must not do as the mechanists and regard everything which is not included in physics and chemistry as being vitalistic or supernatural. On the contrary, the forms of organization and motion of matter may be very varied. To deny this variety is to indulge in oversimplification.

According to the dialectical materialist view, matter is in constant motion and proceeds through a series of stages of development. In the course of this progress there arise ever newer, more complicated and more highly evolved forms of motion of matter having new properties which were not previously present. There can be no doubt that, for a long period after the formation of our planet, there was no life on it. Obviously, all the things which existed on it at that time simply obeyed the laws of physics and chemistry. However, in the process of development of matter on the Earth, the first and most primitive organisms arose, that is to say, life came into being as a qualitatively new form of motion. When this happened, the old laws of physics and chemistry naturally continued to operate, but now they were supplemented by new and more complicated biological laws which had not operated before.

Thus, life is material in nature but its properties are not limited to those of matter in general. Only living beings possess it. This is a special form of the movement of matter, qualitatively different from that of the inorganic world, and the organism has specific biological properties and ways of behaving and does not merely follow the rules governing inorganic nature. Therefore, a dialectical materialist will even formulate the problem of understanding life in a different way from a mechanist. For the latter it consists in a more complete explanation of life in terms of physics and chemistry. For the dialectical materialist, on the other hand, the important thing about understanding life is the establishment of its qualitative difference from other forms of matter, i.e., that difference which obliges us to regard life as a special form of the motion of matter.

ATTEMPTS TO FORMULATE DEFINITIONS OF LIFE

This difference has found, and still finds, a greater or lesser reflection in the definitions of life formulated and expounded by the scientists and thinkers of last century and those of our own time. It is just in their setting out of this difference between the living and the non-living that one can perceive the objective and essential value of these definitions in spite of their absolute contradictoriness and their astounding variety.

At the beginning of his remarkable book, *Leçons sur les phénomènes de la vie communs aux animaux et aux végétaux* (1878/1879), Claude Bernard produces a large number of definitions of life which had been made before that time, but he does so simply in order to show that, in general, any *a priori* definition of life is always chimerical and scientifically unprofitable. However, he also believes that life can be understood completely if approached *a posteriori*, by establishing the characteristic features which differentiate living creatures from non-living bodies. This is certainly not easy and, in doing it we are beset with considerable difficulties and doubts, but all the time we are getting closer to solving our problem.

In an American encyclopedia of 1944 it is stated that there is no single satisfactory definition of life, for, while some include too many phenomena, others suffer from too narrow limitations.

We believe that this is because, in most cases, people try to characterize life as a single point while it is, in fact, a long line, comprising the whole of that section of the development of matter lying between the origin of life on the Earth and our own time, and including among its manifestations the most primitive organisms as well as more highly developed plants and animals, especially man. With the appearance of man, however, there arises a new social form of motion of matter which is more complicated and highly evolved than life and which is characterized by its own peculiar features and by the special laws of development of human society.

It is therefore completely wrong to try to characterize the "line of life" simply on the basis of one point, whether that point lies at the beginning, the middle or the end. In fact, if we try to define life in terms of the characteristics which arose at the very beginning of its emergence on the Earth, we have to exclude from among its features, not merely consciousness, but also respiration, which obviously did not take place among the earliest organisms. On the other hand, if we define life on the basis of phenomena which are typical only of the more highly developed living things, we shall risk relegating the anaerobic bacteria, as well as many primitive organisms, to the category of non-living bodies belonging to inorganic nature.

When Engels made his remarkable definition of life as the "mode of existence of albuminous bodies" he immediately made reservations, indicating the incompleteness of the definition. "Our definition of life" he wrote, "is naturally very inadequate, inasmuch as, far from including *all* the phenomena of life, it has to be limited to those which are the most common and the simplest. From a scientific standpoint all definitions are of little value. In order to gain a really exhaustive knowledge of what life is, we should have to go through all the forms in which it appears, from the lowest up to the highest."

Thus for an exhaustive understanding of life it is necessary to have an understanding of the whole gamut of its characteristic features, starting with those extremely elementary ones, with which the first living beings were endowed, and finishing with the most complicated manifestations of higher nervous activity in animals and man, in which the biological stage of the development of matter culminates. Among this multitude of features characteristic of life, manifested at the very outset of its development and becoming more complicated in the course of its further evolution and increasing complexity, special mention must be made of that clearly defined, specific interaction between the organism and its environment which runs, like a red thread, along the "line of life" and is a characteristic of all living things, the lowest as well as the highest, but which is absent from the objects of the inorganic world.

THE SPECIFIC INTERACTION BETWEEN THE ORGANISM AND ITS ENVIRONMENT

An organism can live and maintain itself only so long as it is continually exchanging material and energy with its environment. As food, drink and gaseous material, various substances of a chemical nature foreign to the organism enter into it. In the organism they undergo far-reaching changes and transformations as a result of which they are converted into the material of the organism itself. That is to say, they are turned into chemical compounds which are, in some degree, similar to those of which the living body is already composed. This is the ascending limb of biological metabolism known as assimilation. In the course of the interaction of substances from outside with the material of the organism, however, the opposite process also occurs continually and is known as dissimilation. The substances of the living body do not remain unchanged. They are broken down fairly quickly to liberate the energy latent in them and the products of their breakdown are discharged into the surrounding medium.

Our bodies flow like rivulets, their material is renewed like water in a stream. This was what the ancient Greek dialectician Heraclitus taught. Certainly the flow, or simply the stream of water emerging from a tap, enables us to understand in their simplest form many of the essential features of such flowing, or open systems as are represented by the particular case of the living body. If the tap is not far open and the pressure in the water cistern remains constant, the external form of the water flowing from the tap remains almost unchanged, as though it were congealed. We know, however, that this form is merely the visible expression of a continual flow of particles of water, which are constantly passing through the stream at a steady rate and emerging from it. If we disturb the relationship between the rates of input and output or the steady process of movement of its particles, the stream, as such, disappears, for the very existence of the stream depends on the steady passage of ever-renewed water molecules through it.

On this analogy the constancy of the external form, and even of the most detailed internal structure of the living being, is merely the visible expression of the constancy of the sequence of processes going on within it as a result of the extremely intricate balancing of the two contrary phenomena already noted, i.e., assimilation and dissimilation. The prolonged existence of a living system in which breakdown and decay are going on all the time is entirely due to this balance. In the place of each molecule or structure which breaks down a new and analogous one appears as a freshly synthesized formation. Thus the organism maintains its form, structure and chemical composition unchanged while its material is continually changing.

Organisms are, thus, not static but stationary or flowing systems. Their ability to exist for a longer or shorter time under given environmental conditions does not depend on their being at rest or unchanging. On the contrary, it depends on the constancy of their movement, i.e., their metabolism.

From a purely chemical point of view, metabolism is merely the sum of a large number of comparatively simple reactions of oxidation, reduction, aldol condensation, hydrolysis, transamination, phosphorylation, cyclization, etc. Each of these reactions can be reproduced, even outside the organism, as there is nothing specifically connected with life about them.

The peculiarity which distinguishes life qualitatively from all other forms of motion of matter (and in particular from inorganic flowing systems) is that, in the living body, the many tens and hundreds of thousands of individual chemical reactions, which, in their sum, make up the metabolism, are not only strictly coordinated in time and space, not merely cooperating harmoniously in a single sequence of self-renewal, but the whole of this sequence is directed in an orderly way towards the continual self-preservation and self-reproduction of the living body as a whole. They are extremely well adapted to solving the problem of the existence of the organism under a given set of environmental conditions.

THE "PURPOSIVENESS" OF THE ORGANIZATION OF LIVING BODIES

This flowing character of the interaction of living bodies with the medium around them, and, what is most important, the amazingly efficient adaptation of the organization of this interaction to the task of self-preservation and self-reproduction of the system under a given set of external conditions, all that which has been referred to by many authors as the "adaptation of form to function" or "purposiveness" in the structure of such a system, is so objectively obvious and makes such a forcible impression on the eyes of those who study living nature that, in one form or another, it figures in the majority of even the most varied definitions of life formulated during the course of many centuries and put forward by members of the most dissimilar schools of philosophy and scientific thought.

The presence in all organisms, without exception, of an adaptation of form to function was noted even by Aristotle who, in his writings, was the first to be able to generalize from the extensive accumulation of biological material which was available at that time. Aristotle designated this specific property of living things as the "entelechy" underlying life or the "principle having its aim within itself."

In one form or another Aristotle's teaching about "entelechy" has left its mark on all idealistic definitions of life. It is reflected in various religious creeds and philosophic teachings and has lasted through many centuries to reach our own twentieth century in the works of Reinke, Driesch, and other contemporary students of vitalism.

In their investigations of life, however, the representatives of the materialist camp naturally could not overlook this specific feature of it. Many of them, following Descartes, defined the vital phenomena of plants and animals merely as responsive reactions of a specifically constructed bodily mechanism to the external influence of the environment. Others saw the orderly direction of metabolism as the specific property which distinguished organisms from non-living things. In this connection Claude Bernard wrote:

"L'édifice organique est le siége d'un perpétuel mouvement nutritif qui ne laisse de repos à aucune partie: chacune, sans cesse ni trére, s'alimente dans le milieu qui l'entoure et y rejette ses dechettes et ses produits. Cette renovation moleculaire est insaisissable pour le regard; mais, comme nous en voyons le debut et le fin, l'entrée et la sortie des substances, nous en concevons les phases intermediaires, et nous nous représentons un courant de matiére qui traverse incessamment l'organisme et le renouvelle dans sa substance en le maintenant dans sa forme.

"L'universalité d'un tel phénoméne chez la plante et chez l'animal et dans toutes leurs parties, sa constance, qui ne souffre pas d'arrét, en font un signe général de la vie, que quelques physiologistes ont employé à sa définition."[2]

"There is one expression which must be applied to all organisms", wrote Engels, "and that is adaptation." Later he puts forward his own definition of life: "Life is the mode of existence of albuminous substances and this mode of existence essentially consists in the constant self-renewal of the chemical constituents of these substances by nutrition and excretion."

In our own times, Perret and later, Bernal, have tried to define life in the following terms which are, perhaps, rather complicated for non-specialists.

"Life is a potentially self-perpetuating open system of linked organic reactions, catalysed stepwise and almost isothermally by complex and specific organic catalysts which are themselves produced by the system".

"The organism represents an entity as a system only in conjunction with the conditions necessary for its life" states a representative of the Michurinist school of biology.

Thus the universal "purposiveness" of the organisation of living beings is an objective and self-evident fact which cannot be ignored by any thoughtful student of nature. The rightness or wrongness of the definition of life advanced by us, and also of many others, depends on what interpretation one gives to the word "purposiveness" and what one believes to be its essential nature and origin.

The idealists see this "purposiveness" as the fulfillment of some predetermined plan of a deity or "universal intellect". The materialists, on the other hand, use the expression (for lack of a better one) as the shortest way of characterizing the direction of the organization of the whole living system towards its self-preservation and self-reproduction under given environmental conditions, as well as to describe the suitability of the structure of the separate parts of the living system to the most efficient and harmonious performance of those vitally necessary functions which the particular part subserves.

The extremely highly developed adaptation of the structure of the individual organs to the performance of their functions and the general "purposiveness" of the whole organization of life is seen to be extremely precise even on a very superficial acquaintance with higher living things. As we have already pointed out, it was noticed a long time ago and found expression in the Aristotelian "entelechy". It had been considered to be essentially mystical and supernatural until Darwin gave a rational, materialistic explanation of the way in which this "purposiveness" arose in higher organisms by means of natural selection.

"Purposiveness" of structure does not, however, manifest itself solely in the more highly organized creatures, it pervades the whole living world from the top to the bottom, right down to the most elementary forms of life. It is necessary to every living body but in the absence of the living body there would be no "purposiveness" under natural conditions. It would therefore be fruitless for us to seek its explanation simply in terms of the laws of the inorganic world, i.e., the laws of physics and chemistry. The "purposiveness" which is characteristic of the organization of all living things can only be understood if one understands the specific interaction between the organism and its environment in terms of the Darwinian principle of natural selection. This new biological law could only arise on the basis of the establishment of life and therefore lifeless, inorganic bodies lack "purposiveness." The striking exception to this rule is the machine.

ATTEMPTS TO TREAT THE ORGANISM AS A MACHINE

It is, of course, impossible to doubt that the principle on which any machine is constructed is that of adapting its structure or external organisation to the performance of the particular specific job which is its work. From this point of view the comparison between the machine and the organism forces itself upon one. In the course of many centuries it has been widely used by many philosophers and scientists, in their attempts to solve the problem of the essential nature of life. The only thing that has changed in these attempts, in the course of the various periods of development of science, has been the opinion as to which of the points common to the organism and the machine ought to be taken into consideration, as being the features most characteristic of life. The way in which the problem was posed and the attempt to describe the organism as a machine of some sort or another has, however, remained essentially unchanged. Undoubtedly, the ideas of each age tend to be expressed in terms of its technology. In his book (1949), N. Wiener very pertinently refers to the seventeenth century and the early part of the eighteenth century as "the age of clocks," the end of the eighteenth century and the whole of the nineteenth century as "the age of steam-engines," and our own times as "the age of communication and control."

In the age of clocks the world was represented by man as a huge mechanism which had been wound up once and for all time. People saw, as the basis of all existence, mechanical motion, the displacement of bodies in space, taking place according to Newton's laws of motion. Life was also discussed from this point of view, as being merely a special kind of mechanical motion. The spontaneous movement of animals and their organs through space may serve as the clearest expression of this. According to the ideas of that time, therefore, the organism is nothing but a "very complicated machine, the structure of which is, nevertheless, completely comprehensible. Its movement depends entirely on its structure and on the pressure and on the collision of particles of matter like the wheels of a water clock" (Descartes). Anatomy therefore occupied the most important place in the study of life at that time.

However, in the next period of the development of science, the age of steam-engines, physiology began, to a greater and greater extent, to aspire to this place and the role of mechanics in the study of life was taken over by energetics.

The prototype of the living creature was now thought of, not as a watch, but as a heat engine. The analogy put forward by Lavoisier between respiration and the burning of fuel was a great step forward. Food is simply the fuel we throw into the furnace of our organism and its importance can therefore be assessed completely in terms of calories. The guiding principles of that time in connection with the understanding of life were those of the conservation and degradation of energy. The first law of thermodynamics, that of the conservation of energy, was found to be universally applicable, both to organisms and to mechanisms.

The second law was a more complicated matter. This is the law which expresses the statistical tendency of nature towards disorder, the tendency to even out energy and thus to devalue it in isolated systems, which is expressed in general terms as the increase of entropy in these systems. If one were to put such a system in uniform conditions and leave it alone, then all the phenomena occurring within it would very soon cease and the system as a whole would come to an end. It would thus attain the unchanging state in which nothing would happen. Physicists call this state "thermodynamic equilibrium" or "maximum entropy."

In organisms, on the other hand, not only does entropy not increase, it may even decrease. Thus, one might say that the fundamental law of physics was a tendency towards disorder or an increase in entropy, while that of biology, on the contrary, was a tendency to increasing organisation or a decrease in entropy.

Some idealist philosophers, such as A. Bergson, defined life as the "struggle against entropy" and even saw, in this contradiction between physics and biology, a reason for accepting the supernatural nature of life.

Now, however, we know that the contradiction is only apparent. Living things can never exist as isolated systems. On the contrary, as we have said above, it is characteristic of living organisms that they constantly interact with their environment and, by virtue of this fact, they must be regarded as "flowing" or "open" systems. The stationary (but not static) state in which they exist is maintained constant, not because they are in a state of "maximum entropy," or because their free energy is at a minimum (as is the case in the thermo-dynamic equilibrium), but because the open system is continually receiving free energy from the medium around it in an amount which compensates for the decrease taking place within the system.

Wiener maintains that the ability to act against the general tendency to the increase of entropy is to be found, not only in organisms, but also in machines which have certain specific ways of interacting with the world outside them. In this way he thinks that machines can create a certain local zone of organization around themselves.

CYBERNETICS

This concept ushers in the third and present period of the history of the problem, the age of communication and control which has superseded the age of steam-engines.

"There is, in electrical engineering," he writes, "a split which is known in Germany as a split between the technique of strong currents and the technique of weak currents, and which we (in the USA and Great Britain) know as the distinction between power and communication engineering. It is this split which separates the age which is just past from that in which we are now living." Communication engineering may make use of currents of any strength and may use enormous motors, but it differs from power engineering in that it is fundamentally interested in the exact reproduction of signals and not in the way in which energy is used.

The energy consumed by an electronic valve may be almost entirely wasted, but nevertheless the valve may be a very effective means of carrying out a necessary operation. Similarly, the efficiency of the work of our nervous systems cannot be calculated simply from the point of view of the rational utilization of the comparatively small amount of energy which reaches the neurons from the blood stream.

Organisms are effectively coupled to the world around them, not merely by their overall metabolism and energy balance, but also by the flow of communications inwards and outwards, the flow of impressions received and actions performed. The extremely well organized and highly differentiated higher nervous activity of man and animals may serve as a particularly clear example of the relationship. Wiener, however, maintains that it is possible to produce a very far-reaching analogy between this activity and the work of contemporary self-regulating machines and auto-mata. Photo-electric elements and other light recep-tors, radiolocating systems, apparatus for the registration of the potential of hydrogen ions, thermometers, manometers, all sorts of microphones, etc. are the equivalents of the sensory organs and serve as mech-anisms for receiving information. The effector organs of the machine may be electric motors, solenoids, heating elements and other similar devices. Between the receptor mechanisms and the effector organs of such machines as, for example, the contemporary quick-acting electronic calculating machine, there are intermediate groups of elements, a central regulating system which might be regarded as being analogous with the brain of animals or man.

The object of this system is to coordinate the incoming messages in such a way as to bring about the desired reaction by the executive organs. As well as the information reaching this central regulating machine from the outside world, it also receives information about the working of the executive organs themselves. This is what is known as "feedback" and it permits recording of the fulfilment or non-fulfilment of its own tasks by the machine itself. "Moreeover," writes Wiener, "the information received by the automat need not necessarily be used at once but may be delayed or stored so as to become available at some future time. This is the analog of memory. Finally, as long as the automaton is running, its very rules of operation are susceptible to some change on the basis of the data which have passed through its receptors, and this is not unlike the process of learning."

Thus, in the transition from the age of the steam engine to the age of communication and control, the prototype of the living thing is becoming the electronic calculating machine, the study of nutrition gives place to the study of the physiology of the central nervous system and energetics is exchanged for cybernetics, which is the scientific study of the reception, transmission, storage, transformation and use of information by regulating apparatus, regardless of whether it is made "of metal or of flesh," i.e., whether it is a machine or an organism.

Like any new branch of knowledge, cybernetics is developing very quickly. In the very short period of its existence it has therefore succeeded in enriching with new ideas and achievements both science and, especially, contemporary technology, in its efforts at maximal automation in the direction of the productive processes. Furthermore, the latest developments in automats and in calculating machines have already advanced so far that the results of experiments planned or already carried out with them may, in many cases, be used in efforts to achieve a rational explanation of the phenomena which take place during the functioning of the nervous system and in many other processes. The understandable attractiveness of these successes as well as the extensive (though little-justified) use of neurophysiological, psychological, and even sociological terminology in cybernetics, has now created a situation in which many contemporary authors are beginning to think that machines which can solve complicated mathematical problems, make translations from one language into another and, in general, carry out many tasks normally considered as brain work, must, in some way, be alive and, therefore, they have come to regard cybernetics as being a fundamentally new and universal road to the understanding of the very essence of life.

Of course, this is wrong. As we have pointed out earlier, attempts have been made, many years ago, to attribute life to machines. The only thing that has changed is the opinion as to what aspect one should concentrate on; movement, energetics, communication or some other property common to organisms and machines which is susceptible to explanation in terms of the laws of physics and chemistry. The basic stimulus which induces investigators to attribute life to machines is always the same. It is as follows. "Purposiveness" in the organization of living things

is what differentiates them in principle from the objects of the inorganic world. Apart from organisms, machines are the only things which show such "purposiveness" in their structure. Furthermore, the work of machines can be completely known in terms of physics and chemistry. This identification of living things with machines was therefore viewed as the one and only way of saving science from the mystical entelechy of the vitalists, the bridge which will carry us over from physics and chemistry to biology.

Of course, we may, and should, try to understand the physical and chemical basis of the various vital phenomena by means of the construction and study of models which will reproduce the same phenomena as occur in organisms but in a simplified form. In doing so, however, we must always remember that we are dealing with models and not confuse them with living things. We must always take into account the differences as well as the similarities between the model and the real thing. Only thus can we avoid very dangerous oversimplification and those mistakes which have always cost mankind dear and which have only been corrected by science at the expense of a tremendous effort. However great the complexity or intricacy of its organization, an electronic calculating machine is still further apart in its nature from a human being than is, for example, the simplest bacterium, although this has not got the differentiated nervous system which the machine imitates so successfully.

Unfortunately this difference is usually slurred over in cybernetic literature. It may be that this is, to some extent, justifiable when we wish to concentrate our attention solely on the general rules of communication and not on any particular systems. However, if the aim of our studies is the understanding of the nature of life, then it is, in principle, impermissible to ignore the difference between organisms and mechanisms.

APPRAISAL OF CONTEMPORARY MECHANISTIC HYPOTHESES

The first difference between machines and living things to strike the eye is the material of which the different systems are made and their actual nature.

Those who hold the theory that living things are machines usually tend to ignore this difference on the ground that the work of a machine depends, essentially, on its structure and not on the material of which it is made. In this connection Jost has written: "We may construct a machine of steel or of brass and this will certainly affect its durability and accuracy but not the nature of the work it does." One might even construct a machine not of metal but of plastic or some other organic material and thus approximate its composition to the chemical composition of a living thing.

Such considerations are, however, radically unsound. The fact that living things are, in Engels' words, "albuminous bodies," that they include in their composition proteins, nucleic acids, lipids, specific carbohydrates, and other multifarious organic compounds is by no means to be regarded as an incidental circumstance of only slight significance. The composition of the living body is the very factor which determines its flowing character. In particular, it is only by understanding the highly specific features of the structure of proteins that we can understand the immediate causes underlying the determinate sequences of individual reactions in metabolism, that is to say, their coordination in time.

Any organic substance can react in very many different ways, it has tremendous chemical possibilities, but outside the living body it is extremely "lazy" or slow about exploiting these possibilities. Inside the living thing, however, organic substances undergo extremely rapid chemical transformation. This is due to the catalytic properties of proteins. If any organic substance is to play a real part in metabolism it must enter into chemical combination with some protein-enzyme, and form with it a particular, very active and unstable intermediate compound. If it does not do so, its chemical potentialities will be realized so slowly that they will be insignificant in the quickly flowing process of life.

Owing to its extreme specificity, each enzyme will only form intermediate compounds with a particular substance (its substrate), and will only catalyze strictly determined individual reactions. The rates at which these reactions take place within the living body may therefore vary greatly, depending, in the first place, on the presence of a set of enzymes, and also on their catalytic activity. This latter can be greatly changed by internal physical and chemical circumstances and also by the action of the external medium. This sort of mobile relationship between the rates of individual biochemical reactions is, in fact, a prerequisite for the determinate sequences and concordances of these reactions in the whole complicated network of metabolism.

This sort of organisation of life may, in some ways, be compared with the organisation of a musical work, such as a symphony, the actual existence of which depends on determinate sequences and concordances of individual sounds. One has only to disturb the sequence and the symphony as such will be destroyed, only disharmony and chaos will remain.

In a similar way, the organisation of life is fundamentally dependent on a regular sequence of metabolic reactions and the form and structure of living bodies are flowing in nature. For this reason, organisms can only exist for any length of time as a result of the continuous accomplishment of chemical transformations, which constitute the essence of living, and the cessation of which would lead to the disruption of the living system and the death of the organism.

In contrast to this, the basic structure of the machine is static. When the machine is working, only the energy source or fuel undergoes chemical change, while the actual structure remains materially unchanged irrespective of whether it is made of metal or of plastic, and the less it is changed (by corrosion, for example), the longer the actual machine will last.

Thus, the actual principle of stability which enables them to exist for a prolonged period is different for organisms and machines. The similarities between them which have been enumerated above are, therefore, only very superficial and, when examined in more detail, they are seen to be purely formal.

We may demonstrate this in the particular case of mechanical movement by an organism. In the muscles of the animal carrying out this movement, the protein fibrils are orientated in a particular way relative to one another. Such a structure, however, cannot be likened in any way to that of a machine. In a machine the structural elements do not play any part whatsoever in the chemical transformation of the energy source. If the component parts of the machine were themselves to undergo chemical transformation during their work, this would, of course, lead quickly to the destruction of the whole mechanism. On the other hand, the elements of construction of the living body, the protein fibrils in this case, themselves take a direct part in the metabolic reactions which serve as the source of that energy which is transformed into mechanical movement. The same may also be said of the comparison of organisms with heat engines in respect of energetics. We now know that the analogy between respiration and combustion is very formal. In combustion the surmounting of the energy of activation, which is necessary for the accomplishment of the oxidative reactions, is done by raising the temperature considerably, whereas in respiration this is not needed. Respiration is based on an enzymic lowering of the energy of activation.

If the transformation of energy took place in the same way in organisms as it does in heat engines, then, at temperatures at which living things can exist, the coefficient of their useful activity would fall to an insignificant fraction of one percent. It is, in fact, amazingly high, considerably higher than that obtained in present-day heat engines. The explanation of this is that the oxidation of sugar, or any other respiratory fuel, in the organism takes place not as a single chemical act, but by a series of individual reactions coordinated in time.

If the oxidation of organic materials in the organism took place suddenly, then the living body would be unable to make rational use of all the energy set free in this way, especially if it was given off in the form of heat. In the oxidation of only 1 mole (180 g) of sugar about 700 kcal are liberated. The instantaneous liberation of this amount of energy would be associated with a sharp rise in temperature, the denaturation of proteins and the destruction of the living body. This same energy effect, which is brought about by the organism under ordinary conditions of low temperature, depends on the fact that, in the process of biological oxidation, sugar is not converted into carbon dioxide and water suddenly, but slowly, by stages. A process of this sort not only gives the possibility of surmounting the energy of activation at ordinary temperatures, it also enables the living body to make rational use of the energy which is gradually set free. Thus, the more highly organized the metabolism, i.e., the better the coordination between the separate reactions comprising it, the higher the coefficient of useful activity.

The principle of evaluating nutrients simply in terms of their content of calories led to many difficulties in its practical application. This principle was only overthrown with great difficulty, as a result of studies of vitamins and essential amino acids, and investigations which showed that, unlike a heat engine, there occurs in the organisms, not only oxidation of energy-providing material, but also transformation of the fundamental protein structures of the living body, which are broken down and resynthesized in the general interaction of the organism with the external medium.

Finally, it must be pointed out that the ways of "overcoming entropy" used by organisms and contemporary mechanisms or automats also differ from one another in principle. As we have shown above, organisms manage to avoid "thermodynamic equilibrium" just because they are open or flowing systems. Recent studies of the thermodynamics of such systems have shown them to be essentially different from classical thermodynamics, which is based on the phenomena observed in closed systems. It gives us a perfectly rational explanation of why it is that, in organisms, entropy not only does not increase, but may even decrease.

According to Wiener, a different principle underlies the ability of contemporary cybernetic automats to counteract the tendency to the increase of entropy and to create zones of organisation around themselves. In order to explain this ability, Wiener uses the same idea as Maxwell used in the form of his demons. According to present-day ideas, however, this "Maxwell's demon" has to be continually obtaining "information" according to which he opens or closes the doors to molecules of high or low rates of movement. For a number of years the overriding wish to identify the organism with a mechanism forced many scientists to ignore all the increasing factual evidence and look for some rigid, unchanging, static structures in the living body so that these structures themselves might be regarded as the specific bearers of life.

At the end of last century it was widely held among biologists that the organization of protoplasm was based on the presence in it of a certain machine-like structure formed of solid and unchanging "beams and braces" interlacing with one another. It was thought that the only thing that prevented us from seeing this structure was the imperfection of our optical methods.

As these methods developed, however, the search for static, "life-determining" structures was first transferred to the realm of colloid chemical formations and then to the realm of intramolecular structure. In this way there arose the concept that the material carriers of life are to be found in the single molecules of heritable substance which have a static, unchanging structure and form a part of the nuclear chromosomes. This concept is associated with the work of T. Morgan and his followers concerning the gene nature of life. According to H. Muller the "living gene molecule" can only undergo change in detail but is essentially so static that it has maintained its internal life-determining structure unchanged throughout the whole development of life on the Earth. This concept of the Morgan school of geneticists was fully expressed in Schrödinger's well-known book *What is Life?* (1944). Schrödinger saw the key to the understanding of life in the fact that the structure which is the only one natural and peculiar to life, namely that of the gene, "is so long-lasting and constant as to border on the miraculous." It is as unchanging as though it were frozen. Thus, according to Schroedinger, the organization of life is based upon the principle of "clockwork," the structure of which remains completely constant at room temperature as well as at absolute zero. In his conclusion Schroedinger writes "Now, I think, few words more are needed to disclose the point of resemblance between a clock-work and an organism. It is simply and solely that the latter also hinges upon a solid—the aperiodic crystal forming the hereditary substance, largely withdrawn from the disorder of heat motion."

The only new thing that has been added to the gene theory up to now is that an attempt has been made to give a chemical reality to the previously rather vague idea of the gene molecule in the form of a suggestion that the living molecule may be a particle of nuclear nucleoprotein or, according to the latest evidence, simply the molecule of deoxyribonucleic acid. It would seem that everything else in the cell is to be regarded as merely the medium for the "living molecule."

From this point of view the capacity for self-reproduction, which is characteristic of living things, is based only on the strictly determinate, static, intra-molecular structure of deoxyribonucleic acid (DNA), on the specific arrangement of purine pyrimidine mononucleotide residues in the polynucleotide chain of DNA. This arrangement represents, in the terminology of cybernetics, the code in which the whole collection of specific characteristics of the living body is stored. The transfer of "hereditary information" may thus be thought of as something like the work of a stamping machine in which the molecule of DNA represents the matrix which always reproduces a single uniform structure. Such hypotheses are very impressive to the protagonists of the machine theory of life and are therefore very widely supported by contemporary physicists as well as biologists. We shall go into this in more detail in the course of further explanation. All that is needed now is to remark briefly on the present state of the problem. The more concrete the biochemical studies of the self-reproduction of living things, the more obvious it becomes that the process is not just bound up with this or that particular substance or a single molecule of it, but is determined by the whole system of organization of the living body which, as we have seen, is flowing in nature and is in no way to be compared with a stamping machine with an unchanging matrix. Speaking on this subject at the Fourth International Congress of Biochemistry, E. Chargaff said "It is even possible that we may be dealing with templates in time rather than in space" by which he meant a definite order in which processes occur in the living thing. In any case when we are not making a formal comparison between the phenomena occurring in living beings and those occurring in machines, but are trying to understand what they are really like, we find, not only similarities, but also a profound difference between the two kinds of system. This difference is not merely fortuitous but forms the very essence of their organization.

In the first place, it is connected with the fact that the "purpose" which inspires a person to create some machine which is necessary to him has nothing in common with the task of self-preservation and self-reproduction which determines the organization of living things. The aims in constructing a watch, a steam locomotive or an automatic device for defence against aerial attack are to tell the time correctly, to transport people and goods and effectively to bring down enemy aircraft. Owing to our present ways of thinking and technical habits which have evolved over many centuries we find it far easier to solve these problems rationally by building static structures out of metal or some solid plastic material, and this is what, in fact, we do.

In this way the actual principles of construction of any machine now in existence reflect the character of the person who made it, his intellectual and techno-logical level, his aims and his methods of solving the problems in front of him.

This also applies fully to the various "cybernetic toys" which are now being made, the point of which is simply to imitate living things, such as Grey Walter's "tortoise," Shannon's "mouse," Ducrocq's "fox," and Ashby's "homeostat," constructions which have been wittily described by Grey Walter as "machines which can serve no useful purpose." The structure of all these necessarily carries the predetermination put there by those who constructed them and P. Cossa was quite right when, in his book, *La cybernétique* (1957), he wrote of them as follows:

"What is inherent in the living thing (adaptability), is not merely the means but the end itself: the preservation of life, the preservation of the continuity of existence by adaptation to the environment. There is nothing like this in the homeostat, it has no inherent ultimate aim. If a living thing, which has had its equilibrium upset, perseveringly tries out, one after another, all possible means of adapting itself to its new environment, this is explained as an effort to survive. If the homeostat tries out its 390,625 combinations one after another it only does so because that is what Ashby wanted of it."[3]

Of course, one may imagine machines of the future which will imitate living things very closely; machines designed as flowing systems in which energy is used by easy stages; they might even be able to reproduce themselves, etc. All the same, the organization of these machines would still reflect the specific task which those who made them had set themselves; they would always bear the marks of their origins.

The insuperable difference in principle between machines and organisms stands out specially clearly when we consider the question of the origin of individual systems. We know that, in its general organization in which the structure is adapted to the performance of particular tasks, a machine develops first in the mind of its creator and not as a real physical system. This idea is then expressed in drawings or plans. These plans usually form the basis for the construction of individual components in accordance with their specifications. These are then assembled and it is only at this stage that the machine appears as a physical object.

The way in which a machine arises is, thus, perfectly clear, but if we try to solve the problem of the origin of living things by analogy with machines, we shall, logically and inevitably, reach an idealistic conclusion.

The book by Schroedinger, which has already been quoted, may serve as a good example of this. In it the author set out to understand life from the point of view of physics, that is, on a purely materialist basis. Nevertheless, in his conclusion, he was forced to characterize life as "the finest masterpiece ever achieved along the lines of the Lord's quantum mechanics," i.e., to put it plainly, he acknowledged the divine origin of life.

There is a difference in form, but not in substance, between this conclusion and those of other attempts to solve the problem of the origin of life which have been made on the basis of purely mechanistic assumptions as to its genetic nature (e.g., those of A. Dauvillier, G. Blum, L. Roka, and others). Essentially, all these attempts arrive at the same explanation. In the primitive and, as yet, lifeless solution of organic material there somehow arose particles of protein, nucleic acid or nucleoprotein and these, suddenly, on their appearance, had an intramolecular structure which was extremely well adapted to the accomplishment of self-reproduction and other vital functions. Thus, there arose the "primary living matrix" which could later be elaborated, but the "purposive" life-determining structure was not necessarily immediately the same as it is today. The question then arises as to what were the natural laws underlying the origin of an intramolecular structure which was adapted to the performance of specific functions. Iron can exist in the elementary form in inorganic nature and, under certain circumstances, may take the form of shapeless lumps, but, as Aristotle wrote, even a sword cannot arise in this way without human intervention, for its structure is suited to the accomplishment of a particular end. In just the same way (as we shall see later) those physical and chemical laws, which were the only ones prevailing in the waters of the primeval ocean, were quite enough to account for the primary formation, in those waters, of high-molecular protein-like polymers and polynucleotides with a more or less irregular arrangement of mononucleotide residues. By themselves, however, these laws are quite insufficient to provide for the possibility of the development of any structures adapted to the performance of particular functions. The supporters of these hypotheses "explain" so to speak, the functional suitability of the structures of their primary matrices as being due to "a lucky chance" or "just pure chance," in which

Dauvillier is justified in seeing "the hand of an eccentric creator." This "hand" does not differ essentially from Schroedinger's "the Lord's quantum mechanics" nor even from St. Augustine's "divine will."

In his book, the English edition of which bears the most intriguing title, *The origins of life* (1957), A. Ducrocq claims to have given a general explanation of life and its origins based on "cybernetic theory" and to have demonstrated the laws of the delicate interaction of forces whereby "conglomerates of atoms are transformed into living aggregates," i.e., living things. On closer acquaintance with this book, however, we find that, in the last analysis, the whole thing amounts to a statement that the chain of DNA, which served as the point of origin for the whole series of living things and which was constructed in a specific way, must have appeared in some improbable way in the solution of organic substances, for its appearance "had a probability which was not nil." In what way is this different from those numerous hypotheses, which we have already mentioned, about the chance origin of life?

Physicists assert that, in principle, it is possible that, by chance, the table on which I am writing might rise up of its own accord into the air owing to the simultaneous orientation of the thermal movement of all its molecules in the same direction. It is hardly likely, however, that anyone would conduct his experimental work, or his practical activities in general, with this possibility in view. Furthermore, the experimental scientist attaches value to those theories which open up possibilities for investigation, but how can one study a phenomenon which, at the best, could only occur once in the whole time of the existence of the Earth? The conception of the chance origin of the "living molecule" is, therefore, completely fruitless from a practical point of view and, as we shall see later, it is also theoretically unsound.

There can be no doubt that the conscious or unconscious attempts to liken the origin of living things to the assembly of a machine also lies at the root of the many contemporary statements to the effect that, in the original solution of organic substances, there were formed various substances which were at once structurally suitable and well adapted to the carrying out of particular vital functions and then, by their combination, they gave rise to the first living body, just as a machine is assembled from separate components the structure of which was already adapted to doing a particular job.

According to these statements the first thing that happened, even before the appearance of the most primitive organism, was the formation of protein-enzymes with their strictly determinate intramolecular structures and their very efficient adaptation to the carrying out of particular catalytic reactions which are very important in metabolism, nucleic acids, which play an essential part in the process of reproduction of organisms and other compounds which are to be found in the very efficient "rationally constructed" organs of living protoplasm as we know it, though this itself only arose secondarily by the combination of the primary compounds.

Such an idea reminds one of the sayings of the ancient Greek philosopher Empedocles who believed that, when living things originally came into being, individual organs were first formed independently of one another—"Thus there grew up a multitude of heads without necks, naked arms wandered around without their shoulders and eyes moved about with no foreheads." Later these unmatched members joined together and in this way the various sorts of animals and people were formed.

IT IS ONLY POSSIBLE TO UNDERSTAND LIFE BY STUDYING ITS ORIGIN AND DEVELOPMENT

From the modern Darwinian point of view the falsity, not to say absurdity, of such a theory is perfectly obvious. Any particular organ can only originate and become perfected as part of the evolutionary development of the organism as a whole.

The specialized and complicated structure of the eye and the hand are adapted to their purpose only when considered in relation to the functions which they carry out. It is impossible, even unthinkable, to take seriously the evolution of an individual organ such as the Empedoclean "eyes with no foreheads" because the very functions which determined their structure would have no meaning under those circumstances. The action of natural selection can, therefore, only affect them as parts of whole living things.

In just the same way enzymes, nucleic acids and so on are only parts of the living body, they are like organs, subserving definite, vitally necessary functions.

Thus the catalytic activity of enzymes or the specific functions of nucleic acids are of no importance to the substances themselves but are only important to the whole living body in which the particular metabolic reactions take place. It follows that, when not a part of such a body, before its formation, they would be quite unable to acquire a "purposive" structure, suited to carrying out their vital functions. It is quite natural and right to suppose that there were successive developments proceeding from simpler to more complicated systems, but, although the individual organs are simpler than the whole organism, we should not be like Empedocles and imagine that animals and people arose by the fusion of individual organs.

Darwin showed the true way in which higher organisms have arisen by the evolution of lower living things which were more simply organized but were still complete systems in themselves.

Similarly it would be wrong to suppose that, in the organically rich waters of the primeval ocean, there arose proteins and nucleic acids with a "purposive" structure extremely accurately and well adapted to the carrying out of particular biological functions and that, later, by their combination, the living body itself was created.

All that we can expect from the action of the laws of physics and chemistry, which were the only ones on the still lifeless Earth, is that there were formed more or less randomly constructed polymers with a haphazard distribution of peptides and mononucleotides and thus having no "purposiveness" or adaptation to the carrying out of particular functions.

These polymers could, however, join together with one another to form complete multimolecular systems though, naturally, these were incomparably simpler than living bodies. It was only as a result of the prolonged evolution of these original systems, their interaction with their environment and their natural selection, that there arose those forms of organization which are peculiar to the living body, namely metabolism, and with it proteins, enzymes, nucleic acids and those other complicated and "purposefully" constructed substances which are characteristic of contemporary organisms. Thus there is not even the most remote similarity between the origin of life and the assembly of a machine.

These two sorts of system show a likeness to each other only if we consider them in their finished state, divorced from their origins. Once we start to deal with this question, the difference between the machine and the organism immediately becomes apparent and it is obvious that the two kinds of system are essentially of a different quality.

This concept is understandable and even simple, for the origin of life and the origin of the machine took place at very widely separated levels of evolutionary development.

We may note the following important stages in this development from the moment of the formation of the Earth to the present day. For the first millions of years of its existence, our planet had no life on it and all the processes occurring on it were subject only to the laws of physics and chemistry. This stage of development may be referred to as inorganic or abiogenic. Life then arose on the Earth and a new biological stage of evolution began. Now new biological laws were added to the old physical and chemical ones and these new laws have now come to the fore and assumed an ever-growing importance in the progressive development of living things. The crowning achievement of this period was the emergence of man heralding the beginning of the third or social stage of evolution. Now even the biological laws have been driven from the foreground and the laws of development of human society have begun to play the leading part in further progress.

It is very important that, with the beginning of each new stage of development, with the origin of a new form of the movement of matter, the tempo of its evolution increases. The abiogenic period of the existence of the Earth lasted for thousands of millions of years, but the decisive progress of biological evolution only required hundreds or perhaps even tens of millions of years for its accomplishment. The whole development of mankind has only lasted a million years. Social transformations have occurred within thousands of years or even centuries and now we can easily notice substantial changes in human society over periods reckoned in decades.

There can hardly have been any significant biological change in the human race since the time of Aristotle, but it is only during the last few hundred years that man has attained hitherto unimaginable power over his environment. He can cover the ground faster than any deer, swim beneath the water better than any fish and fly through the air incomparably faster and further than any bird. But this is not because he has grown wings or fins and gills during that time. The powers acquired by mankind are not the result of biological but of social development. In particular, machines, which play such an outstanding part in man's conquest of the forces of nature are the fruit of this development, for man could only create them by all-round mastery of the experiences accumulated by his forebears over many centuries, only, in fact, on the basis of the communal life of mankind.

Thus machines are not merely inorganic systems operating in accordance with no laws other than those of physics and chemistry. They are, in origin, not biological but higher, i.e., social forms of the motion of matter. We can therefore only understand their real nature through studying their origin. We shall now discuss some other examples so that this may become clearer to the reader.

On the banks of great rivers, which have worn down thick sedimentary formations, one may find stones made of calcite which are commonly called "devil's fingers" because of their queer shape, which certainly does remind one of the shape of a finger except that it is sharpened at one end into a cone. In old times people believed that these objects were formed by lightning striking sand and even their scientific name—belemnites—is derived from this supposition as to their origin. If this were the case, they should always be associated with mineral formations of the abiogenic, inorganic world. In fact, however, it has been shown that belemnites are the fossilized remains of rostra, which are parts of the insides of molluscs, and that these are characteristic of a particular group of cephalopod molluscs which lived in the Jurassic and Cretaceous periods and were completely extinct by the beginning of the Tertiary period. On the surface of some belemnites one may even find traces of the blood vessels of the mantle, or soft envelope, of the body of the mollusc which once enclosed the belemnite. Thus belemnites, taken on their own without reference to their origin, are clearly completely lacking in life. From the point of view of their chemical composition and also from that of their characteristic physical properties they appear to be objects of the inorganic world. Belemnites, however, could not be formed in that world as a result of the elementary forces of inorganic nature alone. For this reason we cannot understand the essential nature of

these objects if we do not know about their biological origin or the history of the development of life on the Earth. In that case they would certainly seem to us to be some miraculous "devil's fingers."

I will now ask my readers to let me indulge in fantasy as this will enable me to present my ideas more clearly.

Let us imagine that people have succeeded in making automatic machines or robots which can not only carry out a lot of work for mankind but can even independently create the energetic conditions necessary for their work, obtain metals and use them to construct components, and from these build new robots like themselves. Then some terrible disaster happened on the Earth, and it destroyed not only all the people but all living things on our planet. The metallic robots, however, remained. They continued to build others like themselves and so, although the old mechanisms gradually wore out, new ones arose and the "race" of robots continued and even, perhaps, increased within limits.

Let us further imagine that all this has already happened on one of the planets of our solar system, on Mars, for example, and that we have landed on that planet. On its waterless and lifeless expanses, we suddenly meet with the robots. Do we have to regard them as living inhabitants of the planet? Of course not. The robots will not represent life but something else. Maybe a very complicated and efficient form of the organization and movement of matter, but still different from life. They are analogous to the belemnites which we have already considered, the only difference being that the belemnites arose in the process of biological development while the robots were based on the higher, social form of the motion of matter.

Life existed in the Jurassic sea and the rostra of cephalopod molluscs played a particular part in it. The life vanished and the belemnites remained, but now they appear to be lifeless objects of the inorganic world. Similarly, automatic machines and, in particular, our imaginary robots, could only develop as offshoots of human (or some similar) society, as the fruit of the social form of organization and movement of matter, and they have played a considerable part in the development of that form of organization. But that form was destroyed, it vanished, and the robots are on their own, not controlled by it. They are completely subject to the laws of physics and chemistry alone.

Nevertheless, just as one cannot understand what a belemnite is if one has no knowledge of life, so it is impossible to grasp the nature of the "Martian robot" without a sufficient acquaintance with the social form of the motion of matter which gave rise to it. This would be true even if one were able to take down the robot into its individual components and reassemble it correctly. Even then there would remain hidden from our understanding those features of the organization of the robot which were purposefully constructed for the solution of problems which those who built them envisaged at some time, but which are completely unknown to us.

When the Lilliputians found a watch in Gulliver's pocket, they were not in a state to understand its nature properly, although, according to Swift, the Lilliputians had a very extensive knowledge of mathematics and mechanics. After prolonged deliberation, they decided that it was a pocket god which Gulliver consulted each time he started to do something.

If some "thinking Martian" were to chance on a watch flying about somewhere in space perhaps he too would be able to take it to pieces and put it together again but there would still be a lot about it that would be incomprehensible to him. And not only will the Martians not understand, but many of my readers will probably not be able to explain why there are only 12 numbers on the faces of ordinary clocks, although the day is divided into 24 hours. This question can only be answered from a good knowledge of the history of human culture and, in particular, of the history of watchmaking.

Similarly, an understanding of the nature of life is impossible without a knowledge of the history of its origin. Usually, however, the nature and origin of life have been regarded, and are even now regarded, as being two completely separate problems. Thus, at the end of last century and the beginning of the present one, the problem of the origin of life was denounced as an accursed and insoluble problem, work on which was unworthy of a serious scientist and was a pure waste of time. People tried to achieve an understanding of the nature of life, which is the main problem of biology at present, primarily in a purely metaphysical way, completely isolated from its origin. In principle, this amounts to their wanting, crudely speaking, to take the living body apart into its component screws and wheels like a watch and then to try to put it together again.

Even Mephistopheles jeered at such an approach in his advice to the young scholar.

"Wer will was Lebendigs erkennen und
    beschreiben,
Sucht erst den Geist herauszutreiben,
Dann hat er die Teile in seiner Hand,
Fehlt leider! nur das geistige Band.
Encheiresin naturae nennt's die Chemie,
Spottet ihrer selbst, und weiss nicht wie."[4]

Of course, a detailed analysis of the substances and phenomena peculiar to contemporary living things is extremely important and absolutely necessary for an understanding of life. That is beyond doubt. The whole question is whether this, by itself, is enough for such an understanding. It is clearly not. Even now, for all our skill in this sort of analysis, we are still very far from being able to point to any way in which life could actually be synthesized although we admit that this synthesis is theoretically perfectly possible.

This is by no means merely because our analysis has, as yet, not been finished, that we still have not found out all the details of the structure of the living body.

"The whole," wrote M. Planck, "is always somewhat different from the sum of the separate parts." It is only possible to understand this whole by knowing it in its maturity and in its development, by studying and reproducing the processes of gradual elaboration and perfection of the more primitive systems which were its precursors.

It is now becoming more and more obvious that a knowledge of the essential nature of life is only possible through a knowledge of its origin. Now, too, this origin no longer seems so puzzling as it did not long ago. We are sketching out in more and more detail the actual ways in which life arose on the Earth. It could only have happened as an integral part of the general historic development of our planet. The facts at our disposal indicate that the origin of life was a gradual process in which organic substances became more and more complicated and formed complete systems which were in a state of continual interaction with the medium surrounding them.

Following the path of the emergence of life in this way we encounter neither the "almighty hand of the Creator" nor machines which made their appearance at a far later stage in the development of matter. We do, however, discover in this way how and why it is that the particular original systems which existed were transformed, in the process of evolution, into those which are characteristic of life instead of into others and how, in that same process of the establishment of life, there arose new biological laws which had not existed before, and also how the "purposiveness" which we notice in all living things came into being.

In this way our knowledge gives us a real understanding of the essential organization of the most primitive forms of life and, on that basis, we can easily follow the further evolution of these forms by applying the precepts of evolutionary theory. We can trace the formation of new features characteristic of highly organized living beings, including man, who is the culmination of the biological stage of the development of matter.

Thus we arrive at the main idea underlying this book which had already been formulated by Heraclitus of Ephesus and was included in the works of Aristotle—"One can only understand the essence of things when one knows their origin and development."

NOTES

This chapter originally appeared as chapter 1 in A. I. Oparin, *Life: Its nature, origin, and development*, trans. Ann Synge, pp. 1–37, New York: Academic Press, 1964.

REFERENCES

1. Bernard, C. (1878/1879). *Leçons sur les phénomènes de la vie communs aux animaux et aux végétaux*. Paris: Librairie J.-B. Baillère et Fils.
2. Cossa, P. (1957). *La cybernétique*. Paris: Masson et Cie.
3. Ducrocq, A. (1957). *The origins of life*, trans. A. Brown. London: Elek Books.
4. Oparin, A. I. (1944). *What is life?* Cambridge, UK: Cambridge University Press.
5. Tolstoï, L. (1886/1887). *O Zhizni [On life]*. Unknown binding.
6. Wiener, N. (1949). *Cybernetics, or control and communication in the animal and in the machine*. New York: Wiley.

ENDNOTES

1 L. Tolstoï is here using the term "scribe" in the bad sense in which it is used in the gospels "scribes and Pharisees."
2 "The organic structure is the seat of perpetual nutritional movement which leaves no part of it at rest: Each part nourishes itself, without rest or pause, from the medium surrounding it and discharges its wastes and products into that same medium. This molecular renewal is not perceptible to the eye, but, as we see the beginning and the end of it, the entry and discharge of substances, we can imagine the intermediate stages and we picture to ourselves a flow of matter, incessantly passing through the organism and renewing its substance while maintaining its form.

The universality of such a phenomenon in plants and animals and in all parts of them, as well as their constancy, which never undergoes arrest, make it a general sign of life which some physiologists have used for its definition."
3 Retranslation from Russian. A. S.
4 Whoso would describe and know aught that's alive
   Seeks first the spirit forth to drive;
   The parts he then hath in his hand,
   But lacks, alas! the spirit-band.
   Encheiresin naturae chemists call it now,
   Mock at themselves and know not how.
   Trans. W. H. van der Smissen

6 · What is the meaning of "life"?

ERNST MAYR

Primitive humans lived close to nature. Every day they were occupied with animals and plants, as gatherers, hunters, or herdsmen. And death—of infants and elders, women in childbirth, men in strife—was forever present. Surely our earliest ancestors must have wrestled with the eternal question, "What is life?"

Perhaps, at first, no clear distinction was made between life in a living organism and a spirit in a nonliving natural object. Most primitive people believed that a spirit might reside in a mountain or a spring as well as in a tree, an animal, or a person. This animistic view of nature eventually waned, but the belief that "something" in a living creature distinguished it from inanimate matter and departed from the body at the moment of death held strong. In ancient Greece this something in humans was referred to as "breath." Later, particularly in the Christian religion, it was called the soul.

By the time of Descartes and the Scientific Revolution, animals (along with mountains, rivers, and trees) had lost their claim to a soul. But a dualistic split between body and soul in human beings continued to be almost universally accepted and is even today still believed by many people. Death was a particularly puzzling problem for a dualist. Why should this soul suddenly either die or leave the body? If the soul left the body, did it go somewhere, such as to some nirvana or heaven? Not until Charles Darwin developed his theory of evolution through natural selection was a scientific, rational explanation for death possible. August Weismann, a follower of Darwin at the end of the nineteenth century, was the first author to explain that a rapid sequence of generations provides the number of new genotypes required to cope permanently with a changing environment. His essay on death and dying was the beginning of a new era in our understanding of the meaning of death.

When biologists and philosophers speak of "life," however, they usually are not referring to life (that is, living) as contrasted with death but rather to life as contrasted with the lifelessness of an inanimate object. To elucidate the nature of this entity called "life" has been one of the major objectives of biology. The problem here is that "life" suggests some "thing"—a substance or force—and for centuries philosophers and biologists have tried to identify this life substance or vital force, to no avail. In reality, the noun "life" is merely a reification of the process of living. It does not exist as an independent entity.[1] One can deal with the process of living scientifically, something one cannot do with the abstraction "life." One can describe, even attempt to define, what living is; one can define what a living organism is; and one can attempt to make a demarcation between living and nonliving. Indeed, one can even attempt to explain how living, as a process, can be the product of molecules that themselves are not living.[2]

What life is, and how one should explain living processes, has been a subject of heated controversy since the sixteenth century. In brief, the situation was this: There was always a camp claiming that living organisms were not really different at all from inanimate matter; sometimes these people were called mechanists, later physicalists. And there was always an opposing camp—called vitalists—claiming instead that living organisms had properties that could not be found in inert matter and that therefore biological theories and concepts could not be reduced to the laws of physics and chemistry. In some periods and at certain intellectual centers the physicalists seemed to be victorious, and in other times and places the vitalists

The Nature of Life, ed. M. A. Bedau and C. E. Cleland. Published by Cambridge University Press. © M. A. Bedau and C. E. Cleland 2010.

seemed to have achieved the upper hand. In this century it has become clear that both camps were partly right and partly wrong.

The physicalists had been right in insisting that there is no metaphysical life component and that at the molecular level life can be explained according to the principles of physics and chemistry. At the same time, the vitalists had been right in asserting that, nevertheless, living organisms are not the same as inert matter but have numerous autonomous characteristics, particularly their historically acquired genetic programs, that are unknown in inanimate matter. Organisms are many-level ordered systems, quite unlike anything found in the inanimate world. The philosophy that eventually incorporated the best principles from both physicalism and vitalism (after discarding the excesses) became known as organicism, and this is the paradigm that is dominant today.

THE PHYSICALISTS

Early beginnings of a natural (as opposed to supernatural) explanation of the world were made in the philosophies of various Greek thinkers, including Plato, Aristotle, Epicurus, and many others. These promising beginnings, however, were largely forgotten in later centuries. The Middle Ages were dominated by a strict adherence to the teachings of the Scriptures, which attributed everything in nature to God and His laws. But medieval thinking, particularly in folklore, was also characterized by a belief in all sorts of occult forces. Eventually, this animistic, magical thinking was reduced, if not eliminated, by a new way of looking at the world that was aptly called "the mechanization of the world picture" (Maier 1938).[3]

The influences leading up to the mechanization of the world picture were manifold. They included not only the Greek philosophers, transmitted to the Western world by the Arabs along with rediscovered original writings, but also technological developments in late medieval and early Renaissance times. There was great fascination with clocks and other automata—and indeed with almost any kind of machine. This eventually culminated in Descartes's claim that all organisms except humans were nothing *but* machines.

Descartes (1596–1650) became the spokesman for the Scientific Revolution, which, with its craving for precision and objectivity, could not accept vague ideas,

immersed in metaphysics and the supernatural, such as souls of animals and plants. By restricting the possession of a soul to humans and by declaring animals to be nothing but automata, Descartes cut the Gordian knot, so to speak. With the mechanization of the animal soul, Descartes completed the mechanization of the world picture.[4]

It is a little difficult to understand why the machine concept of organisms could have had such long-lasting popularity. After all, no machine has ever built itself, replicated itself, programmed itself, or been able to procure its own energy. The similarity between an organism and a machine is exceedingly superficial. Yet the concept did not die out completely until well into this century.

The success of Galileo, Kepler, and Newton in using mathematics to reinforce their explanations of the cosmos also contributed to the mechanization of the world picture. Galileo (1623) succinctly captured the prestige of mathematics in the Renaissance when he said that the book of nature "cannot be understood unless one first learns to comprehend the language and read the letters in which it is composed. It is written in the language of mathematics, and its characters are triangles, circles, and other geometric figures without which it is humanly impossible to understand a single word of it; without these one wanders about in a dark labyrinth."

The rapid development of physics shortly thereafter carried the Scientific Revolution a step further, turning the more general mechanicism of the early period into a more specific physicalism, based on a set of concrete laws about the workings of both the heavens and the Earth.[5]

The physicalist movement had the enormous merit of refuting much of the magical thinking that had generally characterized the preceding centuries. Its greatest achievement perhaps was providing a natural explanation of physical phenomena and eliminating much of the reliance on the supernatural that was previously accepted by virtually everybody. If mechanicism, and particularly its outgrowth into physicalism, went too far in some respects, this was inevitable for an energetic new movement. Yet because of its one-sidedness and its failure to explain any of the phenomena and processes particular to living organisms, physicalism induced a rebellion. This countermovement is usually described under the umbrella term vitalism.

From Galileo to modern times there has been a seesawing in biology between strictly mechanistic and more vitalistic explanations of life. Eventually, Cartesianism reached its culmination in the publication of de La Mettrie's *L'homme machine* (1748). Next followed a vigorous flowering of vitalism, particularly in France and in Germany, but further triumphs of physics and chemistry in the mid-nineteenth century inspired yet another physicalist resurgence in biology. It was largely confined to Germany, perhaps not surprisingly so, since nowhere else did biology flourish in the nineteenth century to the extent it did in Germany.

THE FLOWERING OF PHYSICALISM

The nineteenth-century physicalist movement arrived in two waves. The first one was a reaction to the quite moderate vitalism adopted by Johannes Müller (1801–1858), who in the 1830s switched from pure physiology to comparative anatomy, and of Justus von Liebig (1803–1873), well known for his incisive critiques which helped to bring the reign of inductivism to an end. It was set in motion by four former students of Müller—Hermann Helmholtz, Emil DuBois-Reymond, Ernst Brücke, and Matthias Schleiden. The second wave, which began around 1865, is identified with the names Carl Ludwig, Julius Sachs, and Jacques Loeb. Undeniably, these physicalists made important contributions to physiology. Helmholtz (along with Claude Bernard in France) deprived "animal heat" of its vitalistic connotation, and DuBois-Reymond dispelled much of the mystery of nerve physiology by offering a physical (electric) explanation of nerve activity. Schleiden advanced the fields of botany and cytology through his insistence that plants consist entirely of cells and that all the highly diverse structural elements of plants are cells or cell products. Helmholtz, DuBois-Reymond, and Ludwig were particularly outstanding in the invention of ever-more sophisticated instruments to record the precise measurements in which they were interested. This permitted them, among other achievements, to rule out the existence of a "vital force" by showing that work could be translated into heat without residue. Every history of physiology written since that time has documented these and other splendid accomplishments.

Yet, the underlying philosophy of this physicalist school was quite naive and could not help but provoke disdain among biologists with a background in natural history. In historical accounts of the many achievements of the physicalists, their naivete when it came to living processes has frequently been ignored. But one cannot understand the vitalists' passionate resistance to the claims of the physicalists unless one is acquainted with the actual explanatory statements the physicalists offered.

It is ironic that the physicalists attacked the vitalists for invoking an unanalyzed "vital force," and yet in their own explanations they used such equally unanalyzed factors as "energy" and "movements." The definitions of life and the descriptions of living processes formulated by the physicalists often consisted of utterly vacuous statements. For example, the physical chemist Wilhelm Ostwald defined a sea urchin as being, like any other piece of matter, "a spatially discrete cohesive sum of quantities of energy." For many physicalists, an unacceptable vitalistic statement became acceptable when vital force was replaced by the equally undefined term "energy." Wilhelm Roux (1895), whose work brought experimental embryology into full flower, stated that development is "the production of diversity owing to the unequal distribution of energy."

Even more fashionable than "energy" was the term "movement" to explain living processes, including developmental and adaptational ones. DuBois-Reymond (1872) wrote that the understanding of nature "consists in explaining all changes in the world as produced by the movement of atoms," that is, "by reducing natural processes to the mechanics of atoms... By showing that the changes in all natural bodies can be explained as a constant sum ... of potential and kinetic energy, nothing in these changes remains to be further explained." His contemporaries did not notice that these assertions were only empty words, without substantial evidence and with precious little explanatory value.

A belief in the importance of the movement of atoms was held not only by the physicalists but even by some of their opponents. For Rudolf von Kölliker (1886)—a Swiss cytologist who recognized that the chromosomes in the nucleus are involved in inheritance and that spermatozoa are cells—development was a strictly physical phenomenon controlled by differences in growth processes: "It is sufficient to postulate the occurrence in the nuclei of regular and typical movements controlled by the structure of the idioplasm."

As exemplified in statements by the botanist Karl Wilhelm von Nägeli (1884), another favorite explanation of the mechanists was to invoke "movements of the smallest parts" to explain "the mechanics of organic life."[6] The effect of a nucleus on the rest of the cell—the cytoplasm—was seen by E. Strasburger, a leading botanist of the time, as "a propagation of molecular movements ... in a manner which might be compared to the transmission of a nervous impulse." Thus it did not involve the transport of material; this notion was, of course, entirely wrong. These physicalists never noticed that their statements about energy and movement did not really explain anything at all. Movements, unless directed, are random, like Brownian motion. Something has to give direction to these movements, and this is exactly what their vitalist opponents always emphasized.

The weakness of a purely physicalist interpretation was particularly obvious in explanations of fertilization. When E. Miescher (a student of his and Ludwig) discovered nucleic acid in 1869, he thought that the function of the spermatozoon was the purely mechanical one of getting cell division going; as a consequence of his physicalist bias, Miescher completely missed the significance of his own discovery. Jacques Loeb claimed that the really crucial agents in fertilization were not the nucleins of the spermatozoon but the ions. One is almost embarrassed when reading Loeb's statement that "Branchipus is a freshwater crustacean which, if raised in concentrated salt solution, becomes smaller and undergoes some other changes. In that case it is called Artemia." The sophistication of the physicalists in chemistry, particularly physical chemistry, was not matched by their biological knowledge. Even Sachs, who studied so diligently the effects of various extrinsic factors on growth and differentiation, never seems to have given any thought to the question why seedlings of different species of plants raised under identical conditions of light, water, and food would give rise to entirely different species.

Perhaps the most uncompromising mechanistic school in modern biology was that of Entwicklungsmechanik, founded in the 1880s by Wilhelm Roux. This school of embryology represented a rebellious reaction to the one-sidedness of the comparative embryologists, who were interested only in phylogenetic questions. Roux's associate, the embryologist Hans Driesch, was at first, if anything, even more mechanistic, but he eventually experienced a complete conversion from an extreme mechanist to an extreme vitalist. This happened when he separated a sea urchin embryo at the two-cell stage into two separate embryos of one cell each and observed that these two embryos did not develop into two half organisms, as his mechanistic theories demanded, but were able to compensate appropriately and develop into somewhat smaller but otherwise perfect larvae.

In due time, the vacuousness and even absurdity of these purely physicalistic explanations of life became apparent to most biologists, who, however, were usually satisfied to adopt the agnostic position that organisms and living processes simply could not be exhaustively explained by reductionist physicalism.

THE VITALISTS

The problem of explaining "life" was the concern of the vitalists from the Scientific Revolution until well into the nineteenth century; it did not really become the subject matter of scientific analysis until the rise of biology after the 1820s. Descartes and his followers had been unable to persuade most students of plants and animals that there were no essential differences between living organisms and inanimate matter. Yet after the rise of physicalism, these naturalists had to take a new look at the nature of life and attempted to advance *scientific* (rather than metaphysical or theological) arguments against Descartes's machine theory of organisms. This requirement led to the birth of the vitalistic school of biology.[7]

The reactions of the vitalists to physicalist explanations were diversified, since the physicalist paradigm itself was composite, not only in what it claimed (that living processes are mechanistic and can be reduced to the laws of physics and chemistry) but also in what it failed to take account of (the differences between living organisms and simple matter, the existence of adaptive but much more complex properties—Kant's Zweckmässigkeit—in animals and plants, and evolutionary explanations). Each of these claims and omissions was criticized by one or the other opponent of physicalism. Some vitalists focused on unexplained vital properties, others on the holistic nature of living creatures, still others on adaptedness or directedness (as in the development of the fertilized egg).

All these opposing arguments to the various aspects of physicalism have traditionally been lumped together as vitalism. In some sense, this is not altogether wrong, because all of the antiphysicalists defended the life-specific properties of living organisms. Yet the label vitalist conceals the heterogeneity of this group.[8] For instance, in Germany some biologists (which Lenoir calls teleomechanists) were willing to explain physiological processes mechanically but insisted that this failed to account for either adaptation or directed processes, such as the development of the fertilized egg. These legitimate questions were raised again and again by distinguished philosophers and biologists from 1790 until the end of the nineteenth century, but they had remarkably little effect on the writings of the leading physicalists such as Ludwig, Sachs, or Loeb.

Vitalism, from its emergence in the seventeenth century, was decidedly an antimovement. It was a rebellion against the mechanistic philosophy of the Scientific Revolution and against physicalism from Galileo to Newton. It passionately resisted the doctrine that the animal is nothing but a machine and that all manifestations of life can be exhaustively explained as matter in motion. But as decisive and convincing as the vitalists were in their rejection of the Cartesian model, they were equally indecisive and unconvincing in their own explanatory endeavors. There was great explanatory diversity but no cohesive theory.

Life, according to one group of vitalists, was connected either with a special substance (which they called protoplasm) not found in inanimate matter, or with a special state of matter (such as the colloidal state), which, it was claimed, the physicochemical sciences were not equipped to analyze. Another subset of vitalists held that there is a special vital force (sometimes called Lebenskraft, Entelechie, or élan vital) distinct from the forces physicists deal with. Some of those who accepted the existence of such a force were also teleologists who believed that life existed for some ultimate purpose. Other authors invoked psychological or mental forces (psychovitalism, psychoLamarckism) to account for aspects of living organisms that the physicalists had failed to explain.

Those who supported the existence of a vital force had highly diverse views of the nature of this force. From about the middle of the seventeenth century on, the vital agent was most frequently characterized as a fluid (not a liquid), in analogy to Newton's gravity and to caloric, phlogiston, and other "imponderable fluids." Gravity was invisible and so was the heat that flowed from a warm to a cold object; hence, it was not considered disturbing or unlikely that the vital fluid was also invisible, even though not necessarily something supernatural. For instance, the influential late eighteenth-century German naturalist J. F. Blumenbach (who wrote extensively on extinction, creation, catastrophes, mutability, and spontaneous generation) considered this vital fluid, though invisible, to be nevertheless very real and subject to scientific study, much as gravity was.[9] The concept of a vital fluid was eventually replaced by that of a vital force. Even such a reputable scientist as Johannes Müller accepted a vital force as indispensable for explaining the otherwise inexplicable manifestations of life.

In England, all the physiologists of the sixteenth, seventeenth, and eighteenth centuries had vitalistic ideas, and vitalism was still strong in the 1800–1840 period in the writings of J. Hunter, J. C. Prichard, and others. In France, where Cartesianism had been particularly powerful, it is not surprising that the vitalists' countermovement was equally vigorous. The outstanding representatives in France were the Montpellier school (a group of vitalistic physicians and physiologists) and the histologist F. X. Bichat. Even Claude Bernard, who studied such functional subjects as the nervous and digestive systems and considered himself an opponent of vitalism, actually supported a number of vitalistic notions. Furthermore, most Larmarckians were rather vitalistic in some of their thinking.

It was in Germany that vitalism had its most extensive flowering and reached its greatest diversity. Georg Ernst Stahl, a late seventeenth-century chemist and physician best known for his phlogiston theory of combustion, was the first great opponent of the mechanists. Perhaps he was more of an animist than a vitalist, but his ideas played a large role in the teaching of the Montpellier school.

The next impetus to the vitalistic movement in Germany was the preformation versus epigenesis controversy, which dominated developmental biology in the second half of the eighteenth century. Preformationists held that the parts of an adult exist in smaller form at the very beginning of development. The epigenesists held that the adult parts appear as products of development but are not present as parts in the beginning. In 1759, when the embryologist Caspar Friedrich Wolff refuted preformation and replaced it by epigenesis, he had to invoke some causal agent that would convert the completely unformed mass of the fertilized egg into the adult of a particular species. He called this agent the *vis essentialis*.

J. F. Blumenbach rejected the vague *vis essentialis* and proposed instead that a specific formative force, *nisus formatives*, plays a decisive role not only in the development of the embryo but also in growth, regeneration, and reproduction. He accepted still other forces, such as irritability and sensibility, as contributing to the maintenance of life. Blumenbach was quite pragmatic about these forces, considering them essentially as labels for observed processes of which he did not know the causes. They were black boxes for him, rather than metaphysical principles.

The branch of German philosophy called Naturphilosophie, advanced by F. W. J. Schelling and his followers early in the nineteenth century, was a decidedly metaphysical vitalism, but the practical philosophies of working biologists such as Wolff, Blumenbach, and eventually Müller were antiphysicalist rather than metaphysical. Müller has been maligned as an unscientific metaphysician, but the accusation is unfair. A collector of butterflies and plants from his boyhood on, he had acquired the naturalist's habit of looking at organisms holistically. This perception was lacking in his students, whose leanings were more toward mathematics and the physical sciences. Müller realized that the slogan "life is a movement of particles" was meaningless and without explanatory value, and his alternative concept of Lebenskraft (vital force), though a failure, was closer to the concept of a genetic program than the shallow physicalist explanations of his rebellious students.[10]

Many of the arguments put forth by the vitalists were intended to explain specific characteristics of organisms which today are explained by the genetic program. They advanced a number of perfectly valid refutations of the machine theory but, owing to the backward state of biological explanation available at that time, were unable to come up with the correct explanation of vital processes that were eventually found during the twentieth century. Consequently, most of the argumentation of the vitalists was negative. From the 1890s on Driesch argued, for example, that physicalism could not explain self-regulation in embryonic structures,

regeneration and reproduction, and psychic phenomena, like memory and intelligence. Yet it is remarkable how often perfectly sensible sentences emerge in Driesch's writings whenever his word "Entelechie" is replaced by the phrase "genetic program." These vitalists not only knew that there was something missing in the mechanistic explanations but they also described in detail the nature of the phenomena and processes the mechanists were unable to explain.[11]

Given the many weaknesses and even contradictions in vitalist explanations, it may seem surprising how widely vitalism was adopted and how long it prevailed. One reason, as we have seen, is that at that time there was simply no other alternative to the reductionist machine theory of life, which, to many biologists, was clearly out of the question. Another reason is that vitalism was strongly supported by several other then-dominant ideologies, including the belief in a cosmic purpose (teleology or finalism). In Germany, Immanuel Kant had a strong influence on vitalism, particularly on the school of teleomechanism, an influence still evident in Driesch's writings. A close connection with finalism is evident in the writings of most vitalists.[12]

In part because of their teleological leanings, the vitalists strongly opposed Darwin's selectionism. Darwin's theory of evolution denied the existence of any cosmic teleology and substituted in its place a "mechanism" for evolutionary change—natural selection: "We see in Darwin's discovery of natural selection in the struggle for existence the most decisive proof for the exclusive validity of mechanically operating causations in the whole realm of biology, and we see in this the definitive demise of all teleological and vitalistic interpretations of organisms" (Haeckel 1866). Selectionism made vitalism superfluous in the realm of adaptation.

Driesch was a rabid anti-Darwinian, as were other vitalists, but his arguments against selection were consistently ridiculous and showed clearly that he did not in the least understand this theory. Darwinism, by supplying a mechanism for evolution while at the same time denying any finalistic or vitalistic view of life, became the foundation of a new paradigm to explain "life."

THE DECLINE OF VITALISM

When vitalism was first proposed and widely adopted, it seemed to provide a reasonable answer to the nagging question, "What is life?" Furthermore, at that time it was a legitimate theoretical alternative not just to the crude mechanicism of the Scientific Revolution but also to nineteenth-century physicalism. Vitalism seemingly explained the manifestations of life far more successfully than the simplistic machine theory of its opponents.

Yet considering how dominant vitalism was in biology and for how long a period it prevailed, it is surprising how rapidly and completely it collapsed. The last support of vitalism as a viable concept in biology, disappeared about 1930. A considerable number of different factors contributed to its downfall.

First, vitalism was more and more often viewed as a metaphysical rather than a scientific concept. It was considered unscientific because the vitalists had no method to test it. By dogmatically asserting the existence of a vital force, the vitalists often impeded the pursuit of a constitutive reductionism that would elucidate the basic functions of living organisms.

Second, the belief that organisms were constructed of a special substance quite different from inanimate matter gradually lost support. That substance, it was believed through most of the nineteenth century, was protoplasm, the cellular material outside the nucleus.[13] Later it was called cytoplasm (a term introduced by von Kölliker). Because protoplasm seemed to have what was called "colloidal" properties, a flourishing branch of chemistry developed: colloidal chemistry. Biochemistry, however, together with electron microscopy, eventually established the true composition of cytoplasm and elucidated the nature of its various components: cellular organelles, membranes, and macromolecules. It was found that there was no special substance "protoplasm," and the word and concept disappeared from the biological literature. The nature of the colloidal state was likewise explained biochemically, and colloidal chemistry ceased to exist. Thus all evidence for a separate category of living substance disappeared, and it became possible to explain the seemingly unique properties of living matter in terms of macromolecules and their organization. The macromolecules, in turn, are composed of the same atoms and small molecules as inanimate matter. Wohler's synthesis in the laboratory of the organic substance urea in 1828 was the first proof of the artificial conversion of inorganic compounds into an organic molecule.

Third, all of the vitalists' attempts to demonstrate the existence of a nonmaterial vital force ended in failure. Once physiological and developmental processes began to be explained in terms of physico-chemical processes at the cellular and molecular level, these explanations left no unexplained residue that would require a vitalistic interpretation. Vitalism simply became superfluous.

Fourth, new biological concepts to explain the phenomena that used to be cited as proof of vitalism were developed. Two advances in particular were crucial for this change. One was the rise of genetics, which ultimately led to the concept of the genetic program. This made it possible to explain all goal-directed living phenomena, at least in principle, as teleonomic processes controlled by genetic programs. Another seemingly teleological phenomenon to be newly interpreted was Kant's Zweckmässigkeit. This reinterpretation was achieved by the second advance, Darwinism. Natural selection made adaptedness possible by making use of the abundant variability of living nature. Thus, two major ideological underpinnings of vitalism—teleology and antiselectionism—were destroyed. Genetics and Darwinism succeeded in providing valid interpretations of the phenomena claimed by the vitalists not to be explicable except by invoking a vital substance or force.

If one were to believe the writings of the physicalists, vitalism was nothing but an impediment to the growth of biology. Vitalism took the phenomena of life, so it was claimed, out of the realm of science and transferred them to the realm of metaphysics. This criticism is indeed justified for the writings of some of the more mystical vitalists, but it is not fair when raised against reputable scientists such as Blumenbach and, even more so, Müller, who specifically articulated all the aspects of life that were left unexplained by the physicalists. That the explanation Müller adopted was a failure does not diminish the merit of his having outlined the problems that still had to be solved.

There are many similar situations in the history of science where unsuitable explanatory schemes were adopted for a clearly visualized problem because the groundwork for the real explanation had not yet been laid. Kant's explanation of evolution by teleology is a famous example. It is probably justifiable to conclude that vitalism was a necessary movement to demonstrate the vacuity of a shallow physicalism in the explanation of life. Indeed, as François Jacob (1973) has rightly stated, the vitalists were largely responsible for the recognition of biology as an autonomous scientific discipline.

Before turning to the organicist paradigm which replaced both vitalism and physicalism, we might note in passing a rather peculiar twentieth-century phenomenon—the development of vitalistic beliefs among physicists. Niels Bohr was apparently the first to suggest that special laws not found in inanimate nature might operate in organisms. He thought of these laws as analogous to the laws of physics except for their being restricted to organisms. Erwin Schrödinger and other physicists supported similar ideas. Francis Crick (1966) devoted a whole book to refuting the vitalistic ideas of the physicists Walter Elsasser and Eugene Wigner. It is curious that a form of vitalism survived in the minds of some reputable physicists long after it had become extinct in the minds of reputable biologists.

A further irony, however, is that many biologists in the post-1925 period believed that the newly discovered principles of physics, such as the relativity theory, Bohr's complementarity principle, quantum mechanics, and Heisenberg's indeterminacy principle, would offer new insight into biological processes. In fact, so far as I can judge, none of these principles of physics applies to biology. In spite of Bohr's searching in biology for evidence of complementarity, and some desperate analogies to establish this, there really is no such thing in biology as that principle. The indeterminacy of Heisenberg is something quite different from any kind of indeterminacy encountered in biology.

Vitalism survived even longer in the writings of philosophers than in the writings of physicists. But so far as I know, there are no vitalists among the group of philosophers of biology who started publishing after 1965. Nor do I know of a single reputable living biologist who still supports straightforward vitalism. The few late twentieth-century biologists who had vitalistic leanings (A. Hardy, S. Wright, A. Portmann) are no longer alive.

THE ORGANICISTS

By about 1920 vitalism seemed to be discredited. The physiologist J. S. Haldane (1931) stated quite rightly that "biologists have almost unanimously abandoned vitalism as an acknowledged belief." At the same time, he also said that a purely mechanistic interpretation cannot account for the coordination that is so characteristic of life. What particularly puzzled Haldane was the orderly sequence of events during development. After showing the invalidity of both the vitalistic and the mechanistic approaches, Haldane stated that "we must find a different theoretical basis of biology, based on the observation that all the phenomena concerned tend towards being so coordinated that they express what is normal for an adult organism."

The demise of vitalism, rather than leading to the victory of mechanism, resulted in a new explanatory system. This new paradigm accepted that processes at the molecular level could be explained exhaustively by physicochemical mechanisms but that such mechanisms played an increasingly smaller, if not negligible, role at higher levels of integration. There they are supplemented or replaced by emerging characteristics of the organized systems. The unique characteristics of living organisms are not due to their composition but rather to their organization. This mode of thinking is now usually referred to as *organicism*. It stresses particularly the characteristics of highly complex ordered systems and the historical nature of the evolved genetic programs in organisms.

According to W. E. Ritter, who coined the term organicism in 1919,[14] "Wholes are so related to their parts that not only does the existence of the whole depend on the orderly cooperation and interdependence of its parts, but the whole exercises a measure of determinative control over its parts" (Ritter and Bailey 1928). J. C. Smuts (1926) explained his own holistic view of organisms as follows: "A whole according to the view here presented is not simple, but composite and consists of parts. Natural wholes, such as organisms, are ... complex or composite, consisting of many parts in active relation and interaction of one kind or another, and the parts may be themselves lesser wholes, such as cells in an organism." His statements were later condensed by other biologists into the concise statement that "a whole is more than the sum of its parts."[15]

Since the 1920s, the terms holism and organicism have been used interchangeably. Perhaps, at first, holism was more frequently used, and the adjective "holistic" is still useful today. But holism is not a strictly biological term, since many inanimate systems are also holistic, as Niels Bohr has pointed out correctly. Therefore, in biology the more restricted term "organicism" is now used more frequently.

It encompasses the recognition that the existence of a genetic program is an important feature of the new paradigm.

The objection of the organicists was not so much to the mechanistic aspects of physicalism as to its reductionism. The physicalists referred to their explanations as mechanistic explanations, which indeed they were, but what characterized them far more was that they were also reductionist explanations. For reductionists, the problem of explanation is in principle resolved as soon as the reduction to the smallest components has been accomplished. They claim that as soon as one has completed the inventory of these components and has determined the function of each of them, it should be an easy task to explain also everything observed at the higher levels of organization.

The organicists demonstrated that this claim is simply not true, because explanatory reductionism is quite unable to explain characteristics of organisms that emerge at higher levels of organization. Curiously, even most mechanists admitted the insufficiency of a purely reductionist explanation. The philosopher Ernest Nagel (1961), for instance, conceded "that there are large sectors of biological study in which physico-chemical explanations play no role at present, and that a number of outstanding biological theories have been successfully exploited which are not physico-chemical in character." Nagel tried to save reductionism by inserting the words "at present," but it was already rather evident that such purely biological concepts as territory, display, predator thwarting, and so on could never be reduced to the terms of chemistry and physics without entirely losing their biological meaning.[16]

The pioneers of holism (for example, E. S. Russell and J. S. Haldane) argued effectively against the reductionist approach and described convincingly how well a holistic approach fits the phenomena of behavior and development. But they failed to explain the actual nature of the holistic phenomena. They were unsuccessful when trying to explain the nature of "the whole" or the integration of parts into the whole. Ritter, Smuts, and other early proponents of holism were equally vague (and somewhat metaphysical) in their explanations. Indeed, some of Smuts's wordings had a rather teleological flavor.[17]

Alex Novikoff (1945), however, spelled out in considerable detail why an explanation of living organisms has to be holistic. "What are wholes on one level become parts on a higher one ... both parts and wholes are material entities, and integration results from the interaction of parts as a consequence of their properties." Holism, since it rejects reduction, "does not regard living organisms as machines made of a multitude of discrete parts (physico-chemical units), removable like pistons of an engine and capable of description without regard to the system from which they are removed." Owing to the interaction of the parts, a description of the isolated parts fails to convey the properties of the system as a whole. It is the organization of these parts that controls the entire system.

There is an integration of the parts at each level, from the cell to tissues, organs, organ systems, and whole organisms. This integration is found at the biochemical level, at the developmental level, and in whole organisms at the behavioral level.[18] All holists agree that no system can be exhaustively explained by the properties of its isolated components. The basis of organicism is the fact that living beings have organization. They are not just piles of characters or molecules, because their function depends entirely on their organization, their mutual interrelations, interactions, and interdependencies.

EMERGENCE

It is now clear that two major pillars in the explanatory framework of modern biology were missing in all the early presentations of holism. One, the concept of the genetic program, was absent because it had not yet been developed. The other missing pillar was the concept of emergence—that in a structured system, new properties emerge at higher levels of integration which could not have been predicted from a knowledge of the lower-level components. This concept was absent because either it had not been thought of or it had been dismissed as unscientific and metaphysical. By eventually incorporating the concepts of the genetic program and of emergence, organicism became antireductionist and yet remained mechanistic.

Jacob (1973) describes emergence this way: "At each level units of relatively well-defined size and almost identical structure associate to form a unit of the level above. Each of these units formed by the integration of sub-units may be given the general name 'integron'. An integron is formed by assembling integrons of the level below it; it takes part in the construction of the integron of the level above." Each integron has new characteristics and capacities not present at any lower level of integration; these can be said to have emerged.[19]

The concept of emergence first received prominence in Lloyd Morgan's book on emergent evolution (1923). Darwinians who adopted emergent evolution nevertheless had some misgivings about it because they were afraid that it was antigradualistic. Indeed, some early emergentists were also saltationists, particularly during the period of Mendelism; that is, they believed that evolution proceeded in large, discontinuous leaps, or saltations. These misgivings have now been overcome, because it is now understood that the population (or species), rather than the gene or the individual, is the unit of evolution; one can have different forms (phenetic discontinuities) within populations—by recombination of existing DNA—while a population as a whole must by necessity evolve gradually. A modern evolutionist would say that the formation of a more complex system, representing the emergence of a new higher level, is strictly a matter of genetic variation and selection. Integrons evolve through natural selection, and at every level they are adapted systems, because they contribute to the fitness of an individual. This in no way conflicts with the principles of Darwinism.

To sum up, organicism is best characterized by the dual belief in the importance of considering the organism as a whole, and at the same time the firm conviction that this wholeness is not to be considered something mysteriously closed to analysis but that it should be studied and analyzed by choosing the right level of analysis. The organicist does not reject analysis but insists that analysis should be continued downward only to the lowest level at which this approach yields relevant new information and new insights. Every system, every integron, loses some of its characteristics when taken apart, and many of the important interactions of components of an organism do not occur at the physicochemical level but at a higher level of integration. And finally, it is the genetic program which controls the development and activities of the organic integrons that emerge at each successively higher level of integration.

THE DISTINGUISHING CHARACTERISTICS OF LIFE

Today, whether one consults working biologists or philosophers of science, there seems to be a consensus on the nature of living organisms. At the molecular level, all—and at the cellular level, most—of their functions obey the laws of physics and chemistry. There is no residue that would require autonomous vitalist principles. Yet, organisms are fundamentally different from inert matter. They are hierarchically ordered systems with many emergent properties never found in inanimate matter; and, most importantly, their activities are governed by genetic programs containing historically acquired information, again something absent in inanimate nature.

As a result, living organisms represent a remarkable form of dualism. This is not a dualism of body and soul, or body and mind, that is, a dualism partly physical and partly metaphysical. The dualism of modern biology is consistently physicochemical, and it arises from the fact that organisms possess both a genotype and a phenotype. The genotype, consisting of nucleic acids, requires for its understanding evolutionary explanations. The phenotype, constructed on the basis of the information provided by the genotype, and consisting of proteins, lipids, and other macromolecules, requires functional (proximate) explanations for its understanding. Such duality is unknown in the inanimate world. Explanations of the genotype and of the phenotype require different kinds of theories.

We may tabulate some of the phenomena that are specific to living beings.

Evolved programs

Organisms are the product of 3.8 billion years of evolution. All their characteristics reflect this history. Development, behavior, and all other activities of living organisms are in part controlled by genetic (and somatic) programs that are the result of the genetic information accumulated throughout the history of life. Historically there has been an unbroken stream from the origin of life and the simplest prokaryotes up to gigantic trees, elephants, whales, and humans.

Chemical properties

Although ultimately living organisms consist of the same atoms as inanimate matter, the kinds of molecules responsible for the development and function of living organisms—nucleic acids, peptides, enzymes, hormones, the components of membranes—are macromolecules not found in inanimate nature. Organic chemistry and biochemistry have shown that all substances found in living organisms can be broken down into simpler inorganic molecules and can, at least in principle, be synthesized in the laboratory.

Regulatory mechanisms

Living systems are characterized by all sorts of control and regulatory mechanisms, including multiple feedback mechanisms, that maintain the steady state of the system, mechanisms of a sort never found in inanimate nature.

Organization

Living organisms are complex, ordered systems. This explains their capacity for regulation and for control of the interaction of the genotype, as well as their developmental and evolutionary constraints.

Teleonomic systems

Living organisms are adapted systems, the result of countless previous generations having been subjected to natural selection. These systems are programmed for teleonomic (goal-directed) activities from embryonic development to the physiological and behavioral activities of the adults.

Limited order of magnitude

The size of living organisms occupies a limited range in the middle world, from the smallest viruses to the largest whales and trees. The basic units of biological organization, cells and cellular components, are very small, which gives organisms great developmental and evolutionary flexibility.

Life cycle

Organisms, at least sexually reproducing ones, go through a definite life cycle beginning with a zygote (fertilized egg) and passing through various embryonic or larval stages until adulthood is reached. The complexities of the life cycle vary from species to species, including in some species an alternation of sexual and asexual generations.

Open systems

Living organisms continuously obtain energy and materials from the external environment and eliminate the end-products of metabolism. Being open systems, they are not subject to the limitations of the second law of thermodynamics.

These properties of living organisms give them a number of capacities not present in inanimate systems:

A capacity for evolution

A capacity for self-replication

A capacity for growth and differentiation via a genetic program

A capacity for metabolism (the binding and releasing of energy)

A capacity for self-regulation, to keep the complex system in steady state (homeostasis, feedback)

A capacity (through perception and sense organs) for response to stimuli from the environment

A capacity for change at two levels, that of the phenotype and that of the genotype.

All these characteristics of living organisms distinguish them categorically from inanimate systems. The gradual recognition of this uniqueness and separateness of the living world has resulted in the branch of science called biology, and has led to a recognition of the autonomy of this science.

NOTES

This chapter originally appeared as chapter 1 in Ernst Mayr, *This is biology: The science of the living world*, pp. 1–23 (notes pp. 273–276), Cambridge, MA: Harvard University Press, 1997.

REFERENCES

1. Blandino, G. (1969). *Theories on the nature of life.* New York: Philosophical Library.
2. Cassirer, E. (1950). *The problem of knowledge: Philosophy, science, and history since Hegel.* New Haven: Yale University Press.
3. Crick, F. (1966). *Of molecules and men.* Seattle: University of Washington Press.

4. de La Mettrie, J. O. (1748). *L'homme machine*. Leyden: Elie Luzac.

5. Dijksterhuis, E. J. (1950). *De mechanisering van het wereldbeeld*. Amsterdam: J. M. Meulenhoff.

6. Dijksterhuis, E. J. (1961). *The mechanization of the world picture*, trans. C. Dikshoorn. Oxford: Clarendon Press.

7. Driesch, H. (1905). *Der Vitalismus als Geschichte und als Lehre*. Leipzig: J. A. Barth.

8. DuBois-Reymond, E. (1860). Gedächtnisrede auf-Johannes Müller. *Abteilung Akademie der Wissenschaften*, **1859**, 25–191.

9. DuBois-Reymond, E. (1872). *Über die Grenzen des naturwissenschaftlichen Erkennen*. Leipzig.

10. Dujardin, F. (1835). Recherches sur les organisms inférieurs. *Annales des Sciences Naturelles-Zoologie et Biologie Animale*, **4**, 343–377.

11. Galileo (1623/1960). Il saggiatore [The assayer]. In *The controversy of the comets of 1618*, trans. S. Drake. Philadelphia: The University of Pennsylvania Press.

12. Ghiselin, M. T. (1974). *The economy of nature and the evolution of sex*. Berkeley: University of California Press.

13. Goudge, T. A. (1961). *The ascent of life*. Toronto: University of Toronto Press.

14. Haeckel, E. (1866). *Generelle Morphologie der Organismen: Allgemeine Grundzüge der organischen Formen-Wissenschaft, mechanisch begründet durch die von Charles Darwin reformirte Descendenz-Theorie (2 vols.)*. Berlin: Georg Reimer.

15. Haldane, J. S. (1931). *The philosophical basis of biology*. London: Hodder & Stoughton.

16. Hall, T. S. (1969). *Ideas of life and matter (2 vols.)*. Chicago: University of Chicago Press.

17. Haraway, D. J. (1976). *Crystals, fabrics, and fields*. Cambridge, UK: Cambridge University Press.

18. Jacob, F. (1973). *The logic of life: A history of heredity*. New York: Pantheon.

19. Lenoir, T. (1982). *The strategy of life*. Dordrecht: D. Reidel.

20. Maier, A. (1938). *Die Mechanisierung des Weltbildes: Forschungen zur Geschichte der Philosophie und der Pädagogik*. Leipzig.

21. McLaughlin, P. (1991). Newtonian biology and Kant's mechanistic concept of causality. In G. Funke (Ed.), *Akten siebenten internationalen Kant Kongress* (pp. 57–66). Bonn: Bouvier.

22. Morgan, L. (1923). *Emergent evolution*. London: William and Norgate.

23. Nagel, E. (1961). *The structure of science: Problems in the logic of scientific explanation*. New York: Harcourt, Brace & World.

24. Novikoff, A. (1945). The concept of integrative levels and biology. *Science*, **101**, 209–215.

25. Redfield, R. (1942). Levels of integration in biological and social sciences. In R. Redfield (Ed.), *Biological symposia VIII* (pp. 163–176). Lancaster, PA: Jacques Cattell Press.

26. Rensch, B. (1968). *Biophilosophy*. Stuttgart: Gustav Fischer.

27. Ritter, W. E., & Bailey, E. W. (1928). The organismal conception: Its place in science and its bearing on philosophy. *University of California Publications in Zoology*, **31**, 307–358.

28. Rosenfield, L. L. (1941). *From beast-machine to man-machine*. New York: Oxford University Press.

29. Roux, W. (1895). *Gesammelte Abhandlungen über Entwicklungsmechanik der Organismen (2 vols.)*. Liepzig: Engelmann.

30. Roux, W. (1915). Das Wesen des Lebens. *Kultur der Degenwart III*, **4**(1), 173–187.

31. Sattler, R. (1986). *Biophilosophy*. Berlin: Springer.

32. Smuts, J. C. (1926). *Holism and evolution*. New York: Viking Press.

33. von Kölliker, R. A. (1886). Das Karyoplasma und die Bererbung. In *Kritik der Weismann'schen Theorie von der Kontinuität des Keimplasma*. Leipzig.

34. von Nägeli, C. W. (1845). Über die gegenwärtige Aufgabe der Naturgeschichte, insbesondere der Botanik. *Zeitschrift für Wissenschaftliche Botanik*, 1 & 2.

35. von Nägeli, K. W. (1884). *Mechanisch-physiologische Theorie der Abstammungslehre*. Leipzig: Oldenbourg.

36. Wilson, E. B. (1925). *The cell in development and heredity* (3rd ed.). New York: Macmillan.

37. Woodger, J. H. (1929). *Biological principles: A critical study*. London: Routledge & Kegan Paul.

ENDNOTES

1 The search becomes even more futile if the words "mind" or "consciousness" are substituted for "life." This substitution was made to facilitate a demarcation between human life and the life of animals, but it turned out to be a poor strategy because there is no definition of either mind or consciousness that would be applicable only to humans and exclude all animals.

2 Many attempts were made in the last century to define living or life in a simple sentence, some of them based on physiology,

others based on genetics, but none of them is completely satisfactory. What has been successful is an ever more correct and ever more complete description of all aspects of living. One might say that "living consists of the activities of self-constructed systems that are controlled by a genetic program." Rensch (1968, p. 54) states, "Living organisms are hierarchically ordered open systems composed prevailingly of organic molecules, consisting normally of sharply delimited individuals composed of cells and of limited timespan." Sattler (1986, p. 228) says that a living system can be defined "as an open system that is self-replicating, self-regulating, exhibits individuality, and feeds on energy from the environment." These statements are more descriptions than definitions; they contain statements that are not necessary, and omit references to the genetic program, perhaps the most characteristic feature of living organisms.

3 The historians Maier (1938) and Dijksterhuis (1950, 1961) have splendidly described the gradual change from the Greeks through the "Dark Ages" to scholastic philosophy and, finally, to the beginnings of the Scientific Revolution, indicated by the names Copernicus, Galileo, and Descartes. These historians have determined the manifold influences on this development and what it had retained from the Greek tradition. This includes, for instance, "the passionate endeavor of classical physical science to trace the immutable everywhere behind the variability of phenomena" (Dijksterhuis 1961, p. 8), that is, essentialism. "The fundamental idea of [Plato's] entire philosophy is that the things perceived by us are only imperfect copies, imitations or reflections of ideal forms or ideas" (Dijksterhuis 1961, p. 13). In the developments of these ideas, Plato evidently had more influence than Aristotle. It was he who "wholeheartedly endorsed the Pythagorean principle ... as the germ of the mathematization of science." "Plato makes the cosmos a living being by investing the world-body with a world-soul" (Dijksterhuis 1961, p. 15).

4 Actually, this is a rather simplistic presentation of the pathway by which Descartes arrived at his conclusion. The story goes back to the Aristotelian teaching, accepted by the scholastic philosophers, that plants have a nutritive soul and animals a sensitive soul, only man having a rational soul. Material substance was credited to the sensitive soul of animals, while the rational soul was immortal. The capacities of the sensitive soul of animals were limited to sensory perceptions and memory. From his discussions it is clear that Descartes understood under (rational) soul the "reflective consciousness of self and of the object of thought." To ascribe the capacity of rational thought to animals would

credit them with an immortal soul, and that for Descartes was an unacceptable proposition, because it meant that their souls would go to heaven. (The atheistic thought that perhaps there was no heaven for human souls either apparently never occurred to Descartes.) Ultimately, Descartes's reasoning was based on scholastic definitions of substance and essence, which precluded the existence of soul in animals, and limited it to thinking, rational humans. This conclusion eliminated the unacceptable possibility of animals having an immortal soul rising to heaven after death (Rosenfield 1941, pp. 21–22). If one could deny a soul to animals, it was self-evident that one could not accept the belief still widespread in Europe in the seventeenth century that a soul, an anima or *vita mundi*, permeated the universe.

5 The word mechanist was used throughout the nineteenth and part of the twentieth centuries in two different meanings. On one hand it referred to the views of those who denied the existence of any supernatural forces. For the Darwinians, for instance, it meant the denial of the existence of any cosmic teleology. However, for others the major meaning of the word mechanist was a belief that there is no difference between organisms and inanimate matter, that there is no such thing as life-specific processes. This was the major meaning of mechanist for the physicalists.

6 Von Nägeli (1845, p. 1) says that it must be postulated of the specific terms used in an explanation "that they are expressed generally, absolutely, and in the form of a movement" Rawitz (fide Roux 1915) defines life as "a special form of molecular movements and all manifestations of life are variants of it."

7 Most existing histories of vitalism are rather one-sided, having been written either by vitalists such as Driesch (1905) or by their opponents, who saw nothing good in it. Perhaps the best account is that of Hall (1969, chaps. 28–35). Blandino's (1969) treatment concentrates on Driesch; and Cassirer (1950) likewise focuses on Driesch, his followers and opponents, Jacob's (1973) concise presentation is well balanced and follows the fate of vitalism from animism up. An even more comprehensive and truly balanced history of vitalism, however, is still lacking.

8 As Lenoir (1982) has correctly pointed out.

9 "In fact, various forms of vitalism represent quite legitimate extensions of the Cartesian program in mechanistic biology with Newtonian means" (McLaughlin 1991).

10 How similar Müller's concept of the Lebenskraft was to the concept of the genetic program may be documented by a few quotations: "[Müller's] Lebenskraft acts in all organs as cause and supreme effector of all phenomena according to a definite plan [program]" (DuBois-Reymond 1860, p. 205). Parts of the

Lebenskraft, "representing the whole, are transmitted at reproduction without incurring any loss to every germ where it can remain dormant until germination" (ibid.). The four principal attributes of the Lebenskraft listed by Müller are indeed characteristics of the genetic program: (1) not being localized in a specific organ, (2) being divisible into a very large number of parts all of which still retain the properties of the whole, (3) disappearing at death without leaving any remnant (there is no departing soul), and (4) acting according to a plan (having teleonomic properties). I have described the beliefs of J. Müller in considerable detail, to correct the whiggish treatments by such physicalists as DuBois-Reymond, who maligned Müller as an unscientific metaphysician.

11 Von Uexküll, B. Dürken, Meyer-Abich, W. E. Agar, R. S. Lillie, J. S. Haldane, E. S. Russell, W. McDougall, DeNouy, and Sinnott, to mention only a few of the numerous early twentieth-century vitalists. Ghiselin (1974) designates W. Cannon, L. Henderson, W. M. Wheeler, and A. N. Whitehead as cryptovitalists.

12 Goudge (1961), Lenoir (1982). An opposition to Darwin's selectionism was also a frequent component of vitalist arguments (Driesch 1905).

13 Beginning with C. R. Wolff (1734–1794) the idea developed that there was a basic, undifferentiated stuff which gave rise to the more formed elements. F. Dujardin (1801–1860) first described (1835) and defined it properly under the name "sarcode." More and more attention was paid to it as microscopy flourished. Purkinje coined the term "protoplasm" in 1840. In 1869 protoplasm was for T. H. Huxley the physical basis of life. Cytoplasm was the term introduced by von Kölliker to designate the cellular material outside the nucleus.

14 Actually, this term had been used in the social sciences all the way back to Comte, although organicism meant something rather different for the sociologists from what it meant for the biologists. Bertalanffy (1952, p. 182) listed some 30 authors who had declared their sympathy for a holistic approach. This list was very incomplete, however, not even including the names of Lloyd Morgan, Smuts, and J. S. Haldane. F. Jacob's (1973) concept of the integron is a particularly well-argued endorsement of organismic thinking.

15 Woodger (1929) gives an impressive list of biologists who endorsed the organicist viewpoint, E. B. Wilson (1925, p. 56), for instance, said "even the most superficial acquaintance with the cell activities shows us that [explaining the cell as a chemical machine] cannot be taken in any crude mechanical sense—the difference between the cell and even the most intricate artificial machine still remains too vast by far to be bridged by present knowledge ... modern investigation has brought ever-increasing recognition of the fact that the cell is an organic system, and one in which we must recognize the existence of some kind of ordered structure or organization." Not surprisingly, holistic thinking was always particularly well represented among the developmental biologists. It was strong in the writings of C. O. Whitman, E. B. Wilson, and F. R. Lillie. Haraway (1976) devotes the major part of an entire book to the organicism of three embryologists, Ross Harrison, Joseph Needham, and Paul Weiss. Interestingly, Harrison considered emergence a metaphysical principle, and therefore he considered Lloyd Morgan a vitalist. Like so many biologists in the post-1925 period, he thought that the newly discovered principles of physics, such as the relativity theory, Bohr's complementarity principle, quantum mechanics, and Heisenberg's indeterminacy principle, applied equally to biology and physics.

16 Nagel (1961) defined a mechanist in biology as "one who believes that all living phenomena can be unequivocally explained in physico-chemical terms, that is, in terms of theories that have been originally developed for the domains of enquiry in which the distinction between the living and non-living plays no role, and that by common consent are classified as belonging to physics and chemistry." Such reduction characterizes all of Nagel's account.

17 For instance: "Holism is a specific tendency, with a definite character, and creative of all characters in the universe, and thus fruitful of results and explanations in regard to the whole course of cosmic development" (1926, p. 100). Not surprisingly, holism as presented by Smuts was widely considered to be a metaphysical concept.

18 The subject of levels of integration was discussed in great detail in a special symposium volume (R. Redfield 1942).

19 One mistake that was made particularly often was to consider each level of integration to be a global phenomenon. This is not what these levels are. Every integron from the molecular to the supraorganismic level is singular. There is no conflict between this interpretation and Novikoff's (1945) statement that "the laws describing the unique properties of each level are qualitatively distinct, and their discovery requires methods of research and analysis appropriate to the particular level," and we would now add, appropriate to the particular integron. A modern evolutionist would say that the formation of a more complex system, representing a new higher level, is strictly a matter of genetic variation and selection. There is no conflict with the principles of Darwinism.

7 · The principles of life (selections)

TIBOR GÁNTI

THE CRITERIA OF LIFE

Living and non-living systems are qualitatively different, i.e., living systems have qualitative properties or groups of qualitative properties which occur exclusively in the living world and cannot be found in the non-living world. In what follows, these common characteristics found in living organisms will be called life criteria, the laws uniting these characteristics will be considered the principle of life, and life itself as a common general abstraction of every kind of living being will be accepted as a philosophical and not a biological category. Life criteria will be dealt with in the present chapter, and the principle of life will be discussed in connection with the organization of the chemoton. Life as a philosophical category will not be dealt with in the present book.

The selection and axiomatically exact formulation of life criteria is of fundamental importance in theoretical biology. As we have seen, classical biology has not been able to solve this problem during its long history of 2000 years. We shall present here a quite new system of life criteria, which entirely differ from the classical "life phenomena." This new system of life criteria was first proposed in the original edition of this book in 1971.* However, in the diverse world of biology it is not easy to select fundamentally common characteristics, since it is impossible to find a research worker who could claim to know every part of the living world with the required precision. Thus, the original system of criteria has been somewhat modified since 1971, and here we present a refined and updated form, including both content and formulation. Nevertheless, even this formulation cannot be considered the final version, either in content or in axiomatic exactness, although it is unlikely that major changes will be necessary in the future.

Life is a function of material systems which are organized in a particular way. Thus life is not the property of a special type of matter in the sense of chemistry, i.e., of a chemical compound (e.g., protein or nucleic acid), but is the property of specially organized systems. Thus it is incorrect to speak about living matter; it is more appropriate to speak about living material systems.

A system is living if, and only if, certain characteristically composed processes (life processes) occur in it. The totality of these processes, i.e., the functioning of the system, results in characteristic phenomena which are well suited to differentiate the living from the non-living.

A system suitable for the occurrence of the processes in question may be either functioning or in a state which is not functioning but is capable of doing so. When this system is in its functioning state it is said to be living; when it is in its non-functioning but functionable state it is capable of living but it is not dead. This latter state corresponds to latent life, clinical death, resting seeds, dried-out micro-organisms, and frozen organisms. This is a state which is non-living but not dead.

Death is an irreversible change which makes the system irreversibly incapable of functioning. Consequently, if life corresponds to the functioning state of these special systems, death corresponds to the state which is incapable of functioning. However, there is also an intermediate state which is capable of functioning but is not actually functioning, i.e., a state which has the capability of life but in which the system is not alive because the special processes are not occurring, but it is not dead either because the processes can be started if suitable circumstances arise.

The Nature of Life, ed. M. A. Bedau and C. E. Cleland. Published by Cambridge University Press. © M. A. Bedau and C. E. Cleland 2010.

The totality of life processes, i.e., the functioning of living systems, produces special phenomena which are appropriate for the general characterization of the living state. Some of these phenomena are present in every living being, without exception, at every minute of its life. Without the constant presence of all these phenomena, the system cannot be living. The presence of all these phenomena is an unavoidable criterion of the living state; therefore they will be called real (actual, absolute) life criteria. However, there is another group of life phenomena whose presence is not a necessary criterion for individual organisms to be in the living state, but which are indispensable to the survival of the living world. These will also be considered life criteria, and they will be called potential life criteria.

REAL (ABSOLUTE) LIFE CRITERIA

A living system must inherently be an individual unit

A system is regarded a unit (a "whole") if its properties cannot be additively composed from the properties of its parts, and if this unit cannot be subdivided into parts carrying the properties of the whole system.

A system forming a unit (unit system) is not a simple union of its elements, but a new entity carrying new qualitative properties compared with the properties of its parts. These new properties are determined by interactions occurring according to the organization of the elements of the system. Only the system as a whole displays the totality of these properties.

Living systems form inherently, i.e., by their very existence as a unit. Thus living systems are inherently unit systems—life always is the property of a unit system. However, the inherent unity of living systems does not exclude the possibility of accessory components, i.e., a unit system does not have to be a minimal system at the same time. Since living systems are genetically determined structurally as well as functionally, the biological unity of the system is also genetically determined. The genetic program also carries information with respect to the living unit.

A living system has to perform metabolism

By metabolism we understand the active or passive entrance of material and energy into the system which transforms them by chemical processes into its own internal constituents. Waste products are also produced, so that the chemical reactions result in a regulated and controlled increase of the inner constituents as well as in the energy supply of the system. The waste products eventually leave the system, either actively or passively. Expressions such as "external" and "internal" do not refer here to some spatial separation, but allude to the question of whether or not the material is an organic part of the internal organization of the living system as a unit system. Thus stored nutrients such as glycogen or starch are considered as external materials even if they are spatially located within the living system.

A living system must be inherently stable

Inherent stability is not identical with either the equilibrium or the stationary state. It is a special organizational state of the system's internal processes, which makes the continuous functioning of the system possible and remains constant despite changes in the external environment. It means that the system as a whole, although continuously reacting via dynamic changes occurring within the living system, always remains the same. Further, it means that despite the permanent chemical transformations occurring in the living system, the system itself does not decompose; rather, it grows if necessary.

Inherent stability is something more than homeostasis, since homeostasis follows from it. Inherent stability is an organizational property—a natural consequence (as we shall see later) of the network of chemical and physical processes taking place in the living system. The dormant seed, the frozen tissue culture, the lyophilized micro-organism and the dried-out protozoon are not in homeostasis or in a stationary state, although all of them satisfy the criterion of inherent stability, since under appropriate circumstances they can live again. Thus a system in a state of inherent stability only displays homeostatic properties whilst functioning; hence this criterion includes that of the homeostasis. Thus a living system with inherent stability is capable of living, but does not display homeostasis in a non-living state. Under appropriate circumstances it can be converted from a non-living to a living state, and then it acquires the property of homeostasis.

The concept of homeostasis was proposed by the great American physiologist, Cannon. Living organisms have a special "inner environment," differing in its state and composition from the external environment. Living systems, whether they are simple cells or complex multicellular organisms, strive to hold this inner environment constant despite external changes. Cannon called the constancy of the inner environment homeostasis. As has already been seen, this constancy of the inner environment is also a property of growing and multiplying systems. Thus homeostasis cannot be identified with equilibrium, the stationary state, Ashby's "cybernetic homeostat," or Lyapunov stability.

The constancy of a living system's inner environment, i.e., homeostasis, can only be maintained by detecting the changes occurring in the external environment and reacting to them with active compensating answers. Thus the mechanism of homeostasis is excitability. This can also be carried out in soft molecular systems, as will be seen in connection with the chemoton theory.

Thus inherent stability as a life criterion includes the criteria of homeostasis and excitability in the most general sense, so it is unnecessary to list them separately.

A living system must have a subsystem carrying information which is useful for the whole system

Everything carries within it the information necessary for its origin, development, and function. However, there are some systems which carry information concerning things and events which are independent of them. Examples of such systems are books, magnetic tapes, punched cards, and discs.

In nature the capacity to carry such "surplus information" can only be found in certain subsystems of living systems such as the genetic substance, brain, immune system, etc. These carry not only information about themselves, but also information about the whole system or even the world outside the system. The presence of information-carrying subsystems is characteristic of every living system without exception, and is an indispensable criterion for the development of the living world.

Information coded in a system becomes real information only if there exists another system capable of reading and using this coded information. Living systems can read and use the information stored in their information-carrying subsystems; moreover; they can copy, i.e., replicate, information during their multiplication. Living beings are characterized not only by their ability to store information, but also by their ability to carry out informational operations.

Processes in living systems must be regulated and controlled

An existential condition of every dynamic, i.e., continuously functioning, system is the regulation of its processes. Living systems as soft dynamic systems also have this property. Regulation in living systems occurs primarily through chemical mechanisms. In fact, it could be claimed that regulation should not be a separate criterion, since neither metabolism nor homeostasis can be realized without regulation of the system's processes, and therefore it is already implied by these criteria.

Nevertheless, regulation itself can only ensure the maintenance and functioning of the system. Unidirectional irreversible processes, such as growth, multiplication, differentiation, development, and evolution, also occur in the living world and these cannot be accomplished by regulation alone. They also require steering, i.e., control according to some program. Control in living systems is also performed by molecular mechanisms.

POTENTIAL LIFE CRITERIA

A living system must be capable of growth and multiplication

Growth and multiplication are among the classical life criteria; their presence is general and indispensable in the living world. However, they are not criteria of the living state itself: some domesticated plants and animals are *ab ovo* incapable of reproduction; castrated animals are unable to reproduce; old animals lose their ability to grow and reproduce without losing their lives. The presence of growth and reproduction is not a criterion for individual life, but it is a condition for the existence of the living world and so it must be included among the potential life criteria.

It is necessary to explain why growth and reproduction are combined in a single criterion. Life criteria must be valid for every individual of the living world on any of the levels of the organizational hierarchy. However, growth and multiplication separated only at a given level of evolution; in prokaryotes and in many eukaryotes growth is merely a part of reproduction. This was clearly demonstrated in Hartmann's experiments: For a period of 130 days, he removed daily one-third of the cytoplasm of an amoeba which normally divided every second day, so that the nucleus was left untouched. The amoeba did not die but neither did it divide, since it never reached the stage of growth necessary for cell division to occur.

The reproduction of multicellular organisms is only indirectly related to their growth, but growth is a direct consequence of the reproduction of their cells. In the case of plants it can be disputed whether their vegetative multiplication should be considered growth, multiplication, or regeneration. Therefore treating the abilities to grow and reproduce as a single criterion is rational practice.

A living system must have the capacity for hereditary change and, furthermore, for evolution, i.e., the property of producing increasingly complex and differentiated forms over a very long series of successive generations

Heredity is the capability of living systems to produce individuals identical or similar to themselves, or else to produce germ cells which ensure the development of such individuals. However, the criteria of reproduction include heredity, and so it would be redundant to consider heredity as a separate criterion.

However, the living world could not have evolved if heredity had followed strict rules, i.e., if the characteristics of the offspring had been always identical with those of the parents. In this case new characteristics could never have arisen in the living systems. Hence the capacity for hereditary change, i.e., the appearance of characteristics in the offspring which were not found in the endless chain of its progenitors, must form a separate criterion. This capacity is an indispensable condition for evolution, but it is not a prerequisite for the individual's living state. Hence it is included in the potential life criteria.

The capacity for hereditary change is a necessary but probably insufficient condition of evolution. A prerequisite of evolution is the possibility that hereditary changes with different (adaptive) "values" can appear. This property can generally be found in the living world, but is not a criterion for individual life. Hence it is included in the potential life criteria.

Living systems must be mortal

Death is undoubtedly characteristic of living systems in the sense that non-living systems cannot die. However, this only means that death and life are connected to form a complementary pair of concepts. But death is indispensable to life in a deeper sense—death ensures the recycling of organic material. Without death exactly the same primaeval cells which consumed the organic substance of the biosphere at the beginning of life would still exist on Earth today.

Thus the death of the individual is indispensable to the living world. It is undoubtedly a life criterion, but it is not a criterion of the living state of the individual life, since the latter may cease without death. For example, in the case of cell division by binary fission and mitosis (because of the semiconservative replication mechanism of DNA) one of the two strands of the DNA molecules of the individual daughter cells comes from the mother, while the other is newly synthesized. The other cell substances are statistically distributed among the two daughter cells, which therefore contain in a statistically equal ratio the original substances of the mother cell and the newly synthesized substances. Thus the result is two cells which are equally young, neither of which can be considered a "mother" or a "daughter" cell. The life of these cells ceases during division in such a manner that no corpse is produced. Thus these cells are potentially immortal, at least in view of the continuity of their generation.

Cessation of the life of the individual without death is not restricted to unicellular organisms; it can also occur at the multicellular level. Simple multicellular animals (e.g., fresh-water polyps, planaria, and earthworms) can be cut into two or more pieces so that each part grows into a whole animal; thus, as in the case of cell division, the original individual ceases to exist but there is no corpse. This behavior is well known in the vegetable kingdom (slip planting, vegetative multiplication).

Thus death is not an absolute but a potential criterion of life, which results from the existence of the living world.

All classical life phenomena (except movement as a simple change of place) are included in the above criteria but in a more exact and better-defined form. Furthermore, they include other criteria involving characteristics which have only been disclosed by molecular biology, such as the regulation and control of information processing and other processes. Hence the life criteria lifted above give a stricter prescription for the "living" condition of a system than the classical life phenomena.

Since potential life criteria do not form a prerequisite of individual life, we consider a system to be living if it satisfies the real (absolute) life criteria, regardless of its concrete material construction or of the quality of the chemical substances forming it.

This definition provides a method of treating the fundamental laws of life completely generally, independent of their actual realizations. Thus this definition is not restricted to proteins, nucleic acids, or even to carbon compounds. If, in the future, exobiology (the science of life outside Earth) discovers non-carbon-based living systems, these will also be included on the basis of our general definition built upon the lift criteria summarized below.

REAL (ABSOLUTE) LIFE CRITERIA

1. Inherent unity
2. Metabolism
3. Inherent stability
4. Information-carrying subsystem
5. Program control

POTENTIAL LIFE CRITERIA

1. Growth and reproduction
2. Capability of hereditary change and evolution
3. Mortality

LEVELS OF LIFE AND DEATH

Introduction

Most biologists think, or at least feel, that the question "What is life?" is only an abstract philosophical question, outside the consideration of modern biology. This point of view is not only erroneous, but is also definitely harmful for the development of the biological sciences. Of course, the nature of life *is* a philosophical question. But in addition to this, it is also a fundamental question in the natural sciences. An answer to this question is an indispensable precondition for the birth of theoretical biology. It is also a very important social question, as a multitude of social problems will remain unsolved without a concrete understanding of the nature of life.

Starting points of the natural sciences are usually some minimal model systems. These are systems which can show, during the evolutionary development of matter, the special properties particular to a given scientific discipline. For example, elementary electronic charges (protons and electrons) became the fundamental starting points for the electrical sciences, molecules for chemistry, elementary cell structures for crystallography, and genes for genetics. Until these discoveries, biology had no such minimal system. However, it is obvious that during the evolution of the organization of matter, material systems had to achieve a certain level or critical degree of complexity. Below this level systems are not alive, but above it they start to show the most general and principal phenomena of life. Gánti proposed such a minimal construction in 1971, which he called the chemoton. In order for the chemoton model to fulfill the role of the minimal system for biology, its organizational principles must be present in every living being and must be absent in non-living systems. If we have a model of such a minimal system, then the question "What is life?" is no longer an abstract philosophical question. It is a basic question of the natural sciences and the answer has a scientific basis; it can be described with exact mathematical methods and its correctness can be verified by concrete experiments. Such a model makes it possible to obtain exact answers for the question of the origin of life, and in the absence of such a model we do not even know what we are searching for.

In addition, human societies have a multitude of problems which cannot be solved without exact knowledge of the nature of life. Some of these are really urgent today: genetic manipulations (especially on human embryos and embryonic tissues), transplantation, abortion, and the question of clinical death. These questions have many legal, ethical, moral, and religious consequences which cannot be addressed without understanding what is living and what is not.

Explanation of some basic concepts

"What is the characteristic feature of life? When is a piece of matter alive?" asked Erwin Schrödinger in his famous booklet *What Is Life?* His answer was: "It is alive if it goes on 'doing something', moving, exchanging material with its environment, and so forth, and if it is doing that for a much longer period than we would expect an inanimate piece of matter to 'keep going' under similar circumstances." This description really characterizes one of the most fundamental features of living things. All living systems—while alive—do something, work, function. A system that does not do anything is not alive. It may be lifeless, which means that it does not live, never was alive, and is not ever going to live. It may be dead, which means that it may have lived earlier, but has lost this capacity. Or it may be in an inactive state, which means that it would be able to live but this function does not work at the given moment (e.g., frozen bacteria or dried seeds). It may start to live again under suitable conditions. Schrödinger's characterization really referred to the most important property of living beings.

However, living things are not the only systems that "do something" and also do it for long periods. For instance, the wind may blow for prolonged periods and rivers do their erosive work continuously. In technology, engines, vehicles, and automata are also able to work without interruption. What is common to these systems is that they are positioned between the higher and the lower potential level of some kind of energy, and, part of the energy which flows through the system is transformed to work.

However, there is an essential difference between the work performed by the wind or a river and that performed by some machinery constructed by humans or by living systems. The work of natural forces is undirected: It is caused by some temporary environmental condition and produces in random changes in the environment. The work of machines and living systems is directed; it is programmed by some inner structure, i.e., by their construction, and produces useful work for the creator (humans) or for the living entity.

To get from random to directed work, the flow of energy must be manipulated along a series of forced trajectories within the system. Mechanical machines are manipulated by mechanical means (e.g., wheels, axles, springs, pendulums), and electronic apparatus is

manipulated by wires, coils, rectifiers, capacitors, etc. The same procedure occurs if the driving force of the machine is chemical energy, as is the case with steam engines or battery-driven electric apparatus. Here, the chemical energy is first transformed into mechanical or electric energy and this is subsequently manipulated in forced trajectories by mechanical or electronic means.

The driving force of living systems is chemical energy. This is also true for plants, in which the energy of the light is initially converted into chemical energy. However, in contrast with the situation for mechanical machines, the energy flow in living systems is manipulated by chemical means. This is the most important difference between mechanically constructed machines and living systems. It is also the source of many special properties of living systems, such as, the capacity for spontaneous formation, growth, and regeneration, but above all the capacity for reproduction. In contrast with manmade technologies, where the machines are based on mechanical or electronic automata, living systems are fundamentally chemical automata. During evolution, the mechanisms of living systems have sometimes been extended using mechanical and electronic components, but their basic structures remain chemical automata. They manipulate the driving energy by chemical methods.

As chemical reactions can proceed with suitable intensity only in the fluid phase (gas or solution), a chemical automaton can only work in the fluid phase, i.e., the continuous presence of some kind of solvent is essential. This is why chemical automata are also known as "fluid automata." The functioning of mechanical automata is restricted to a rigorous geometrical order of their parts, and the functioning of electronic automata is also restricted to some geometric arrangement of their components. The functioning of the fluid automata is largely independent of any kind of geometrical order. It works even if the solution is stirred, or if half of it is poured into another container, etc. Compared with mechanical and electrical automata, i.e., the "hard automata," these properties provide living systems with highly favourable possibilities. One of these is the capacity for reproduction—autocatalytic systems are well known in chemistry. In the following sections, we shall consider living systems as fluid machines. Higher levels of evolution produced living systems with hard and partially hard (so-called "soft") components for the manipulation of the flow of mechanical and electrical energy. Nevertheless, because the fundamental functioning of these systems depends on chemical manipulation, we shall refer to them as soft automata.

The minimal system of life: the chemoton

There are some examples of the existence and use of fluid automata in human technology, but these are mostly known as chemical reactions (or reaction systems) and not as fluid automata. Examples of such fluid automata include the well-known families of oscillating chemical reactions or the biochemical systems used in the chemical industry to produce sugar phosphates. But the fundamental unit (i.e., the minimal system) of biology must have some specific properties:

- it must function under the direction of a program
- it must reproduce itself
- it and its progeny must be separate from the environment.

Gánti first described such a model in 1971 and he continued to develop the idea in further publications. He named these fluid automata chemotons.

A chemoton consists of three different autocatalytic (i.e., reproductive) fluid automata, which are connected with each other stoichiometrically. The first is the metabolic subsystem, which is a reaction network (optionally complicated) of chemical compounds with mostly low molecular weight. This must be able to produce not only all the compounds needed to reproduce itself, but also the compounds needed to reproduce the other two subsystems. The second subsystem is a two-dimensional fluid membrane, which has the capacity for autocatalytic growth using the compounds produced by the first subsystem. The third subsystem is a reaction system which is able to produce macromolecules by template polycondensation using the compounds synthesized by the metabolic subsystem. The byproducts of the polycondensation are also needed for the formation of the components of the membrane. In this way, the third subsystem is able to control the working of the other two solely by stoichiometrical coupling. As they work, the three fluid automata become a unified chemical supersystem through the forced stoichiometrical connections. This means that they are unable to function without each other, but the supersystem formed by their cooperation can function. It can even proliferate in a rigorous geometrical sense, as has been shown by Gánti in *The Chemoton Theory*.

The chemical construction of the chemotons is illustrated by the chemoton model (Fig. 7.1), which is the most elementary simplified description of chemotons. Presumably it cannot be realized in this form, but this simple model demonstrates all the indispensable stoichiometrical connections which must be present in every kind of chemoton construction. Kinetic analysis of the elementary chemical reactions allows us to perform an exact numerical investigation of the workings of the chemotons using a computer. These investigations have shown that chemotons behave as living beings, i.e., they satisfy all the criteria which characterize living systems in general (life criteria).

The chemoton model is an abstract model. By using it we can understand how it is possible to organize a chemical supersystem from several autocatalytic subsystems, which are directed by a central program, and which can reproduce itself. The fact that it is an abstract system means that its components are not restricted to particular chemical compounds. However, they must have certain stoichiometric capabilities and, they must be able to produce certain compounds, which are important for the whole system. Any of the simplified subsystems of the model can be replaced by a complex real chemical network if it is able to ensure the required stoichiometric connections. Our working group has constructed such a realistic system, consisting of more than 100 different organic components. Some of these components are known to be materials which existed prebiotically; the existence of others under such conditions is possible. This realistic model (at least stoichiometrically) could work as a program-controlled reproductive chemical supersystem, using well-known chemical compounds, which existed prebiotically, as nutrients. Máté Hidvégi constructed the major part of the chemical network and Eörs Szathmáry performed the stoichiometrical calculations. Of course, this realistic prebiotic model is not the only possible one; there are many other possible solutions. Because there are no chemical restrictions on the realization of the model, it could even consist of silicon compounds. The only prerequisite is that the silicon chemistry should be able to produce reaction systems suitable for the stoichiometric criteria of the

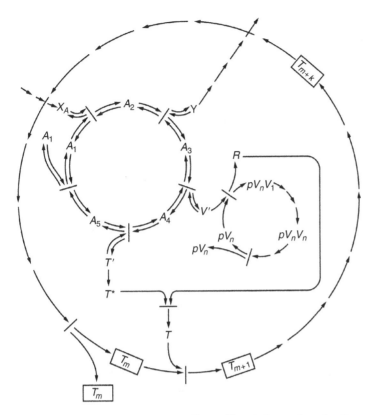

Fig. 7.1. Minimum model of chemotons. Three self-producing systems are coupled together stoichiometrically: cycle A → 2A, template polycondensation pV_n → $2pV_n$, and membrane formation T_m → $2T_m$. This coupling results in a proliferating program-controlled fluid automaton, known as the chemoton.

chemoton model. Therefore the chemoton model is able to represent any type of living system, even extraterrestrial ones.

The chemoton model is designed from the bottom up, starting from the direction of the chemical reactions. Linkage of three autocatalytic subsystems has resulted in a supersystem which is controlled by a program and is able to reproduce itself. Appropriate investigations have shown that this construction is able to satisfy the criteria for life. It is probable that chemotons are the simplest systems possible which show the properties of life. Consequently, the chemoton model, as the simplest system capable of satisfying the criteria for life, can be regarded as the minimal system of biology.

The chemoton model has a further significant property. In the model only stoichiometrical connections are defined, i.e., only the quality and quantity of the chemical components and the pathways of their formation, transformation and decomposition are described. The model does not contain any prescription or restriction on the speed of the chemical reactions in the system. Therefore, it remains valid whether the reaction rates are determined exclusively by the concentrations of the components or are influenced by catalytic effects, or even if the processes are accelerated by complicated enzymatic systems, as happens in the living world today.

Life at the prokaryotic level

If we are searching for the secret of life—trying to define the most fundamental differences between the animate and the inanimate—then we have to investigate the simplest and most primitive level of life existing in the world today. According to our present knowledge, this is represented by the prokaryotes. The world of the prokaryotes is now extremely diversified, as they have developed to their present level of diversity over a period of billions of years. Even so, they have common properties which are the fundamental characteristics of their being.

The simplest known prokaryotic organisms are the thermoplasmas and the mycoplasmas, which are completely fluid organisms bounded by two-dimensional fluid membranes. There are two subsystems inside the boundary membrane—the cytoplasm and the genetic material. The cytoplasm is essentially the metabolic subsystem, consisting of small-molecule metabolic networks which produce all the chemical components needed for the working and reproduction of the total system. These components include the building blocks for membrane formation as well as for the reproduction of the genetic materials. The genetic material contains the genetic information for the whole system. Therefore, the structural bases of the myco- and thermoplasmas correspond very well to the chemoton model. They have the three subsystems required—the metabolic, the membranous, and the informational. These three subsystems have no solid components; everything, even the boundary (membrane) subsystem, is in the fluid state. However, they may contain additional subsystems which have been developed during evolution. Such subsystems include, for example, the protein-synthesizing subsystem or, in some cases, the flagella which are semi-solid structures.

The world of the prokaryotes is very diverse. They include many different metabolic types, ranging from the sulphur bacteria to the methane-producing bacteria, from the obligate anaerobes to the aerobes, and from halophilic to hyperthermophilic organisms. They also have an enormous variety of morphological properties. They can be formless, spherical, stick- or thread-like, or spiral. They can be motionless or fast-moving. Some are surrounded by fluid membranes only; others have solid cell walls or even a slimy shell enclosing their cell walls. But the fundamental construction of the three subsystems is present in every one of them. During evolution several additional parts and subsystems have been added to the fundamental construction without disturbing the coordinated functioning of the three original subsystems.

The chemoton model can be considered as a suitable representation of living prokaryotic systems. We can state that, at the level of the prokaryotes, only those systems which are based on the fundamental properties of stoichiometric organization, defined by the chemoton model, are alive. Although many different types of metabolism are known, which is what makes it possible for prokaryotes to live in a variety of extreme environmental conditions, each of them must have an autocatalytic metabolic network to produce the raw materials for membrane formation and to reproduce the information subsystem, i.e., the genetic subsystem. It is also true that both the boundary membrane and the information subsystem (genetic apparatus) are autocatalytic and self-reproducing, and that their cooperation results in a chemical supersystem, where the coordinated functioning comes primarily from the stoichiometric connections between the three different subsystems. Everything else, including the chemical kinetic relations, is based on these stoichiometric connections.

The existence and cooperation of these three subsystems is the prior condition for the presence of life at the prokaryotic level. If any of them is absent, the system is no longer alive. In the case of viruses, only the information subsystem and, for more complex viruses, the boundary membrane are present—the metabolic subsystem is missing. Viruses cannot function by themselves—they cannot grow and they cannot multiply. They can reproduce only by entering cells and forcing them to make copies to produce identical viruses.

Similarly, one may make lysates from prokaryotes, i.e., material in which the cell membranes have been destroyed. These lysates can be used to synthesize biochemicals, because the metabolic subsystem still works for a time. However, these are no longer living systems; they do not have the organization of the chemotons and they cannot grow or proliferate. Their functions are not directed by an information subsystem, but only by the general rules of chemistry.

Thus, with the help of the chemoton model we can define clearly the difference between living and non-living systems at the prokaryotic level.

Life at the animal level

At least two different kinds of life exist. I would like to repeat and emphasize this: We can identify at least two different kinds of life. I must emphasize that this statement is not a hypothesis but the everyday experience of biology and medicine. However, to the best of my knowledge, nobody has yet recognized, or at least stated, this fact. Because this important fact has not been clearly stated, the concept of the two kinds of life is confused with other phenomena in both scientific and everyday applications. We know that after the death of an animal its organs, tissues, and cells remain alive for a time. This is why human organs can be transplanted from a dead to a living person, and why it is possible to produce tissue and cell cultures. If we consider death as nothing other than the irreversible end of a life, then logically we must conclude that the life of a human or an animal and the life of its organs, tissues or cells are not the same.

There is also a difference in construction between the two kinds of life, i.e., the life of an animal and that of its cells. Life at the prokaryotic level is characteristic of systems which are organized directly from chemical processes into living entities. Life at the metazoan level is characteristic of systems which have been organized into living entities directly from cells, which are also living entities. These systems, where the matter has a higher hierarchical organization than in the cells, use chemical as well as mechanical and electronic processes for energy utilization.

Life itself is always the property of an entity. This means that, for example, it is impossible to cut a prokaryotic cell or an animal in half in such a way that both halves remain alive. There are some exceptional cases (e.g., annelids) which can be explained.

Perhaps we could define life at the prokaryotic level, which we have characterized using the chemoton model, as primary life. Life at the metazoan level, i.e., the life of those "biological supersystems" whose elementary units are living cells themselves, can be defined as secondary life. It is significant that the three-subsystem structure defined by the chemoton model is also characteristic of systems having secondary life. Each animal entity has a subsystem governing the geometrical structure (skin, bones), each has a subsystem governing the metabolic processes, i.e., the supply and transformation of nutrients for the whole animal (digestive tract, secretory organs, blood circulation), and each has an information and control subsystem (the nervous system) to ensure the coordinated working of the organs under given internal and environmental conditions. Naturally, as with prokaryotes,

several additional parts or subsystems may be added to the three fundamental subsystems to satisfy special requirements or living conditions. It seems that the three-subsystem organization of the chemoton model, which leads to the phenomena designated as life at the prokaryotic level, also leads to similar phenomena at higher levels of the organizational hierarchy of matter. Such phenomena have been designated as "life" in several languages during historical times.

Each living animal represents two kinds of life at each moment of its existence: the primary life of its cells and the secondary life of its body. The secondary life is the real life of the animal. However, the ovum, which contains all the information about a living animal, does not show or contain any phenomena related to its secondary life. Therefore, we must conclude that the life of an animal does not start in its ovum or at fertilization, but begins during embryonic development as the information subsystem (the nervous system) begins to control the functioning of the two other subsystems. In the same way, the animal dies at the moment that its information subsystem irreversibly ceases to control the other two subsystems. This is why, in medical practice, assigning the moment of death to the irreversible cessation of brain function is correct. And it is also why considering the fertilized ovum as a living human being is incorrect. Human life only starts at the moment when the central nervous system begins to control the complete collection of embryonic cells during the development of the embryo. Laboratories working on artificial insemination discard fertilized but unused eggs from time to time, but this practice cannot be regarded as a massacre.

Between and beyond

The prokaryotes and the metazoans are only two parts of the living world. A considerable part of the living world lies between them: the enormous variety of eukaryotic cells and fungi, and the whole world of plants. So far, we have not investigated these parts of the living world from our point of view. Nevertheless, we can draw some general conclusions on the basis of our present knowledge.

The organization of eukaryotic cells is in accordance with the chemoton model. It is not a contradiction that their information (subsystem (genetic material)) is in a special unit (nucleus) covered with a membrane; the chemoton model defines only the stoichiometric connections and says nothing about the geometrical structures. The stoichiometric requirements are satisfied in eukaryotic cells in exactly the same way as they are satisfied in the prokaryotes. The presence of the cell organelles with their own information subsystems (chloroplasts, mitochondria) could cause us some problems, but their presence is well explained by endosymbiotic theory. Eukaryotic cells do not appear to have any type of secondary life which can be terminated without the termination (death) of the primary life of the cell.

In the multicellular organisms, an interesting situation exists in the case of the fungi and of the plants. They have two subsystems, one governing the geometrical structure, and the other governing the metabolic processes. The connection between the two subsystems corresponds to the requirements of the chemoton model. In these subsystems, just as in the case of the animals, the building components are not direct chemical reactions but cells with their own (primary) life. We would expect these organisms to have some sort of secondary life. However, experience shows that such a secondary life does not exist in plants and fungi: It is impossible to kill them in the way that animals can be killed. Animals cannot be cut in two so that that both parts remain alive; plants can be cut into many pieces, each of which is still living. Some plants are propagated in this way in agriculture (vegetative proliferation). This is because no secondary information subsystem has evolved in plants and fungi. All information about plants and fungi is stored in the DNA of their cells and no new information subsystem has developed above all level. Therefore the life of plants and fungi is no more than the coordinated collaboration of the primary life of their cells. Human experience mirrors this fact: Several languages have words to name the act of killing people or animals, but none for killing plants or fungi. In the same way, some oriental religions care for the life of animals but not for the life of the plants.

What is at the level above the organization of animal life? In the same way as eukaryotic cells, as living units, are the building elements of a system, developing a higher level of the organizational hierarchy of matter, it is possible that cooperative functional organization can develop between multicellular organisms. This is again novel; even higher levels of evolutionary organization may be created using multicellular organisms as building elements.

Evolution has already started to proceed in this way. As a first experiment it produced families, and subsequently troops, herds and hordes evolved from several groups of animals. Perhaps the insect societies represent the highest level in this evolutionary process. For example, an ant-hill shows such a high level of organization that even it own metabolic processes appear in it: ants bring food into the ant-hill raise fungi in it, breed plant lice, remove the waste materials, etc. All these activities equate to a metabolic subsystem on this level of the organizational hierarchy. Geometrical separation is also present, as the ant-hill is their well-defined territory. We can see that two of the three subsystems of the chemoton model are already present in the ant-hill as an entity. The question is whether they have any information or control subsystems. Several languages speak about the "life" of the ant-hills, but this is not proof of the existence of some form of tertiary life above the primary and secondary levels.

With the appearance of humans, supra-individual systems, which clearly possess the organization characterized by the three subsystems of the chemoton model, have evolved. These are the nation states, developed during human history. Their metabolic processes are industry, agriculture, traffic, trade, etc., they have well-defined borders, and they also have an informational-controlling subsystem in the form of jurisdiction and the executive authorities. Could it be possible that the formation of nation states during history is nothing but the most recent experiment of evolution to create systems with tertiary life, in addition to primary and secondary lives? Who knows?

NOTES

This chapter originally appeared with notes and commentary by James Griesemer and Eörs Szathmary as section 3.5 and chapter 1 in Tibor Gánti, *The principles of life*, edited by James Griesemer and Eörs Szathmary, pp. 1–10 and 74–80, Oxford: Oxford University Press, 2003.

* Ed. note: Gánti T. (1971). *Az élet princípiuma* [*The principles of life*]. Budapest: Gondolat.

Section II
The origin and extent of natural life

The chapters in this section discuss contemporary scientific views about two issues: how the familiar forms of life that exist on Earth originated from nonliving material by physico-chemical processes, and how familiar life forms might differ from other forms of life that naturally exist. Our scientific knowledge of natural life is limited to current life on Earth, which includes microbes (true bacteria, bacteria-like archea, and protists, e.g., paramecia) as well as larger, more familiar fungi (e.g., mushrooms), plants, and animals. For better or worse, our paradigm of life is founded upon these organisms because they represent the only examples of life of which we can be certain. Our understanding of life in general, including the possibilities for artificial life and extraterrestrial life, is strongly influenced by what biologists and biochemists have discovered about life on Earth. Moreover, by asking specific biological questions, deeper philosophical ones inevitably emerge. The import of these chapters is thus not limited to the topic of this section. Taken together, these chapters bring important and sometimes underappreciated discoveries in biology and biochemistry to bear on some of the oldest questions about life.

Philosophical questions inevitably arise as soon as one starts probing the origin and extent of natural life. Both scientific projects raise general philosophical questions about life, such as the following: What presuppositions about life are embedded in different views about how life arose or what forms could exist? What weight should we place on those presuppositions? Which (if any) of these presuppositions are truly essential to any form of life that exists anywhere, and which represent physical and chemical contingencies at the time of the origin of familiar life, presumably on early Earth? Indeed, from what standpoint can we possibly determine what is accidental to life and what is

not? Are there steps we can take to free our view of life from being biased by the accidents of our epistemic circumstances?

Additional philosophical questions arise in the context of specific scientific paradigms for the nature, origin, and extent of life. The earliest biologists were naturalists. Their observations convinced them that life is distinguished from nonliving objects by abstract functional characteristics, especially the capacity to (1) self-organize and maintain self-organization for an extended period of time against internal and external perturbations and (2) reproduce and transmit to progeny heritable modifications; in this, they followed Aristotle (Ch. 1). As the writings of René Descartes and Immanuel Kant in Section I illustrate, with the development of classical physics the emphasis shifted to how the functional characteristics of life could be realized in a Newtonian physical system. Unfortunately, classical physics did not prove very helpful in illuminating this issue. It was not until the development of biochemistry and molecular biology in the twentieth century that biologists began making substantial progress on understanding the material basis of life. They discovered that the high-level functional characteristics identified by their predecessors as essential to life are realized through a complex cooperative arrangement centrally involving proteins and nucleic acids. Questions about the origin and extent of life in the universe were reframed in terms of biochemistry and molecular biology; Alexander Oparin (Ch. 5) and Ernst Mayr (Ch. 6) played leading roles in this process. This view is reflected in the contemporary writings of Orgel (Ch. 8) and Shapiro (Ch. 9) on the origin of life, and Pace (Ch. 11) and Benner, Ricardo, and Carrigan (Ch. 12) on the molecular possibilities for life elsewhere.

The Nature of Life, ed. M. A. Bedau and C. E. Cleland. Published by Cambridge University Press. © M. A. Bedau and C. E. Cleland 2010.

The molecular approach to the origin and extent of life gave rise to new philosophical questions, such as the following. Is the transition from a nonliving ensemble of molecules to a primitive living thing gradual or abrupt? Can an account of the nature of life be given independently of an account of the origin of life? The latter question is particularly pressing in the context of the historical character of Darwin's theory of evolution by natural selection. (We return to these issues below.) An even bigger philosophical worry is the assumption that an understanding of the general nature of life, wherever and whenever it might be found, lies at the level of molecules and biochemical processes, as opposed to higher-level organizational and functional properties. This assumption precludes the possibility of certain forms of artificial life, e.g., purely informational software systems, by fiat; see Section III. But even supposing that life is fundamentally biochemical, there is the further question of whether its distinctive organizational and functional characteristics can be fully explained in terms of, or reduced to, biochemistry. This raises the question of whether some of the higher-level properties of life are *emergent*, i.e., whether they represent novel autonomous properties arising spontaneously at certain levels of molecular complexity. Some scientists and philosophers think so. The problem is making good sense of the concept of emergence; see Bedau and Humphreys (2008) for a collection of articles on this issue.

Some discussions of the biochemical nature of life in this section are quite technical, but nevertheless well worth the effort. A brief review of basic biochemistry and molecular biology provides useful background information for the scientifically unsophisticated. Life as we know it on Earth today involves a complex coordination of two types of very large, carbon-containing molecules: proteins and nucleic acids. Proteins supply the bulk of the structural material for building organismal bodies as well as the catalytic (enzymatic) material for powering and maintaining them. Nucleic acids, on the other hand, store (DNA or RNA) and process (RNA) the hereditary information required for reproduction and for synthesizing the enormous quantity and variety of proteins required by an organism during its lifetime. The crucial process of coordinating these two functions, i.e., translating the hereditary material stored in nucleic acids into proteins for use in growth, maintenance, and repair, is handled by ribosomes,

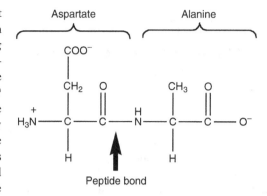

Fig. II.1. Schematic diagram showing a peptide bond between two amino acids, aspartate and alanine, from Shelley Copley; reprinted by permission of Shelley Copley.

minuscule but intricately structured molecular devices (composed of both protein and RNA) found in large numbers in all cells. In short, all known life on Earth utilizes the same macromolecules (proteins and nucleic acids) and core molecular architecture (ribosomes) to realize the abstract functional characteristics held up since Aristotle as essential to life.

The molecular similarities among contemporary organisms on Earth do not end, however, at the level of proteins and nucleic acids. A protein typically consists of 50 to 1000 amino acids joined together by peptide bonds into a long chain or polymer; Fig. II.1 depicts two amino acids joined together by a peptide bond. Amino acids are fairly complex molecules consisting of two functional groups, a basic amino group ($-NH_2$) and an acidic group ($-COOH$). There are over 100 different types of amino acids, and yet life on Earth (from the simplest bacterium to the most majestic redwood tree or gifted human being) constructs its proteins primarily from the same 20. Like many three-dimensional objects, amino acids are asymmetric and thus have the geometric property of handedness, which chemists refer to as "chirality." By convention one of these chiral arrangements is labeled "L" (a.k.a. left-handed) and the other "D" (a.k.a. right-handed). Known life on Earth utilizes only L-amino acids in its proteins, but there are no known biochemical reasons to think that life could not utilize D-amino acids instead, or for that matter, an alternative suite of amino acids (Benner, Ricardo, and Carrigan, Ch. 12).

Nucleic acids are long polymers made up of nucleotides, as opposed to amino acids. A nucleotide consists

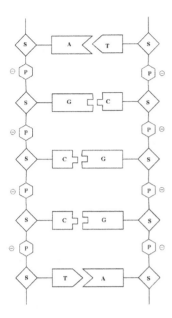

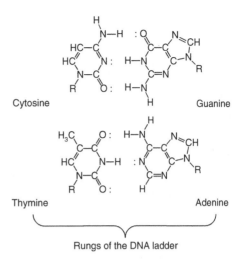

Fig. II.3. Schematic diagram showing the molecular structure of two rungs (paired nucleotide bases) of a DNA molecule from Steven A. Benner, *Life, the Universe and the Scientific Method*, reprinted as Fig. II.3 in "Introduction to Section II: The origin and extent of natural life." Copyright © 2008, Steven A. Benner. Reprinted by permission of Steven A. Benner.

Fig. II.2. Schematic diagram illustrating the structure of an unwound portion of a DNA molecule from Steven A. Benner, *Life, the Universe and the Scientific Method*, reprinted as Figure II.2 in "Introduction to Section II: The origin and extent of natural life." Copyright © 2008, Steven A. Benner. Reprinted by permission of Steven A. Benner.

of three molecular subunits, a phosphate unit bonded to a sugar unit (ribose in RNA and deoxyribose in DNA) that, in turn, is attached to a base unit. With the exception of some viruses, which use RNA, DNA is the repository of hereditary information. RNA supervises the intricate process of translating the information encoded on DNA into proteins for immediate use by the organism. Each strand of DNA's famous double helix consists of a sequence of individual nucleotides covalently bonded into a long chain by sugar (S)-phosphate (P) units (Fig. II.2). Hydrogen bonds between their respective nucleotide bases hold the two strands together (Fig. II.3). DNA utilizes four standard bases—adenine (A), thymine (T), guanine (G), and cytosine (C)—to encode hereditary information. These bases pair off in a complementary pattern; C pairs to G and A pairs to T. RNA utilizes uracil (U) instead of thymine, and is typically single-stranded, though it contains folded regions that allow complementary bases on the same strand to bind together. As Benner, Ricardo, and Carrigan (Ch. 12) mention, however, biochemists have identified seven additional bases that

could be but are not used by known life on Earth. Because of its complementary base-pairing scheme, each strand of DNA is a template for reconstructing the other half. The complementary double-stranded structure enables a simple mechanism for replicating DNA molecules, as well as providing an important source of redundancy in case one strand is damaged.

Hereditary information is encoded on a single strand of DNA (the "coding" strand) by sequences of three consecutive bases. Each triplet of bases (codon) designates a specific amino acid or, alternatively, the initiation or termination of the construction of a chain of amino acids constituting a protein. With some minor exceptions, all known life on Earth utilizes the same genetic code, with the same triplets of bases encoding the same amino acids. Yet there is little reason to suppose that a different coding schema could not have been used. The information on the coding strand of DNA is transcribed onto a special (messenger) RNA molecule and transported to a ribosome for translation into a protein. It is unlikely that the earliest life on Earth employed ribosomes for translating hereditary information into structural and catalytic material because they are incredibly complex molecular "machines" composed of both protein and RNA. The origin of the ribosome is one of the great mysteries of the evolution of life on Earth.

As the chapters in this section illustrate, the functional, molecular, and biochemical similarities among contemporary Earth organisms provide the springboard for scientific speculation about both how life emerges from nonliving material under natural conditions and how physics and chemistry constrain the possibilities for natural life anywhere in the universe. They also raise new philosophical questions. As we suggested above, one of these questions concerns whether and, if so, how an understanding of the nature of life presupposes an understanding of life's origin. Such a connection was explicitly proposed by Oparin (Ch. 5), and many contemporary scientists implicitly accept this view. One example is Pace (Ch. 11), an advocate of the RNA world theory of the origin of life, who contends that all life in the universe is likely to utilize an organic genetic molecule closely resembling the nucleic acids of familiar life.

But this view can be questioned. In the first place, it is not true in general that an understanding of the nature of something presupposes an understanding of how to make it. As an example, the chemical composition of quartz (SiO_2) was known long before anyone knew how natural volcanic processes produced it. But perhaps the nature of biological systems is different from that of chemical substances such as quartz. One can envision an evolutionary argument to the effect that, unlike inanimate chemical compounds, life is an historical entity and its origin is essential to its nature. In a similar vein, Tibor Gánti (Ch. 7) identifies the problem of understanding the nature of life with that of fully characterizing minimal life, and argues that the earliest form of life on Earth exemplifies minimal life.

Proper treatment of these arguments requires care. Someone might maintain that metabolism is an (or even the most) essential property of life and yet still accept a genes-first account of the origin of life, on the grounds that chemical systems require genetic information in order to generate and maintain the kind of self-organization required for metabolism (see Orgel, Ch. 8). On such a view, the most austere version of the RNA world (a community of evolving, self-replicating, naked RNA molecules) might count as a crucial stage in the emergence of life but not life itself. So, it would be a mistake to assume that evidence for a genes- or metabolism-first theory of the origin of life also counts as evidence for a genetic or metabolic definition or

explanation of life (Section IV contains more on definitions and explanations of life). These arguments call for careful examination, clarification, and evaluation on the part of the reader.

In Chapter 8, biologist Leslie Orgel identifies two bottlenecks in our understanding of the emergence of life on Earth: (1) the source of the basic molecular building blocks of life, and (2) the identity and origin of the first biological polymer constructed from them. The first is not as serious as the second because there are three highly plausible, empirically supported conjectures about how the early Earth was supplied with small organic molecules such as amino acids and nucleotide bases: (i) synthesis in a "prebiotic soup," (ii) delivery by meteorites (comets and asteroids), and (iii) synthesis by hydrothermal processes in deep sea volcanic vents. All three processes could have played a role.

For Orgel, the hard problem is the origin of the first biological macromolecule. Like many contemporary biologists, he believes that hereditary information is required to organize and sustain the long sequences of chemical reactions required for metabolism; see also Pace (Ch. 11), Benner, Ricardo, and Carrigan (Ch. 12), and the National Research Council report (Ch. 15). He thus embraces a genes-first account of the origin of life. The question is what kind of polymer constituted the first genetic structures. Along with most scientists currently working on the origin of life, Orgel endorses the RNA world theory. Proposed in the 1960s by Woese (1967), Orgel (1968), and Crick (1968), the RNA world theory holds that the first life on Earth consisted of communities of small, self-replicating RNA molecules. Orgel concedes, however, that the abiotic (non-biological) synthesis of RNA oligomers (very small polymers) from nucleotides, under geochemical conditions that are likely to have occurred on the early Earth, poses serious problems. He discusses some alternative genetic structures that might have preceded RNA, namely p-RNA and PNA, but admits that they do not seem much easier to synthesize abiotically under natural conditions. Orgel concludes that we have yet to hit upon the original genetic polymer, and that it might have been substantially different from the nucleic acids used by contemporary Earth life. Nevertheless, he believes that it was organic; he briefly discusses and dismisses Cairns-Smith's radical suggestion that the earliest genetic structures were inorganic clay crystals (Cairns-Smith, 1982), which provided a catalytic template for the construction of RNA oligomers and were eventually replaced by them in a "genetic takeover." But regardless of the identity of the first genetic polymer, Orgel, like many biologists, is committed to the idea that the RNA world represents a crucial developmental stage in the evolution of life as we know it on Earth today.

In a sustained, multifaceted attack on the chemical plausibility of the RNA world, biochemist Robert Shapiro argues in Chapter 9 that the abiotic synthesis of nucleotides and RNA oligomers from basic chemical building blocks under natural (vs. laboratory) conditions is so improbable that it ought to be dismissed as mere fantasy. In light of these considerations, he defends a metabolism-first approach to the emergence of life. On Shapiro's view the earliest life consisted of self-organizing networks of small organic molecules (supplied by meteorites, hydrothermal processes, etc.) coupled to an external energy source (most likely, geochemical processes). Thus, chemical self-organization (of the sort required by life) does not presuppose a genetic structure, although as he points out, such a chemical system could be construed as having a "compositional genome."

As mentioned in the introduction to Section I, Oparin (see Ch. 5) is the father of metabolism-first theories of the origin of life. The metabolism-first approach is not as popular today as the genes-first approach, perhaps because its defenders have not rallied around a specific version, as proponents of genes-first accounts have with the RNA world. A number of quite different versions have been proposed in recent years, including Stuart Kauffman's "far from equilibrium reaction systems" (Ch. 30), the "lipid world" (Segré *et al.*, 2001), the "iron-pyrite world" (Wächtershäuser, 1992), de Duve's autocatalytic metabolic systems (de Duve, 1991), and the "double-origin theory" (Dyson, 1999). Despite their striking differences, Shapiro believes that all metabolism-first theories are committed to the following five conditions: (1) some kind of boundary for selecting, concentrating, and protecting biomolecules and their building blocks, (2) an energy source to drive the molecular organization process, (3) a mechanism for coupling the release of energy to the molecular organization process, (4) a chemical network permitting adaptation and evolution, and (5) a chemical network that grows and reproduces.

Shapiro's five conditions reveal that he is at least implicitly committed to the importance of both of the functional characteristics that have been attributed to life since the time of Aristotle, despite the fact that he believes that metabolism (qua self-organizing chemical processes involving small organic molecules) preceded the synthesis of genetic polymers. In general, it is illuminating to examine whether metabolism-first views of the origin of life like Shapiro's and genes-first views like Orgel's presuppose different fundamental conceptions of what life is. On Shapiro's view, there is little reason to think that all life in the universe shares the same basic chemistry as familiar life; any collection of small molecules, organic or otherwise, capable of meeting his five conditions qualifies as life. He concludes that there might be many different chemical pathways to life, depending upon the environment, and that life is probably common in the universe.

Shapiro attributes the surprisingly common view among scientists (seriously entertained by the National Research Council in Ch. 15) that life on Earth represents a "happy accident" to the widespread acceptance of the highly improbable RNA world. In Chapter 10 philosopher Iris Fry points out that there is not much difference between the happy accident view and the creationist account of the origin of life. Both remove the origin of life from the realm of scientific investigation. She argues that scientists ought to explicitly commit themselves to a continuity thesis that "there is no unbridgeable gap between inorganic and living physical systems, and under suitable physical circumstances the emergence of life is highly probable." The continuity thesis represents a philosophical commitment similar to philosophical naturalism, but specifically directed at the origin of life. Fry argues that the continuity thesis has enormous heuristic value insofar as it demands that scientists fill crucial epistemological gaps in their theories of the origin of life with specific chemical and physical processes, rather than appealing to "scientific miracles" to explain them away. Different theories of the origin of life will of course face different challenges, and hence require correspondingly different chemical and physical solutions. But genuine scientific progress on the question of how life emerges from nonliving chemicals requires specific, gap-filling, physico-chemical mechanisms. Failure to employ the continuity thesis in one's research diminishes its scientific status, for the theory with the fewest physico-chemical gaps provides the best *scientific* explanation for the origin of life. The remainder of Fry's chapter shows how a variety of different theories of the origin of life, from genes-first to metabolism-first approaches, deploy the continuity thesis in explaining the emergence of life. Unlike Shapiro, she does not view the RNA world as so implausible as to violate the continuity thesis, but she does concur that it faces some serious problems. Fry's chapter is noteworthy for canvassing a variety of different theories of the origin of life, highlighting their strengths and weakness, and identifying problems that have yet to be solved. It is clear that she favors hybrid theories, involving the coevolution of proteins and nucleic acids.

In Chapter 11 microbiologist Norman Pace explicitly endorses a genetic definition of life, which he explains in terms of the capacity of a chemical system for self-replication and (Darwinian) evolution by natural selection. Pace uses this definition in his arguments for the universality of our basic biochemistry. He contends that there are not any other plausible chemical options for building physical systems capable of reproducing, adapting, and evolving by means of natural selection. According to Pace, all natural life, wherever it is found in the universe, consists of complex, organic (carbon-containing) macromolecules made up of repeating simple subunits. Many of these subunits (e.g., amino acids, 5-carbon sugars, and nucleotide bases) are likely to be very similar, perhaps even the same, as in our biomolecules. Pace believes that chemical differences will be found among organisms descended from alternative origins of life but that they will exist at higher levels, in the structure of the macromolecules assembled from the basic subunits, the mechanisms (e.g., ribosomes) utilized in assembling the macromolecules, and the specific biochemical pathways in which the macromolecules participate.

Because he thinks that all life closely resembles familiar Earth life in its basic biochemistry, Pace believes that Earth life forms, especially microbes (which exhibit the greatest metabolic diversity and environmental tolerance), provide a good model for thinking about extraterrestrial life. Based upon what we know about the range of conditions tolerated by the "toughest" and most versatile form of Earth life, one can infer that the minimal requirements for a "habitable" environment include, among other things, CO_2 as a carbon source and water as a solvent. As a consequence, instrumentation for searching for extraterrestrial life, which is likely to be microbial, should focus on identifying the basic building blocks of familiar life, e.g., biotic amino acids and nucleotide bases. He also recommends using gene sequencing techniques because he believes that this (differences in codon sequences) is the level at which life elsewhere in the universe is most likely to differ from us. Interestingly, Pace believes that any life elsewhere in our solar system is likely to share a common ancestor with us, having been transported around the solar system by meteors.

Like Pace, biochemist Steven Benner and colleagues (Ch. 12) ground their discussion of the chemical possibilities for life on a genetic definition, more specifically, NASA's "chemical Darwinian definition," which amounts to the same thing as Pace's definition. Benner, Ricardo, and Carrigan are not, however, convinced that the chemical options for building a system satisfying this definition are as limited as Pace believes. They propose that the definition entails a hierarchy of general requirements for chemical life, listed in decreasing order of importance: (1) thermodynamic disequilibrium (life requires an energy source); (2) chemical bonding (to hold together the subunits of biological macromolecules); (3) isolation (to select, concentrate, and protect delicate biomolecules); (4) carbon-like scaffolding (to build the complex polymers required for the structural, catalytic, and hereditary properties of life); (5) energetic patterns in metabolism (that exploit chemical disequilibria via multi-step, downhill chemical reactions running close to equilibrium); (6) solvent (to speed up rates of chemical reactions by separating compounds into reactive parts, transporting them to other locations, and bringing them into contact).

Benner and colleagues subsequently explore different chemical possibilities for meeting these requirements under a wide range of environmental conditions found outside the Earth, including Venus, Saturn's moon Titan, gas giants like Jupiter, terrestrial planets ejected from solar systems, and even gases in interstellar space. They conclude that (under the right conditions) life might use silicon instead of carbon as scaffolding for its biomolecules, and that life could exist in a surprisingly wide range of environments, exploiting almost any thermodynamic disequilibrium as an energy source and using a variety of solvents other than water. According to Benner and colleagues, although life in Earth-like environments undoubtedly uses carbon as the scaffold and water as the solvent, the only "absolute" requirements for life (as defined by the chemical Darwinian definition) are thermo-dynamic disequilibrium and temperatures consistent with chemical bonding. So while Benner and colle-agues agree with Pace about the intrinsic nature of life, they nonetheless disagree about the chemical possibi-lities for life elsewhere in the universe.

Microbiologist Kenneth Nealson has been exten-sively involved in the design of life-detection equip-ment for NASA space missions. Like Pace and Benner et al. (and many other scientists), Nealson thinks that the first step in theorizing about or searching for extraterrestrial life is to formulate a general definition of life; see Cleland and Chyba (Ch. 26) for an argu-ment against the definitional approach. Unlike Pace and Benner et al., however, he does not endorse a specific definition. Instead, he bases his discussions of life-detection strategies in Chapter 13 on what biologists have learned in recent years about Earth's microbial world.

Nealson and Pace both believe that the vast majo-rity of life in the universe is microbial. Microbes are not only ubiquitous on Earth today, but represent the oldest, most diverse, and longest living form of life on Earth. There is evidence that they emerged as early as 3.8 billion years ago, and the fossil record indicates that they remained the only form of life for more than two billion years. Microbes are also remark-able for their metabolic diversity and environmental tolerance. Unlike animals and plants, which are limited to organic matter and light, microbes metabolize a wide range of chemicals, including carbon monoxide, ferrous iron, hydrogen sulfide, and hydrogen gas. Microbes are also much tougher, thriving under con-ditions of pH, temperature, salinity, radiation, and dryness that no plant or animal could tolerate. In short, microbes have taught us that life can exploit a much greater variety of energetic resources and toler-ate a much wider range of chemical and physical conditions than previously thought. This explains why Nealson and many other scientists believe that microbial life is likely to be the most common form of life in the universe, and hence that it ought to be the focus of any search for extraterrestrial life. Unlike Pace, however, Nealson is open to the possibility of microbes with exotic chemistries. In order to minimize Earth-centricity, he recommends searching for struc-tures with constant elemental compositions that are chemically out of equilibrium with their surrounding

environments, rather than hunting for specific elem-ents or molecules. This suggests that Nealson con-siders metabolism (interpreted generally in terms of the maintenance of self-organization against environ-mental perturbations) to be a universal, perhaps the most essential, property of living systems.

Philosopher Carol Cleland and biochemist Shelley Copley argue in Chapter 14 that the Earth itself is a good place to look for "alien" forms of microbial life; see also Davies and Lineweaver (2005) and Cleland (2007). The assumption that all life on Earth today shares the same basic molecular architecture and bio-chemistry is part of the paradigm of modern biology. But as Cleland and Copley point out, there is little theoretical or empirical support for this widely held assumption. Scientists know that life could have been at least modestly different at the molecular level (utili-zed different amino acids to build its proteins, differ-ent nucleotide bases in its DNA, etc.) and it is clear that abiotic processes (meteorites, hydrothermal pro-cesses, etc.) supplied the early Earth with alternative molecular building blocks for life. If the emergence of life is highly probable under the right chemical and physical conditions, as many scientists believe, then it seems likely that the early Earth hosted multiple origins of life, some of which produced chemical vari-ations on life as we know it. It is often claimed that even if this were the case, descendents of an alternative origin of life would long ago have been either elimin-ated by our form of life, in the ruthless Darwinian competition for resources, or amalgamated into a single form of life (familiar Earth life) by lateral gene transfer. The authors of the National Research Council report (Ch. 15), for example, sympathize with this view. Cleland and Copley reject this claim, arguing that it conflicts with what microbiologists have dis-covered in recent years about the complexity and diversity of microbial communities. Another argument that is often advanced against the possibility of alter-native forms of microbial life is that we would have already detected them with the sophisticated methods used by contemporary microbiologists. As Cleland and Copley point out, however, the tools used to explore the microbial world (microscopy with mol-ecular staining techniques, cultivation, and sophisti-cated genomic methods such as PCR amplification of rRNA genes) could not detect even a modestly differ-ent form of microbial life. Given these considerations, and the profound philosophical and scientific importance that such a discovery would represent, they contend that the hypothesis that the contemporary Earth contains a "shadow biosphere" of unfamiliar forms of microbial life is worthy of serious scientific consideration.

The capstone piece in this section is the short but wide-ranging introduction of the National Research Council's report *The Limits of Organic Life in Planetary Systems* (Ch. 15). This chapter provides an overview of the latest scientific thinking about the possibilities and limits of natural and unnatural life—or, as the authors term it, "weird" life. Like Cleland and Chyba (Ch. 26), the National Research Council eschews endorsing a definition of life. Instead they list some universal characteristics of known Earth life and discuss which ones seem most likely to characterize all life. The authors are very sympathetic to the idea that Darwinian evolution is essential to life, a view defended by Dawkins (Ch. 29), and shared by many biologists. Some of their arguments for universal Darwinism, however, seem to confuse issues about the nature of life with issues about the origin of life (the necessity of a genetic system for achieving and maintaining metabolic processes). Like Nealson and Benner *et al.*, and most authors in other sections of this book, the Council seeks a conception of life that avoids Earth-centricity. Scientists know of important ways in which carbon-based life might differ from familiar Earth life, and they can conceive of life based on a scaffolding of silicon rather than carbon. But these possibilities are inferred from what we know about the molecular biology and biochemistry of familiar Earth life, so they might not exhaust the full range of possibilities for life. The committee closes by identifying some problems that must be solved if we are to truly understand the limits of chemical life in the universe.

The National Research Council's chapter on weird life includes an interesting discussion of robotic and computer-based, purely informational forms of life (these are respectively the "hard" and "soft" artificial life discussed in Section III). They argue that, if it exists, robotic and purely informational "life" would almost certainly be designed by some other pre-existing form of intelligent life, such as humans; this reflects the view, common among many biologists, that natural selection is the fundamental mechanism for the evolution of the first life and intelligence in the universe. Thus, they conclude that robotic or purely informational forms of life are not relevant to life's *origin*. However, the Council concedes that their conclusion might merely reflect the narrowness of our current preconceptions about life. For extensive discussion of whether computational life could nevertheless be illuminating about life's *nature*, see the material in Section III.

REFERENCES

Bedau, M. A. & Humphreys, P. (Eds.) (2008). *Emergence: Contemporary readings in philosophy and science.* Cambridge: MIT Press.

Benner, S. A. (2008). *Life, the universe, and the scientific method.* Steven A. Benner Foundation Press.

Cairns-Smith, A. G. (1982). *Genetic takeover and the mineral origins of life.* Cambridge: Cambridge University Press.

Cleland, C. E. (2007). Epistemological issues in the study of microbial life: Alternative terran biospheres? *Studies in History and Philosophy of Biological and Biomedical Science*, 38, 847–861.

Crick, F. H. C. (1968). The origin of the genetic code. *Journal of Molecular Biology*, 38, 367–379.

Davies, P. C. W. & Lineweaver, C. H. (2005). Finding a second sample of life on earth. *Astrobiology*, 5, 154–163.

de Duve, C. (1991). *Blue print for a cell: The nature and origin of life.* Burlington, NC: Neil Patterson Publishers.

Dyson, F. (1999). *Origins of Life.* Cambridge: Cambridge University Press.

Orgel, L. E. (1968). Evolution of the genetic apparatus. *Journal of Molecular Biology*, 38, 381–393.

Segré, D., Ben-Eli, D., Deamer, D. W., & Lancet, D. (2001). The lipid world. *Origin of Life and Evolution of the Biosphere*, 31, 119–145.

Wächtershäuser, G. (1992). Groundworks for an evolutionary biochemistry: The iron–sulfur world. *Progress in Biophysics and Molecular Biology*, 58, 85–201.

Woese, C. R. (1967). *The genetic code: The molecular basis for genetic expression.* New York: Harper and Row.

8 · The origin of life: a review of facts and speculation

LESLIE E. ORGEL

The problem of the origin of life on the Earth has much in common with a well-constructed detective story. There is no shortage of clues pointing to the way in which the crime, the contamination of the pristine environment of the early Earth, was committed. On the contrary, there are far too many clues and far too many suspects. It would be hard to find two investigators who agree on even the broad outline of the events that occurred so long ago and made possible the subsequent evolution of life in all its variety. Here, I outline two of the main questions and some of the conflicting evidence that has been used in attempts to answer them. First, however, I summarize the few areas where there is fairly general agreement.

The Earth is slightly more than 4.5 billion years old. For the first half billion years or so after its formation, it was impacted by objects large enough to evaporate the oceans and sterilize the surface.[1,2] Well-preserved microfossils of organisms that have morphologies similar to those of modern blue-green algae, and date back about 3.5 billion years, have been found,[3] and indirect but persuasive evidence supports the proposal that life was present 3.8 billion years ago.[4] Life, therefore, originated on or was transported to the Earth at some point within a window of a few hundred million years that opened about four billion years ago. The majority of workers in the field reject the hypothesis that life was transported to the Earth from somewhere else in the galaxy and take it for granted that life began *de novo* on the early Earth.

The uniformity of biochemistry in all living organisms argues strongly that all modern organisms descend from a last-common ancestor (LCA). Detailed analysis of protein sequences suggests that the LCA had a complexity comparable to that of a simple modern bacterium and lived 3.2–3.8 billion years ago.[5] If we knew the stages by which the LCA evolved from abiotic components present on the primitive Earth, we would have a complete account of the origin of life. In practice, the most ambitious studies of the origins of life address much simpler questions. Here, I discuss two of them. What were the sources of the small organic molecules that made up the first self-replicating systems? How did biological organization evolve from an abiotic supply of small organic molecules?

ABIOTIC SYNTHESIS OF SMALL ORGANIC MOLECULES

Miller, a graduate student who was working with Harold Urey, began the modern era in the study of the origin of life at a time when most people believed that the atmosphere of the early Earth was strongly reducing. Miller[6] subjected a mixture of methane, ammonia, and hydrogen to an electric discharge and led the products into liquid water. He showed that a substantial percentage of the carbon in the gas mixture was incorporated into a relatively small group of simple organic molecules and that several of the naturally occurring amino acids were prominent among these products. This was a surprising result; organic chemists would have expected a much less tractable product mixture. The Urey–Miller experiments were widely accepted as a model of prebiotic synthesis of amino acids by the action of lightning.

Miller and his coworkers went on to study electric-discharge synthesis of amino acids in greater detail.[7,8] Using more powerful analytical techniques, they identified many more amino acids—some, but by no means all, of which occur in living organisms. They also

The Nature of Life, ed. M. A. Bedau and C. E. Cleland. Published by Cambridge University Press. © M. A. Bedau and C. E. Cleland 2010.

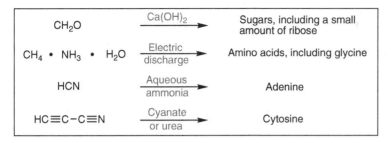

Fig. 8.1. Early prebiotic syntheses of biomonomers.

showed that a major synthetic route to the amino acids is through the Strecker reaction—that is, from aldehydes, hydrogen cyanide, and ammonia. Glycine, for example, is formed from formaldehyde, cyanide, and ammonia—all of which can be detected among the products formed in the electric-discharge reaction.

In the years following the Urey–Miller experiments, the synthesis of biologically interesting molecules from products that could be obtained from a reducing gas mixture became the principle aim of prebiotic chemistry (Fig. 8.1). Remarkably, Oro[9] was able to synthesize adenine from hydrogen cyanide and ammonia. Somewhat later, Sanchez, Ferris and I[10] showed that cyanoacetylene is a major product of the action of an electric discharge on a mixture of methane and nitrogen and that cyanoacetylene is a plausible source of the pyrimidine bases uracil and cytosine. This new information, together with previous studies that showed that sugars are formed readily from formaldehyde,[11,12] convinced many students of the origins of life that they understood the first stage in the appearance of life on the Earth: the formation of a prebiotic soup of biomonomers.

As in any good detective story, however, the principal suspect, the reducing atmosphere, has an alibi. Recent studies have convinced most workers concerned with the atmosphere of the early Earth that it could never have been strongly reducing.[13] If this is true, Miller's experiments, and most other early studies of prebiotic chemistry, are irrelevant. I believe that the dismissal of the reducing atmosphere is premature, because we do not completely understand the early history of the Earth's atmosphere. It is hard to believe that the ease with which sugars, amino acids, purines and pyrimidines are formed under reducing-atmosphere conditions is either a coincidence or a false clue planted by a malicious creator.

Many of those who dismiss the possibility of a reducing atmosphere believe that the crime was an outside job. A substantial proportion of the meteorites that fall on the Earth belong to a class known as carbonaceous chondrites.[14] These are particularly interesting because they contain a significant amount of organic carbon and because some of the standard amino acids and nucleic-acid bases are present. Could the prebiotic soup have originated in preformed organic material brought to the Earth by meteorites and comets?

Supporters of the impact theory have argued convincingly that sufficient organic carbon must have been present in the meteorites and comets that reached the surface of the early Earth to have stocked an abundant soup. However, would this material have survived the intense heating that accompanies the entry of large bodies into the atmosphere and their subsequent collisions with the surface of the Earth? The results of theoretical calculations depend strongly on assumptions made about the composition and density of the atmosphere, the distribution of sizes of the impacting objects, etc.[15] The impact theory is probably the most popular at present, but nobody has proved that impacts were the most important sources of prebiotic organic compounds.

The newest suspects are the deep-sea vents, submarine cracks in the Earth's surface where superheated water rich in transition-metal ions and hydrogen sulfide mixes abruptly with cold sea water. These vents are sites of abundant biological activity, much of it independent of solar energy. Wächtershäuser[16,17] has proposed a scenario for the origin of life that might fit such an environment. He hypothesizes that the reaction between iron(II) sulfide and hydrogen sulfide (a reaction that yields pyrites (FeS_2) and hydrogen) could provide the free energy necessary for reduction of carbon dioxide to molecules capable of supporting the origin of life. He asserts that life originated on the surface of iron sulfides as a result of such chemistry. The assumptions that complex metabolic cycles self-organize on the surface and that the significant products never escape from the surface are essential parts of this theory; in Wächtershäuser's opinion, there never was a prebiotic soup!

Stetter and colleagues[18] have confirmed the novel suggestion that hydrogen sulfide, in the presence of iron(II) sulfide, acts as a reducing agent. They have reduced, for example, acetylene to ethane, and mercaptoacetic acid to acetic acid, but they have not reported reduction of CO_2. However, in a new study, Wächtershäuser and Huber[19] have shown that FeS spiked with NiS reduces carbon monoxide. Given that carbon monoxide might well have been present in large amounts in the gases escaping from the vents, Wächtershäuser's findings could well prove important. If metal sulfides can be shown to catalyze the synthesis of a sufficient variety of organic molecules from carbon monoxide, the vent theory of the origins of biomonomers will become very attractive.

In summary, there are three main contending theories of the prebiotic origin of biomonomers (not to mention several other less popular options). No theory is compelling, and none can be rejected out of hand. Perhaps it is time for a conspiracy theory; more than one of the sources of organic molecules discussed above may have collaborated to make possible the origin of life.

SELF-ORGANIZATION

There is no general agreement about the source of prebiotic organic molecules on the early Earth, but there are several plausible theories, each backed by some experimental data. The situation with regard to the evolution of a self-replicating system is less satisfactory; there are at least as many suspects, but there are virtually no experimental data.

The fairly general acceptance of the hypothesis that there was once an RNA world (i.e., a self-contained biological world in which RNA molecules functioned both as genetic materials and as enzyme-like catalysts) has changed the direction of research into the origins of life.[20] The central puzzle is now seen to be the origin of the RNA world. Two specific, but

intertwined, questions are central to the debate. Was RNA the first genetic material or was it preceded by one or more simpler genetic materials? How much self-organization of reaction sequences is possible in the absence of a genetic material? I shall concentrate on the first question.

The assumption that a polymer that doubled as a genetic material and as a source of enzyme-like catalytic activity once existed profoundly changes the goals of prebiotic synthesis. The central issue becomes the synthesis of the first genetic monomers: nucleotides or whatever preceded them. The synthesis of amino acids, coenzymes, etc. becomes a side issue, because there is no reason to believe that they were ever synthesized abiotically; some or all of them might have been introduced as direct or indirect consequences of the enzyme-like activities of RNA or its precursor(s). Supporters of the hypothesis that RNA was the first genetic material must explain where the nucleotides came from and how they self-organized. Those who believe in a simpler precursor have the difficult task of identifying such a precursor, but they hope that explaining monomer synthesis will then be simpler.

Returning to the idiom of the detective story, accumulating evidence suggests that RNA, a prime suspect, could have completed the difficult task of organizing itself into a self-contained replicating system. It has proved possible to isolate sequences that catalyze a wide variety of organic reactions from pools of random RNA.[21,22] As regards the origin of the RNA world, the most important reactions are those in which a preformed template-RNA strand catalyzes the synthesis of its complement from monomers or short oligomers. Eklund and coworkers[23] have isolated catalysts for the ligation of short oligonucleotides surprisingly easily, and the catalysts carry out ligation with adequate specificity. These molecules are the RNA equivalents of the RNA and DNA ligases. Considerable progress has also been made in selecting RNA equivalents of RNA polymerases.[24]

If the RNA world evolved de novo, it must have depended initially on an abiotic source of activated nucleotides. However, oxidation–reduction, methylation, oligosaccharide synthesis, etc., supported by nucleotide-containing coenzyme, probably became part of the chemistry of the RNA world before the invention of protein synthesis. Unfortunately, we cannot say just how complex the RNA world could have been until we know more about the range of reactions that can be catalyzed by ribozymes. It seems likely that RNA could have catalyzed most of the steps involved in the synthesis of nucleotides,[25] and possibly the coupling of redox reactions to the synthesis of phosphodiesters and peptides, but this remains to be demonstrated experimentally.

The experiments on the selection of ribozymes that catalyze nucleic acid replication (discussed above) use as inputs pools of RNA molecules synthesized by enzymes. Recently, Ferris and coworkers[26,27] have made considerable progress in the assembly of RNA oligomers from monomers, using an abundant clay mineral, montmorillonite, as a catalyst. The substrates that they use, nucleoside 5'-phosphorimidazolides, were probably not prebiotic molecules, but the experiments do indicate that the use of minerals as adsorbents and catalysts could allow the accumulation of long oligo-nucleotides once suitable activated monomers are available. We have shown that, using activated mono-mers, non-enzymatic copying of a wide range of oligo-nucleotide sequences is possible[28] and have obtained similar, but less extensive, results for ligation of short oligomers.

An optimist could propose the following scenario. First, activated mononucleotides oligomerize on mont-morillonite or an equivalent mineral. Next, copying of longer templates, using monomers or short oligomers as substrates, leads to the accumulation of a library of dsRNA molecules. Finally, an RNA double helix, one of whose strands has generalized RNA-polymerase activity, dissociates; the polymerase strand copies its complement to produce a second polymerase molecule, which copies the first to produce a second comple-ment, and so on. The RNA world could therefore have arisen from a pool of activated nucleotides.[29] All that would have been needed is a pool of activated nucleotides!

Nucleotides are complicated molecules. The syn-thesis of sugars from formaldehyde gives a complex mixture, in which ribose is always a minor component. The formation of a nucleoside from a base and a sugar is not an easy reaction and, at least for pyrimidine nucleosides, has not been achieved under prebiotic conditions; the phosphorylation of nucleosides tends to give a complex mixture of products.[30] The inhibi-tion of the template-directed reactions on D-templates by L-substrates is a further difficulty.[31] It is almost

inconceivable that nucleic acid replication could have got started, unless there is a much simpler mechanism for the prebiotic synthesis of nucleotides. Eschenmoser and his colleagues[32] have had considerable success in generating ribose 2,4-diphosphate in a potentially pre-biotic reaction from glycolaldehyde monophosphate and formaldehyde. Direct prebiotic synthesis of nucle-otides by novel chemistry is therefore not hopeless. Nonetheless, it is more likely that some organized form of chemistry preceded the RNA world. This leads us to a discussion of genetic takeover.

Cairns-Smith,[33] long before the argument became popular, emphasized how improbable it is that a mol-ecule as high tech as RNA could have appeared *de novo* on the primitive Earth. He proposed that the first form of life was a self-replicating clay. He suggested that the synthesis of organic molecules became part of the competitive strategy of the clay world and that the inorganic genome was taken over by one of its organic creations. Cairns-Smith's postulate of an inorganic life form has failed to gather any experimental support. The idea lives on in the limbo of uninvestigated hypotheses. However, Cairns-Smith also contemplated the possibility that RNA was preceded by one or more linear organic genomes.[34] This idea has taken root, but its implications have not always been appreciated.

If RNA was not the first genetic material, bio-chemistry might provide no clues to the origins of life. Presumably, the biological world that immediately preceded the RNA world already had the capacity to synthesize nucleotides. This should help us to for-mulate hypotheses about its chemical characteristics. However, if there were two or more worlds before the RNA world, the original chemistry might have left no trace in contemporary biochemistry. In that case, the chemistry of the origins of life is unlikely to be discovered without investigating in detail all the chemistry that might have occurred on the primitive Earth—whether or not that chemistry has any relation to biochemistry. This gloomy prospect has not pre-vented discussion of alternative genetic systems.

The only potentially informational systems, other than nucleic acids, that have been discovered are closely related to nucleic acids. Eschenmoser[35] has undertaken a systematic study of the properties of nucleic acid analogs in which ribose is replaced by another sugar or in which the furanose form of ribose is replaced by the pyranose form (Fig. 8.2b).

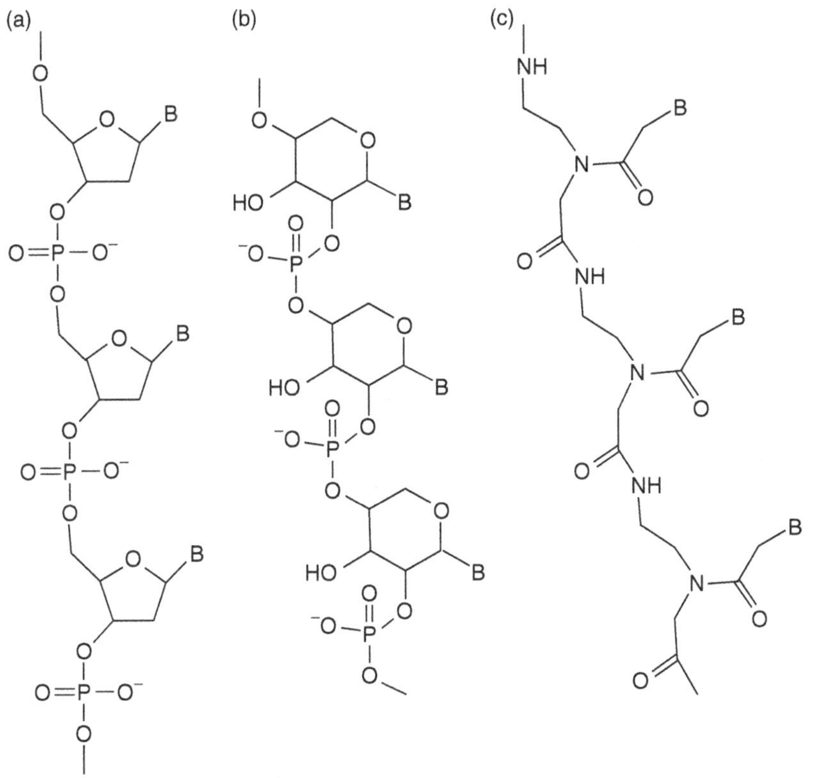

Fig. 8.2. DNA and potentially informational oligonucleotide analogs. (a) DNA. (b) Pyranosyl analog of RNA. (c) Peptide nucleic acid.

Strikingly, polynucleotides based on the pyranosyl isomer of ribose (p-RNA) form Watson–Crick-paired double helices that are more stable than RNA, and p-RNAs are less likely than the corresponding RNAs to form multiple-strand competing structures.[35] Furthermore, the helices twist much more gradually than those in the standard nucleic acids, which should make it easier to separate strands during replication. Pyranosyl RNA seems to be an excellent choice as a genetic system; in some ways, it might be an improvement on the standard nucleic acids. However, prebiotic synthesis of pyranosyl nucleotides is not likely to prove much easier than synthesis of the standard isomers, although a route through ribose 2,4-diphosphate is being explored by Eschenmoser and his colleagues.

Peptide nucleic acid (PNA) is another nucleic acid analog that has been studied extensively (Fig. 8.2c). It was synthesized by Nielsen and colleagues[36] during work on antisense RNA. PNA is an uncharged, achiral analog of RNA or DNA; the ribose–phosphate backbone of the nucleic acid is replaced by a backbone held together by amide bonds. PNA forms very stable double helices with complementary RNA or DNA.[36,37] We have shown that information can be transferred from PNA to RNA, and vice versa, in template-directed reactions[38,39] and that PNA–DNA chimeras form readily on either DNA or PNA templates.[40] Thus, a transition from a PNA world to an RNA world is possible. Nonetheless, I think it unlikely that PNA was ever important on the early Earth, because PNA monomers cyclize when they are activated; this would make oligomer formation very difficult under prebiotic conditions.

The studies described above suggest that there are many ways of linking together nucleotide bases into chains that can form Watson–Crick double helices. Perhaps a structure of this kind will be discovered that can be synthesized easily under prebiotic conditions. If so, it would be a strong candidate for the very first genetic material. However, another possibility remains to be explored: the first genetic material might not have involved nucleoside bases. Two or more very simple molecules could have the pairing properties needed to form a genetic polymer—a positively charged and a negatively charged amino acid, for example. However, it is not clear that stable structures of this kind exist. RNA is clearly adapted to double-helix formation: its constrained backbone permits simultaneous base pairing and stacking. It is unlikely that much simpler molecules could substitute for the nucleotides. Perhaps some other interaction between the chains can stabilize a double helix in the absence of base stacking; binding to a mineral surface might supply the necessary constraints, but this remains to be demonstrated. In the absence of experimental evidence, little useful can be said.

The above discussion reveals a very large gap between the complexity of molecules that are readily synthesized in simulations of the chemistry of the early Earth and the molecules that are known to form potentially replicating informational structures. Several authors have therefore proposed that metabolism came before genetics.[41–43] They have suggested that substantial organization of reaction sequences can occur in the absence of a genetic polymer and, hence, that the first genetic polymer probably appeared in an already-specialized biochemical environment. Because it is hard to envisage a chemical cycle that produces β-D-nucleotides, this theory would fit best if a simpler genetic system preceded RNA.

There is no agreement on the extent to which metabolism could develop independently of a genetic material. In my opinion, there is no basis in known chemistry for the belief that long sequences of reactions can organize spontaneously—and every reason to believe that they cannot. The problem of achieving sufficient specificity, whether in aqueous solution or on the surface of a mineral, is so severe that the chance of closing a cycle of reactions as complex as the reverse citric acid cycle, for example, is negligible. The same, I believe, is true for simpler cycles involving small molecules that might be relevant to the origins of life and also for peptide-based cycles.

CONCLUSION/OUTLOOK

In summary, there are several tenable theories about the origin of organic material on the primitive Earth, but in no case is the supporting evidence compelling.

Similarly, several alternative scenarios might account for the self-organization of a self-replicating entity from prebiotic organic material, but all of those that are well formulated are based on hypothetical chemical syntheses that are problematic. Returning to our detective story, we must conclude that we have identified some important suspects and, in each case, we have some ideas about the method they might have used. However, we are very far from knowing whodunit. The only certainty is that there will be a rational solution.

This review has necessarily been highly selective. I have neglected important aspects of prebiotic chemistry (e.g., the origin of chirality, the organic chemistry of solar bodies other than the Earth, and the formation of membranes). The best source for such material is the journal *Origins of Life and Evolution of the Biosphere*, particularly those issues that contain the papers presented at meetings of the International Society for the Study of the Origin of Life.

ACKNOWLEDGMENTS

This work was supported by NASA (grant number NAG5–4118) and NASA NSCORT/EXOBIOLOGY (grant number NAG5–4546), I thank Aubrey R. Hill, Jr for technical assistance and Bernice Walker for manuscript preparation.

NOTES

This chapter originally appeared in *Trends in Biochemical Sciences* 23 (1998), 491–495.

REFERENCES

1. Sleep, N. H., Zahnle, K. J., Kasting, J. F., & Morowitz, H. J. (1989). Annihilation of ecosystems by large asteroid impacts on early Earth. *Nature*, 342, 139–142.
2. Chyba, C. F. (1993). The violent environment of the origin of life: Progress and uncertainties. *Geochimica et Cosmochimica Acta*, 57, 3351–3358.
3. Schopf, J. W. (1993). *The Earth's earliest biosphere: Its origin and evolution*. Princeton, NJ: Princeton University Press.
4. Mojzsis, S. J., Arrhenius, G., McKeegan, K. D., Harrison, T. M., Nutman, A. P., & Friend, C. R. (1996). Evidence for life on Earth before 3,800 million years ago. *Nature*, 384, 55–59.

5. Feng, D.-F., Cho, G., & Doolittle, R. F. (1997). Determining divergence times with a protein clock: Update and reevaluation. *Proceedings of the National Academy of Sciences*, **94**, 13,028–13,033.

6. Miller, S. L. (1953). A production of amino acids under possible primitive Earth conditions. *Science*, **117**, 528–529.

7. Ring, D., Wolman, Y., Friedmann, N., & Miller, S. L. (1972). Prebiotic synthesis of hydrophobic and protein amino acids. *Proceedings of the National Academy of Sciences*, **69**, 765–768.

8. Wolman, Y., Haverland, H., & Miller, S. L. (1972). Nonprotein amino acids from spark discharges and their comparison with the Murchison meteorite amino acids. *Proceedings of the National Academy of Sciences*, **69**, 809–811.

9. Oro, J. (1960). Synthesis of adenine from ammonium cyanide. *Biochemical and Biophysical Research Communications*, **2**, 407–412.

10. Ferris, J. P., Sanchez, A., & Orgel, L. E. (1968). Studies in prebiotic synthesis III. Synthesis of pyrimidines from cyanoacetylene and cyanate. *Journal of Molecular Biology*, **33**, 693–704.

11. Butlerow, A. (1861). Formation synthétique d'une substance sucrée. *Comptes Rendus de l' Académie des Sciences*, **53**, 145–147.

12. Butlerow, A. (1861). Bildung einer zuckerartigen Substanz durch Synthese. *Justus Liebigs Annalen der Chemie*, **120**, 295–298.

13. Kasting, J. F. (1993). Earth's early atmosphere. *Science*, **259**, 920–926.

14. Cronin, J. R., Pizzarella, S., & Crukshank, D. P. (1988). Organic matter in carbonaceous chondrites, planetary satellites, asteroids and comets. In J. F. Kerridge & M. S. Matthew (Eds.), *Meteorites and the early solar system* (pp. 819–857). Tuscan, AZ: University of Arizona Press.

15. Chyba, C. & Sagan, C. (1992). Endogenous production, exogenous delivery and impact-shock synthesis of organic molecules: An inventory for the origins of life. *Nature*, **355**, 125–132.

16. Wächtershäuser, G. (1988). Before enzymes and templates: Theory of surface metabolism. *Microbiological Reviews*, **52**, 452–484.

17. Wächtershäuser, G. (1992). Groundworks for an evolutionary biochemistry: The iron-sulfur world. *Progress in Biophysics and Molecular Biology*, **58**, 85–201.

18. Blöchl, E., Keller, M., Wächtershäuser, G., & Stetter, K. O. (1992). Reactions depending on iron sulfide and

linking geochemistry with biochemistry. *Proceedings of the National Academy of Sciences*, **89**, 8117–8120.

19. Huber, C. & Wächtershäuser, G. (1997). Activated acetic acid by carbon fixation on (Fe,Ni)S under primordial conditions. *Science*, **276**, 245–247.

20. Gesteland, R. F. & Atkins, J. F. (1993). *The RNA world: The nature of modern RNA suggests a prebiotic world*. Long Island, NY: Cold Spring Harbor Laboratory Press.

21. Pan, T. (1997). Novel and variant ribozymes obtained through in vitro selection. *Current Opinion in Chemical Biology*, **1**, 17–25.

22. Breaker, R. R. (1997). DNA aptamers and DNA enzymes. *Current Opinion in Chemical Biology*, **1**, 26–31.

23. Eklund, E. H., Szostak, J. W., & Battel, D. P. (1995). Structurally complex and highly active RNA ligases derived from random RNA sequences. *Science*, **269**, 364–370.

24. Eklund, E. H. & Bartel, D. P. (1996). RNA-catalysed RNA polymerization using nucleoside triphosphates. *Nature*, **382**, 373–376.

25. Unrau, P. J. & Bartel, D. P. (1998). RNA-catalysed nucleotide synthesis. *Nature*, **395**, 260–263.

26. Ferris, J. P. & Ertem, G. (1993). Montmorillonite catalysis of RNA oligomer formation in aqueous solution: A model for the prebiotic formation of RNA. *Journal of the American Chemical Society*, **115**, 12,270–12,275.

27. Kawamura, K. & Ferris, J. P. (1994). Kinetic and mechanistic analysis of dinucleotide and oligonucleotide formation from the 5'-phosphorimidazolide of adenosine on NA+-montmorillonite. *Journal of the American Chemical Society*, **119**, 7564–7572.

28. Hill, A. R., Jr., Wu, T., & Orgel, L. E. (1993). The limits of template-directed synthesis with nucleoside-5'-phosphoro(2-methyl)imidazolides. *Origins of Life and Evolution of the Biosphere*, **23**, 285–290.

29. Orgel, L. E. (1994). The origin of life on Earth. *Scientific American*, **271**, 52–61.

30. Ferris, J. P. (1987). Prebiotic synthesis: Problems and challenges. *Cold Spring Harbor Symposia on Quantitative Biology*, **LII**, 29–39.

31. Joyce, G. F., Visser, G. M., van Boeckel, A. A., van Boom, J. H., Orgel, L. E., & van Westrenen, J. (1984). *Nature*, **310**, 602–604.

32. Müller, D., Pitsch, S., Kittaka, A., *et al.* (1990). Chimie von a-Aminonitrilen. Aldomerisierung von Glycolaldehyd-phosphat zu racemischen Hexose-2,4,6-triphosphaten und (in Gegenwart von

Formaldehyd) racemischen Pentose-2,4-diphosphaten: Rac-Allose-2,4,6-triphosphat und rac-Ribose-2,4-diphosphat sind die Reaktionshauptprodukte. *Helvetica Chimica Acta*, **73**, 1410–1463.

33. Cairns-Smith, A. G. (1982). *Genetic takeover and the mineral origins of life*. Cambridge, UK: Cambridge University Press.

34. Cairns-Smith, A. G. & Davis, C. J. (1997). The design of novel replicating polymers. In R. Duncan and M. Weston-Smith (Eds.), *Encyclopaedia of ignorance* (pp. 397–403). New York: Pergamon Press.

35. Eschenmoser, A. (1997). Towards a chemical etiology of nucleic acid structure. *Origins of Life and Evolution of the Biospheres*, **27**, 535–553.

36. Egholm, M., Buchardt, O., Nielsen, P. E., & Berg, R. H. (1992). Peptide of nucleic acids (PNA): Oligonucleotide analogs with an achiral peptide backbone. *Journal of the American Chemical Society*, **114**, 1895–1897.

37. Egholm, M., Buchardt, O., Christensen, L., *et al.* (1993). PNA hybridizes to complementary oligonucleotides obeying the Watson-Crick hydrogen-bonding rules. *Nature*, **365**, 566–568.

38. Schmidt, J. G., Nielsen, P. E., & Orgel, L. E. (1997). Information transfer from DNA to peptide nucleic acids by template-directed syntheses. *Nucleic Acids Research*, **25**, 4792–4796.

39. Schmidt, J. G., Nielsen, P. E., & Orgel, L. E. (1997). Information transfer from peptide nucleic acids to RNA by template-directed synthesis. *Nucleic Acids Research*, **25**, 4797–4802.

40. Koppitz, M., Nielsen, P. E., & Orgel, L. E. (1998). Formation of oligonucleotide-PNA-chimeras by template-directed ligation. *Journal of the American Chemical Society*, **120**, 4563–4569.

41. Kauffman, S. A. (1986). Autocatalytic sets of proteins. *Journal of Theoretical Biology*, **119**, 1–24.

42. Wächtershäuser, G. (1988). Before enzymes and templates: Theory of surface metabolism. *Microbiological Reviews*, **52**, 452–484.

43. de Duve, C. (1991). *Blueprint for a cell: The nature and origin of life*. Burlington, NC: Neil Patterson.

9 · Small molecule interactions were central to the origin of life

ROBERT SHAPIRO

A SIMPLER ORIGIN FOR LIFE

Extraordinary discoveries inspire extraordinary claims. Thus, James Watson reported that immediately after he and Francis Crick uncovered the structure of DNA, Crick "winged into the Eagle [pub] to tell everyone within hearing that we had discovered the secret of life." Their structure—an elegant double helix—almost merited such enthusiasm. Its proportions permitted information storage in a language in which four chemicals, called bases, played the same role as 26 letters do in the English language.

Further, the information was stored in two long chains, each of which specified the contents of its partner. This arrangement suggested a mechanism for reproduction: The two strands of the DNA double helix parted company, and new DNA building blocks that carry the bases, called nucleotides, lined up along the separated strands and linked up. Two double helices now existed in place of one, each a replica of the original.

The Watson–Crick structure triggered an avalanche of discoveries about the way living cells function today. These insights also stimulated speculations about life's origins. Nobel laureate H. J. Muller wrote that the gene material was "living material, the present-day representative of the first life," which Carl Sagan visualized as "a primitive free-living naked gene situated in a dilute solution of organic matter." (In this context, "organic" specifies compounds containing bound carbon atoms, both those present in life and those playing no part in life.) Many different definitions of life have been proposed. Muller's remark would be in accord with what has been called the NASA definition: Life is a self-sustained chemical system capable of undergoing Darwinian evolution.

Richard Dawkins elaborated on this image of the earliest living entity in his book *The Selfish Gene:* "At some point a particularly remarkable molecule was formed by accident. We will call it the *Replicator*. It may not have been the biggest or the most complex molecule around, but it had the extraordinary property of being able to create copies of itself." When Dawkins wrote these words 30 years ago, DNA was the most likely candidate for this role. Later, researchers turned to other possible molecules as the earliest replicator, but I and others think that this replicator-first model of the origin of life is fundamentally flawed. We prefer an alternative idea that seems much more plausible.

WHEN RNA RULED THE WORLD

COMPLICATIONS to the DNA-first theory soon set in. DNA replication cannot proceed without the assistance of a number of proteins—members of a family of large molecules that are chemically very different from DNA. Both are constructed by linking subunits together to form a long chain, but whereas DNA is made of nucleotides, proteins are made of amino acids. Proteins are the handymen of the living cell. Enzymes, proteins' most famous subclass, act as expediters, speeding up chemical processes that would otherwise take place too slowly to be of use to life. Proteins used by cells today are built following instructions encoded in DNA.

The above account brings to mind the old riddle: Which came first, the chicken or the egg? DNA holds the recipe for protein construction. Yet that information cannot be retrieved or copied without the assistance of proteins. Which large molecule, then, appeared first—proteins (the chicken) or DNA (the egg)?

The Nature of Life, ed. M. A. Bedau and C. E. Cleland. Published by Cambridge University Press. © M. A. Bedau and C. E. Cleland 2010.

A possible solution appeared when attention shifted to a new champion—RNA. This versatile class of molecule is, like DNA, assembled of nucleotide building blocks but plays many roles in our cells. Certain RNAs ferry information from DNA to ribosomes, structures (which themselves are largely built of other kinds of RNA) that construct proteins. In carrying out its various duties, RNA can take the form of a double helix that resembles DNA or of a folded single strand, much like a protein.

In the early 1980s scientists discovered ribozymes, enzyme like substances made of RNA. A simple solution to the chicken-and-egg riddle now appeared to fall into place: Life began with the appearance of the first self-copying RNA molecule. In a germinal 1986 article, Nobel Laureate Walter Gilbert wrote in the journal *Nature*: "One can contemplate an RNA world, containing only RNA molecules that serve to catalyze the synthesis of themselves ... The first step of evolution proceeds then by RNA molecules performing the catalytic activities necessary to assemble themselves from a nucleotide soup." In this vision, the first self-replicating RNA that emerged from nonliving matter carried out the various functions now executed by RNA, DNA, and proteins.

A number of additional clues support the idea that RNA appeared before proteins and DNA in the evolution of life. For example, many small molecules, called co-factors, play a role in enzyme-catalyzed reactions. These co-factors often carry an attached RNA nucleotide with no obvious function. Such structures have been considered "molecular fossils," relics descended from the time when RNA alone, without DNA or proteins, ruled the biochemical world.

This clue and others, however, support only the conclusion that RNA preceded DNA and proteins; they provide no information about the origin of life, which may have involved stages prior to the RNA world in which other living entities ruled supreme. Confusingly, researchers use the term "RNA world" to refer to both notions. Here I will use the term "RNA first" for the claim that RNA was involved in the origin of life, to distinguish it from the assertion that RNA merely arose before DNA and proteins.

THE SOUP KETTLE IS EMPTY

The RNA-first hypothesis faces a tremendously challenging question: How did that first self-replicating RNA arise? Enormous obstacles block Gilbert's picture of RNA forming in a nonliving nucleotide soup.

RNA's building blocks, nucleotides, are complex substances as organic molecules go. Each contains a sugar, a phosphate, and one of four nitrogen-containing bases as sub-subunits. Thus, each RNA nucleotide contains nine or 10 carbon atoms, numerous nitrogen and oxygen atoms, and the phosphate group, all connected in a precise three-dimensional pattern. Many alternative ways exist for making those connections, yielding thousands of plausible nucleotides that could readily join in place of the standard ones but that are not represented in RNA. That number is itself dwarfed by the hundreds of thousands to millions of stable organic molecules of similar size that are not nucleotides.

The idea that suitable nucleotides might nonetheless form draws inspiration from a well-known experiment published in 1953 by Stanley L. Miller. He applied a spark discharge to a mixture of simple gases that were then thought to represent the atmosphere of the early Earth and saw that amino acids formed. Amino acids have also been identified in the Murchison meteorite, which fell in Australia in 1969. Nature has apparently been generous in providing a supply of these particular building blocks. By extrapolation of these results, some writers have presumed that *all* life's building blocks could be formed with ease in Miller-type experiments and were present in meteorites. This is not the case.

Amino acids, such as those produced in experiments like Miller's, are far less complex than nucleotides. Their defining features are an amino group (a nitrogen and two hydrogens) and a carboxylic acid group (a carbon, two oxygens and a hydrogen), both attached to the same carbon. The simplest of the 20 amino acids used to build natural proteins contains only two carbon atoms. Seventeen of the set contain six or fewer carbons. The amino acids and other substances that were prominent in the Miller experiment contained two and three carbon atoms. In contrast, no nucleotides of any kind have been reported as products of spark-discharge experiments or in studies of meteorites. Apparently, inanimate nature has a bias toward the formation of molecules made of fewer rather than greater numbers of carbon atoms and thus shows no partiality in favor of creating the nucleotides required by our kind of life.

To rescue the RNA-first concept from this otherwise lethal defect, its advocates have created a discipline called prebiotic synthesis. They have attempted to show that RNA and its components can be prepared in their laboratories in a sequence of carefully controlled reactions, using what they consider to be relevant conditions and starting materials.

The Web version of this article, available at www.sciam.com/ontheweb, goes into more detail about the shortcomings of prebiotic synthesis research. The problems bring the following analogy to mind. Consider a golfer who, having played a ball through an 18-hole course, then assumes that the ball could also play itself around the course in his absence. He had demonstrated the possibility of the event; it was only necessary to presume that some combination of natural forces (earthquakes, winds, tornadoes, and floods, for example) could produce the same result, given enough time. No physical law need be broken for spontaneous RNA formation to happen, but the chances against it are immense.

Some chemists have suggested that a simpler replicator molecule similar to RNA arose first and governed life in a "pre-RNA world" (Fig. 9.1). Presumably this first replicator would also have the catalytic capabilities of RNA. Because no trace of this hypothetical primal replicator and catalyst has been recognized so far in modern biology, RNA must have completely taken over all its functions at some point after its emergence.

Yet, even if nature could have provided a primordial soup of suitable building blocks, whether nucleotides or a simpler substitute, their spontaneous assembly into a replicator involves implausibilities that dwarf those required for the preparation of the soup. Let us presume that the soup of building blocks has somehow been assembled, under conditions that favor their connection into chains. They would be accompanied by hordes of defective units, the inclusion of which in a nascent chain would ruin its ability to act as a replicator. The simplest kind of flawed unit would have only one "arm" available for connection to a building block, rather than the two needed to support further growth of the chain.

An indifferent nature would theoretically combine units at random, producing an immense variety of short, terminated chains, rather than the much longer one of uniform backbone geometry needed to support

replicator and catalytic functions. The probability of this latter process succeeding is so vanishingly small that its happening even once anywhere in the visible universe would count as a piece of exceptionally good luck.

LIFE WITH SMALL MOLECULES

Nobel Laureate Christian de Duve has called for "a rejection of improbabilities so incommensurably high that they can only be called miracles, phenomena that fall outside the scope of scientific inquiry." DNA, RNA, proteins, and other elaborate large molecules must then be set aside as participants in the origin of life. Inanimate nature instead provides us with a variety of mixtures of small molecules with which to work.

Fortunately, an alternative group of theories that can employ these materials has existed for decades. The theories use a thermodynamic, rather than a genetic, definition of life, under a scheme put forth by Sagan in the *Encyclopedia Britannica*: a localized region that increases in order (decreases in entropy) through cycles driven by an energy flow would be considered alive. This small-molecule approach is rooted in the ideas of Soviet biochemist Alexander Oparin. Origin-of-life proposals of this type differ in specific details; here I will list five common requirements (and add some ideas of my own) (Fig. 9.2).

A boundary is needed to separate life from non-life

Life is distinguished by its great degree of organization, yet the second law of thermodynamics requires that the universe move in a direction in which disorder, or entropy, increases. A loophole, however, allows entropy to decrease in a limited area, provided that a greater increase occurs outside the area. When living cells grow and multiply, they convert chemical energy or radiation to heat. The released heat increases the entropy of the environment, compensating for the decrease in living systems. The boundary maintains this division of the world into pockets of life and the nonliving environment in which they must sustain themselves.

Today sophisticated double-layered cell membranes, made of chemicals classified as lipids, separate living cells from their environment. When life began, some natural feature probably served the same

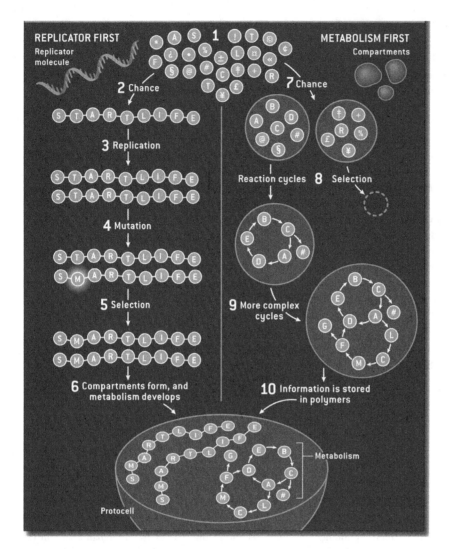

Fig. 9.1. Replicator vs. metabolism

Scientific theories of the origin of life largely fall into two rival camps: replicator-first and metabolism-first. Both models must start from molecules formed by nonbiological chemical processes, represented here by balls labeled with symbols (1). In the replicator-first model, some of these compounds join together in a chain, by chance forming a molecule—perhaps some kind of RNA—capable of reproducing itself (2). The molecule makes many copies of itself (3), sometimes forming mutant versions that are also capable of replicating (4). Mutant replicators that are better adapted to the conditions supplant earlier versions (5). Eventually this evolutionary process must lead to the development of compartments (like cells) and metabolism, in which smaller molecules use energy to perform useful processes (6). Metabolism first starts off with the spontaneous formation of compartments (7). Some compartments contain mixtures of the starting compounds that undergo cycles of reactions (8), which over time become more complicated (9). Finally, the system must make the leap to storing information in polymers (10).

purpose. In support of this idea, David W. Deamer of the University of California, Santa Cruz, has observed membrane like structures in meteorites. Other proposals have suggested natural boundaries not used by life today, such as iron sulfide membranes, rock surfaces (in which electrostatic interactions segregate selected molecules from their environment), small ponds and aerosols.

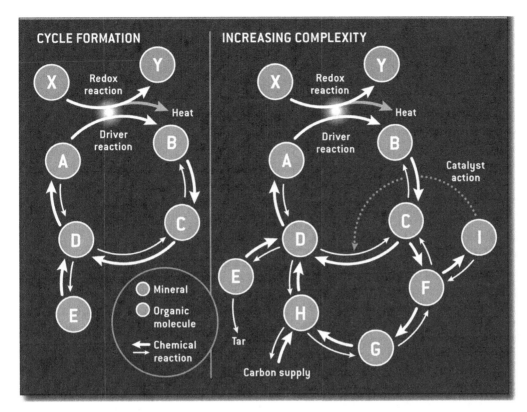

Fig. 9.2. Evolution of chemical networks

The metabolism-first hypothesis requires the formation of a network of chemical reactions that increases in complexity and adapts to changes in the environment. CYCLE FORMATION: An energy source (here the so-called redox reaction converting mineral X to mineral Y) couples to a reaction that converts the organic molecule A to molecule B. Further reactions (B to C, C to D....) form a cycle back to A. Reactions involving molecular species outside the cycle (E) will tend to draw more material into the cycle.

INCREASING COMPLEXITY: If a change in conditions inhibits a reaction in the cycle (for example, C to D), then other paths can be explored. Here a bypass has been found by which C is converted to D through intermediates F, G and H. Another solution would be the incorporation into the reaction network of a catalyst (I) whose action (*dotted line*) unblocks the C to D transformation. To survive, the evolving network must draw in carbon-containing materials from the environment more rapidly than it loses them by diffusion and side reactions, such as the formation of tars that settle out of the solution.

An energy source is needed to drive the organization process

We consume carbohydrates and fats, combining them with oxygen that we inhale, to keep ourselves alive. Microorganisms are more versatile and can use minerals in place of the food or the oxygen. In either case, the transformations that are involved are called redox reactions. They entail the transfer of electrons from an electron-rich (or reduced) substance to an electron-poor (or oxidized) one. Plants can capture solar energy directly and adapt it for the functions of life. Other forms of energy are used by cells in specialized circumstances—for example, differences in acidity on opposite sides of a membrane. Yet others, such as radioactivity and abrupt temperature differences, might be used by life elsewhere in the universe.

A coupling mechanism must link the release of energy to the organization process that produces and sustains life

The release of energy does not necessarily produce a useful result. Chemical energy is released when gasoline is burned within the cylinders of an automobile, but the vehicle will not move unless that energy is used to turn the wheels. A mechanical connection, or coupling, is required. Every day, in our own cells, each of us degrades pounds of a nucleotide called ATP. The energy released by this reaction serves to drive processes necessary for our biochemistry that would otherwise proceed too slowly or not at all. Linkage is achieved when the reactions share a common intermediate, and the process is sped up by the intervention of an enzyme. One assumption of the small-molecule approach is that coupled reactions and primitive catalysts sufficient to get life started exist in nature.

A chemical network must be formed to permit adaptation and evolution

We come now to the heart of the matter. Imagine, for example, that an energetically favorable redox reaction of a mineral drives the conversion of an organic chemical, A, to another one, B, within a compartment. I call this key transformation a driver reaction, because it serves as the engine that mobilizes the organization process. If B simply reconverts back to A or escapes from the compartment, we would not be on a path that leads to increased organization. In contrast, if a multistep chemical pathway—say, B to C to D to A—reconverts B to A, then the steps in that circular process (or cycle) would be favored to continue operating because they replenish the supply of A, allowing the continuing useful discharge of energy by the mineral reaction (Fig. 9.2).

Branch reactions will occur as well, such as molecules converting back and forth between D and another chemical, E, that lies outside the ABCD cycle. Because the cycle is driven, the E-to-D reaction is favored, moving material into the cycle and maximizing the energy release that accompanies the driver reaction.

The cycle could also adapt to changing circumstances. As a child, I was fascinated by the way in which water, released from a leaky hydrant, would find a path downhill to the nearest sewer. If falling leaves or dropped refuse blocked that path, the water would back up until another route was found around the obstacle. In the same way, if a change in the acidity or in some other environmental circumstance should hinder a step in the pathway from B to A, material would back up until another route was found. Additional changes of this type would convert the original cycle into a network. This trial-and-

error exploration of the chemical "landscape" might also turn up compounds that could catalyze important steps in the cycle, increasing the efficiency with which the network used the energy source.

The network must grow and reproduce

To survive and grow, the network must gain material faster than it loses it. Diffusion of network materials out of the compartment into the external world is favored by entropy and will occur to some extent. Some side reactions may produce gases, which escape, or form tars, which will drop out of solution. If these processes together should exceed the rate at which the network gains material, then it would be extinguished. Exhaustion of the external fuel would have the same effect. We can imagine, on the early earth, a situation where many start-ups of this type occur, involving many alternative driver reactions and external energy sources. Finally, a particularly hardy one would take root and sustain itself.

A system of reproduction must eventually develop. If our network is housed in a lipid membrane, physical forces may split it after it has grown enough. (Freeman Dyson of the Institute for Advanced Study in Princeton, NJ has described such a system as a "garbage bag world" in contrast to the "neat and beautiful scene" of the RNA world.) A system that functions in a compartment within a rock may overflow into adjacent compartments. Whatever the mechanism may be, this dispersal into separated units protects the system from total extinction by a local destructive event. Once independent units were established, they could evolve in different ways and compete with one another for raw materials; we would have made the transition from life that emerges from non-living matter through the action of an available energy source to life that adapts to its environment by Darwinian evolution.

CHANGING THE PARADIGM

Systems of the type I have described usually have been classified under the heading "metabolism first," which implies that they do not contain a mechanism for heredity. In other words, they contain no obvious molecule or structure that allows the information stored in them (their heredity) to be duplicated and passed on to their descendants. Yet a collection of small items holds the same information as a list that describes the items. For example, my wife gives me a shopping list for the supermarket; the collection of grocery items that I return with contains the same information as the list. Doron Lancet of the Weizmann Institute of Science in Rehovot, Israel, has given the name "compositional genome" to heredity stored in small molecules, rather than a list such as DNA or RNA.

The small-molecule approach to the origin of life makes several demands on nature (a compartment, an external energy supply, a driver reaction coupled to that supply, a chemical network that includes that reaction, and a simple mechanism of reproduction). These requirements are general in nature, however, and are immensely more probable than the elaborate multistep pathways needed to form a molecule that is a replicator.

Over the years, many theoretical papers have advanced particular metabolism-first schemes, but relatively little experimental work has been presented in support of them. In those cases where experiments have been published, they have usually served to demonstrate the plausibility of individual steps in a proposed cycle. The greatest amount of new data has perhaps come from Günter Wächtershäuser and his colleagues at Munich Technical University. They have demonstrated parts of a cycle involving the combination and separation of amino acids in the presence of metal sulfide catalysts. The energetic driving force for the transformations is supplied by the oxidation of carbon monoxide to carbon dioxide. The researchers have not yet demonstrated the operation of a complete cycle or its ability to sustain itself and undergo further evolution. A "smoking gun" experiment displaying those three features is needed to establish the validity of the small-molecule approach.

The principal initial task is the identification of candidate driver reactions—small-molecule transformations (A to B in the preceding example) that are coupled to an abundant external energy source (such as the oxidation of carbon monoxide or a mineral). Once a plausible driver reaction has been identified, there should be no need to specify the rest of the system in advance. The selected components (including the energy source), plus a mixture of other small

molecules normally produced by natural processes (and likely to have been abundant on the early Earth), could be combined in a suitable reaction vessel (Fig. 9.3). If an evolving network were established, we would expect the concentration of the participants in the network to increase and alter with time. New catalysts that increased the rate of key reactions might appear, whereas irrelevant materials would decrease in quantity. The reactor would need an input device (to allow replenishment of the energy supply and raw materials) and an outlet (to permit removal of waste products and chemicals that were not part of the network).

In such experiments, failures would be easily identified. The energy might be dissipated without producing any significant changes in the concentrations of the other chemicals, or the chemicals might be converted to a tar, which would clog the apparatus. A success might demonstrate the initial steps on the road to life. These steps need not duplicate those that took place on the early earth. It is more important that the general principle be demonstrated and made available for further investigation. Many potential paths to life may exist, with the choice dictated by the local environment.

An understanding of the initial steps leading to life would not reveal the specific events that led to the familiar DNA–RNA–protein-based organisms of today. Still, because we know that evolution does not anticipate future events, we can presume that nucleotides first appeared in metabolism to serve some other purpose, perhaps as catalysts or as containers for the storage of chemical energy (the nucleotide ATP continues to serve this function today). Some chance event or circumstance may have led to the connection of nucleotides to form RNA. The most obvious function of modern RNA is to serve as a structural element that assists in the formation of bonds between amino acids in the synthesis of proteins. The first RNAs may have served the same purpose, but without any preference for specific amino acids. Many further steps in evolution would be needed to "invent" the elaborate mechanisms for replication and specific protein synthesis that we observe in life today.

If the general small-molecule paradigm were confirmed, then our expectations of the place of life in the universe would change. A highly improbable start for life, as in the RNA-first scenario, implies a universe in

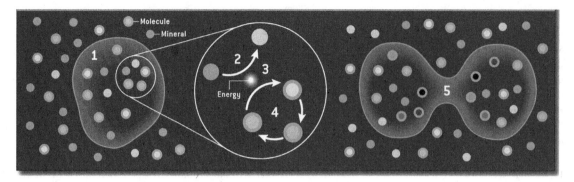

Fig. 9.3. Five requirements for metabolism first
At least five processes must occur for small molecules to achieve a kind of life—here defined as the creation of greater order in localized regions by chemical cycles driven by an energy flow. First, something must create a boundary to separate the living region from the nonliving environment (1). A source of energy must be available, here depicted as a mineral undergoing a heat-producing reaction (2). The released energy must drive a chemical reaction (3). A network of chemical reactions must form and increase in complexity to permit adaptation and evolution (4). Finally, the network of reactions must draw material into itself faster than it loses material, and the compartments must reproduce (5). No information-storing molecule (such as RNA or DNA) is required; heredity is stored in the identity and concentration of the compounds in the network.

which we are alone. In the words of biochemist Jacques Monod, "the universe was not pregnant with life nor the biosphere with man. Our number came up in the Monte Carlo game."

The small-molecule alternative, however, is in harmony with the views of biologist Stuart Kauffman: "If this is all true, life is vastly more probable than we have supposed. Not only are we at home in the universe, but we are far more likely to share it with as yet unknown companions."

NOTES

This chapter originally appeared in the June issue of *Scientific American* (2007), 47–53.

10 · Are the different hypotheses on the emergence of life as different as they seem?

IRIS FRY

INTRODUCTION

The scientific field devoted to the study of the origin of life is characterized by its high degree of controversy. Disagreement among the active participants in research seems to run very deep and relate to the most basic features of the various models proposed. This probably stems from the relative novelty of the field, from its interdisciplinary nature which brings together scientists who use different methodological and theoretical categories, and above all, from the inevitable speculative dimension of the subject. Despite this pronounced division, there is a most basic common denominator which binds together all researchers of the emergence of life. This common element, which I will coin "the continuity thesis," is the assumption that there is no unbridgeable gap between inorganic matter and living systems, and that *under suitable physical conditions the emergence of life is highly probable*. It is the adoption of the "continuity thesis," concerning the philosophical dimension of life-matter relationship, which has turned the origin of life into a legitimate scientific question, and which constitutes a necessary condition for any scientific research in this domain.

The "thesis" is reflected in the different researchers' attempts to base the emergence of the biological features of self-organized systems upon universal physical principles and the physico-chemical properties of these systems.[1] I claim that the continuity thesis, which is a *philosophical presupposition*, though strengthened by research in the field, is not derived from it. Evidently, empirical data along with other factors contributed to the historical establishment of the naturalistic world-view, of which the continuity thesis forms a part. However, within this world-view the "thesis" is now a logical, necessary corollary.

My discussion will focus on the "continuity thesis," as assumed and applied currently and during the last several decades within an evolutionary context, and will touch only peripherally on the history of the subject. The crucial question, how to account for both relatively simple physical objects and complex living systems within one theoretical framework, gave rise throughout the ages to two major answers—basically, the two possible answers within a non-evolutionary context. The "mechanical answer," still not aware of the enormous complexity of life, claimed that inanimate and animate matter are essentially the same, and hence the production of the latter from the former requires only mechanical means.[2] The vitalistic answer, which overlapped during most eras with the theological position, regarded life as an absolutely distinct category. The connection between it and inorganic matter could be achieved only through the intervention of some transcendent agent or principle (Toulmin & Goodfield 1982). As regards the problem of the origin of life, these two answers were manifested in the different explanations provided for the doctrine of spontaneous generation, and in the controversies on this issue starting in the seventeenth and eighteenth centuries (Farley 1974).[3]

Alexander Oparin, in his pioneering studies on the origin of life starting in the 1920s, was among the first to suggest a third alternative, which, based on the evolutionary paradigm, rejected the traditional mechanistic as well as the vitalistic answer. He stated that "the development of life was an integral part of the general development of matter" (Oparin 1972, p. vii), emphasizing, nevertheless, the unique features of the organization of biological systems (Oparin 1961 [1938], p. 25). Following Oparin's studies, the continuity thesis, which embodied the logic of the evolutionary

The Nature of Life, ed. M. A. Bedau and C. E. Cleland. Published by Cambridge University Press. © M. A. Bedau and C. E. Cleland 2010.

paradigm, was adopted in many theoretical texts. Both the gradual development of biological organization through consecutive stages, and the deterministic nature of the process were, and are currently, pointed out (Lehninger 1970, p. 771; Farley 1974, p. 183; Keosian 1974, pp. 285–293; Eigen 1971, p. 519; Wicken 1987; de Duve 1991, p. 211).[4]

My objectives in this chapter are the following: first, to call attention to the existence of the continuity thesis, the proponents of which I group under *"the law camp"*; *second*, to discuss the views expressed by the opponents of the thesis *within the scientific community*, who comprise *"the almost miracle camp"*; *third*, to point out the heuristic advantage in the explicit adoption and use of the continuity thesis by researchers in the origin-of-life field. Let me elaborate. The purpose of this chapter is not only to draw attention to a unifying philosophical principle which underlies the more explicit controversies among the various origin-of-life theories (important as this "unification" is on the background of seemingly irreconcilable differences). As against this purpose, the objection might be raised that, in fact, there is no point in discussing the continuity thesis, since developments in science have made it self-evident and hence trivial. My claim is that quite a few biologists still entertain the view that the emergence of life was the result of a "happy accident." The notion that life arose following a highly improbable, in fact, almost a miraculous event (see Monod 1974, but also, Mayr 1982; Crick 1981), is definitely not obsolete. The question of "chance and necessity," which is still frequently debated within the origin-of-life field and in the wider literature on evolution,[5] is highly relevant to the issues raised by the continuity thesis.

It is indeed trivial to point out that a creationist position on the origin of life is anti-scientific. It is, on the other hand, far from trivial to draw the philosophical connection between such an explicit position and its implicit counterparts within the scientific community. This connection, which I will attempt to portray, has two important aspects. First, based on historical and philosophical analysis it can be shown that the "happy accident" hypothesis precludes the possibility of scientific research of the question, how life emerged. Second, the "almost miracle" point of view is based on the traditional mechanistic view of matter as passive, and as devoid of any capability of self-organization.

This conception thus equates the material processes which led to the emergence of life with the "random shuffling of molecules" (Hoyle & Wickramasinghe 1981, p. 3). The continuity thesis, on the other hand, is translated in the various current origin-of-life theories into specific suggestions of mechanisms of self-organization. This brings me to my third objective. Though sharing the common notion that we are not dealing here with a mysterious gap, the different theories do differ in emphasis and particularly in awareness of the significance of the continuity thesis. I claim, relying on a few scenarios suggested in the field, that a strong, self-aware adoption of the continuity thesis serves as an incentive to new research ideas.

As a last introductory remark, a general caveat should be added here. My discussion of certain aspects in origin-of-life models will not deal with their plausibility or validity, but rather with the philosophical presuppositions that are instrumental in their shaping. Neither do I presume to give a detailed, up to date picture of the whole scene, but rather to indicate the main philosophical trends. It will thus be instructive to start by discussing a few significant differences between philosophical and empirical considerations.

ON THE "CONTINUITY THESIS" AND VARIOUS "PRINCIPLES OF CONTINUITY"

Distinction should be made between the *"continuity thesis"* used in this paper and other references in the literature on the origin of life to some kind of *"principle of continuity."* Put shortly, the different "principles of continuity" are all methodological rules which already presuppose the philosophical assumption of the "continuity thesis." I claim that the continuity thesis is an epistemologically necessary condition which makes research on the origin of life possible. As such, it forms a *negative heuristic*, eliminating from the domain of possible scientific research in this field any theory which does not share its assumptions. It can act, however, also as a positive guidance for research—as a *positive heuristic*.[6] In this capacity, the thesis gives rise to methodological and empirical claims concerning specific scenarios and mechanisms of the emergence of life, among them different versions of "the principle of continuity."

I will mention now briefly a few of the "principles of continuity" to be found in the literature. Back in

1968, Leslie Orgel declared himself to be guided by a "Principle of Continuity which requires that each stage in evolution develops continuously from the previous one." Orgel made it clear that what he had in mind was a methodological principle, the need for which stems, according to him, from the fact that "it is very difficult to see how a totally different biological organization could have undergone a continuous transition to a nucleic acid–protein system with which we are familiar" (Orgel 1968). Quoting Orgel's formulation of the Principle of Continuity, Weiner and Maizels rely on the same methodological assumption when they put forward their "molecular fossils" notion (Weiner & Maizels 1991). In fact, this notion serves to explain Orgel's claim that certain features of the contemporary genetic system emerged very early in biological evolution (Orgel 1968). A molecular fossil is any molecule or function which became "frozen in time" because its change would entail an impossibly large number of simultaneous changes in the system. Thus, Weiner and Maizels conclude that molecular fossils "give us clues about the origin and history of life" (Weiner & Maizels 1991, p. 53).

Another expression of the same methodological rule is to be found in the notion of "biochemical orthogenesis," originally suggested in 1949 by Marcel Florkin, one of the pioneers of comparative biochemistry and molecular evolution. This is pointed out by Christian de Duve in his *Blueprint for a Cell* where, based on "biochemical orthogenesis," he advances the idea that primitive peptide catalysts were responsible at a very early stage of biogenesis for the gradual development of protometabolism (de Duve 1991, p. 135). The principle of continuity is discussed in considerable length by Harold Morowitz, especially in his *Beginnings of Cellular Life* (Morowitz 1992). It is in this discussion that we can discern the connection and, yet, the differences between the principle of continuity and the philosophical "continuity thesis." Morowitz emphasizes the methodological nature of "his" principle which states "that for any postulated stage in biogenesis there must be a continuous path backward to the prebiotic state of the Earth and forward to modern organisms" (p. 27). He relies positively on the principle of continuity in order to probe deeper into prebiotic processes based on our knowledge of contemporary cells. In addition, he uses the principle negatively, as a critique of other theories.

As is to be expected, all the different versions of the principle of continuity enumerated here presuppose unquestionably the "continuity thesis" as defined in this chapter. Yet, the philosophical and the methodological do not completely overlap. This is, for example, the case of the disagreement between Morowitz and Cairns-Smith. On the basis of his principle of continuity, Morowitz rejects the clay theory of Cairns-Smith and the theory of pyrite surface suggested by Wächtershauser. Since no clay or pyrite structures or vestiges of them exist in contemporary cells, their introduction, he says, which violates the principle of continuity, "needlessly complicates origins of life theory" (pp. 90–91). On the other hand, Cairns-Smith, *adhering to the continuity thesis*, claims that he was led to suggest an unconventional scenario based on inorganic clay chemistry, since the gap between "the simplest conceivable version of organisms as we know them, and components that the Earth might reasonably have been able to generate ... is enormous," and since nevertheless, Cairns-Smith believes, it was bridged through natural causes (Cairns-Smith 1985, p. 4, p. 8).

The various principles of continuity might indeed push forward the experimental investigation of the emergence of life; as such, they do represent the heuristic advantage of the continuity thesis. However, the *decision to* adopt the continuity thesis is a *philosophical* one, and does not depend on methodological considerations. Moreover, this decision does not depend on the success of a *specific* experimental program, nor can it be revoked on the basis of its failure.

MOROWITZ'S FORMULATION OF THE CONTINUITY THESIS

Morowitz does formulate his version of the "continuity thesis" (not identified by any name), which he rightly presents as one of three possible answers to the question, *how life emerged?* The first possible answer, says Morowitz, is that the origin of life is "the result of a divine act that lies beyond the laws of science." According to the second answer, the origin of life is understandable within the laws of physics. However, "since it is the result of so many random events, the nature of life is essentially unique" (Morowitz 1992, p. 2). The third answer, which corresponds to the continuity thesis presented here, states that "the origin of life is a deterministic event, the result of the operation of the laws of nature on a physical chemical system of a certain type. This system evolves in time, is governed by physical principles, and eventually gives rise to living forms." Morowitz indeed points out that while the first two approaches lie outside experimental science, the third option is a necessary condition for a scientific study of the origin of life. Such a study, he makes clear, requires that the event under investigation be "largely deterministic within the scope of ordinary physics and chemistry" (p. 3).

Morowitz's formulation of the third answer touches on the most essential element of the emergence-of-life process indicated by the continuity thesis—its *deterministic nature*. Though, as put by Morowitz, the details of the evolution of living systems from physico-chemical systems need not be totally deterministic in every aspect, "the overall behavior follows in a predictable way" (p. 3). It is similarly important to note that the essential characteristic of the second option, the seemingly within-natural-processes answer, is its emphasis on the *random character* of life's emergence. This option necessarily leads to the conclusion that the emergence of life was a highly improbable event, in fact, a singularity. As will be pointed out shortly, there is a strong philosophical connection between the theological and the "random" hypotheses—both come to view the emergence of life as a sort of miracle. This connection is referred to by Cairns-Smith, who says that "there is a temptation, in any case, to suppose that if the origin of life was not actually supernatural it was at least some very extraordinary event, an event of low probability, a statistical leap across a great divide. That way a trace of magic can be held onto" (Cairns-Smith 1985, p. 2).

ON PANSPERMIA THEORIES AND THE CONTINUITY THESIS

Screening the six possible combinations of answers to the questions, *how and where life emerged*, Morowitz reaches the conclusion that only one combination, that of a deterministic event on Earth, is "of major interest from the viewpoint of present-day experimental science" (Morowitz 1992, p. 4). Because of his lack of distinction between empirical and philosophical considerations, Morowitz misses some important differences. Most researchers today opt for terrestrial or extraterrestrial scenarios, based on empirical arguments

which have nothing to do with their position on the continuity thesis. On the other hand, several scientists who advocated panspermia theories at the turn of the century presupposed the *eternal existence of life in the cosmos* and, hence, denied the continuity thesis (for a thorough treatment of the history of the subject, see Kamminga 1982).

Distinction should be made between *empirical arguments* that deal with the question of *the actual probability of appropriate physical conditions* for the emergence of life on Earth or elsewhere in the universe, and *philosophical arguments* dealing with *the probability of the emergence of life, whatever the physical conditions are*. The formal structure of the two kinds of arguments is similar. However, while the empirical contentions rest on astrophysical and geophysical data, the philosophical ones are based on a general *a priori* conception of the nature of matter and life. As will be pointed out later, the conception of matter as passive and devoid of any possibility of self-organization leads to the conclusion that the probability of life's emergence is extremely small. As for the empirical arguments, some people conclude, based on empirical assumptions and data, that appropriate biogenic conditions probably arose only on Earth. The demands of carbon-chemistry, it is claimed, put stringent requirements both on the type of planet that can support life, and on the environment in which early chemical development can take place. Hence, there are many critical factors that might reduce the probability of life elsewhere (Rood & Trefil 1981).

On the other hand, following several new developments, there is a renewed interest in panspermia scenarios that question the probability of appropriate biogenic conditions on Earth. Recent data indicate that the "time window" during which life could have formed on Earth was much shorter than previously assumed—only about half a billion years at most (Joyce 1989, p. 219). While these data prompt some researchers to suggest that the processes of chemical evolution and the origin of terrestrial life happened very quickly (Joyce 1989, p. 219; Morowitz 1992, p. 32), others reject altogether the possibility of a terrestrial origin of life. They doubt whether the standard model of a chemical evolution initiated by Oparin and Haldane is valid, and question the possibility that such a short time interval would be sufficient for the prebiological stage. The conclusion is that at least certain stages of the process by which life originated occurred in space (see Goldanskii & Kuzmin 1989; Gribbin 1993).

It is interesting to discern both empirical and philosophical considerations mixed up in Francis Crick's ideas on the emergence of life. In some of his writings on the subject Crick deals with the difficult empirical question, how complex organic molecules could have been synthesized on the primitive Earth, assuming that the primordial atmosphere was not as reducing as was assumed before. Thus, Crick raises the possibility that "planets elsewhere in the universe may have had more reducing atmospheres, and thus have on them a more favorable prebiotic soup" (Crick 1981, p. 79). Yet, it appears that Crick is not ready to consider a probable scenario for the emergence of life even in those sites where the environmental conditions seem more promising. An honest man, he claims, armed with all the knowledge available to us now, could only state that "in some sense, the origin of life appears at the moment to be *almost a miracle*, so many are the conditions which would have had to been satisfied to get it going" (p. 88, emphasis mine). The basic philosophical assumption underlying the "almost miracle" notion becomes apparent once we learn that, for Crick, the emergence of life was "*a happy accident*," which only the passage of millions of years helped to bring about (p. 39).

First in 1973 with Leslie Orgel, and then in 1981, in his book *Life Itself*, Crick presented his "*directed panspermia*" theory on the emergence of life, according to which, life, in the form of anerobic bacteria, was sent to the Earth in a spaceship by an advanced civilization, just when conditions for life on our planet became appropriate (Crick & Orgel 1973; Crick 1981). The directed panspermia hypothesis was supposed to call the attention of the scientific community to the enormous difficulties faced by any current conventional scenario for the emergence of life. However, combined with Crick's idea that the origin of life was a very extraordinary event, it seems that Crick's story about bacteria intentionally sent to Earth in an unmanned spacecraft indeed holds onto "a trace of magic" referred to by Cairns-Smith. Crick's directed panspermia theory is definitely "anti-experimental," to use Morowitz's terms. However, from a philosophical perspective, its main drawback is not its panspermic dimension but rather its "randomistic character"— and hence, its inability to deal scientifically with the

question, how biological organization emerged. Relegating the question to another planet obviously does not solve the problem, which cannot be solved as long as the "happy accident" notion is entertained.

THE "ALMOST MIRACLE CAMP"

An exposition of the philosophical rival of the continuity thesis will help to focus our attention on the most relevant issues involved in the controversy. While the continuity thesis assumes the stepwise, probable emergence of biological organization, the "almost miracle camp" regards the emergence of this organization as a real puzzle. We are dealing here, says Jacques Monod, with one of the most pronounced representatives of this position, not so much with a "problem" but with "a veritable enigma" (Monod 1974, p. 135). The assumption that unites Monod and other members of the miracle camp is that the physical "mechanism" responsible for the emergence of a self-organized system is, in fact, its *chance association from its constituents*. The enigma, as will be shown shortly, arises because the probability of this chance association is virtually zero. Monod's "chance hypothesis" applies already to the formation of the macromolecules that constitute the living cell. The first specific sequence of amino acids that arose on earth, thinks Monod, was the result of a purely random process (p. 96). Monod reaches this conclusion based on the examination of sequences of amino acids (and, by implication, the sequences of nucleotides in the genes that encode these proteins) in existing proteins.

The sequences are *entirely random* in the sense that "even knowing the exact order of 199 residues in a protein containing 200, it would be impossible to formulate any rule, theoretical or empirical, enabling us to predict the nature of one residue not yet verified by analysis" (p. 96). The existent amino-acid sequences in today's proteins are supposed to reflect the original formation of the first protein on Earth. The random law of assembly means, according to Monod, that there were no physico-chemical constraints whatsoever that directed the formation of the first protein—"*no regularity, special features, or restrictive characteristic*" (p. 96). Hence, "random" in the sense used here means first, the *noninstructed* (noninformed) formation of the first proteins, and second, *the equal chances for each amino acid that participated in the synthesis.*[7] Monod emphasizes, in addition to his "molecular roulette" conception, that each specific sequence is reproduced in each organism by a highly accurate mechanism which guarantees the invariance of the structure, and on the basis of this invariance arises in each organism, through the translation of the linear structure of proteins into the "teleonomic inter-actions of the globular structure," the order and func-tionality which characterize the living system. Monod thus concludes that "a *totally* blind process can by definition lead to anything" (p. 96).

Monod is not bothered (unjustifiably, as we'll presently see) with the question of the prebiotic chance formation of macromolecules—proteins and nucleic acids. He is struck by the real puzzle of the emergence of a primitive cell—of a primitive linkage between proteins and nucleic acids. The *a priori* probability of the *random* formation of such as system (random for-mation being the only feasible physical process he is ready to consider), he acknowledges, was "virtually zero" (p. 136). However, not only is the probability of the random emergence of a single-cell organism com-parable, as was rightly claimed by Fred Hoyle, to the assemblage of a 747 by a tornado whirling through a junkyard, but even the formation of a particular macromolecule poses a deep problem. The hypothesis that a primitive polypeptide or polynucleotide arose "by chance" assumes that all the possible alternative sequences are physically equivalent.

The reciprocal of the number of all possible com-binations gives the probability of the appearance of a specific, particular sequence. Calculations for both polypeptides and polynucleotides prove that the number of possible variants is beyond our realistic grasp. In the case of a polymer which corresponds to a single gene containing 1000 nucleotides—the number of possible random variants produced is 10,602 (Eigen 1992, p. 10). The chance construction of the smallest catalytically active protein molecule that consists of at least 100 amino acid residues involves more than 10,130 variants (Küppers 1990, p. 60). Based on these figures, the chance formation of a specific polymer is beyond belief. These facts have led researchers in the origin-of-life field to suggest alternatives, to be discussed later, to the "chance formation" of macromolecules in prebiotic conditions. As for Monod's puzzle—it seems that the entire matter of the universe will not suffice for the purpose of the random construction of the most primitive organism (for some relevant calculations, see Eigen 1993).

Organisms, however, do exist on Earth. Not willing to consider any other physical option than chance, Monod is, thus, left with a real enigma. Under his assumptions he has to envisage the whole process of evolution as a huge lottery. Luckily for us, he says, "our number came up in the Monte Carlo game" (Monod 1974, p. 137).[8] However, based on the "virtually zero" *a priori* probability of the emergence of life, we are dealing here, according to Monod, with a unique and unrepeatable event. We cannot exclude the possibility, he says, that "the decisive event occurred *only once*" (p. 136).

The direct bearing of this conclusion on the pos-sibility of the scientific investigation of the origin of life should be made clear. Monod himself admitted how distasteful to him as a scientist was his "casino" conclusion, for he acknowledged that "science can neither say nor do anything about a unique occurrence. It can only consider events which form a class, whose *a priori* probability, however faint, is definite" (p. 136). Karl Popper, based on his agreement with Monod that life emerged from inanimate matter by an extremely improbable combination of chance circumstances, has to admit that the origin of life becomes "an impene-trable barrier to science and a residue to all attempts to reduce biology to chemistry and physics" (Popper 1974, p. 270; 1982, p. 148). The last point raised by Popper—*the possibility of the reduction of biology to physico-chemistry*—is highly relevant to our discussion. Examination of some of the "virtually-zero-probability" claims reveals their preoccupation with the question, whether life and inanimate matter are continuous, and with the question of the autonomy of biology.

One of the most notable examples is to be found in Ernst Mayr's views. Mayr, the renown evolutionist, is one of the vocal proponents of the "autonomy of biol-ogy" tenet (see Mayr 1982). His distinction between physical and biological systems, and hence between the practice of biology and the physical sciences, as well as his insistence on the scientific legitimacy of the bio-logical way of doing science, are fully justified. How-ever, Mayr misapplies the "autonomy of biology" notion when he asserts that "a full realization of the *near impossibility* of an origin of life brings home the point how improbable this event was" (p. 584). Such claims, in fact, raise a barrier between the physical world and the emergence of biological organization.

Association between notions that the subject of biology is the "whole living unit" and that a gap exists between physics and biology, tend to appear in discussions on the improbability of the origin of life (see Mora 1965, p. 46, p. 57). Opinions in this vein create the impression that the distinction between the "law camp" and the "almost miracle camp" is, in fact, the distinction between "reductionists" and "holists" who insist on treating the organism as a "whole living unit." This impression is strengthened not only by the "autonomists" themselves but also by some of their opponents. According to Manfred Eigen, the two philosophical options for the explanation of the emergence of biological information are reduction, or the assumption of *vis vitalis* (Eigen 1992, p. 122). However, at the same time, Eigen does claim that genetic information represents a quality that "far transcends chemistry" (p. 124). There is clearly a need for a more rigorous definition of "reduction."

The framework of this chapter does not allow me to get into a serious discussion of the intricate subject of reductionism. The point I would like to make here, stated briefly, is that though living systems, whose structure and function are based on information, manifest novel features not to be found in unorganized systems, their emergence and properties can be explained on the basis of physical principles. I thus believe that the dichotomy between "reductionistic" and "organismic" attitudes can be transcended. The claim, that it is possible to offer a scientific evaluation of the evolution of life which is organismic in nature, and at the same time based on universal physical principles, is one of the central themes of the new thermodynamic evolutionary paradigm (see, for example, Weber *et al.* 1988).

The basic philosophical issue here is the question already posed by Kant—*can we explain, on causal materialistic terms, the production of an organized whole from its parts*, taking into account that in an organized whole, parts and whole are reciprocally dependent. That this is indeed the issue becomes clear when Kant's basic notion of the organism as a vicious circle—"a cause and effect of itself" is compared to Monod's and Mayr's real source of puzzlement and despair of solution—the circular interdependence of proteins and nucleic acids in the organism (Monod 1974, p. 135; Mayr 1982, p. 583). According to Kant, mechanistic, materialistic explanations can account

only for the production of an *aggregate from its individual parts*. Our discursive reason can fathom the formation of an integrated whole only as a result of external design (Kant 1987 [1790], pp. 288–294). This poses a difficult dilemma as regards biological organization, in which evidently every part and process makes sense only in terms of the functional whole. However, in distinction from a designed artifact, the different parts of an organism produce each other seemingly according to an *inner plan of design* (p. 253). Thus, Kant concludes that, while we can obviously account for the production of artifacts, we cannot give a causal explanation of the origin of biological organization.[9]

This last conclusion is based not only on Kant's conception of the structure of our understanding. What is highly relevant to current views on the question of the origin of life is Kant's notion of *matter as basically inert, as being by definition lifeless* (p. 276). Within this conception, structures can form through mechanical associations, while the idea of physical principles of self-organization, which involves a different evaluation of whole-parts relationship, is a contradiction in terms. This is why the very possibility of the spontaneous production of life from inorganic matter seems to Kant absurd (Kant 1987, p. 311). (Monod and Mayr, on their part, wonder how and when did the circle of protein and nucleic acids close, evaluating the probability of this event as "virtually zero".) Following these considerations, Kant addresses the philosophical enigma: How to conduct a science of biology while acknowledging that organization cannot be formed without prior organization? In distinction from physics, the science of organisms is limited, he says. It has to assume an *original organization as given*, and to pursue scientific, mechanistic procedures from this starting point (p. 311).[10]

Kant's philosophical evaluation of the status of biological organization reveals the connection between the claim for the autonomy of biology and the rejection or suspension of the scientific study of the emergence of organization. Kant declared categorically that never could another Newton arise that would explain in terms of natural laws the production of "a mere blade of grass" (p. 282). Biologists and chemists who claim today that the origin of life borders on the miraculous, probably would not like to share this prophecy. Yet, in actuality, they suspend the scientific study of the origin of biological organization and create a barrier between biological evolution and the preceding stages of evolution, as well as between physics and biology.

The position of Monod, Mayr, and others who assume that life arose as a "happy accident" (Crick 1981), entails a serious philosophical choice. This is made clear by the following argument presented by Christian de Duve. The *a priori* answer to the question how life originated, he says, is that *"unless one adopts a creationist view ...* life arose through the succession of an enormous number of small steps, almost each of which, given the condition at the time, *had a very high probability of happening."* This assumption, he adds, *has to be made* simply because *the alternative amounts to a miracle,* that falls outside the scope of scientific inquiry (de Duve 1991, p. 112, emphasis mine). We encounter here, in an explicit way, the *resolution* to adopt the continuity thesis in order to be able to *choose* the scientific course. "Viewing each step as highly likely, if not bound, to happen under the conditions that prevailed follows from the fact that the number of individual steps must have been very large. Let the probability of each step be even moderately low—say 50%—and the combined probability, which is the product of the individual probabilities, soon reaches levels that border on the miraculous (10^{300} for as few as 1000 steps)"[11] (p. 112).

That the philosophical choice is indeed one between a natural science point of view and a creationist one becomes obvious when the "virtually zero probability" argument is carried to its logical conclusion, as done by the astronomers Fred Hoyle and Chandra Wickramasinghe (Hoyle & Wickramasinghe 1981). Both Monod and Hoyle equate the production of life through natural causes with series of chance events. According to Monod, since (a) life arose through natural causes, (b) natural causes in this case means "no regularity, special feature, or restrictive characteristic" put on the chance assembly of molecules (no physical principles of self-organization), and (c) the probability of chance encounter producing such enormously complex result is virtually zero—Monod's unavoidable conclusion is that science faces an enigma with which it cannot deal.

Hoyle's reasoning, starting from the same assumption follows a different path. Since (a) life springing from natural causes is nothing but "the random shuffling of simple organic molecules," and (b) the chance of biochemical systems being thus

formed is "exceedingly minute, to a point indeed where it is insensibly different from zero" (Hoyle & Wickramasinghe 1981, pp. 2–3), Hoyle and Wickramasinghe reject the chance option, and consequently, the natural-causes avenue altogether, and opt instead for the introduction of purpose into nature. Based on their calculation, they claim that a random shuffling of amino acids would have as little chance as one part in 1,040,000 of producing the original enzymes, thus, "any theory with a probability of being correct that is larger than one part in $10^{40,000}$ must be judged superior to random shuffling. The theory that life was assembled by an intelligence has, we believe, a probability vastly higher than one part in $10^{40,000}$ of being the correct explanation" (pp. 129–130, emphasis mine; see also, Hoyle & Wickramasinghe 1993, p. 2).

On the face of it, nothing could be more opposed than Hoyle's views to Monod's "principle of objectivity" which is "the systematic or axiomatic denial that scientific knowledge can be obtained on the basis of theories that involve, explicitly or not, a teleological principle" (Monod 1974, p. 357). Nothing, it seems, could be stronger anathema to Ernst Mayr, who vehemently denies any association of teleology with the process of evolution (Mayr 1974) than Hoyle's "outrageous notion" of purpose. Yet, ideas have their own, sometimes independent, logic. Regarding, as do Monod and Mayr, the physical appearance of a primordial living entity as highly improbable (as does Crick who sidesteps the issue of the origin of organization through his hypothesis of "directed panspermia"), amounts to questioning the sufficiency of physical principles of self-organization to produce life. This leaves the door open to Hoyle's teleological option.

As already indicated by Kant, in explaining natural phenomena we subject the contingent to the rule of a law. When we lack the appropriate mechanistic laws, he claims, we use teleological explanations in order to make the contingent lawful (Kant 1987 [1790], p. 287). Thus, there is no genuine dichotomy between purpose, or some notion of predestination in evolution, and the option of chance or Monod's "lottery." The two horns of the faulty dilemma—chance and telos—converge philosophically. In addition, both of them imply the end of scientific investigation. My analysis so far substantiates the claim that the *genuine dichotomy* is between *the continuity thesis,* according to which life arose from inanimate matter through probable physical mechanisms of self-organization, and *the "almost miracle" thesis*, which regards the origin of life as a highly improbable event.

THE "LAW CAMP"

The focus of the discussion so far has been on the question whether the origin of life was based on chance events. Our analysis has pointed to the different philosophical positions on this focal point. It was indicated that due to the enormous complexity of the most primitive system considered to be alive, the chance hypothesis amounts to the "improbability hypothesis." Current experimental and theoretical research constitutes an alternative to this position. The philosophical spirit underlying this alternative is well represented in the words of J. D. Bernal, who in response to miracle-camp claims has said: "the question, could life have originated by a chance occurrence of atoms, clearly leads to a negative answer. This answer, combined with the knowledge that life is actually here, leads to the conclusion that some sequences other than chance occurrences must have led to the appearances of life" (Bernal 1965, p. 53). Generally speaking, it might be said that the differences among the various current theories in the origin-of-life field concern different suggestions for "sequences other than chance occurrences" that will account for life's emergence.

Though one of my objectives in this chapter is to draw attention to the role of the continuity thesis as a necessary epistemological condition in *all* emergence-of-life theories, very significant divisions and points of contentions cannot be ignored. Perhaps the deepest division concerns the question, *what constituted a primitive living system*. This division has long historical roots in the different biological traditions that stem from the disciplines of biochemistry and genetics. In 1936 Oparin formulated his classical theory of the origin of life within the tradition of biochemistry based on the central concept of metabolism (Kamminga 1986). The development of modern genetics, based on advances in cytology and the revival of Mendel's laws, brought about the assumption that the genetic functions—replication and mutation—are more fundamental to life than any other biological function. As regards the question of the origin of life, the debate between these two conceptions was formulated in terms of the question, whether life had arisen in the

form of a "living molecule" capable of replication, or in the form of a polymolecular system of a certain amount of structural and functional heterogeneity (Kamminga 1986, p. 7).

Following the discoveries of molecular biology, which introduced new terms into the discussion, the previous debate is paraphrased in the question, which came first, a self-replicating system of comprised nucleic acids or a metabolic system consisting of proteins? This is a "chicken and egg" question since, in existing organisms proteins can form only on the basis of information stored in nucleic acids, and nucleic acids can replicate and be translated only with the aid of proteins. Several current texts in the field discuss the replication versus metabolism, or the nucleic acids versus protein controversy. Known among them is Freeman Dyson's *Origins of life* (1985) which attributes equal significance to replication and metabolism as the characteristics of life, but nevertheless assumes, for various empirical and logical reasons, the emergence of metabolism first and replication second in the course of biogenesis (Dyson 1985). Another line of division, parallel in certain respects to the nucleic-acid–protein division, concerns the question, whether compartmentation into some sort of protocell and the separation of the evolving system from its environment preceded the development of replication (Fox & Dose 1972), or whether the formation of compartments was a relatively late event. As will be pointed out later, there is a growing tendency to think in terms of "compromise solutions" (see, for example, Wicken 1987; de Duve 1991; Joyce 1989, p. 222; see also Dyson 1985, p. 34).

Starting more than 20 years ago and despite severe difficulties, the notion of replication-first is the dominant one in the field. This might be due to historical reasons.[11] In addition, there are strong chemical reasons why many biochemists adopt the notion of an "RNA world," according to which there was a time, before the origin of protein synthesis, when life was based entirely on RNA (for extensive survey of the subject, see Joyce 1989; Gesteland & Atkins 1993). There is no doubt that the discovery in the early 1980s that RNA molecules have enzymatic properties and are thus the only molecules known to function as both genotype and phenotype caused molecular biologists to take the notion of an "RNA world" seriously (Joyce 1989, p. 217).

In order to substantiate my claim about the fundamental role of the continuity thesis in the study of the emergence of life, we should examine whether indeed current theories in the field, each on its own terms, embody the search for "sequences other than chance occurrences" supposed to bridge the gap between inorganic molecules and primordial living systems. Since Manfred Eigen's is one of the most influential theories within the replication-first conception, some of its basic tenets should be examined. Analyzing Eigen's theory in light of the continuity thesis poses an interesting challenge, since in several of his texts, especially the earlier ones, Eigen based his scenario on the preliminary occurrence of random events. Thus, we'll have to clarify the notion of "random events" in this case and see whether this theory can be reconciled with the "law camp."

MANFRED EIGEN'S THEORY OF MOLECULAR SELF-ORGANIZATION

In 1971 Manfred Eigen published an article which dealt with the evolution of biological macromolecules, claiming that this evolution can be explained only on the basis of the prevalence of the Darwinian principle of natural selection, not only in organic evolution, but also in the transition phase between inanimate and animate systems. Self-producing macromolecules such as RNA or DNA, in a suitable environment, said Eigen, exhibit a behavior which might be called Darwinian (Eigen 1971). Following the rather imprecise replication of the first oligonucleotides, a population of closely similar but not identical variant macromolecules—the "*quasispecies*"—is formed. Molecular evolution within the quasispecies, which results in the emergence of optimal self-replication, Eigen claims, is the first necessary step on the way to the formation of a living system (Eigen 1992, p. 126). The following stage in Eigen's model, necessary in order to secure the development of functional linkage between nucleic acids and proteins, involves the integration of several quasispecies into a cooperative system (Eigen & Schuster 1977, p. 563).

In his 1971 paper, Eigen claimed that evolution started from a random matrix. In the beginning, he said, "there must have been a molecular chaos, without any functional aggregation among the immense variety of chemical species. Thus, the self-organization of

matter we associate with the 'origin-of-life' must have started from random events" (Eigen 1971, p. 467). Eigen was criticized for his "randomistic attitude" (see for example, Fox 1984, p. 17), and his model, coined "*the random replicator*," was described as based on the highly improbable, accidental formation of one or more replicating RNA molecules (Shapiro 1986, p. 164; p. 166). Considering these allegations, can Eigen's position be reconciled with the "law camp"? In my view, the answer is positive even as regards Eigen's "randomistic" texts. Moreover, in his later work, there is a clear shift toward emphasis on directedness in the evolution of life and away from the conception of statistical randomness (Eigen 1992, p. 29).

It is my claim that when Eigen's "randomistic" aspect comes to the fore, it is not due to a "chance camp" attitude but rather to a misplaced focus of attention. In many of his texts, Eigen, erroneously I believe, tends to consider the prebiotic phase as non-problematic, and to simply assume that the "right environment" was ready (Eigen & Schuster 1978, p. 346). However, it should be made clear that Eigen's conception of "randomness" is far from "statistical randomness." Unlike Monod, who ruled out the contribution of physicochemical constraints in the formation of prebiotic proteins (Monod 1974, p. 95), Eigen does acknowledge the presence of atomic, molecular, or even supramolecular non-accidental structural constraints (Eigen, 1971, p. 467, note 1; Eigen *et al.* 1981, p. 82). It should be pointed out that the notion of prebiotic physical selection, based on thermodynamic criteria, operating in the production of particular sequences both of RNA and proteins, is considered essential by many investigators (see, for example, de Duve 1991, pp. 141–142). Wicken, for example, enumerates several factors instrumental in these selection processes: homogeneous optical activity, high folding capabilities of certain protein sequences, hydrophobic tension with water as a factor in relative stability of "certain macromolecules and microspheres" (see Wicken 1987, pp. 108–109). The decisive importance of such physical constraints in the prebiotic environment is indicated in one of Eigen's later discussions of the possible sources of nucleic acid and protein chirality (Eigen 1992, p. 35).

In 1971, though postulating a prebiotic molecular chaos, Eigen adds in qualification, that randomness is restricted to functional aspects. (Eigen 1971, p. 467, note 1). In 1981, Eigen does acknowledge the importance of some "functional molecules" to the "chemistry of a prebiotic soup," though clarifying that only self-replication and natural selection could bring about evolution of function (Eigen *et al.* 1981, p. 82). The role of primitive catalysis, mentioned in his previous works, is emphasized in 1992, when Eigen's answer to the question, *how did the first self-reproducing molecules originate?*, speaks of the role of catalysis, albeit weak, which was probably performed by proteins, "no doubt ... the first on the scene" (Eigen 1992, pp. 29–32). Thus, not only does Eigen postulate a prebiotic scene in which various physical and chemical constraints favored certain evolutionary directions, he stresses the importance of substances with catalytic ability—mainly proteinoids, but also ribozymes and metal ions—which could overcome the huge improbability involved in a Monod-like "chance scenario" (pp. 32–33).[12]

Eigen's rejection of a "chance scenario" is even more pronounced when it comes to the question, how did optimal genes—RNA sequences capable of self-replication with relatively high degree of fidelity— arise. *Genetic information, according to Eigen, could not have evolved on the basis of chance.* Rather, the generation of a superior mutant—in terms of copying fidelity, stability and especially replication rate—in a population of self-replicating macromolecules must have been *the result of a determined causal chain of events* (p. 25). This conclusion is reached through the *quantitative and detailed analysis*, not possible in classical genetics, but made available now by molecular biology techniques, of the distribution of sequences in a population of macromolecules in the quasispecies.[13] The ramifications Eigen draws from the quasispecies conception involve a *reformulation of the classical neo-Darwinian point of view* which relies heavily on the completely random character of mutations, and consequently on the entirely opportunistic nature of evolutionary development (see pp. 22–30). The quasispecies theory denies the classical assumption that "each mutant appears with a probability that is independent of whether it is a superior, a neutral or an inferior variant" (p. 23). In the process of molecular evolution, according to Eigen, selection is not aimed at the individual wild type that produces chance mutations in a completely random way. In fact, in a quasispecies the "wild type" never emerges as a single winner, and the object of selective evaluation is rather the quasispecies as a whole, the entire ensemble of coexisting mutants (p. 27).

Eigen's quantitative formulation of the theory of molecular self-organization is based on the concept of *the sequence space*—an imaginary space, in which the number of points is equal to the total number of all possible sequences. In this space, each nucleotide sequence occupies a unique position. The positions are arranged to reflect the sequence kinship between each and every possible variant. The processes of molecular evolution and the generation of more superior mutants can be portrayed as a "walk" through the "fitness, or value topography" of the sequence space. A decisive factor in the structure of the quasispecies has to do with the existence, within any replicating population, of many *neutral mutants*, whose fitness— and their ability to replicate—is only slightly or not at all different from that of the wild type (see Kimura 1983). As a result of the structure of the quasispecies and the fact that the fitness of the entire population is evaluated as a whole, the numbers within the population of specific mutants depend not only on their degree of kinship with the wild type (as they do in the classical model), but on their specific rate of replication and their distribution in the "sequence space" (Eigen 1992, pp. 24–27).

Neutral or nearly neutral mutants are far better represented than others, because of their relatively efficient self-replication. In addition, their relative numbers are affected by the fitness of their neighbors in the sequence space. Not only does an efficient mutant replicate itself, it also arises by the erroneous replication of its efficient neighbors. Thus, most mutants arise in the "mountain regions" of the value landscape, close to the peak of optimization (pp. 27–28). This dynamic leads to a shift in population numbers. Thus, a bias is introduced in the development of the quasispecies as a whole and evolution seems to be directed towards the production of the superior mutant. "*This directedness of the evolution process*," Eigen says, "*is perhaps the clearest expression of the present-day paradigmatic change in the established Darwinian world-picture*" (p. 29).

The paradigmatic change which points to a deterministic element in the evolution of molecular self-organization enables Eigen to substantiate his claim that natural selection is the physical principle of order operating in the emergence of life. Otherwise, within the classical interpretation of natural selection, according to which the production of mutants is completely random, the problem of the emergence of complexity could not be solved (p. 23).[14] The reinterpreted principle of natural selection is the antithesis to the (in all likelihood physically impossible) notion that genetic information was produced through a purely random synthesis (p. 11).

Some features of the following stages in Eigen's model will be considered shortly, however, we might return now to our original question, whether Eigen's ideas may be reconciled with the "law camp." Based on the above presentation of Eigen's view, there is no doubt that the answer is definitely positive. Since Eigen was quite often criticized for his "random replicator" theory, his emphasis on the deterministic aspect of the continuity thesis carries a special weight. It should be remembered here that on several occasions Eigen expressed the view that life is probably a universal phenomenon. "The evolution of life, if it is based on a derivable physical principle [the principle of natural selection], must be considered *an inevitable process* ... not only in principle but also sufficiently probable in a realistic span of time. It requires appropriate environmental conditions (which are not fulfilled everywhere) and their maintenance. These conditions have existed on earth and must still exist on many planets in the universe" (Eigen 1971, p. 519).

ON SOME PROBLEMS IN EIGEN'S THEORY AND THE RNA-WORLD THEORIES

Let us recapitulate: At the beginning of this chapter I have described the *continuity thesis* which grounds the emergence and evolution of life on universal physical principles, claiming that the emergence of life cannot be due to blind chance. I have then stated that one of my purposes is to demonstrate that this thesis is at the basis of the various models suggested in the origin-of-life field. Eigen's theory of the evolution of genetic information in populations of self-replicating macromolecules, despite its seemingly chance-like appearance, proved to be a decisive case in point. It is time now to point out the presence of the continuity thesis in other origin-of-life theories that differ from Eigen's in some of their basic claims.

As already stated, the notion that a self-replicating apparatus based on RNA was the first chemical "world" on the way to life is quite prevalent in the field. Recently, a growing number of origin-of-life

chemists have started to doubt this conception. Though agreeing that RNA has probably played an important role in the very early history of life, they raise the question, what preceded the appearance of RNA self-replication. In a comprehensive review of this problem, Gerald Joyce, a leading contributor to the field, discusses several objections to the "RNA first" notion. The first deals with the obvious advantage proteins have over RNA in terms of catalytic power, even when the discovery of catalytic RNA—of ribozymes—is taken into account (Joyce 1989, p. 218).[15] The second, more serious objection, has to do with the tremendous difficulties, in fact, the unlikelihood of the prebiotic synthesis of RNA (p. 219).

Joyce methodically examines all the possible chemical pathways that could have led to the synthesis of the necessary building blocks of RNA, relying on data provided by geochemistry, prebiotic chemistry and nucleic acid biochemistry. He focuses on the main stumbling blocks in this synthesis (pp. 220–221), and, in addition, mentions several possible cross-inhibitions and isomeric interferences among the various building blocks, (assuming that they could have been formed), that would have made the synthesis of RNA unimaginable. The most reasonable interpretation, Joyce concludes, is that "life did not start with RNA. The RNA world came into existence after many of the problems associated with the prebiotic synthesis and template directed replication of RNA had been solved" (p. 222).

There are three ways, according to Joyce, in which the difficulties of a preliminary RNA world could have been reduced. *First*, life did not originate with self-replication. Rather as a result of *chemical evolution*, non-instructed processes gradually changed the chemical environment, mainly through *chemical ordering of complex peptide structures, and possibly the formation of microspheres or membranous vesicles*. This stage made it easier for the later appearance of self-replication. *Second*, the development of a primitive self-replicating system, e.g., a mineral surface, as suggested by Cairns-Smith, served as a template for the formation of additional mineral layers. According to Cairns-Smith, these *clay "organisms"* first synthesized RNA molecules that served as catalysts in the mineral worlds Eventually, the RNA catalysts "took over" the genetic role (Cairns-Smith 1985). *Third*, the chemistry of the prebiotic environment made possible the appearance of self-replicating *RNA-like* systems easier to synthesize, e.g., genetic material based on purines alone, or including glycerol instead of ribose. Yet, Joyce mentions several obstacles that every chemical self-replication faces, all of them having to do with the need for catalytic activity. Thus, the three alternatives can be reduced to the first option—Joyce concludes that a period of chemical evolution was needed so that a genetic system based on some simple RNA-like molecules would eventually be able to arise (Joyce 1989, pp. 222–223).

Though Eigen does rely on catalytic proteins for the preliminary phase of his model, the option of chemical evolution that attributes a primary first role to catalytic proteins and not to self-replicating macromolecules serves as an interesting alternative to Eigen's model. One of the most difficult problems facing theories that postulate self-replication as the first phase of life is the requirement for almost faultless copying. The replication of prebiotic oligomers is supposed to involve enough errors to give natural selection raw material to work on. However, too many errors will destroy all information. Freeman Dyson considers the narrow straits between which replication had to maneuver as unrealistic in a prebiotic scenario. He views *error-tolerance* as the primary requirement for a model of an emerging molecular population (Dyson 1985, p. 73). This is one of the reasons why he prefers an Oparin-like model, according to which life first arose as a metabolic enclosed system comprised of a population of catalytically active oligopeptides, each rearranging each other's structure. Dyson's ideas on the origin of life are based, as well, on his description of the "essential characteristic of living cells" as "*homeostasis*, the ability to maintain a steady and more-or-less constant chemical balance in a changing environment" (p. 61). He envisions the development of replication at a later stage, made possible only in the context of the evolution of the genetic code, ribosomes and chromosomes (p. 74).

How does Eigen cope with the complication that arises in a system in which the first replicating molecules were supposed to manage without informed enzymes, and hence had to put up with high error rates? This error rate limited the sequence size of the replicating molecules to short oligonucleotides (100 nucleotides or fewer). In order to improve replication, clearly a primitive translation mechanism into proteins had to come into existence. However, this goal could not be achieved with the postulated short oligonucleotides. *Cooperation* is Eigen's answer to this "catch 22 of the origin of life" (Maynard Smith 1986, p. 118). The breakthrough in molecular evolution which enabled the development of sophisticated living systems, he claims, must have been brought about by an integration of several self-reproducing units into a cooperative system.

A mechanism capable of such integration, which will assure the emergence of functional linkages between nucleic acids and proteins, can be provided only by the class of the *hypercycles* (Eigen and Schuster 1977, p. 563). Eigen's basic unit of the hypercycle consists of a feedback loop in which some of the translation products of an RNA sequence has a beneficial effect upon the reproducibility of their *own* particular genotype (Eigen 1992, p. 41).[16] However, in order to mobilize the information content of several RNA sequences (several genes), which cannot be combined into a continuous molecular chain because of the error-threshold relationship, Eigen postulates a more complex hypercycle (p. 41). Thus, he envisages the existence of cyclically arranged series of self-replicative units in which an intermediate product of one cycle is employed by the next one in the series (Eigen *et al.* 1981, p. 89).

However, it has to be noted that Eigen's hypercycle model presupposes what is to be explained in the first place (de Duve 1991, p. 187). In order to make the hypercycle concept feasible, Eigen has to assume the existence of "more or less specific mechanisms for the translation of RNA sequences into protein molecules" (Eigen 1992, p. 41). According to Christian de Duve, the neglect to specify how a crude mechanism of translation came into place, similarly to the neglect to specify how the first RNA molecules arose, both stem from not giving detailed attention to the catalytic activities of oligopeptides, without which the RNA world could not have materialized (de Duve 1991, p. 133).

A more serious attack on Eigen's hypercycle notion, based on chemical and philosophical considerations, is voiced by Jeffrey Wicken (Wicken 1987). Wicken does consider the replicative quasispecies to be "the most empirically supported and theoretically justifiable aspect of molecular Darwinism" (p. 107). However, he claims that RNA sequences could not begin to serve as templates for protein synthesis in a quasispecies setting, which promotes selection *for better replication*. Information for making functional proteins, on the other hand, Wicken points out, is burdensome to bare replicators. "The evolutionary genesis and stabilization of genetic information [*information for proteins*] require that it has a functional referent to which it is useful" (p. 104). According to Wicken, the referent to which information for making catalytic proteins is useful might have been a *catalytic whole*, a protoorganization, e.g., a microsphere postulated by Sidney Fox (p. 104), or, one might add, a "Dyson metabolic unit." Similar to Dyson, who mentioned the development of some ribosome organization as a prerequisite for the emergence of translation, Wicken refers to several experiments which indicate the formation of particles from lysine-rich proteinoids which are catalytically active, with acidic proteinoids and nucleic acids. These articles might be regarded as model ribosomes in both structure and function (p. 105).

It seems as if the two points of view—the "replicator first" and the "cell first"—are very far apart. This seems to be the case, especially when the question of compartmentation is considered. There is a famous quotation by Eigen to the effect that organization into cells was surely postponed as long as possible (Eigen *et al.* 1981, p. 107). On the other hand, following Oparin for whom the formation of a coacervate—a droplet separated from its environment—was a necessary condition for the emergence of life, several researchers in the field continue to present cell-first claims. The most vocal of them is Sidney Fox, whose proteinoid microspheres, produced by stirring heat-polymerized amino acids with water, were mentioned before (Fox & Dose 1972). Another proponent of early compartmentation, on energetic grounds, is Harold Morowitz, who says that "the necessity of thermodynamically isolating a subsystem is an irreducible condition of life ... elaborate evolutionary development in an unpartitioned, aqueous environment ... places an extraordinary burden to counter diffusion and the other dissipative consequences of the Second Law of Thermodynamics" (Morowitz 1992, p. 8). For Morowitz, it is the closure of a phospolipid bilayer membrane into a vesicle that represents "a discrete transition from nonlife to life" (p. 9).

In fact, Eigen's emphasis on "hypercycle first, enclosure second" is characteristic of his earlier work.

More recently he is saying that "both forms of organization (hypercyclic coupling and enclosure) *are needed at once*" (Eigen 1992, p. 109). He points out several chemical reasons why protocells could not have been formed before the self-organization of the hypercycle. At the sametime, hypercycles could not be generated without "breaking up the homogeneity of the soup" (p. 43). Eigen's concomitant hypercycling and enclosure does not seem to be that different from Wicken's suggestion that "a more realistic possibility is that life emerged through the coevolution of nucleic acids and proteins within catalytic microspheres" (Wicken 1987, p. 106). (For a similar hypothesis that speaks about the development of hypercycles within a "Fox's microsphere" or a "Morowitz's vesicle," see Weber *et al.* 1989, p. 385; Bresch *et al.* 1980). We have already mentioned Joyce's ideas about a period of chemical evolution in which proteins were structured into enclosed units, as a more probable stage leading to the RNA world. In my view, these models are all examples of *compromise solutions* that reflect the need to overcome the presumed improbability of certain stages on which one-sided models are based. These compromise solutions are, thus, another manifestation of *the continuity thesis* at work.

THE CONTINUITY THESIS IN CELL-FIRST THEORIES

We have already discussed the continuity thesis as formulated by Harold Morowitz. Sidney Fox and his associates, for many years the major promoters of the cell-first notion, regard the continuity thesis as their basic philosophical framework. Based on experiments in simulated chemical evolution, Fox and his group concluded that biogenesis was not governed by chance events, but was rather "constrained and directed" by the physico-chemical properties of the reacting materials (Fox & Dose 1972). Whereas for Eigen, the principle of order which combines the physical and the biological realms is the principle of natural selection, Fox emphasizes the importance of non-Darwinian selection at the molecular level, which functions on the thermodynamic criteria of stability.

First, he claims, the prebiotic synthesis of amino acids is not statistically random. Second, these amino acids contain the instructions for their own arrangements in polymers, which are "highly nonrandom in type" (Fox 1980, p. 579). Third, these polymers which display catalytic activity, on contact with water undergo self-assembly to form microspheres—a process internally controlled by selective intermolecular forces (Fox & Dose 1972, p. 242). Fox and his associates adopt Oparin's claim that "the development of the first forms of life on earth was not a solitary 'happy event'... but an event whose repetition was an integral part of the general development of matter" (Oparin 1972, p. vii). They thus predict the possible production of amino acids, proteins and microspheres on other celestial bodies, when the needed precursors, notably carbon compounds and water, are present (Fox & Dose 1972, pp. 338–339). Fox's model is severely criticized by many in the origin-of-life field, and some of his central contentions are doubted, notably that the formation of proteinoids represents a process of self-ordering. The reality of his scenario and the biogenic relevance of his microspheres are questioned as well (Shapiro 1986, pp. 191–205). Yet, we have seen that both Eigen's replication model and Wicken's organismic, thermodynamic model rely on some important aspects of Fox's theory. For our discussion here, the crucial point is the important role played by the continuity thesis in this theory.

Recently, a new evolutionary paradigm based on non-equilibrium thermodynamics attributes to the Second Law of thermodynamics a central, positive role in evolution. There is a very strong correlation between the continuity thesis as portrayed in this paper and the basic tenets of the new "paradigm." *First*, there is the explicit insistence on the law-governed, continuous nature of all phases of evolution. Biological phenomena are considered not only as consistent with physical laws. "Some of their most fundamental characteristics follow directly from such laws" (Hull 1988, p. 3). *Second*, there is emphasis not only on the continuity of physical and biological systems, but also on the uniqueness of the latter. Jeffrey Wicken, among the founders of the new paradigm, explicitly states his aim to formulate the principles of continuity that connect life with prelife, and his belief that thermodynamics provides the needed conceptual connective tissue. Combining "physicalist" and "organismic" points of view, he defines life on the basis of general physical principles as an example of a "dissipative structure"—"a system that maintains a high degree of internal order by dissipating entropy

to its surroundings." However, the living system is unique in that it involves information, and hence organization—a term Wicken exclusively applies to informed systems, biological and social (Wicken 1987, p. 32). As was already pointed out in Wicken's criticism of Eigen, the generation of information is possible, he asserts, only in an organismic context.

It is Wicken's claim that *dissipation through structuring* is an evolutionary first principle. All phases of evolution, starting from cosmic origins and including both the emergence and the evolution of life, are causally connected with the second law of thermodynamics (p. 5). Contrary to widespread notions, he claims, the formation of structures of growing complexity does not contradict the Second Law. In fact, since "putting smaller entities together to form larger entities will generate entropy through the conversion of potential energy to heat" (p. 72), the process of structuring, e.g., of producing diverse chemical compounds in the prebiotic phase, was promoted by this law. At this stage, selection "chose" on the basis of thermodynamic stability. As evolution entered its biotic phase, nonequilibrium thermodynamics became more important in selection (p. 109). Indeed, according to the new paradigm, "biological systems are stabilized far from equilibrium *by way of* self-organizing, autocatalytic structures that serve as pathways for the dissipation of unusable energy and material ... [thus] *entropy production and organization are positively correlated*" (Weber *et al*. 1989, p. 375, emphasis mine). This is why biological evolution, contrary to traditional wisdom, is seen as an entropic process, in which despite the universal tendency to deplete thermodynamic potential, there is building up of structure (Wicken 1987, p. 72).

The dichotomy of "physicalism" versus "organicism" is not the only one transcended in Wicken's theory and in the new "paradigm." Chance and necessity are "fused" as well on the basis of thermodynamics. Wicken's concept of the thermodynamics of evolution speaks of "microscopic chance" within the framework of macroscopic law (p. 129). Since one can predict the direction processes take on the basis of entropy production, "thermodynamics allows us to ask *why* processes occur, in an entirely materialistic way" (p. 57). In their review *Consequences of Nonequilibrium Thermodynamics for Darwinism*, Depew and Weber maintain that nonequilibrium models can provide "a set of principles showing why the evolution of biological systems is *something to be expected*, rather than something that needs to be explained against a theoretical background that does not strongly anticipate it" (Depew and Weber 1988, p. 318, emphasis mine).

Mention should be made here of another theoretical attempt along similar philosophical lines to the thermodynamic paradigm. Stuart Kauffman and his colleagues, who study mathematical models for certain complex biological systems, have recently suggested the existence of a key principle that has shaped the development of life—"*spontaneous self-organization: the tendency of complex dynamical systems to fall into an ordered state without any selective pressure whatsoever.*" The combination of spontaneous self-organization and the molding action of natural selection, they claim, is responsible for the fact that "evolution is not just a series of accidents" (Waldrop 1990, p. 1543; Kauffman 1991, p. 78). Based on mathematical analysis and a computer model, Kauffman, his colleague Doyne Farmer and associates suggest an origin-of-life scenario in which an "autocatalytic set" made of catalytic oligopeptides and involving a primitive metabolism, was the initial phase (Waldrop 1990, p. 1543; see also Kauffman 1993).

In summary, all the theories examined here, their differences notwithstanding, postulate as an alternative to "chance occurrences" some physical dynamics of self-organization responsible for the emergence of life. In this, they in fact, follow the original ideas of Alexander Oparin and J. B. S. Haldane in their papers of 1924 and 1929, respectively, which initiated the modern study of the origin of life. Both papers contain the basic idea that life on earth has been preceded by long chemical evolution of organic compounds, and the notion of intermediate forms on the way to the primitive first cells. Both postulate a primordial environment entirely different from the earth today, and attempt to portray a specific scenario for the early stages of prebiotic evolution. Yet, there are important differences between the two papers. Whereas Oparin developed his ideas within the context of the then highly popular biochemistry of colloids, Haldane was drawn to the origin-of-life subject via the discovery of viruses. Haldane speaks on the "first living or half-living things" as "only capable of reproduction," compares the prebiotic medium with its variety of molecules to the host cell of present viruses, and

postulates a long delay until the enclosure into a cell (Haldane 1967 [1929], p. 247).

In distinction from Oparin, who is the first "cell-first" proponent, Haldane's might be called a "naked gene" scenario. Still, they are both fathers of the *Oparin–Haldane hypothesis* whose breakthrough importance lies in its *philosophical tenets*. The Oparin–Haldane scenario is often equated with the now much criticized notion that organic compounds could have accumulated in the "primordial soup" under reducing atmospheric conditions. However, *the major joint philosophical contribution of Oparin and Haldane's ideas was the establishment of the continuity thesis*.

THE HEURISTIC VALUE OF THE CONTINUITY THESIS

My analysis so far has indicated that scientists in the field are trying, each according to his chemical or physical inclination, to sustain the philosophical continuity thesis with specific materialistic statements and details. Several "principles of continuity" were mentioned at the beginning of the paper as an example of the positive heuristic of the continuity thesis. However, as far as one can tell from the texts examined, in most cases the thesis is assumed implicitly. I claim that being self-aware of the epistemological role of the thesis results in a deliberate effort to devise more probable models for bridging the gap between life and inorganic matter. The philosophical tenets of the "thermodynamics school" explicitly express this motivation. A good example of the heuristic advantage of the continuity thesis is provided by the work of Christian de Duve.

As already noted, de Duve believes it necessary to assume that almost all successive steps in the emergence of life had a very high probability of happening. The only alternative to this assumption is the anti-scientific notion of a miraculous origin (de Duve 1991, p. 112). The continuity thesis which de Duve deliberately chooses to adopt is used by him in order to examine whether the proposed models in the field are acceptable as they stand, or whether they need to be improved. According to his reconstruction, emerging life went through four main successive "worlds": the primeval prebiotic, the thioester, the RNA, and the DNA worlds. De Duve's contribution to the script— *the insertion of the thioester world*, is essential, he thinks, "because I cannot accept the view of an RNA world arising through purely random chemistry" (pp. 112–113).

Though de Duve points out that this random chemistry is highly determined by physical and chemical factors, he, however, feels that this physico-chemical determination of the products in the prebiotic phase is not enough in order to sustain the continuity thesis with "materialistic flesh." According to him, without additional help of both a catalytic and energetic nature the "prebiotic broth" would have remained sterile. This is where the thioester bond becomes essential (p. 113). So what could the thioesters do? Their first function, says de Duve, was to support the assembly of multimers, notably from amino acids, with the help of the energy derived from the thioester bond. The multimerization process, though random and undirected, yielded a minute subset of multimers that survived, among them a number of crude catalysts (pp. 113–114). De Duve describes many other functions of the thioester bond and their consequences for the possibility of establishing a *"protometabolic network,"* "Having inaugurated electron transfer, phosphorylation, group transfer, and energy coupling, all thanks to the thioester bond, *protometabolism could now exploit the full catalytic potential of the multimer population"* (p. 115, emphasis mine). Following the stages of the formation of AMP and other mononucleotides, the rise of ATP, of adenylyl transfer, the making of coenzymes, finally the first oligonucleotides were assembled. "Emerging life had arrived at the threshold of the RNA world" (p. 115).

An additional clarification of terms will further indicate de Duve's alliance with the continuity thesis. De Duve emphasizes that the deterministic aspect of his model does not relate to the question, how probable in the universe are the conditions that "breed life." It rather relates to *the probability of the outcome of these conditions*—to the fact that the different stages of the process, *once it takes place*, are determined by the intrinsic properties of the materials involved (p. 214). Randomness in this process, which implies only being noninformed, is not statistical randomness. When de Duve speaks of "random multimerization" of small peptides from amino acids, he does not mean that the composition of the resulting mixture would be a simple statistical function of the relative abundances of the different building blocks. Rather, different

physical and chemical constraints come into play and result in a *strong deterministic control* of the composition of the population of peptides synthesized and preserved (pp. 141–143). Summing up the key elements of his model, de Duve claims that several of these elements—building the primary building blocks by abiotic synthesis, formation of catalytic multimers by random (undirected) assembly of these building blocks, the formation of thioesters—are achieved *"only by exploiting randomness fully and leaving nothing to chance"* (p. 213). Only after all the main lines of the blueprint, continues de Duve, had been fulfilled "would complexity reach a degree sufficient for chance to begin playing a role and bring in diversity and unpredictability" (p. 213).[17]

Thus, de Duve's model is a specific expression of his general claim that *"were it [the emergence of life] not an obligatory manifestation of the combinatorial properties of matter, it could not possibly have arisen naturally"* (p. 217). Can such a position be described as *"predestinistic,"* using the term suggested by Robert Shapiro in his all-embracing criticism of current origin-of-life models? According to Shapiro, the replication-first, as well as the cell-first theorists are "predestinists" who believe "that the laws of the universe contain a built-in bias that favors the production of chemicals vital to biochemistry and ultimately to human life" (Shapiro 1986, p. 108). Shapiro finds fault with the specific "predestinist" attempts of "prebiotic chemists to devise series of plausible reactions" that demonstrate how chemistry changed into biochemistry within a certain scenario (p. 176). In my view, and as a conclusion of this chapter, what Shapiro describes and condemns is exactly the recipe for a scientific study of the emergence of life. Based on the continuity thesis as the rational alternative adopted by science against the alternative of miracle or design, different models are conceived which direct the specific experimental or theoretical search.

CONCLUSION

This chapter was devoted to the discussion of what I have coined "the continuity thesis." This thesis states that the development of life from matter is a gradual process to be explained on the basis of physical principles. The thesis rejects the "chance camp" notion, expressed by several prominent scientists, that the gap between inanimate matter and life was bridged by a unique, miraculous event. I described the continuity thesis as a philosophical presupposition that unites researchers of the origin of life, and that forms the basis for the "law camp." Surveying several models suggested in the field, e.g., replication-first and cell-first theories, I pointed out the presence of the continuity thesis in all of them, despite their differences. The assumption that life emerged from matter based on physical mechanisms of self-organization is, I claimed, not a "passive ingredient" of all these theories. When acknowledged, this assumption can serve as guidance to devise more probable scenarios. This was exemplified in the work of Christian de Duve.

Scientists quite often tend to deny any relevance of philosophical considerations to their specific work. Based on the analysis carried out in this paper, the importance of philosophical assumptions and arguments in the study of the emergence of life field cannot be doubted. Philosophy, in this case, goes to the core—to the very "right of existence" of this scientific endeavor.

NOTES

This chapter originally appeared in *Biology and Philosophy* 10(4) (1995), 389–417.

REFERENCES

1. Bartel, D. P. & Szostak, J. W. (1993). Isolation of new ribozymes from a large pool of random sequences. *Science*, **261**, 1411–1418.
2. Bernal, J. D. (1965). Discussion. In S. W. Fox (Ed.), *The origin of prebiological systems and of their molecular matrices* (pp. 65–88). New York: Academic Press.
3. Bresch, C., Neisert, H., & Harnasch, D. (1980). Hypercycles, parasites and packages. *Journal of Theoretical Biology*, **85**, 399–405.
4. Cairns-Smith, A. G. (1985). *Seven clues to the origin of life.* Cambridge, UK: Cambridge University Press.
5. Crick, F. (1981). *Life itself.* New York: Simon & Schuster.
6. Crick, F. & Orgel, L. E. (1973). Directed panspermia. *Icarus*, **19**, 341–346.
7. Dawkins, R. (1986). *The blind watchmaker.* London: Penguin.
8. de Duve, C. (1991). *Blueprint for a cell.* Burlington, NC: Neil Patterson.
9. Depew, D. J. & Weber, B. H. (1988). Consequences of nonequilibrium thermodynamics for the Darwinian tradition. In B. H. Webber, D. J. Depew, and J. D. Smith (Eds.), *Entropy, information, and evolution* (pp. 317–354). Cambridge, MA: MIT Press.
10. Dyson, F. (1985). *Origins of life.* Cambridge, UK: Cambridge University Press.
11. Eigen, M. (1971). Self-organization of matter and the evolution of biological macromolecules. *Naturwissenschaften*, **58**, 465–523.
12. Eigen, M. (1992). *Steps towards life.* Oxford: Oxford University Press.
13. Eigen, M. (1993). Viral quasispecies. *Scientific American*, July, 42–49.
14. Eigen, M. & Schuster, P. (1977). The hypercycle, part A: The emergence of the hypercycle. *Naturwissenschaften*, **64**, 541–565.
15. Eigen, M. & Schuster, P. (1978). The hypercycle, part C: The realistic hypercycle. *Naturwissenschaften*, **65**, 341–369.
16. Farley, J. (1974). *The spontaneous generation controversy from Descartes to Oparin.* Baltimore: John Hopkins University Press.
17. Fox, S. W. (1980). Life from an orderly cosmos. *Naturwissenschaften*, **67**, 576–581.
18. Fox, S. W. (1984). Proteinoid experiments and evolutionary theory. In M. W. Ho and P. T. Saunders (Eds.), *Beyond neo-Darwinism* (pp. 15–60). London: Academic Press.
19. Fox, S. W. & Dose, K. (1972). *Molecular evolution and the origin of life.* San Francisco: W. H. Freeman.
20. Gesteland, R. F. & Atkins, J. F. (Eds.) (1993). *The RNA world.* Cold Spring Harbor, NY: Cold Spring Harbor Press.
21. Goldlanskii, V. L. & Kuzmin, V. V. (1989). Spontaneous breaking of mirror symmetry in nature and the origin of life. *Soviet Physics Uspekhi*, **32**, 1–29.
22. Gould, S. J. (1989). *Wonderful life.* New York: W. W. Norton.
23. Gribbin, J. (1993). *In the beginning.* Boston: Little, Brown & Company.
24. Haldane, J. S. (1921). *Mechanism, life and personality.* London: John Murray.
25. Haldane, J. B. S. (1967). The origins of life. In J. D. Bernal, *The origin of life* (pp. 242–249). London: Wiedenfeld & Nicholson.
26. Hoyle, F. & Wickramasinghe, N. C. (1981). *Evolution from space.* London: Dent & Sons.

27. Hoyle, F. & Wickramasinghe, N. C. (1993). *Our place in the cosmos.* London: J. M. Dent.

28. Hull, D. L. (1988). Introduction. In B. H. Weber, D. J. Depew, and J. D. Smith (Eds.), *Entropy, information, and evolution* (pp. 1–8). Cambridge, MA: MIT Press.

29. Joyce, G. F. (1989). RNA evolution and the origins of life. *Nature,* **338**, 217–224.

30. Kamminga, H. (1982). Life from space—a history of panspermia. *Vistas in Astronomy,* **267**, 67–86.

31. Kamminga, H. (1986). The protoplasm and the gene. In A. G. Cairns-Smith and H. Hartman (Eds.), *Clay minerals and the origin of life.* Cambridge, UK: Cambridge University Press.

32. Kant, I. (1987). *Critique of judgment, trans.* W. S. Pluhar. Indianapolis: Hackett.

33. Kauffman, S. A. (1991). Antichaos and adaptation. *Scientific American,* August, 78–84.

34. Kauffman, S. A. (1993). *The origins of order: Self-organization and selection in evolution.* Oxford: Oxford University Press.

35. Keosian, J. (1974). Life's beginnings—origin or evolution. *Origins of Life,* **5**, 285–293.

36. Kimura, M. (1983). *The neutral theory of molecular evolution.* Cambridge, UK: Cambridge University Press.

37. Küppers, B.-O. (1990). *Information and the origin of life.* Cambridge, MA: MIT Press.

38. Lehninger, A. L. (1970). *Biochemistry.* New York: Worth.

39. Maynard Smith, J. (1986). *The problems of biology.* Oxford: Oxford University Press.

40. Mayr, E. (1974). Teleological and teleonomic, a new analysis. *Boston Studies in the Philosophy of Science,* **14**, 91–117.

41. Mayr, E. (1982). *The growth of biological thought.* Cambridge, MA: Harvard University Press.

42. Monod, J. (1974). *Chance and necessity.* Glasgow: Collins Publishing; Fontana Books.

43. Mora, P. T. (1965). The folly of probability. In S. W. Fox (Ed.), *The origin of prebiological systems and of their molecular matrices* (pp. 39–52). New York: Academic Press.

44. Morowitz, H. J. (1992). *Beginnings of cellular life.* New Haven, CT: Yale University Press.

45. Oparin, A. I. (1961). *The origin of life, trans.* S. Morgolis. New York: Macmillan.

46. Oparin, A. I. (1967). *Genesis and evolutionary development of life, trans.* E. Maas. New York: Academic Press.

47. Oparin, A. I. (1972). Forward. In S. W. Fox and K. Dose, *Molecular evolution.* San Francisco: W. H. Freeman.

48. Orgel, L. E. (1968). Evolution of the genetic apparatus. *Journal of Molecular Biology,* **38**, 381–393.

49. Popper, K. (1974). Reduction and the incompleteness of science. In F. Ayala and T. Dobzhansky (Eds.), *Studies in the philosophy of biology.* Berkeley: University of California Press.

50. Popper, K. (1982). *The open universe: An argument for indeterminism.* Totowa, NJ: Rowman & Littlefield.

51. Rood, R. T. & Trefil, J. S. (1981). *Are we alone?* New York: Charles Scribner's Sons.

52. Shapiro, R. (1986). *Origins.* Toronto: Bantam Books.

53. Toulmin, S. & Goodfield, J. (1982). *The architecture of matter.* Chicago: University of Chicago Press.

54. Waldrop, M. M. (1990). Spontaneous order, evolution and life. *Science,* **247**, 1543–1545.

55. Weber, B. H., Depew, D. J., & Smith, J. D. (Eds.) (1988). *Entropy, information, and evolution.* Cambridge, MA: MIT Press.

56. Weber, B. H., Depew, D. J., Dyke, C., *et al.* (1989). Evolution in thermodynamic perspective: An ecological approach. *Biology and Philosophy,* **4**, 373–406.

57. Weiner, A. M. & Maizels, N. (1991). The genomic tag model for the origin of protein synthesis. In S. Osawa and T. Honjo (Eds.), *Evolution of life* (pp. 51–65). Tokyo: Springer-Verlag.

58. Wicken, J. (1987). *Evolution, thermodynamics and information.* New York: Stanford University Press.

ENDNOTES

1 Not surprisingly, as has been recently commented by Manfred Eigen, Darwin had some thoughts on a sort of "continuity thesis." While he realized that the question of the origin of life was unanswerable in his day, he still contemplated, in a letter to Nathaniel Wallich (1881), the possibility of a "principle of continuity" which "renders it probable that the principle of life [natural selection] will hereafter be shown to be a part, or consequence, of some general law" (see Eigen 1992, p. 17).

2 A mirror image of this position was the one that saw all matter as equally alive (Toulmin and Goodfield 1982, p. 317). On this view, categories of organic development are more basic to nature than mechanical categories (see, for example, Haldane 1921, pp. 100–101).

3 The belief in the "spontaneous generation" of life—the sudden, immediate appearance of fully formed organisms from inanimate matter—was justified in different periods either on the basis of the action of mechanical principles

(e.g., the atomists, Descartes), or by relying on the intervention of a life-donating spirit, or of God, in the implementation of form in crude matter (e.g., Plotinus, Thomas Aquinas) (see Oparin 1967, pp. 9–40).

4 "Determinism" here means the complete determination of the phenomena that led to the emergence of life by the physico-chemical conditions prevailing at the scene. Hence, under the same conditions, the emergence of life is reproducible (see de Duve 1991, p. 211, note 1). Notice should be taken that we are discussing here the evolutionary phase concerning the formation of life, and not the subsequent stages of biological evolution. The "deterministic requirement" applies mainly to the transition stage, and has, for various reasons, to be relaxed in later evolutionary stages (see de Duve 1991, pp. 211–217; Gould, 1989, p. 289).

5 Just to cite a few examples, Küppers 1990; Gould 1989; de Duve 1991; Wicken 1987; Dawkins 1986.

6 I wish to thank one of the reviewers of a previous version of this paper for his suggestion to apply Imre Lakatos's "positive and negative heuristic" terms, in the context of my analysis of the continuity thesis.

7 As will be shown later, the term "random" as used by Christian de Duve, for example, in the context of the emergence of life, is not equated *with statistical randomness*, but rather with a noninformed and strictly deterministic processes affected by *kinetic and steric factors* (de Duve 1991, p. 187).

8 Monod regards the "lottery image" as a sobering response against the ideologies, notably dialectical materialism, which claim to be founded upon science, but in fact rely on "animistic projection." The alternative view to the idea that evolution of the biosphere is part of cosmic evolution, culminating in man, he says, is to portray evolution as basically a random process (Mono, 1974, pp. 39–40).

9 Kant's *critical* solution to the difficult question of the original formation of biological organization is that we have to view the organism *as if* (*als ob*) it was externally designed (Kant, 1987, pp. 280–283).

10 This is why Kant praises Johann Friedrich Blumenbach, the German biologist (1752–1848), known for his embryological research (and for his anthropological studies), who on the one hand based his embryological explanations on the original existence of "organized matter," and on the other, "left an unmistakable share to natural mechanism" (Kant 1987 [1790], p. 311).

11 Dyson attributes this fact to the impact of Max Delbrück's bacteriophages experiments in the late 1930s and 1940s, on Erwin Schrödinger and his successors in molecular biology (Dyson 1985, pp. 3–5).

12 Eigen's anti-chance position can serve as a good argument against the scenario advocated by another arch-selectionist, Richard Dawkins, both in his *The Selfish Gene* (1976) and in his *The Blind Watchmaker* (Dawkins, 1986). In the later book, Dawkins says that we have to assume "a single-step chance event in the origin of cumulative selection itself" (p. 140). Similarly to Monod, Dawkins considers this "chance event" as the only alternative to *Design*. Though considering Cairns-Smith's clay theory as an example of a mechanism capable of bringing about life, in the end Dawkins relies on the enormity of time and space in the universe to make even the most miraculous event possible (pp. 141–166).

13 Eigen's model of the quasispecies is mathematically limited to sufficiently large populations (10^{10}–10^{12}), in which information content does not exceed 10^3–10^6 nucleotides. These limits apply to populations of molecular replicators, viruses and microorganisms (Eigen 1992, p. 82). The predictions of the mathematical theory, Eigen claims, have been confirmed with RNA viruses and with several laboratory systems (p. 29).

14 This is due to the fact that the way toward a superior mutant is not uninterrupted. If, as assumed in classical genetics, the system has to try out all possible mutations in order to find an advantage, the origin of complexity—of optimal genes—cannot be explained (Eigen 1992, pp. 22–23).

15 It should be added that even when a central role is allotted to ribozymes in the RNA world, based on recent discoveries of RNA acting as a replicating enzyme of itself, it becomes obvious that RNA replicase could not arise from a random pool of RNA sequences, and that non-random reactions and the participation of cofactors were needed (Bartel and Szostak 1993).

16 Such a hypercycle, claims Eigen, has been recently shown to be operating in the mechanism of infection of host cells by viruses (Eigen, 1992, p. 42).

17 A distinction between the strictly deterministic nature of the origin of life phase, and the entirely contingent nature of the following biological evolution is made by S. J. Gould, who considers the origin of life on earth to be "virtually inevitable, given the chemical composition of early oceans and atmospheres, and the physical principles of self-organizing systems" (Gould 1989, p. 289).

11 · The universal nature of biochemistry

NORMAN R. PACE

People have long speculated about the possibility of life in settings other than Earth. Only in the past few centuries, however, have we been able to conceive of the specific nature of such settings: other planets around our own Sun and solar systems similar to our own elsewhere in the physical universe. Speculation on the nature of life elsewhere often has paid little heed to constraints imposed by the nature of biochemistry, however. A century of fanciful science fiction has resulted not only in social enthusiasm for the quest for extraterrestrial life, but also in fanciful notions of the chemical and physical forms that life can take, what the nature of life can be. Since the time of the Viking missions to Mars, in the mid-1970s, our view of life's diversity on Earth has expanded significantly, and we have a better understanding of the extreme conditions that limit life. Consequently, our search for extant life elsewhere in the solar system can now be conducted with broader perspective than before.

How can life be detected regardless of its nature and origin? Considering the recent spectacular advances in observational astronomy, it seems likely that the first sign of life elsewhere will be the spectroscopic detection of co-occurring non-equilibrium gases, for instance oxygen and methane, in the atmosphere of a planet around some distant star. Co-occurrence of such gases would indicate that they are replenished, perhaps most readily explained by the influence of life.[1] By observation of oxygen and methane, Earth could possibly be seen as a home for life even from distant galaxies. Other potential habitats for life in this solar system, such as Mars and Europa, however, are not so obvious. The search for life on those bodies will be conducted at the level of analytical chemistry. As we undertake the detection of extraterrestrial life, it is instructive to try to put constraints on what the nature of life can be. These constraints, the requirements for life, tell us where and how to look for life, and the forms that it can take.

WHAT IS LIFE?

An early question that needs to be confronted, indeed a question that in the last analysis requires definition, is: What is life? Most biologists would agree that self-replication, genetic continuity, is a fundamental trait of the life process. Systems that generally would be deemed nonbiological can exhibit a sort of self-replication, however.[2] Examples would be the growth of a crystal lattice or a propagating clay structure. Crystals and clays propagate, unquestionably, but life they are not. There is no locus of genetic continuity, no organism. Such systems do not evolve, do not change in genetic ways to meet new challenges. Consequently, the definition of life should include the capacity for evolution as well as self-replication. Indeed, the mechanism of evolution—natural selection—is a consequence of the necessarily competing drives for self-replication that are manifest in all organisms. The definition based on those processes, then, would be that life is any self-replicating, evolving system.

The processes of self-replication and evolution are not reliably detectable, even in the terrestrial setting. Consequently, in the practical search for life elsewhere we need to incorporate information on the nature of the chemistries that can provide the basis for self-replication and evolution. Considering the properties of molecules likely to be needed to replicate and evolve, it is predictable that life that we encounter anywhere in the universe will be composed of organic chemicals that follow the same general principles as

The Nature of Life, ed. M. A. Bedau and C. E. Cleland. Published by Cambridge University Press. © M. A. Bedau and C. E. Cleland 2010.

our own organic-based terrestrial life. The operational definition of life then becomes: Life is a self-replicating, evolving system expected to be based on organic chemistry.

WHY ORGANIC CHEMISTRY?

The basic drive of life is to make more of itself. The chemical reactions required for the faithful propagation of a free-living organism necessarily require high degrees of specificity in the interactions of the molecules that carry out the propagation. Such specificity requires information, in the form of complex molecular structure—large molecules. The molecules that serve terrestrial organisms typically are very large, proteins and RNAs with molecular weights of thousands to millions of daltons, or even larger as in the case of genetic DNA. It is predictable that life, wherever we encounter it, will be composed of macromolecules.

Only two of the natural atoms, carbon and silicon, are known to serve as the backbones of molecules sufficiently large to carry biological information. Thought on the chemistry of life generally has focused on carbon as unique.[3] As the structural basis for life, one of carbon's important features is that unlike silicon it can readily engage in the formation of chemical bonds with many other atoms, thereby allowing for the chemical versatility required to conduct the reactions of biological metabolism and propagation. The various organic functional groups, composed of hydrogen, oxygen, nitrogen, phosphorus, sulfur, and a host of metals, such as iron, magnesium, and zinc, provide the enormous diversity of chemical reactions necessarily catalyzed by a living organism. Silicon, in contrast, interacts with only a few other atoms, and the large silicon molecules are monotonous compared with the combinatorial universe of organic macromolecules.

Life also must capture energy and transform that energy into the chemistry of replication. The electronic properties of carbon, unlike silicon, readily allow the formation of double or even triple bonds with other atoms. These chemical bonds allow the capture and delocalization of electronic energy. Some carbon-containing compounds, therefore, can be highly polarized and thereby capture "resonance energy" and transform this chemical energy to do work or to produce new chemicals in a catalytic manner.

The potential polarizability of organic compounds also contributes to the specificity of intermolecular interactions, because ionic and van der Waals complementarities can shift to mesh with or to repulse one another. Finally, it is critical that organic reactions, in contrast to silicon-based reactions, are broadly amenable to aqueous conditions. Several of its properties indicate that water is likely to be the milieu for life anywhere in the universe.[2]

The likelihood that life throughout the universe is probably carbon-based is encouraged by the fact that carbon is one of the most abundant of the higher elements. Astronomical studies find complex organic compounds strewn throughout interstellar space. Moreover, the common occurrence of carbonaceous meteorites testifies to an organic-rich origin for our own solar system. If life indeed depends on the properties of carbon, then life is expected to occur only in association with second- or later-generation stars. This is because carbon is formed only in the hearts of former stars, so far as we know.

THE UNIVERSAL NATURE OF BIOCHEMISTRY

Life as we know it builds simple organic molecules that are used as building blocks for large molecules. Amino acids are used to construct the long chains of proteins; simple sugars combine with the purine and pyrimidine bases and phosphate to construct the nucleic acids. It seems logical that the evolution of any organic-based life form would similarly result in the construction of complex molecules as repeating structures of simple subunits. Indeed, it seems likely that the basic building blocks of life anywhere will be similar to our own, in the generality if not in the detail. Thus, the 20 common amino acids are the simplest carbon structures imaginable that can deliver the functional groups used in life, with properties such as repeating structure (the peptide unit), reactivity with water, and intrinsic chirality. Moreover, amino acids are formed readily from simple organic compounds and occur in extraterrestrial bodies such as meteorites, so are likely to form in any setting that results in the development of chemical complexity necessary for life.

Similarly, the five-carbon sugars used in nucleic acids are likely to be repeated themes, perhaps in part because they are the smallest sugars that can cyclize and thereby confer spatial orientation on other molecules, for instance the purines and pyrimidines that comprise the genetic information of terrestrial organisms. Further, because of the unique abilities of purines and pyrimidines to interact with one another with particular specificity, these subunits, too, or something very similar to them, are likely to be common to life wherever it occurs. Differences in evolutionary systems likely will lie at the higher-order levels: the structures of the large molecules assembled from the simple units, and the mechanisms by which they are assembled and in which they participate.

Themes that are probably common to life everywhere extend beyond the building blocks. Energy transformation is a critical issue. The processes of life require the capture of adequate energy, from physical or chemical processes, to conduct the chemical transformations requisite for life. Based on thermodynamics there are only two such energy-capturing processes that can support "primary productivity," the synthesis of biological materials from inorganic carbon dioxide. One process, termed lithotrophy, involves the oxidation and concomitant reduction of geochemical compounds. For instance, methanogenic organisms gain energy for growth by the use of hydrogen (H_2) as a source of high-energy electrons, which are transferred to carbon dioxide (CO_2), forming methane (CH_4). Other microbes might use hydrogen sulfide (H_2S) as an energy source, respiring with oxygen (O_2), to produce sulfuric acid (H_2SO_4). It is thought that the earliest life on Earth relied on lithotrophic metabolism.

The second general process for obtaining energy, photosynthesis, captures light energy and converts it into energetic electrons that can be used to accomplish biochemical tasks. Photosynthesis arose early in the history of terrestrial life and probably drives most primary productivity on Earth today. The contribution of lithotrophy to terrestrial primary productivity remains unknown, however, because there currently is little information on such organisms that may be distributed throughout the Earth's crust, wherever the physical conditions permit.

Although terrestrial life and life that might arise independently of Earth are expected to use many similar, if not identical, building blocks, they also are expected to have some biochemical qualities that are unique. This expectation is based on the fact that different evolutionary lines of terrestrial evolution also have engendered novelties unique to those lines. Thus, the biochemistry of methanogenesis arose uniquely in Archaea, whereas the property of chlorophyll-based photosynthesis was invented among the phylogenetic domain Bacteria (below). The cytoskeleton, which is probably a requirement for large and complex cell structure, arose in the eukaryotes. Considering the variety of Earth's life, novelty, as well as commonality, must be expected elsewhere.

THE PHYSICAL LIMITS OF LIFE: THE HABITABLE ZONE

Thought on where in the solar system life might occur was limited historically by the belief that life relies ultimately on light and warmth from the Sun and, therefore, is restricted to the surfaces of planets. The inner boundary of the "habitable zone" in our solar system was considered as approximately between Earth and Venus, not so close to the sun as to be too hot for life. The outer boundary was considered to lie between Mars and Jupiter, not so far from the sun that the surface of a body would necessarily be frozen or receive too little light for efficient photosynthesis. Light probably is not directly required for life to arise, however, except as it may be involved in the formation of organic compounds during the accretion of a planetary system. On the other hand, the biological use of light energy, photosynthesis, may be a prerequisite for persistence of planetary life over billions of years. The reason for this conjecture is that light provides a continuous and relatively inexhaustible source of energy. Life that depends only on chemical energy inevitably will fail as resources diminish and cannot be renewed.

Nonetheless, we know that life occurs in Earth's crust, away from the direct influence of light, and that many organisms have metabolisms that function independently of light. Thus, the outer boundary of the potentially habitable zone extends into the far reaches of the solar system, to any rocky body with internal heating, regardless of its distance from the Sun. (I specify "rocky" body to accommodate chemicals expected to be required for metabolism (below).) Life can persist in the absence of light by using inorganic energy sources, as do lithotrophic organisms, or organic sources deposited in planetary interiors during their accretion, as do heterotrophs.[4] Therefore, rather than proximity to the Sun, it seems more useful to define the habitable zone for life in terms of the chemical and physical conditions that are expected to be required for life. Our view of life's possible extremes currently is limited to the extremes of terrestrial life. Considering the intrinsic fragility of complex organic systems, coupled with the powerful force of natural selection, I venture that the physical limits of life are likely to be about the same anywhere in the universe. The window of chemical and physical settings that permit life are broad, however. Some important considerations are the following.

Chemical setting

Although the general energy requirement of life is a state of chemical disequilibrium, in which some oxidation–reduction reaction can occur, the specific thermodynamic requirements of biological energy-gathering strategies constrain the sites where life can occur. For example, a setting for lithotrophic organisms requires the occurrence of an appropriate mix of oxidized and reduced chemicals. Photosynthetic organisms require sufficient light of appropriate frequency. The light must be sufficiently energetic to support biosynthesis, but not so energetic as to be chemically destructive. These considerations constrain photosynthesis-based life to the spectral zone of about 300–1500 nm in wavelength. (Terrestrial photosynthesis is limited to about 400–1200 nm.) Beyond the requirements for energy metabolism and CO_2 as a carbon source, terrestrial life requires only a few elements: H, N, P, O, S, and the suite of metals.

Physical setting

Physical constraints on life include temperature, pressure, and volume. The extreme diversity of terrestrial life probably provides an analog for life's diversity anywhere.

Temperature is a critical factor for life. Temperatures must be sufficiently high that reactions can occur, but not so high that complex and relatively fragile biomolecules are destroyed. Moreover, because life probably depends universally on water, the temperature must be in a range for water to have the properties necessary for solute transfer. Water can be stabilized against boiling by pressure, but at too-low temperatures, water becomes crystalline and inconsistent with transport. Currently, the upper temperature record for culturable microbes is 112–113 °C, held by hyperthermophilic archaeons of the genera *Pyrolobus* and *Pyrodictium*.[5] Even the spectacularly durable bacterial endospore does not survive extended heating beyond ca. 120 °C. The lower-temperature boundary for life is not established, but microbes are recoverable from ice, and growth of organisms has been detected in ice to –20 °C.[6] The physical properties of ice can allow solute diffusion at temperatures much lower than the freezing point.[7] Thus, if based on aqueous organic chemistry, the temperature span for life anywhere in the universe is likely to be less than 200 °C, within roughly –50° to 150 °C.

Pressure and volume

The pressure required of a setting for life is probably limited at the lower end only by the vapor pressure required to maintain water or ice. An upper limit for pressure tolerance is unknown. Organisms on the terrestrial seafloor experience pressure over 1000 atmospheres, and microbes recovered from deep oil wells are exposed to far higher pressures. The upper pressure limit for life probably is determined mainly by the effect of the pressure in reducing volume for occupancy. Life can be remarkably small, however. It is estimated that cells only a few hundred nanometers in diameter can contain all of the components considered necessary for life.[8]

The expected commonality of chemistry in life's processes assists in life detection because it predicts that terrestrial types of biochemicals are useful targets for analysis even in an extraterrestrial setting. On the other hand, the expected similarity of terrestrial and potentially alien life complicates the interpretation of positive chemical tests for biochemicals. Thus, analyses of simple terrestrial-like biochemical compounds might not distinguish between a signal of life, on one hand, and an abiotically derived organic chemical, or between an alien life form and a terrestrial contaminant. Distinction between organisms with different evolutionary origins may require analysis of macromolecules and genes. Particularly, the nature and detail of the genetic information would be telling.

A GENETIC SIGNATURE OF TERRESTRIAL LIFE

All life on Earth is genetically related through an evolutionary past that extends beyond 3.8 billion years ago. We see this relatedness in the many common structural and mechanistic features that make up all cells. The relationships between different terrestrial life forms are quantitatively explicit in the now-emerging maps of the course of evolution, phylogenetic trees based on DNA sequences. Even if potentially alien organisms were to present the same biochemistry as seen in terrestrial organisms, genetic sequences could provide criteria to distinguish them if they are of different evolutionary sources.

The gene sequence-based overview of terrestrial biological diversity is embodied in phylogenetic trees, relatedness diagrams such as that shown in Fig. 11.1.[9] The construction of a phylogenetic tree is conceptually simple. The number of differences between pairs of corresponding sequences from different organisms is taken to be some measure of the "evolutionary distance" that separates them. Pair-wise differences between the sequences of many organisms can be used to infer maps of the evolutionary paths that led to the modern-day sequences. The phylogenetic tree shown in Fig. 11.1 is based on small-subunit ribosomal RNA (rRNA) gene sequences, but the same topology results from comparing sequences of any other genes involved in the nucleic acid-based information-processing system of cells, the core of genetic continuity. On the other hand, phylogenetic trees based on metabolic genes, those involved in manipulation of small molecules and in interaction with the environment, sometimes are not congruent with the rRNA topology. Such genes do not offer any consistent alternative to the rRNA tree, however. Consequently, patterns that deviate from the rRNA tree probably are best interpreted to reflect lateral transfers of genes or even the intermixing of genomes in the course of evolution.[10] The genome of any particular organism is comprised of genes derived evolutionarily from both vertical and lateral transmission.

Phylogenetic trees are rough maps of the evolution and diversification of life on Earth. From the standpoint of both sequence divergence and complexity, most of Earth's life is seen to be microbial in nature, which is surely what we need to expect of life that might occur elsewhere in our solar system. Some of the conclusions that can be drawn from the molecular trees verify previously advanced biological hypotheses. For instance, the molecular trees confirm what was once a hypothesis, that the major organelles, mitochondria and chloroplasts, were derived from bacteria, *proteobacteria* and *cyanobacteria*, respectively. The biochemical trait of oxygenic photosynthesis arose with the cyanobacterial radiation (indicated in Fig. 11.1 by the lines leading to *Synechococcus, Gloeobacter,* and chloroplast). Because the cyanobacterial radiation is peripheral in the tree, early life must have been anaerobic and most of bacterial diversification must have happened before the availability of oxygen.

Other conclusions from the molecular trees clarify relationships among terrestrial life. The molecular trees show that Earth's main lines of descent fall into three relatedness groups, the "domains" of Archaea, Bacteria, and Eucarya (eukaryotes). The point of origin of the lines leading to the modern domains cannot be determined by using rRNA gene sequences alone. Comparison of gene family sequences such as the protein synthesis factors Ef-Tu and Ef-G, that diverged before the last common ancestor of life, however, indicate an origin deep on the bacterial line as shown in Fig. 11.1. This relationship means that Archaea and Eucarya shared common ancestry subsequent to the separation of their common ancestor from Bacteria. Biochemical properties of the organisms are consistent with this conclusion. For example, the transcription and translation machineries of modern-day representatives of Archaea and Eucarya are far more similar to one another than either is to corresponding functions in Bacteria. This result shows that the eucaryal nuclear line of descent is not a relatively recent derivative of symbiosis, rather, is as old as the line of Archaea. This result also indicates that the common textbook presentation of life as divided into two categories, prokaryote and eukaryote, is incomplete. Rather, terrestrial life is of three kinds: archaeal, bacterial, and eucaryal, distinct from one another in fundamental ways.

Gene sequences that are common to all organisms are incisive signatures of terrestrial origin. This is because organisms with independent origins are unlikely to have evolved identical genetic sequences, even if the chemical structures of the subunits that comprise the genetic information were identical. Thus,

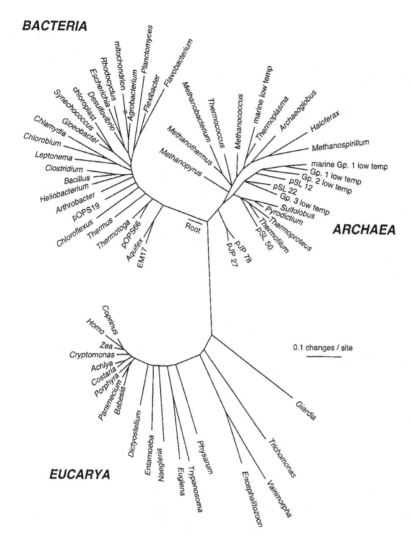

Fig. 11.1. Universal phylogenetic tree based on small-subunit rRNA sequences. Sixty-four rRNA sequences representative of all known phylogenetic domains were aligned, and a tree was constructed by using the program FASTDNAML, correcting for multiple and back mutations. That tree was modified to the composite one shown by trimming lineages and adjusting branch points to incorporate results of other analyses. Evolutionary distance (sequence difference) between the species shown is read along line segments. The scale bar corresponds to 0.1 changes per nucleotide. (Reproduced with permission from ref. 9 (Copyright 1997, American Association for the Advancement of Science).)

in gene sequences we can recognize terrestrial life, distinguish it from life derived from a different evolutionary origin even in the face of substantial biochemical similarity. This would become a significant issue if life—or its remains—were discovered on another body in the solar system.

Because planetary systems are formed by accretion, I think it unlikely that life on another body in the solar system arose independently of terrestrial life. It is now clear from meteorite studies that bodies can be transported from one planet to another, for instance from Mars to Earth, without excessive heating that would sterilize microbial organisms.[11] Although such transfer events are now rare, they must have been far more frequent during the accretion of the planets. Large-scale infall, blasting ejecta throughout the forming solar system, probably extended until at least about 4 billion years ago and so probably overlapped with the processes that resulted in the origin of life. In principle, life, regardless of where it arose, could have survived interplanetary transport and seeded the solar system wherever conditions occur that are permissible to life. So, if we go to Mars or Europa and find living creatures there, and read their rRNA genes, we should not be surprised if the sequences fall into our own relatedness group, as articulated in the tree of life.

My research activities are supported by the National Institutes of Health, National Science Foundation, and the National Aeronautics and Space Administration Astrobiology Institute.

NOTES

This chapter originally appeared in *Proceedings of the National Academy of Sciences* 98(3) (2001), 805–808.

REFERENCES

1. Committee on Planetary and Lunar Exploration, Space Studies Board, Commission on Physical Sciences, Mathematics and Applications, & National Research Council (1990). *Strategy for the detection and study of other planetary systems and extrasolar planetary materials: 1990–2000.* Washington, DC: National Academy Press.

2. Feinberg, G. & Shapiro, R. (1980). *Life beyond Earth: The intelligent Earthling's guide to life in the universe.* New York: Morrow.

3. Miller, S. L. & Orgel, L. E. (1974). *The origins of life on Earth.* Englewood Cliffs, NJ: Prentice-Hall.

4. Gold, T. (1998). *The deep, hot biosphere.* New York: Springer.

5. Stetter, K. (1999). Extremophiles and their adaptation to hot environments. *FEBS Letters*, **452**, 22–25.

6. Rivkina, E. M., Friedmann, E. L, McKay, C. P., & Gilichinsky, D. A. (2000). Metabolic activity of permafrost bacteria below the freezing point. *Applied and Environmental Microbiology*, **66**, 3230–3233.

7. Price, B. P. (2000). A habitat for psychrophiles in deep Antarctic ice. *Proceedings of the National Academy of Sciences*, **97**, 1247–1251.

8. Commission on Physical Sciences, Mathematics and Applications, & the Space Studies Board (1999). *Size limits of very small microorganisms: Proceedings of a workshop.* Washington DC: National Academy Press.

9. Pace, N. R. (1997). A molecular view of microbial diversity and the biosphere. *Science*, **276**, 734–740.

10. Woese, C. R. (2000). Interpreting the universal phylogenetic tree. *Proceedings of the National Academy of Sciences*, **97**, 8392–8396.

11. McKay, D. S., Gibson, E. K., Jr., Thomas-Kerpta, K. L., Vali, H., Romanek, C. S., & Zare, R. N. (1996). Search for past life on Mars: Possible relic biogenetic activity in Martian meteorite ALH84001. *Science*, **273**, 924–930.

12 • Is there a common chemical model for life in the universe?

STEVEN A. BENNER, ALONSO RICARDO,
AND MATTHEW A. CARRIGAN

INTRODUCTION

A simple "We don't know" is often the best answer for some questions, perhaps even for the title question, suggested by this issue's editors of *Current Opinion in Chemical Biology*. Because we have direct knowledge of life only on Earth, and as known life on Earth descended from a single ancestor, we have only one data point from which to extrapolate statements about the chemistry of life generally. Until life is encountered elsewhere, or aliens contact us, we will not have an independent second dataset. We may not even then, if the alien life itself shares an ancestor with life on Earth.

We can, of course, conceive of alternative chemical solutions to specific challenges presented to living systems.[1] We can then test their plausibility by synthesizing, in the laboratory, unnatural organic molecules that represent the alternatives, and seeing if they behave suitably. Some of these alternatives come from simple "Why?" or "Why not?" questions. For example: why are 20 standard amino acids used in Terran proteins? Experiments with unnatural amino acids (using the natural ribosome to incorporate them into proteins) have expanded the amino acid repertoire of proteins.[2–7] These experiments find no reason to exclude alternative sets of amino acids from hypothetical proteins in hypothetical alien life forms.

Similar questions can be asked about DNA. For example, why does Terran genetics use ribose and deoxyribose? Why not glycerol, a hexose or a tetrose? Again, synthetic chemists have made DNA analogs using each, and studied their behavior.[8–10] These studies underlie a rational discussion of the etiology and design of DNA.

Some questions reach farther. Is a three-biopolymer system (DNA-RNA-proteins) essential for life? Why not two biopolymers, or just one? Is water necessary? Is carbon essential? Why not silicon? Here, direct experiments are more difficult to conceive; few have actually been attempted.

Nevertheless, a discussion of "Why?" and "Why not?" questions posed by Terran biochemistry can start with a trialectic of three classes of explanations.[11] The first class is functional, and hypothesizes that a chemical structure is found in life because it is the best solution to a particular biological problem. Under the Darwinian paradigm, "best" means "confers most fitness", or (at least) more fitness than unused alternatives. This explanation implies that the living system had access to alternative solutions, individuals chose among these, and those who chose poorly failed to survive.

The second class of explanation is historical. The universe of chemical possibilities is huge. For example, the number of different proteins 100 amino acids long, built from combinations of the natural 20 amino acids, is larger than the number of atoms in the cosmos. Life on Earth certainly did not have time to sample all possible sequences to find the best. What exists in modern Terran life must therefore reflect some contingencies, chance events in history that led to one choice over another, whether or not the choice was optimal.

Explanations of this class are difficult to evaluate because we know little about the hypersurface relating the sequence of a protein to its ability to confer fitness.[11] If that hypersurface is relatively smooth, then optimal solutions might be reached via Darwinian search processes, regardless of where contingencies began the search. If the surface is rugged, however, then many structures of Terran life are locally optimal, and suboptimal with respect to the universe of

The Nature of Life, ed. M. A. Bedau and C. E. Cleland. Published by Cambridge University Press. © M. A. Bedau and C. E. Cleland 2010.

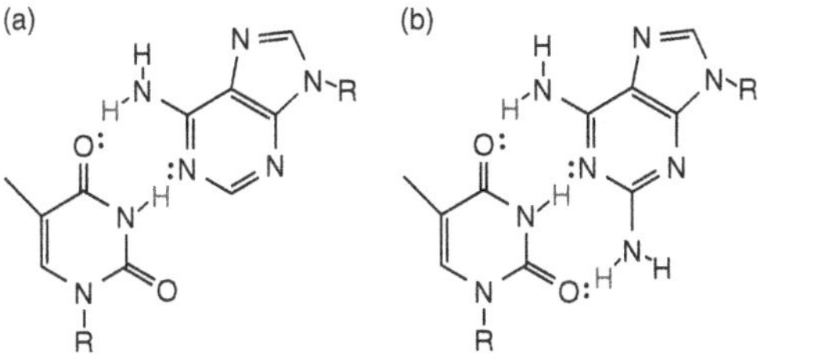

Current Opinion in Chemical Biology

Fig. 12.1. The thymine–adenine T · A base pair (a) is joined by two hydrogen bonds. Why is aminoadenine (b) not used instead of adenine, to give a stronger base pair joined by three hydrogen bonds?

possibilities. Darwinian mechanisms cannot easily find a globally optimal solution on a rugged hypersurface.

The third class draws on the concept of vestigiality. Here, it is recognized that some features of contemporary Terran biochemistry may have emerged via selective pressures that no longer exist. The contemporary feature may therefore not represent optimization for the modern world, but rather be a vestige of optimization in an ancient world. Such models explain, for example, the human appendix, and many of the chemical details found in modern life (such as the RNA component of many cofactors).

These three classes of explanations are illustrated by possible answers to a question asked by many biochemistry students: why does DNA use adenine, which forms only two hydrogen bonds with thymidine? Aminoadenine gives stronger pairing.[12] (Fig. 12.1). A functional explanation might argue that genomes are optimal when they have both a weak (A*T) and strong (G*C) pair. A historical explanation suggests that adenine was arbitrarily chosen over other available candidates by a contingent accident, and conserved because it was too difficult to later replace without losing fitness. A vestigiality explanation holds that adenine, not aminoadenine, is made prebiotically from ammonium cyanide.[13] Here, adenine's prebiotic availability made it better for starting life, even though aminoadenine might be preferable now.

It is useful to construct these alternative explanations as hypotheses, even if no experiments can presently test them. The act of constructing the trialectic allows scientists to appreciate how accessible alternative explanations are. This, in turn, provides a level of discipline to a discussion where the "We don't know" answer is probably most appropriate. This article reviews lightly some of the chemical constraints on life, and some possible chemistries of alternative life forms.

WHAT IS LIFE?

To decide whether life has a common chemical plan, we must decide what life is.[14,15] A panel assembled by NASA in 1994 was one of many groups to ponder this question. The panel defined life as a "chemical system capable of Darwinian evolution."[16] This definition, which follows an earlier definition by Sagan,[17] will be used here.

This definition contrasts with many others that have been proposed, and avoids many of their pitfalls. For example, some definitions of life confuse "life" with the concept of being "alive." Thus, asking if an entity can move, eat, metabolize or reproduce might ask whether it is alive. But an individual male rabbit (for example) that is alive cannot (alone) support Darwinian evolution, and therefore is not "life."[18]

Further, many efforts to define "life" fall afoul of the fact that no non-trivial term can be defined to philosophical completeness.[19] The general difficulty of defining terms, theoretically or operationally, was one of the discoveries of twentieth century philosophy, and is not unique to the definition of "life". It is impossible, for example, to define "water" in a complete way. We can say that water is "dihydrogen oxide," but are we speaking of a water molecule, water as a substance, or water operationally? And what is "hydrogen"? Any effort to deal with the definition of life at this level encounters analogous questions, which can easily be paralyzing.

The phrase "Darwinian evolution" carries baggage from 150 years of discussion and elaboration. It makes specific reference to a process that involves a molecular system (DNA on Earth) that is replicated imperfectly, where the imperfections are themselves heritable. Therefore, Darwinian evolution implies more than reproduction, a trait that ranks high in many definitions of life.

The panel's definition also avoids confusion from many non-living systems that reproduce themselves. For example, a crystal of sodium chlorate can be powdered and used to seed the growth of other crystals;[20] the crystal thereby reproduces. Features of the crystal, such as its chirality, can be passed to descendants.[20] The replication is imperfect; a real crystal of sodium chlorate contains defects. To specify all of the defects would require enormous information, easily the amount of information in the human genome. But the information in these defects is not itself heritable. Therefore, the crystal of sodium chlorate cannot support Darwinian evolution. Therefore, a sodium chlorate system is not life.[21]

The NASA panel's definition of life is interesting for other reasons. First, it provides information about what forms of life were believed to be possible, not just conceivable. As fans of Star Trek know, forms of life that are *not* chemical systems capable of Darwinian evolution are easily conceivable. Besides aliens resembling Hollywood actors with prostheses, the Enterprise has encountered conceptual aliens that do not fit the panel's definition. The nanites that infected the Enterprise computer in Episode 50 of Star Trek: The Next Generation ("Evolution") are informational in essence; their Darwinian evolution is not tied to an informational molecule, like DNA (although they require a chemical matrix to survive). The Crystalline Entity of Episodes 18 ("Home Soil") and 104 ("Silicon Avatar") appears to be chemical, but not obviously Darwinian. The Calamarain (Episode 51: "Deja Q") are a conceptual life form that is purely energy, not evidently requiring matter. And Q (Episode 1: "Encounter at Farpoint") appears to be neither matter nor energy, flitting instead in and out of the Continuum without the apparent need of either.

If we were to encounter any of these other conjectural entities during a real, not conceptual, trek through the stars, we would be forced to concede that they represent living systems. We would be obligated to change our definition of life. We do not change it now simply because we do not believe that the weirder life conceived in the Star Trek scripting room is possible outside of that room.

Nanites and the fictional android Data are examples of artificial life. We do not doubt that androids can be created, including androids who (note the pronoun) desire to be human. We do not, however, believe that Data could have arisen spontaneously, without a creator that had already emerged by Darwinian process (as is indeed the case with Data). Hence, we might regard Data as a biosignature, or even agree that he is alive, even if we do not regard him as a form of life. For the same reason, a computer holding

nanites would be evidence that a life form existed to create it; the computer is a biosignature, and the nanites are an artificial life form, something requiring natural life to emerge.

THE BASIC REQUIREMENTS FOR LIFE

The NASA panel's definition is also useful because it makes direct reference to chemical principles, which we understand well from Terran science. Chemical principles generate a hierarchy of requirements needed for life under the definition. These have a good chance of being common for life in the cosmos, at least the life falling within the definition. We list them from the most stringent (and most likely to be universal) to the less stringent.

THERMODYNAMIC DISEQUILIBRIUM

To the extent that Darwinian evolution is a progressive process, and to the extent that life actually does something, life requires an environment that is not at thermodynamic equilibrium. This statement is almost certainly true for all life, including artificial life.

The statement does not imply a universal chemical metabolism, even on Earth. Wherever a thermodynamic disequilibrium exists on Earth, organisms appear to have adapted a different metabolism to exploit it. But these organisms all require the disequilibrium.

The requirement for an environment that is at disequilibrium is often paraphrased as a requirement for an "energy source" or "high energy compounds." While technically true, this locution can create problems. For example, de Duve speaks of life emerging in acidic environments at high temperatures, because "energy rich" thioesters form spontaneously in such environments.[22] Indeed they do. But in that environment, thioesters are not "high energy" (they form spontaneously). Likewise, in an environment poor in water but rich in ADP and inorganic phosphate, the formation of ATP is spontaneous, and ATP has a lower "energy" than ADP and inorganic phosphate.

Thermodynamic disequilibrium is easy to find in the cosmos. Almost any environment, in the vicinity of a nuclear fusion reaction, such as that occurring in our Sun, will not be at thermodynamic equilibrium. Indeed, unshielded exposure to the energy efflux of the Sun creates a problem of having too much energy, as those with sunburn appreciate.

For most of human history, a star was regarded as the *only* way to obtain thermodynamic non-equilibrium (Genesis 1.3 and the Popol Vuh, for example). Robert Heinlein suggested that the ultimate punishment from an intergalactic court is to deprive a miscreant planet of its sun.[23] In this view, a universal feature of life is a food chain that begins with an autotrophic organism that converts photons from a star into chemical disequilibrium.

But other disequilibria are available within planets. For example, heavier atomic nuclei, left from a supernova, are not at thermodynamic equilibrium. Decay of these nuclei is a powerful source of planetary not-at-equilibrium environments. Radioactive decay drives tectonics and volcanism on Earth. These create non-equilibrium environment in many areas,[24] such as black smokers on the ocean floor.[25] The consequent energetic disequilibrium supports life despite the absence of direct solar energy.

Recognizing that geothermal energy is easily placed at the bottom of a food chain, Stevenson noted that small, Earth-like planets that support life from geothermal energy might be ejected from forming solar systems.[26] They could then travel the galaxy, carrying life completely independent of a star. Depending on the frequency of these ejections, such planets might hold the vast majority of life in our galaxy, all living without a sun on the decay of radioactive nuclei left over from past supernova explosions.

BONDING

A weaker constraint on life requires that it be based on covalent bonds. The concept is imperfect; the distinction between covalent and non-covalent bonding itself reflects the language of organisms living at 310 K. There is, in fact, a range of chemical bonds, from the very strong, surviving for significant times even at temperatures as high as 1000 K, to the very weak, existing only near absolute zero.

Terran life uses various types of bonding. The carbon–carbon covalent bond, typically $400 \, kJ \, mol^{-1}$ with respect to homolytic fragmentation, is the strongest of these. A hydrogen bond is worth only $20 \, kJ \, mol^{-1}$. It is nevertheless strong enough to be important in Terran life at 273–373 K. The hydrogen bond gives water its most distinctive physical properties, including its high boiling point, the large range in temperature over which it is a liquid, and its increase in volume upon freezing. In addition, the hydrogen bond is largely responsible for the hydrophobic effect, and the fact that water and oil form separate phases at 310 K.

Most of the universe does not lie between 273 and 373 K, however, the range where covalent bonds are necessary and hydrogen bonds are useful. At lower temperatures, biochemistry may well be dominated by non-covalent bonds. The bonding that supports information transfer needed for Darwinian evolution might universally require bonding sufficiently strong at the ambient temperature to be stable for some appropriate time. In water at 273–373 K, a combination of covalent bonding, hydrogen bonding and the hydrophobic effect meets this requirement. In different solvents or temperatures, different bonding may be better suited.

ISOLATION WITHIN THE ENVIRONMENT

Selfishness is essential to Darwinian evolution. A Darwinian cycle can proceed only if it replicates itself in preference to others. An enormous literature discusses isolation in the context of parasitism and altruism.[27] Much of the recent literature on molecular systems to achieve isolation focus on liposomes and other membrane-like structures that resemble lipid bilayers found throughout Terran life.[28–31] An interesting result from the Szostak laboratory connects mineralogy to liposome formation.[32] This type of compartmentalization is based on liquid/liquid immiscibility and the hydrophobic effect. This need not be the only way to achieve isolation, of course. Isolation can be achieved on a two-dimensional surface, where a living system has a fractile 2+ dimensionality.[33]

CARBON-LIKE SCAFFOLDING

The stability of the carbon–carbon bond at 273–373 K has made it the first choice element to scaffold biomolecules. Even in Terran biochemistry, other elements are required. Hydrogen is needed for many reasons; at the very least, it terminates C–C chains. Heteroatoms

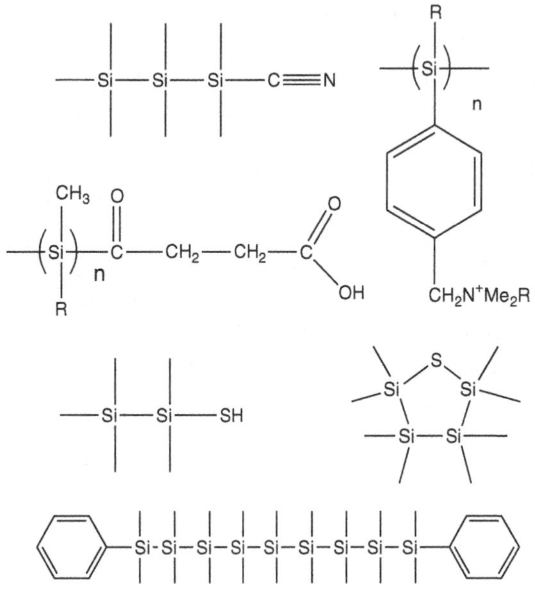

Current Opinion in Chemical Biology

Fig. 12.2. Structures of some synthetic polysilanes that have been described in the literature. There is little question that polymeric diversity is possible using silicon, rather than carbon, as a scaffolding.

(atoms that are neither carbon nor hydrogen) determine the reactivity of carbon-scaffolded biomolecules. In Terran life, these are oxygen, nitrogen and, to a lesser extent, sulfur, phosphorus, selenium, and an occasional halogen.

While carbon is widely regarded as the only scaffolding element,[34,35] silicon is a frequently mentioned second choice (see the original Star Trek Episode 26, "The Devil in the Dark"). Like carbon, silicon can form four bonds. The silicon–silicon bond is weaker than the carbon–carbon bond,[36] but not excessively (typically ca. $300 \, kJ \, mol^{-1}$, compared with $400 \, kJ \, mol^{-1}$ for a typical carbon–carbon bond).

Many compounds built on a scaffolding of silicon–silicon bonds have in fact been made. Oligosilanes having up to 26 consecutive Si–Si bonds are known; these support a variety of functionalized and non-functionalized side chains (Fig. 12.2). These structures resemble those from Terran membranes. These have been targets of chemists because of their ability to form sigma-conjugated polysilanes.[37] The electronic

properties of these compounds are similar to those of the carbon-containing pi-conjugated system.

Oligosilanes generally have alkyl side chains. These are generally soluble in non-polar solvents. Shorter polysilanes are known containing reactive ends (cyanide, hydroxyl or phenol) that allow further modifications and solubility in polar solvents. For example, phenolic oligosilanes dissolve in alcohol, ether acetone and slightly alkaline water. Oligosilanes carrying carboxylic acid groups are soluble in water.[38,39] Amphiphilic oligosilanes self-aggregate in water, creating vesicles and micelles.[40] Oligosilanes can also be chiral (pace the statement on the NASA astrobiology website http://nai.arc.nasa.gov/astrobio/feat_ questions/silicon_life.cfm).

The silicon–oxygen–silicon (silicone) unit also makes scaffolds, found today in superballs and breast implants. Again, silicon and oxygen alone cannot generate particularly interesting chemistry. For this, the Si–O- units must have side chains that include carbon, hydrogen, nitrogen, oxygen and other heteroatoms. The life would nevertheless be called silicon-based, just as we call Terran life carbon-based, even though it depends on other atoms as much as on carbon.

The reactivities of silicon and carbon differ in some notable ways. First, its position in the third row of the periodic table means that silicon carries low energy d orbitals. Therefore, the associate mechanism for reaction at silicon, where an incoming nucleophile forms a bond before a leaving group leaves, is available to silicon, but not to carbon. This makes many compounds containing silicon more reactive than their carbon analogs. Whether this difference in reactivity is advantageous or disadvantageous depends on perspective. The greater reactivity of silicon compared with carbon may be an advantage in cold environments.[1]

The weaker criterion is that life requires scaffolding based on carbon or silicon. There is no third choice if covalent scaffolding is viewed as a requirement. Weirder forms of life based on ionic bonding, or metallic bonding, are conceivable in principle, but not otherwise at this time. Nevertheless, it is instructive to review the writings of Cairns-Smith and Hartman,[41,42] who have considered issues relating to information transfer in systems that are neither carbon-based nor dependent on covalent bonding.

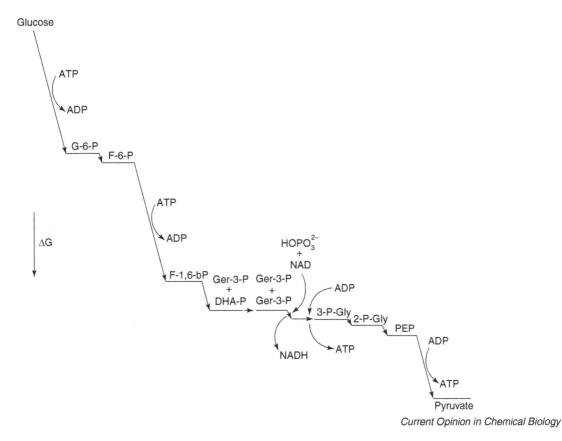

Fig. 12.3. Physiological free energy diagram for the glycolysis reaction for one of the two Ger-3-P generated. G, glucose; G-6-P, glucose-6-phosphate; F-6-P, fructose-6-phosphate; F-1,6-bP, fructose-1,6-bisphosphate; Ger-3-P, glyceraldehyde- 3-phosphate; DHA-P, dihydroxyacetone-P; 3-P-Gly, 3-phosphoglycerate; 2-P-Gly, 2-phosphoglycerate; PEP, phosphoenolpyruvate. Vertical height is proportional to the change in free energy (ΔG).

Current Opinion in Chemical Biology

ENERGETIC PATTERNS IN METABOLISM

For a metabolic sequence to convert substrates into products, the two must be in disequilibrium. When they are not, the overall reaction must be coupled to another reaction where substrates and products are at disequilibrium. This is frequently phrased by textbooks as a requirement that a metabolic pathway must be energetically downhill, or coupled to an energy source.

Many Terran metabolic multistep pathways are run close to equilibrium for internal steps (Fig. 12.3), with the first step being energetically downhill, and a site of regulation. Required, however, is that the last step be energetically downhill, thereby pulling the reaction to completion.[43] This feature may be universal in metabolic pathways, simply because it exploits most economically a surrounding chemical disequilibrium. Having large free energy drops at every step in a pathway is wasteful.

This characteristic energetic relationship between a set of compounds that are intermediates in an evolved metabolism may be a universal biosignature. If an inventory of the small molecules in a suspected living system (on Saturn's largest moon, Titan, for example) reveals this characteristic energetic relationship, this may be evidence of Darwinian evolution acting to create an optimal metabolism.

THE SOLVENT

A liquid phase facilitates chemical reactions, something known empirically for centuries. As a solvent, a liquid allows dissolved reactants to encounter each other at rates higher than the rates of encounter between species in a solid. Chemical reactions can take place in the gas and solid phases as well, of course. But each of these has disadvantages relative to the liquid phase.

In the gas phase, chemistry is limited to molecules that are sufficiently volatile to deliver adequate amounts of material to the gas phase at moderate temperatures, and/or to molecules sufficiently stable to survive higher temperatures where vapor pressures are higher. Obviously, if volume is abundant, pressures are low, and time scales are long, even low concentrations of biomolecules might support a biosystem in the gas phase. It is even conceivable that in the vacuum of interstellar space, life exists through molecules at high dilution reacting in the gas phase. The possible disadvantage of this is, of course, the difficulties in holding together the components of the interstellar life form. One advantage would be that it need not be encumbered by the lifetime of a planetary system. Indeed, one can imagine life in the gas phase associated with a galaxy and its energy flux for nearly the age of the universe.

Likewise, species can diffuse through solids to give chemical reactions.[44] Solid-phase diffusion is slow, however. Nevertheless, given cosmic lengths of time, and the input of energy via high energy particles, a biochemistry able to support Darwinian evolution can be conceived.[45] A weird life form might reside in solids of the Oort cloud (a large sphere of cosmic debris surrounding the solar system, and the origin of comets) living in deeply frozen water, obtaining energy occasionally from the trail of free radicals left behind by ionizing radiation, and carrying out only a few metabolic transformations per millennium.

Solvents like water, but not water

Liquid ammonia is a possible solvent for life, an alternative to water. Indeed, water and ammonia are analogous. Ammonia, like water, dissolves many organic compounds. Preparative organic reactions are done in ammonia in the laboratory. Ammonia, like water, is liquid over a wide range of temperatures (195–240 K at 1 atm). The liquid range is even broader at higher pressure. For example, at 60 atm ammonia is liquid at 196–371 K. Further, liquid ammonia may be abundant in the solar system. It exists, for example, in the clouds of Jupiter.

The weaker hydrogen-bonding potential of ammonia is often considered undesirable in some discussions of ammonia as a biosolvent. Ammonia has three hydrogen bond donors and only one hydrogen-bonding acceptor, whereas water has two of each. This imbalance is used to explain the lower boiling point of ammonia compared with water.

The increased ability of ammonia to dissolve hydrophobic organic molecules (again compared with water) suggests an increased difficulty in using the hydrophobic effect to generate compartmentalization in ammonia, relative to water. This, in turn, implies that liposomes, a compartment that works in water, generally will not work in liquid ammonia.

Hydrophobic phase separation is possible in ammonia, however, albeit at lower temperatures. For example, Brunner reported that liquid ammonia and hydrocarbons form two phases, where the hydrocarbon chain contains 1–36 CH_2 units.[46] Different hydrocarbons become miscible with ammonia at different temperatures and pressures. Thus, phase separation useful for isolation would be conceivable in liquid ammonia at temperatures well below its boiling point at standard pressures.

The greater basicity of liquid ammonia must also be considered. The acid and base in water are H_3O^+ and HO^-. In ammonia, NH_4^+ and NH_2^- are the acid and base, respectively. H_3O^+, with a pK_a of –1.7, is ca. 11 orders of magnitude stronger an acid than NH_4^+, with a pK_a of 9.2 (in water). Likewise, NH_2^- is about 15 orders of magnitude stronger as a base than HO^-.

Ammonia would not support the chemistry found in Terran life, of course. Terran life exploits compounds containing the C=O carbonyl unit. In ammonia, carbonyl compounds are converted to compounds containing the corresponding C=N unit. Nevertheless, hypothetical, reactions that exploit a C=N unit in ammonia can be proposed in analogy to the metabolic biochemistry that exploits the C=O unit in Terran metabolism in water (Fig. 12.4).[47] Given this adjustment, metabolism in liquid ammonia is easily conceivable.

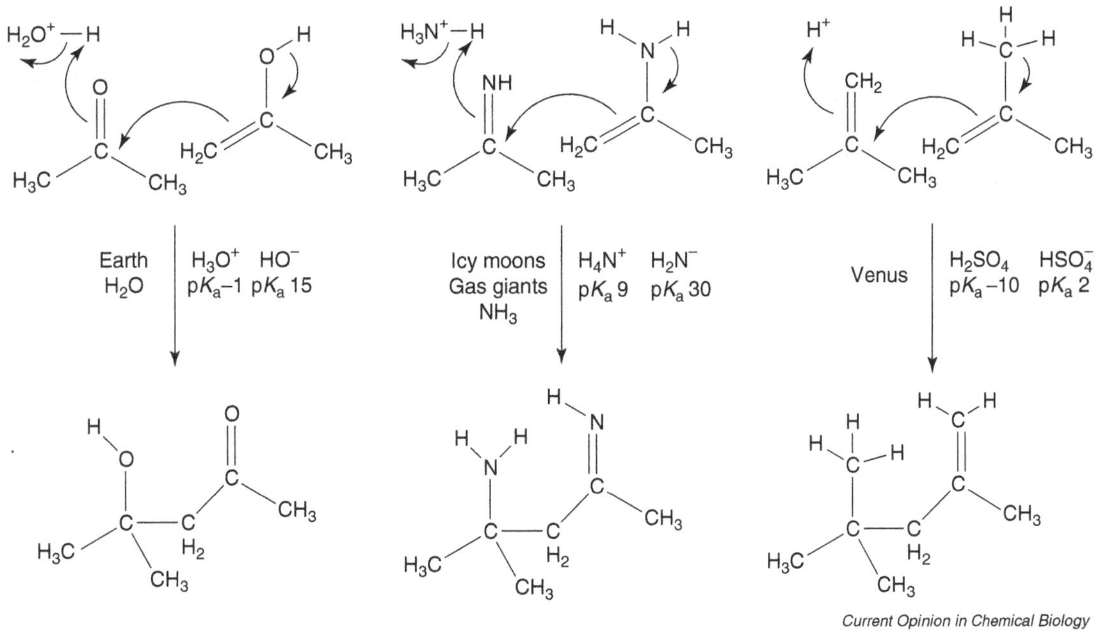

Fig. 12.4. Different functional groups, but analogous mechanisms, could be used to form new carbon–carbon bonds in different solvents. In water, the C=O unit would provide the necessary reactivity. In ammonia, the C=N unit would provide the necessary reactivity. In sulfuric acid, the C=O unit is sufficient to provide the necessary reactivity.

Current Opinion in Chemical Biology

Ammonia is not the only polar solvent that might serve as an alternative to water. For example, sulfuric acid is a reasonably good solvent that supports chemical reactivity.[48] Sulfuric acid is known above Venus.[49] Here, three cloud layers at 40–70 km are composed mostly of aerosols of sulfuric acid, ca. 80% in the upper layer and 98% in the lower layer.[50] The temperature (ca. 310 K at ca. 50 km altitude, at ca. 1.5 atm) is consistent with stable carbon–carbon covalent bonds.

Many authors have discussed the possibility of life on Venus in its acidic environment,[51,52] replacing earlier views that Venus might be covered by swamps,[53] or by hot seas at the poles.[54] The surface temperature of Venus is approximately 740 K. Sagan and Morowitz even considered organisms that float above the hot surface using hydrogen "float bladders"[55] analogous to those found in Terran aquatic organisms.[56] Schulze-Makuch et al. argued for sample return from the Venusian atmosphere to address the possibility of life there.[57]

Metabolic hypotheses are not in short supply for the hypothetical life in acidic aerosols. In strong acid, the C=O bond is reactive as a base, and can support a metabolism as an analog of the C=O unit (Fig. 12.4). This type of chemical reactivity is exemplified in some Terran biochemistry. For example, acid-based reactions of the C=O unit are used by plants when they synthesize fragrant molecules.[58,59] Nor are sources of energy in short supply in the Venusian atmosphere. For example, a Venusian metabolism might exploit the relatively high flux of ultraviolet radiation in the Venusian clouds.[60]

Formamide is a third solvent of biological interest. Formamide, formed by the reaction of hydrogen cyanide with water, is polar like water. In formamide, however, many species that are thermodynamically unstable in water with respect to hydrolysis, are spontaneously synthesized. This includes ATP (from ADP and inorganic phosphate), nucleosides (from ribose borates[61] and nucleobases), peptides (from amino acids), and even oligoribonucleotides.[62–64] Formamide is itself hydrolyzed, meaning that it persists only in a relatively dry environment, such as a desert. As desert environments have recently been proposed as being potential sites for the prebiotic synthesis of ribose,[61] they may hold formamide as well.

Hydrocarbons: Non-polar solvents for biology?

There is no need to focus on polar solvents, like water, when considering possible habitats for life. Hydrocarbons, ranging from the smallest (methane) to higher homologs (ethane, propane, butane etc.) are abundant in the solar system.

Methane, ethane, propane, butane, pentane and hexane have boiling points of ca. 109, 184, 231, 273, 309 and 349 K, respectively, at standard pressure. Thus, oceans of ethane may exist on Titan. At a mean surface temperature estimated to be 95 K, methane (which freezes at 90 K) would be liquid, implying that oceans of methane could cover the surface of Titan.

Many discussions of life on Titan have considered the possibility that water, normally frozen at the ambient temperature, might remain liquid following heating by impacts.[65] Life in this aqueous environment would be subject to the same constraints and opportunities as life in water. Water droplets in hydrocarbon solvents are, in addition, convenient cellular compartments for evolution, as Tawfik and Griffiths have shown in the laboratory.[66] An emulsion of water droplets in oil is obtainable by simple shaking. This could easily be a model for how life on Titan achieves Darwinian isolation.

But why not use the hydrocarbons that are naturally liquid on Titan as a solvent for life directly? Broad empirical experience shows that organic reactivity in hydrocarbon solvents is no less versatile than in water. Indeed, many Terran enzymes are believed to catalyze reactions by having an active site that is not water-like.

Further, hydrogen bonding is difficult to use in the assembly of supramolecular structures in water. In ethane as a solvent, a hypothetical form of life would be able to use hydrogen bonding more; these would have the strength appropriate for the low temperature. Further, hydrocarbons with polar groups can be hydrocarbon-phobic; acetonitrile and hexane, for example, form two phases. One can conceive of liquid/liquid phase separation in bulk hydrocarbon that could achieve Darwinian isolation.

The reactivity of water means that it destroys hydrolytically unstable organic species. Thus, a hypothetical form of life living in a Titan hydrocarbon ocean would not need to worry as much about the hydrolytic deamination of its nucleobases, and would be able to guide reactivity more easily than life in water.

This is understood by preparative organic chemists, who prefer non-aqueous solvents to water as media for running organic reactions in the laboratory. For example, in a recent issue of the *Journal of Organic Chemistry*, chemists used a solvent other than water to run their reactions over 80% of the time. Chemists avoid water as a solvent because it itself is reactive, presenting both a nucleophilic oxygen and an acidic hydrogen at 55 molar concentrations. Thus, in many senses, hydrocarbon solvents are better than water for managing complex organic chemical reactivity.

Thus, as an environment, Titan certainly meets all of the stringent criteria outlined above for life. Titan is not at thermodynamic equilibrium. It has abundant carbon-containing molecules and heteroatoms. Titan's temperature is low enough to permit a wide range of bonding, covalent and non-covalent. Titan undoubtedly offers other resources believed to be useful for catalysis necessary for life, including metals and surfaces.

This makes inescapable the conclusion that if life is an intrinsic property of chemical reactivity, life should exist on Titan. Indeed, for life *not* to exist on Titan, we would have to argue that life is *not* an intrinsic property of the reactivity of carbon-containing molecules under conditions where they are stable. Rather, we would need to believe that either life is scarce in these conditions, or that there is something special, and better, about the environment that Earth presents (including its water).

Solvents that are not water and not hydrocarbon

The most abundant compound in the solar system is dihydrogen, the principal component (86%) of the upper regions of the gas giants, Jupiter, Saturn, Uranus and Neptune. The other principal component of the outer regions of the giant planets is helium (0.14%). Minor components, including methane (2×10^{-3}%), water (6×10^{-4}%), ammonia (2.5×10^{-4}%), and hydrogen sulfide (7×10^{-5}%) make up the rest.

But is dihydrogen a liquid? The physical properties of a substance are described by a phase diagram that relates the state of a material (solid of various types, liquid or gas) to temperature and pressure. A line typically extends across the phase diagram. Above this line, the substance is a gas; below the line, the substance is a liquid. Typically, however, the line

Table 12.1. *Critical temperature and pressure for selected substances*

Liquid	Critical temperature (K)	Critical pressure (atm)
Hydrogen	33.3	12.8
Neon	44.4	26.3
Nitrogen	126	33.5
Argon	151	48.5
Methane	191	45.8
Ethane	305	48.2
Carbon dioxide	305	72.9
Ammonia	406	112
Water	647	218

ends at a critical point. Above the critical point, the substance is a supercritical fluid, neither liquid nor gas. Table 12.1 shows the critical temperatures and pressures for some substances common in the solar system.

The properties of supercritical fluids are generally different than those of the regular fluids. For example, supercritical water is relatively non-polar and acidic. Further, the properties of a supercritical fluid, such as its density and viscosity, change dramatically with changing pressure and temperature as the critical point is approached.

The changing solvation properties of supercritical fluids near their critical points is widely used in industry.[67] For example, supercritical carbon dioxide, having a critical temperature of 304.2 K and pressure of 73.8 atm, is used to decaffeinate coffee. Supercritical water below the Earth's surface leads to the formation of many of the attractive crystals that are used in jewelry.

Little is known about the behavior of organic molecules in supercritical dihydrogen as a solvent. In the 1950s and 1960s, various laboratories studied the solubility of organic molecules (e.g., naphthalene) in compressed gases, including dihydrogen and helium.[68,69] None of the environments examined in the laboratory explored high pressures and temperatures, however.

Throughout most of the volume of gas giant planets where molecular dihydrogen is stable, it is a supercritical fluid. For most of the volume, however, the temperature is too high for stable carbon–carbon covalent bonding. We may, however, define two radii for each of the gas giants. The first is the radius where dihydrogen becomes supercritical. The second is where the temperature rises to a point where organic molecules are no longer stable; for this discussion, this is chosen to be 500 K. If the second radius is smaller than the first, then the gas giant has a "habitable zone" for life in supercritical dihydrogen. If the second radius is larger than the first, however, then the planet has no habitable zone.

If such a zone exists on Jupiter, it is narrow. Where the temperature is 300 K (clearly suitable for organic molecules), the pressure (ca. 8 atm) is still subcritical. At about 200 km down, the temperature rises above 500 K, approaching the upper limit where carbon–carbon bonds are stable.[70] For Saturn, Uranus and Neptune, however, the habitable zone appears to be thicker (relative to the planetary radius). On Saturn, the temperature is ca. 300 K when dihydrogen becomes supercritical. On Uranus and Neptune, the temperature when dihydrogen becomes supercritical is only 160 K; organic molecules are stable at this temperature.

The atmospheres of these planets convect. To survive on Jupiter, hypothetical life based on carbon–carbon covalent bonds would need to avoid being moved by convection to positions in the atmosphere where they can no longer survive. This is, of course, true for life in any fluid environment, even in Terran oceans. Sagan and Salpeter presented a detailed discussion of what might be necessary for a "floater" to remain stable in Jupiter's atmosphere.[71]

LIFE IN TERRAN-LIKE ENVIRONMENTS

Weird life might exist in weird environments. But we might also ask whether a common chemical model is found in life in Terran-like environments. Here, carbon is the scaffold and water is the solvent. By constraining our discussion to these environments, we fall into line with most literature, which is weighted heavily towards these as being required for life. Thus, in its search for signs of life on Mars, NASA has chosen to follow the water. Given what we know, this seems to be a sensible strategy.

WATER AS THE SOLVENT

Water certainly appears to be required for Terran life, which appears to live only under conditions where water is a liquid. Thus, the extremophiles that live above 373 K on the ocean floor do so only because high pressures there keep water liquid. Bacteria in the Antarctic ice pack presumably require melting to grow.[72–74]

As a test for the hypothesis that water is required for life, places on Earth that lack water might be searched to see if life has evolved to fill those environments. McKay and his colleagues have searched for life carefully in the Atacama desert, one of the driest spots on Earth.[75] They have concluded that life here is very scarce; assays using PCR failed to detect typical Terran biosignatures, and the number of cultures arising from samples from the driest area is at or near background.

Virtually every environment on Earth that has been examined, from deep in mine shafts to deep in oceans, in environments dominated by reductants as strong as hydrogen sulfide and methane to oxidants as strong as dioxygen, seems to hold life that has evolved from the universal ancestor of all life on Earth. As long as water is available, life finds a way to exploit whatever thermodynamic disequilibria exists.[76–78]

This experience suggests that Terran life needs water. But we can still ask whether water is specifically needed for life, or whether the Atacama desert example simply shows that *some* solvent is needed (as no other solvent is present in the Atacama). Once again, we have little experimental information to support an opinion one way or the other. But the properties of water are well known, and we can at least ask whether there is a common chemical model for life in water.

ADVANTAGES AND DISADVANTAGES OF WATER

Most of the literature in this area comments on properties of water that make it well suited for life. We can construct for each a "good news–bad news" dialectic.

Water, for example, expands when it freezes. This is useful to process rocks to make soils. This expansion also means that ice floats, permitting surface ice to insulate liquid water beneath, which can harbor life on an otherwise freezing planet. Ammonia and methane are both denser as solids than as liquids.

They therefore shrink when they freeze, exposing the liquid surface to further freezing.

Another feature of liquid water is the hydrophobic effect. This rubric captures the empirical observation that water and oil do not mix. This is a manifestation of the fact that water forms stronger hydrogen bonds to other water molecules than to oily molecules. The hydrophobic effect is key to the formation of membranes, which in turn support isolation strategies in Terran life. Likewise, when proteins fold, they put their hydrophobic amino acids inside, away from water.

On the other hand, water engages in undesired reactions as well. Thus, cytidine hydrolytically deaminates to give uridine with a half-life of ca. 70 years in water at 300 K.[79] Adenosine hydrolytically deaminates (to inosine), and guanosine hydrolytically deaminates to xanthosine at only slightly slower rates. As a consequence, Terran DNA in water must be continuously repaired. The toxicity of water creates special problems for the prebiotic chemistry, as repair mechanisms presumably require a living system.

UNIVERSAL CONSTRAINTS FOR LIFE IN WATER

Water also constrains the structure of carbon-containing biomolecules and their subsequent metabolism. At the very least, for life to exploit the power of water as a solvent, many biomolecules must be soluble in water. This is the simplest explanation for the prevalence of hydroxyl groups in organic molecules central to biological metabolism. Glucose has five, and dissolves well in water.

Charge also helps dissolution in water. Many intermediates in Terran metabolism are phosphorylated, giving them charges that increase solubility in water. The citric acid cycle is based on tri- and dicarboxylic acids. These are ionized at physiological pH, making them soluble in water.

DNA, the molecule at the center of Darwinian evolution on Earth, shows the importance of being soluble.[80] The DNA duplex, where a polyanion binds another polyanion, appears to disregard Coulomb's law. One might think (and indeed, many have thought)[81,82] that the duplex would be more stable if one strand were uncharged, or polycationic.[83,84]

Many efforts have been made to create non-ionic analogs of DNA and RNA.[82] For example, replacing

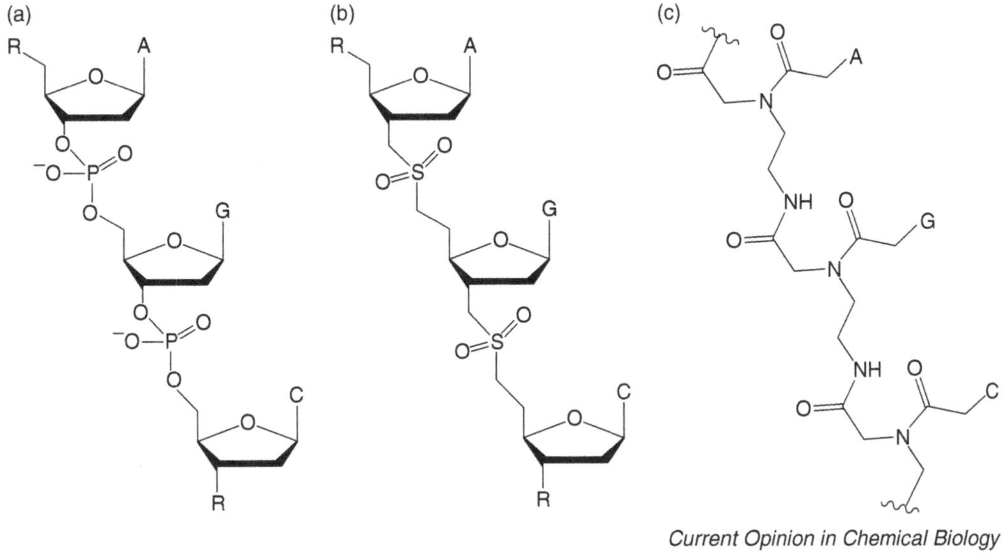

Current Opinion in Chemical Biology

Fig. 12.5. Replacing (a) the anionic phosphate diester linker with (b) the uncharged dimethylenesulfone linker generates DNA and RNA analogs that are rough isosteres of the phosphate analog. (c) Peptide nucleic acid (PNA) is a DNA/RNA mimic in which the phosphate–sugar backbone has been replaced by uncharged N-(2-aminoethyl)glycine linkages with the nucleobases attached through methylene carbonyl linkages to the glycine amino group.

the anionic phosphate diester linker with the uncharged dimethylenesulfone linker generates DNA and RNA analogs that are rough disasters of the phosphate analog (Fig. 12.5).[85–87] Short sulfone-linked DNA analogs (SNAs) display molecular recognition of the Watson–Crick type.[88] In longer oligosulfones, however, the loss of the repeating charge damages rule-based molecular recognition.[86,89] Further, SNAs differing by only one nucleobase displayed different levels of solubility, aggregation, folding and chemical reactivity.[90,91]

These results suggest three hypotheses for why charged phosphate linkages are important to molecular recognition in DNA. First, the repeating charges in the backbone force inter-strand interactions away from the backbone, causing the strands to contact at the Watson–Crick edge of the heterocycles (Fig. 12.6). Without the polyanionic backbone, inter-strand contacts can be anywhere.[92] Further, the repeating charges in the backbone keep DNA strands from folding. A flexible polyanion is more likely to adopt an extended conformation suitable for templating than a flexible neutral polymer, which is more likely to fold.[93]

Last, the repeating backbone charges allow DNA to support Darwinian evolution. As noted above,

replication is not sufficient for a genetic molecule to support Darwinian evolution. The Darwinian system must also generate inexact replicates, descendants whose chemical structures are different from those of their parents. These differences must then be replicable themselves. While self-replicating systems are well known in chemistry, those that generate inexact copies, with the inexactness itself replicable, are not.[94] Indeed, small changes in molecular structure often lead to large changes in the physical properties of a system. This means that inexact replicates need not retain the general physico-chemical properties of their ancestors, in particular, properties that are essential for replication.

In DNA, the polyanionic backbone dominates the physical properties of DNA. Replacing one nucleobase by another, therefore, has only a second-order impact on the physical behavior of the molecule. This allows nucleobases to be replaced during Darwinian evolution without losing properties essential for replication.

For this reason, a repeating charge may be a universal structure feature of any genetic molecule that supports Darwinian evolution in water.[85] Polycationic backbones may be as satisfactory as polyanionic

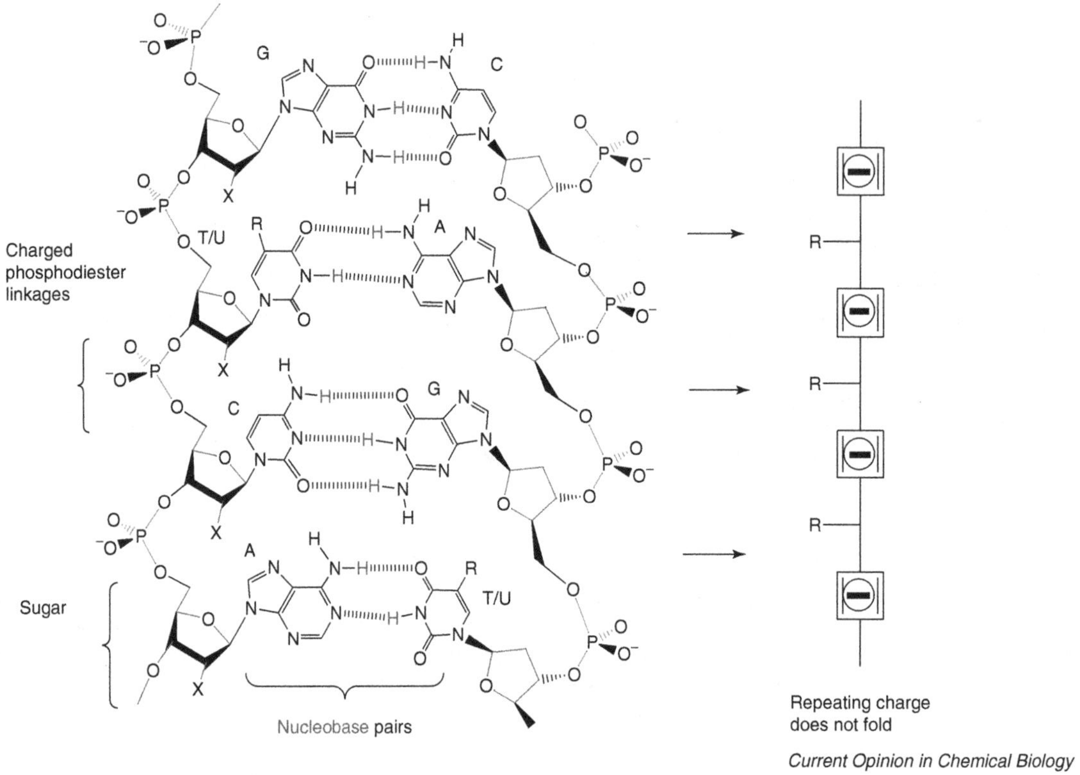

Fig. 12.6. The repeating charges in the backbone force inter-strand interactions away from the backbone, causing the strands to contact at the Watson–Crick edge of the heterocycles. Without the polyanionic backbone, inter-strand contacts can be anywhere.

backbones, however.[82,83] Thus, if life is detected in water on other planets, their genetics are likely to be based on polyanionic or polycationic backbones, even if their nucleobases differ from those found on Earth. This structural feature can be easily detected by simple analytical devices.

A REPEATING DIPOLE AS A UNIVERSAL STRUCTURAL FEATURE OF CATALYSTS IN WATER

As noted above, the specific 20 amino acids that are common in standard Terran proteins need not be universal. But what about the amide backbone? Unlike DNA and RNA, where the repeating element is a monopole (a charge), the repeating element of a polypeptide chain is a dipole. This is ideal for folding; the positive ends of one dipole interact favorably with the negative end of another dipole. By comparison, a biopolymer based on a repeating ester linkage would not fold via backbone-backbone interactions (Fig. 12.7). As folding is almost certainly required for efficient catalysis, one might expect repeating dipoles as found in polyamides to be found throughout the galaxy in biospheres that are based on water.

Amino acids appear to be products of prebiotic synthesis. Further, they are found in carbonaceous meteorites, both as *a* amino acids that are standard in Terran life, and in the non-standard *a* -methyl amino acids found by Cronin and Pizzarello.[95]

The uniqueness of bonding between carbon, oxygen and nitrogen makes it difficult to conceive of an alternative backbone that contains a repeating dipole. Selecting from the third row of the periodic table, repeating sulfonamide certainly would be one possibility (Fig. 12.7).[96] Repeating phosphonamides, where the negative charge is blocked, would be another alternative.[97]

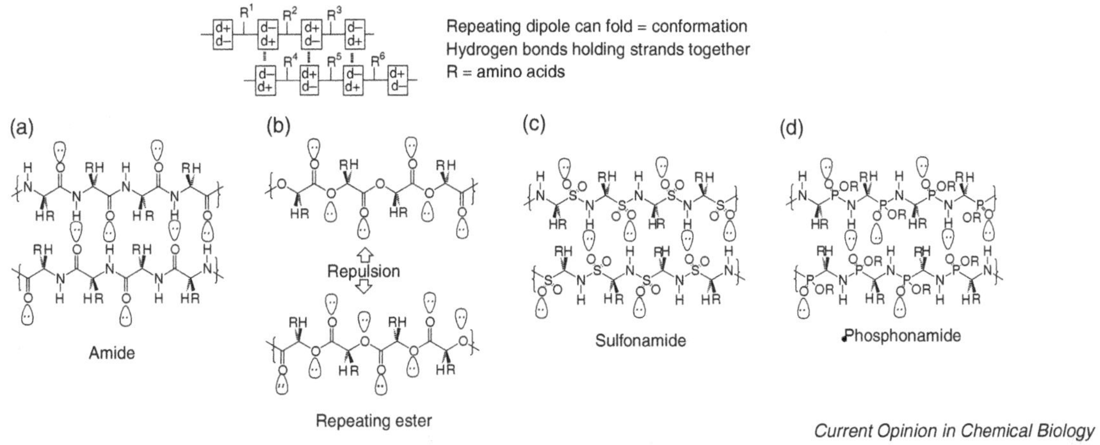

Repeating dipole can fold = conformation
Hydrogen bonds holding strands together
R = amino acids

(a) Amide

(b) Repulsion / Repeating ester

(c) Sulfonamide

(d) Phosphonamide

Current Opinion in Chemical Biology

Fig. 12.7. A repeating dipole (amide) in polypeptides is ideal for folding in water. (a) The polyamide found in natural proteins. (b) A polyester linkage by contrast will not fold. If we select elements from the third row of the periodic table, (c) repeating sulfonamide certainly would be one possibility and (d) a repeating phosphonamide in which the charge of the phosphate has been cancelled (as in a phosphosester) will be another.

UNIVERSAL STRUCTURAL FEATURES OF METABOLITES IN WATER

The strength of the carbon–carbon bond that makes it valuable to support covalent bonding becomes a disadvantage when it comes to metabolism. Bonds between two carbon atoms do not break at temperatures where water is liquid unless they have reason to do so. In particular, a reactant must be structured so that the electron pair that forms the bond between two carbon atoms has a place to go other than on to the carbon itself. In modern Terran metabolism, the electrons generally move, directly or indirectly, to an oxygen atom that begins the reaction doubly bonded to a carbon atom. The $C{=}O$ (carbonyl) group is central to all of contemporary Terran metabolism.

Other atoms can be doubly bonded to carbon and generate similar reactivity. For example, nitrogen doubly bonded to carbon supports much of the same chemistry. In water, the $C{=}N$ (imine) unit is rapidly hydrolyzed to give an amine and a carbonyl-containing product containing a $C{=}O$ at the corresponding position. Therefore, in modern Terran biochemistry, the imine is a transient species in many chemical reactions, including those that occur in an active site of the enzyme. Pyridoxal chemistry is a good example of this.

UNIVERSAL PROBLEMS IN WATER

Many organic molecules that are nearly inescapably parts of a carbon-based metabolism display problematic reactivity in water. Life that does not escape these molecules must deal with their reactivity.

For example, carbon dioxide is a key to carbon cycling in any moderately oxidizing environment. Water and carbon dioxide are a problematic pair, however. The carbon of carbon dioxide is a good electrophilic center. But carbon dioxide itself is poorly soluble in water (0.88 v/v at 293 K and 1 atm), and dissolves at pH 7 primarily in the form of the bicarbonate anion. This, however, has its electrophilic center shielding by the anionic carboxylate group. This means that bicarbonate is intrinsically unreactive as an electrophilic. Thus, the metabolism of carbon dioxide is caught in a conundrum. The reactive form is insoluble; the soluble form is unreactive.

Terran metabolism has worked hard to manage this conundrum. The reactivity of the biotin cofactor was discussed nearly three decades ago in this context.[98] Biotin is metabolically expensive, however, and cannot be used to manage carbon dioxide and its problematic reactivity in large amounts. The enzyme ribulose bisphosphate carboxylase attempts to manage the problem without biotin. But here, the problematic reactivity of carbon dioxide competes with the

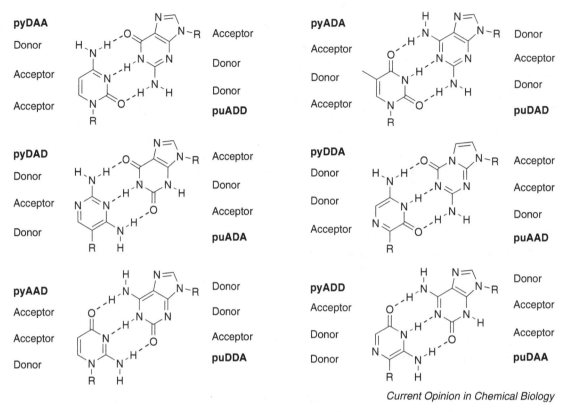

Current Opinion in Chemical Biology

Fig. 12.8. Expanded genetic alphabet. Following the rules of size complementarity (large purines pair with small pyrimidines) and hydrogen bonding complementarity (hydrogen bond donors from one nucleobase pair with hydrogen bond acceptors from another), the four Terran nucleobases are a minority of the 12 that can form mutually exclusive pairs that fit the Watson–Crick base pairing geometry.

problematic reactivity of dioxygen. Even in highly advanced plants, a sizable fraction of the substrate intended to capture carbon dioxide is destroyed through reaction with dioxygen.[99] Terran life, and nearly a billion years, has not found a compelling solution to this problem, which may be universal. Indeed, if we encounter non-Terran carbon-based life, it will be interesting to see how they have come to manage the unfortunate properties of carbon dioxide.

DNA NUCLEOBASES AS AN ILLUSTRATION OF THE TRIALECTIC

The four standard Terran nucleobases pair following simple rules (A pairs with T, G pairs with C). These, in turn, follow from two rules of complementarity: size complementarity (large purines pair with small pyrimidines), and hydrogen bonding complementarity (hydrogen bond donors pair with hydrogen bond acceptors; Fig. 12.8). This perspective shows that 12 letters forming six mutually exclusive base pairs are possible within the Watson–Crick geometry, obeying size and hydrogen bonding complementarity.[100] Representative examples of each of these have been synthesized, and base pairing is indeed observed following an expanded set of Watson–Crick base pairing rules. The extra letters in an artificial genetic alphabet are useful. They are now used in FDA-approved clinical diagnostics tests that monitor the load of HIV and hepatitis C viruses in ca. 400,000 patients annually.[101] An expanded genetic alphabet also supports the polymerase chain reaction,[102] thereby enabling Darwinian evolution with a non-Terran genetic molecule.

Given that these two rules of complementarity have manipulative value in creating artificial genetic systems under standard Terran aqueous conditions, they might also under extreme conditions. Consider, for example, the problem of genetics at low pH. Adenine is half protonated at pH 3.88; cytidine is half-protonated at pH 4.56. Protonation changes the hydrogen bonding patterns of these nucleobases, and disrupts the Watson–Crick pairing that holds the double helix together. Therefore, these nucleobases will not sustain life at low pH, such as in the Venusian atmosphere.

Reflecting this, modern Terran organisms that live at low pH (pHs as low as 0 are known[103]) pump protons to maintain an intracellular pH well above the pK_a of the protonated nucleobases. They could, however, have altered the structure of the nucleobases to alter their basicity. For example, 5-nitrocytidine has a pK_a that should be considerably lower than that of cytidine.

Why have Terran organisms that live at low pH chosen to pump protons rather than alter their DNA nucleobases? A functional explanation would hold that the fitness lost through a requirement to pump protons is less than the fitness lost through the biosynthesis of alternative nucleobases. An explanation based on vestigiality accepts that replacing cytidine by 5-nitrocytidine might indeed generate an organism that is more fit, but that the life that invaded the low pH environments had cytidine, not 5-nitrocytidine, together with the metabolism to make and degrade it, the polymerases to copy it, and the molecular biology to use it. Changing all of this was not possible using Darwinian processes.

An analogous discussion extends to the ribose in Terran RNA.[104] Evidence for an "RNA world," an episode of life on Earth where RNA was the only genetically encoded component of biological catalysts,[105] is found in the ribosome,[106] in catalytic RNA molecules,[107-109] and in RNA fragments from contemporary metabolism.[110] Many have doubted, however, that the first forms of life used RNA as its genetic material.[111] This doubt arises in part because of the chemical instability of ribose in water.[112]

Ribose can be made abiologically. When incubated with technical calcium hydroxide, formaldehyde and glycolaldehyde (both in the interstellar medium[113,114]) are converted into pentoses, including ribose, in a transformation known as the formose reaction.[115,116]

Unfortunately, the ribose rapidly suffers degradation under these conditions. Threose, explored as an alternative to ribose,[117] does also.

To address the stability problem, ribose phosphates have been proposed.[118] Further, borates stabilize ribose under conditions where it is formed.[61] Boron is not a common element on Earth. Yet, borate is excluded from many silicate minerals and enriched in igneous melts. Further, boron is enriched in the crust over its concentration in the mantle. Thus, sediments can contain as much as 100–150 ppm boron,[119] and its appearance in subduction volcanoes is taken as evidence of their origin. Boron also appears in igneous rocks on the surface as tourmalines. Alkaline borate minerals are well known on Earth. Serpentinization of olivine in mantle rocks generates alkaline solutions.[120] Weathering processes deliver borate to aqueous solution, the oceans, and to evaporite minerals.[121]

The prebiotic synthesis of sugars is now being examined with renewed interest. Non-racemic amino acid mixtures and zinc–proline complexes act as asymmetric catalyst during the condensation of glycolaldehyde and formaldehyde.[122,123] Ribose has been shown to be selectively precipitated by derivatization reaction with cyanamide. When the racemic mixtures of D-, and L-ribose were reacted with cyanamide, pure crystals containing homochiral domains were observed.[124] These recent discoveries make plausible the vestigial role of ribose in the origins of life.

NON-TERRAN LIFE ON EARTH

Does an alternative biosphere exist on Earth? This question is not as absurd as it might seem. Just 50 years ago, the Archaeal kingdom was not recognized. Life deep in the ocean was not known. With over 90% of microorganisms not culturable, we cannot exclude the possibility that Earth still carries a form of life that we do not already know of. Indeed, as this manuscript was being written, a new macroscopic, multicellular worm was discovered feeding on whale carcasses two miles below the Pacific Ocean.[125]

This question can be approached from two extremes. First, we can ask whether a life form that many believe was certainly possible still remains on Earth. The strongest current view for the natural history of life suggests that it began as part of an RNA world, a chemical system capable of Darwinian evolution that used RNA as the sole genetically encoded component of biological catalysis.[105,126] We might ask: do RNA organisms still exist on Earth?

It is not clear that we would have found them if they were here. First, virtually all of the "universal" probes for life target the ribosome, a molecule that makes encoded proteins. The ribosome is, almost by definition, not present in RNA organisms.

Where on Earth might environments be found where RNA organisms have a selective advantage, and therefore may have survived the 2 billion years of competition with protein based life? Three thoughts come to mind. First, the encoded biopolymer of RNA organisms does not require sulfur; proteins do. RNA organisms may therefore survive in an environment depleted in sulfur.

Alternatively, ca. 70% of the volume of a typical protein-based three polymer microorganism is consumed by the translation machinery needed to make proteins. This machinery is not needed by an RNA organism. Therefore, RNA organisms can be much smaller than protein-based organisms.[127] This suggests we might look for RNA organisms on Earth by looking for environments that are space-constrained. Many minerals have pores that are smaller than one micron across. These might hold smaller RNA organisms.

RNA is also easier to denature and then refold, compared with proteins. This might create a niche for an RNA organism in environments that cycle between very high and very low temperatures.

CONCLUSION

This review suggests that life might exist in a wide range of environments. These include non-aqueous solvent systems at low temperatures (such as found on Saturn's moon Titan). Life may even exist in more exotic environments, such as the supercritical dihydrogen–helium mixtures found on gas giants. We propose that the only absolute requirements are a thermodynamic disequilibrium and temperatures consistent with chemical bonding. Weaker desiderata include a solvent system, the availability of elements such as carbon, hydrogen, oxygen and nitrogen, certain thermodynamic features available to metabolic pathways, and the opportunity for isolation. If we constrain life to water, more specific criteria can be proposed,

including soluble metabolites, genetic materials with repeating charges, and a well-defined temperature range.

ACKNOWLEDGMENTS

The authors are indebted to support from the NASA Exobiology program (NAG5–12362) and the National Science Foundation (CHE-0213575).

NOTES

This chapter originally appeared in *Current Opinion in Chemical Biology* 8 (2004), 672–689.

REFERENCES

1. Bains, W. (2004). Many chemistries could be used to build living systems. *Astrobiology*, **4**, 137–167.
2. Hecht, S. M., Alford, B. L., Kuroda, Y., & Kitano, S. (1978). Chemical aminoacylation of transfer-RNAs. *Journal of Biological Chemistry*, **253**, 4517–4520.
3. Chin, J. W., Cropp, T. A., Anderson, J. C, Mukherji, M., Shang, Z. W., & Schultz, P. G. (2003). An expanded eukaryotic genetic code. *Science*, **301**, 964–967.
4. Noren, C. J., Anthony-Cahill, S. J., Griffith, M. C, & Schultz, P. G. (1989). A general method for site-specific incorporation of unnatural amino acids into proteins. *Science*, **244**, 182–188.
5. Baldini, G., Martoglio, B., Schachenmann, A., Zugliani, C., & Brunner, J. (1988). Mischarging *Escherichia coli* tRNAPhe with L-4'-[3-(trifluoromethyl)-3H-diazirin-3-yl]phenylalanine, a photoactivatable analog of phenylalanine. *Biochemistry*, **27**, 7951–7959.
6. Bain, J. D., Diala, E. S., Glabe, C. G., Dix, T. A., & Chamberlin, A. R. (1989). Biosyntheic site-specific incorporation of a non-natural amino acid into a polypeptide. *Journal of American Chemical Society*, **111**, 8013–8014.
7. Hohsaka, T. & Masahiko, S. M. (2002). Incorporation of non-natural amino acids into proteins. *Current Opinion in Chemical Biology*, **6**, 809–815.
8. Schneider, K. C & Benner, S. A. (1990). Oligonucleotides containing flexible nucleoside analogs. *Journal of the American Chemical Society*, **112**, 453–455.
9. Augustyns, K., Vanaerschot, A., & Herdewijn, P., (1992). Synthesis of l-(2,4-dideoxy-beta-D-eiythro-hexopyranosyl)

thymine and its incorporation into oligonucleotides. *Bioorganic and Medicinal Chemistry Letters*, **2**, 945–948.

10. Eschenmoser, A. (1999). Chemical etiology of nucleic acid structure. *Science*, **284**, 2118–2124.

11. Benner, S. A. & Ellington, A. D. (1988). Interpreting the behavior of enzymes: Purpose or pedigree? *CRC Critical Reviews in Bioengineering*, **23**, 369–426.

12. Geyer, C. R., Battersby, T. R., & Benner, S. A. (2003). Nucleobase pairing in expanded Watson–Crick-like genetic information systems: The nucleobases. *Structure*, **11**, 1485–1498.

13. Oró, J. (1960). Synthesis of adenine from ammonium cyanide. *Biochemical and Biophysical Research Communications*, **2**, 407–412.

14. Cleland, C. E. & Chyba, C. F. (2002). Defining 'life.' *Origins of Life and Evolution of the Biosphere*, **32**, 387–393.

15. Ruiz-Mirazo, K., Pereto, J., & Moreno, A. (2004). A universal definition of life: Autonomy and open-ended evolution. *Origins of Life and Evolution of the Biosphere*, **34**, 323–346.

16. Deamer, D. W. & Fleiscchaker, G. R. (Eds.) (1994). *Origins of life: The central concepts*. Boston: Jones & Bartlett.

17. Sagan, C. (1970). Life. In *The encyclopedia Britannica*. London: William Benton.

18. Koshland, D. E., Jr. (2002). The seven pillars of life. *Science*, **295**, 2215–2216.

19. Schwartz, S. P. (1977). Introduction. In S. P. Schwarz (Ed.), *Naming, necessity, and natural kands*. Ithaca, NY: Cornell University Press.

20. Kondepudi, D. K., Kauffinan, R. J., & Singh, N. (1990). Chiral symmetry-breaking in sodium-chlorate crystallization. *Science*, **250**, 975–976.

21. Arrhenius, G. (2003). Crystals and life. *Helvetica Chimica Acta*, **86**, 1569–1586.

22. de Duve, C. (1991). *Blueprint for a cell: The nature and origin of life*. Burlington, NC: Neil Patterson.

23. Heinlein, R. (1983). *Have space suit, will travel Del Ray*, VA: Del Ray Books.

24. Kelley, D. S., Karson, J. A., Blackman, D. K., *et al.* (2001). AT3–60 Shipboard Party: An off-axis hydrothermal vent field near the Mid-Atlantic ridge at 30° N. *Nature*, **412**, 145–149.

25. Corliss, J. B., Dymond, J., Gordon, L. I., *et al.* (1979). Submarine thermal springs on the Galapagos Rift. *Science*, **203**, 1073–1083.

26. Stevenson, D. (1999). Life-sustaining planets in interstellar space? *Nature*, **400**, 32.

27. Dawkins, R. (1989). *The selfish gene* (2nd ed.). Oxford: Oxford University Press.

28. Chen, I. A. & Szostak, J. W. (2003). Membrane growth can generate a transmembrane pH gradient in fatty acid vesicles. *Proceedings of the National Academy of Sciences*, **101**, 7965–7970.

29. Luisi, P. L., Walde, P., & Oberholzer, T. (1999). Lipid vesicles as possible intermediates in the origin of life. *Current Opinion in Colloid and Interface Science*, **4**, 33–39.

30. Szostak, J. W., Bartel, D. P., & Luisi, P. L. (2001). Synthesizing life. *Nature*, **409**, 387–390.

31. Deamer, D., Dworkin, J. P., Stanford, S. A., Bernstein, M. P., & Allamandola, L. J. (2002). The first cell membranes. *Astrobiology*, **2**, 371–381.

32. Hanczyc, M. M L, Fujikawa, S. M., & Szostak, J. W. (2003). Experimental models of primitive cellular compartments: Encapsulation, growth, and division. *Science*, **302**, 618–622.

33. Wächtershäuser, G. (1990). Evolution of the first metabolic cycles. *Proceedings of the National Academy of Sciences*, **87**, 200–204.

34. Pace, N. (2001). The universal nature of biochemistry. *Proceedings of the National Academy of Sciences*, **98**, 805–808.

35. Miller, S. L. & Orgel, L. E. (1974). *The origins of life on the Earth*, Englewood Cliffs, NJ: Prentice-Hall.

36. Walsh, R. (1981). Bond dissociation energy values in silicon-containing compounds and some of their implications. *Accounts of Chemical Research*, **14**, 246–252.

37. Maxka, J., Huang, L. M., & West, R. (1991). Synthesis and NMR spectroscopy of permethylpolysilane oligomers $Me(SiMe_2)_{10}Me$, $Me(SiMe_2)_{16}Me$, and $Me(SiMe_2)_{22}Me$. *Organometallics*, **10**, 656–659.

38. Hayase, S., Horiguchi, R., Onishi, Y., & Ushirogouchi, T. (1989). Syntheses of polysilanes with functional groups 2: Polysilanes with carboxylic acids. *Macromolecules*, **22**, 2933–2938.

39. Hayase, S. (1995). Polysilanes with functional groups. *Endeavor*, **19**, 125–131.

40. Sanji, T., Kitayama, F., & Sakurai, H. (1999). Self-assembled micelles of amphiphilic polysilane block copolymers. *Macromolecules*, **32**, 5718–5720.

41. Cairns-Smith, A. G. (1966). The origin of life and the nature of the primitive gene. *Journal of Theoretical Biology*, **10**, 53–88.

42. Cairns-Smith, A. G. & Hartman, H. (1986). *Clay minerals and the origin of life*. Cambridge, UK: Cambridge University Press.

43. Voet, D. & Voet, J. (2004). *Biochemistry*. Hoboken, NJ: J. Wiley & Sons.

44. Huang, B. & Walsh, J. J. (1998). Solid-phase polymerization mechanism of poly(ethyleneterephthalate) affected by gas flow velocity and particle size. *Polymer*, 39, 6991–6999.

45. Goldanskii, V. I. (1996). Nontraditional mechanisms of solid-phase astrochemical reactions. *Kinetics and Catalysis*, 37, 608–614.

46. Brunner, E. (1988). Fluid mixtures at high pressures VI: Phase separation and critical phenomena in 18(n-alkane + ammonia) and 4(n-alkane + methanol) mixtures. *Journal of Chemical Thermodynamics*, 20, 1397–1409.

47. Haldane, J. B. S. (1954). The origins of life. *New Biology*, 16, 12–27.

48. Olah, G. A., Salem, G., Staral, J. S., & Ho, T. L. (1978). Preparative carbocation chemistry 13: Preparation of carbocations from hydrocarbons via hydrogen abstraction with nitrosonium hexafluorophosphate and sodium nitrite trifluoromethanesulfonic acid. *Journal of Organic Chemistry*, 43, 173–175.

49. Kolodner, M. A. & Steffes, P. G, (1998). The microwave absorption and abundance of sulfuric acid vapor in the Venus atmosphere based on new laboratory measurements. *Icarus*, 132, 151–169.

50. Schulze-Makuch, D., Grinspoon, D. H., Abbas, O., Irwin, L. N., & Bullock, M. A. (2004). A sulfur-based survival strategy for putative phototropic life in the Venusian atmosphere. *Astrobiology*, 4, 11–18.

51. Cockell, C. S. (1999). Life on Venus. *Planetary Space Science*, 47, 1487–1501.

52. Colin, J. & Kasting, J. F. (1992). Venus: A search for clues to early biological possibilities. In G. Carle, D. Schwartz, & J. Huntington (Eds.), *Exobiology in solar system exploration* (NASA special publication 512) (pp. 45–65). Moffett Field, CA: NASA, Ames Research Center.

53. Arrhenius, S. (1918). *The destinies of the stars*. New York: Putnam.

54. Seckbach, J. & Libby, W. F. (1970). Vegetative life on Venus? Or investigations with alga which grow under pure CO$_2$ in hot acid media at elevated pressures. *Space Life Science*, 2, 121–143.

55. Sagan, C. & Morowitz, H. (1967). Life in the clouds of Venus. *Nature*, 215, 1259–1260.

56. Hoar, W. S. & Randall, D. J. (1981). Fish physiology. In J. E. Webb, J. A. Wallwork, and J. H. Elgood (Eds.), *Guide to living fishes* (6 vols., 1969–1971). London: Macmillan.

57. Schulze-Makuch, D., Irwin, L. N., & Irwin, T. (2002). Astrobiological relevance and feasibility of a sample collection mission to the atmosphere of Venus. In H. Lacoste (Ed.), *Proceedings of the first European workshop on exo-astrobiology* (pp. 247–250). Noordwijk, Netherlands: ESA Publication Division.

58. Kreuzweiser, J., Schnitzler, J. P., & Steinbrecher, R. (1999). Biosynthesis of organic compounds emitted by plants. *Plant Biology*, 1, 149–159.

59. Weyerstah, P. (2000). Synthesis of compounds isolated from essential oils. In V. Lanzotti and O. Taglialatela-Scafati (Eds.), *Flavour and fragrance chemistry: Proceedings of the Phytochemical Society of Europe* (vol. 46, pp. 57–66). Dordrecht: Kluwer Academic Publishers.

60. Schulze-Makuch, D. & Irwin, L. N. (2004). *Life in the universe: Expectations and constraints*. Berlin: Springer-Verlag.

61. Ricardo, A., Carrigan, M. A., Olcott, A. N., & Benner, S. A. (2004). Borate minerals stabilize ribose. *Science*, 303, 196.

62. Schoffstall, A. M. (1976). Prebiotic phosphorylation of nucleosides in formamide. *Origins of Life and Evolution of the Biosphere*, 7, 399–412.

63. Schoffstall, A. M., Barto, R. J., & Ramo, D. L. (1982). Nucleoside and deoxynucleoside phosphorylation in formamide solutions. *Origins of Life and Evolution of the Biosphere*, 12, 143–151.

64. Schoffstall, A. M. & Liang, E. M. (1985). Phosphorylation mechanisms in chemical evolution. *Origins of Life and Evolution of the Biosphere*, 15, 141–150.

65. Sagan, C, Thompson, W. R., & Khare, B. N. (1992). Titan: A laboratory for prebiological organic chemistry. *Accounts of Chemical Research*, 25, 286–292.

66. Tawfik, D. S., & Griffiths, A. D. (1998). Man-made cell-like compartments for molecular evolution. *Nature Biotechnology*, 16, 652–656.

67. Lu, B. C. Y., Zhang, D., & Sheng, W. (1990). Solubility enhancement in supercritical solvents. *Pure Applied Chemistry*, 62, 2277–2285.

68. Robertson, W. W. & Reynolds, R. E. (1958). Effects of hydrostatic pressure on the intensity of the singlet-triplet transition of 1-chloronaphthalene in ethyl iodide. *Journal of Chemical Physics*, 29, 138–141.

69. King, A. D., Jr. & Robertson, W. W. (1962). Solubility of naphthalene in compressed gases. *Journal of Chemical Physics*, 37, 1453–1455.

70. West, R. A. (1999). Atmospheres of the giant planets. In L.-A. McFadden, P. R. Weissman and T. V. Johnson (Eds.) *Encyclopedia of the solar system* (vol. *2*, pp. 383–402). New York: Academic Press.

71. Sagan, C. & Salpeter, E. E. (1976). Particles, environments, and possible ecologies in the Jovian atmosphere. *Astrophysics Journal*, **32**, 737–755.

72. Layboum-Parry, J. (2002). Survival mechanisms in Antarctic lakes. *Philosophical Transactions of the Royal Society of London Series B: Biological Sciences*, **357**, 863–869.

73. Junge, K., Eicken, H., & Deming, J. W. (2003). Motility of *Colwellia psychrerythraea* strain 34H at subzero temperatures. *Applied Environmental Microbiology*, **69**, 4282–4284.

74. Junge, K., Eicken, H., & Deming, J. W. (2004). Bacterial activity at −2 to −20 degrees C in Arctic wintertime sea ice. *Applied Environmental Microbiology*, **70**, 550–557.

75. Navarro-Gonzalez, R., Rainey, F. A., Molina, P., *et al.* (2003). Mars-like soils in the Atacama Desert, Chile, and the dry limit of microbial life. *Science*, **302**, 1018–1021.

76. Gold, T. (1992). The deep, hot biosphere. *Proceedings of the National Academy of Sciences*, **89**, 6045–6049.

77. Pedersen, K. (1993). The deep subterranean biosphere. *Earth-Science Review*, **34**, 243–260.

78. Stevens, T. O. (1997). Subsurface microbiology and the evolution of the biosphere. In P. S. Amy and D. L. Halderman (Eds.), *Microbiology of the terrestrial deep subsurface: Microbiology of extreme and unusual environments* (pp. 205–224). Boca Raton, FL: Chemical Rubber Company Press.

79. Frick, L., MacNeela, J. P., & Wolfenden, R. (1987). Transition state stabilization by deaminases: Rates of nonenzymatic hydrolysis of adenosine and cytidine. *Bioorganic Chemistry*, **15**, 100–108.

80. Westheimer, F. H (1987). Why Nature chose phosphates. *Science*, **235**, 1173–1178.

81. Jayaraman, K., McParland, K. B., Miller, P., & Tso, P. O. P. (1981). Non-ionic oligonucleoside methylphosphonates 4: Selective-inhibition of *Escherichia coli* protein-synthesis and growth by non-ionic oligonucleotides complementary to the 3′ end of 16S ribosomal-RNA. *Proceedings of the National Academy of Sciences*, **78**, 1537–1541.

82. Miller, P. S., McParland, K. B., Jayaraman, K., & Tso, P. O. P. (1981). Biochemical and biological effects of nonionic nucleic acid methylphosphonates. *Biochemistry*, **20**, 1874–1880.

83. Reddy, P. M. & Braice, T. C. (2003), Solid-phase synthesis of positively charged deoxynucleic guanidine (DNG) oligonucleotide mixed sequences. *Bioorganic and Medicinal Chemistry Letters*, **13**, 1281–1285.

84. Linkletter, B. A., Szabo, I. E., & Bruice, T. C. (2001). Solid-phase synthesis of oligopurine deoxynucleic guanidine (DNG) and analysis of binding with DNA oligomers. *Nucleic Acids Research*, **29**, 2370–2376.

85. Benner, S. A. & Hutter, D. (2002). Phosphates, DNA, and the search for nonterran life. A second generation model for genetic molecules. *Bioorganic Chemistry*, **30**, 62–80.

86. Huang, Z., Schneider, K. C., & Benner, S. A. (1991). Building blocks for oligonucleotide analogs with dimethylene-sulfide, -sulfoxide and -sulfone groups replacing phosphodiester linkages. *Journal of Organic Chemistry*, **56**, 3869–3882.

87. Huang, Z., Schneider, K. C., & Benner, S. A. (1993). Oligonucleotide analogs with dimethylene-sulfide, -sulfoxide and -sulfone groups replacing phosphodiester linkages. *Methods in Molecular Biology*, **20**, 315–353.

88. Roughton, A. L., Portmann, S., Benner, S. A., & Egli, M. (1995). Crystal structure of a dimethylene-sulfone-linked ribodinucleotide analog. *Journal of the American Chemical Society*, **117**, 7249–7250.

89. Richert, C., Roughton, A. L., & Benner, S. A. (1996). Nonionic analogs of RNA with dimethylene sulfone bridges. *Journal of the American Chemical Society*, **118**, 4518–4531.

90. Schmidt, J. G., Eschgfaeller, B., & Benner, S. A. (2003). A direct synthesis of nucleoside analogs homologated at the 3′ and 5′ positions. *Helvetica ChimicaActa*, **86**, 2957–2997.

91. Eschgfaeller, B., Schmidt, J. G., & Benner, S. A. (2003). Synthesis and properties of oligodeoxynucleotide analogs with bis(methylene) sulfone-bridges. *Helvetica Chimica Acta*, **86**, 2937–2956.

92. Steinbeck, C. & Richert, C. (1998). The role of ionic backbones in RNA structure: An unusually stable non-Watson-Crick duplex of a nonionic analog in an apolar medium. *Journal of the American Chemical Society*, **120**, 11,576–11,580.

93. Flory, P. J. (1953). *Principles of polymer chemistry*. Ithaca, NY: Cornell University Press.

94. Lee, D. H., Granja, J. R., Martinez, J. A., Severin, K., & Ghadiri, M. R. (1996). A self-replicating peptide. *Nature*, **382**, 525–528.

95. Cronin, J. R. & Pizzarello, S. (1986). Amino-acids of the Murchison meteorite III: Seven carbon acyclic primary alpha-amino alkanoic acids. *Geochimica et Cosmochimica Acta*, **50**, 2419–2427.

96. Ahn, J.-M, Boyle, N. A., MacDonald, M. T., & Janda, K. D. (2002). Peptidomimetics and peptide backbone modifications. *Mini Reviews in Medicinal Chemistry*, **2**, 463–473.

97. Yamauchi, K., Mitsuda, Y., & Kinoshita, M. (1975). Peptides containing aminophosphonic acids III. The synthesis of tripeptide analogs containing aminomethylphosphonic acid. *Bulletin of the Chemical Society of Japan*, **48**, 3285–3286.

98. Visser, C. M. & Kellog, R. M. (1978). Biotin: Its place in evolution. *Journal of Molecular Evolution*, **11**, 171–178.

99. Ogren, W. L. & Bowes, G. (1972). Oxygen inhibition and other properties of soybean ribulose 1,5-diphosphate carboxylase. *Journal of Biological Chemistry*, **247**, 2171–2176.

100. Benner, S. A. (2004). Understanding nucleic acids using synthetic chemistry. *Accounts of Chemical Research*, **37**, 794–797.

101. Elbeik, T., Surtihadi, J., Destree, M., *et al.* (2004). Multicenter evaluation of the performance characteristics of the Bayer VERS ANT HCV RNA 3.0 assay (bDNA). *Journal of Clinical Microbiology*, **42**, 563–569.

102. Sismour, A. M., Lutz, S., Park, J. H., *et al.* (2004). PCR amplification of DNA containing non-standard base pairs by variants of reverse transcriptase from human immunodeficiency virus-1. *Nucleic Acids Research*, **32**, 728–735.

103. Edwards, K. J., Bond, P. L., Gihring, T. M., & Banfield, J. F. (2000). An archaeal iron-oxidizing extreme acidophile important in acid mine drainage. *Science*, **287**, 1796–1799.

104. Freier, W. M. & Altmann, K. H. (1997). The ups and downs of nucleic acid duplex stability: Structure-stability studies on chemically-modified DNA:RNA duplexes. *Nucleic Acids Research*, **25**, 4429.

105. Gilbert, W. (1986). The RNA world. *Nature*, **319**, 818.

106. Ban, N., Nissen, P., Hansen, J., Moore, P. B., & Steitz, T. A. (2000). The complete atomic structure of the large ribosomal subunit at 2.4 angstrom resolution. *Science*, **289**, 905–920.

107. Kruger, K., Grabowski, P. J., Zaug, A. J., Sands, J., Gottschling, D. E., & Cech, T. R. (1982). Self-splicing RNA: Autoexcision and autocyclization of the ribosomal RNA intervening sequence of Tetrahymena. *Cell*, **31**, 147–157.

108. Guerrier-Takada, C, Bardiner, K., Marsh, T., Pace, N., & Altaian, S, (1983). The RNA moiety of ribonuclease P is the catalytic subunit of the enzyme. *Cell*, **35**, 849–857.

109. Battel, D. P. & Szostak, J. W. (1993). Isolation of new ribozymes from a large pool of random sequences. *Science*, **261**, 1411–1418.

110. Benner, S. A., Ellington, A. D., & Tauer, A. (1989). Modem metabolism as a palimpsest of the RNA world. *Proceedings of the National Academy of Sciences*, **86**, 7054–7058.

111. Shapiro, R. (1988). Prebiotic ribose synthesis: A critical analysis. *Origins of Life and Evolution of the Biosphere*, **18**, 71–85.

112. Larralde, R., Robertson, M. P., & Miller, S. L. (1995). Rates of decomposition of ribose and other sugars: Implications for chemical evolution. *Proceedings of the National Academy of Sciences*, **92**, 8158–8160.

113. Hollis, J. M., Lovas, F. J., & Jewell, P. R. (2000). Interstellar glycolaldehyde: The first sugar. *Astrophysics Journal*, **540**, L107–L110.

114. Hollis, J. M., Vogel, S. N., Snyder, L. E., Jewell, P. R., & Lovas, R. J. (2001). The spatial scale of glycolaldehyde in the galactic center. *Astrophysics Journal*, **554**, L81–L85.

115. Butlerow, A. (1861). Bildung einer zuckerartingen Substanz durch Synthese. *Annalen*, **120**, 295–298.

116. Breslow, R. (1959). On the mechanism of the formose reaction. *Tetrahedron Letters*, **21**, 22–26.

117. Schöning, K. U., Scholz, P., Guntha, S., Wu, X., Krishnamurthy, R., & Eschenmoser, A. (2000). Chemical etiology of nucleic acid structure: The α-threofuranosyl-(3′–2′) oligonucleotide system. *Science*, **290**, 1347–1351.

118. Krishnamurthy, R., Arrhenius, G., & Eschenmoser, A. (1999). Formation of glycolaldehyde phosphate from glycolaldehyde in aqueous solution. *Origins of Life and Evolution of the Biosphere*, **29**, 333–354.

119. Ryan, J. G., Leeman, W. P., Morris, J. D., & Langmuir, C. H. (1996). The boron systematics of intraplate lavas: Implications for crust and mantle evolution. *Geochimica et Cosmochimica Acta*, **60**, 415–422.

120. Moody, J. B. (1976). Serpentinization. *Lithos*, **9**, 125–138.

121. Kawakami, T. (2001). Tourmaline breakdown in the migmatite zone of the Ryoke Metamorphic

Belt, SW Japan. *Journal of Metamorphic Geology*, **19**, 61–75.

122. Pizzarello, S. & Webber, A. L. (2004). Prebiotic amino acids as asymmetric catalysts. *Science*, **303**, 1151.

123. Kofoed, J., Machuquerio, M., Reymond, J. L., & Darbre, T. (2004). Zinc-proline catalyzed pathway for the formation of sugars. *Chemical Communications*, **13**, 1540–1541.

124. Springsteen, G. & Joyce, G. F. (2004). Selective derivatization and sequestration of ribose from a prebiotic mix. *American Chemical Society*, **126**, 9578–9583.

125. Rouse, G. W., Goffredi, S. K., & Vrijenhoek, R. C. (2004). Osedax: Bone-eating marine worms with dwarf males. *Science*, **305**, 668–671.

126. Watson, J. D., Hopkins, N. H., Roberts, J. W., Steitz, J. A., & Weiner, A. M. (1987). *Molecular biology of the gene* (4th ed., p. 1115). Menlo Park, CA: Benjamin Cummings.

127. Benner, S. A. (1999). How small can a microorganism be? In *Size limits of very small organisms: Proceedings of a workshop* (pp. 126–135). Washington DC: National Academy Press.

13 · Searching for life in the universe: lessons from Earth

KENNETH H. NEALSON

Within the next decade, NASA, in conjunction with colleagues from several nations, will embark on one of the most exciting missions yet undertaken in the exploration of space: the return of samples from Mars. The mission, which has been moved back several years, will almost certainly have an architecture similar to that originally planned for the Mars 03/05 mission (shown in Fig. 13.1). That is, it will have two separate launches, each of which will conduct experiments on the surface of Mars, retrieve and store samples, and put these collected samples into Mars orbit in two separate sample canisters. Subsequently, a Mars orbiter (sample return vehicle) will be employed to retrieve the orbiting sample canisters, and return them to Earth for analysis. Probably by 2011 we should have on Earth the samples from the surface of Mars on the order of 1–2 kg for scientific study. These samples will add an immense amount to our knowledge of the solar system, and of Earth itself, and will also be carefully scrutinized for the presence of indicators of present or past life on the red planet. Given the absence of any obvious features on Mars that suggest life, and the negative results (with regard to life) obtained in the Viking mission of the 1970s, a skeptical observer might well ask, "Why send such a mission?"

In essence, the results of recent scientific inquiry in several areas provide the basis for increased optimism for finding life elsewhere in the universe, and if we are going to launch such a search our nearby neighbor, Mars, is a reasonable first step. With regard to new knowledge: (1) we have learned things about the universe that have made the search for extraterrestrial life much more reasonable; and (2) our understanding of life on Earth has changed dramatically, altering our view of our own planet's biota, and therefore of the possibility that life might exist on places previously regarded as too hostile for life. While the astronomical discoveries (new suns, new planets, new information about the structure and history of Mars, the purported ocean on Europa, etc.) are important, it is the "lessons from the Earth" that I will concentrate on here.

As the director of the Center for Life Detection at the Jet Propulsion Laboratory, I was delighted to be asked to contribute to this symposium on science and religion.* The connection with the purported conflict between these two areas is not at all obvious for a microbiologist interested in the early evolution of the Earth and its biota. From my perspective, both of these endeavors were possible only after humankind developed to a point where we had the honor and pleasure to practice science and to contemplate our own destiny, ethics, and morality. Thus, for most of the Earth's history our planet was dominated by life that had nothing to do with either science or religion— it really could not have cared less! To this end, I believe it is appropriate to begin my presentation with our current view of life's history on our own planet, keeping in mind that any search for life elsewhere must be framed by our knowledge of the history of life on Earth.

In its early life-compatible stages, the Earth was still a fairly hostile spot for life as we know it now. It was warm, lacked oxygen in its atmosphere, and, because of the lack of oxygen, had little ozone to protect the planet from harmful ultraviolet light. Yet, it was in such an environment that life arose and left its earliest records. From recent studies of the Issua formation in Greenland, traces of metabolic activity (carbon metabolism) indicate that life existed on Earth as early as 3.8 Ga (billion years ago). This suggests that the invention of life took place rather rapidly, roughly within 200 million years of when the planet cooled and thereby became hospitable for carbon-based life.

The Nature of Life, ed. M. A. Bedau and C. E. Cleland. Published by Cambridge University Press. © M. A. Bedau and C. E. Cleland 2010.

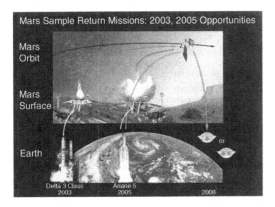

Fig. 13.1. Architecture of a Mars Sample Return (MSR) Mission. The original architecture of the MSR as envisioned in 2000 shows the general plan and time scale for a joint mission concept between the U.S. and France. As originally planned, the mission consisted of two launches, one in 2003 with a lander and rover, and a second one in 2005 with a similar lander, rover, and orbiter (Earth Return Vehicle or ERV). After landing on the Mars surface, the rover will collect samples (consisting of cores of both rocks and soil) and return these samples to a small cache in a rocket on the lander. These samples will be placed in a small sphere (orbiting cylinder or OS) and put into Mars Orbit, so that at the end of this phase, there will be two small "satellites" orbiting Mars. These will be retrieved by the ERV, and returned to Earth for analysis. The dates of this mission have been moved backwards, with a projected launch date of late in the decade.

These discoveries have triggered speculation about life in general (e.g., the problems associated with the invention of complex living systems), as well as about the possibility that similar living systems might have evolved on other planets. For example, it is generally agreed that in the early period of planetary development, and up until about 3.5 Ga, Mars and Earth may have shared similar planetary conditions. This has led many to posit that life might have had adequate time, and the proper conditions, to develop on early Mars as well as here. Subsequently, however, Mars lost its magnetosphere, hydrosphere, and most of its atmosphere, making the surface of Mars, by Earthly standards, an extremely hostile environment. While the current conditions of high UV light, absence of liquid water, and sub-freezing temperatures suggest that extant surface life would be precluded, the possibility that it may have once existed cannot be excluded on the basis of our knowledge of the history of the planet.

Most of the evidence of the earliest life on Earth is in the form of chemical tracers: there are few truly ancient fossils. This absence of a fossil record implies that simple, unicellular life dominated the early Earth, a fact consistent with what we know of unicellular life on the planet today, and of conditions necessary for fossil preservation. In fact, until about 2 Ga (billion years ago), there was little oxygen on the planet, and the development of complex eukaryotic cells, which live via oxygenic respiration, was probably not possible.

On the basis of the study of ancient soils, it is believed that oxygen first appeared and rose rapidly in the atmosphere approximately 2 Ga (Fig. 13.2) and that it was only upon this rise that the development of eukaryotic cells was possible. The Cambrian explosion of species and complex multicellular eukaryotes did not occur until approximately 500 million years ago, when oxygen reached current levels. From that point onward, the Earth began to take on what we would find a familiar appearance: occupied by plants, animals, and fungi.

However, even before the rise of oxygen, Earth was teeming with microbial life—this is the perspective that must be kept in mind when searching for life on other planets of unknown evolutionary age. Indeed, other planets could be in any of these stages, and the search for life cannot simply assume that a given stage of life or planetary evolution will have been reached. One should also keep in mind that planets evolve and life evolves, and that the interplay between these alters both: the evolution of Earth (on which the presence of life has made a strong impact) has had a drastic impact on the subsequent evolution of life. The oxygen we breathe is a product of the early evolution of photosynthesis, which supplies it. Without this innovation, the planet might well be alive, but its life would look much different from what we see today.

But what is life on Earth really like today—how well do we know the life on our own planet? The answer to this question is probably, "Very poorly," especially if we consider the full complement of life on Earth. To do so, we must take into account not only the "higher" organisms, but also the single-celled, anucleate organisms collectively referred to as prokaryotes. These organisms need to be factored into any discussion of the nature of life on Earth, both past and present. To illustrate the way in which prokaryotes are viewed now in comparison to just a few years ago,

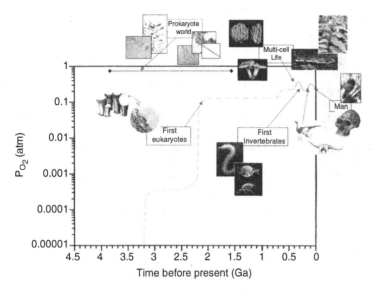

Fig. 13.2. Evolution of life on Earth as related to the appearance of oxygen in the atmosphere. The plot of oxygen versus time is modeled after the data of Rye and Holland (1998), who have proposed this pattern as the most likely based on studies of ancient soils (paleosols). The pictures of organisms are meant to emphasize that the early Earth was colonized by simple organisms, probably prokaryotic in nature, and that complex organisms (multicellular large creatures) did not appear until the oxygen levels were near to what they are now, approximately 500 million years ago.

I offer the following quotation from Dr. E. O. Wilson, the noted expert on animal evolution, from his book, *The Naturalist:*

> If I could do it all over again and relive my vision of the 21st century, I would be a microbial ecologist. Ten billion bacteria live in a gram of ordinary soil, a mere pinch held between the thumb and the forefinger. They represent thousands of species, almost none of which are known to science. Into that world I would go with the aid of modern microscopy and molecular analysis.

Dr. Wilson has made immense contributions to our understanding of macroscopic life on Earth, and in this quote he expresses the opinion that it is now time to move such thinking to the microbial level. To expand on these thoughts, I would point out that while microbial biomass is thought to account for 50% or more of the Earth's biomass, we have been able to culture and characterize only a small percentage of these prokaryotes, and thus know almost nothing about nearly half of our Earthly biota. This is a sad state of affairs for a society that claims to be ready to embark on a search for life in the Universe!

In addition to our new insights about the early appearance of Earthly life, a number of other biological findings have changed our perceptions of life on Earth. These new developments, which must be factored into any search for extraterrestrial life, involve the nature and diversity of life on Earth (e.g., the very definition of Earthly life), as well as new insights into the toughness and tenacity of life as we know it.

With regard to our view of the nature of life on Earth, major changes have occurred in the past two decades. We have moved from a peculiarly eukaryo-centric view of life to one that openly admits that the small, single-celled creatures that were once ignored play a vitally important role in the metabolism of our planet. The view of life that most of us learned from our biology teachers is that commonly referred to as the five kingdoms (Fig. 13.3). It was derived through the work of Linnaeus and others in the mid-1700s.

This classification scheme relied upon observation of the visible features of organisms to give each a name (e.g., *Homo sapiens* for humans), and to group organisms of similar appearance together. The diagram in Fig. 13.3 is called a phylogenetic tree; these trees are used to illustrate the evolutionary progression that

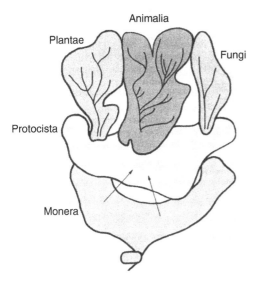

Fig. 13.3. The five kingdoms of life. This view represents Earthly life as composed of five kingdoms, four of which are eukaryotic, and the fifth, the monera (what we now call prokaryotes), at the base of the tree. This view of life is based on structural and functional analyses of organisms, and a proposed evolutionary line in which life moved from simple to more complex, and smaller to larger.

may have occurred to result in the extant organisms (e.g., to answer the question of which organisms preceded which in time).

Largely because of the nature of the tools available (human eye, hand lens, and later, simple microscopes), it is not surprising that such trees were dominated by the macroscopic, many-celled eukaryotes such as the fungi, plants, and animals. The tiny eukaryotic protists (amoebae, paramecia, giardia, etc.), being visible but not understood, were relegated to the next-to-the-bottom rung of the ladder, while the prokaryotes (bacteria) were handily put at the bottom, where they could be acknowledged, but not seriously so. This entire approach was reasonable at the time, in the sense that structural diversity was driving classification, and the single-celled, anucleate prokaryotes, as they are called, have little that is comparable with the structurally and behaviorally diverse larger organisms, collectively referred to as the eukaryotes.

This view of the biosphere has dramatically changed in the last decade with the advent of molecular taxonomy and phylogeny. The basic idea behind this approach is that there are some molecules

common to all Earthly life (16S ribosomal RNA, for example), and that, if one could sequence such molecules and compare the sequences, it might be possible to use this chemical information to compare all life, even that which can be seen only with a microscope. While the germ of this idea is actually decades old, its demonstration was realized only recently with new development of techniques in sequencing of nucleic acids, and the use of this information for organismic comparisons.

The work pioneered by Dr. Karl Woese of the University of Illinois has changed the way we look at life on Earth (Woese 1987, 1994; Pace 1996). From the point of view of the prokaryotes, which lack features that can be used to compare them to each other or to the eukaryotes, this molecular methodology allowed one, for the first time, to have a sense of the phylogeny (a natural history which had been previously lacking) of the various groups (Olsen, Woese, and Overbeek 1994; Stahl 1993). Not only could the prokaryotes be compared to each other, but also because the eukaryotes also contained these same molecules, the comparisons could encompass all of the five kingdoms. The results of this approach were quite dramatic: the four eukaryotic kingdoms were found to be quite homogeneous, while the prokaryotes were found to be very diverse, and thus were expanded to two separate kingdoms, referred to as Bacteria and Archaea (Fig. 13.4).

A quick glance at the molecular tree reveals that the major genetic variation among the eukaryotes is seen in the single-celled protists, while the three dominant kingdoms (plants, animals, and fungi) are actually clustered at the end of the eukaryotic assemblage, and display only a modicum of genetic diversity (Fig. 13.4). Apparently, it is possible to achieve structural and behavioral diversity (traits that have appeared only in the last 500 million years) while remaining genetically rather homogeneous. Given that multicellular eukaryotes evolved only recently, and that for nearly 3 billion years the prokaryotes dominated the surface of the Earth, one should not be surprised that the bulk of the apparent genetic diversity on the planet resides in this group.

Another notable feature of life on Earth is that of its toughness and tenacity. In order to consider this issue, we should briefly return to the discussion of prokaryotes and eukaryotes. Some of the key properties used to distinguish the prokaryotes from their

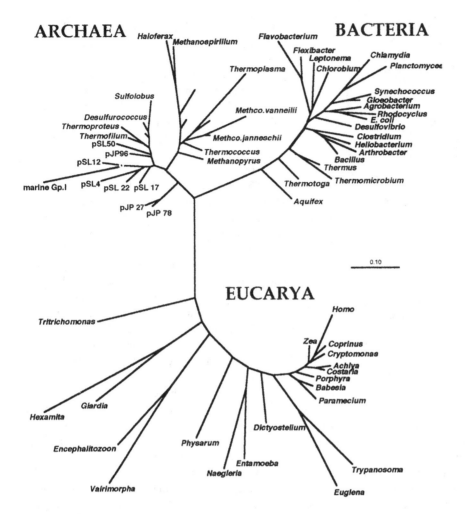

Fig. 13.4. The three kingdoms of life. On the basis of sequence analyses of the 16S ribosomal RNA gene, which is contained in all Earthly organisms, it is possible to construct a molecular phylogeny that can quantitatively compare all organisms, even those that can not be cultivated in the laboratory. Such an approach has yielded quite a different view of life, as shown here, in which the major kingdoms of life (animals, plants, fungi and protocists) shown in Figure 13.3, are grouped into one kingdom, and the prokaryotes are expanded into two kingdoms, called the Archaea and the Bacteria.

more complex eukaryotic cohorts are shown in Table 13.1. The eukaryotes are defined by the presence of a nucleus and nuclear membrane (*eu* = true; *karyon* = nucleus), and in general are characterized by complex structures, complex behavioral features, and simple metabolism. Their metabolism is primarily via oxygen-based respiration of organic carbon, and the sizeable energy yields from these processes are used to support their complex structural and behavioral investments. Basically, plants make organic carbon via photosynthesis and animals eat the plants (and other animals), leading to the kind of complex communities we easily recognize under the general heading of trophic levels or predator–prey cycles. The very existence of complex structures (both intracellular organelles, and multicellular tissues and organs) renders the eukaryotes sensitive to environmental extremes often easily tolerated by their structurally simple prokaryotic relatives (e.g., above 50 °C it is unusual to find functional eukaryotes).

Table 13.1. *Properties of the prokaryotes and eukaryotes. The small anucleate organisms known as prokaryotes share some properties that allow us to group them into functional domains that are quite different from their eukaryotic counterparts*

Key properties	
Prokaryotes	Eukaryotes
Small Size (1–2 μm) (high S/V ratio) favors chemistry	Larger Size Cells (10–25 μm) complex structures multicells/tissues
Rigid Cell Wall requires transport extracellular enzymes	Flexible Cell Walls phagocytosis particle (organism) uptake
Metabolic Diversity Alternate energy sources (light, organics, inorganics) Alternate oxidants (O$_2$, metals, CO$_2$, etc.)	Metabolic Specialization O$_2$ respiration organic C as fuel

On the other hand, the prokaryotes are the environmental "tough guys," tolerant to many environmental extremes of pH, temperature, salinity, radiation, and dryness. I refer to these organisms as the sundials of the living world—tough, simple, effective, and nearly indestructible. Some of the fundamental properties that distinguish them from the eukaryotes are shown in Table 13.1. First, they are small—they have optimized their surface to volume ratio so as to maximize chemistry. On the average, for the same amount of biomass, a prokaryote may have 10–100 times more surface area. Thus, for a human, whose body mass may include a few percent (by mass) bacteria (as gut symbionts), the bacteria make up somewhere between 24 and 76% of the effective surface area! In environments like lakes and oceans, where bacterial biomass is thought to be approximately 50% of the total, the bacteria compose 91 to 99% of the active surface area, and in anoxic environments, where the biomass is primarily prokaryotic, the active surface areas are virtually entirely prokaryotic. In essence, if you want to know about environmental chemistry, you must look to the prokaryotes!

Prokaryotes have rigid cell walls, which preclude life as predators. They are restricted to life as chemists,

and do their metabolism via transport and chemistry. This is in marked contrast to the eukaryotes, which are capable of engulfing (by a process called phagocytosis) other cells, and thus engaging in biology. In essence, the prokaryotes spurn life as biologists in order to optimize their skills as chemists. The full effect of such evolution is now easily visible through the genomic analyses of prokaryotes, where, in general, high percentages of the structural genes are involved with membrane and transport processes.

In many cases, up to 25% or more of the total genome deals with the interface between the cell surface and the environment and is involved with uptake, transport, or metabolism of environmental chemicals. In eukaryotes on the other hand, much of the DNA is devoted to the more biological concerns such as development, regulation, and differentiation. Finally, the prokaryotes are metabolically very diverse, while the eukaryotes are quite restricted in their abilities. The prokaryotes have developed a metabolic repertoire that allows them to utilize almost any energetically useful chemical available that is abundant on the Earth (Nealson, 1997a,b, 1999). Being opportunists, these ingenious chemists have simply harvested every worthwhile corner of the chemical market, learning to utilize organic and inorganic energy sources of nearly all kinds. Let us look, for example, at the major sources of energy available on Earth today, as shown in Fig. 13.5.

On the left one sees the potential energy sources, ranked from the most energy-rich at the top to the least energy-rich on the bottom. On the right are the oxidants that can be used to "burn" these fuels, with the best oxidant (oxygen) at the bottom, and the worst one (carbon dioxide) towards the top. Since a fuel needs to be "burned" to yield energy, we can estimate the amount of energy available simply by connecting a given fuel with an oxidant. If the arrow connecting any given so-called redox pair slopes downwards, it indicates that energy is available from this combination, and there is almost certain to be one or more microorganisms capable of using this combination. In marked contrast, the eukaryotes utilize only a few organic carbon compounds, and only molecular oxygen as the oxidant—they sacrifice diversity for high-energy yield, while the prokaryotes occupy the diverse, lower-energy habitats.

But what about the toughness of prokaryotes? The word "extremophile" has crept into our vocabulary in

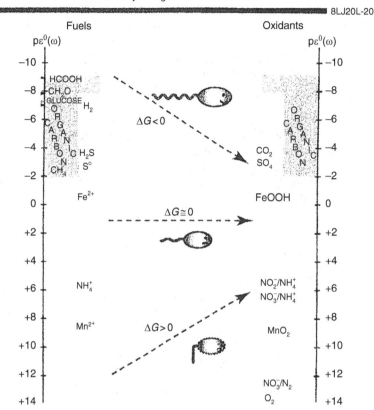

Fig. 13.5. Sources of energy and oxidants on Earth. Some of the energy sources available to organisms on Earth are shown on the *left*, with the most energy-rich at the top and the least energy-rich at the bottom. On the *right* are shown the available oxidants for the burning of these biological fuels. The fuels and oxidants commonly used by the eukaryotes are glucose and oxygen, while those available to the prokaryotes are HCOOH, CH_2O, H_2, H_2S, CH_4, S_0, Fe^{++}, NH_4^+, Mn^{++}, and CO_2, SO_4, FeOOH, NO_2^-/NH_4^+, NO_3^-/NH_4^+, MnO_2, NO_3^-/N_2 respectively.

the past decade: invented to describe organisms that are resistant to, and even thrive in, extreme conditions. As shown in Fig. 13.6, these extremophiles can be resistant to chemical (pH, salinity), physical (temperature, dryness), or nutritional extremes, and it is seldom in nature that an organism encounters just one extreme. For example, under high temperatures, it is common to find anoxic conditions, as the solubility of oxygen is very low in hot water. Furthermore, due to high evaporation rates, warm systems are often associated with high salinity. Thus, desert ponds are often of high pH and salinity, as evaporating water and the minerals trapped there interact to produce extreme conditions.

The most notorious extremophiles are perhaps those associated with high-temperature environments—hyperthermophilic bacteria capable of growth at 100 °C and above with a maximum temperature of about 115 °C, well above the boiling point of water. These organisms can only be grown under pressure where the water is stable, and will freeze to death at temperatures as high as 80 °C, temperatures that would result in severe burns to humans. Other bacteria are known that live in saturated salt brines, and at pH values as low as −1 and as high as +11.

One of the strategies of life that often emerges when things get tough is that of an endolithic lifestyle—the ability to associate with rocks, either on or

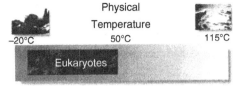

Physical
Temperature

-20°C   50°C   115°C

Eukaryotes

Psychrophiles Mesophiles Thermophiles Hyperthermophiles

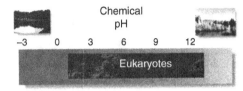

Chemical
pH

-3   0   3   6   9   12

Eukaryotes

Extreme Acidophiles Acidophiles Neutralophiles Alkalophiles

Extremophiles

Metabolic

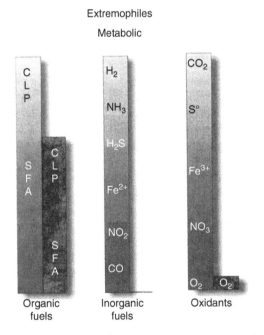

Organic   Inorganic   Oxidants
fuels   fuels

Fig. 13.6. Extremophiles. This figure illustrates the ranges of environments encountered by various types of extremophiles, and to make the point that extreme conditions can be physical, chemical, or metabolic. It is seldom that one encounters just one of these conditions at a time.

just under the surface. In California's alkaline Mono Lake, for example, we can see that the tufa mounds that dominate the alkaline lake, and which appear to be dead, are actually teeming with life (Fig. 13.7)

(Sun *et al.* 1999; Sun and Nealson 1999). A few millimeters under the rock surface are populations of cyanobacteria that hide from the intense sunlight, positioning themselves for optimum growth in the now-filtered light. A similar situation occurs in many hot and cold desert rocks and soils, where the photosynthetic microbes are found under the surfaces of rock layers (Friedmann 1982, 1993).

In addition to the physical and chemical extremes noted above, I would like to point out another property of prokaryotes, referred to here as nutritional extremophily. Given that eukaryotes are almost entirely limited to growth on organic carbon with oxygen as the oxidant, any set of conditions in which organic carbon or oxygen is absent is a potential life-threatening situation. For the prokaryotes, however, such environments are simply opportunities to continue living but with a different nutritional system.

While it cannot be said with certainty *when* such metabolic diversity arose on Earth, its very existence forces one who is hunting for life to include such "extreme" habitats in the search, and to broaden the definition of life to include metabolic abilities that a few years ago might have been summarily dismissed. The ability to grow on energy sources such as carbon monoxide, ferrous iron, hydrogen sulfide, or hydrogen gas implies that Bacteria could inhabit worlds not previously considered as candidates by most scientists seeking extraterrestrial life, and must be included in any emerging search strategies.

A final point regarding the prokaryotes relates to their tenacity and ability to survive for long periods of time. There are many examples of bacteria being revived after long-term storage, but perhaps none any more dramatic than those from the Antarctic permafrost, where soils that have been permanently frozen for 3 million years or more have yielded copious numbers of living bacteria. Dr. David Gillichinsky and his colleagues from Puschino, Russia, have been drilling in such sites for many years now, and a number of organisms have been "revived" from their carefully collected samples. It is not unusual to find 10^6 to 10^7 (1 to 10 million) viable bacteria from each gram of permafrost (Shi *et al.* 1997). These are not cold-loving (psychrophilic) bacteria that have adapted to these freezing conditions, but simply mesophilic organisms that have been trapped within this icy storage facility for millions of years.

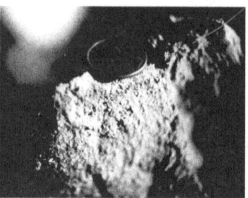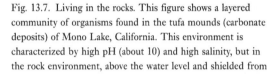

Fig. 13.7. Living in the rocks. This figure shows a layered community of organisms found in the tufa mounds (carbonate deposits) of Mono Lake, California. This environment is characterized by high pH (about 10) and high salinity, but in the rock environment, above the water level and shielded from the sun by the carbonate rocks, is a healthy and ubiquitous population of photosynthetic cyanobacteria, living in harmony with many other prokaryotic species.

So, as we are poised to proceed to other celestial bodies in search of life, we find that our definition of habitability is quite different from what we adhered to just a few years ago. In response to this, we must: (*a*) consider that the physical and chemical conditions tolerant to life are broader than we once thought; (*b*) examine the potential energy sources available and look carefully for life forms utilizing any such energy; and, (*c*) be prepared for subtle, single-celled life that may not be obvious at first glance, even looking in places where life might have been preserved as dormant forms.

Given these "lessons from the Earth," is it possible to design a strategy that will allow us to search for life with some degree of confidence? The success of any search will depend upon the ability to define the general features of life, to develop methods for measuring such general features, and to employ these methods for remote sensing, for *in situ* studies, and for the analysis of returned samples (Fig. 13.8).

Studies of Earthly life suggest that metabolic evolution, one of the keys to life's becoming a global phenomenon, was already in full swing more than 500 million years ago. Most of the Earth's geology, and many of its atmospheric properties, which we still see today, were in place by that time. If we want to search for life elsewhere, we must keep in mind that there is no guarantee that a particular planet will have evolved to the same advanced stages we have on Earth—a historical perspective is thus key to developing a strategy for life detection. To put this another way, we must know the early history of a planet in order to frame the search for life properly.

Since Earth is the only place where we are certain that life exists, it will serve as our laboratory for the development of the search strategy. The overall strategy is still in its early stage of definition, but a general idea of it involves three phases: (*a*) the development of non-Earth-centric biosignatures for life detection; (*b*) the testing of these biosignatures on Earthly samples to see just how good they are; and, (*c*) the eventual use of these biosignatures and tests for the analyses of extraterrestrial samples.

For most biologists this entire process is a new endeavor, asking new questions. It is rare that a biologist is handed a rock and asked: "Is it alive?" or, "From this sample, can you prove whether there was ever life on Earth?" Yet that is what we will be faced with in a few years when samples are returned from Mars. In fact, if another planet was teeming with life, as is Earth, this would not be a difficult task, even with very old rocks. It would be relatively easy to tell that Earth was (and is) alive from almost any distance, and especially so if samples were available for detailed physical and chemical analysis. However, if the signs

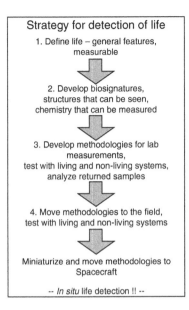

Strategy for detection of life

1. Define life – general features, measurable

2. Develop biosignatures, structures that can be seen, chemistry that can be measured

3. Develop methodologies for lab measurements, test with living and non-living systems, analyze returned samples

4. Move methodologies to the field, test with living and non-living systems

Miniaturize and move methodologies to Spacecraft

-- *In situ* life detection !! --

Fig. 13.8. Searching for life in the universe. Searching for life when we do not know what it looks like may represent one of the truly great challenges facing human scientists. One must move to fundamental definitions of life, and develop biosignatures to aid in the search—biosignatures that are not dependent upon known earthly molecules, but which would never miss earthly life if it were encountered.

of life are subtle or unfamiliar, then the task becomes much more difficult.

This difficulty is demonstrated by the present controversy surrounding the now famous Mars meteorite, ALH 84001. Four years ago, this 4.5-billion-year-old rock was reported to contain evidence for life on Mars. But even now, after extensive research, the jury is still out as to whether the evidence is convincing. The problems stem from many fronts, including the age of the sample, the difficulties in separating indigenous signals from those due to Earth contamination, and the very definition of life and how one proves that it is (or was) present. What this meteorite really has taught us is that we have a lot to learn about how to distinguish life from non-life.

Biologists may not even be the right group of scientists to answer the question, "Is there life in this rock?" As a biology student, I was never asked such a question. Rather I was given a frog and asked, "How does it work?" or "What is it made of?" These days, the questions have changed to "What genes are there,

and how do they function?" but the general problem remains—biologists study life they already know how to detect, they do not seek to detect life they do not know about. This is a question inherently interdisciplinary in nature, and perhaps best suited to those who are willing to define life in general terms that would include all life on Earth, but not exclude life made of different types of molecules.

As a group, we biologists should be extremely well suited to detect life as we know it primarily because we understand its chemistry so well. There are molecules that can be detected at very high sensitivity, allowing us to find a single bacterium in a liter of water. However, if such key indicator molecules are not there, the search becomes much more difficult, and the likelihood that life elsewhere would contain the same key molecules certainly cannot be depended upon. The problem then takes on a different aspect: Because we rely on Earth-centric indicators of life, we biologists may unwittingly be the least well suited to detect life that might differ in its chemical make-up.

To this end, our astrobiology group is focusing on what we feel are two fundamental properties of all life, structure and chemical composition, both of which can be detected and measured. Historically, structures are the paleontologists' keys to recognition of past life on Earth. It is structures that characterize life as we know it and that should be expected to characterize any new forms of life we encounter. We do not know in advance the nature of the structures or the size scales over which to search, but we do expect there to be structural elements associated with any life forms. When one is hunting in a new spot, dependency on known structures has a number of potential traps, including the fact that one might discard structures simply because they are unfamiliar. It will be important to remain open-minded about the types and sizes of structures found in samples from new sites.

While we believe that life will be linked to some structural elements, structure alone will not prove the existence of life. However, coupling structural analysis with the determination of chemical content may well provide a tool for strongly inferring the presence of life. On Earth, life is carbon-based, with a peculiar and remarkably constant elemental composition (hydrogen, nitrogen, phosphorous, oxygen, carbon, etc.), which is remarkably out of equilibrium with the crustal elemental abundance of our planet. In other words, there are more or less of some elements than would be present if there were no life on Earth. Life is, almost by definition, a source of negative entropy: a structure composed of groups of chemical monomers and polymers that are not predicted to be present on thermodynamic grounds, given the abundance of chemicals in the atmosphere and crust of the Earth. The exact nature of these chemicals is not so important as the fact that they are grossly out of equilibrium with their surrounding geological environment. So, if methods were available for analysis of the chemistry of structures at the proper size scale(s), then the possibility of detecting extant (or even extinct) life would be greatly increased. While there are other properties of life that may be measurable (such as replication, evolution, and energetic exchange with the environment), and which may leave traces in the geological record, we believe that if life does or did exist, then it will be detectable by the existence of structures and their distinctive chemistries.

Ultimately, we would like to have samples from many places in our solar system and beyond, but realistically, we will probably need to make measurements remotely and *in situ,* and be satisfied with these as our indicators of life. As our ability to measure structures and chemistry improves, the possibility of answering the question of whether life does or does not exist beyond Earth will improve as well. A strategy for exploration, sample collection and return, and finally, sample analysis will be needed. Given the number of other solar systems already known to exist, and the emerging numbers of planets around faraway stars, it seems unlikely that life will not be found elsewhere. Development of the proper strategy, and definition of those conditions that do and do not support life, will be key to the ultimate discovery of extraterrestrial life. With the proper strategy and approach, the question seems to be not one of whether, but when.

NOTES

This chapter originally appeared in *Annals of the New York Academy of Sciences* 950(1) (2001), 241–258.

REFERENCES

1. Friedmann, E. (1982). Endolithic microorganisms in the Antarctic cold desert. *Science*, **215**, 1045–1053.
2. Friedmann, E. (1993). *Antarctic microbiology* (1st ed.). New York: Wiley Interscience.
3. Glombek, M. P. (1999). A message from warmer times. *Science*, **283**, 1470–1471.
4. Klein, H. P. (1979). The Viking mission and the search for life on Mars. *Reviews of Geophysics*, **17**, 1655–1662.
5. Knoll, A. H. (1992). The early evolution of eukaryotes: A global perspective. *Science*, **256**, 622–627.
6. Margulis, L. (1981). *Symbiosis in cell evolution.* San Francisco: W. H. Freeman.
7. McKay, C. P., Friedmann, E. I., Wharton, R. A., & Davies, W. L. (1992). History of water on Mars: A biological perspective. *Advances in Space Research*, **12**, 231–238.
8. Mojzsis, S. J., Arrhenius, G., McKeegan, K. D., Harrison, T. M., Nutman, A. P., & Friend, C. R. L. (1996). Evidence for life on Earth before 3,800 million years ago. *Nature*, **384**, 55–59.
9. Nealson, K. H. (1997a). Sediment bacteria: Who's there, what are they doing, and what's new? *Annual Review of Earth and Planetary Science*, **25**, 403–434.
10. Nealson, K. H. (1997b). The limits of life on Earth and searching for life on Mars. *Journal of Geophysical Research*, **102**, 23,675–23,686.
11. Nealson, K. H. (1999). Post-Viking microbiology: New approaches, new data, new insights. *Origins of Life and Evolution of the Biosphere*, **29**, 73–93.
12. Olsen, G. J., Woese, C. R., & Overbeek, R. (1994). The winds of evolutionary change: Breathing new life into microbiology. *Journal of Bacteriology*, **176**, 1–6.
13. Pace, N. (1996). New perspective on the natural microbial world: Molecular microbial ecology. *American Society of Microbiology News*, **62**, 463–470.
14. Rye, R. & Holland, H. D. (1998). Paleosols and the evolution of atmospheric oxygen: A critical review. *American Journal of Science*, **298**, 621–672.
15. Schidlowski, M., Appel, P. W. U., Eichmann, R., & Junge, C. E. (1979). Carbon isotope geochemistry of the $3.7 \times 10$9-yr-old Isua sediments, West Greenland: Implications for the Archaean carbon and oxygen cycles. *Geochimica et Cosmochimica Acta*, **43**, 189–199.

16. Schopf, J. W. & Klein, C. (1992). *The Proterozoic biosphere.* Cambridge, UK: Cambridge University Press.

17. Shi, T., Reves, R. H., Gilichinsky, D. A., & Friedmann, E. (1997). Characterization of viable bacteria from Siberian permafrost by 16S rDNA sequencing. *Microbial Ecology,* **33,** 169–179.

18. Stahl, D. A. (1993). The natural history of microorganisms. *American Society of Microbiology News, Series B,* **59,** 609–613.

19. Sun, H., Nealson, K. H., & Venkateswaren, K. (1999). Alkaliphilic and alkalotolerant actinomycetes isolated from Mono Lake tufa. *Abstracts from the 99th annual meeting of the American Society of Microbiologists.* Washington DC: ASM Press.

20. Sun, H. & Nealson, K. H. (1999). Endolithic cyanobacteria in Mono Lake. *Abstracts from the 99th annual meeting of the American Society of Microbiologists.* Washington DC: ASM Press.

21. Woese, C. R. (1987). Bacterial evolution. *Microbiology Review,* **51,** 221–271.

22. Woese, C. R. (1994). There must be a prokaryote somewhere: Microbiology's search for itself. *Microbiology Review,* **58,** 1–9.

ENDNOTE

* [Editor's Note]: Here the author is referring to the symposium that led to the original publication of the chapter.

14 · The possibility of alternative microbial life on Earth

CAROL E. CLELAND AND SHELLEY D. COPLEY

INTRODUCTION

Finding a form of life that differs in its molecular architecture and biochemistry from life as we know it would be profoundly important both from a scientific and a philosophical perspective. There is compelling evidence that life as we know it on Earth today shares a last universal common ancestor (LUCA; Woese 1967, 2004). It is unlikely that LUCA was the earliest form of life on Earth since it was already quite sophisticated, having nucleic acids and proteins, as well as complex metabolic processes. In short, life as we know it represents a single example of a fairly advanced stage of life. One cannot safely generalize from a single example to all life, wherever or whenever it may be found. Indeed, in the absence of additional examples of life we are in a position analogous to that of a zoologist trying to formulate a theory of mammals based only upon their experience with zebras. It is unlikely that she will focus on their mammary glands since they are characteristic only of the females. Yet the mammary glands tell us more about what it means to be a mammal than the ubiquitous stripes seen in both male and female zebras. Finding a form of life having a different molecular architecture and biochemistry would help us to understand the nature of life in general—the processes that led to its emergence and the various forms it may take, whether on the early Earth or elsewhere in the Universe. Furthermore, it would have profound philosophical implications for our understanding of our place in the Universe.

When scientists speak of searching for "life as we *don't* know it," they typically have in mind extraterrestrial life. Considerable attention has been given to the question of what life might look like in other places in the Universe and how we might detect its presence with the aid of remote and *in situ* robotic devices. There is, however, another possibility that is rarely considered, and that is that the contemporary Earth itself might host forms of life differing at the molecular level in fundamental ways from life as we currently know it.

Discussions of the origin of life on Earth have appeared in the literature over a period of many decades (for an excellent overview, see Fry 2000). While most researchers have assumed that life originated only once on Earth, a few pioneers have considered the idea of multiple origins of life: Shapiro (1986) has suggested that familiar life may have originated more than once on Earth. Sleep *et al.* (1989) have suggested that familiar life may have emerged and been extinguished several times on the early Earth during the period of "heavy bombardment." Wächtershäuser (1992) has suggested that primitive surface metabolists preceded cellular life and might even persist in habitats that cannot be occupied by heterotrophs. Here we discuss in detail a possibility that has received little attention. We suggest that if life originated more than once on Earth, it may have produced proto-organisms differing at the molecular level in fundamental ways from the forerunners of our form of life and moreover, that microbial descendants of some of these proto-organisms may still be with us on Earth today, as yet unrecognized for what they represent. We argue that this idea, which is contrary to the paradigm that all life on Earth today descends from a common ancestor, should not be dismissed. Davies & Lineweaver have discussed this idea, although with little biological and chemical detail, in a recent paper attempting to quantify the likelihood of the emergence of multiple forms of life on Earth (Davies & Lineweaver 2005) (*vide infra*).

The Nature of Life, ed. M. A. Bedau and C. E. Cleland. Published by Cambridge University Press. © M. A. Bedau and C. E. Cleland 2010.

WHAT MIGHT MODESTLY DIFFERENT LIFE LOOK LIKE?

Life as we know it on Earth today shares a number of fundamental characteristics at the molecular level. It contains catalytic and structural macromolecules made of protein, and genetic material made of nucleic acids. It is clear that proteins and DNA are remarkably well suited for their particular functions, and many alternative structures that have been considered fall short in terms of providing suitable structures for these functions. However, it is also clear that some of the molecular building blocks of proteins and nucleic acids could have been different. Indeed, it is an open question as to whether all life (wherever it may be found) is constructed of proteins and nucleic acids. This question is difficult to answer outside the context of a general theory of living systems, something that we currently lack. We do not explore the possibility here of forms of life that differ radically at the molecular level because, as discussed below, detection of even modestly different life forms poses a tremendous challenge.

Familiar life utilizes nucleic acids to store its hereditary information. DNA is well suited for this function for a number of reasons. First, it is double-stranded, and the resulting redundancy provides the correct sequence information in case of damage to one strand of DNA that must be repaired. The polyanionic backbone causes DNA to adopt an extended structure that facilitates replication. Importantly, this extended structure is quite insensitive to the exact sequence of bases in the DNA (Benner & Hutter 2002). Finally, the interaction between the two complementary strands that is mediated by hydrogen bonding interactions between the Watson–Crick faces of the bases is strong enough to provide molecular recognition and structural integrity, but not so strong that the strands cannot be easily separated to allow replication. Much effort has been invested into the exploration of alternative structures for a genetic polymer. The possibility of alternative backbone structures or alternative sugars has been explored, but with limited success in terms of reproducing the ability of DNA to form an extended double-stranded structure regardless of the identity of the bases in the polymer (Miller *et al.* 1981; Huang *et al.* 1993; Richert *et al.* 1996; Eschenmoser 1999; Benner & Hutter 2002; Reddy & Bruice 2003). However, the identity of the

Table 14.1. *Number of distinct codons available for various combinations of base pairs and codon sizes*

Number of base pairs	Number of positions in codon			
	2	3	4	5
1	4	8	16	32
2	16	64	256	1024
3	36	216	1296	7776
4	64	512	4096	32,768

bases used in DNA is a characteristic that might have been substantially different. Benner and co-workers (Piccirilli *et al.* 1990; Benner 1994, 2004; Benner & Switzer 1999; Geyer *et al.* 2003) have explored the possibility of different base pairs, and have shown that a number of alternative base pairs can be accommodated in duplex DNA. In addition, life as we know it employs a triplet genetic code, although the code is not universal—there are some variations in codon assignments, particularly in mitochondria and ciliates. The possibility of codes that utilize a different number of bases or different sizes of codons can be considered (see Table 14.1). However, if we assume that approximately 20 amino acids are required to create good protein structures, then most of the possible codes listed in Table 14.1 either have too little coding capacity or far too much (a situation that would probably introduce too much complexity into the process of translation). Only a triplet code using four bases and a doublet code using six bases have coding capacities in the right range.

Extant life on Earth uses proteins for the majority of structural and catalytic functions. Proteins are particularly suited for these functions because of the structural properties of polymers of amino acids. The polyamide backbone of proteins is neutral, unlike that of nucleic acids, and thus the polymer is able to fold into globular structures. The planarity of the amide functionalities in the backbone restricts rotation around the C–N bond, thus providing some restrictions on the number of conformations that can be adopted. Furthermore, the repeating pattern of hydrogen bond donors and acceptors in the backbone allows interactions along the strand that promote the formation of stable secondary structures, such as alpha helices and beta sheets. The linkage between amino acids is quite stable, but not infinitely so, and it can be relatively easily hydrolyzed by enzymes to allow the turnover of proteins within cells. This propitious combination of properties is conferred by the amide bonds linking the amino acids in the polymer; polymers linked by ester, thioester, ether or C−C bonds would lack one or more these properties.

Life as we know it builds its proteins primarily from the same 20 amino acids.[1] Yet there are many other amino acids that could have been utilized. While it is important that the collection of amino acids used in proteins includes a sufficient number of small, large, hydrophilic, hydrophobic, and charged amino acids, the exact identities of the amino acids in each of these classes may not be critical. Moreover, the amino acids utilized for protein synthesis by familiar life are all L-amino acids, and there is no reason to think that D-amino acids could not have been utilized instead. Indeed, proteins that have been chemically synthesized from D-amino acids fold correctly and are functional (Milton *et al.* 1992; Zawadzke & Berg 1992; Fitzgerald *et al.* 1995; Canne *et al.* 1997).

Given that alternative combinations of bases in DNA and amino acids in proteins might have been chosen, why does the form of life with which we are familiar construct its proteins and nucleic acids out of the particular combination of molecular building blocks that it does? Given our current understanding of chemistry and molecular biology, the best explanation is that these building blocks resulted from chemical and physical contingencies present on the early Earth. Had circumstances been relevantly different, so would life on Earth. This suggests an intriguing possibility. Perhaps a number of different locations on the early Earth were conducive to the formation of life. Locations that have been proposed as important sites for the pre-biotic chemical reactions that provided the building blocks of life include hydrothermal vents (Holm & Andersson 1998; Martin & Russell 2003), mineral surfaces (Wächtershäuser 1988; Cairns-Smith *et al.* 1992; Cody 2004) and organic aerosol particles (Dobson *et al.* 2000). In addition, a variety of amino acids were deposited on Earth from meteorites derived from both asteroids (Oró *et al.* 1971; Anders 1989; Glavin *et al.* 1999; Botta & Bada 2002) and comets (Chyba *et al.* 1990). Racemic

mixtures of 70 amino acids, only 8 of which are utilized by life on Earth today, have been identified in meteorites (Anderson & Haack 2005). A novel, and as yet untested, theory proposes that the synthesis of amino acids might have been catalysed on the primitive Earth by dinucleotides, once conditions favorable for their production arose (Copley *et al.* 2005). It is unlikely that the chemical conditions in these incipient "cradles of life" were identical. Thus, the building blocks available for life, as well as the stability of critical intermediates and the types of reactions that might have been catalysed, would certainly have differed in different locations. If conditions conducive to the emergence of life were present at a number of different locations, then corresponding differences in biomolecules might have arisen in the earliest Terran life forms, perhaps communities of self-replicating RNA molecules (Gilbert 1986; Joyce 2002) or vesicles containing self-reproducing populations of molecules (Oparin 1957; Dobson *et al.* 2000). The hypothesis that the early Earth hosted multiple, alternative origins of life is thus compatible with our current chemical and biological understanding of the nature of familiar life. As we discuss below, some of these alternative types of life might still exist on Earth.

It is, of course, possible that the origin of life is an exceedingly improbable affair, and that life either originated only once on Earth or originated elsewhere and was brought to Earth in a meteorite. But it is important to bear in mind that ignorance concerning how life actually originated on Earth does not provide support for the claim that the origin of life is a cosmic coincidence of some sort; ignorance cannot support a knowledge claim of any sort except perhaps for the trivial claim that we simply do not know. Besides, to the extent that science operates under the guiding principle that natural phenomena are explicable in terms of natural processes, appeal to unnatural occurrences, whether cosmic coincidences or supernatural creation, is self-defeating. If, like other natural phenomena, life is the product of natural processes operating under certain kinds of chemical and physical constraints, then it seems more likely than not that the early Earth hosted more than one origin of life. Some of these separate origins might have produced primitive organisms differing in their basic molecular building blocks in some of the ways discussed above.

COULD ALTERNATIVE LIFE CO-EXIST WITH FAMILIAR LIFE ON EARTH?

If there were alternative origins of life on Earth, it seems clear that they did not give rise to proto-organisms that evolved into large organisms such as higher plants and animals. However, there is little reason to suppose that the processes of evolution inevitably produce large organisms. Microbes are the most abundant form of life on Earth. In most cases, they multiply more rapidly than large organisms, allowing them to evolve more rapidly in response to changing environmental conditions. Microbes exploit more energy resources than multicellular organisms. Some of them photosynthesize, others metabolize organic material and others metabolize inorganic material such as ammonia, hydrogen sulphide and iron. They prosper under an astonishingly wide range of environmental conditions, being found in highly acidic streams, boiling hot springs, several kilometres beneath the Earth's crust and in the coldest regions of Antarctica (Rothschild & Mancinelli 2001). In other words, the biological diversity of the microbes is much greater than that of large multicellular organisms. Indeed, organisms such as higher plants and animals seem to be the exception rather than the rule on Earth. This may also be true of the Universe as a whole; Ward & Brownlee (2000) argue that microbial forms of life are probably very common in the Universe, but that large complex organisms are not. The point is the absence of large complex descendents of alternative forms of early life does not count as evidence that alternative life forms did not exist early in Earth's history, or that they could not persist today.

It might be argued that our form of life is so aggressive and evolutionarily robust that any form of alternative life would have been eliminated long ago. This argument does not bear up under consideration of the structure and dynamics of microbial communities. Although small in number, rare microbes successfully compete in environments swarming with common microbes (McCaig et al. 1999; Spear et al. 2005; Walker et al. 2005). Indeed, they typically participate with other organisms in an interdependent biological system, producing or utilizing material that is utilized, produced or ignored by other microbes. There is little reason to suppose that the microbial descendents of an alternative origin of life could not participate in such a system. For example, a microbe that used only D-amino acids for protein synthesis could survive quite well in a milieu containing L-amino acids simply by having a suite of racemases to convert the abundant L-amino acids to D-amino acids. Bacteria typically have such racemases to generate the D-amino acids used for peptidoglycan synthesis (Yoshimura & Esaki 2003).

There are also at least two plausible alternative evolutionary scenarios. Those forms of novel life that differed the most from familiar life (e.g., where the proteins utilized a very different suite of amino acids) might have had an evolutionary edge. Familiar life would have found them the most difficult to metabolize, and hence the poorest source of nutrition. Such micro-organisms might not only have survived, but gone on to evolve their own independent, interlocking ecological system of predator–prey relations. Another way in which novel forms of early Earth life might have survived is by becoming adapted to environments that are less hospitable to familiar microbial life. In short, rather than being eliminated, novel forms of early life might have evolved in such a way as to remove themselves from competition with familiar life.

Another hypothetical objection that might be raised against the possibility of novel forms of life, descended from an alternative origin of life, has to do with lateral gene transfer. Lateral gene transfer contrasts with vertical ("normal") transfer, which is what happens when genes are transferred from parent to offspring. Lateral gene transfer involves the transfer of genetic material from one organism to another without replication or reproduction. This material can be incorporated into the recipient's genome and passed on to its offspring. Lateral gene transfer is known to have played a significant role in the evolution of microbes. Indeed, many microbiologists believe that the earliest life consisted of a community of proto-organisms that shared genetic material (Woese 1998). If this were the case then (the argument goes) primitive microbes deriving from different origins of life would have been amalgamated into this homogeneous pool of primitive proto-organisms, which subsequently evolved into familiar life.

However, this scenario glosses over some serious problems. Lateral gene transfer as we know it today presupposes significant similarities in the genomes of the microbes involved. Familiar microbes could not incorporate pieces of a genome utilizing alternative base pairs, not to mention different numbers of bases, into their genomes or vice versa. If microbes deriving from alternative origins of life exchanged biomolecules they must have done it before the complex cooperative arrangement between proteins and nucleic acids that characterize familiar life was worked out. Indeed, exchanges among proto-organisms may have been indiscriminate, involving precursor biomolecules of all kinds. In other words, it is not at all clear that a community of diverse proto-organisms deriving from alternative origins of life could have hybridized into a single form of life that evolved into life as we know it. Given our limited understanding of the origin and early development of life, we cannot dismiss the possibility that familiar life arose from a fortuitous mixture of chemicals and that fortuitous mixtures of different chemicals produced alternative forms of microbial life.

Of course none of this proves that such organisms ever existed, let alone still exist. The point is only that many of the arguments that are commonly advanced against their possibility do not hold up well under close scrutiny. How likely is it that they exist? Answering this question is difficult because we know of only one form of life, and we do not yet understand the mechanism by which it emerged. Indeed, Davies & Lineweaver (2005) have recently argued that an alternative origin of life on Earth is not only possible, but also highly probable; they calculate the probability to be 90 percent. We believe their calculation is based upon questionable assumptions. Their argument is based upon the assignment of a 50 percent probability for the emergence of life over a 100 million year period. We simply cannot assign such a probability based upon the single data point we have. However, the possibility of alternative life forms cannot be ruled out on the basis of our current knowledge of chemistry and biology.

LIMITATIONS TO CURRENT TECHNOLOGIES

The possibility that an early alternative life form could have evolved into microbes that either co-exist with familiar life as part of a single, unified biosphere or exist in an independent, parallel biosphere should not be discounted out of hand for a very simple reason: our current technology would not allow us to detect an alternative form of life.[2] We have three major tools with which to explore the microbial world. The first is microscopy. Unfortunately, the morphology of most non-eukaryotic microbes provides little insight into their phylogenetic classification or metabolic capabilities, and we are unlikely to be able to distinguish between normal life and alternative life just by looking. Moreover, molecular biology has taught us that superficial similarities in morphology can hide important differences in molecular architecture and biochemistry. The Archaea provide a particularly salient example. Most Archaea look pretty much like bacteria under a microscope. However, the Archaea are genetically and biochemically more different from bacteria than they are from eukaryotes. Indeed, the discovery that the Archaea are so different from bacteria revolutionized biological taxonomy, with the five kingdoms of familiar life (animals, plants, fungi, protists and monera (bacteria)) being replaced by three domains of life (Archaea, Eubacteria, and Eukarya; Woese et al. 1990). The moral is that morphology does not allow one to eliminate the possibility of *shadow microbes*, i.e., microbes that resemble Archaea and Eubacteria in their gross morphology, but differ from them in fundamental ways at the molecular level. This possibility is underscored by the fact that evolutionary pressures can produce similar adaptations from different biological building blocks. It is probable that conditions on the early Earth favored the development of a morphology along the lines of the Archaea and the Eubacteria, just as conditions on Earth later favoured the independent development of wings in insects, birds and bats. It would thus be a mistake to conclude that every microbe that resembles a bacterium under a microscope is an Archaeaon or a Eubacterium.

The power of microscopy has been expanded by the use of stains for specific cellular components such as nucleic acids (see Fig. 14.1) and lipids, and particularly by fluorescence *in situ* hybridization (FISH) techniques, in which oligonucleotide probes targeting specific genes are used to identify specific organisms in an environmental sample. However, these approaches do not allow us to conclude that all cells visible under a microscope are representatives of familiar life. Stains such as 4′,6-diamidino-2-phenylindole (DAPI) intercalate into double-stranded nucleic acids, and would probably stain nucleic acids in a different life form that contained different bases or a different

Fig. 14.1. DAPI-stained microbes in a water sample taken from a flooded area in New Orleans after Hurricane Katrina (1000× magnification). Photo courtesy of Mari Rodriguez and Mark Hernandez.

backbone in its genetic material. Stains for lipids cannot distinguish between familiar and alternative life forms, since the presence of a lipid membrane reveals nothing about the components of the genetic material and proteins enclosed within. Finally, FISH as it is usually performed is not at all useful, as it can only be used to identify cells containing a gene complementary to the probe being used. If an alternative form of life uses a different type of genetic material, oligonucleotide probes will not hybridize. Thus, microscopy, even combined with standard molecular tools such as FISH, cannot eliminate the possibility that alternative life forms, even those that are not very different from known life, are present in natural microbial populations.

A second tool that has provided us with most of our information about the genetic composition and physiological properties of microbes is cultivation. By growing large quantities of a single microbe, we can determine the chemical components of its genetic and structural materials, the composition of its membrane, the types of metabolic processes it uses to obtain nutrients and energy, and, assuming that its genetic material is DNA, the sequence of its genome. However, we can currently culture less than 1% of the microbes that can be visualized by microscopy (Pace 1997). Efforts to improve cultivation techniques are bearing fruit (Leadbetter 2003), and this situation may improve in the next few years. However, difficulties in cultivation certainly limit our ability to detect alternative forms of microbial life, particularly since they would be more likely than familiar life to require

growth conditions that we might not expect. Without being able to culture a shadow microbe, it would be difficult to determine, for example, that it utilized different bases in its DNA or different amino acids in its proteins.

Finally, PCR amplification of 16S rRNA from environmental samples has provided an extraordinarily powerful tool for identifying non-culturable components of microbial communities (Pace 1997). Unfortunately, however, DNA amplification is not useful for detecting novel forms of microbial life. The process requires "universal primers" capable of supporting amplification of 16S rRNA from one of the three domains of familiar life. Their effectiveness in amplifying microbial DNA depends upon whether it contains coding regions that are sufficiently similar to those on the DNA of familiar life. This poses a serious problem for the prospect of identifying shadow microbes by means of DNA amplification. Even if an alternative life form had DNA as we know it, its ribosomal RNA (assuming that it has ribosomes) might be so different from those of familiar life that it could not be amplified by PCR. Moreover, if the backbone, sugars, or bases of the genetic material were different in an alternative form of life, its genetic material could not be amplified using PCR.

QUESTIONING THE PARADIGM OF LIFE ON EARTH

The paradigm for modern biology includes the assumption that life as we know it is the only form of life on Earth. We owe our understanding of the role played by paradigms in science to the work of Thomas Kuhn (1970). As Kuhn argued, scientific research is conducted within the confines of a paradigm. In addition to theories, paradigms include methods, instrumentation and subsidiary assumptions concerning a particular subject matter. Even though our theoretical understanding of life does not preclude alternative forms of microbial Terran life, the supposition that they do not exist is tacitly incorporated into the paradigm of modern biology.

Paradigms are invaluable tools for scientific research. They facilitate the construction of hypotheses, the design of experiments and the interpretation of results. However, as Kuhn discussed, paradigms sometimes act as blinkers, hindering the exploration of nature by discouraging certain avenues of exploration and biasing the way in which results are interpreted. As a result, important scientific discoveries, and the theoretical advances that wait upon them, may be delayed for many years. Kuhn (1970, pp. 115–117) illustrated this point with several examples from astronomy. William Herschel's discovery of the planet Uranus is particularly salient for our purposes. Between 1690 and 1781 some of Europe's most eminent astronomers reported seeing a star in positions that we now know were occupied by Uranus. Twelve years later Herschel observed the same object with a newly developed, more powerful telescope, and what he saw stumped him. Under higher magnification, it appeared disc shaped, which was highly unusual for a star. Further investigation revealed that the mysterious object moved among (rather than with) the stars. Herschel concluded that he had discovered a new comet. However, as subsequent investigations revealed, the orbit of the object did not conform to that of a comet. After several more months of investigation, another astronomer ventured that the orbit was planetary. Thus, what had been taken to be a star was discovered to be something quite different, namely, a planet. The discovery of Uranus was rapidly followed by the discovery of numerous smaller objects having planetary orbits. Kuhn speculates that the minor paradigm change imposed upon astronomers by the discovery of Uranus prepared them to see objects (namely, asteroids) that they had not seen before but that had nonetheless been there all along. In this light, it is instructive to consider some analogous examples from the biological sciences. These cases resemble the cases discussed by Kuhn insofar as they involve discoveries that were astonishing at the time but nevertheless represented phenomena that had been present all along, unrecognized because they conflicted with a reigning paradigm.

In the Middle Ages infectious diseases were attributed to such things as bad air, supernatural influences, and humoral imbalances, in conjunction with the constitution of the body. The foundations for the modern paradigm for infectious disease were laid by Louis Pasteur, Robert Koch and others towards the end of the nineteenth century (Reid 1974; Madigan & Martinko 2006). Koch, who first identified a bacterium (*Bacillus anthracis*) as the cause of anthrax, developed techniques for culturing and propagating bacteria, and

for determining that a particular bacterium is the cause of a particular disease. Koch's new paradigm (the germ theory of disease) was powerful but it was unable to establish that bacteria caused all infectious diseases. In particular, his techniques were unable to identify viruses, which were far too small to be seen using the technology available at the time and which could not be cultured in isolation. Yet experimental work strongly suggested that infectious agents of some sort were involved in the transmission of the diseases concerned. The mystery was finally solved in the early twentieth century by a combination of two new technologies. Electron microscopy allowed the visualization of the extremely tiny viral particles, and cultivation in the presence of cells (in particular, in eggs) allowed the propagation of viruses. In more recent times, our understanding of the causes of infectious disease has been shaken again by the discovery of prions, proteinaceous infection particles that cause diseases such as scrapie and bovine spongiform encephalopathy (Prusiner 1998). Prions are simply proteins, and the idea that a protein could transmit an infectious disease was so revolutionary that a Nobel Prize was awarded to Stanley Prusiner for the discovery of prions in 1997.

The discovery of the Archaea provides a particularly salient example since it involved the discovery of a previously unsuspected form of microbial life, which resulted in the overthrow of a dominant biological paradigm (Woese 2004), Prior to 1977, scientists believed that living organisms fell into two categories—bacteria and eukaryotes. This way of thinking originated from microscopic studies, which revealed that bacteria were small and contained no membrane-bound organelles, while eukaryotic cells were larger and contained several membrane-bound organelles, including the nucleus, mitochondria and endoplasmic reticulum. The demise of this paradigm started modestly in the 1960s, when Carl Woese began to sequence ribosomal RNA in order to generate phylogenetic trees based on molecular characteristics. As a large database of rRNA sequences became available, it became evident that the rRNA sequences of some microbes clustered together to the exclusion of bacteria and eukaryotes. This group was initially called the Archae-bacteria, but the name was later changed to Archaea, as it was realized that these organisms are fundamentally different from bacteria and, in fact, constitute the third domain of life. The Archaea resemble bacteria in terms of morphology, transcribe RNA using machinery that is more similar to that of eukaryotes than that of bacteria, and have a cell wall structure that is markedly different from both bacteria and eukaryotes. The discovery of Archaea required the development of molecular techniques because Archaea and bacteria look similar under a light microscope. Indeed, what are now understood to be telling chemical differences in the cell membranes of the Archaea and bacteria were originally interpreted as mere adaptations to what was perceived to be extreme environments. In the words of Brock (1978): "The fact that *Sulfolobus* and *Thermoplasma* have similar lipids is of interest, but almost certainly this can be explained by convergent evolution. This hypothesis is strengthened by the fact that *Halobacterium*, another quite different organism, also has lipids similar to the two acidophilic thermophiles." Consequently, the presence of a third domain of life was completely unexpected. Biologists had stumbled across a new form of microbial life without recognizing that they had done so.

A final example is the discovery of catalytic RNA, which upset what is fondly known as the "Central Dogma." The Central Dogma posits that DNA is transcribed into mRNA, which is then translated into proteins. Proteins were believed to carry out the interesting catalytic, structural and regulatory functions required for life. Although Carl Woese had speculated, as early as 1967, that nucleotides might catalyse chemical reactions (Woese 1967), this idea was not given serious consideration. RNA was seen as simply the intermediary between DNA and protein. This paradigm was upset by the unexpected and independent discoveries by Tom Cech (Kruger *et al.* 1982) and Sidney Altman (Guerrier-Takada *et al.* 1983) that RNA could catalyse chemical reactions. Cech and Altman shared the Nobel Prize in 1989 for this discovery. More recently, important roles of RNA in regulating gene expression have been discovered, requiring yet another remodelling of the Central Dogma paradigm. In summary, some of the most lauded work in scientific history has upset the paradigm prevailing at the time. Yet, we continue to operate in the framework of paradigms because they are so useful. Paradigms can be upset by the emergence of new technology that allows exploration in a new way, or by recognition that the results of an experiment do not fit the paradigm and are so compelling that revision of the paradigm is necessary. As discussed above, exploration of the microbial world has continued to yield new and unexpected discoveries. It is not unreasonable to think that this process will continue as we develop more sophisticated methods and tools for probing the invisible world of microbial life.

CONCLUSIONS

The possibility of microbial descendents of alternative origins of life on Earth cannot be dismissed based on current knowledge. The fact that we have not discovered any does not mean they do not exist, since the tools that we currently use to explore the microbial world could not detect them if they existed. Furthermore, arguments to the effect that alternative microbes could not co-exist with familiar life are belied by what we know of the complexity and diversity of microbial communities. If such microbes exist, there is little doubt that they cast heretofore unrecognized physical and chemical shadows (so to speak) upon our familiar biosphere, and hence could be detected with the right tools. The challenge, of course, is to develop methods for recognizing these elusive chemical and physical traces. However, even if shadow microbes do not exist on Earth today, the development of such tools would be an invaluable contribution to the search for unfamiliar forms of microbial life on other planets and moons.

The discovery of a shadow Terran biosphere would have profound scientific and philosophical ramifications. It is clear that life as we know it on Earth has a common origin, which means that we are currently limited to a single sample of life. One cannot generalize on the basis of a single sample. In order to formulate a truly general theory of living systems, we need examples of unfamiliar forms of life. Although we have good theoretical reasons for believing that life on Earth could have been at least modestly different in its biochemistry and molecular architecture, we do not know how different it could have been. It is important that we do not artificially constrain our thinking about the origin of life on Earth and the possibilities for extraterrestrial life on the basis of a limited and possibly very misleading example of life. Indeed, a dedicated search for shadow microbes might produce surprising results, providing us with unexpectedly novel forms of microscopic life. Given that the possibility of alternative forms of life on Earth cannot be discounted and the profound importance such a discovery would represent, we believe that a dedicated search for them ought to be seriously considered.

Finding an alternative form of life on Earth poses an enormous technical challenge. First, we cannot predict the most likely place for finding an alternative life form on Earth. With no knowledge of the biology of such a life form, we cannot predict whether it would more probably be found in rich ecosystems with much microbial diversity, or in extreme conditions, where only a few types of familiar microbes thrive. Indeed, we have often been surprised by finding familiar microbes in unexpected places; for example, the discovery of abundant *Mycobacterium* species in an endolithic community from a highly acidic silica rock sample from Yellowstone's Norris Geyser Basin (Walker *et al.* 2005) was unexpected, as *Mycobacteria* are generally found at only very low levels in environmental samples. Second, the problem resembles finding the proverbial needle in a haystack. The extent of microbial diversity is staggering. A recent study estimates that soil carrying 2×10^9 cells g^{-1} can contain nearly 10^7 species, with 99.9% of the species present at levels of less than 10^5 cells g^{-1} (Gans *et al.* 2005). Even identifying the presence of a rare eubactenum is challenging under these circumstances. Finally, as described above, our current methods are inadequate for detecting forms of life that do not have DNA containing the four canonical bases.

It is worthwhile considering what new methods would allow us to identify and characterize an alternative life form. Studies of the unique biology of an alternative life form would require a sample large enough for the chemical analysis of its constituent macromolecules. Growth in a pure culture would be the optimal way to accomplish this. However, we do not have the methods for high-throughput screening of cultivation conditions that could ensure the growth of a pure culture of an alternative microbe from an environmental sample. Thus, initial efforts might best be directed toward detecting potential alternative life forms *in situ*. The most expedient way to detect an alternative form of life might be to develop reagents that can distinguish typical DNA from other genetic materials differing either in the backbone or the nature of the bases. Such reagents could be used to stain environmental samples to look for cells that do not bind the reagent, always keeping in mind the possibility that an unusual cell wall structure might lead to an inadequate permeabilization of the cell and a consequent lack of staining. For example, antibodies against DNA should be able to discriminate between typical DNA and alternative nucleic acids containing a different type of backbone; this could easily be tested using synthetic analogues of DNA, An alternative approach would be to develop reagents that would recognize alternative backbone structures. Although the creation of such reagents is certainly feasible, this approach would be a fishing expedition limited by our ability to predict what backbone structures might be found in alternative life forms. However, an advantage to this approach would be that a fluorescent probe that does not stain normal cells could be used to collect stained cells using fluorescence-activated cell sorting (FACS).

A different methodology would be required to detect alternative life forms utilizing a different suite of bases in DNA. Antibodies that recognize the bases of DNA as well as the backbone could be used. Antibodies that recognize alternative bases in a DNA context could also be developed and used to stain environmental samples to look for cells that utilize different bases in DNA. Again, fluorescent probes might be used to sort out cells binding the probe for additional analyses. Alternatively, antibodies against DNA containing non-standard bases could be immobilized and used to capture DNA containing unusual bases from bulk DNA isolated from environmental samples, possibly in amounts sufficient for chemical characterization. Each of these proposed methods poses technical challenges, particularly the daunting signal-to-noise problems inherent in trying to detect a rare microbe in a large and diverse population. However, a search for shadow microbes on Earth should be considered because finding an alternative form of life would be one of the greatest scientific discoveries of all time.

ACKNOWLEDGMENTS

Carol Cleland thanks researchers at the Centro de Astrobiología (Spain) and Susan Jones (Department of History, University of Colorado, Boulder) for helpful discussions. We thank Lynn Rothschild for helpful comments on an earlier version of this paper. This work was supported in part by a NASA grant to the University of Colorado's Astrobiology Center.

NOTES

This chapter originally appeared in *International Journal of Astrobiology* 4 (2005), 165–173.

REFERENCES

1. Anders, E. (1989). Pre-biotic organic matter from comets and asteroids. *Nature*, **342**, 255–257.
2. Andersen, A. C. & Haack, H. (2005). Carbonaceous chondrites: Tracers of the prebiotic chemical evolution of the solar system. *International Journal of Astrobiology*, **4**, 12–17.
3. Benner, S. A. (1994). Expanding the genetic lexicon: Incorporating non-standard amino acids into proteins by ribosome-based synthesis. *Trends in Biotechnology*, **12**, 158–163.
4. Benner, S. A. (2004). Understanding nucleic acids using synthetic chemistry. *Accounts of Chemical Research*, **37**, 784–797.
5. Benner, S. A. & Hutter, D. (2002). Phosphates, DNA, and the search for nonterrean life: A second generation model for genetic molecules. *Bioorganic Chemistry*, **30**, 62–80.
6. Benner, S. A. & Switzer, C. Y. (1999). Chance and necessity in biomolecular chemistry: Is life as we know it universal? In H. Frauenfelder, J. Deisenhofer, and P. J. Wolynes (Eds.), *Simplicity and complexity in proteins and nucleic acids* (pp. 339–363). Berlin: Dahlem University Press.
7. Benner, S. A., Ricardo, A., & Carrigan, M. A. (2004). Is there a common chemical model for life in the universe? *Current Opinion in Chemical Biology*, **8**, 672–689.
8. Botta, O. & Bada, J. (2002). Extraterrestrial organic compounds in meteorites. *Surveys in Geophysics*, **23**, 411–467.
9. Brock, T. D. (1978). *Thermophilic microorganisms and life at high temperatures* (p. 178). New York: Springer-Verlag.
10. Cairns-Smith, A. G., Hall, A. J., & Russell, M. J. (1992). Mineral theories of the origin of life and an iron-sulfide example. *Origin of Life and Evolution of the Biosphere*, **22**, 161–180.
11. Canne, L. E., Figliozzi, G. M., Robson, B., *et al.* (1997). The total chemical synthesis of L- and D-superoxide dismutase. *Protein Engineering*, **10**, 23.
12. Chyba, C. F., Thomas, P. J., Brookshaw, L., & Sagan, C. (1990). Cometary delivery of organic molecules to the early Earth. *Science*, **249**, 366–373.

13. Cody, G. (2004). Transition metal sulfides and the origin of metabolism. *Annual Review of Earth and Planetary Sciences*, **32**, 569–599.
14. Copley, S. D., Smith, E., & Morowitz, H. J. (2005). A mechanism for the association of amino acids and their codons and the origin of the genetic code. *Proceedings of the National Academy of Sciences*, **102**, 4442–4447.
15. Davies, P. C. W. & Lineweaver, C. H. (2005). Finding a second sample of life on Earth. *Astrobiology*, **5**, 154–163.
16. Dobson, C. M., Ellison, G. B., Tuck, A. F., & Vaida, V. (2000). Atmospheric aerosols are prebiotic chemical reactors. *Proceedings of the National Academy of Sciences*, **97**, 11,864–11,868.
17. Eschenmoser, A. (1999). Chemical etiology of nucleic acid structure. *Science*, **284**, 2118–2123.
18. Fitzgerald, M. C., Chernushevich, I., Standing, K. G., Kent, S. B. H., & Whitman, C. P. (1995). Total chemical synthesis and catalytic properties of the enzyme enantiomers L- and D-4-oxalocrotonate tautomerase. *Journal of the American Chemical Society*, **117**, 11,075–11,080.
19. Fry, I. (2000). *The emergence of life on Earth*. New Brunswick, NJ: Rutgers University Press.
20. Gans, J., Wolinsky, M., & Dunbar, J. (2005). Computational improvements reveal great bacterial diversity and high metal toxicity in soil. *Science*, **309**, 1387–1390.
21. Geyer, C. R., Battersby, T. R., & Benner, S. A. (2003). Nucleobase pairing in expanded Watson–Crick-like genetic information systems. *Structure*, **11**, 1495–1498.
22. Gilbert, W. (1986). Origin of life: The RNA world. *Nature*, **319**, 618.
23. Glavin, D. P., Bada, J. L., Bringon, K. L., & McDonald, G. D. (1999). Amino acids in the Martian meteorite Nakhla. *Proceedings of the National Academy of Sciences*, **96**, 8835–8838.
24. Guerrier-Takada, C., Gardiner, K., Marsh, T., Pace, N., & Altman, S. (1983). The RNA moiety of ribonuclease P is the catalytic subunit of the enzyme. *Cell*, **35**, 849–857.
25. Holm, N. G. & Anderson, E. M. (1998). Hydrothermal systems. In A. Brack (Ed.), *The molecular origins of life: Assembling pieces of the puzzle*. Cambridge, UK: Cambridge University Press.
26. Huang, Z., Schneider, K. C., & Benner, S. A. (1993). Oligonucleotide analogs with dimethyl-sulfide, -sulfoxide and -sulfone groups replacing phosphodiester linkages. In S. Agrawal (Ed.), *Methods in Molecular*

Biology (pp. 315–353). Totowa, NJ: Rowman & Littlefield.

27. Joyce, G. (2002). The antiquity of RNA-based evolution. *Nature*, **418**, 214–221.

28. Kruger, K., Grabowski, P. J., Zuang, A. J., Sands, J., Gottschling, D. E., & Cech, T. R. (1982). Self-splicing RNA: Autoexcision and autocyclization of the ribosomal RNA intervening sequence of Tetrahymena. *Cell*, **31**, 147–157.

29. Kuhn, T. (1970). *The structure of scientific revolutions*. Chicago: University of Chicago Press.

30. Leadbetter, J. R. (2003). Cultivation of recalcitrant microbes: Cells are alive, well, and revealing their secrets in the 21st century laboratory. *Current Opinion in Microbiology*, **6**, 276–281.

31. Madigan, M. T. & Martinko, J. M. (2006). *Brock biology of microorganisms* (11th ed.). Upper Saddle Riber, NJ: Pearson Prentice-Hall

32. Martin, W. & Russell, M. (2003). On the origin of cells: A hypothesis for the evolutionary transitions from abiotic geochemistry to chemoautotrophic prokaryotes, and from prokaryotes to nucleated cells. *Philosophical Transactions of the Royal Society of London Series B: Biological Sciences*, **358**, 59–85.

33. McCaig, A. E., Glover, L. A., & Prosser, J. I. (1999). Molecular analysis of bacterial community structure and diversity in unimproved upland grass pastures. *Applied Environmental Microbiology*, **65**, 1721–1730.

34. Miller, P. S., McParland, K. B., Jayaraman, J., & Tso, P. O. P. (1981). Biochemical and biological effects of nonionic nucleic acid methylphosphonates. *Biochemistry*, **20**, 1874–1880.

35. Milton, R. C. L., Milton, S. C. F., & Kent, S. B. H. (1992). Total chemical synthesis of a D-enzyme: The enantiomers of HIV-1 protease show demonstration of reciprocal chiral substrate specificity. *Science*, **245**, 1445–1448.

36. Oparin, A. I. (1957). *The origin of life on Earth* (3rd ed.). Edinburgh: Oliver & Boyd.

37. Pace, N. R. (1997). A molecular view of microbial diversity and the biosphere. *Science*, **274**, 734–740.

38. Piccirilli, J. A., Krauch, T., Moroney, E. E., & Benner, S. A. (1990). Extending the genetic alphabet: Enzymatic incorporation of a new base pair into DNA and RNA. *Nature*, **343**, 33–37.

39. Prusiner, S. B. (1998). Prions. *Proceedings of the National Academy of Sciences*, **95**, 13,363–13,383.

40. Reddy, P. M. & Bruice, T. C. (2003). Solid-phase synthesis of positively charged deoxynucleic guanidine (DNG) oligonucleotide mixed sequences. *Bioorganic and Medicinal Chemistry Letters*, **13**, 1281–1285.

41. Reid, R. (1974). *Microbes and men*. London: British Broadcasting Corporation.

42. Richert, C., Roughton, A. L., & Benner, S. A. (1996). Nonionic analogs of RNA with diethyl sulfone bridges. *Journal of the American Chemical Society*, **118**, 4518–4531.

43. Rothschild, L. J. & Mancinelli, R. L. (2001). Life in extreme environments. *Nature*, **409**, 1092–1101.

44. Shapiro, R. (1986). *Origins: A skeptic's guide to the creation of life on Earth* (pp. 293–295). Toronto: Bantam Books.

45. Sleep, N. H., Zahnle, K. J., Kastings, J. F., & Morowitz, H. J. (1989). Annihilation of ecosystems by large asteroid impacts on the early Earth. *Nature*, **342**, 139–142.

46. Spear, J. R., Walker, J. J., McCollum, T. M., & Pace, N. R. (2005). Hydrogen and bioenergetics in the Yellowstone geothermal ecosystem. *Proceedings of the National Academy of Sciences*, **102**, 2555–2560.

47. Wächtershäuser, G. (1988). Before enzymes and templates: Theory of surface metabolism. *Microbiological Review*, **52**, 452–484.

48. Wächtershäuser, G. (1992). Groundworks for an evolutionary biochemistry: The iron-sulfur world. *Progress in Biophysics and Molecular Biology*, **58**, 85–201.

49. Walker, J. J., Spear, J. R., & Pace, N. (2005). Geobiology of a microbial endolithic community in the Yellowstone geothermal environment. *Nature*, **434**, 1011–1014.

50. Ward, P. D. & Brownlee, D. (2000). *Rare Earth*. New York: Springer-Verlag.

51. Woese, C. R. (1967). *The genetic code: The molecular basis for genetic expression* (pp. 186–189). New York: Harper & Row.

52. Woese, C. R. (2004). The archaeal concept and the world it lives in: A retrospective. *Photosynthesis Research*, **80**, 361–372.

53. Woese, C. R., Kandler, O., & Wheelis, M. L. (1990). Towards a natural system of organisms: Proposal for the domains Archaea, Bacteria, and Eucarya. *Proceedings of the National Academy of Sciences*, **87**, 4576–4579.

54. Yoshimura, T. & Esaki, N. (2003). Amino acid racemases: Functions and mechanisms. *Journal of Bioscience and Bioengineering*, **96**, 103–109.

55. Zawadzke, L. E. & Berg, J. M. (1992). A racemic protein. *Journal of the American Chemical Society*, **114**, 4002–4003.

ENDNOTES

1 Seleno-cysteine is found in a small number of enzymes. It is incorporated during protein synthesis at the ribosome using a tRNA that recognizes what would normally be a stop codon. Seleno-methionine is incorporated into proteins randomly in place of methionine. Post-translational modifications of some amino acids in proteins occur in specific cases; examples include the formation of dehydroalanine from serine and γ-carboxy-glutamate from glutamate.

2 Davies & Lineweaver mention this point, but do not discuss it in detail.

15 · Introduction to the limits of organic life in planetary systems

NATIONAL RESEARCH COUNCIL OF THE NATIONAL ACADEMIES

THE SEARCH FOR LIFE IN THE COSMOS

The National Aeronautics and Space Administration (NASA) has long given high priority to missions that ask whether extraterrestrial life might exist in the solar system and beyond. That priority reflects public interest, which was enhanced in the mid 1990s when fragments of Mars delivered to Earth as meteorites were shown to contain small structures reminiscent of microbial life.

The proper interpretation of those structures remains controversial, but it is certain that nothing would alter our view of humanity and our position in the cosmos more than the discovery of alien life. Nothing would contribute more to NASA's goal of exploring the cosmos, or to inspiring and educating the next generation of students in the hard sciences and engineering, than a search for alien life. Nothing would be more unfortunate than to expend considerable resources in the search for alien life and then not recognize it if it is encountered.

The search for life in the cosmos begins with our understanding of life on Earth. This understanding has grown enormously over the past century. It is now clear that, although terran life is conveniently categorized into millions of species, studies of the molecular structure of the biosphere show that all organisms that have been examined have a common ancestry. There is no reason to believe, or even to suspect, that life arose on Earth more than once, or that it had biomolecular structures that differed greatly from those shared by the terran life that we know.

Our only example of life has been quite successful in dominating the planet. Earth itself presents a variety of environments, some extreme by human standards.

One lesson learned from studies of terran biochemistry and its environmental range on Earth is that the life we know requires liquid water. Wherever a source of energy is found on Earth with liquid water, life of the standard variety is present.

That observation has already helped to guide NASA missions through the directive to "follow the water" in searching for life in the solar system. Environments where liquid water might be or might have been present are high on the list of locales planned for NASA missions. Excitement runs high when sites are found where the geology indicates with near certainty the past presence of liquid water in substantial amounts.

As pragmatic as the strategy is, scientists and laypeople alike have asked whether it might be parochial, or "terracentric." As Carl Sagan noted, it is not surprising that carbon-based organisms breathing oxygen and composed of 60 percent water would conclude that life must be based on carbon and water and metabolize free oxygen (Sagan 1973).

The depth and breadth of our knowledge of terran chemistry tempts us to focus on carbon because terran life is based on carbon, and organic chemistry as we know it emerged from nineteenth-century natural-product chemistry based on the isolation of compounds from nature. If terran life had provided silicon-based molecules, then our knowledge of silicon-based chemistry would now be advanced.

The natural tendency toward terracentricity requires that we make an effort to broaden our ideas of where life is possible and what forms it might take. Furthermore, basic principles of chemistry warn us against terracentricity. It is easy to conceive of chemical reactions that might support life involving non-carbon compounds, occurring in solvents other

The Nature of Life, ed. M. A. Bedau and C. E. Cleland. Published by Cambridge University Press. © M. A. Bedau and C. E. Cleland 2010.

than water, or involving oxidation–reduction reactions without dioxygen. Furthermore, there are reactions that are not redox. For example, life could get energy from $NaOH + HCl$; the reaction goes fast abiotically, but an organism could send tendrils into the acid and the base and live off the gradient. An organism could get energy from supersaturated solution. It could get relative humidity from evaporating water. It is easy to conceive of alien life in environments quite different from the surface of a rocky planet. The public has become aware of those ideas through science fiction and non-fiction, such as Peter Ward's *Life as We Do Not Know It*. (Ward 2005)

The public and the scientific community have become interested in authoritative perspectives on the possibility of life in environments in the solar system very much different from the ones that support life on Earth and life supported by "weird" chemistry in exotic solvents and exploiting exotic metabolisms. To NASA those ideas would help to guide missions throughout the solar system and permit them to recognize alien life if it is encountered, however it is structured. Given the inevitability of human missions to Mars and other locales potentially inhabited by alien life, an understanding of the scope of life will improve researchers' chance to study such life before a human presence contaminates it or, through ignorance or inaction, destroys it.

In broadest outline, this report shows that the committee found no compelling reason for life being limited to water as a solvent, even if it is constrained to use carbon as the scaffolding element for most of its biomolecules. In water, varied molecular structures are conceivable that could (in principle) support life, but it would be sufficiently different from life on Earth that it would be overlooked by unsophisticated life-detection tools. Evidence suggests that Darwinian processes require water, or a solvent like water, if they are supported by organic biopolymers (such as DNA). Furthermore, although macromolecules using silicon are known, there are few suggestions as to how they might have emerged spontaneously to support a biosphere.

DEFINING THE SCOPE OF THE PROBLEM

For generations the definition of life has eluded scientists and philosophers. (Many have come to recognize that the concept of "definition" itself is difficult to define (Cleland & Chyba 2002).) We can, however, list characteristics of the one example of life that we know—life on Earth:

- It is chemical in essence; terran living systems contain molecular species that undergo chemical transformations (metabolism) under the direction of molecules (enzyme catalysts) whose structures are inherited, and heritable information is itself carried by molecules.
- To have directed chemical transformations, terran living systems exploit a thermodynamic disequilibrium.
- The biomolecules that terran life uses to support metabolism, build structures, manage energy, and transfer information take advantage of the covalent bonding properties of carbon, hydrogen, nitrogen, oxygen, phosphorus, and sulfur and the ability of heteroatoms, primarily oxygen and nitrogen, to modulate the reactivity of hydrocarbons.
- Terran biomolecules interact with water to be soluble (or not) or to react (or not) in a way that confers fitness on a host organism. The biomolecules found in terran life appear to have molecular structures that create properties specifically suited to the demands imposed by water.
- Living systems that have emerged on Earth have done so by a process of random variation in the structure of inherited biomolecules, on which was superimposed natural selection to achieve fitness. These are the central elements of the Darwinian paradigm.

Various published definitions of life understandably incorporate those features, given that we are the life form defining it. Indeed, because the chemical structures of terran biomolecular systems all appear to have arisen through Darwinian processes, it is hardly surprising that some of the more thoughtful definitions of life hold that it is a "chemical system capable of Darwinian evolution" (Joyce *et al.* 1994).

IS EVOLUTION AN ESSENTIAL FEATURE OF LIFE?

Many of the definitions of life include phrases like *undergoes Darwinian evolution*. The implication is that phenotypic changes and adaptation are necessary to exploit unstable environmental conditions and to function optimally in the environment. Evolutionary changes have even been suggested for the hypothesized "clay crystal life" of Cairns-Smith (Cairns-Smith, 1982), referring to randomly occurring errors in crystal structure during crystal growth as analogous to mutations. Would a self-replicating chemical system capable of chemical transformations in the environment be considered life? If self-replicating chemical compounds are not life, replication by itself is not sufficient as a defining characteristic of life. Likewise, the ability to undergo Darwinian evolution, a process that results in heritable changes in a population, is also not sufficient to define life if we consider minerals that are capable of reproducing errors in their crystal structure to be equivalent to evolution. Although that property of clays may have been vital in the origin of life and particularly in the prebiotic synthesis of organic macromolecules and as catalysts for metabolic reactions, can the perpetuation of "mistakes" in crystal structure result in the selection of a "more fit" crystal structure? It is important to emphasize that evolution is not simply reproducing mutations (mistakes in clays), but also selecting variants that are functionally more fit.

The canonical characteristics of life are an inherent capacity to adapt to changing environmental conditions and to interact with other living organisms (and, at least on Earth, also with viruses) (Brown 2003; Martin et al. 2003; Ochman, Lawrence, and Groisman 2000; Woese 2002). Natural selection is the key to evolution and the main reason that Darwinian evolution persists as a characteristic of many definitions of life. The only alternative to evolution for producing diversity would be to have environmental conditions that continuously create different life forms or similar life forms with random and frequent "mistakes" in the synthesis of chemical templates used for replication or metabolism. Such mistakes would be equivalent to mutations and could lead to traits that gave some selective advantage in an existing community or in exploiting new habitats. That random process could lead to life forms that undergo a form of evolution without a master information macromolecule, such as DNA or RNA. It is difficult to imagine such life forms as able to "evolve" into complex structures unless other mechanisms, such as symbiosis or cell-cell fusion, are available.

Evolution is the key mechanism of heritable changes in a population. However, although mutation and natural selection are important processes, they are not the only mechanisms for acquiring new genes. It is

understood that lateral gene transfer is one of the most important and one of the earliest mechanisms for creating diversity and possibly for building genomes with the requisite information to result in free-living cells (Martin et al. 2003). Lateral gene transfer is also one of the mechanisms to align genes from different sources into complex functional activities, such as magnetotaxis and dissimilatory sulfate reduction (Grünberg et al. 2001; Mussmann et al. 2005; Mazel 2006). It is possible that this mechanism was important in the evolution of metabolic and biosynthetic pathways and other physiological traits that may have evolved only once even though they are present in a wide variety of organisms. Coevolution of two or more species is also a hallmark of evolution manifested in many ways, from insect–plant interactions to the involvement of hundreds of species of bacteria in the nutrition of ruminant animals. Organisms and the environment also coevolve, depending on the dominant characteristics of the environment and the availability of carbon and energy sources.

If the ability to undergo Darwinian evolution is a canonical trait of life no matter how different a life form is from Earth life, are there properties of evolving extraterrestrial organisms that would be detectable as positive signs of life? Evolution provides organisms the opportunity to exploit new and changing environments, and one piece of evidence for the cosmic ubiquity of evolution is that on Earth life occupies all available habitats and even creates new ones as a consequence of metabolism. Another hallmark of evolution is the ability of organisms to coevolve with other organisms and to form permanent and obligatory associations. It is highly probable that an inevitable consequence of evolution is the elimination of radically different biochemical lineages of life that may have formed during the earliest period of the evolution of life. Extant Earth life is the result of either selection of the most fit lineage or homogenization of some or all of the different lineages into a common ancestral community that developed into the current three major lineages (domains). All have a common biochemistry based on presumably the most "fit" molecular information strategies and energy-yielding pathways among a potpourri of possibilities.

Thus, one of the apparent generalizations that can be drawn from extant Earth life, and the explanation for the development of a "unity of biochemistry" in all organisms, is that lateral gene transfer is an ancient and efficient mechanism for rapidly creating diversity and complexity. Lateral gene transfer is also an efficient mechanism for selecting the genes that are most "fit" for specific proteins and transferring them into diverse groups of organisms. The results are the addition of genes and the replacement of less-fit genes that have similar functions. Natural selection based solely on mutation is probably not an adequate mechanism for evolving complexity. More important, lateral gene transfer and endosymbiosis are probably the most obvious mechanisms for creating complex genomes that could lead to free-living cells and complex cellular communities in the short geological interval between life's origin and the establishment of autotrophic CO_2 fixation about 3.8 billion years ago and microbial sulfate reduction 3.47 billion years ago on the basis of isotope data (Rosing 1999; Shen, Buick and Canfield 2001; Shidlowski 1988). An important implication of the existence of viruses or virus-like entities during the early evolution of cellular organisms is that their genomes may have been the source of most genetic innovations because of their rapid replication, high rates of mutation due to replication errors, and gene insertions from diverse host cells (Claverie 2006; Forterre 2006; Koonin and Martin 2005).

Is evolution an essential feature of life? Cells are more than the information encoded in their genomes; they are part of a highly integrated biological and geochemical system in whose creation and maintenance they have participated. The unity of biochemistry among all Earth's organisms emphasizes the ability of organisms to interact with other organisms to form coevolving communities, to acquire and transmit new genes, to use old genes in new ways, to exploit new habitats, and, most important, to evolve mechanisms to help to control their own evolution. Those characteristics would probably be present in extraterrestrial life even if it had a separate origin and a unified biochemistry different from that of Earth life.

BRIEF CONSIDERATIONS OF POSSIBLE LIFE FORMS OUTSIDE THE SCOPE OF THIS REPORT

As discussed in the literature (Ricardo and Carrigan 2004), chemical models of non-Earth-centric life reveal much about what the scientific community considers possible, particularly regarding ways in which systems organize matter and energy to generate life. Thus, truly "weird" life might utilize an element other than carbon for its scaffolding. Less weird, but still alien to human biological experience, would be a life form that does not exploit thermodynamic disequilibria that are largely chemical. Weirder would be a life form that does not exploit water as its liquid milieu. Still weirder would be a life form that exists in the solid or gas phase (Allamandola and Hudgins 2003). In a different direction, yet also outside the scope of life that most communities think possible, would be a life form that lacks a history of Darwinian evolution.

Some features of terran life are almost certainly universal, however. In particular, the requirement for thermodynamic disequilibrium is so deeply rooted in our understanding of physics and chemistry that it is not disputable as a requirement for life. Other criteria are not absolute. Terran biology contains clear examples of the use of non-chemical energy; photosynthesis is the best known, although energy from light is soon converted to chemical energy. Silicon, in some environments, can conceivably support the scaffolding of large molecules. This report explicitly considers non-aqueous environments.

Even Darwinian evolution is presumably not an absolute. For example, depending on how human civilization applies gene therapy, our particular form of life could be able to evolve via Lamarckian,[1] as opposed to Darwinian, processes. Humankind will be able to perceive and solve problems in human biology without needing to select among random events, thus sparing the species the need to remove unavoidable genetic defects through the death of individuals. That will make the human biosphere no less living, even to those who make Darwinian evolution central in their concept of life.

Likewise, we can easily conceive of robots that are self-reproducing or computer-based processes that grow and replicate (Adami and Wilke 2004; Rosing 1999; Shen, Buick, and Canfield 2001; Shidlowski 1988). Here, information transfer is not based on a specific molecular replication but on a replication involving information on a matrix. Whether such entities will be called life remains to be seen.

What is clear is that the scientific community does not believe that Lamarckian, robotic, or informational "life" could have arisen spontaneously from inanimate matter. At the very least, its matrix would have to be constructed initially by a chemical Darwinian life form arising from processes similar to those seen on Earth. Again, it is not clear whether those views are constrained by our inability to conceive broadly from what we know or whether they reflect true constraints on the processes by which life might emerge in natural history.

Those thoughts introduce a subsidiary theme of this report. It is conceivable that chemistry, structure, or environments able to *support* life were not suited for the *initiation* of life. For example, Earth can support life today, but prevailing views hold that life could not have originated in an atmosphere that is as oxidizing as Earth's today. If that is true, the surface of Earth would be an environment that is habitable but not able to give rise to life.

STRATEGIES TO MITIGATE ANTHROPOCENTRICITY

We have only one example of biomolecular structures that solve problems posed by requirements for life, and the human mind finds it difficult to create ideas truly different from what it already knows. It is thus difficult for us to imagine how life might look in planetary environments very different from what we find on Earth. Recognizing that difficulty, the committee chose to embrace it. The committee exploited a strategy that began with characterization of the terran life that humankind has known well, first because of its macroscopic visibility and then through microscopic observation that began in earnest four centuries ago. This, of course, is like life that is associated with humankind. As the next step in the strategic process, the committee assembled a set of observations about life that is considered exotic when compared with human-like life. Exploration of Earth has taken researchers to environments that human-like organisms find extreme, to the highest temperatures at which liquid water is possible, to the lowest temperatures at which water is liquid, to the depths of the ocean where pressures are high, to extremes of acidity and alkalinity, to places where the energy flux is too high for human-like life to survive, to locales where thermodynamic disequilibria are too scarce to support human-like life, and to locations where the chemical environment is toxic to human-like life.

The committee then asked, can we identify environments on Earth where Darwinian processes that

exploit human-like biochemistry cannot exploit available thermodynamic disequilibria? The answer is an only slightly qualified no. It appears that wherever the thermodynamic minimum for life is met on Earth and water is found, life is found. Furthermore, the life that is found appears to be descendant from an ancestral life form that also served as the ancestor of humankind (perhaps we would not necessarily have recognized it if its ancestry were otherwise) and exploits fundamentally human-like biochemistry.

The committee then reviewed evidence of abiotic processes that manipulate organic material in a planetary environment. It asked whether the molecules that we see in contemporary terran life might be understood as the inevitable consequences of abiotic reactivity. Although signatures of such predecessor reactivity can be adumbrated within contemporary biochemistry, they are generally faint (Benner, Ellington, and Tauer 1989). Some 4 billion years of biological evolution have attached a strong Darwinian signature to whatever went before; hypotheses regarding evidence of our inanimate ancestry within modern biostructures are the subject of intense dispute.

If life originated first on Earth, it was long ago when conditions on the surface of this planet were very different from what they are today. We do not know what those conditions were, and we may never know. Furthermore, the organisms around today are all highly evolved descendants of the first life forms and probably contributed long ago to the demise of their less fit, more primitive competitors. The historical slate has been wiped clean both geologically and biologically. Finally, because life forms replicate, singular events can have enormous impacts on future developments. Life does not have to be a probable outcome of spontaneous physicochemical processes, although it may well be. Arguments based on probability are not as powerful in this sphere as they usually are in the physical sciences.

The committee surveyed the inventory of environments in the solar system and asked which non-Earth ones might be suited to life of the terran type. Such locales are few, unless there are laws not now understood that could govern the early stages of the self-organization of biochemical structures and processes that could lead inevitably to evolving life forms (Kauffman 1995). Subsurface Mars and the putative sub-ice oceans of the Galilean satellites are the only locales in the solar system (other than Earth itself) that are clearly compatible with terran biochemistry.

The committee's survey made clear, however, that most locales in the solar system are at thermodynamic disequilibrium—an absolute requirement for chemical life. Furthermore, many locales that have thermodynamic disequilibrium also have solvents in liquid form and environments where the covalent bonds between carbon and other lighter elements are stable. Those are weaker requirements for life, but the three together would appear, perhaps simplistically, to be sufficient for life. The committee asked whether it could conceive of biochemistry adapted to those exotic environments, much as human-like biochemistry is adapted to terran environments. Few detailed hypotheses are available; the committee reviewed what is known, or might be speculated, and considered research directions that might expand or constrain understanding about the possibility of life in such exotic environments.

Finally, the committee considered more exotic solutions to problems that must be solved to create the emergent properties that we agree characterize life. It considered a hierarchy of "weirdness":

- Is the linear dimensionality of biological molecules essential? Or can a monomer collection or two-dimensional molecules support Darwinian evolution?
- Must a standard liquid of some kind serve as the matrix for life? Can a supercritical fluid serve as well? Can life exist in the gas phase? In solid bodies, including ice?
- Must the information content of a living system be held in a polymer? If so, must it be a standard bio-polymer? Or can the information to support life be placed in a mineral form or in a matrix that is not molecularly related to Darwinian processes?
- Are Darwinian processes and their inherent struggle to the death essential for living systems? Can altruistic processes that do not require death and extinctions and their associated molecular structures support the development of complex life?

NOTES

This chapter originally appeared in the National Academy of Sciences, *The limits of organic life in planetary systems*, pp. 5–10, Washington D. C.: National Academies Press, 2007. This work was composed with the combined efforts of the Committee on the Limits of Organic Life in Planetary Systems, the Space Studies Board, and the Board on Life Sciences. The full text is available at http://www.nap.edu/catalog/11919.html (accessed July, 2008). Members of the Committee on the Limits of Organic Life in Planetary Systems: John A. Baross (Chair), Steven A. Benner, George D. Cody, Shelley D. Copley, Norman R. Pace, James H. Scott, Robert Shapiro, Mitchell L. Sogin, Jeffrey L. Stein, Roger Summons, and Jack W. Szostak. Members of the Space Studies Board: Lennard A. Fisk (Chair), A. Thomas Young (Vice Chair), Spiro K. Antiochos, Daniel N. Baker, Steven J. Battel, Charles L. Bennett, Judith A. Curry, Jack D. Farmer, Jack D. Fellows, Jacqueline N. Hewitt, Tamara E. Jernigan, Klaus Keil, Berrien Moore III, Kenneth H. Nealson, Norman P. Neureiter, Suzanne Oparil, James Pawelczyk, Harvey D. Tananbaum, Richard H. Truly, Joseph F. Veverka, Warren M. Washington, Gary P. Zank and Marcia S. Smith. Members of the Board on Life Sciences: Keith Yamamoto (Chair), Ann M. Arvin, Jeffrey L. Bennetzen, Ruth Berkelman, Deborah Blum, R. Alta Charo, Jeffrey L. Dangl, Paul R. Ehrlich, Mark D. Fitzsimmons, Jo Handelsman, Ed Garlow, Kenneth H. Keller, Randall Murch, Gregory A. Petsko, Muriel E. Poston, James Reichman, Marc T. Tessier-Lavigne, James Tiedje Terry L. Yates and Frances Sharples.

REFERENCES

1. Adami, C. & Wilke, C. O. (2004). Experiments in digital life. *Artificial Life*, **10**, 117–122.
2. Allamandola, L. J. & Hudgins, D. M. (2003). From interstellar polycyclic aromatic hydrocarbons and ice to astrobiology. In V. Pirronello and J. Krelowski (Eds.), *Proceedings of the NATO ASI, solid state astrochemistry* (pp. 1–54). Dordrecht: Kluwer.
3. Benner, S. A., Ellington, A. D., & Tauer, A. (1989). Modern metabolism as a palimpsest of the RNA World. *Proceedings of the National Academy of Sciences*, **86**, 7054–7058.
4. Brown, J. R. (2003). Ancient horizontal gene transfer. *Nature Reviews Genetics*, **4**, 121–132.
5. Cairns-Smith, A. G. (1982). *Genetic takeover and the mineral origins of life*. Cambridge, UK: Cambridge University Press.
6. Claverie, J. M. (2006). Viruses take center stage in cellular evolution. *Genome Biology*, **7**, 110.

7. Cleland, C. E. & Chyba, C. F. (2002). Defining 'Life'. *Origins of Life and Evolution of the Biosphere*, 32, 387–393.

8. Forterre, P. (2006). Three RNA cells for ribosomal lineages and three DNA viruses to replicate their genomes: A hypothesis for the origin of cellular domain. *Proceedings of the National Academy of Sciences*, 103, 3669–3674.

9. Grünberg, K., Wawer, C., Tebo, B. M., & Schüler, D. (2001). A large gene cluster encoding several magnetosome proteins is conserved in different species of magnetotactic bacteria. *Applied Environmental Microbiology*, 67, 4573–4582.

10. Joyce, G. F., Young, R., Chang, S., *et al.* (1994). In D. W. Deamer and G. R. Fleischaker (Eds.), *The origins of life: The central concepts*. Boston: Jones & Bartlett.

11. Kauffman, S. A. (1995). *At home in the universe: The search for laws of self organization and complexity*. New York: Oxford University Press.

12. Koonin, E. V. & Martin, W. (2005). On the origin of genomes and cells within inorganic compartments. *Trends in Genetics*, 21, 647–654.

13. Martin, W., Rotte, C., Hoffmeister, M., *et al.* (2003). Early cell evolution, eukaryotes, anoxia, sulfide, oxygen, fungi first (?), and a tree of genomes revisited. *IUBMB Life*, 55, 193–204.

14. Mazel, D. (2006). Integrons: Agents of bacterial evolution. *Nature Review of Microbiology*, 4, 608–620.

15. Mussmann, M., Richter, M., Lombardot, T., *et al.* (2005). Clustered genes related to sulfate respiration in uncultured prokaryotes support the theory of their concomitant horizontal transfer. *Journal of Bacteriology*, 187, 7126–7127.

16. Ochman, H., Lawrence, J. G., & Groisman, E. S. (2000). Lateral gene transfer and the nature of bacterial innovation. *Nature*, 405, 299–304.

17. Ricardo, A. & Carrigan, M. A. (2004). Is there a common chemical model for life in the universe? *Current Opinion in Chemical Biology*, 8, 672–689.

18. Rosing, M. T. (1999). ^{13}C-depleted carbon microparticles in >3700-Ma sea-floor sedimentary rocks from West Greenland. *Science*, 283, 674–676.

19. Sagan, C. (1973). Extraterrestrial life. In C. Sagan (Ed.), *Communication with extraterrestrial intelligence* (CETI) (pp. 42–67). Cambridge, MA: MIT Press.

20. Shen, Y., Buick, R., & Canfield, D. E. (2001). Isotopic evidence for microbial sulfate reduction in the early Archaean era. *Nature*, 410, 77–81.

21. Shidlowski, M. A. (1988). A 3800-million-year isotopic record of life from carbon in sedimentary rocks. *Nature*, 333, 313–318.

22. Ward, P. (2005). *Life as we do not know it*. New York: Viking.

23. Woese, C. R. (2002). On the evolution of cells. *Proceedings of the National Academy of Sciences*, 99, 8742–8747.

ENDNOTE

1 Lamarck recognized a similar principle of evolution referred to as "inheritance of acquired characters," stating that variations in characteristics seen in organisms were acquired in response to the environment.

Section III
Artificial life and synthetic biology

As the discussions in Sections I and II underscore, natural life as we know it on Earth today is an extremely complex phenomenon, having distinctive functional characteristics at the abstract level and comprising specific chemical elements, molecules, and biochemical pathways at the physical level. Its complex, multi-leveled, interconnected character poses problems for biologists trying to generalize to all forms of life: Which features of our only unequivocal example of life are fundamental to all life? Artificial life (also known as "ALife") is an interdisciplinary study that aims to circumvent this difficulty by artificially synthesizing life-like processes both in isolation and together at different levels of analysis. This allows researchers to better explore questions such as whether the stuff that comprises life is essential to it. Underlying research in ALife is the view that one of the best ways to understand something is to learn how to make it. The term "artificial" refers to the fact that humans are involved in the construction process, and indicates that the results might be quite unlike natural forms of life. There is a lively debate among ALife researchers over what types of artificial life qualify as genuine examples of life, instead of mere simulations or models of life. The question is, when is a simulation or model of life complete enough to be the real thing?

Artificial life research can be subdivided into three broad categories, reflecting three different types of synthetic methods (Bedau 2003, 2007). "Soft" artificial life creates computer simulations or other purely informational constructions that exhibit lifelike behavior. Soft artificial life overlaps somewhat with the new science of systems biology, which systematically studies complex interactions in biological systems with the aid of computational models. However, whereas systems biology aims to capture quantitative details about specific actual biological systems, soft artificial life focuses more on understanding qualitative aspects of broader classes of hypothetical living systems. Stuart Kauffman (Ch. 30) describes two specific examples of soft artificial life. Some researchers believe that soft artificial life is essential for making progress on understanding the nature of life because there is no way to investigate certain theoretical scenarios without the aid of computer simulations. One question about soft artificial life concerns the extent to which the informational processes that it focuses upon—those that it identifies as essential to life—are inferred from what we know about familiar natural life on Earth. As with all attempts to generalize from a limited example, there is the worry that known natural forms of life may be unrepresentative of life's full potential in important but unknown ways. Some of these ways might include characteristics at the abstract, informational level as well as at the physical level of chemical elements, biomolecules, and biochemical processes. Also, as Elliott Sober (Ch. 16) and Margaret Boden (Ch. 18) discuss, it is not clear that important functional features of life (e.g., metabolism) can be treated independently of their physical embodiments.

"Hard" artificial life typically implements lifelike systems in hardware made from silicon, steel, and plastic. Hard artificial life is a branch of robotics research that is motivated specifically by the hypothesis that the best way to create robots with the same capacities for autonomous adaptive behavior found in natural life forms is to make devices that operate on the same abstract principles as living systems. Recent hard artificial life achievements include the first widely available commercial robotic domestic vacuum, Roomba® (Brooks, 2002), and the walking robots designed by an artificial form of evolution and fabricated by automated manufacturing processes, described by Lipson and

The Nature of Life, ed. M. A. Bedau and C. E. Cleland. Published by Cambridge University Press. © M. A. Bedau and C. E. Cleland 2010.

Pollack (Ch. 19). Such robots are perhaps the clearest contemporary example of what a Cartesian mechanistic conception of life would look like today (recall Descartes, Ch. 2).

"Wet" artificial life uses the resources found in chemical and biological laboratories to synthesize lifelike systems in a test tube. Wet artificial life overlaps with the new science of synthetic biology, which aims to design and construct new biological parts, devices, and systems, and to re-design existing, natural biological systems for useful purposes. However, most synthetic biology tends to be "top-down" and amounts to modifying existing forms of life, while wet artificial life is the "bottom-up" form of synthetic biology that aims to make new forms of life from scratch. These new forms of life might differ in important ways from any forms of life that exist naturally in the universe, and they might be created by processes that bear little resemblance to the natural process by which life first arose on Earth. In this context it is worth noting that some researchers contend that one cannot separate a theory of the nature of life from a theory of the origin of life. Alexander Oparin (Ch. 5), for instance, would reject the claim that life could be created by processes differing greatly from those that produced the first life on Earth. Wet artificial life might produce a credible counterexample to Oparin's point of view.

Unfortunately, the terminology "wet artificial life" and "synthetic biology" is somewhat confusing. One source of confusion is that wet artificial life and bottom-up synthetic biology are essentially the same activity. In addition, wet artificial life could naturally be called "synthetic biology" because of the central role of synthetic methodologies in artificial life. Adding to the confusion are headlines about synthetic biology employing the phrase "artificial life."

Artificial life has a checkered reputation, and many natural scientists are quick to dismiss it. This is partly because of the amount of hype that accompanied its early history (somewhat like the hype in the early history of artificial intelligence). But another reason is that many people mistakenly equate artificial life with *soft* artificial life. Once one realizes how wet artificial life overlaps with synthetic biology, it becomes clear that these research programs are relevant to biology and could even illuminate the nature of life. Some researchers believe that work in these areas provides a much more precise and detailed account of the minimal molecular requirements of simple cellular life, thereby supporting the view that life can be captured in terms of the materials out of which it is composed. At the other extreme, if one can accept the possibility that fundamental principles governing living systems could be abstracted away from their chemical embodiments explored in molecular biology, wet artificial life, and synthetic biology, then one can begin to appreciate how hard and soft artificial life could shed important light on how living systems operate. This provides a much more abstract, functional perspective on the nature of life.

These possibilities help to explain why the advent of synthetic biology and especially of soft, hard, and wet artificial life is fuelling interest in the nature of life. To learn more about the full extent of life as it could be, it is helpful to explore a diverse range of hypothetical possibilities for life, particularly given that we currently have access to only the life forms discovered to date on Earth. Artificial life and synthetic biology are continually producing new kinds of systems that challenge our preconceptions about life, and thus continually renewing the question of what kinds of systems deserve to be thought of as alive. The existence of seemingly autonomous agents made in hard artificial life raises the question whether a device made only of plastic, silicon, and steel could ever literally be alive, as Lipson and Pollack imply (Ch. 19). Further scientific and technological advances in hard artificial life will produce robots with new and richer kinds of autonomous, adaptive, goal-directed behavior, including perhaps the capacities for self-repair and self-reproduction. This could make the traditional presumption that living systems could not be built from plastic, silicon, and steel seem more and more parochial, and our uncertainty about how exactly to demarcate living things could grow.

In a similar way, the current race (wet artificial life) to synthesize a minimal artificial cell or protocell in a test tube (Chs. 20 and 21; see also Szostak *et al.* 2001; Rasmussen *et al.* 2004; Rasmussen *et al.* 2008) confronts us with important questions about life. This race requires some agreement about which chemical features it is important to achieve, and those chemical features must extend beyond the specific chemical details found in familiar life forms. Deamer (Ch. 20) and Luisi and colleagues (Ch. 21) propose different lists of what they take to be key general chemical properties to be realized in bottom-up synthetic biology and wet artificial life. Many of them have already been achieved, and as progress continues in these areas, we will continually confront the question of whether various proposed chemical features seem upon reflection to be crucial for living systems. Our actual experience with real chemical embodiments of those properties will trump any prior purely hypothetical speculation about their significance for life. The social and ethical implications of creating new forms of life in the laboratory will also increase the need for understanding what life is (Garfield *et al.* 2007; Bedau & Parke 2009).

Soft artificial life in particular has focused attention on a very specific question about the nature of life: Does the stuff of which it is composed matter? Some people think that a soft artificial life computer system could amount to more than a *simulation* of life. They believe that appropriately designed informational computer systems could be *literally* alive (Langton 1989; Ray 1992). The claim that software systems could be literally alive is known as the thesis of "strong" (soft) artificial life. This thesis is analogous to the thesis of "strong" artificial intelligence, according to which a properly programmed ordinary digital computer could literally be intelligent, at least as intelligent as an ordinary human being. Similarly, strong (soft) artificial life is the view that a properly programmed ordinary digital computer (or more accurately, the informational structures being implemented upon it) could literally be alive, as alive as any bird, beetle, or bacterium. Strong forms of artificial intelligence and artificial life contrast with the relatively uncontroversial "weak" claims that artificial intelligence and artificial life software systems can teach us a lot about intelligence and life. Many of the philosophical examinations of soft artificial life are preoccupied with the strong thesis; see, e.g., the discussions in this book by Sober (Ch. 16), Lange (Ch. 17), and Boden (Ch. 18). (There are analogous "strong" theses about hard and wet artificial life, discussed below.)

The constructions produced in artificial life and bottom-up synthetic biology could potentially increase the actual examples of forms of life. They also call special attention to the complexity of the question about the nature of life. The attempt to make new forms of life cannot get started without some preconception about the crucial properties of living systems. The chapters in Sections I and II provide many examples of such preconceptions about life, such as

that living systems must involve a metabolism (e.g., Schrödinger (Ch. 4), Oparin (Ch. 5), and many others) or that living systems must be composed of macromolecules (e.g., Mayr (Ch. 6), Pace (Ch. 11), and many others). The constructions of artificial life and synthetic biology can challenge and help us to revise some of these preconceptions, as discussed by Sober (Ch. 16), Lange (Ch. 17), and Boden (Ch. 18), as well as in Section IV by Bedau (Ch. 31) and Kauffman (Ch. 30). This raises the question of whether there is any firm foundation from which to adjudicate whether and how our preconceptions about life should be revised. It even raises the specter that life has no determinate nature, as Keller stresses (Ch. 22).

The strong thesis of soft artificial life is the central focus of Chapter 16, in which philosopher Elliott Sober examines the analogy between the philosophy of mind and the philosophy of life. The central premise for strong artificial intelligence is the supposed multiple realizability of mental states, that is, the assumption that mental states could be possessed by creatures that are constructed out of a very wide range of different kinds of materials. This multiplicity of physical realizations of mental states makes it impossible to identify mental states with any specific physical properties. One consequence is to make it more plausible that a suitably programmed computer could have genuine mental states—the thesis of strong artificial intelligence. The philosophical theory typically associated with strong artificial intelligence is called "functionalism," and it holds that possession of mental states is determined not by an entity's specific material composition but by the functional organization of that material. On this view, mental states can be realized in a vast multiplicity of materials because the essential properties of mental states are functional, not material.

Functionalism about the mind is analogous to the broadly Aristotelian perspective on life (Ch. 1), according to which the essential properties of living systems are their functional capacities rather than their material composition. To determine the plausibility of functionalism about life, Sober investigates to what extent fundamental biological properties are also multiply realizable. He concludes that some biological properties such as fitness or predator and prey behavior have quite heterogeneous instances, while others, such as the property of being hereditary material, are significantly constrained by the laws of chemistry and physics. The discussions in Chapters 11 and 12 provide detailed scientific investigations into these sorts of constraints. The physical and chemical constraints on any possible life forms provide reason to question functionalism with respect to life, but they do not necessarily refute it. For all forms of functionalism will allow chemistry and physics to set some constraints on the kinds of materials that could realize the functional properties in question. The critical issue for functionalism about life is the extent to which the essential characteristics of life are independent of its material composition. Modest versions of functionalism require that only some of the essential features of life be fully independent of the stuff of which it is composed, whereas a fully functionalist account requires that all of them be independent; the latter position supports the position of strong (soft) ALife. Different versions of functionalism will demand different degrees of stringency, and the pros and cons of the alternatives remain open questions.

One further issue that Sober raises specifically concerns computer realizations of life. Computer systems that seem to mimic the fundamental properties of living systems are the main motivation for the strong thesis of soft artificial life. Sober points out that some biological processes, like digestion and metabolism, seem non-computational on the surface, and this raises doubts that any purely computational system could literally be alive. For these and related reasons, Sober concludes that the prospects for strong soft artificial life are dim. It is an open question whether these kinds of arguments have any force against hard or wet artificial life or bottom-up synthetic biology.

The field of artificial life presupposes that the distinction between life and nonlife is an important topic for scientific investigation, but some would disagree. Sober, for example, questions whether biologists really care about this distinction (Ch. 16). Philosopher Marc Lange takes up this question in Chapter 17 and addresses two central issues: what to make of the familiar hallmarks of life, and what to conclude about the admitted difficulty of defining life. Many authors have stressed the distinctive hallmarks of living systems, such as the ability to reproduce, grow, respond to the environment, and evolve, and those hallmarks are sometimes used to characterize or define the living. However, as Lange reminds us, there are well-known counterexamples to such characterizations

(see also Ch. 26). An important open question about the nature of life is what to make of the status of these hallmarks. Are they pretheoretic posits that any valid conception of life must respect? If so, for what reason? Are they just common preconceptions that scientific progress could overturn? If so, on what basis? One can sense the tension that these questions generate when these hallmarks are discussed (see, e.g., Chs. 6, 7, 15, and 24). Lange makes an important point about the hallmarks of life. Rather than asking merely whether a system exhibits a certain hallmark, Lange points out that we should ask whether the system exhibits the hallmark for "the same reasons" that paradigmatic living systems do so. If not, then the presence of the hallmarks might be irrelevant to whether the system is alive.

The difficulty of defining life is sometimes taken to imply that the life/nonlife distinction does little or no useful work in biology (e.g., Chs. 16 and 28). Lange counters that worries about the definition of life should be separated from concerns with the scientific legitimacy and usefulness of the concept of life. He points out that scientific concepts often have vague boundaries, so the vagaries of life do not show that the concept is unscientific, useless, or bogus in some other way. Lange concludes that, even if biologists cannot define away life or reduce it to some other independently intelligible properties, life could nevertheless still be a natural kind, and in fact the property of being alive could explain why living systems exhibit life's characteristic hallmarks.

The idea that metabolism is a characteristic property of living systems can be traced back to Aristotle's capacity for self-nutrition (Ch. 1), and it still commands wide assent. Erwin Schrödinger, for example, argues in Chapter 4 that metabolism is needed to enable complex ordered living systems to survive in the face of the second law of thermodynamics, and the investigations into physical and chemical constraints on all possible living systems by Pace (Ch. 11) and Benner and colleagues (Ch. 12) presume that all life forms will need a metabolism. In Chapter 18 philosopher Margaret Boden evaluates both soft and hard artificial life from the perspective of the role that they afford to metabolism. Boden distinguishes three grades of metabolism. $Metabolism_1$ is the property of those systems that depend on energy to survive. $Metabolism_2$ applies to those systems that use energy packets to power the behavior of their body. $Metabolism_3$ is the property of using energy packets to maintain a body and power its behavior. Boden makes this distinction in order to focus attention on the third grade, for a central premise in her argument is that metabolism$_3$ plays an ineliminable role in any acceptable account of genuine life. Her other central premise is that both soft and hard artificial life abstract away from metabolism$_3$. From these premises, it follows directly that neither soft nor hard artificial life can produce genuine instances of life; in other words, the strong thesis of soft and hard artificial life is false (though the strong thesis of wet artificial life is not threatened). Boden's argument highlights the role of metabolism in life, and it prompts us to examine anew the nature and status of the requirement that life forms have a metabolism. Even if we agree that all familiar life forms have a metabolism, why think that this must hold for any possible form of life? Is the idea that life must involve a metabolism a discovery about nature, or does it merely reflect a contingent choice about our use of the word 'life'?

The strong connection between life and metabolism defended by Boden is implicitly rejected by computer scientists Hod Lipson and Jordan Pollack in Chapter 19. This chapter describes the automated design and manufacture of robotic "life forms" made out of plastic, silicon, and steel—nicely illustrating some of the achievements of hard artificial life. Lipson and Pollack emphasize that familiar biological life forms control the means of their own reproduction (a point stressed by Kant (Ch. 3), among others), in contrast with the human–controlled process by which robots are usually constructed. Lipson and Pollack take significant strides toward the autonomous reproduction of robots by coupling a genetic algorithm for autonomously testing and improving robot design with an automated evolutionary process for fabricating the robots. Their robots were designed simply to move across the floor, but it was crucial that they actually functioned in the real world. They were constructed out of three kinds of parts: plastic struts connected with ball joints, linear actuators propelled backward and forward in a straight line by means of an electric motor, and an artificial neural network that controlled the linear actuators.

The bodies (connected struts and actuators) and brains (neural networks) of these robots were designed by an evolutionary process analogous to the evolution by natural selection that has shaped biological life.

Lipson and Pollack's evolutionary process created successive generations of possible robot designs. Each design in a population was tested with a computer simulation to determine how rapidly it could move across the floor. The best designs were then "mutated" to produce the next generation of robot designs. After repeating this process of selection and mutation for a number of generations, the best resulting designs were fabricated with automated rapid manufacturing technology, by printing layer after layer of plastic parts. The resulting robots are well designed and possess the teleological aspects that characterize life in general (see the chapters by Aristotle (Ch. 1), Kant (Ch. 3), Dawkins (Ch. 29), Oparin (Ch. 5), and Mayr (Ch. 6), among others). Although Lipson and Pollack have not yet fully automated the process by which these robots reproduce themselves, they are well on the way toward this goal. This vivid possibility raises in sharp relief the question of whether a population of fully autonomously acting and reproducing robots would literally be alive, in contrast to what Boden argues.

The current state of wet artificial life and bottom-up synthetic biology is reviewed in Chapters 20 and 21. Biophysicist David Deamer concludes in Chapter 20 that "the prospect of life in a test tube has always seemed just around the corner, and the corner is closer than ever." This conclusion rests on an analysis of twelve chemical mechanisms that one could argue must be integrated for even the simplest microorganism to emerge in the laboratory. According to Deamer, even the simplest forms of life exhibit a complex set of 12 interconnected molecular mechanisms, and the fact that life requires these 12 complex chemical steps explains why life is so hard to define. Examples of Deamer's 12 chemical requirements for life include the requirement that boundary membranes self-assemble from soap-like molecules to form microscopic cell-like compartments, the requirement that energy be captured by the membranes either from light and a pigment system or from chemical energy (or both), the requirement that macromolecules be encapsulated in the compartments but that smaller molecules be able to cross the membrane barrier to provide nutrients and chemical energy for primitive metabolism, and the requirement that the macromolecules grow by polymerization of the nutrient molecules.

For the most part, each of Deamer's requirements for life has been realized in the laboratory. The main remaining challenge is to achieve them not singly but in concert. Chemically integrating the requirements for life might be somewhat easier than Deamer suggests, for some of the items on the list might be optional. At least, this is the presumption behind the design of the so-called "Los Alamos bug" (Rasmussen *et al.*, 2003), which can lack boundary membranes, lack genes for protein synthesis, and lack proteins (and so typical catalysts) altogether. In this way, the attempt to create new forms of life in the laboratory prompts us to rethink the nature of life.

One of the pioneers of wet artificial life and bottom-up synthetic biology is the Italian chemist Pier Luigi Luisi. In Chapter 21, Luisi and colleagues review the recent history of wet artificial life in more detail, focusing on attempts to create a minimal cell in the laboratory. Luisi's team's efforts are informed by the theory of autopoiesis developed by Maturana and Varela (1980), and their achievements can be seen as steps toward the chemical realization of something analogous to Gánti's chemoton (Ch. 7). A "minimal cell" is supposed to be the simplest possible form of life. One simplification is to shrink the genome as much as possible; the chapter outlines a very primitive and feeble form of life that would require only 45–50 genes. At the more complex extreme are schemes that involve complex molecules (often catalysts, including especially polymerases) and complex cellular substructures (especially ribosomes) extracted from existing forms of life (typically, bacteria).

Luisi's minimal cell design is motivated by a quite common view of life, according to which life essentially involves the concomitance of three processes: self-maintenance or metabolism, self-reproduction, and evolvability. Luisi recounts the key bottom-up chemical achievements to date that approximate minimal cells (what Luisi calls "limping" life). First, lots of experimental work has demonstrated that liposomes (microscopic spherical containers made from a bilayer membrane, that spontaneously self-assemble in aqueous solutions from amphiphilic molecules) can function as containers for complex biological molecular reactions (Rasmussen *et al.* 2008). These liposomes grow spontaneously from amphiphilic precursors, and they spontaneously multiply. The biochemical reactions that have been produced inside liposomes include the synthesis of RNA and proteins, including critical pore-forming proteins and polymerase

enzymes. Some of these results depend on commercial cell-extract kits that are complex chemical black boxes containing myriad molecular constituents of simple living cells.

The work in wet artificial life and bottom-up synthetic biology described by Deamer and Luisi is interesting in part because the prospect of making new forms of life from scratch in the laboratory is closer than many people realize (for a recent overview of the field, see Rasmussen *et al.* 2008). As we get closer to this goal and finally reach it, and as more and more different kinds of artificial life forms arise in the laboratory, our preconceptions about the nature of life will come under increasing pressure.

Most views about the nature of life presuppose that life is a natural kind, that is, a category of objects that is unified by a real objective similarity. The debate is over exactly which properties define that category. Reflection on artificial life and synthetic biology has led historian and philosopher Evelyn Fox Keller in Chapter 22 to challenge this presupposition. She argues that life is not a natural but a "human kind," that is, a category of objects that is unified merely by human stipulation or convention. Now, the distinction between natural and human kinds is a difficult and open philosophical question, and some people are skeptical about the very notion of natural kinds. Keller's skepticism is presumably not global, but rather aims to distinguish the concept of life from certain other central scientific concepts that are genuine natural kinds (e.g., the element oxygen, the substance water, and the chemical categories of acids and bases).

Keller's perspective clashes with the views presented in most of the chapters in this book. Whether growing out of astrobiology and the attempt to understand the origin of life, or motivated by artificial life and synthetic biology, recent philosophical and scientific interest in the nature of life mostly assumes that the distinction between life and nonlife is objective and independent of human interests. Keller marshals a number of considerations to support the claim that life is a human kind. One is that the belief that there is a defining property of living beings that is absent from all nonliving things is a contingent product of human history, which she argues can be dated to the nineteenth century. Another is the fact that all attempts to find the essence of life have failed. Furthermore, the more we learn about life, the less stable our view of the nature of life becomes. In addition, attempts to bridge the gap between nonlife and life in the laboratory tend to dissolve the boundary between them. When you articulate the different steps between life and nonlife, you discover not a sharp, black and white dividing line but a gradual transition among many shades of gray. Lange (Ch. 17) and Cleland and Chyba (Ch. 26) would not, however, view this as a problem, since, as they point out, few natural categories have crisp boundaries; most kind terms (natural and non-natural) are vague and subject to borderline cases. Moreover, according to Keller, the scientific and technological progress of artificial life and synthetic biology violate our older taxonomies and require us to invent new concepts. If we choose to use the old word "life" to refer to these new concepts, it is, on her view, merely a human stipulation or convention. No question about nature hangs on whether artificial life software systems or robots or artificial cells created in a laboratory are called "alive," so the question of whether they are "really" alive is meaningless. But as Cleland and Chyba (Ch. 26) point out, the history of science reveals that new scientific theories frequently change old taxonomies and concepts, and this is not just a matter of human stipulation or convention. Many of the problems that Keller cites might reflect little more than the inadequacy of our current concept of life, rather than an objective fact about the disunity of life.

In addition to the chapters in this section, artificial life and synthetic biology are discussed in the other sections of this collection, including in Chapters 15, 28, 30, and 31. Those chapters fit thematically within the central concerns of this section, and anyone interested in what artificial life and synthetic biology can teach us about the nature of life (or vice versa) should read those chapters along with the ones in this section.

REFERENCES

Bedau, M. A. (2003). Artificial life: Organization, adaptation, and complexity from the bottom up. *Trends in Cognitive Science*, 7, 505–512.

Bedau, M. A. (2007). Artificial life. In M. Matthen & C. Stephens (Eds.), *Handbook of the philosophy of biology* (pp. 585–603). Amsterdam: Elsevier.

Bedau, M. A. & Parke, E. C. (Eds.) (2009). *The ethics of protocells: Moral and social implications of creating life in the laboratory.* Cambridge: MIT Press.

Garfield, M. S., Endy, D., Epstein, G. L., & Friedman, R. M. (2007). *Synthetic genomics: Options for governance.* Available on the web at http://www.jcvi.org/cms/fileadmin/site/research/ projects/synthetic-genomics-report/synthetic-genomics-report.pdf (accessed May 2008).

Langton, C. G. (1989). Artificial life. In C. G. Langton (Ed.), *Artificial life* (Santa Fe Institute studies in the sciences of complexity, proceedings vol. IV) (pp. 1–47). Redwood City, CA: Addison-Wesley.

Maturana, H., & Varela, F. (1980). *Autopoiesis and cognition: The realization of the living.* Boston: D. Reidel.

Rasmussen, S., Chen, L., Nilsson, M., & Shigeaki, A. (2003). Bridging nonliving and living matter. *Artificial Life*, 9, 269–316.

Rasmussen, S., Chen, L., Deamer, D., *et al.* (2004). Transitions from nonliving to living matter. *Science*, 303, 963–965.

Rasmussen, S., Bedau, M. A., Chen, L., *et al.* (Eds.) (2008). *Protocells: Bridging nonliving and living matter.* Cambridge: MIT Press.

Ray, T. S. (1992). An approach to the synthesis of life. In C. Langton, C. Taylor, J. D. Farmer & S. Rasmussen (Eds.), *Artificial Life II* (pp. 371–408). Redwood City, CA: Addison-Wesley.

Szostak, J. W., Bartel, D. P., & Luisi, P. L. (2001). Synthesizing life. *Nature*, 409, 387–390.

16 · Learning from functionalism: prospects for strong artificial life

ELLIOTT SOBER

TWO USES FOR COMPUTERS

There are two quite different roles that computers might play in biological theorizing. Mathematical models of biological processes are often analytically intractable. When this is so, computers can be used to get a feel for the model's dynamics. You plug in a variety of initial condition values and allow the rules of transition to apply themselves (often iteratively); then you see what the outputs are.

Computers are used here as aids to the theorist. They are like pencil and paper or a sliderule. They help you think. The models being investigated are about life. But there is no need to view the computers that help you investigate these models as alive themselves. Computers can be applied to calculate what will happen when a bridge is stressed, but the computer is not itself a bridge.

Population geneticists have used computers in this way since the 1960s. Many participants in the Artificial Life research program are doing the same thing. I see nothing controversial about this use of computers. By their fruits shall ye know them. This part of the AL research program will stand or fall with the interest of the models investigated. When it is obvious before-hand what the model's dynamics will be, the results provided by computer simulation will be somewhat uninteresting. When the model is very unrealistic, computer investigation of its properties may also fail to be interesting. However, when a model is realistic enough and the results of computer simulation are surprising enough, no one can deny the pay-off.

I just mentioned that computers may help us understand bridges, even though a computer is not a bridge. However, the second part of the artificial life research program is interested in the idea that computers are instances of biological processes. Here the computer is said to be alive, or to exemplify various properties that we think of as characteristic of life.

This second aspect of the artificial life research program needs to be clearly separated from the first. It is relatively uncontroversial that computers can be tools for investigating life; in contrast, it is rather controversial to suggest that computers are or can be alive. Neither of these ideas entails the other; they are distinct.

Shifting now from AL to AI, again we can discern two possible uses of computers. First, there is the use of computers as tools for investigating psychological models that are too mathematically complicated to be analytically solved. This is the idea of computers as tools for understanding the mind. Second and separate, there is the idea that computers are minds. This latter idea, roughly, is what usually goes by the name of strong AI.

Strong AI has attracted a great deal of attention, both in the form of advocacy and in the form of attack. The idea that computers, like paper and pencil, are tools for understanding psychological models has not been criticized much, nor should it have been. Here, as in the case of AL, by their fruits shall ye know them.

So in both AI and AL, the idea that computers are tools for investigating a theory is quite different from the idea that computers are part of the subject-matter of the theory. I'll have little more to say about the tool idea, although I'll occasionally harp on the importance of not confusing it with the subject-matter idea. Table 16.1 depicts the parallelism between AI and AL; it also should help keep the tool idea and the subject matter idea properly separated: Of course, "strong" does not mean plausible or well defended; "strong" means daring and "weak" means modest.

The Nature of Life, ed. M. A. Bedau and C. E. Cleland. Published by Cambridge University Press. © M. A. Bedau and C. E. Cleland 2010.

Table 16.1. *Parallelism between AI and AL*

	Psychology	Biology
Computers are tools for investigating	weak AI	weak AL
Computers are part of the subject matter of	strong AI	strong AL

Where did the idea that computers could be part of the subject matter of psychological and biological theories come from? Recent philosophy has discussed the psychological issues a great deal, the biological problem almost not at all. Is it possible to build a case for living computers that parallels the arguments for thinking computers? How far the analogy can be pushed is what I want to determine.

THE PROBLEM OF MIND AND THE PROBLEM OF LIFE

The mind/body problem has witnessed a succession of vanguard theories. In the early 1950s, Ryle[12] and Wittgenstein[18] advocated forms of logical behaviorism. This is the idea that the meaning of mentalistic terms can be specified purely in terms of behavior. Ryle attacked the Myth of the Ghost in the Machine, which included the idea that mental states are inner causes of outward behavior.

In the mid 1950s to mid 1960s, behaviorism itself came under attack from mind/brain identity theorists. Australian materialists like Place[10] and Smart[14] maintained that mental states are inner causes. But more than that, they argued that mental properties would turn out to be identical with physical properties. Whereas logical behaviorists usually argued for their view in an *a priori* fashion, identity theorists said they were formulating an empirical thesis that would be borne out by the future development of science.

The identity theory can be divided into two claims. They claimed that each mental object is a physical thing. They also claimed that each mental property is a physical property. In the first category fall such items as minds, beliefs, memory traces, and afterimages. The second category includes believing that snow is white and feeling pain. This division

may seem a bit artificial—why bother to make separate claims about my belief that snow is white and the property of believing that snow is white? In a moment, the point of this division will become plain.

Beginning in the mid 1960s the identity theory was challenged by a view that philosophers called functionalism. Hilary Putnam,[11] Jerry Fodor,[6] and Daniel Dennett[3] argued that psychological properties are multiply realizable. If this is correct, then the identity theory must be rejected.

To understand what multiple realizability means, it is useful to consider an analogy. Consider mousetraps. Each of them is a physical object. Some are made of wire, wood, and cheese. Others are made of plastic and poison. Still others are constituted by bunches of philosophers scurrying around the room armed with inverted wastepaper baskets.

What do all these mousetraps have in common? Well, they are all made of matter. But more specifically, what properties do they share that are unique to them? If mousetraps are multiply realizable, then there is no physical property that all mousetraps, and only mousetraps, possess.

Each mousetrap is a physical thing, but the property of being a mousetrap is not a physical property. Here I am putting to work the distinction between object and property that I mentioned before.

Just as there are many physical ways to build a mousetrap, so, functionalists claimed, there are many physical ways to build a mind. Ours happens to be made of DNA and neurons. But perhaps computers could have minds. And perhaps there could be organisms in other species or in other galaxies that have minds, but whose physical organization is quite different from the one we exemplify. Each mind is a physical thing, but the property of having a mind is not a physical property.

Dualism is a theory that I have not mentioned. It claims that minds are made of an immaterial substance. Identity theorists reject dualism. So do functionalists. The relationship between these three theories can be represented by saying how each theory answers a pair of questions (see Table 16.2).

I mentioned earlier that identity theorists thought of themselves as advancing an empirical thesis about the nature of the mind. What about functionalism? Is it an empirical claim? Here one finds a division between two styles of argument. Sometimes functionalists

Table 16.2. *Three theories of mind*

| | | Are mental __ physical? | |
	Dualism	Functionalism	Identity theory
Objects	No	Yes	Yes
Properties	No	No	Yes

appear to think that the meaning of mentalistic terminology guarantees that the identity theory must be false. This *a priori* tendency within functionalism notwithstanding, I prefer the version of that theory that advances an empirical thesis. It is an empirical question how many different physical ways a thinking thing can be built. If the number is enormously large, then functionalism's critique of the identity theory will turn out to be right. If the number is very small (or even one), then the identity theory will be correct. Perhaps the design constraints dictated by psychology are satisfiable only within a very narrow range of physical systems; perhaps the constraints are so demanding that this range is reduced to a single type of physical system. This idea cannot be dismissed out of hand.

If philosophers during the same period of time had been as interested in the problem of life as they were in the problem of mind, they might have formulated biological analogs of the identity theory and functionalism. However, a biological analog of mind/body dualism does not have to be invented—it existed in the form of vitalism. Dualists claim that beings with minds possess an immaterial ingredient. Vitalists claim that living things differ from inanimate objects because the former contain an immaterial substance—an *élan vital*.

Had the problem of life recapitulated the problem of mind, the triumphs of molecular biology might then have been interpreted as evidence for an identity theory, according to which each biological property is identical with some physical property. Finally, the progression might have been completed with an analog of functionalism. Although each living thing is a material object, biological properties cannot be identified with physical ones.

Actually, this functionalist idea has been espoused by biologists, although not in the context of trying to recapitulate the structure of the mind/body problem. Thus, Fisher[5] says that "fitness, although measured by a uniform method, is qualitatively different for every different organism." Recent philosophers of biology have made the same point by arguing that an organism's fitness is the upshot of its physical properties even though fitness is not itself a physical property. What do a fit cockroach and a fit zebra have in common? Not any physical property, any more than a wood and wire mousetrap must have something physical in common with a human mouse catcher. Fitness is multiply realizable.

There are many other biologically interesting properties and processes that appear to have the same characteristic. Many of them involve abstracting away from physical details. For example, consider Lewontin's[7] characterization of what it takes for a set of objects to evolve by natural selection. A necessary and sufficient condition is heritable variation in fitness. The objects must vary in their capacity to stay alive and to have offspring. If an object and its offspring resemble each other, the system will evolve, with fitter characteristics increasing in frequency and less-fit traits declining.

This abstract skeleton leaves open what the objects are that participate in a selection process. Darwin thought of them as organisms within a single population. Group selectionists have thought of the objects as groups or species or communities. The objects may also be gametes or strands of DNA, as in the phenomena of meiotic drive and junk DNA.

Outside the biological hierarchy, it is quite possible that cultural objects should change in frequency because they display heritable variation in fitness. If some ideas are more contagious than others, they may spread through the population of thinkers. Evolutionary models of science exploit this idea. Another example is the economic theory of the firm; this describes businesses as prospering or going bankrupt according to their efficiency.

Other examples of biological properties that are multiply realizable are not far to seek. Predatory/prey theories, for example, abstract away from the physical details that distinguish lions and antelopes from spiders and flies.

I mentioned before that functionalism in the philosophy of mind is best seen as an empirical thesis about the degree to which the psychological characteristics of a system constrain the system's physical realization. The same holds for the analog of functionalism as a thesis about biological properties. It may be somewhat obvious that some biological properties—like the property of being a predator—place relatively few constraints on the physical characteristics a system must possess. But for others, it may be much less obvious.

Consider, for example, the fact that DNA and RNA are structures by which organisms transmit characteristics from parents to offspring. Let us call them hereditary mechanisms. It is a substantive question of biology and chemistry whether other molecules could play the role of a hereditary mechanism. Perhaps other physical mechanisms could easily do the trick; perhaps not. This cannot be judged *a priori*, but requires a substantive scientific argument.

WHAT ARE PSYCHOLOGICAL PROPERTIES, IF THEY ARE NOT PHYSICAL?

In discussing functionalism's criticism of the mind/brain identity theory, I mainly emphasized functionalism's negative thesis. This is a claim about what psychological states are not; they aren't physical. But this leaves functionalism's positive proposal unstated. If psychological properties aren't physical, what are they, then?

Functionalists have constructed a variety of answers to this question. One prominent idea is that psychological states are computational and representational. Of course, the interest and plausibility of this thesis depends on what "computational" and "representational" are said to mean. If a functionalist theory entails that desk calculators and photoelectric eyes have beliefs about the world, it presumably has given too permissive an interpretation of these concepts. On the other hand, if functionalism's critique of the identity theory is right, then we must not demand that a system be physically just like us for it to have a psychology. In other words, the problem has been to construct a positive theory that avoids, as Ned Block[1] once put it, both chauvinism and liberalism. Chauvinistic theories are too narrow, while liberal theories are too broad, in their proposals for how the domain of psychology is to be characterized.

BEHAVIORISM AND THE TURING TEST

Functionalists claim that psychological theories can be formulated by abstracting away from the physical details that distinguish one thinking system from another. The question is: How much abstracting should one indulge in?

One extreme proposal is that a system has a mind, no matter what is going on inside it, if its behavior is indistinguishable from some other system that obviously has a mind. This is basically the idea behind the Turing Test. Human beings have minds, so a machine does too, if its behavior is indistinguishable from human behavior. In elaborating this idea, Turing was careful that "irrelevant" cues not provide a tip off. Computers don't look like people, but Turing judged this fact to be irrelevant to the question of whether they think. To control for this distracting detail, Turing demands that the machine be placed behind a screen and its behavior standardized. The behavior is to take the form of printed messages on a tape.

Besides intentionally ignoring the fact that computers don't look like people, this procedure also assumes that thinking is quite separate from doing. If intelligence requires manipulating physical objects in the environment, then the notion of behavior deployed in the Turing Test will be too meager. Turing's idea is that intelligence is a property of pure cogitation, so to speak. Behavior limited to verbal communication is enough.

So the Turing Test represents one possible solution to the functionalist's problem. Not only does thought not require a physical structure like the one our brains possess. In addition, there is no independently specifiable internal constraint of any kind. The only requirement is an external one, specified by the imitation game.

Most functionalists regard this test as too crude. Unfortunately, it seems vulnerable to both type-1 and type-2 errors. That is, a thing that doesn't think can be mistakenly judged to have a mind and a thinking thing can be judged to lack a mind by this procedure (see Table 16.3).

In discussions of AI, there has been little attention to type-2 errors. Yet, it seems clear that human beings with minds can imitate the behavior of mindless computers and so fool the interrogator into thinking that they lack minds. Of perhaps more serious concern is the possibility that a machine might have a mind, but have beliefs and desires so different from those of any human being that interrogators would quickly realize that they were not talking to a human being. The machine would flunk the Turing Test, because it cannot imitate human response patterns; it is another

Table 16.3. *Errors in the Turing Test*

	The subject thinks	The subject does not think
The subject passes	ok	type-1
The subject does not pass	type-2	ok

matter to conclude that the machine, therefore, does not have a mind at all.

The possibility of type-2 error, though real, has not been the focus of attention. Rather, in order to overcome the presumption that machines can't think, researchers in AI have been concerned to construct devices that pass the Turing Test. The question this raises is whether the test is vulnerable to type-1 errors.

One example that displays the possibility of type-1 error is due to Ned Block.[2] Suppose we could write down a tree structure in which every possible conversation that is 5 hours or less in duration is mapped out. We might trim this tree by only recording what an "intelligent" respondent might say to an interrogator, leaving open whether the interrogator is "intelligent" or not. This structure would be enormously large, larger than any current computer would be able to store. But let's ignore that limitation and suppose we put this tree structure into a computer.

By following this tree structure, the machine would interact with its interrogator in a way indistinguishable from the way in which an intelligent human being would do so. Yet the fact that the machine simply makes its way through this simple tree structure strongly suggests that the machine has no mental states at all.

One might object that the machine could be made to fail the Turing Test if the conversation pressed on beyond five hours. This is right, but now suppose that the tree is augmented in size, so that it encompasses all sensible conversations that are ten hours or less in duration. In principle, the time limit on the tree might be set at any finite size—four score and ten years if you like.

Block draws the moral that thinking is not fully captured by the Turing Test. What is wrong with this branching structure is not the behaviors it produces but how it produces them. Intelligence is not just the ability to answer questions in a way indistinguishable

from that of an intelligent person. To call a behavior "intelligent" is to comment on how it is produced. Block concludes, rightly I think, that the Turing Test is overly behavioristic.

In saying this, I am not denying that the Turing Test is useful. Obviously, behavioral evidence can be telling. If one wants to know where the weaknesses are in a simulation, one might try to discover where the outputs mimic and where they do not. But it is one thing for the Turing Test to provide fallible evidence about intelligence, something quite different for the test to define what it is to have a mind.

A large measure of what is right about Searle's[13] much-discussed paper "Minds, Brains, and Programs" reduces to this very point. In Searle's Chinese room example, someone who speaks no Chinese is placed in a room, equipped with a manual, each of whose entries maps a story in Chinese (S) and a question in Chinese about that story (Q) onto an answer in Chinese to that story (A). That is, the man in the room has a set of rules, each with the form

$$S + Q \to A.$$

Chinese stories are sent into the room along with questions about those stories, also written in Chinese. The person in the room finds the $S + Q$ pairing on his list, writes down the answer onto which the pair is mapped, and sends that message out of the room. The input/output behavior of the room is precisely what one would expect of someone who understands Chinese stories and wishes to provide intelligent answers (in Chinese) to questions about them. Although the system will pass the Turing Test, Searle concludes that the system that executes this behavior understands nothing of Chinese.

Searle does not specify exactly how the person in the room takes an input and produces an output. The details of the answer manual are left rather vague. But Block's example suggests that this makes all the difference. If the program is just a brute-force pairing of stories and questions on the one hand and answers on the other, there is little inclination to think that executing the program has anything to do with understanding Chinese. But if the manual more closely approximates what Chinese speakers do when they answer questions about stories, our verdict might change. Understanding isn't definable in terms of the ability to answer questions; in addition, how one obtains the answers must be taken into account.

Searle considers one elaboration of this suggestion in the section of his paper called "The Brain Simulator Reply (Berkeley and MIT)."[13] Suppose we simulate "the actual sequence of neuron firings at the synapses of the brain of a native Chinese speaker when he understands stories in Chinese and gives answers to them." A system constructed in this way would not just duplicate the stimulus/response pairings exemplified by someone who understands Chinese; in addition, such a system would closely replicate the internal processes mediating the Chinese speaker's input/output connections.

I think that Searle's reply to this objection is question-begging. He says "the problem with the brain simulator is that it is simulating the wrong things about the brain. As long as it simulates only the formal structure of the sequence of neuron firings at the synapses, it won't have simulated what matters about the brain, namely its causal properties, its ability to produce intentional states."

Although Searle does not have much of an argument here, it is important to recognize that the denial of his thesis is far from trivial. It is a conjecture that might or might not be right that the on/off states of a neuron, plus its network of connections with other neurons, exhaust what is relevant about neurons that allows them to form intentional systems. This is a meager list of neuronal properties, and so the conjecture that it suffices for some psychological characteristic is a very strong one. Neurons have plenty of other characteristics; the claim that they are all irrelevant as far as psychology goes may or may not be true.

There is another way in which Searle's argument recognizes something true and important, but, I think, misinterprets it. Intentionality crucially involves the relationship of "aboutness." Beliefs and desires are about things in the world outside the mind. How does the state of an organism end up being about one object, rather than about another? What explains why some states have intentionality whereas others do not? One plausible philosophical proposal is that intentionality involves a causal connection between the world and the organism. Crudely put, the reason my term "cat" refers to cats is that real cats are related to my use of that term in some specific causal way. Working out what the causal path must be has been a difficult task for the causal theory of reference. But leaving that issue aside, the claim that some sort of causal relation is necessary, though perhaps not sufficient, for at least some of the concepts we possess, has some plausibility.

If this world/mind relationship is crucial for intentionality, then it is clear why the formal manipulation of symbols can't, by itself, suffice for intentionality. Such formal manipulation is purely internal to the system, but part of what makes a system have intentionality involves how the system and its states are related to the world external to the system.

Although conceding this point may conflict with some pronouncements by exponents of strong AI, it does not show that a thinking thing must be made of neuronal material. Consistent with the idea that intentionality requires a specific causal connection with the external world is the possibility that a silicon-chip computer could be placed in an environment and acquire intentional states by way of its interactions with the environment.

It is unclear what the acquisition process must be like, if the system is to end up with intentionality. But this lack of clarity is not specific to the question of whether computers could think; it also applies to the nature of human intentionality. Suppose that human beings acquire concepts like *cat* and *house;* that is, suppose these concepts are not innate. Suppose further that people normally acquire these concepts by causally interacting with real cats and houses. The question I wish to raise is whether human beings could acquire these concepts by having a neural implant performed at birth that suitably rewired their brains. A person with an implant would grow up and feel about the world the way any of us do who acquired the concepts by more normal means. If artificial interventions in human brains can endow various states with intentionality, it is hard to see why artificial interventions into silicon computers can't also do the trick.

I have tried to extract two lessons from Searle's argument. First, there is the idea that the Turing Test is overly behavioristic. The ability to mimic intelligent behavior is not sufficient for having intelligence. Second, there is the idea that intentionality—aboutness—may involve a world/mind relationship of some specifiable sort. If this is right, then the fact that a machine (or brain) executes some particular program is not sufficient for it to have intentionality. Neither of these conclusions shows that silicon computers couldn't have minds.

What significance do these points have for the artificial life research program? Can the biological properties of an organism be defined purely in terms of environment/behavior pairings? If a biological property has this characteristic, the Turing Test will work better for it than the test works for the psychological property that Turing intended to describe. Let's consider photosynthesis as an example of a biological process. Arguably, a system engages in photosynthesis if it can harness the Sun's energy to convert water and CO_2 into simple organic compounds (principally CH_2O). Plants (usually) do this in their chloroplasts, but this is just one way to do the trick. If this is right, either for photosynthesis or for other biological properties, then the first lesson I drew about the Turing Test in the context of philosophy of mind may actually provide a disanalogy between AI and AL. Behaviorism is a mistake in psychology, but it may be the right view to take about many biological properties.

The second lesson I extracted from Searle's argument concerned intentionality as a relationship between a mental state and something outside itself. It is the relation of aboutness. I argued that the execution of a program cannot be the whole answer to the question of what intentionality is.

Many biological properties and processes involve relationships between an organism (or part of an organism) and something outside itself. An organism reproduces when it makes a baby. A plant photosynthesizes when it is related to a light source in an appropriate way. A predator eats other organisms. Although a computer might replicate aspects of such processes that occur inside the system of interest, computers will not actually reproduce or photosynthesize or eat unless they are related to things outside themselves in the right ways. These processes involve actions—interactions with the environment; the computations that go on inside the skin are only part of the story. Here, then, is an analogy between AI and AL.

THE DANGER OF GOING TOO FAR

Functionalism says that psychological and biological properties can be abstracted from the physical details concerning how those properties are realized. The main problem for functionalism is to say how much abstraction is permissible. A persistent danger for functionalist theories is that they err on the side of

being too liberal. This danger is especially pressing when the mathematical structure of a process is confused with its empirical content. This confusion can lead one to say that a system has a mind or is alive (or has some more specific constellation of psychological or biological properties) when it does not.

A simple example of how this fallacy proceeds may be instructive. Consider the Hardy–Weinberg Law in population genetics. It says what frequencies the diploid genotypes at a locus will exhibit, when there is random mating, equal numbers of males and females, and no selection or mutation. It is, so to speak, a "zero-force law"—it describes what happens in a population if no evolutionary forces are at work (see Sober[16]). If p is the frequency of the A allele and q is the frequency of a (where $p + q = 1$), then in the circumstances just described, the frequencies of AA, Aa and aa are p^2, $2pq$, and q^2, respectively.

Consider another physical realization of the simple mathematical idea involved in the Hardy–Weinberg Law. A shoe manufacturer produces brown shoes and black shoes. By accident, the assembly line has not kept the shoes together in pairs, but has dumped all the left shoes into one pile and all the right shoes into another. The shoe manufacturer wants to know what the result will be if a machine randomly samples from the two piles and assembles pairs of shoes. If p is the frequency of black shoes and q is the frequency of brown ones in each pile, then the expected frequency of the three possible pairs will be p^2, $2pq$, and q^2. Many other examples of this mathematical sort could be described.

Suppose we applied the Hardy–Weinberg Law to a population of *Drosophila*. These fruit flies are biological objects; they are alive and the Hardy–Weinberg Law describes an important fact about how they reproduce. As just noted, the same mathematical structure can be applied to shoes. But shoes are not alive; the process by which the machine forms pairs by random sampling is not a biological one.

I wish to introduce a piece of terminology: the Shoe/Fly Fallacy is the mistaken piece of reasoning embodied in the following argument:

Flies are alive.
Flies are described by law L.
Shoes are described by law L.
Hence, shoes are alive.

A variant of this argument focuses on a specific biological property—like reproduction—rather than on the generic property of "being alive."

Functionalist theories abstract away from physical details. They go too far—confusing mathematical form with biological (or psychological) subject matter—when they commit the Shoe/Fly Fallacy. The result is an overly liberal conception of life (or mind).

The idea of the Shoe/Fly Fallacy is a useful corrective against overhasty claims that a particular artificial system is alive or exhibits some range of biological characteristics. If one is tempted to make such claims, one should try to describe a system that has the relevant formal characteristics but is clearly not alive. The Popperian attitude of attempting to falsify is a useful one.

Consider, for example, the recent phenomenon of computer viruses. Are these alive? These cybernetic entities can make their way into a host computer and take over some of the computer's memory. They also can "reproduce" themselves and undergo "mutation." Are these mere metaphors, or should we conclude that computer viruses are alive?

To use the idea of the Shoe/Fly Fallacy to help answer this question, let's consider another, similar system that is not alive. Consider a successful chain letter. Because of its characteristics, it is attractive to its "hosts" (i.e., to the individuals who receive them). These hosts then make copies of the letters and send them to others. Copying errors occur, so the letters mutate.

I don't see any reason to say that the letters are alive. Rather, they are related to host individuals in such a way that more and more copies of the letters are produced. If computer viruses do no more than chain letters do, then computer viruses are not alive either.

Note the "if" in the last sentence. I do not claim that computer viruses, or something like them, cannot be alive. Rather, I say that the idea of the Shoe/Fly Fallacy provides a convenient format for approaching such questions in a suitably skeptical and Popperian manner.

FROM HUMAN COGNITION TO AL, FROM BIOLOGY TO AL

One of the very attractive features of AI is that it capitalizes on an independently plausible thesis about human psychology. A fruitful research program about human cognition is based on the idea that cognition involves computational manipulations of representations. Perhaps some representations obtain their intentionality via a connection between mind and world. Once in place, these representations give rise to other representations by way of processes that exploit the formal (internal) properties of the representations. To the degree that computers can be built that form and manipulate representations, to that degree will they possess cognitive states.

Of course, very simple mechanical devices form and manipulate representations. Gas gauges in cars and thermostats are examples, but it seems entirely wrong to say that they think. Perhaps the computational view can explain why this is true by further specifying which kinds of representations and which kinds of computational manipulations are needed for cognition.

Can the analogous case be made for the artificial life research program? Can an independent case be made for the idea that biological processes in naturally occurring organisms involve the formation and manipulation of representations? If this point can be defended for naturally occurring organisms, then to the degree that computers can form and manipulate representations in the right ways, to that degree will the AL research program appear plausible.

Of course, human and animal psychology is part of human and animal biology. So it is trivially true that some biological properties involve the formation and manipulation of representations. But this allows AL to be no more than AL. The question, therefore, should be whether life processes other than psychological ones involve computations on representations.

For some biological processes, the idea that they essentially involve the formation and manipulation of representations appears plausible. Thanks to our understanding of DNA, we can see ontogenesis and reproduction as processes that involve representations. It is natural to view an organism's genome as a set of instructions for constructing the organism's phenotype. This idea becomes most plausible when the phenotypic traits of interest are relatively invariant over changes in the environment. The idea that genome represents phenotype must not run afoul of the fact that phenotypes are the result of a gene/environment interaction. By the same token, blueprints of buildings don't determine the character of a building in every detail. The building materials available in the environment and the skills of workers also play a role. But this does not stop us from thinking of the blueprint as a set of instructions.

In conceding that computers could exemplify biological characteristics like development and reproduction, I am not saying that any computers now do. In particular, computers that merely manipulate theories about growth and reproduction are not themselves participants in those processes. Again, the point is that a description of a bridge is not a bridge.

What of other biological properties and processes? Are they essentially computational? What of digestion? Does it involve the formation and manipulation of representations? Arguably not. Digestion operates on food particles; its function is to extract energy from the environment that the system can use. The digestive process, *per se*, is not computational.

This is not to deny that in some systems digestion may be influenced by computational processes. In human beings, if you are in a bad mood, this can cause indigestion. If the mental state is understood computationally, then digestion in this instance is influenced by computational processes. But this effect comes from without. This example does not undermine the claim that digestion is not itself a computational process.

Although I admit that this claim about digestion may be wrong, it is important that one not refute it by trivializing the concepts of representation and computation. Digestion works by breaking down food particles into various constituents. The process can be described in terms of a set of procedures that the digestive system follows. Isn't it a short step, then, to describing digestion as the execution of a program? Since the program is a representation, won't one thereby have provided a computational theory of digestion?

To see what is wrong with this argument, we need to use a distinction that Kant once drew in a quite different connection. It is the distinction between following a rule and acting in accordance with a rule. When a system follows a rule, it consults a representation; the character of this representation guides the system's behavior. On the other hand, when a system acts in accordance with a rule, it does not consult a representation; rather, it merely behaves as if it had consulted the rule.

The planets move in ellipses around the sun. Are they following a rule or are they just acting in accordance with a rule? Surely only the latter. No representation guides their behavior. This is why it isn't possible to provide a computational theory of planetary motion without trivializing the ideas of computation and representation. I conjecture that the same may be true of many biological processes; perhaps digestion is a plausible example.

In saying that digestion is not a computational process, I am not saying that computers can't digest. Planetary motion is not a computational process, but this does not mean that computers can't move in elliptical orbits. Computers can do lots of things that have nothing to do with the fact that they are computers. They can be doorstops. Maybe some of them can digest food. But this has nothing to do with whether a computational model of digestion will be correct.

I have focused on various biological processes and asked whether computers can instantiate or participate in them. But what about the umbrella question? Can computers be alive? If Turing's question was whether computers can think, shouldn't the parallel question focus on what it is to be alive, rather than on more fine-grained concepts like reproduction, growth, selection, and digestion?

I have left this question for last because it is the fuzziest. The problem is that biology seems to have little to tell us about what it is to be alive. This is not to deny that lots of detailed knowledge is available concerning various living systems. But it is hard to see which biological theories really tell us about the nature of life. Don't be misled by the fact that biology has lots to say about the characteristics of terrestrial life. The point is that there is little in the way of a principled answer to the question of which features of terrestrial life are required for being alive and which are accidental.

Actually, the situation is not so different in cognitive science. Psychologists and others have lots to tell us about this or that psychological process. They can provide information about the psychologies of particular systems. But psychologists do not seem to take up the question "What is the nature of mind?" where this question is understood in a suitable, nonchauvinistic way.

Perhaps you are thinking that these are the very questions a philosopher should be able to answer. If the sciences ignore questions of such generality, then it is up to philosophy to answer them. I am skeptical about this. Although philosophers may help clarify the implications of various scientific theories, I really doubt that a purely philosophical answer to these questions is possible. So if the sciences in question do not address them, we are pretty much out of luck.

On the other hand, I can't see that it matters much. If a machine can be built that exemplifies various biological processes and properties, why should it still be interesting to say whether it is alive? This question should not preoccupy AL any more than the parallel question should be a hang up for AI. If a machine can perceive, remember, desire, and believe, what remains of the question of whether it has a mind? If a machine can extract energy from its environment, grow, repair damage to its body, and reproduce, what remains of the issue of whether it is "really" alive?

Again, it is important to not lose sight of the if's in the previous two sentences. I am not saying that it is unimportant to ask what the nature of mind or the nature of life is. Rather, I am suggesting that these general questions be approached by focusing on more specific psychological and biological properties. I believe that this strategy makes the general questions more tractable; in addition, I cannot see that the general questions retain much interest after the more specific ones are answered.

CONCLUDING REMARKS

Functionalism, both in the study of mind and the study of life, is a liberating doctrine. It leads us to view human cognition and terrestrial organisms as examples of mind and life. To understand mind and life, we must abstract away from physical details. The problem is to do this without going too far.

One advantage that AL has over AI is that terrestrial life is in many ways far better understood than the human mind. The AL theorist can often exploit rather detailed knowledge of the way life processes are implemented in naturally occurring organisms; even though the goal is to generalize away from these examples, real knowledge of the base cases can provide a great theoretical advantage. Theorists in AI are usually not so lucky. Human cognition is not at all well understood, so the goal of providing a (more) general theory of intelligence cannot exploit a detailed knowledge of the base cases.

An immediate corollary of the functionalist thesis of multiple realizability is that biological and psychological problems are not to be solved by considering physical theories. Existing quantum mechanics is not the answer, nor do biological and psychological phenomena show that some present physical theory is inadequate. Functionalism decouples physics and the special sciences. This does not mean that functionalism is the correct view to take for each and every biological problem; perhaps some biological problems are physical problems in disguise. The point is that, if one is a functionalist about some biological process, one should not look to physics for much theoretical help.

The Turing Test embodies a behavioristic criterion of adequacy. It is not plausible for psychological characteristics, though it may be correct for a number of biological ones. However, for virtually all biological processes, the behaviors required must be rather different from outputs printed on a tape. A desktop computer that is running a question/answer program is not reproducing, developing, evolving, or digesting. It does none of these things, even when the program describes the processes of reproduction, development, digestion, or evolution.

It is sometimes suggested that, when a computer simulation is detailed enough, it then becomes plausible to say that the computer is an instance of the objects and processes that it simulates. A computer simulation of a bridge can be treated as a bridge, when there are simulated people on it and a simulated river flowing underneath. By now I hope it is obvious why I regard this suggestion as mistaken. The problem with computer simulations is not that they are simplified representations, but that they are representations. Even a complete description of a bridge—one faithful in every detail—would still be a very different object from a real bridge.

Perhaps any subject matter can be provided with a computer model. This merely means that a description of the dynamics can be encoded in some computer language. It does not follow from this that all processes are computational. Reproduction is a computational process because it involves the transformation of representations. Digestion does not seem to have this characteristic. The AL research program has plenty going for it; there is no need for overstatement.

NOTES

This chapter originally appeared in Christopher G. Langton, Charles Taylor, J. Doyne Farmer, and Steen Rasmussen (Eds.), *Artificial life II*, pp. 749–765, Boulder, CO; Oxford: Westview Press, 2003.

REFERENCES

1. Block, N. (1975). Troubles with functionalism. In W. Savage (Ed.), *Perception and cognition: Issues in the foundations of psychology* (Minnesota studies in the philosophy of science, volume IX) (pp. 261–326). Minneapolis: University of Minnesota Press.

2. Block, N. (1981). Psychologism and behaviorism. *Philosophical Review*, 90, 5–43.

3. Dennett, D. (1978). *Brainstorms*. Cambridge, MA: MIT Press.

4. Dretske, F. (1985). Machines and the mental. *Proceedings and Addresses of the American Philosophical Associations*, 59, 23–33.

5. Fisher, R. (1958). *The genetical theory of natural selection*. New York: Dover.

6. Fodor, J. (1968). *Psychological explanation*. New York: Random House.

7. Lewontin, R. (1970). The units of selection. *Annual Review of Ecology and Systematics*, 1, 1–14.

8. Pattee, H. (1989). Simulations, realizations, and theories of life. In C. G. Langton (Ed.), *Artificial life* (Santa Fe Institute studies in the sciences of complexity, proceedings volume VI) (pp. 63–77). Redwood City, CA: Addison-Wesley.

9. Penrose, R. (1990). *The emperor's new mind*. Oxford: Oxford University Press.

10. Place, U. T. (1956). Is consciousness a brain process? *British Journal of Psychology*, 47, 44–50.

11. Putnam, H. (1975). The nature of mental states. In H. Putnam, *Mind, language, and reality* (pp. 429–440). Cambridge, UK: Cambridge University Press.

12. Ryle, G. (1949). *The concept of mind*. New York: Barnes & Noble.

13. Searle, J. (1980). Minds, brains, and programs. *Behavior and Brain Sciences*, 3, 417–457.

14. Smart, J. J. C. (1959). Sensations and brain processes. *Philosophical Review*, 68, 141–156.

15. Sober, E. (1985). Methodological behaviorism, evolution, and game theory. In J. Fetzer (Ed.), *Sociobiology and epistemology* (pp. 181–200). Dordrecht: Reidel.

16. Sober, E. (1984). *The nature of selection*. Cambridge, MA: MIT Press.

17. Turing, A. (1950). Computing machinery and intelligence. *Mind*, 59, 433–460.

18. Wittgenstein, L. (1953). *Philosophical investigations*. Oxford: Basil Blackwell.

17 · Life, "artificial life," and scientific explanation

MARC LANGE

INTRODUCTION

What is it for an entity to be a *living thing*? This question has recently been treated with benign neglect by philosophers. (It goes entirely unmentioned in current philosophy of biology texts—e.g., Rosenberg 1985; Sober 1993.) In contrast, it is now being much discussed by certain biologists and computer scientists. They deny that the distinction between life and nonlife is an artificial one; on their view, it is an important subject for scientific investigation. They have organized themselves into a new field, "artificial life," with its own conferences (e.g., Langton 1989; Langton *et al.* 1991), its own journal (Volume 1 of *Artificial Life*, dated 1994, has been published by MIT Press), and even its own popularizations (e.g., Levy 1992; Emmeche 1994).

"Alife" researchers believe that it is possible to create and to study *digital organisms*, each of which consists of a machine-language program *living* in a computer's memory. When the program is executed, it is sometimes copied to another location in memory. The program can arrange for its "copy" to include certain new elements, permitting variation and selection to occur. "Evolving digital organisms will compete for access to the limited resources of memory space and CPU time, and evolution will generate adaptations for the more agile access to and the more efficient use of these resources" (Ray 1994, p. 185). While many advocates of "strong-AL" maintain (for reasons I will mention) that currently there are no living software entities, they argue that "it is not hard to imagine computer viruses of the future that will be just as alive as biological viruses" (Farmer & Belin 1991, p. 820) and that certain kinds of self-replicating patterns of instructions would be alive not merely as biological

viruses are, but in the fullest sense. C. G. Langton, a prominent Alife researcher, expresses a view common in the field:

> The ultimate goal of the study of artificial life would be to create "life" in some other medium, ideally a *virtual* medium where the essence of life has been abstracted from the details of its implementation in any particular model. We would like to build models that are so life-like that they cease to become *models* of life and become *examples* of life themselves.
>
> (1986, p. 147)

Obviously, any argument that certain self-replicating segments of machine code are living—not in some metaphorical sense but in the literal sense used in biology—presupposes that some such notion *is* used in biology. But it is not evident that there is any work in biology to be performed by a distinction between life and nonlife. Furthermore, any argument that certain pieces of software are alive, in the literal biological sense, presupposes some rough sense of what "life" is. But it is notoriously difficult to explicate "vitality." Finally, the strong-AL claim faces arguments (made, e.g., by Elliott Sober (1991)) that no entities consisting of software could possibly be alive in the biological sense.

In this paper, I will examine all of these matters in the context of the renewed discussion of "life" in the scientific literature. I will try to understand what work (if any) this concept may perform in biology—especially in connection with the scientific explanation of certain phenomena—and what (if anything) turns on scientific disputes over this concept's utility. My purpose is not to judge the plausibility of any particular theories advanced in connection with

The Nature of Life, ed. M. A. Bedau and C. E. Cleland. Published by Cambridge University Press. © M. A. Bedau and C. E. Cleland 2010.

"artificial life." Rather, it is to investigate whether the concept of "vitality" might play some role in a rational reconstruction of biology, and (if so) to identify what role that might be. Occasionally, it will prove helpful to try to understand the function that this concept performs in the Alife research program. It will also be illuminating to consider some episodes from the history of biology that do not involve vitalism, but in which scientists nevertheless put the concept of "vitality" to work. It will also sometimes be useful to compare "life" to various other concepts used in science.

"SIGNS OF LIFE" AND HOW NOT TO SHOW THAT "LIFE" PERFORMS NO WORK IN BIOLOGY

How do advocates of strong-AL argue that certain self-replicating segments of machine code would be alive? Langton summarizes the argument:

> Artificial Life will force us to rethink what it is to be "alive." The fact is we have no commonly agreed upon definition of "the living state." When asked for a definition, biologists will often point to a long list of characteristic behaviors and features shared by most living things ... However, as most such lists are constituted of strictly behavioral criteria, it is quite possible that we will soon be able to exhibit computer processes that exhibit all of the behaviors on such a list ... [M]any researchers in the field of Artificial Life believe in the Strong Claim ... that any definition or list of criteria broad enough to include all known biological life will also include certain classes of computer processes, which, therefore, will have to be considered to be "actually" alive.
>
> (1991, p. 19f.)

Attributes typically mentioned by biologists (e.g., Pirie 1938, p. 14; Curtis 1977, pp. 14f.) and Alife researchers (e.g., Farmer & Belin 1991, p. 18) as "signs of life" include the capacities to reproduce, to grow, to move, to evolve, to respond to the environment, to remain stable under small environmental changes ("homeostasis"), and to convert matter and energy from the environment into the organizational pattern and activities of the organism ("metabolism"). Similar lists have been given for a very long time, but their status is

obscure, since their presentation is invariably followed by certain provisos: that some non-living things possess one or more of these attributes, that presumably an entity could possess all of these properties and nevertheless not be alive, and that some living things lack some of these properties. There are canonical examples: a whirlpool or ocean wave assimilates surrounding matter into its form; crystals, clouds, and fires grow whereas many mature organisms do not; iron rusts in response to being surrounded by oxygen; bubbles are stable under various small environmental changes; mules and neutered pets cannot reproduce; pepsin in the stomach catalyzes the reaction that produces pepsin from pepsinogen; Penrose's (1959) plywood blocks reproduce; fires, ocean waves, and planets move; and so on.

One way to avoid having to explicate vitality's relation to the "signs of life" is to deny that biologists are (or ought to be) interested in any property over and above the conjunction of these signs. Sober writes:

> If a machine [or perhaps more to the point, a segment of machine code—ML] can be built that exemplifies various biological processes and properties, why should it still be interesting to say whether it is alive? ... If a machine [or a segment of machine code—ML] can extract energy from its environment, grow, repair damage to its body, and reproduce, what remains of the issue of whether it is "really" alive? ... I cannot see that the general questions retain much interest after the more specific ones are answered.
>
> (1991, p. 763)

Some eminent biologists apparently agree with Sober, though others do not. Immunologist P. B. Medawar, Nobel laureate in 1960, says that discussions of what it is to be alive "are felt to mark a low level in biological conversation" (1977, p. 7), whereas geneticist Joshua Lederberg, Nobel laureate in 1958, writes:

> An important aim of theoretical biology is an abstract definition of life. Our only consensus so far is that such a definition must be arbitrary [since] life has gradually evolved from inanimate matter ...
>
> (1960, p. 394)

Lederberg's remark reminds us that one cannot show that the distinction between life and nonlife has no place in modern biology simply by noting that biologists recognize a continuum between "living" and "non-living" entities, an intermediate region inhabited by, e.g., viruses, prions, and various entities figuring in theories of "the origin of life." Many important scientific concepts have vague boundaries—even the various chemical elements, the paradigm examples of "natural kinds." Although the difference between an atomic number of 7 and an atomic number of 8 appears to be discrete, consider what happens when an alpha particle is absorbed by a nitrogen atom and a proton emitted; as Sober elsewhere (1992, p. 353) notes, there may well be no particular moment when the nitrogen atom becomes an oxygen atom. There may be no need to specify some precise moment of transmutation, just as presumably "nothing turns on whether a virus is described as a living organism or not" (Medawar & Medawar, 1977, p. 7; cf. Pirie 1938, p. 22). That there are intermediate cases in connection with which the concept "living" is not useful does not show that this concept is idle in biology, but only that an adequate rational reconstruction of its role must permit such cases.

It is often said that biologists cannot presently *define* "life." Yet, of course, they know what the word means—at least in the ordinary sense that enables them to use it properly—and dictionaries have satisfactory entries for it. Biologists cannot presently define "life" only in the sense that they cannot presently "define it away;" they cannot presently reduce vitality entirely to other, independently intelligible properties. This inability does not suggest that the concept performs no work in biology. The concept "electron" performs work in physics, it is used by physicists, they know (in the ordinary sense) what it means, and they know some properties that electrons possess, though they cannot define "electron-hood" away. Natural philosophers regarded "water" as a natural kind and knew how to recognize it long before they knew that water was H_2O and presumably before they had any proposal for a reductive definition of "water."

Moreover, "life" may perform work in biology, and one goal of biology may be to *understand* vitality, without one goal of biology being to present a definition (in the reductive sense) of "vitality," any more than one goal of biology is to define "ostrich" or one goal of physics is to define "electron." (For that matter, in order for the distinction between life and non-life to perform work in science, an *understanding* of vitality need not be a *goal* of science. It is possible that an

understanding of vitality is valuable to science as a means, not as an end in itself. I will return to this point.) I find it very curious that theoretical biologists, whenever they mention the concept "life," immediately turn to various possible definitions of "life" and bemoan their inadequacy. It is unclear to me why biologists should think that their failure to define "life" reductively makes this concept suspect or its use unwarranted. As I have mentioned (and as further examples will illustrate), this is not the attitude that scientists take towards other concepts.

Perhaps biologists worry that to use "life" without the prospect of defining it reductively commits them to vitalism. But it should go without saying that in order for the distinction between living and non-living things to figure in our rational reconstruction of research in biology, scientists need not believe that living things may violate the laws governing inanimate matter, nor must they consider seriously the possibility that some special constituent (like Bergson's "élan vital" or Driesch's "entelechy") endows living things with their vitality. In order to recognize the distinction between nitrogen and oxygen as a natural one, a scientist need not deny that nitrogen and oxygen are made of the same types of fundamental particles. The distinction between nitrogen and oxygen is nevertheless an ontological one between natural kinds.

I have just explained several ways that the distinction between life and nonlife cannot be shown to be idle in modern biology. But I have come no closer to explicating the relation between the "signs of life" and vitality, nor have I explained the role that attributions of vitality might play in biology. By the same token, I just referred to a scientist as recognizing the distinction between nitrogen and oxygen as a "natural" one. Presumably, this attitude leads the scientist to regard certain uses of the categories "nitrogen" and "oxygen" as appropriate. But in what ways is the scientist thereby prepared to put these categories to work? Do scientists use "life" in analogous ways?

VITALITY AS EXPLAINING "SIGNS OF LIFE"

As we have seen, Sober argues that it is empty to attribute vitality to software entities, insofar as it goes beyond attributions of various "signs of life." Sober also argues that there is a fundamental obstacle to any software entity's genuinely displaying any of these signs. Even if a software entity possesses a property with the same "mathematical structure" (Sober 1991, p. 759) as a "sign of life," it may well be impossible, even in principle, for that property to have the "empirical content" (ibid.) of the biological property. In that event, while the software entity may in a certain respect *simulate* a living creature, it does not genuinely possess the relevant sign of life. Here is his argument:

> Many biological properties and processes involve relations between an organism (or part of an organism) and something outside itself. An organism reproduces when it makes a baby. A plant photosynthesizes when it is related to a light source in an appropriate way. A predator eats other organisms. Although a computer [or software entity—ML] might replicate aspects of such processes that occur inside the system of interest, computers [or software entities—ML] will not actually reproduce or photosynthesize or eat unless they are related to things outside themselves in the right ways.
>
> (1991, p. 759)

But one segment of machine-language code can bear various relations to things outside of it (e.g., other programs, the CPU) that are isomorphic to various relations borne by a biological entity to other things. Sober's point, then, appears to be not so much that it is the right *relation* to outside things that is essential in order for an entity to display a sign of life, as that it is the right *outside things*. Intuitively, this seems more plausible in the case of photosynthesis—a light source is essential—than in the case of predation or reproduction; it is not clear why another software entity would not do as prey or as progeny, at least without begging the question. (Of course, there are certain activities that require a relation to the right outside things; drinking, for instance, presumably requires liquids. But to my knowledge, drinking and eating are not ever listed as "signs of life." I think that this is because they are essentially connected with food, nourishment, health, etc.—in short, they *presuppose* vitality, whereas no "sign of life" is a sufficient condition for vitality. A *"sign* of life" is a strictly behavioral property; eating and drinking are not strictly behavioral, since they presuppose vitality and so amount to more than mere

"ingesting" or "taking in.") Indeed, Sober's argument presupposes that in order to display a certain sign of life, an entity needs to figure in a relation involving not merely a certain "mathematical structure" but also a certain "empirical content." On what principle can Sober defend this claim? His resources are meager, once he precludes grounding the content of a given sign of life in some concept of "being alive" over and above displaying these various signs. If the "signs of life" are an arbitrary collection, then there is really no importance to the distinction between simulating and genuinely displaying one of them, whereas if there is some point to drawing this distinction, then that is presumably because the genuine sign, but not its simulation, bears some relation to vitality, which is important because vitality has some significance over and above its "signs."

How might this be? The "signs of life," while neither individually necessary nor jointly sufficient for something to be living, bear a special relation to vitality: Certain things display a given "sign of life" *because they are alive* (while certain other, non-living things display a given "sign of life" for some other reason). In other words, that a given thing is living *explains* why in certain circumstances it can reproduce, metabolize, move, and so on. This distinguishes the "signs of life" from other properties, including their simulations. And, I will argue, it is in connection with such explanations that the concept "life" performs its work in biology.

More precisely, I will argue that this is a function that the concept of "vitality" has often been called upon to perform, and I will argue that so far as we can presently tell, it might well turn out to play this role in the correct biological theory. Whether the distinction between living and nonliving things is in fact important for biologists to draw is an empirical question. The answer is for science to discover, just as science has ascertained that the distinction between nitrogen and oxygen is important and that certain of an object's properties can be explained by the fact that the object is composed of copper. My purpose is not to discover whether some thing's vitality can indeed explain its exhibiting a given "sign of life." Rather, I will try to explicate what it would take for the concept of "vitality" to be capable of performing this function. I will argue that for "vitality" to be so capable not only is consistent with our current understanding of living things, but also makes sense of the way in which the "signs of life" function in biology.

To see this, it is useful to begin by examining some examples from the history of science. Admittedly, it is difficult to find examples in which scientists offer something's vitality as an explanation of its exhibiting various "signs of life." Obviously, familiar animals and plants were understood to be living long before any kind of explicit scientific research was undertaken. Consequently, it was not thought worth saying that if, e.g., snakes are living creatures, then that fact would explain why they can move; I would be astonished if we found this consideration being advanced explicitly in the literature as, say, evidence that snakes are alive. If scientists wrote about explaining the locomotion of snakes, it was in the course of trying to give a different explanation of this phenomenon, e.g., an explanation involving the snake's musculature. (There are many ways of explaining the fact that snakes can move, just as there are many ways of explaining the fact that a given wire can conduct electricity, e.g., by noting that it is copper or by presenting the wire's molecular structure and describing how electrons pass through it. Different conversational contexts call for different explanations of the same fact.) But in the history of science, we can find some disputes in which scientists explicitly discussed whether some entity lives and what its vitality might explain (and in which the entity was not thought to lie in the intermediate range where drawing any line between life and nonlife is unilluminating). I will consider two examples: astronomical bodies and fungi.

That the celestial bodies are alive was a common view in antiquity and was much debated through Galileo's time. One argument often given for their vitality was that it explains their capacity to move, one of the "signs of life." Plato (*Statesman* 269c4–270a8, *Timaeus* 38e5, 40a8–b1) so argues in general terms, and while Aristotle gives several different explanations of the capacity of stars and planets to move, scholastic natural philosophers often took him as having attributed this capacity to their vitality, especially in *De Caelo* 285a29–30, 292a18–21, and 292b1–2 ("We must then think of the action of the stars as similar to that of animals and plants"). An "inference to the best explanation" appears in Cicero's summary (*De Natura Deorum* 2.16.14) of Aristotle's argument in *De Philosophic* (now lost):

Aristotle, indeed, is entitled to praise for having laid down that everything which moves does so either by nature, necessity, or choice. Moreover, the sun, moon, and all the stars move, but things which move by nature are carried either downwards by their weight or upwards by their lightness, neither of which happens to the stars, since their course is directed around in a circle. And it certainly cannot be said that it is some greater force which makes the stars move in a way contrary to nature, for what greater force can there be? It remains, therefore, to conclude that the movement of the stars is voluntary . . .

(Scott 1991, p. 24)

Although non-living things can move, vitality can explain movement, and if no other explanation is promising, an object's movement is evidence of its vitality. This argument was very often made by Hellenistic and scholastic philosophers (see Grant 1994, p. 71). For example, Ptolemy (*Planetary Hypotheses* 2.12; see Taub 1993, p. 117ff.) argues that the explanation of planetary motion is analogous to the explanation of the motion of birds. One of the two or three most widely discussed questions about the heavens during the five centuries following the year 1200 was whether the stars and planets are alive (Grant 1994, pp. 236, 703), and the popular phrase among advocates of their vitality was that they move "just as a bird flies through the air or a fish swims in water" (Grant 1994, pp. 274, 348); the reason that astronomical bodies can move is similar to the reason that birds and fish can move.

Conversely, those who held that astronomical bodies are not alive argued (setting theological arguments aside) either that they lack various "signs of life," so that there is nothing to be explained by positing their vitality or that we can best explain the "signs of life" they display without positing that they are living. Aquinas, Oresme, Galileo, and others argued (e.g., Grant 1994, p. 473ff.) that astronomical bodies do not ingest, grow, or reproduce. (Some advocates of the vitality of astronomical bodies—e.g., Franciscus Bonae Spei (Grant 1994, p. 483)—replied that the capacity to grow and to ingest, though "signs of life," are fundamentally just instances of "self-motion" and since astronomical bodies (like birds) certainly do move without external compulsion, that they fail to grow and to ingest is not strong evidence against their vitality.) It was also commonly argued that since astronomical bodies are homogeneous, they have no organs of sensation (according, e.g., to Aquinas (Grant 1994, p. 475)) and they lack the internal organization that constitutes a sign of life (according, e.g., to Franciscus de Oviedo (Grant 1994, p. 481ff)).

Likewise, from ancient times until the mid-nineteenth century, natural philosophers debated whether fungi are living creatures, and these arguments concerned whether fungi display the familiar "signs of life" and, if so, whether the best explanation is that they are livings. It was long unclear, for instance, whether fungi grow; Pliny, for example, believed that they may spring into existence full size and may not be alive (*Naturalis Historia*, Bk. XIX, Sect. 11). Since for a very long time nothing seed-like was observed to be associated with fungi, it was believed that they do not reproduce. Some (e.g., Bock in 1552, Bauhin in 1623, Medicus in 1788) thought that they derive from moisture and are merely "superfluous moisture" (Ainsworth 1976, pp. 13, 18); the lore, common in Shakespeare's time, that they result from the condensation of moisture rising from the earth in low-lying damp areas, as near water mills, is the origin of the term "mildew" (Large 1962, p. 18). Alternatively, many classical authors (e.g., Juvenal and Plutarch—see Houghton 1885) believed fungi to result from lightning or thunder. Many who believed that fungi display none of the familiar "signs of fife" were especially impressed by the "fungus-stone," which appeared to be a rock that, whenever watered, would produce an edible fungus. (Though it resembles a large grey stone—to cut it, one must use a saw—it is actually the particularly hard sclerotium of the fungus *Polyporus tub-eraster*) Widely discussed since ancient times (see Ramsbottom 1931/1932; 1938/1939, p. 311), the fungus-stone led many (even as late as Villemet in 1784) to classify fungi in the mineral kingdom (Ainsworth 1976, p. 32). On the other hand, some scientists posited the vitality of fungi to explain their internal organization (e.g., Seyffert in 1744—see Ramsbottom 1938/1939, p. 313), growth (e.g., Builliard in 1793—see Ainsworth 1976, p. 32), and capacity for motion (e.g., the explosion of the "mortier" of *Lycoperdon car-pobolus*, discussed by Durande in 1785 as a sign of life and later understood as functioning to disperse spores—see Ramsbottom 1938/1939, p. 311f).

(Here, briefly, is one further historical example. In 1827, the English botanist Robert Brown observed through a microscope that grains of pollen suspended in water perform ceaseless erratic motions. Having experimental evidence that the motions arise neither from currents in the water nor from its gradual evaporation, Brown (1828) attributed them to the vitality of the pollen grains—until he noticed that dust particles not derived from living things exhibit the same motions (later termed "Brownian"). Brown (1829, p. 164f) discussed other cases in which some scientist has employed "very ingenious reasoning . . . to show that, to account for the motions [of an object], the least improbable conjecture is to suppose the [object] animated.")

These examples suggest that "It is living" has sometimes been considered able to explain why some object exhibits a certain "sign of life." Accordingly, to understand the role in science that might be played by the concept "life," we must understand the value that those who offer these purported explanations believe them to have. How these "explanations" might be informative is made especially puzzling by two considerations. First, I have already noted that scientists expect there to be explanations of these same phenomena that do not invoke vitality, as when a snake's capacity to move is explained by accounts of its anatomy and physiology. Of what possible use, then, is an explanation that instead invokes "life"? Second, scientists offer these explanations without knowing what vitality involves over and above the various "signs of life." In that circumstance, it is not at all evident how appeal to an object's "vitality" has any power to explain its exhibiting various signs of life.

THE VALUE OF EXPLANATIONS THAT INVOKE VITALITY

Earlier, I noted several respects in which using "It is living" to explain why an object displays a certain "sign of life" resembles using "It is copper" to explain why an object is electrically conductive. Neither explanation describes the object's structure or specifies how that structure results in the phenomenon being explained. Both "It is living" and "It is copper" were regarded as explanatory long before anything about the relevant structures was known. Now I will pursue this resemblance; I suggest that if "It is living" is indeed explanatory, then it helps us to understand why an object displays some sign of life in just the way that "It is copper" helps us to understand why an object is electrically conductive.[1]

In what way is that? Each of these explanations informs us that the phenomenon being explained results entirely from the nature of the given object. One reason that this explanation is useful is because it informs us, regarding certain facts about the object's present situation, that these facts are irrelevant to the phenomenon being explained. For instance, if we are told that the object is electrically conductive because it is copper, then we learn that it would still have been electrically conductive even if it had, e.g., been a different shape or been in a different location or been painted a different color—if we already knew that it would still have been copper under those circumstances. Moreover, we learn that it might no longer be electrically conductive were it to participate in a chemical reaction. Likewise, by attributing, e.g., an object's motility to its vitality, we convey that certain external circumstances are irrelevant to its capacity to move, whereas the proper functioning of its bodily organs is relevant. This is brought out by examples in which the object's vitality has no explanatory relevance. For instance, that Leo the Lion is a living thing does not explain why Leo at a certain moment moves in one direction rather than another, since his direction is sensitive to external circumstances. Similarly, that Leo is a living thing does not explain why he can move downward when released from a flying airplane, since Leo would still fall even if he were not a living thing; the proper functioning of his bodily organs is irrelevant.

Of course, the power of "Object a is copper" to explain why object a is electrically conductive cannot be due entirely to the fact that "Because it is copper" tells us with regard to certain facts that they are relevant (in some sense cashed out in terms of counterfactuals) and regarding certain other facts that they are irrelevant to a's electrical conductivity. After all, if we want to know why someone who took opium fell asleep, and we are told that it is because opium has dormative virtue, we learn something about what is relevant and what is irrelevant to the fact being explained—e.g., that the person's wearing a blue shirt is irrelevant. Yet it is standardly held that a claim ascribing a dispositional property (e.g., "Opium has dormative

virtue") cannot help to explain a manifestation of that disposition.[2] Likewise, that "Object a is a living thing" can point the way towards a genuine explanation of the object's exhibiting a given sign of life, and can even in certain contexts stand in for that explanation (by telling us everything we wanted to know when we asked why the object exhibits that sign of life), does not show that "Object a is living" is itself explanatory.

What would it be for "It is living" to have the same sort of explanatory power as "It is copper"? To recognize the full significance of saying "Object a is electrically conductive because it is copper," we must remember that "Because it is copper" answers many other why-questions, both about the given object ("Why is a thermally conductive?") and about other objects ("Why is object b electrically conductive?"). By saying "Object a is electrically conductive because it is copper," we convey that when a's electrical conductivity is explained in more detail, that explanation will share something important with the detailed explanations of a's thermal conductivity and b's electrical conductivity. Long before scientists knew much about the structure of copper objects or how their structure makes them excellent conductors of electricity and heat, scientists presumed that there are properties that any copper object has in common with all others non-accidentally, i.e., in virtue of their being copper, and that their electrical and thermal conductivity can be explained by these properties.

Of course, these various explanations may have important differences; the mechanism by which copper conducts electricity is different in important ways from the mechanism by which copper conducts heat. Nevertheless, copper's structure is crucial to both phenomena. Scientists initially did not know much about this structure. Yet they regarded "It is copper" as explanatory; they would not have done so had they not believed that there are important similarities among these various detailed explanations. To put the point very roughly: It is explanatory to classify something as "copper"—this term denotes a natural kind rather than an artificial category—precisely because there are such similarities, i.e., because it turns out to be heuristically useful for scientists to presume that in some important respects, these various explanations are similar. In other words, scientists learn more about the world more quickly when they expect these explanations to be similar in some important respects—and when, consequently, they regard their discoveries about why, e.g., *a* is electrically conductive as inductively confirming something similar about why *b* is thermally conductive—than they would by regarding these explanatory tasks as unrelated.

I believe that some scientists expect the concept "living thing" to perform the same kind of work as "copper." That is, they expect that any two living things display their "signs of life" for importantly similar reasons.[3] Scientists who believe that living things constitute a natural kind recognize that the similarity among living things in the reasons they display various signs of life must lie at a fairly deep level. Recall, for example, that those who viewed astronomical bodies as alive held that they move "just as a bird flies through the air or a fish swims in water;" they held, in other words, that the reason astronomical bodies can move is similar to the reason birds and fish can move. To see how abstract these scientists expected the similarity to be, let us set the astronomical bodies aside momentarily. Though the details of how birds and fish manage to move were as yet unknown, these scientists undoubtedly recognized that the reason that birds can move through air must differ in some significant respects from the reason that fish can move through water. Nevertheless, these scientists apparently expected there to be certain fundamental respects in which these explanations are similar—and similar to explanations of the capacity of astronomical bodies to move.

Science has ascertained that copper things constitute a natural kind. Whether living things constitute a natural kind is likewise a question for scientific research to answer. It remains an open question; scientists currently disagree on this issue. Lederberg, for instance, apparently believes that living creatures probably display their signs of life for importantly similar reasons. In contrast, Medawar is impressed by how little has been derived from research (e.g., by vitalists) guided by the presupposition that living things constitute a natural kind, and rightly sees the past sterility of this approach as some evidence against it.

Alife researchers are motivated by the thought that artificial "software organisms" display various "signs of life" for fundamentally the same reasons as natural "wet-ware" organisms do; these scientists are prepared to inductively project their eventual discoveries about "artificial life" onto natural organisms.

Indeed, many Alife researchers believe that they have already begun to elucidate one important similarity in the reasons that various living things exhibit various "signs of life." Alife researchers currently describe this similarity by referring to the "parallel," "distributed," "decentralized," "bottom-up," "self-organized," "emergent" way in which "signs of life" arise in living things. These terms are often not fully explicated, but the general thought seems roughly to be that a living entity displays a "sign of life" not as the result of the organizing activity of some central coordinator functioning to realize an established plan, but as the result of local interactions among many autonomous components, each component governed by the same simple laws. My purpose here is not to evaluate the actual prospects for ultimately creating or simulating life by developing software that is so organized. I wish only to understand the work that any such research program intends the concept of "life" to perform, as well as to identify one sense in which living things, though different in many respects, might display signs of life for "the same reason."

Those who pursue this research program do not regard vitality as a dispositional property—i.e., as like dormative virtue, water solubility, and water absorbency. A dispositional claim (e.g., "This cube is water soluble") is a very minimal description; in regarding a certain property (D-ness) as dispositional, a scientist holds that objects possessing this property may differ fundamentally in the microstructures and mechanisms whereby they manifest this disposition. In other words, the scientist holds that all physically possible D objects are similar in virtue of being D's *only* in how they would respond to a particular kind of stimulus under certain circumstances (e.g., only in that they would dissolve were they placed in water under standard conditions). For instance, an ionic molecule (such as sodium chloride) is soluble in water for a very different reason from a covalent molecule (such as sucrose). Likewise, one fabric is water absorbent because water can be incorporated into each of its fibers, whereas another fabric is water absorbent because water can be held in place between its fibers; both are water absorbent, but their similarity in possessing this dispositional property does not result from any similarity in microstructure or mechanism. Scientists who regard a given property D-ness as dispositional expect there to be no law-statements of the form "All D's are ..."; roughly speaking, they are not prepared to inductively project their eventual discoveries about the microstructure or mechanism responsible for one entity's D-ness onto each other possible D entity.[4] In contrast, Alife researchers regard living organisms as constituting a natural kind, like copper objects. They do not regard "It is a living thing" as a minimal description. They expect (though are not certain) that any two living things display their signs of life for (in some fundamental sense) the same reason. Consequently, they are willing to take certain discoveries they make about the mechanism responsible for one living thing's exhibiting a given sign of life as inductively confirming a lawlike hypothesis about all living things—that is (roughly speaking), as confirming, for any other living thing, something similar about why it displays its signs of life. If these scientists are correct, then vitality contrasts with dispositional properties in its capacity to figure in scientific explanations.[5]

Living things may constitute a natural kind. If so, then as I have explained, co-classifying various objects as "living" would be useful as a means to learning more about them. It again must be emphasized that the goal of such research is to learn more about living things, not necessarily to define "life" reductively. Perhaps this is easier to see in connection with other concepts in colloquial use whose utility in science is not yet clear and whose role, if they turn out to be useful, would be analogous to that of "life." Consider, for instance, the concept "storm." Jupiter's Great Red Spot and Neptune's Great Dark Spot are often categorized as "storms." One might ask: Is this merely a loose analogy, or are they really storms? Some scientists believe that although these phenomena display some of the same features as terrestrial storms, the reasons that the Great Red Spot and the Great Dark Spot exhibit these features are not similar to the reasons that these features appear in terrestrial storms—and, furthermore, that there are many, fundamentally different kinds of terrestrial "storms," bearing only superficial resemblances. These scientists see nothing at issue in asking whether the Great Red Spot is "really" a storm. Other scientists disagree. They believe that the Great Red Spot and the Great Dark Spot display various features for reasons that are, in certain profound (and largely unknown) respects, similar to the reasons that terrestrial storms exhibit the

same features. These scientists believe that much can ultimately be learned about terrestrial storms by studying these phenomena in the atmospheres of other planets. If this turns out to be true, then the category "storm" is a natural kind and all of these phenomena really are "storms." But there is no particular reason to regard this research as aiming to *define* "storm" in other terms (though a definition may result); the research may aim simply to discover the natural laws concerning storms. In thinking that we can better understand terrestrial storms by inductively projecting from storms that involve very different material constituents and exist under very different circumstances, these astronomers are similar to Alife researchers.

Some may be inclined to say that, even if all of the entities towards the living end of the spectrum exhibit many signs of life as "emergent" phenomena, to exhibit many signs of life for this reason cannot be part of *what it is* to be a living thing. For, it might be insisted, we can conceive of an entity that is intuitively living but exhibits many signs of life for another reason entirely. I am not willing to endorse an analysis of "vitality" in terms of "emergence"; it is far too early to say whether this research program will flourish or even to tell whether its notion of "emergence" is well defined. But I do not believe that the sort of analysis of "vitality" that this research program aims to provide is ruled out by our ability to conceive of a living thing that exhibits many signs of life for "non-emergent" reasons. In the same way, a chemist in the late nineteenth century, who accepted the atomic theory before the discovery of the proton, would have said (after being told something about these hypothetical "protons") that he can imagine a gold atom that does not contain 79 protons. This would not have been taken as precluding an analysis of "gold" as the element each atom of which contains 79 protons. We can best understand various episodes from the history of science if we regard scientists as prepared to count some thing as "gold" if and only if it possesses the same fundamental structure as certain paradigmatic examples of gold. We can (and nineteenth-century chemists could) conceive of various different accounts of this fundamental structure. Likewise, I have suggested that we can best understand various episodes from the history of science if we regard scientists as prepared to count some thing as "living" if and only if it exhibits many "signs of life" for the same fundamental reason as certain paradigmatic examples of living things do. And we can conceive of various different accounts of this fundamental reason.

LIFE AS INTERESTING OVER AND ABOVE THE "SIGNS OF LIFE"

It may be that most things displaying a given "sign of life" (e.g., growth) do not display any of the other "signs." But that some thing is living may count as considerable evidence that it exhibits a given sign of life, even if it has never yet been observed to do so. Of course, this evidence would not be conclusive—a living thing need not display every "sign of life"— but this evidence may justify scientists at least in looking more carefully for indications that the object displays further "signs of life." For instance, naturalists in the sixteenth, seventeenth, and eighteenth centuries who believed there to be no direct evidence (as yet) that fungi reproduce, but believed that fungi are living creatures, believed for this reason that fungi probably reproduce, and so that if one looked carefully, one would probably discover their seeds. Though spores were first observed (by della Porta) in 1588, and were observed (by Micheli) in 1729 to be the means by which fungi reproduce, a great many scientists failed in attempts to replicate these observations. Hence, we find that as late as 1785, Durande argued that since fungi exhibit various other "signs of life," fungi presumably have seeds, though they are too small to have yet been seen (Ramsbottom 1938/1939, pp. 312, 316f).

If a scientist regards the living things as constituting a natural kind, then she regards the hypothesis that some thing displays a given "sign of life" as less well confirmed by its displaying some other "sign of life" than by its displaying that other "sign of life" *for the reason that living things do.*[6] That is not to say that she regards some thing's exhibiting a given "sign of life" for the same reason as living things do (perhaps, that is, as an "emergent" phenomenon) as *ensuring* that it is alive. A computer virus may display a given sign of life (e.g., homeostasis, as when it adjusts to different operating systems, compensates for insufficient storage, or defeats anti-virus mechanisms) for the reason that living things do. Yet a computer virus, like a biological virus, displays so few other signs of life that at best, it falls into the same intermediate region between life

and nonlife as biological viruses do. (And contemporary computer viruses do not display some of the signs of life that even biological viruses exhibit. For instance, there are currently no computer viruses that evolve other than by programmers deliberately altering them or by their being programmed to make some slight alteration when copied in order to enable their progeny to avoid detection. Contemporary computer viruses do not evolve "naturally" as biological viruses do. Apparently, it is for reasons such as this that in the passage I quoted earlier, Farmer and Belin deny that current computer viruses are just as alive as biological viruses. See Spafford 1994.) The fact that an entity exhibits a given sign of life for the same reason as living things do is insufficient to make that entity alive. But (if living things constitute a natural kind) this fact contributes to the thing's vitality, i.e., makes it closer to being alive.

I have said that the hypothesis that some thing displays a given "sign of life" may be less well confirmed by its displaying some other "sign of life" than by its displaying that other "sign of life" for the reason that living things do. One might reply that this difference in confirmatory value would not show that scientists care about a thing's vitality over and above its displaying many "signs of life," since the hypothesis that some thing displays a given "sign of life" is less well confirmed by its displaying some other "sign of life" than by its displaying many other "signs of life." However, that it displays many other "signs of life" is not much evidence that it displays the given "sign of life" if we deny that it displays these other "signs of life" for the same reason as living things do.

Scientists are interested not merely in whether a thing displays a given "sign of life," but in whether it does so for the reason that living things do. As we have seen, scientists are interested in whether an object's displaying a given "sign of life" contributes to its vitality not only because scientists are interested in discovering which other signs of life it displays, but also because scientists are interested in explaining why it displays the given sign of life. Earlier, we saw Sober argue that software entities may simulate certain "signs of life" but cannot genuinely display them. I argued that if (as Sober says) science were not interested in an object's vitality over and above the "signs of life" it exhibits, then science would have no principled basis for drawing the distinction that Sober invokes: between an object's displaying and its simulating a given "sign of life." Those scientists who believe that living things constitute a natural kind are justified in distinguishing between an object's displaying a given sign of life in a manner that contributes to its vitality from its displaying a given sign of life in another fashion. For instance, when Alife researchers argue that a software entity does not simulate an activity of some living things, but actually performs the very same activity, their argument is that the reason for the entity's performances is fundamentally similar to the explanation of the analogous performances by familiar living things. Consider one example: To assist animal behaviorists in understanding how the birds in a flock manage to fly together in a coordinated fashion, a computer program was recently written (Reynolds 1987; see also Emmeche 1994, p. 88ff) that moves bird-shaped figures around on a computer screen in accordance with certain simple rules determining how the velocity of each figure at each moment is to be changed according to the positions and velocities of the other figures near it a moment before. The result is that the figures perform the turning and wheeling behavior characteristic of flocks of birds (and schools of fish and herds of mammals). Alife researchers have argued that these figures (while obviously not living) are flocking—duplicating, not imitating what birds do—and their argument (Langton 1989, p. 32f) is that the coordinated behavior of these figures is explained in the same "decentralized, bottom-up" way as the organization of flocks of birds. There is no lead bird; the collective behavior emerges from many local interactions among autonomous birds, each bird behaving according to the same simple rules.

Hence, if (as Sober says) science takes no interest in whether some thing is alive over and above whether it performs a certain list of activities, then each of the performances on the list must consist not merely of displaying a given "sign of life," but of doing so *for a certain reason*. But this is just exhibiting the given "sign of life" for the same (perhaps unknown) reason as various living things do. In other words, the activities on *this* list are not presently intelligible independently of the concept of being a living creature; this concept is used in distinguishing a genuine performance of one of these activities from a simulation.

To conclude: Advocates of strong-AL should not feel obliged to offer a *definition* of "life" in order to justify holding that some software entity is alive. In addition, they should recognize that (contrary to Langton (1991, 19f), a passage I quoted earlier) their arguments for the vitality of some artificial "software organism" actually turn not so much on the *number* of "signs of life" it exhibits, as on the *reason* that it exhibits whichever signs it exhibits: for the same reason that paradigmatic living things do. Whether there is such a common reason is for science to discover. While certain capacities (e.g., to photosynthesize) cannot be exhibited by software, no particular sign of life is necessary for vitality. The relation between the "signs of life" and vitality has remained obscure, since all have recognized that the signs are neither individually necessary nor jointly sufficient for vitality. I have proposed a novel rational reconstruction of this relation: according to many scientists, the living things constitute a natural kind and vitality sometimes explains why an entity exhibits various "signs of life." I presently see little reason to believe that it is in principle impossible for a segment of machine code to exhibit various signs of life because it is alive. The view that biologists are interested not in distinguishing life from nonlife, but merely in identifying which "signs of life" various entities exhibit, ignores the obstacles to specifying the "signs" in which biologists are interested without appealing to the concept of "life." And it overlooks the work, evident in various episodes from the history of science, that the concept of "life" has been called upon to perform in scientific explanations.

NOTES

This chapter originally appeared in *Philosophy of Science* 63(2) (1996), 225–244.

REFERENCES

1. Ainsworth, G. C. (1976). *Introduction to the history of mycology.* Cambridge, UK: Cambridge University Press.
2. Brown, R. (1828). A brief account of microscopical observations made in the months of June, July and August 1827, on the particles contained in the pollen of plants. *Philosophical Magazine*, **4**, 161–173.
3. Brown, R. (1829). Additional remarks on active molecules. *Philosophical Magazine*, **6**, 161–166.

4. Curtis, H. (1977). *Invitation to biology* (2nd ed.). New York: Worth.

5. Emmeche, C. (1994). *The garden in the machine: The emerging science of artificial life.* Princeton: Princeton University Press.

6. Farmer, J. D. & Belin, A. d'A. (1991). Artificial life: The coming evolution. In C. G. Langton, C. Taylor, J. D. Farmer, and S. Rasmussen (Eds.), *Artificial life II* (Santa Fe Institute studies in the sciences of complexity, proceedings vol. **X**) (pp. 815–840). Redwood City, CA: Addison-Wesley.

7. Grant, E. (1994). *Planets, stars, and orbs.* Cambridge, UK: Cambridge University Press.

8. Houghton, W. (1885). Notices of fungi in Greek and Latin authors. *Annals and Magazine of Natural History Series 5*, 15, 22–49.

9. Lange, M. (1994). Dispositions and scientific explanation. *Pacific Philosophical Quarterly*, **75**, 108–132.

10. Lange, M. (1996). Inductive confirmation, counterfactual conditionals, and laws of nature. *Philosophical Studies*, **85**(1), 1–36.

11. Langton, C. G. (1986). Studying artificial life with cellular automata. *Physica D*, **22**, 120–149.

12. Langton, C. G. (1989). Artificial life. In Langton, C. G. (Ed.), *Artificial life* (Santa Fe Institute studies in the sciences of complexity, proceedings vol. **IV**) (pp. 1–47). Redwood City, CA: Addison-Wesley.

13. Langton, C. G. (1991). Introduction. In C. G. Langton, C. Taylor, J. D. Fanner, and S. Rasmussen (Eds.), *Artificial life II* (Santa Fe Institute studies in the sciences of complexity, proceedings vol. **X**) (pp. 3–23). Redwood City, CA: Addison-Wesley.

14. Langton, C. G., Taylor, C., Farmer, J. D., & Rasmussen, S. (Eds.) (1991). *Artificial life II* (Santa Fe Institute studies in the sciences of complexity, proceedings vol. **X**). Redwood City, CA: Addison-Wesley.

15. Large, E. C. (1962). *The advance of the fungi.* New York: Dover.

16. Lederberg, J. (1960). Exobiology: Approaches to life beyond the Earth. *Nature*, **132**, 393–400.

17. Levy, S. (1992). *Artificial life.* New York: Vintage.

18. Medawar, P. B. & Medawar, J. S. (1977). *The life science.* New York: Harper & Row.

19. Penrose, L. S. (1959). Self-reproducing machines. *Scientific American*, **200**(6), 105–113.

20. Pirie, N. W. (1938). The meaninglessness of the terms 'life' and 'living.' In J. Needham and D. E. Green (Eds.), *Perspectives in biochemistry* (pp. 11–22). Cambridge, UK: Cambridge University Press.

21. Ramsbottom, J. (1931/1932). The fungus-stone. *Proceedings of the Linnean Society* (London), **144**, 76–79.

22. Ramsbottom, J. (1938/1939). The expanding knowledge of mycology since Linnaeus. *Proceedings of the Linnean Society* (London), **151**, 280–367.

23. Ray, T. (1994). An evolutionary approach to synthetic biology. *Artificial Life*, **1**, 170–209.

24. Reynolds, C. (1987). Flocks, herds, and schools: A distributed behavioral model. *Computer Graphics*, **21**(4), 25–34.

25. Rosenberg, A. (1985). *The structure of biological science.* Cambridge, UK: Cambridge University Press.

26. Scott, A. (1991). *Origen and the life of the stars.* Oxford: Oxford University Press.

27. Sober, E. (1991). Learning from functionalism— Prospects for strong artificial life. In C. G. Langton, C. Taylor, J. D. Farmer, and S. Rasmussen (Eds.), *Artificial life II* (Santa Fe Institute studies in the sciences of complexity, proceedings vol. **X**) (pp. 749–765). Redwood City, CA: Addison-Wesley.

28. Sober, E. (1992). Evolution, population thinking, and essentialism. In M. Ereshefsky (Ed.), *The units of evolution* (pp. 247–278). Cambridge, MA: MIT Press/ Bradford Books.

29. Sober, E. (1993). *Philosophy of biology.* Boulder: Westview.

30. Spafford, E. (1994). Computer viruses as artificial life. *Artificial Life*, **1**, 249–265.

31. Taub, L. (1993). *Ptolemy's universe.* Chicago; La Salle, IL: Open Court.

32. Wolfram, S. (1985). Undecidability and intractability in theoretical physics. *Physical Review Letters*, **54**, 735–738.

ENDNOTES

1 One dissimilarity between "It is living" and "It is copper" is that it is a natural law that all copper bodies are electrically conductive, whereas for many signs of life (e.g., motility) it is not even true—much less a law of nature—that all living things display that sign of life. If Fa explains Ga only if "All F's are G" is physically necessary, then the fact that object *a* is living cannot explain why object *a* displays certain signs of life. But I do not believe that Fa explains Ga only if Fa physically necessitates Ga. The precise relation between natural law and scientific explanation has never been properly explicated. Though I cannot pursue this *topic* here, perhaps it will help

if I express my belief that Fa's explanatory relevance to Ga derives from the truth-values of various counterfactual conditionals (e.g., perhaps from the falsehood of "Had Fa not obtained, then Ga would still have obtained"). Since laws of nature support counterfactuals, the fact that it is a natural law that all F's are G may be responsible for the truth-values of some of these various counterfactuals. But even if it is not physically necessary that all F's are G, it is possible for these various counterfactuals to have the truth-values that enable Fa to explain Ga. So a counterfactual account of scientific explanation could allow the fact that object *a* is *living* to explain why object *a* displays certain signs of life, and in the same way allow the fact that object *b* is copper to explain why object *b* is electrically conductive, even though it is not physically necessary that a living thing display those signs of life but it is physically necessary that a copper body be electrically conductive. For some defense of a counterfactual approach to scientific explanation, see Lange (1994).

2   For a more careful discussion of this standard view, references to its adherents, and an account of why it is correct, see Lange (1994).

3   By these "reasons," I do not mean to include *historical* reasons, e.g., reasons having to do with natural selection. I mean something to do with the *mechanism* by which living things manage to exhibit their signs of life. (I will shortly give an example.) Participation in natural selection is instead commonly regarded as a "sign of life," one that biological viruses exhibit but (as I will shortly explain) computer viruses presently do not.

4   It has often been held that not all confirmation of "All F's are G" is *inductive* confirmation, and that we can confirm "All F's are G" *inductively* only if we believe that this regularity may be physically necessary. But it requires considerable care to express the intuitive difference between inductive confirmation and confirmation simpliciter in precise terms (e.g., in terms of the probability calculus). I do this in Lange (forthcoming), and I show there that we cannot confirm "All F's are G" inductively if we believe that this clam is an accidental generalization if it is true.

5   To justify this claim properly, I would need to offer a conception of scientific explanation that reveals precisely why a dispositional claim is otiose in a scientific explanation of a manifestation of that disposition. I propose such an account in Lange (1994). There I cash out and defend the rough characterization I have just given of the distinction between dispositional and categorical properties.

6   To learn whether some "software organism" displays a certain sign of life, wouldn't we do much better to ask the programmers than to observe whether it displays certain other signs of life? (I am grateful to an anonymous referee for posing this question.) My first response is to note that the sequence of instructions of constituting a "software organism" may not have been designed by a programmer. Rather, it may have resulted from "unnatural selection": a random population of strings of computer code was produced; its members were graded on some criterion; the worst were erased; the best were "mated" to each other, while random mutations and crossings-over were introduced; and the process was repeated on the new population. Sophisticated computer programs have been generated by such a "genetic algorithm." The objection might then become that in order to ascertain whether some "software organism" displays a certain sign of life, we would do much better to examine the sequence of instructions constituting the "organism" than to note whether it displays certain other signs of life. But the same could be said of a biological organism; if we knew enough molecular biology, we could ascertain whether some biological organism exhibits a certain sign of life by examining its genetic endowment. This fact takes nothing away from the confirmatory value of observing some thing to display a given sign of life when we have no access to the instructions or know very little molecular biology. Finally, even if we know the sequence of instructions, it may be that we can establish for certain whether the program exhibits a given sign of life only by running it (or by effectively simulating it) (see, e.g., Wolfram 1985). To confirm the *prediction* that the software entity displays a given sign of life, we may find it useful to observe that it exhibits various other signs of life.

18 · Alien life: how would we know?

MARGARET A. BODEN

INTRODUCTION

To recognize alien life, we would need to know what we mean by *alien*, and what we mean by *life*. Let us take the first thing first. We would see life as alien, I suggest, if it were discovered on a different planet, if it involved a fundamentally different biochemistry on planet Earth or if it was artificially generated by artificial intelligence/artificial life (AI/A-life) research (robots, perhaps?).

In particular, we would see it as alien if it consisted of purely *virtual* organisms: creatures existing only in computer memory, and manifested on the VDU screen. Most talk about life, alien or otherwise, assumes some physical object (some "body")—perhaps microscopically small—existing as a material thing. But this is not applicable to purely virtual organisms. The claim that such cyber-creatures could properly be regarded as *alive* is the claim that "strong A-life" is possible. (Strong A-life is so-called by analogy with strong AI (Searle 1980).) And that claim cannot be assessed without considering what, in general, we would count as *life*.

That is a notoriously difficult question. In this paper, I will concentrate on one of the commonly listed criteria of life: metabolism. As we shall see, metabolism (in the sense used by biologists) is a form of biochemical fine-tuning. It characterizes all known life, as a matter of fact—and, I will argue, as a matter of necessity too. If that is right, then something that does not metabolize cannot properly be regarded as being alive. So if virtual creatures are not fine-tuned in this sort of way, they cannot be counted as *living* things. Whatever marvels the realms of alien life may hold, cyber-organisms are not among them.

To address these questions, we must consider carefully just what is meant by "metabolism". The first thing to note is that metabolism concerns *the role of matter/energy in organisms considered as physically existing things*. It is not an abstract functionalist concept, divorced from the specific material realities. By contrast, the other features typically mentioned in definitions of life—self-organization, emergence, autonomy, growth, development, reproduction, adaptation, responsiveness and (sometimes) evolution—can arguably be glossed in functionalist, informational terms.

The core concept of self-organization, for example, involves the emergence (and maintenance) of order, out of an origin that is ordered to a lesser degree. It concerns not mere superficial change, but fundamental structural development. The development is spontaneous, or autonomous, in that it results from the intrinsic character of the system (often in interaction with the environment), rather than being imposed on it by some external force or designer.

Similarly abstract definitions can be given of the other items on the list. Thus, emergence is the appearance of novel properties that seem (at least at first sight) to be inexplicable in terms of earlier stages or lower-level components. Growth is increase in quantity; development is autonomous structural change leading to a higher degree of order; adaptation is an improved response to the environment by means of structural and/or behavioral change (which may be heritable); reproduction is self-copying; and evolution is adaptive change by means of reproduction, heredity, variation, and selection.

It is because no comparable definition can be given for metabolism that the concept is problematic for strong A-life. A-life in general is a functionalist enterprise; that is, A-life researchers typically think of vital phenomena in terms of information and computation, not matter or energy. For example, John von Neumann defined the general requirements of reproduction in logical–computational terms, and pointed out that copying errors (an informational notion) could result in adaptive evolution (Burks 1966, 1970). Similarly, in the "Call for Papers" for the first conference identifying "Artificial Life" as a unitary project, Christopher Langton said: "The ultimate goal of A-life is to extract the logical form of living systems" (Levy 1992, p. 113).

Of course, none of these A-life researchers doubts that living things are material entities of some sort. In other words, life is not pure information. Langton makes this explicit in his statement that life is "a property of the organization of matter, rather than a property of the matter which is so organized" (Langton 1989, p. 2). So far, then, the question "Are matter and energy essential to life?" seems to be answered with a guarded "yes". Some matter is organized, somehow. But the nature of the material stuff is philosophically irrelevant to the status of the physical system as a living thing. It could, for example, be silicon. And nothing can (or need) be said about the general type of physicochemical processes that must be going on, except that they are organized in the relevant ways.

However, Langton (1986) also says: "The ultimate goal of the study of artificial life would be to create 'life' in some other medium, ideally [sic] a virtual medium where the essence of life has been abstracted from the details of its implementation in any particular model".

Such "life" would inhabit cyber-space, a virtual world of informational processes grounded in computers. The virtual creatures would be defined in purely informational terms, as strings of bits or computer instructions. But their activity (the execution of the instructions)—without which, they could not be regarded even as candidates for life—would require the computer. So, like biological creatures, they would have some physical existence: namely, the material ground, in computer memory, of the relevant information processing. The "matter which is organized" would be the stuff of which the relevant computers are constructed—which might be almost anything. As Langton's word "ideally" makes clear, the molecules and physicochemical processes involved would be of no concern to the A-life functionalist. The only interesting properties of the virtual creatures, *qua* living things, would be abstract, informational ones.

This claim of Langton's is disputed even by many A-life researchers, so his "ultimate goal" cannot be ascribed to A-life work in general. (It follows that A-life as a whole could not be dismissed merely because one rejected strong A-life; similar remarks apply to weak and strong AI, as Searle allows.) Nevertheless, Langton is not alone in making such claims.

One of the A-life researchers who agree with him is Thomas Ray, an ecologist specializing in tropical forests. Indeed, Ray goes even further than Langton: he believes he has already implemented primitive forms of real—albeit virtual—life. His computer models of co-evolution in the virtual world "Tierra" have led to the foundation of the "Digital Reserve" (Ray 1992, 1994). This is a virtual memory space spread across a worldwide network of computers, which allow their spare capacity to be used at idle times. Tierra is one example (another is described in the section on strong metabolism and strong A-life) of A-life work described by its proponent as the creation of actual, if primitive, life-forms.

The creatures (Ray's word) inhabiting the Digital Reserve, like those within Tierra itself, are strings of self-replicating computer code. They can mate (exchange genetic instructions), mutate, compete and evolve. For example, some code-strings evolve which lack the instructions responsible for self-replication, but which can "parasitize" the code of other creatures in order to replicate themselves. This is a successful evolutionary strategy because fitness is defined in terms of access to computer memory—and a "species" with shorter strings can fit more individuals into a given memory space.

The creatures let loose in the Digital Reserve move from one computer to another in their search for unused memory space. (Because they are implemented in a virtual computer, simulated by some actual computer, the software creatures cannot "escape" into computers not on the Reserve network, nor infest the everyday workings of those that are included.) Ray insists that the Digital Reserve is an experiment in the creation of new forms of life.

Those, like Langton and Ray, who regard strong A-life as a real possibility defend their counterintuitive view by making two interconnected claims. First, that the virtuality is limited: computers, after all, are material things, and need energy in order to function. Secondly, that the criteria for life are essentially abstract, or functionalist, saying nothing whatever about the nature of its (admittedly necessary) material grounding. To show that they are mistaken, one must show that at least one of these claims is false.

Since the first claim is indisputable, the focus falls on the second. I suggested, above, that all but one of the items on the typical list of vital properties can indeed be viewed as abstract, informational concepts. The one obvious exception is metabolism. The proponent of strong A-life must therefore show that virtual systems can genuinely metabolize. (The alternative strategy—dropping metabolism from the list of vital criteria—is discussed, and rejected, in a later section.)

In the next section, I distinguish three senses of metabolism: The first two (weaker) senses are found in the arguments of some proponents of strong A-life, for on each of these interpretations some A-life artifacts would count as genuinely alive. The third, strongest, sense is not. It is drawn rather from biology, and posits a form of bodily identity which (I shall argue in the section on strong metabolism and strong A-life) is not attained by virtual creatures.

Irrespective of questions concerning A-life, the strong sense of metabolism is more interesting than is sometimes thought. Besides referring to the biochemical processes (whatever they may be) that maintain the growth and function of an organism, it denotes various general properties that those processes *must necessarily* possess. Life on Mars or Alpha Centauri, then, would have those properties too.

THREE CONCEPTS OF METABOLISM

What, exactly, is metabolism? It locates life in the physical world (no angels on pinheads). But it does not denote mere materiality. A volcano is a material thing, and so is a grain of sand, but neither of these metabolizes. Rather, metabolism—in the minimal sense of the term—denotes energy dependency, as a condition for the existence and persistence of the living thing.

If energy dependency were all there was to it, then strong A-life would be possible. For, as both Langton and Ray are quick to point out, virtual life satisfies this criterion. Strong A-life is utterly dependent on energy. Electrical power is needed to execute the information processes that define "this" creature, or "that" one. Pull the plugs on the computers, stop the electrons inside from jumping, and cyber-space is not merely

emptied, but destroyed. Strong A-life, having once existed, would have died.

However, "metabolism" is normally used to mean more than mere energy dependency. Two further senses of the term can be distinguished, each associated with notions of using, collecting, spending, storing and budgeting energy. These activities—a form of "fine-tuning"—are characteristic of life. (Active volcanoes involve huge amounts of energy, without which they would not exist. But they do not use it, collect it, store it or even spend it, except in a weakly metaphorical sense—and they certainly do not budget it.)

A second (stronger) sense of metabolism supplements mere energy dependency with the idea of individual energy packets used to power the activities of the creature, its physical existence being taken for granted. Each living system has assigned to it, or collects for itself, a finite amount of energy. This is used up as it engages in its various activities. When the individual's energy is spent, either because it is no longer available in the environment or because the system can no longer collect or use it, the energy-dependent behavior must cease and the creature dies.

Some very early efforts in A-life (around mid century) already involved the idea—and the reality—of individual energy packets. Grey Walter's (1950, 1951) mechanical "tortoises," Elmer and Elsie, were simple robots that used their energy to engage in physical behavior. They moved around the floor by means of electric power, every so often abandoning their current activity in order to recharge their batteries. The second definition of metabolism would also cover those more recent A-life robots that are broadly comparable to Grey Walter's tortoises, some of which even have distinct energy stores devoted to different types of activity. Such robots could, therefore, be termed alive insofar as this (second-sense) criterion is concerned.

But A-life robots are not germane to the question of whether strong A-life is possible. For "strong A-life" does not refer indiscriminately to just any A-life artifacts, including robots and physical systems grounded in exotic biochemistries. Rather, it refers to virtual creatures inhabiting virtual worlds.

As remarked above, virtual creatures exist only in computer memory, manifested to the observer on the VDU screen. They "exist" in the sense that they consist of a particular (perhaps continuously varying) distribution of electric charges at various (perhaps widely scattered) locations inside the machine (these locations may change, as the relevant instructions are swapped from one part of the machine to another for execution or storage). In this sense, then, they may be said to have physical existence. But that is not to say that they have bodies (see below). Nor is it to say (which is required for the second sense of metabolism) that they store and budget real energy so as to engage in their activities and to continue their physical existence.

Many virtual creatures are intended by their human creators as computer simulations of real life. That is, their manifest behavior on the VDU screen (caused by the underlying electronic processes in computer memory) has some systematic relation to, or isomorphism with, certain features of living organisms. And some of these model metabolism (understood in the second sense), at least in a crude manner.

So examples abound of programs that simulate individual animals with distinct energy levels, raised by eating and rest, and reduced by activities such as food-seeking, fighting and mating. A few of these even assign different sub-packets of energy to various drives, so that at a particular time a creature might have energy available to mate, but not to fight. (For a very early example, where a simulated rat has to choose between seeking warmth and food, see Doran (1968).)

However, the "packets" and "sub-packets" here are not actual identifiable energy sources or energy stores, but mere simulations of these. At any given time, the program may dictate that the creature will seek food, but this merely means that some numerical variable has fallen below a threshold value, so triggering the food-seeking instructions. Certainly, energy is needed to execute the instructions. But this comes, via electric plug or battery, from the general undiscriminated energy source on which the whole program is passively dependent. If the program simulates more than one creature, this energy source is equally available to all, given the relevant program instructions. Metabolism in the first sense is achieved, but in the second sense it is merely modeled.

Suppose that separate energy sources, distinct real energy packets, were to be supplied (in the computer) for each simulated creature. What then? The second sense of metabolism would have been satisfied. If our concept of life involved this sense of the term, strong A-life would be conceivable.

However, one must note two important features of the second definition, as given above. First, it speaks of the creature's "physical existence," not of the creature's "body"—nor even of its being a "unitary" physical system. Second, and crucially, it speaks of that physical existence being taken for granted.

Clearly, then, the second sense of metabolism is not the biologist's concept of it. For no biologist ignores the fact that an organism's physical existence is an integrated material system, or body. (Apparent exceptions include slime moulds, within whose life-cycle the multicellular organism splits into many unicellular "amoebae," which later coalesce into a multicellular creature. But at every point, even at the amoebic stage, there is one or more integrated material system. Whether one chooses to call a reconstituted multicellular structure the "same" organism, or body, is not important here.)

Furthermore, no biologist takes the existence of a creature's body for granted. On the contrary, one of the prime puzzles of biology is to explain how living bodies come into existence, and how they are maintained until the organism dies. We therefore need a third, still stronger, definition of metabolism if we are to capture what biologists normally mean by the term.

The third sense of metabolism refers to the use, and budgeting, of energy for bodily construction and maintenance, as well as for behavior. Metabolism, in other words, is more than mere material self-organization. That occurs (for instance) in the Belousov–Zhabotinsky reaction, where mixing two liquids results in the spontaneous emergence of order (visible whorls and circles) – but no-one would speak of life here: too many of the other vital properties listed at the start of this chapter are missing.

Rather, metabolism is a type of material self-organization which, unlike the Belousov–Zhabotinsky reaction, involves the autonomous use of matter and energy in building, growing, developing and maintaining the bodily fabric of a living thing. (For present purposes, we may apply the term "body" to plants as well as animals.)

The matter is needed as the stuff of which the body is made. And the energy is needed to organize this matter, and new matter appropriated during the lifetime, into something that persists in its existence despite changes in external conditions. Metabolism, in this strong sense, both generates and maintains the distinction between the physical matter of the individual organism and that of other things, whether living or not.

Metabolism in this third sense *necessarily* involves closely interlocking biochemical processes. In other words, it necessarily involves a form of "fine-tuning".

There are several (inter-related) reasons for this. A multicellular organism must, and a unicellular organism may, sometimes grow. (I take it that multicellular organisms start off as unicellular ones; normally, this is a single spore or fertilized egg, but multicellular slime moulds "grow" by the aggregation of many unicellular creatures.) And even a unicellular organism must (sometimes) repair damage. Since living matter cannot be created from nothing, growth and repair require that new molecules be synthesized by the organism—which molecules themselves make up the organism. Moreover, the living system (subject, like every physical thing, to the second law of thermodynamics) continuously tends to disorder and the dissipation of energy. Hence metabolism must involve continual energy intake from the environment.

The simplest conceivable living things might take their energy directly from the environment whenever they needed it. (They would satisfy only the first sense of metabolism, not the third.) Perhaps the very earliest organisms actually did this. But this would leave them vulnerable to situations in which no "new" energy was immediately available. (Analogously, computers that rely on the plug in the wall are vulnerable to power cuts.) If, by chance, the organism became able to store even small amounts of excess energy for later use, its viability—and Darwinian fitness—would be enormously increased. Once evolution got started, this fact would be reflected in the evolution of metabolism.

Inevitably, then, all metabolic systems (other than the very earliest, perhaps) must not only exchange energy with the outside world but also perform internal energy budgeting. Excess energy is stored, so that reliance on direct energy collection is avoided. If (which is likely) the inputted energy cannot be conveniently stored in its initial form, it must be changed into some other form. In other words, living organisms must convert external energy into some substance ("currency") that can be used to provide energy for any of the many different processes going on inside the organism. This is "the first fundamental law of bio-energetics".

Apparently, only three convertible energy currencies—one of which is ATP, or adenosine tri-phosphate—are used by terrestrial life (Moran *et al.* 1997, paragraph 4.2). This *may* be an evolutionary accident, ATP having turned up (due to mutation) early in the game and having been retained ever since. In other words, it is not known (so my informants tell me) whether some other substance *could*, in principle, fulfil the same role. If not, then "alien" life on Mars would be ATP-based too. If so, then perhaps it would not. It would show metabolic fine-tuning, to be sure. But the basic chemicals playing the core biophysical tunes would be different.

Additional, purely internal, energy exchanges are required as the collected energy is first converted into substances suitable for storage and then, on the breakdown of those substances, released for use. Very likely, these processes will produce waste materials, which have to be neutralized and/or excreted by still other processes. In short, metabolism necessarily involves a nice equilibrium between anabolism and catabolism, requiring a complex biochemistry to effect these vital functions.

Bodily maintenance is normally continuous. But the underlying metabolic processes are more active at some times—of the day, year and life-cycle—than at others. Sometimes, they are drastically slowed down, or (perhaps) even temporarily suspended. In hibernating animals, for instance, metabolism is kept to a minimum: respiration and excretion occur at a very low rate. Even in the case of seeds or spores frozen, or entombed, for centuries, some minimal metabolic activity may have been going on.

But what if it has not? It is not clear that this strong concept of metabolism assumes that active self-maintenance must be absolutely continuous, allowing no interruptions whatsoever. If biochemical research were to show that metabolism is occasionally interrupted, in highly abnormal conditions (such as freezing), so be it. Indeed, we already speak of "suspended animation": a spore may be currently inactive, but if it retains the potential to metabolize in suitable conditions we do not regard it as "dead".

STRONG METABOLISM AND STRONG A-LIFE

The previous section showed that the first sense of metabolism is satisfied by all A-life systems, and that the second could conceivably be satisfied by certain types of A-life simulation. But what of the third, strongest, sense? Could this be found in any A-life creatures, so allowing us to regard them as living things? If so, would these creatures necessarily be robots, or could they also include virtual life?

A-life robots as currently envisaged do not fit the bill. These are typically "situated" robots, engineered (or evolved) to respond directly to environmental cues. Some do not look at all life-like (Cliff *et al.* 1993). Others resemble insects in their physical form, and may have control systems closely modeled on insect neuroanatomy (Brooks 1986, 1991; Beer 1990).

Certainly, such robots are in a significant sense autonomous, especially if they have been automatically evolved over many thousands of generations (Boden 2000a). And they undoubtedly consume real energy as they make their way around their physical environment. Unlike classical robots, they are embedded in the world, in the sense that they react directly to it rather than by means of a complex internal world model. But being embedded does not necessitate being (truly) embodied. Earlier I argued that a body is not a mere lump of matter, but the physical aspect of a living system, created and maintained as a functional unity by an autonomous metabolism. If that is right, then these robots do not have bodies.

Conceivably, some future A-life robots might be self-regulating material systems, based on some familiar or exotic biochemistry. Just how exotic their biochemistry might be is unclear, however.

In principle, it need not even be carbon based. It may be the case, however, that carbon is the only element capable of forming the wide range of stable yet complex molecular structures that seem to be necessary for life. And Eric Drexler (1989) has argued that even utterly alien (non-carbon) biochemistries would have to share certain *relational* properties with ours. They would have to employ general diffusion, not channels devoted to specific molecules; molecular shape-matching, not assembly by precise positioning; topological, not geometric, structures; and adaptive, not inert, components. In effect, Drexler is offering a functionalist characterization of biochemistry (the chemistry of metabolism), which can perhaps be instantiated in many different ways. Metabolism has also been characterized in even more abstract, thermodynamic, terms (Moreno & Ruiz-Mirazo 1999).

Whatever the details, artefacts grounded in exotic biochemistries might well merit the ascription of life: not strong A-life confined to cyber-space, but real, metabolizing, life. There is nothing in A-life at present that promises such alien creatures. (It is conceivable, however, that human biochemists have already created artificial life-forms—though not robots—without realizing it, by unwittingly "creating the conditions under which [metabolizing systems] form themselves" (Zeleny 1977, p. 27).) In any event, such artifacts are irrelevant to our main question. If novel robots and biochemistries were to be engineered or artificially evolved, they would count as *successful* A-life rather than *strong* A-life. The question thus remains as to whether the third sense of metabolism rules out strong A-life.

Metabolism in this strong sense, as we have seen, involves material embodiment—embodiment, not mere physical existence. It also requires a complex equilibrium of biochemical processes of certain definable types. It cannot be adequately modelled by a system freely helping itself to electricity by plug or battery, or even by assigning notional "parcels" of computer power to distinct functions within the program. Virtual creatures might have individual energy packets, and some form of energy budgeting, but these would be pale simulations of the real thing. Even "biochemical" A-life models are excluded from the realm of the living, if they are confined to cyber-space.

This forbids us from regarding as truly living things a "species" of A-life that has recently attracted considerable attention—and whose main designer Steve Grand insists that its virtual denizens are primitive forms of life (Grand, personal communication). I am thinking of the cyber-beings conjured up by running "Creatures," a computer game, or more accurately a computer world, built by the use of A-life techniques (Grand *et al.* 1996). It is a far richer virtual world than that of other computerized "pets"—such as "Dogz," "Catz" and the electronic Tamagochi chick that the user must rest, exercise and clean. What is of special interest here is that Creatures includes a (crude) model of metabolism, as well as of behavior.

The human user of Creatures can hatch, nurture, aid, teach and evolve apparently cuddly little VDU creatures called norns. Up to ten norns can co-exist in the virtual world (future increases in computer power will make larger populations possible), but even one solitary individual will keep the person quite busy.

One of the user's tasks is to ensure that all the norns can find food when they are hungry, and to help them learn to eat the right food and avoid poisons. Another is to teach them to respond to simple linguistic inputs (proper names, categories, and commands), with different norns receiving different lessons. Yet another is to help them learn to cooperate in various simple ways. In addition, the user must protect them—and teach them to protect themselves—from grendels, predatory creatures also present in the virtual world. The human can evolve new norns likely to combine preferred features of appearance and behavior, since mating two individuals results in (random) recombinations of their "genes".

A norn's genes determine its outward appearance and the initial state of its unique neural-network "brain" (at birth, 1000 neurones and 5000 synapses), with the specific connection weights changing with the individual's experience. The genes also determine its idiosyncratic "metabolism". Each creature's behavior is significantly influenced by its (simulated) biochemistry. This models global features such as widespread information flow in the brain, hormonal modulations within the body, the norn's basic metabolism, and the state of its immune system.

The virtual biochemistry is defined in terms of four types of biochemical object. First, there are 255 different "chemicals," each of which can be present in differing concentrations. (These are not identified with specific biochemical molecules: the functions of the 255 substances are assigned randomly.) Second, various biochemical "reactions" are represented. These include fusion, transformation, exponential decay and catalysis (of transformation and of breakdown). Third and fourth, there are a number of "emitter" and "receptor" chemicals, representing various processes in the brain and body (for example, activity in the sense organs). Taken together, these (abstract, functional) biochemical categories are used to build feedback paths modelling phenomena such as reinforcement learning, drive reduction, synaptic atrophy, glucose metabolism, toxins and the production of antibodies.

This general architecture offers significant potential for theoretically interesting advances in A-life modeling. Its largely untapped complexity, including

its ability to model global features of information processing, makes it a promising test-bed. It could be developed, for example, by incorporating recent AI ideas on the computational architecture underlying motivation and emotion (Sloman 1990; Beaudoin 1994; Wright et al. 1996), which as yet have been modeled only in very preliminary ways (Wright 1997).

Even now, without such additions, Creatures is undeniably seductive. All but the most hard-headed of users spontaneously address the norns as though they were alive, and some mourn the demise of individuals (each of whose "life history" is unique) despite being able to hatch others at the touch of a button.

For all that, Creatures is a simulation of life, not a realization of it. There is no actual glucose, and no actual chemical transformation; the system is not even a chemically plausible model of specific molecular processes. Moreover, the simulated metabolism is concerned with controlling the norns' behavior, not with building or maintaining its bodily fabric. (Still less does it regulate the VDU creature's underlying, electronic, physical existence.)

Admittedly, the "foods" and "poisons" are associated with simulated metabolites and metabolic processes. At present, however, these affect the norns' behavioral, not bodily, integrity. They do not froth at the mouth when ingesting poison; and they do not have "hearts" that stop beating, or "flesh" that rots without oxygen. Certainly, some future development of Creatures might include a much richer metabolic simulation. The user might even be able to help a favourite norn to acquire a suntan, or to feed and exercise so as to develop its "biceps". Nevertheless, there would be no real metabolism, no real body—and no real life.

What if the "foods" were to be associated with real energy, which was used only to run the electronic processes underlying the VDU manifestation of the individual norn?

This would be an example of the type of A-life system discussed above (in relation to the second sense of metabolism), in which the creature's continuing physical existence depends upon its being able to commandeer specific packets of real energy. In such a case, since the norns can evolve, they might even evolve new ways of attracting real energy and of using it (for instance) to repair their electronic grounding when damaged. Nevertheless, the points remarked above still stand: this imaginary scenario concerns the creature's physical existence, not its metabolically integrated body, and it takes that physical existence for granted. The construction of the computer, and of the parts/ processes within it that constitute the norn's material being, was effected by artificial construction, not by autonomous metabolism.

In short, if we regard metabolism (in the third, biological, sense) as—literally—vital, we must reject the claim that norns, and their cyber-cousins, are simple forms of life. Even energy-gobbling and self-repairing norns, evolved without human direction, would not metabolize in this strong sense.

CAN WE DROP METABOLISM?

Someone might suggest at this point that we adopt a weaker sense of metabolism when defining life, or that we drop the criterion of metabolism altogether. In that event, some of the virtual artefacts envisaged by Langton, Ray or Grand *could* properly be regarded as alive. Such suggestions cannot be instantly dismissed as absurd. For one cannot define life, define metabolism and conclude that strong A-life is—or is not—possible in a way that will immediately convince everyone. In contrast, the concept of life is negotiable.

There are two reasons for this. First, there is no universally agreed definition of life. It is not even obvious that what one should do, in this situation, is to *try* to justify (*a priori*) a list of necessary and sufficient conditions, since our everyday concept may not name what philosophers call a "natural kind". I noted one example of definitional disagreement in the introduction, where I remarked that evolution is "sometimes"—which is to say, not always—added to the typical list of vital properties. Indeed, it is regarded as "the" fundamental criterion by many biologists, and by some philosophers—such as Mark Bedau (1996). But there are problems lurking here.

Taking evolution (or, in Bedau's terminology, "supple adaptation") to be essential has several philosophical difficulties, as Bedau himself admits. One is that creationist biology becomes logically incoherent, not just empirically false. Another is that evolving populations, rather than individual organisms, must be taken as the paradigm case of life. This conflicts with ordinary usage, where we say that the lion—not the lion lineage—is "alive". It also sits uneasily with

the concept of metabolism: we saw earlier that even the weakest sense of this term is defined with reference to the physical maintenance of individual things. (By the same token, including metabolism in the list of vital criteria underscores our usual assumption that individual organisms are paradigms of life.) Nevertheless, Bedau argues that evolution is so important in theoretical biology that it should be regarded as the very essence of life. Others, by contrast, argue that evolution—and reproduction, too—is merely a secondary feature of life, and that one can envisage living things incapable of either (see below). Secondly, the concept is negotiable because *even if* everyone today defined "life" in the same way, they might tomorrow have good reason for defining it differently. Scientific discoveries might lead to an (*a posteriori*) theoretical identification of the real essence of life, and hence to a change in the way that non-scientists use the term.

The suggestion that evolution be taken as essential, for example, dates from the development of modern biology. Before Darwin's theoretical work, perhaps even before the twentieth-century synthesis of Darwinism and genetics, it would have been unreasonable to propose this—*even though* many of his predecessors believed that, as a matter of fact (not explanatory theory), living things had somehow evolved. Again, one of the research aims of A-life is to study "life as it could be," not merely "life as we know it" (Langton 1989, p. 2). This might eventually lead to a different, more inclusive, definition. Indeed, one new "essential" vital property has already been suggested: Langton (1990, 1992) conjectures that all living things satisfy a narrow range of numerical values of the "lambda parameter," a simple statistical measure of the degree of order and novelty in a system. It is not obvious that this sort of discovery is impossible. In short, with scientific advance, the list of vital properties can change.

It might appear, then, that the possibility of strong A-life hangs on mere definitional fiat. Given that there are several senses of metabolism, why not simply choose the weakest, or the strongest, so as to allow or disallow strong A-life respectively? More radically, why not drop metabolism entirely? If we can consider adding evolution, surely we can consider dropping metabolism? We could retain a commitment to physicalism. And metabolism would still be recognized as a universal characteristic of the sort of (biological) life we happen to know about. But it would no longer be seen as essential.

To see the situation in this way is to confuse fiat with negotiation. I said, above, that the concept of life is negotiable, not that it can be defined just anyhow. Both scientific and philosophical judgment must be involved in favouring one definition rather than another. And both types of judgment imply that to drop metabolism from the concept of life would not be a sensible move. That is, the analogy we are asked to draw here—between adding evolution and dropping metabolism—is too weak to be persuasive.

There are strong scientific reasons for adding evolution to the definition of life, even for making it the most fundamental criterion. Specifically, evolutionary theory has enormous explanatory and integrative power, interconnecting all (or most) biological phenomena. Even in molecular biology and genetics, evolutionary explanations provide many insights. And most biologists who resist the reductionist approach of molecular biology, taking the form of whole organs and organisms as their explanandum, see it as not merely universal, but fundamental.

A minority do not. For instance, Brian Goodwin (Goodwin 1990; Webster & Goodwin 1996, part 2) and Stuart Kauffman (1992) argue that biological self-organization is a more fundamental explanatory concept than evolution—and that the two processes can sometimes pull in different directions (see also Wheeler 1997). But even these theoretical mavericks allow that Darwinian evolution selects, and so (superficially) shapes, the range of living things that survive, given the (deeper, wider) potentialities afforded by self-organization. In short, all serious biologists—I do not include creationists—acknowledge that evolution has considerable explanatory force. This is why Bedau is willing to accept the admittedly counter-intuitive implications of taking evolution to be necessary.

That is not to say that everyone will judge the strong reasons for adding evolution to the definition to be strong enough. In particular, those who stress metabolism as a criterion are likely to insist that we should continue to take individual creatures, not evolved species, as the paradigm of life.

Consider, for example, the argument of the biologists Humberto Maturana and Francisco Varela (1980, pp. 105–107). Their definition of life as "autopoiesis in the physical space" is broadly equivalent to the third

sense of metabolism defined previously (broadly, but not exactly: see Boden 2000b). They remark that the concept of evolution *logically presupposes* the existence of some identifiable unity—that is, of a living thing self-generated and self-sustained by autopoiesis. Evolution, therefore, cannot be a defining criterion of life.

Their refusal to regard evolution as essential is not a merely semantic point, following trivially from their preferred definition of life. Rather, it is a biological hypothesis. They point out that a living, self-organizing, cell could conceivably be incapable of reproduction. Even if it could be split (either accidentally or autonomously) into two autopoietic halves, there might be no self-copying involved. Self-copying requires some relation of particulate heredity between the mother and daughter systems. Furthermore, without such (digital) heredity, there can be no evolution (Maynard Smith 1996, p. 117). So the first living things might not have been capable of evolution.

My own view is that to regard evolution as an essential criterion of life is unwise. For the reasons I have outlined above, it would be better regarded as a universal characteristic, though admittedly one offering enormous explanatory power. Because of this explanatory power, it is not surprising that many biologists take evolution to be a defining property. But this definition, interpreted strictly, generates too many counterintuitive—and biologically paradoxical—implications. That is, I do not find Bedau's arguments compelling. Even so, one must allow that he and others like him have a respectable case to make.

The same cannot be said of someone who proposes to *drop* metabolism as a defining criterion of life. There is no persuasive argument for rejecting our intuitions about its necessity. We have just seen that metabolism is even more fundamental than evolution, since non-reproducing organisms are conceivable and may once have lived. And the second section showed that metabolism, in the third sense, is essential for self-organizing bodily creatures that take in energy from their environment. Or rather, it is essential if that energy is not always immediately available, and it is useful if the energy is not always immediately needed. As for explanatory power, metabolism provides this. Biochemists have identified a host of specific molecular reactions involving general types of metabolic relation (such as breakdown and catalysis), and satisfying general principles concerning the storage and budgeting of energy (the "laws of bio-energetics" mentioned when discussing concepts of metabolism).

In short, scientific advance in biology and bio-chemistry reinforces our everyday assumption that metabolism is crucial, while also enriching the concept considerably.

To outweigh this combination of scientific theory and everyday usage, powerful countervailing considerations would be needed. But none exists. The only reason for proposing that we drop metabolism from our concept of life is to allow a strictly functionalist-informational account of life in general, and A-life in particular. The same applies in respect of suggestions that we weaken the notion of metabolism, abandoning the third interpretation and substituting mere energy dependency (with or without individual energy packets). The only purpose of this recommendation is to allow virtual beings, which have physical existence but no body, to count as life. These question-begging proposals have no independent grounds to buttress them.

Significantly, it is even difficult to imagine what such independent grounds could be like. Perhaps some future science might discover strange wispy clouds, distributed over a large space yet somehow identifiable as (one or more) unitary individuals, and having causal properties analogous to those of living things—but lacking metabolism? In that case, we would have to think again. The concept of life remains negotiable, after all.

However, this futuristic scenario is well-nigh unintelligible. What are these "causal properties analogous to those of living things" that do not require bodily unity? And how, in the absence of metabolism, could the clouds satisfy any self-organizing principle of living unity? The fact that science fiction writers (including the cosmologist Fred Hoyle) have sometimes asked us to consider such ideas does not prove that, carefully considered, they make sense.

Similar remarks apply to the speculative idea of a "cosmic computer" (or "computers") distributed across the atmosphere, supposedly supporting information-processes that "evolve and adapt" much as Ray's virtual creatures do. Many philosophers argue that life is a necessary ground of cognition. If that is so, then nothing can be regarded as intelligent which is not also alive. And if life requires some metabolizing bodily unity, then the "cosmic computer" is irredeemably suspect.

The argument of this chapter suggests that such science-fictional ideas are not just implausible, but irredeemably incoherent. Without independent grounds for doing so, we should not drop metabolism from the concept of life. Nor should we weaken our (third) interpretation of it. On the contrary, we should acknowledge it as a fundamental requisite of the sort of fine-tuned self-organization that is characteristic of—indeed, necessary for—life.

In summary, metabolism is necessary, so strong A-life is impossible. Alien organisms, whatever their habitats may be, do not exist in cyber-space.

NOTES

This chapter originally appeared in *International Journal of Astrobiology* 2(2) (2003), 121–129.

REFERENCES

1. Beaudoin, L. P. (1994). Goal-processing in autonomous agents. Ph.D. thesis, School of Computer Science, University of Birmingham, Birmingham.
2. Bedau, M. A. (1996). The nature of life. In M. A. Boden (Ed.), *The philosophy of artificial life* (pp. 332–357). Oxford: Oxford University Press.
3. Beer, R. D. (1990). *Intelligence as adaptive behavior: An experiment in computational neuroethology.* New York: Academic Press.
4. Boden, M. A. (2000a). Life and cognition. In J. Branquinho (Ed.), *The foundations of cognitive science at the end of the century* (pp. 11–22). Oxford: Oxford University Press.
5. Boden, M. A. (2000b). Autopoiesis and life. *Cognitive Science Quarterly*, **1**, 115–143.
6. Brooks, R. A. (1991). Intelligence without representation. *Artificial Intelligence*, **47**, 139–159.
7. Burks, A. W. (1966). *Theory of self-reproducing automata.* Urbana, IL: University of Illinois Press.
8. Burks, A. W. (1970). *Essays on cellular automata.* Urbana, IL: University of Illinois Press.
9. Cliff, D., Harvey, I., & Husbands, P. (1993). Explorations in evolutionary robotics. *Adaptive Behavior*, **2**, 71–108.
10. Doran, J. E. (1968). Experiments with a pleasure-seeking automaton. In D. Michie (Ed.), *Machine intelligence III* (pp. 195–216). Edinburgh: Edinburgh University Press.
11. Drexler, K. E. (1989). Biological and nanomechanical systems: Contrasts in evolutionary complexity.

In C. G. Langton (Ed.), *Artificial life* (Santa Fe Institute studies in the sciences of complexity, proceedings vol. **IV**) (pp. 501–519). Redwood City, CA: Addison-Wesley.

12. Goodwin, B. C. (1990). Structuralism in biology. *Science Progress*, **74**, 227–244.

13. Grand, S., Cliff, D., & Malhotra, A. (1996). *Creatures: Artificial life autonomous software agents for home entertainment*. Research report CSRP 434. Brighton: University of Sussex School of Cognitive and Computing Sciences. Available at ftp://ftp.cogs.susx.ac.Uk/pub/reports/csrp/csrp434.ps.Z (accessed August 2008).

14. Grey Walter, W. (1950). An imitation of life. *Scientific American*, **182**, 42–45.

15. Grey Walter, W. (1951). A machine that learns. *Scientific American*, **185**, 60–63.

16. Kauffman, S. A. (1992). *The origins of order: Self-organization and selection in evolution*. Oxford: Oxford University Press.

17. Langton, C. G. (1986). Studying artificial life with cellular automata. *Physica D*, **22**, 1120–1149.

18. Langton, C. G. (1989). Artificial life. In C. G. Langton (Ed.), *Artificial life* (Santa Fe Institute studies in the sciences of complexity, proceedings vol. **IV**) (pp. 1–47). Redwood City, CA: Addison-Wesley. Reprinted, with revisions, in M. A. Boden (Ed.), *The philosophy of artificial life* (pp. 39–94). Oxford: Oxford University Press.

19. Langton, C. G. (1990). Computation at the edge of chaos: Phase-transitions and emergent computation. *Physica D*, **42**, 12–37.

20. Langton, C. G. (1992). Life at the edge of chaos. In C. G. Langton, C. Taylor, J. D. Farmer, and S. Rasmussen (Eds.), *Artificial life II* (Santa Fe Institute studies in the sciences of complexity, proceedings vol. **X**) (pp. 41–91). Redwood City, CA: Addison-Wesley.

21. Levy, S. (1992). *Artificial life: The quest for a new creation*. New York: Pantheon.

22. Maturana, H. R. & Varela, F. J. (1980). *Autopoiesis and cognition: The realization of the living*. London: Reidel.

23. Maynard Smith, J. (1996). Evolution: Natural and artificial. In M. A. Boden (Ed.), *The philosophy of artificial life* (pp. 173–178). Oxford: Oxford University Press.

24. Moran, F., Moreno, A., Montero, F., & Minch, E. (1997). Further steps towards a realistic description of the essence of life. In C. G. Langton and T. Shimohara (Eds.), *Artificial life V: Proceedings of the 5th international workshop on the synthesis and simulation of living systems* (pp. 255–263). Cambridge, MA: MIT Press.

25. Moreno, A. & Ruiz-Mirazo, K. (1999). Metabolism and the problem of its universalization. *BioSystems*, **49**, 45–61.

26. Ray, T. S. (1992). An approach to the synthesis of life. In C. G. Langton, C Taylor, J. D. Farmer, & S. Rasmussen (Eds.), *Artificial life II* (Santa Fe Institute studies in the sciences of complexity, proceedings vol. **X**) (pp. 371–408). Redwood City, CA: Addison-Wesley.

27. Ray, T. S. (1994). An evolutionary approach to synthetic biology: Zen and the art of creating life. *Artificial Life*, **1**, 179–210.

28. Searle, J. R. (1980). Minds, brains, programs. *Behavioral Brain Science*, **3**, 417–457.

29. Sloman, A. (1990). Motives, mechanisms, emotions. In M. A. Boden (Ed.), *The philosophy of artificial intelligence* (pp. 231–247). Oxford: Oxford University Press.

30. Webster, G. & Goodwin, B. C. (1996). *Form and transformation: Generative and relational principles in biology*. Cambridge, UK: Cambridge University Press.

31. Wheeler, M. (1997). Cognition's coming home: The reunion of life and mind. In P. Husbands and I. Harvey (Eds.), *4th European conference on artificial life* (pp. 10–19). Cambridge, MA: MIT Press.

32. Wright, I. P. (1997). Emotional agents. Ph.D. thesis, School of Computer Science, University of Birmingham, Birmingham.

33. Zeleny, M. (1977). Self-organization of living systems: A formal model of autopoiesis. *International Journal of General Systems*, **4**, 13–22.

19 · Automatic design and manufacture of robotic life forms

HOD LIPSON AND JORDAN P. POLLACK

In the field of artificial life, "life as it could be" is examined on the basis of understanding the principles, and simulating the mechanisms, of real biological forms.[4] Just as aeroplanes use the same principles as birds, but have fixed wings, artificial lifeforms may share the same principles, but not the same implementation in chemistry. Stored energy, autonomous movement, and even animal communication are replicated in toys using batteries, motors, and computer chips.

Our central claim is that to realize artificial life, full autonomy must be attained not only at the level of power and behavior (the goal of robotics, today[5]), but also at the levels of design and fabrication. Only then can we expect synthetic creatures to sustain their own evolution. We thus seek automatically designed and constructed physical artifacts that are functional in the real world, diverse in architecture (possibly each slightly different), and automatically producible with short turnaround time, at low cost and in large quantities. So far these requirements have not been met.

The experiments described here use evolutionary computation for design, and additive fabrication for reproduction. The evolutionary process operates on a population of candidate robots, each composed of some repertoire of building blocks. The evolutionary process iteratively selects fitter machines, creates offspring by adding, modifying and removing building blocks using a set of operators, and replaces them into the population (see Methods section). Evolutionary computation has been applied to many engineering problems.[6] However, studies in the field of evolutionary robotics reported to date involve either entirely virtual worlds,[2,3] or, when applied in reality, adaptation of only the control level of manually designed and constructed robots.[7-9] These robots have a predominantly fixed architecture, although Lund et al.[10] evolved partial aspects of the morphology, Thompson[11] evolved physical electric circuits for control only, and we evolved static Lego structures, but had to manually construct the resultant designs.[12] Other works involving real robots make use of high-level building blocks comprising significant pre-programmed knowledge.[13] Similarly, additive fabrication technology has been developing in terms of materials and mechanical fidelity[14] but has not been placed under the control of an evolutionary process.

Our approach is based on the use of only elementary building blocks and operators in both the design and fabrication process. As building blocks are more elementary, any inductive bias associated with them is minimized, and at the same time architectural flexibility is maximized. Similarly, use of elementary building blocks in the fabrication process allows the latter to be more systematic and versatile. As a theoretical extreme, if we could use only atoms as building blocks, laws of physics as constraints and nanomanipulation for fabrication, the versatility of the manufacturable design space would be maximized. Earlier reported work that used higher level components and limited architectures (such as only tree structures[2,3]) resulted in expedited convergence to acceptable solutions, but at the expense of truncating the design space. Furthermore, these design spaces did not consider manufacturability.

The design space that we used consisted of bars and actuators as building blocks of structure and artificial neurons as building blocks of control. Bars connected with free joints can potentially form trusses—that can represent arbitrary rigid, flexible and articulated structures—as well as multiple detached structures, and emulate revolving, linear and planar joints at various levels of hierarchy. Similarly, artificial neurons can connect to

The Nature of Life, ed. M. A. Bedau and C. E. Cleland. Published by Cambridge University Press. © M. A. Bedau and C. E. Cleland 2010.

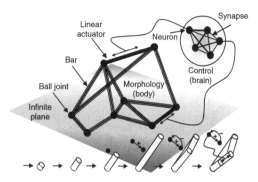

Fig. 19.1. Schematic illustration of an evolvable robot. Bars connect to each other to form arbitrary trusses; by changing the number of bars and the way they connect, the structural behavior of the truss is modified—some substructures may become rigid, while others may become articulated. Neurons connect to each other via synapses to form arbitrary recurrent neural networks. By changing the synapse weights and the activation threshold of the neuron, the behavior of the neuron is modified. By changing the number of neurons and their connectivity, the behavior of the network is modified. Also,we allow neurons to connect to bars: in the same way that a real neuron governs the contraction of muscle tissue, the artificial neuron signal will control the length of the bar by means of a linear actuator. All these changes can be brought about by mutational operators. A sequence of operators will construct a robot and its controller from scratch by adding modifying and removing building blocks. The sequence at the bottom of the image illustrates an arbitrary progression of operators that create a small bar, elongate it and split it. Simultaneously, other operators create a neuron, and another neuron, connect them in a loop, and eventually connect one of the neurons to one of the bars. The bar is now an actuator. Because no sensors were used, these robots can only generate patterns and actions, but cannot directly react to their environment.

create arbitrary control architectures such as feed-for-ward and recurrent nets, state machines and multiple independent controllers (like multiple ganglia). Additive fabrication, where structure is generated layer by layer, allows for the automatic generation of arbitrarily complex physical structures and the rapid construction of physically different bodies, including any that are composed of our building blocks. A schematic illustration of a possible architecture is shown in Fig. 19.1. The bars connect to each other through ball-and-socket joints, neurons can connect to other neurons through synaptic connections, and neurons can connect to bars. In the latter case, the length of the bar is governed by the output

of the neuron by means of a linear actuator. No sensors were used.

Starting with a population of 200 machines that were composed initially of zero bars and zero neurons, we conducted evolution in simulation. The fitness of a machine was determined by its locomotion ability— the net distance that its centre of mass moved on an infinite plane in a fixed duration. The process itera-tively selected fitter machines, created offspring by adding, modifying and removing building blocks, and replaced them into the population (see Methods). This process typically continued for 300 to 600 generations. Both body (morphology) and brain (control) were thus co-evolved simultaneously.

The simulator that we used for evaluating fitness (see Methods) supported quasi-static motion in which each frame is statically stable. This kind of motion is simpler to transfer reliably into reality, yet is rich enough to support low-momentum locomotion. Typically, several tens of generations passed before the first movement occurred. For example, at a min-imum, a neural network generating varying output must assemble and connect to an actuator for any motion at all (see the sequence in Fig. 19.1 for an example). Various patterns of evolutionary dynamics emerged, some of which are reminiscent of natural phylogenetic trees. Figure 19.2 presents examples of extreme cases of convergence, speciation and massive extinction. A sample instance of an entire generation thinned down to unique individuals is shown in Fig. 19.3.

Selected (virtual) robots out of those with winning performance were then automatically converted into physical objects: their bodies, represented only as points and lines, were first expanded into solid models with ball joints and accommodations for linear motors according to the evolved designs (Fig. 19.4a). This "solidifying" stage was performed by an automatic program that combined predesigned components describing a generic bar, ball joint, and actuator. The virtual solid bodies were then "materialized" using commercial rapid prototyping technology (Fig. 19.4b). This machine used a temperature-controlled head to extrude thermoplastic material layer by layer, so that the arbitrarily evolved morphology emerged as a solid three-dimensional structure without tooling or human intervention. The entire pre-assembled machine was fabricated as a single unit, with fine plastic supports connecting between moving parts (Fig. 19.4c); these

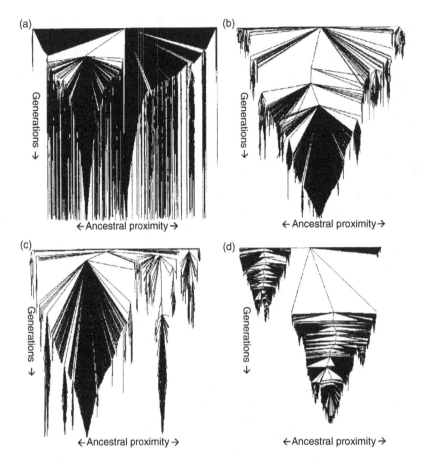

Fig. 19.2. Phylogenetic trees of several different evolutionary runs. Each node in the tree represents an individual and links represent parent–child relationships. The vertical axis represents generations, and the horizontal axis represents ancestral proximity in terms of the hops along the tree necessary to get from one individual to another. All trees originate at a common root denoting an empty robot with zero bars and actuators. Trees exhibit various degrees of divergence and speciation: (a) extreme divergence, resulting from niching methods[22]; (b) extreme convergence, resulting from fitness-proportionate selection; (c) intermediate level of divergence, typical of earlier stages of fitness-proportionate selection; and (d) massive extinction under fitness-proportionate selection. The trees are thinned, and depict several hundred generations each.

supports broke away at first motion. The resulting structures contained complex joints that would be difficult to design or manufacture using traditional methods (Figs. 19.4d and 19.5). Standard stepper motors were then snapped in, and the evolved neural network was executed on a microcontroller to activate the motors. The physical machines (three to date) then faithfully reproduced their virtual ancestors' behavior in reality (see Table 19.1).

In spite of the relatively simple task and environment (locomotion over an infinite horizontal plane), surprisingly different and elaborate solutions were evolved. Machines typically contained around 20 building blocks, sometimes with significant redundancy (perhaps to make mutation less likely to be catastrophic[15]). Not less surprising was the fact that some (for example, Fig. 19.5b) exhibited symmetry, which was neither specified nor rewarded for anywhere in the code; a possible explanation is that symmetric machines are more likely to move in a straight line, consequently covering a greater net distance and acquiring more fitness. Similarly, successful designs

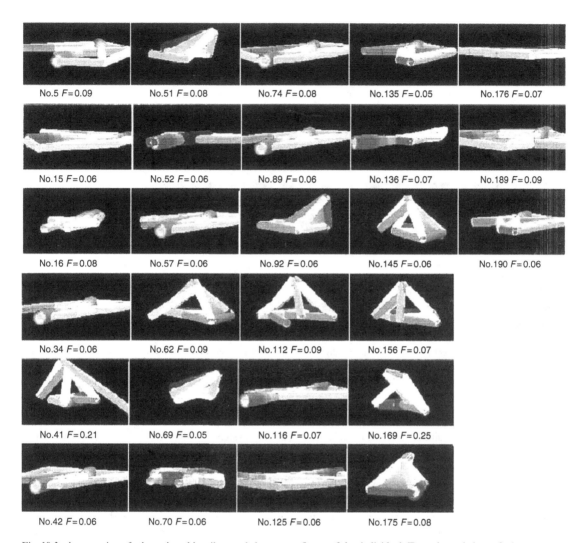

No.5 F=0.09   No.51 F=0.08   No.74 F=0.08   No.135 F=0.05   No.176 F=0.07

No.15 F=0.06   No.52 F=0.06   No.89 F=0.06   No.136 F=0.07   No.189 F=0.09

No.16 F=0.08   No.57 F=0.06   No.92 F=0.06   No.145 F=0.06   No.190 F=0.06

No.34 F=0.06   No.62 F=0.09   No.112 F=0.09   No.156 F=0.07

No.41 F=0.21   No.69 F=0.05   No.116 F=0.07   No.169 F=0.25

No.42 F=0.06   No.70 F=0.06   No.125 F=0.06   No.175 F=0.08

Fig. 19.3. A generation of robots. An arbitrarily sampled instance of an entire generation, thinned down to show only significantly different individuals. The caption under each image provides an arbitrary index number (used for reference) and the fitness of that individual. Two subpopulations of robots are observable, each with its own variations: one flat on the ground, and the other containing some elevated structure.

appear to be robust in the sense that changes to bar lengths would not significantly hamper their mobility. Three samples are shown and described in detail in Fig. 19.5, exploiting principles of ratcheting (Fig. 19.5a), anti-phase synchronization (Fig. 19.5b) and dragging (Fig. 19.5c). Others (not shown) used a sort of a crawling bipedalism, where a body resting on the floor is advanced using alternating thrusts of left and right "limbs". Some mechanisms used sliding articulated components to produce crab-like sideways motion. Other machines used a balancing mechanism to shift a friction point from side to side and advance by oscillatory motion. Table 19.1 compares the performances of three physical machines to their virtual ancestors. We note that, although overall distance travelled in the second and third cases does not match, in all cases the physical motion was achieved using corresponding mechanical and control principles. The difference in distance results

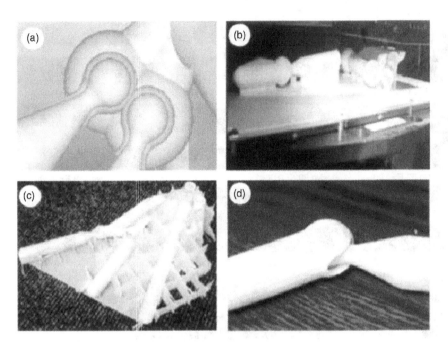

Fig. 19.4. Physical embodiment process. (a) Automatically 'fleshed' joints in virtual space; (b) a physical replication process in a rapid prototyping machine that builds the three-dimensional morphology layer after layer; (c) pre-assembled body in mid print with discardable support structure; (d) a close-up image of a joint printed as a single unit. The ball is printed inside the socket.

from the limbs slipping on the surface, implying that the friction model used in the simulation was not realistic.

Although both the machines and the task that we describe here are fairly simple compared with the products of human teams of engineers (and with the products of biological evolution), we have demonstrated a robotic "bootstrap" process, in which automatically designed electromechanical systems have been manufactured robotically. We have carefully minimized human intervention in both the design and the fabrication stages. Apart from snapping in the motors, the only human work was in informing the simulation about the "universe" that could be manufactured.

Without reference to specific organic chemistry, life is an autonomous design process that is in control of a complex set of chemical "factories," allowing the generation and testing of physical entities that exploit the properties of the medium of their own construction. Using a different medium, namely off-the-shelf rapid manufacturing, and evolutionary design in simulation, we have made progress towards replicating this autonomy of design and manufacture. This is, to our knowledge, the first time any artificial evolution system has been connected to an automatic physical

construction system. Taken together, our evolutionary design system, "solidification" process, and rapid prototyping machine form a primitive replicating robot. Although there are many, many further steps before this technology could become dangerous,[16] we believe that if indeed artificial systems are to ultimately interact and integrate with reality, they cannot remain virtual; it is crucial that they cross the simulation-reality gap to learn, to evolve,[17] and to affect the physical world directly.[18] Eventually, the evolutionary process must accept feedback from the live performance of its products.

Future work is needed primarily in understanding how more complex modular structures might self-organize, and how these complex structures may transfer into reality under control of the evolutionary process. Technological advances in micro-electro-mechanic systems (MEMS), nanofabrication, and multi-material rapid prototyping that can embed circuits[19] and actuators[20] in bulk material, together with higher-fidelity physical simulation and an increased understanding of evolutionary computational processes, may pave the way for the self-sustaining progress that Moravec has termed "escape velocity."[21]

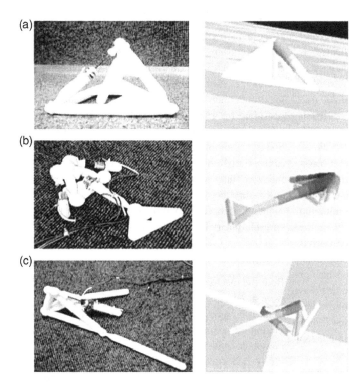

Fig. 19.5. Three resulting robots. Real robots (left); simulated robots (right). (a) A tetrahedral mechanism that produces hinge-like motion and advances by pushing the central bar against the floor. (b) This surprisingly symmetric machine uses a seven-neuron network to drive the center actuator in perfect anti-phase with the two synchronized side limb actuators. While the upper two limbs push, the central body is retracted, and vice versa.

(c) This mechanism has an elevated body, from which it pushes an actuator down directly onto the floor to create ratcheting motion. It has a few redundant bars dragged on the floor, which might be contributing to its stability. Print times are 22, 12, and 18 hours, respectively. These machines perform in reality in the same way that they perform in simulation. Motion videos of these robots and others are available: see Supplementary Information.

Table 19.1. *Results*

Machine	Distance traveled (cm)	
	Virtual	Physical
Tetrahedron (Fig. 19.5a)	38.5	38.4 (35)
Arrow (Fig. 19.5b)	59.6	22.5 (18)
Pusher (Fig. 19.5c)	85.1	23.4 (15)

Comparison of the performance of physical machines versus their virtual origin. Values are the net distance that the center of mass of each machine traveled over 12 cycles of neural network. Distances given in the column headed "physical" are compensated for scale reduction (actual distance is shown in parentheses). The mismatch in the last two rows is primarily due to the slipping of limbs on the surface.

METHODS

Robot representation

A robot was represented by a string of integers and floating-point numbers that describe bars, neurons and their connectivity, as follows:

robot:= $\langle$vertices$\rangle$ $\langle$bars$\rangle$ $\langle$neurons$\rangle$ $\langle$actuators$\rangle$

vertex:= $\langle x, y, z\rangle$

bar:= $\langle$vertex 1 index, vertex 2 index, relaxed length, stiffness$\rangle$

neuron:= $\langle$threshold, synapse coefficients of connections to all neurons$\rangle$

actuator:= $\langle$bar index, neuron index, bar range$\rangle$

Evolution process

Experiments were performed using version 1.2 of GOLEM (Genetically Organized Lifelike Electro

Mechanics), which is available at <http://www.demo.cs.brandeis.edu/golem>. We performed a simulated evolutionary process: the fitness function was defined as the net euclidean distance that the centre-of-mass of an individual moves over a fixed number (12) of cycles of its neural control. We started with a population of 200 null (empty) individuals. Each experiment used a different random seed. Individuals were then selected, mutated, and replaced into the population in steady state as follows: the selection functions we tried were random, fitness-proportionate or rank-proportionate. The mutation operators used to generate an offspring were independently applied with the following probabilities: a small mutation in length of bar or neuron synaptic weight (0.1), the removal or addition of a small dangling bar or unconnected neuron (0.01), split vertex into two and add a small bar, or split bar into two and add vertex (0.03), attach or detach neuron to bar (0.03). At least one mutation was applied. The mutations took place on the symbolic representation of the phenotype. After mutation, a new fitness was assigned to the individual by means of a simulation of the mechanics and the control (see details below). The offspring was inserted into the population by replacing an existing individual. The replacement functions we tried chose individuals to replace either randomly, in inverse-proportion to their fitness, or using similarity-proportionate criteria (deterministic crowding[22]). Various permutations of selection-replacement methods are possible; the results we report here were obtained using fitness-proportionate selection and random replacement. However, using rank selection instead of fitness-proportionate selection, or using random selection with fitness-proportionate replacement yielded equivalent results. The process continued for 300 to 600 generations (approximately 10^5 evaluations overall). The process was performed both serially and in parallel (on a 16-processor computer). On parallel computers we noticed an inherent bias towards simplicity: simpler machines could complete their evaluation sooner and consequently reproduce more quickly than complex machines (this could be avoided with a generational implementation).

Our evolutionary simulation was based on evolutionary strategies[23] and evolutionary programming,[24] because it directly manipulated continuous valued representations and used only elementary operators of mutation. Alternatively, we could have used genetic algorithms[25] and genetic programming[26] that introduce crossover operators that are sensitive to the structure of the machines, which might change the rate of evolution and lead to replicated structures. We did not form a morphological grammar from which the body is developed,[27] but evolved directly on the symbolic representation of the phenotype. And, instead of separating body (morphology) and brain (control) into separate populations, or providing for a "neonatal" stage that might allow us to select for brains that are able to learn to control their bodies, we simply applied selection to bodies and brains as integrated units. This simplified experimental set-up followed our focus on completing the simulation and reality loop, but we anticipate that the many techniques that have been developed in evolutionary and co-evolutionary learning[28-30] will enrich our results.

Simulation

Both the mechanics and the neural control of a machine were simulated concurrently. The mechanics were simulated using quasi-static motion, where each frame of the motion was assumed to be statically stable. This kind of motion is simple to simulate and easy to induce in reality, yet is rich enough to support various kinds of low-momentum motion like crawling and walking (but not jumping). The model consisted of ball-joined cylindrical bars with true diameters. Each frame was solved by relaxation: an energy term was defined, taking into account elasticity of the bars, potential gravitational energy, and penetration energy of collision and contact. The degrees of freedom of the model (vertex coordinates) were then adjusted iteratively according to their derivatives to minimize the energy term, and the energy was recalculated. Static friction was also modelled. The use of relaxation permitted handling singularities (for example, snap-through buckling) and under-constrained cases (like a dangling bar). Noise was added to ensure the system does not converge to unstable equilibrium points, and to cover the simulation–reality gap. The material properties modelled correspond to the properties of the rapid prototyping material (modulus of elasticity, $E = 0.896$ GPa; specific density, $\rho = 1,000$ kg m^{-3}; yield stress, $\sigma_{\text{yield}} = 19$ MPa). The neural network was simulated in discrete synchronized cycles. In each cycle, actuator lengths were modified in small increments not larger than 1 cm.

NOTES

This chapter originally appeared in *Nature* 406 (2000), 974–978.

REFERENCES

1. Langton, C. G. (Ed.) (1989). *Artificial life* (Santa Fe Institute studies in the sciences of complexity, proceedings vol. **IV**). Redwood City, CA: Addison-Wesley.

2. Sims, K. (1994). Evolving 3D morphology and behavior by competition. In R. Brooks and P. Maes (Eds.), *Artificial life IV* (proceedings of the 4th international workshop on the synthesis and simulation of living systems) (pp. 28–35). Cambridge, MA: MIT Press.

3. Komosinski, M. & Ulatowski, S. (1999). Framsticks: Towards a simulation of a nature-like world, creatures and evolution. In D. Floreano, J.-D. Nicoud, and F. Mondada (Eds.). *Advances in artificial life* (proceedings of the 5th European conference, ECAL) (pp. 261–265). Berlin: Springer.

4. Smith, J. M. (1992). Byte-sized evolution. *Nature*, **355**, 772–773.

5. Swinson, M. (1998). Mobile autonomous robot software. *DARPA Report BAA-99–09*. Arlington, Virginia: DARPA.

6. Bentley, P. (Ed.) (1999). *Evolutionary design by computers*. San Francisco: Morgan Kaufmann.

7. Floreano, D. & Mondada, F. (1994). Automatic creation of an autonomous agent: Genetic evolution of a neural-network driven robot. In D. Cliff, P. Husbands, J.-A. Meyer, and S. W. Wilson (Eds.), *From animals to animats III* (pp. 421–430). Cambridge, MA: MIT Press.

8. Husbands, P. & Meyer, J.-A. (1998). *Evolutionary robotics*. Berlin: Springer.

9. Nolfi, S. (1992). Evolving non-trivial behaviors on real-robots: A garbage collecting robot. *Robotics and Autonomous Systems*, **22**, 187–198.

10. Lund, H., Hallam, J., & Lee, W. (1996). A hybrid GP/GA approach for co-evolving controllers and robot bodies to achieve fitness-specified tasks. In *Proceedings of IEEE, 3rd international conference on evolutionary computation* (pp. 384–389). Piscataway, NJ: IEEE Press.

11. Thompson, A. (1997). Artificial evolution in the physical world. In T. Gomi (Ed.), *Evolutionary robotics: From intelligent robotics to artificial life (ER '97)* (pp. 101–125). Ontario: AAI Books.

12. Funes, P. & Pollack, J. (1998). Evolutionary body building: Adaptive physical designs for robots. *Artificial Life*, **4**, 337–357.

13. Leger, C. (1999). Automated synthesis and optimization of robot configurations: An evolutionary approach. Thesis, Carnegie Mellon University, Pittsburg.

14. Kochan, A. (1997). Rapid prototyping trends. *Rapid Prototyping Journal*, **3**, 150–152.

15. Lenski, R. E., Ofria, C, Collier, T., & Adami, C. (1999). Genome complexity, robustness and genetic interactions in digital organisms. *Nature*, **400**, 661–664.

16. Joy, B. (2000). Why the future doesn't need us. *WIRED Magazine*, **8**(4), 238–264.

17. Watson, R. A., Ficici, S. G., & Pollack, J. B. (1999). Embodied evolution: Embodying an evolutionary algorithm in a population of robots. In P. Angeline (Ed.), *1999 congress on evolutionary computation* (pp. 335–342). New York: IEEE Press.

18. Beer, R. D. (1990). *Intelligence as adaptive behavior*. New York: Academic Press.

19. Ziemelis, K. (1998). Putting it on plastic. *Nature*, **393**, 619–620.

20. Baughman, R. H., Cui, C., Zakhidov, A. A., *et al.* (1999). Carbon nanotube actuators. *Science*, **284**, 1340–1344.

21. Moravec, H. (1999). *Robot—From mere machine to transcendent mind*. Oxford: Oxford University Press.

22. Mahfoud, S. W. (1995). *Niching methods for genetic algorithms*. Thesis, University of Illinois, Urbana-Champaign.

23. Rechenberg, I. (1973). *Evolutionsstrategie: Optimierung technischer Systeme nach Prinzipien der biologischen Evolution*. Stuttgart: Frommann-Holzboog.

24. Fogel, L. J., Owens, A. J., & Walsh, M. J. (1966). *Artificial intelligence through simulated evolution*. New York: Wiley.

25. Holland, J. (1975). *Adaptation in natural and artificial systems*. Ann Arbor, MI: University of Michigan Press.

26. Koza, J. (1992). *Genetic programming*. Cambridge, MA: MIT Press.

27. Gruau, F. & Quatramaran, K. (1997). Cellular encoding for interactive evolutionary robotics. In P. Husbands and I. Harvey (Eds.), *Fourth European conference on artificial life* (pp. 368–377). Cambridge, MA: MIT Press.

28. Chellapilla, K. & Fogel, D. (1999). Evolution, neural networks, games and intelligence. In *Proceedings of 1987 IEEE conference on computer design, VLSI in computers and processors* (pp. 1471–1496). Washington DC: IEEE Computer Society Press.

29. Hillis, W. D. (1992). Co-evolving parasites improves simulated evolution as an optimizing procedure. In C. G. Langton, C. Taylor, J. D. Farmer, and S. Rasmussen (Eds.), *Artificial life II* (Santa Fe Institute studies in the sciences of complexity, proceedings vol. X) (pp. 313–322). Redwood City, CA: Addison-Wesley.

30. Pollack, J. B. & Blair, A. D. (1998). Co-evolution in the successful learning of backgammon strategy. *Machine Learning*, **32**, 225–240.

20 · A giant step towards artificial life?

DAVID DEAMER

Step by step, the components of an artificial form of cellular life are being assembled by researchers. Lipid vesicles the size of small bacteria can be prepared and under certain conditions are able to grow and divide, then grow again. Polymerase enzymes encapsulated in the vesicles can synthesize RNA from externally added substrates. Most recently, the entire translation apparatus, including ribosomes, has been captured in vesicles. Substantial amounts of proteins were produced, including green fluorescent protein used as a marker for protein synthesis. Can we now assemble a living cell? Not quite yet because no one has produced a polymerase that can be reproduced along with growth of the other molecular components required by life. But we are closer than ever before.

INTRODUCTION

Evidence from phylogenetic analysis suggests that microorganisms resembling today's bacteria were the first forms of cellular life. Fossilized traces of their existence have been found in Australian rocks at least 3.5 billion years old, and isotopic signatures from Greenland suggest that life might have existed even earlier, around 3.8 billion years ago.[1] In the time since life's beginnings, the machinery of life has become advanced. For instance, when researchers knocked out genes in one of the simplest known bacterial species, they reached a limit of ~265–350 genes that appear to be an absolute requirement for contemporary microbial cells.[2]

Yet life did not spring into existence with a full complement of 300+ genes, ribosomes, membrane transport systems, metabolism and the DNA $\rightarrow$ RNA $\rightarrow$ protein information transfer that dominates

all life today. There must have been something simpler, a kind of scaffold life that was left behind in the evolutionary rubble. Can we deduce the nature of that scaffold, then test it experimentally? One possibility is the RNA World concept, which arose from the discovery of catalytic RNA molecules called ribozymes. The fact that genetic information and a catalytic site can be combined in the same molecule means that RNA by itself could have functioned in the same way that DNA, RNA and protein do today. The RNA World idea was greatly strengthened when it was discovered that the catalytic core of ribosomes is not composed of protein, but instead is a small piece of RNA machinery. This is convincing evidence that RNA probably came first, and was then overlaid by more complex and efficient protein machinery.[3]

Another approach to understanding the origin of life is now underway. Instead of subtracting genes from an existing organism, researchers are attempting to incorporate one or a few genes into microscopic lipid vesicles to produce molecular systems that display the properties of life. Those properties might then provide clues to the process by which life began in a natural setting on the early Earth.

DEFINING ARTIFICIAL LIFE

What would such a system do? We can answer this question by listing the steps that would be required for a microorganism to emerge as the first cellular life form on the early Earth:

(i) Boundary membranes self-assemble from soap-like molecules to form microscopic cell-like compartments.

The Nature of Life, ed. M. A. Bedau and C. E. Cleland. Published by Cambridge University Press. © M. A. Bedau and C. E. Cleland 2010.

(ii) Energy is captured by the membranes either from light and a pigment system, or from chemical energy, or both.

(iii) Ion concentration gradients are maintained across the membranes and can serve as a major source of metabolic energy.

(iv) Macromolecules are encapsulated in the compartments but smaller molecules can cross the membrane barrier to provide nutrients and chemical energy for primitive metabolism.

(v) The macromolecules grow by polymerization of the nutrient molecules.

(vi) Macromolecular catalysts evolve that speed the growth process.

(vii) The macromolecular catalysts themselves are reproduced during growth.

(viii) Genetic information is encoded in the sequence of monomers in one set of polymers.

(ix) The information is used to direct the growth of catalytic polymers.

(x) The membrane-bounded system of macromolecules can divide into smaller structures that continue to grow.

(xi) Genetic information is passed between generations by duplicating the gene sequences and sharing them among daughter cells.

(xii) Occasional mistakes (mutations) are made during replication or transmission of information so that the system can evolve through natural selection.

Looking down this list, one is struck by the complexity of even the simplest form of life. This is why it has been so difficult to "define" life in the usual sense of a definition—that is, boiled down to a few sentences in a dictionary. Life is a complex system that cannot be captured in a few sentences, so perhaps a list of its observed properties is the best we can ever hope to do.

LIFE IN THE LABORATORY

One is also struck by the fact that all but one of the functions in the list have now been reproduced in the laboratory. It was shown 40 years ago that lipid vesicles self-assemble into bilayer membranes that maintain ion gradients.[4] The vesicles can grow by addition of lipids, and can even be made to divide by imposing shear forces, after which the vesicles grow again.[5] If

bacteriorhodopsin is in the bilayer, light energy can be captured as a proton gradient. If an ATP synthase is in the membrane, the photosynthetic system can couple the energy of the proton gradient to the synthesis of a pyrophosphate bond between ADP and phosphate so that ATP is produced.[6] Macromolecules such as proteins and nucleic acids can easily be encapsulated.[7]

And finally, macromolecules such as RNA[8] and proteins can be synthesized inside lipid vesicles. The first attempt to assemble a translation system in vesicles was made by Oberholzer et al.[9] However, only small amounts of peptide were synthesized, mainly because the lipid bilayer was impermeable to amino acids. This limited ribosomal translation to the small number of amino acids that were encapsulated within the vesicles. Yu et al.[10] and Nomura et al.[11] improved the yield substantially by using larger vesicles, demonstrating that GFP can be synthesized by an encapsulated translation system. Most recently, Noireaux and Libchaber[12] published an elegant solution to the dilemma of bilayer impermeability. They broke open bacteria and captured samples of the bacterial cytoplasm in lipid vesicles. The samples included ribosomes, tRNAs and other components required for protein synthesis. The researchers then carefully chose two genes to translate, one for GFP and a second for a pore-forming protein called α-hemolysin. If the system worked as planned, the GFP would accumulate in the vesicles as a visual marker for protein synthesis and the hemolysin would allow externally added "nutrients" in the form of amino acids and ATP— the universal energy source required for protein synthesis—to cross the membrane barrier and supply the translation process with energy and monomers (Fig. 20.1). The system worked as predicted. The vesicles began to glow with the classic green fluorescence of GFP and nutrient transport through the hemolysin pore allowed synthesis to continue for as long as four days.

The next obvious step is to incorporate both a gene transcription system and protein synthesis in lipid vesicles. This was reported last year by Ishikawa et al.[13] who managed to assemble a two-stage genetic network in liposomes, in which the gene for an RNA polymerase was expressed first, and the polymerase was then used to produce mRNA required for GFP synthesis.

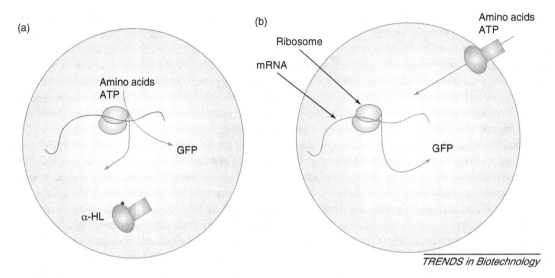

Fig. 20.1. Translation in a microenvironment. In (a) amino acids and ATP encapsulated in the vesicle volume are used to make small amounts of GFP and α-hemolysin (α-HL). The membrane prevents access to external amino acids and ATP, so the system will quickly exhaust the "nutrients" trapped in the vesicle. However, because the α-HL is a pore-forming protein, it migrates to the membrane and assembles into a heptamer with a pore large enough to permit amino acids and ATP to enter the vesicle. Now the system can synthesize significant amounts of new protein and the vesicle begins to glow green from the accumulation of GFP (b). Something similar must have happened on the pathway to the origin of cellular life so that the first cells had access to nutrients available in the environment.

A SECOND ORIGIN OF LIFE?

Can we now synthesize life? All that remains to be done, it seems, is to add up the individual processes, integrate them into a complete system and we will have a second origin of life, 3.8 billion years after the first origin but this time in a laboratory setting. However, would this system really be alive? This question brings us back to the one exception mentioned earlier. Everything in the system grows and reproduces except the catalytic macromolecules themselves, the polymerase enzymes or ribosomes. Every other part of the system can grow and reproduce, but the catalysts get left behind.

This is the final challenge: to encapsulate a system of macromolecules that can make more of themselves, a molecular version of von Neumann's replicating machine.[14] Is there any hope? The answer is yes, a faint glimmer of hope. Using a technique developed for selection and molecular evolution of RNA, David Bartel and co-workers[15] have produced a ribozyme that can grow by copying a sequence of bases in its own structure. To date, the polymerization has only been able to copy a short string of nucleotides, but this is a good start. If a ribozyme system can be found that catalyzes its own complete synthesis using genetic information encoded in its structure, it could rightly be claimed to have the essential property of self-replication. Given such a ribozyme, it would not be difficult to imagine its incorporation into lipid vesicles to produce a simple cell that would have the basic properties of the living state. The prospect of life in a test tube has always seemed just around the corner, and the corner is closer than ever.

NOTES

This chapter originally appeared in *Trends in Biotechnology* 23(7) (2005), 336–338.

REFERENCES

1. Mojzsis, S. J., Arrhenius, G., McKeegan, K. D., Harrison, T. M., Nutman, A. P., & Friend, C. R. L. (1996). Evidence

for life on Earth before 3,800 million years ago. *Nature*, **384**, 55–59.

2. Hutchison, C, III, Peterson, S. N., Gill, S. R., *et al.* (1999). Global transposon mutagenesis and a minimal Mycoplasma genome. *Science*, **286**, 2165–2169.

3. Hoang, L., Fredrick, K., & Noller, H. F. (2004). Creating ribosomes with an all-RNA 30S subunit P site. *Proceedings of the National Academy of Sciences*, **101**, 12,439–12,443.

4. Bangham, A. D., Standish, M. M., & Miller, N. (1968). Cation permeability of phospholipids model membranes: Effect of narcotics. *Nature*, **208**, 1295–1297.

5. Hanczyc, M. M., Fujikawa, S. M., & Szostak, J. W. (2003). Experimental models of primitive cellular compartments: Encapsulation, growth and division. *Science*, **302**, 618–622.

6. Racker, E. & Stoeckenius, W. (1974). Reconstitution of purple membrane vesicles catalyzing light-driven proton uptake and adenosine triphosphate formation. *Journal of Biological Chemistry*, **249**, 662–663.

7. Shew, R. & Deamer, D. W. (1985). A novel method for encapsulation of macromolecules in liposomes. *Biochimica et Biophysica Acta*, **816**, 1–8.

8. Chakrabarti, A., Breaker, R., Joyce, G. F., & Deamer, D. W. (1994). RNA synthesis by a liposome-encapsulated polymerase. *Journal of Molecular Evolution*, **39**, 555–559.

9. Oberholzer, T., Nierhaus, K. H., & Luisi, P. L. (1999). Protein expression in liposomes. *Biochemical and Biophysical Research Communications*, **261**, 238–241.

10. Yu, W., Sato, K., Wakabayashi, M., *et al.* (2001). Synthesis of functional protein in liposome. *Journal of Bioscience and Bioengineering*, **92**, 590–593.

11. Nomura, S., Tsumoto, K., Hamada, T., Akiyoshi, K., Nakatani, Y., & Yoshikawa, K. (2003). Gene expression within cell-sized lipid vesicles. *ChemBioChem*, **4**, 1172–1175.

12. Noireaux, V. & Libchaber, A. (2004). A vesicle bioreactor as a step toward an artificial cell assembly. *Proceedings of the National Academy of Sciences*, **101**, 17,669–17,674.

13. Ishikawa, K., Sato, K., Shima, Y., Urabe, I., & Yomo, T. (2004). Expression of a cascading genetic network within liposomes. *FEBS Letters*, **576**, 387–390.

14. von Neumann, J. (1966). *Theory of self-reproducing automata*. Chicago: University of Illinois Press.

15. Johnston, W. K., Unrau, P. J., Lawrence, M. S., Glasner, M. E., & Bartel, D. P. (2001). RNA-catalyzed RNA polymerization: Accurate and general RNA-templated primer extension. *Science*, **292**, 1319–1325.

21 · Approaches to semi-synthetic minimal cells: a review

PIER LUIGI LUISI, FRANCESCA FERRI,
AND PASQUALE STANO

THE NOTION OF MINIMAL CELL

The simplest living cells existing on Earth have several hundred genes, with hundreds of expressed proteins, which, more or less simultaneously, catalyse hundreds of reactions within the same tiny compartment—an amazing enormous complexity.

This picture elicits the question of whether or not such complexity is really essential for life, or whether or not cellular life might be possible with a much smaller number of components. This question is also borne out of considerations on early cells, which could not have been as complex. The enormous complexity of modern cells is probably the result of billions of years of evolution in which a series of defence and security mechanisms, redundancies and metabolic loops (which, in highly permissive conditions, were probably not necessary) was developed. These considerations led to the notion of minimal cell, now broadly defined as a cell having the minimal and sufficient number of components to be considered alive. This automatically precedes the next fundamental, but complex, question, "What does 'alive': mean?" One may choose quite a general definition, defining life at a cellular level as the concomitance of three basic properties: self-maintenance (metabolism), self-reproduction and evolvability (Fig. 21.1).

Evolvability is a Darwinian notion. As such, it refers to populations rather than individual cells. Consequently, one should take into consideration an entire family of minimal cells in the stream of environmental pressure and corresponding genetic evolution.

The trilogy defining cellular life may not be perfectly implemented, particularly in synthetic constructs, and several kinds of approximations to cellular life can be envisaged. For example, we may have protocells capable of self-maintenance but not of self-reproduction, or

vice versa. Or we might have protocells in which self-reproduction is active for only a few generations, or systems that are not capable of evolvability. In any given type of minimal cell (i.e., one with all three attributes), there may be quite different ways of implementation and sophistication. So, clearly, the term "minimal cell" depicts large families of possibilities and not simply one particular construct. The idea that the minimal forms of life are not univocally defined, and correspond rather to a large family, is not new in the field of the origin of life and early evolution. However, it is important to keep in mind that we are not simply considering theoretical possibilities, but something new: a synthetic biology approach and the particular methodology of experimental implementation.

The question on minimal cell has been considered for many years. One should recall, in particular, the work of Morowitz,[35] who estimated that the size of a minimal cell should be about one tenth smaller than *Mycoplasma genitalium*, based on enzymatic components of primary metabolism. Earlier insights of significance in the field were provided by Jay and Gilbert,[19] Woese[67], and Dyson.[7] More recently, reviews by Pohorille and Deamer,[50] Luisi,[28] and Oberholzer and Luisi[42] have sharpened the question and have brought it to the perspective of modern molecular tools. In fact, the last years have seen a significant revival of interest in the field of the minimal cell. In this chapter, we wish to review this work, emphasizing experimental aspects. Over the last few years, many theoretical approaches to minimal forms of life have been presented in the literature, but they will not be discussed in this review. This is not out of lack of interest in them, but out of the desire to focus this review on the art of synthetic biology of minimal cells. The idea of writing a review on the subject was also prompted by

The Nature of Life, ed. M. A. Bedau and C. E. Cleland. Published by Cambridge University Press. © M. A. Bedau and C. E. Cleland 2010.

The notion of "minimal cell"

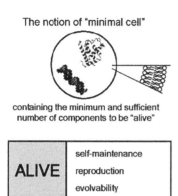

containing the minimum and sufficient
number of components to be "alive"

ALIVE	self-maintenance
	reproduction
	evolvability

Fig. 21.1. The notion of minimal cell. As explained in the text, this definition does not identify one particular structure, but is rather a descriptive term for a wide variety of minimal cells.

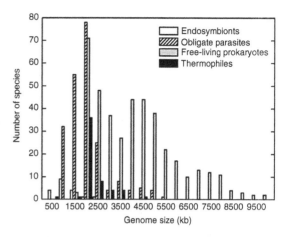

Fig. 21.2. Prokaryotic genome size distribution (N = 641). Genome sizes, complete proteomes and the number of open reading frames were all retrieved from the National Center for Bio-technology Information (http://www.ncbi.nlm.nih.gov) (Adapted from Islas et al.).[17]

the rise in interest that the field of minimal cell has been witnessing over the last few years, as documented, for example, by two international meetings on the subject held last year.[a]

To put this work into a more concrete perspective, it is useful to first look at the smallest unicellular organisms on Earth, focusing on the notion of minimal genome.

THE MINIMAL GENOME

Figure 21.2 compares genome size distributions calculated in a series of assumptions[17] of free-living prokaryotes, obligate parasites, thermophiles, and endosymbionts. DNA contents of free-living prokaryotes can vary over a tenfold range, from 1450 kb for *Halomonas halmophila* to 9,700 kb for *Azospirillum lipoferum* Sp59b. In comparison, consider that *Escherichia coli* K-12 has a genome size of ca. 4640 kb, and *Bacillus subtilis* has a genome size of 4200 kb.

Classification of endosymbionts as a separate group shows that their DNA content may be significantly smaller; the smallest sizes are then those of *M. genitalium* and *Buchnera*, with a value that confirms the predictions of Shimkets,[55] who states that the minimum genome size for a living organism should approximately be around 600 kb. It is argued that these two organisms have undergone massive gene losses and that their limited encoding capacities are due to their adaptation to highly permissive intracellular environments provided by the hosts.[17]

WHAT DO THESE FIGURES MEAN IN TERMS OF MINIMAL GENE NUMBERS?

Table 21.1, also taken from Islas et al.,[17] reports the number of coding regions in some small genomes. The table also gives an account of redundant genes, amounting to an average of 6%–20% of the whole genome. How can one work with the data of Table 21.1 to envisage further simplifications of the genome?

Gil et al.,[13] in Valencia, asked this question and arrived at the smaller number of 206 genes, basing the figure on their work with *Buchnera* spp. and other organisms. The results are given in Table 21.2.

Notice that the figures provided by Gil et al. are close to those obtained by other authors based on different considerations, as summarized in Table 21.3, which reports the most salient data relative to minimal genome calculations and observations. In fact, the question of the minimal genome has been considered, for example, by Mushegian,[36] Shimkets,[55] Mushegian and Koonin,[37] Kolisnychenko et al.,[20] and Koonin.[21, 22] In particular, Mushegian and Koonin[37] calculated an inventory of 256 genes, which represents the amount of DNA required to sustain a modern type of minimal cell in permissible conditions. This number, as indicated later by Koonin,[21,22] is quite similar to the values of viable minimal genome sizes inferred by site-directed

Table 21.1. *Genetic redundancies in small genomes of endosymbionts and obligate parasites*[a]

Proteome	Genome size (kb)	Number of ORFs	Number of redundant sequences	Redundancy (%)
Mycoplasma genitalium	580	480	52	10.83
Mycoplasma pneumoniae	816	688	134	19.48
Buchnera sp. APS	640	574	67	11.67
Ureaplasma urealyticum	751	611	105	17.18
Chlamydia trachomatis	1000	895	60	6.70
Chlamydia muridarum	1000	920	60	6.52
Chlamydophila pneumoniae J138	1200	1,070	148	13.83
Rickettsia prowazekii	1100	834	49	5.88
Rickettsia conorii	1200	1,366	189	13.84
Treponema pallidum	1100	1,031	78	7.57

ORFs Open reading frames.
[a]Genome sizes, complete proteomes and the number of ORFs were all retrieved from the National Center for Biotechnology Information. (http://www.ncbi.nlm.nih.gov).

Table 21.2. *Core of a minimal bacterial gene set*[a]

DNA metabolism	16
Basic replication machinery	13
DNA repair, restriction and modification	3
RNA metabolism	106
Basic transcription machinery	8
Translation: aminoacyl-tRNA synthesis	21
Translation: tRNA maturation and modification	6
Translation: ribosomal proteins	50
Translation: ribosome function, maturation and modification	7
Translation factors	12
RNA degradation	2
Protein processing, folding and secretion	15
Protein post-translational modification	2
Protein folding	5
Protein translocation and secretion	5
Protein turnover	3
Cellular processes	5
Energetic and intermediary metabolism	56
Poorly characterized	8
Total	343

[a]Courtesy of Professor A. Moya (Institut Cavanilles de Biodiversitat i Biologia Evolutiva, Universitat de València).

gene disruptions in *B. subtilis*[18] and transposon-mediated mutagenesis knockouts in *M. genitalium* and *Mycoplasma pneumoniae*.[15] Concerning this last work, one may recall that the notion of the "minimal genome" is approached in quite a different way by Hutchinson *et al*. In a study carried out at the Institute for Genomic Research in Rockville, MD, Hutchinson *et al*. knocked out genes from a *M. genitalium* bacterium one by one, and they estimated that of the 480 protein-coding regions, about 265–350 are essential in laboratory growth conditions, including about 100 genes of unknown functions.[15]

Taking a step further, the idea was to remove the original genetic material from the bacterium and to insert the synthetic one to see whether it works or not.[70] This approach had already been used by Cello *et al*.[5] at Stony Brook to create an infectious poliovirus that is much simpler than a bacterium.

We have reached the number of 200–300 genes as the minimal genome. This is a considerable simplification of the initial number, but it still corresponds to a formidable complexity, which, once again, induces the question of whether and how it can further go down.

FURTHER SPECULATIONS

Obviously, only speculations can help us at this point. Imagine a kind of theoretical knockdown of the

Table 21.3. *Works on the minimal genome*

Description of the system	Main goal and results	References
The complete nucleotide sequence (580,070 bp) of the *M. genitalium* genome has been determined by whole-genome random sequencing and assembly	Only 470 predicted coding regions were identified (genes required for DNA replication, transcription and translation; DNA repair; cellular transport; and energy metabolism)	Fraser *et al.*[9]
Site-directed gene disruption in *B. subtilis*	Values of viable minimal genome size were inferred	Itaya[18]
The 468 predicted *M. genitalium* protein sequences were compared with 1703 protein sequences encoded by the other completely sequenced small bacterial genome, that of *Haemophilus influenzae*	A minimal self-sufficient gene set: the 256 genes that are conserved in Gram-positive and Gram-negative bacteria are almost certainly essential for cellular function	Mushegian and Koonin[37]
Computational analysis (quantification of gene content, of gene family expansion and of orthologous gene conservation, as well as their displacement)	A set close to 300 genes was estimated as the minimal set sufficient for cellular life	Mushegian[36]
Global transposon mutagenesis was used to identify non-essential genes in *Mycoplasma* genome	265–350 of the 480 protein-coding genes of *M. genitalium* are essential in laboratory growth conditions, including about 100 genes of unknown function	Hutchinson *et al.*[15]
Several theoretical and experimental studies are reviewed	The concept of minimal gene set	Koonin[21]
The article focuses on the notion of a DNA minimal cell	The conceptual background of the minimal genome is discussed	Luisi *et al.*[31]
Full-length poliovirus cDNA was synthesized by assembling oligonucleotides of plus and minus strand polarity	It is possible to create an infectious poliovirus, which is much simpler than a bacterium, by a synthetic approach	Cello *et al.*[5]
A technique for precise genomic surgery was developed and applied to delete the largest K-islands of *E. coli*, which are identified by comparative genomics as recent horizontal acquisitions to the genome	Twelve K-islands were successfully deleted, resulting in an 8.1% reduction in genome size, a 9.3% reduction of gene count and elimination of 24 of 44 transposable elements of *E. coli*; the goal was to construct a maximally reduced *E. coli* strain to serve as a better model organism	Kolisnychenko *et al.*[20]
Physical mapping of *Buchnera* genomes obtained from five aphid lineages	They suggest that the *Buchnera* genome still experiences a reductive process towards a minimum set of genes necessary for its symbiotic lifestyle	Gil *et al.*[12]
Computational and experimental methods on comparative genomics	60 proteins are common to all cellular life; a core of 500–600 genes should represent the gene set of the last universal common ancestor	Koonin[22]

(Continued)

Table 21.3. (*cont.*)

Description of the system	Main goal and results	References
Buchnera and other organism genomes were compared	206 genes were identified as the core of a minimal bacterial gene set	Gil *et al.*[13]
Comparative genomics	Estimates of the size of minimal gene complement were performed to infer the primary biological functions required for a sustainable, reproducible cell today and throughout evolutionary times	Islas *et al.*[17]

genome that simultaneously reduces cellular complexity and part of non-essential functions.[31]

The first pit stop of this intellectual game is to imagine that a cell without enzymes (then the corresponding genes) needed to synthesize low molecular weight compounds—assuming that low molecular weight compounds, including nucleotides and amino acids, were available in the surrounding medium and were able to permeate into the cell membrane. This would be an entirely permeable minimal cell. Further simplifications[31] finally bring us to a cell that is able to perform protein and lipid biosyntheses through a modern ribosomal system, but is limited to a rather restricted number of enzymes (see Table 21.4). This cell would have ca. 25 genes for the entire DNA/RNA synthetic machinery, ca. 120 genes for the entire protein synthesis (including RNA synthesis and 55 ribosomal proteins) and 4 genes for membrane synthesis—which brings us to a total of about 150 genes, somewhat less than Gil *et al.*'s previously introduced figure of 206.

Thanks to the outside supply of substrates, such a cell should be capable of self-maintenance and self-reproduction, including replication of membrane components. However, it would neither synthesize low molecular weight compounds nor have redundancies for its own defence and security (in fact, all self-repair mechanisms are missing). Furthermore, cell division would simply be due to a physically based statistical process.

There is, however, no proof that this theoretical construct would be viable, but this also goes for Gil *et al.*'s 206 genes. It is nevertheless instructive to take these theoretical knockdown experiments further, with the next victims being ribosomal proteins. Can we take

them out? Some indications suggest that ribosomal proteins may not be essential for protein synthesis,[69] and there are other suggestions about an ancient and simpler translation system.[4,38]

Of course, this sort of discussion takes us directly into the scenario of early cells at the origin of life; in fact, some claim that the first ribosomes consisted of rRNA associated simply with basic peptides.[66] If we accept this and take out the 55 genes for ribosomal proteins and some other enzymes, we would then have a number of genes around 110.

FURTHER REDUCTIONS

A large portion of foreseen genes corresponds to RNA and DNA polymerases. A number of data[10,24,25,58] suggest that a simplified replicating enzymatic repertoire, as well as a simplified version of protein synthesis, might be possible. In particular, the idea that a single polymerase could play multiple roles as a DNA polymerase, transcriptase and primase is conceivable in very early cells.[31]

The game could go on by assuming that, at the time of early cells, not all "our" 20 amino acids were involved and that a lower number of amino acids would reduce the number of aminoacyl-tRNA synthetases and tRNA genes.

All these considerations may help to decrease the number of genes down to a number of, say, 45–50 genes (see Table 21.4 for a living, although certainly limping, minimal cell).[31]

This number is significantly lower than the one proposed by Professor Moya in Table 21.2, but is of course based on a higher degree of speculation. Many

Table 21.4. *A hypothetical list of gene products, sorted by functional category, that defines the minimal cell according to the definitions used in this chapter*

Gene product	Number of genes		
	Minimal DNA cell[a]	"Simple ribosome" cell	Extremely reduced cell
DNA/RNA metabolism			
DNA polymerase III	4[b]	4[b]	1
DNA-dependent RNA polymerase	3[c]	3[c]	1
DNA primase	1	1	
DNA ligase	1	1	1
Helicases	2–3	2–3	1
DNA gyrase	2[d]	2[d]	1
Single-stranded DNA-binding proteins	1	1	1
Chromosomal replication initiator	1	1	
DNA topoisomerases I and IV	1+2[d]	1+2[d]	1
ATP-dependent RNA helicase	1	1	
Transcription elongation factor	1	1	
RNase (III, P)	2	2	
DNase (endo/exo)	1	1	
Ribonucleotide reductase	1	1	1
Protein biosynthesis/ translational apparatus			
Ribosomal proteins	51	0	0
Ribosomal RNA	1[e]	1[e]	1[e] (Self-splicing)
Aminoacyl-tRNA synthetases	24	24	14[f]
Protein factors required for biosynthesis and membrane protein synthesis	9–12[g]	9–12[g]	3
tRNA	33	33	16[h]
Lipid metabolism			
Acyltransferase 'plsX'	1	1	1
Acyltransferase 'plsC'	1	1	1
PG synthase	1	1	1
Acyl carrier protein	1	1	1
Total	146–150	105–107	46

[a]Based on *M. genitalium*.

[b]Subunits a, b, y and tau.

[c]Subunits a, b and b'.

[d]Subunits a and b.

[e]One operon with three functions (rRNA).

[f]Assuming a reduced code.

[g]Including a possible limited potential to synthesize membrane proteins.

[h]Assuming the third base to be irrelevant.

authors would doubt that a cell with only 45–50 genes would be able to work. But again, the consideration goes on to early cells and to the consideration that the first cells could have not started with dozens of genes from the very beginning in the same compartment. This last consideration permits a logical link with the notion of compartments.

Suppose that these 45–50 macromolecules, or their precursors, developed first in solution (i.e., let us forget for a moment the possibility of compartments). Then, to start cellular life, compartmentation should have come later on, and one would then have to assume a simultaneous entrapment of all these different genes in the same vesicle. This could indeed be regarded as highly improbable; in fact, a scenario in which the complexity of cellular life evolved from within the compartment is more reasonable—a situation where the 45 (or 206) macromolecules were produced and evolved from a much smaller group of components from inside the photocell.

Until now, we have speculated on 'normal' protein/DNA/RNA cells—the ones we know in nature. In a further speculative leap, we could ask the question, "What about a theoretical RNA cell?" Let us briefly consider this question before proceeding further with the usual cells.

THE MINIMAL RNA CELL

One of the simplest constructs that responds to the criteria of evolvability, self-maintenance and reproduction is the so-called "RNA cell" (Fig. 21.3). This purely theoretical object, developed by Szostak et al.,[60] represents a synthesis of RNA and compartment models.

In this case, the combined "genetic" and catalysing properties of ribozymes play a central role. The RNA cell consists of a vesicle containing two ribozymes: one with replicase activity and the other with catalysing activity for the synthesis of membrane components. The first ribozyme is capable of replicating itself, and the second ribozyme is replicated by the first one. At the same time, a precursor is transformed into a membrane-forming compound, allowing the growth and subsequent division of the parent vesicle. In this way, a concerted core-and-shell reproduction of the entire construct may be obtained.

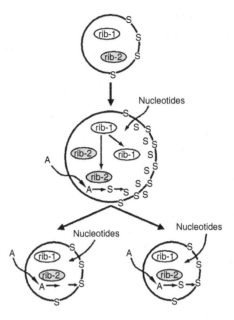

Fig. 21.3. The RNA cell, containing two ribozymes. Rib-1 is an RNA replicase capable of reproducing itself and making copies of Rib-2. Rib-2 is capable of synthesizing the cell membrane by converting precursor A to membrane-forming S. All necessary low molecular weight components required for macromolecular synthesis are provided from the surrounding medium and are capable of permeating the membrane. (Adapted from Luisi et al.)[31] For the sake of simplicity, an ideal cell division is represented in this figure, where all core components are equally shared between new vesicles.

As mentioned previously, this is a hypothetical scheme based on not-yet-existing ribozymes and a series of additional assumptions (e.g., full permeability of the membrane to precursor A and nucleotides, both present in large excesses in the environment), or the assumption that the cell divides, distributing both kinds of ribozyme to the daughters (so that, in each cell, there are always first- and second-type ribozymes).

In conclusion, the construct of Fig. 21.3, although quite exciting for its simplicity, remains a theoretical model, considering that the two ribozymes are still non-existent. In addition, the RNA cell, in a realistic scenario, must eventually evolve into the DNA/protein cell. Despite all these limits, the RNA cell is very interesting for one insight: it shows that, at least theoretically, cellular life can be implemented by a limited number of RNA genes.

TOWARDS THE CONSTRUCTION OF THE MINIMAL PROTEIN/DNA CELL: SETTING THE STAGE

Going back to the discussion on theoretical and practical backgrounds for the achievement of a minimal protein/DNA cell, it is certainly more complex than a minimal RNA cell, while, at the same time, it is more realistic and accessible from an experimental point of view, since all the ingredients exist. In fact, as already mentioned, self-replicating ribozymes, although fascinating objects, are not available (and it is questionable whether they ever will be), whereas genes and enzymes of a protein/DNA are available. In particular, the question is whether the construction of the corresponding minimal cell is possible with present laboratory tools.

Traditionally, people working in the area of prebiotic chemistry have been pursuing the so-called bottom-up approach, based on the notion that a continuous and spontaneous increase of molecular complexity transformed inanimate matter into the first self-reproducing cellular entities. For a number of reasons, this approach has not yet been successful, and another approach to the construction of the minimal living cell has been proposed in the last few years (indicated in Fig. 21.4). We use extant nucleic acids and enzymes and insert them into a vesicle, thus reconstructing the minimal living cell.

While the term "bottom–up" is recognized and accepted, this alternative route to the minimal cell is less clear and could give rise to different interpretations. The term "top–down" has been used to indicate the use of extant cellular components (DNA and enzymes) to build simple cellular constructs. However, such terminology could be misunderstood, since, in a way, this is also a bottom-up approach, in the sense that it goes in the direction of increasing complexity (the cell) starting from single components (DNA and enzymes). Moreover, there are different possible interpretations of the terms "top-down" and "bottom-up" in the literature, and we believe that, to avoid confusion, the term 'reconstruction' is perhaps more appropriate in minimal cell studies, making it clear that, in this procedure, one does not necessarily reach the construction of an extant cell or something that exists on Earth. Since they do not exist in our biological life, the term "artificial cell" may be used.[50] This is

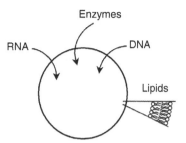

Fig. 21.4. The semi-synthetic approach to the construction of the minimal cell.

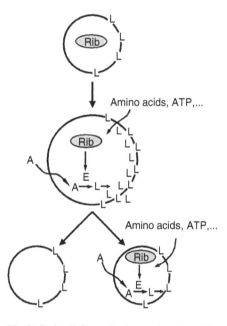

Fig. 21.5. A cell that makes its own boundary. The complete set of biomacromolecules needed to perform protein synthesis (genes, RNA polymerases and ribosomes) is indicated as Rib. The product of this synthesis (indicated as E) is the complete set of enzymes for lipid (L) synthesis. After growth and division, some of the "new" vesicles might undergo "death by dilution."

acceptable; however, since, generally, extant enzymes and genes are utilized, the term "semi-artificial cell" might be considered more appropriate.

Having clarified this, the next point is to set the stage of the experimental approach. We need a cell-like compartment, with vesicles (liposomes when they are constituted by lipids) being the preferred candidates. Figure 21.5 suggests that the incorporation of components into vesicles is the most obvious way to start.

In fact, as already well known, several attempts have been carried out in this direction. In all these studies, as we will see, a reaction is supposed to take place in the inner water pool of the vesicles, but to begin with, one must also consider that the membrane surface can also function as a reaction site, particularly if hydrophobic compounds are used. However, this has not been studied in detail yet.

This review aims to provide basic information that is limited to experimental approaches—a choice that implies neglecting the many theoretical models of minimal life provided by computer scientists and theoreticians of complexity.

PRELIMINARIES: REACTIVITY IN VESICLES

Two areas of study are preliminary to the utilization of liposomes as cellular models. The first considers possible analogies between vesicles and cellular membranes in terms of physico-chemical properties, such as stability, permeability and self-reproduction, to see whether (and to what extent) vesicles are close to cellular structures. The second area of inquiry considers the use of vesicles as hosts for complex molecular biological reactions to see if vesicles can indeed support the biochemistry of cellular life.

Concerning the first area of study, it has been shown that vesicles are capable of multiplying themselves at the expense of surfactant precursors;[1,32,64] in certain conditions, this may happen with the retention of the original size distribution (the so-called matrix effect; see Bloechliger et al.,[3] Lonchin et al.,[26] and Rasi et al.[51]). Again, it is not the aim of this article to review all these data, while it is important here to keep in mind that one of the most critical mechanisms of living cells can be simulated by vesicles based only on physical and chemical properties (i.e., without the use of sophisticated biochemical machineries). This consideration is relevant if one focuses on the prebiotic scenario.

Another important preliminary physico-chemical property is membrane permeability to solutes. Here things are more complicated, as vesicles and liposomes offer considerable resistance to the uptake of simple biochemicals in their water pool. This is particularly true with phospholipid membranes, which are commonly used as models for modern membrane bilayers.

Note, however, that phospholipids are relatively modern compounds; most likely, the first membranes and vesicles were constituted by surfactants, which could offer higher permeability (although possibly less stability) by virtue of their presumable chemically heterogeneous composition. It is reasonable, in fact, that early cells might have been somehow more "permissive" in terms of boundary properties and functions.

The use of membrane channels offers a possibility, but has, until now, met only modest success. An exception has been the use of α-hemolysin by Noireaux and Libchaber.[39] This approach has been quite successful. On the other hand, however, one should consider that, in more general terms, α-hemolysin pores are unselective and bidirectional and are therefore characterized by low specificity and are not very efficient due to gradient dissipation. In addition, α-hemolysin cannot be considered a primeval protein, however precisely these features of low selectivity represent a scenario where the first unspecific protocells were developed. The main problem is how to bring together the high local concentration needed in the water pool of liposomes. This difficulty might be partially circumvented if two or more liposomes, each containing a given substrate, could fuse together to produce liposomes containing all reagents. In fact, fusion of vesicles is becoming an active area of research, and interesting results have been already obtained.[33,47,57,61]

Fusion of compartments can also be achieved by utilizing water-in-oil emulsion. Actually, in this way, as we will see further on in detail, protein synthesis could be obtained by mixing compartments containing various ingredients for synthesis.

Concerning the area of biochemical reactions in liposomes, a large amount of experimental work (mainly studies in which liposomes have been used as host systems for molecular biology reactions) has paved the way for significant developments (Table 21.5).

For example, biosynthesis of poly(A) (a model for RNA) was reported independently by two groups.[6,64] In both cases, polynucleotide phosphorylase was entrapped in vesicles, and the synthesis of poly(A), which remained in the aqueous core of such vesicles, was observed. In one case,[64] internal poly(A) synthesis proceeded simultaneously through the reproduction of vesicle shells due to external addition of a membranogenic precursor (oleic anhydride).

Table 21.5. *Molecular biology reactions in liposomes*

Description of the system	Main goal and results	References
Enzymatic poly(A) synthesis	Polynucleotide phosphorylase producing poly(A) from ADP	Chakrabarti et al.[6]
Enzymatic poly(A) synthesis	Poly(A) is produced inside simultaneously with the (uncoupled) self-reproduction of vesicles	Walde et al.[64,65]
Oleate vesicles containing the enzyme Qβ replicase, an RNA template and ribonucleotides; the water-insoluble oleic anhydride was added externally	A first approach to a synthetic minimal cell: the replication of an RNA template proceeded simultaneously with the self-replication of the vesicles	Oberholzer et al.[43,44]
POPC liposomes containing all different reagents necessary to carry out a PCR reaction	DNA amplification by the PCR inside the liposomes; a significant amount of DNA was produced	Oberholzer et al.[43,44]
POPC liposomes incorporating the ribosomal complex with the other components necessary for protein expression	Ribosomal synthesis of polypeptides can be carried out in liposomes; synthesis of poly(Phe) was monitored by quantification of ^{14}C-labelled products	Oberholzer et al.[45]
T7 DNA within cell-sized giant vesicles formed by natural swelling of phospholipid films	Transcription of DNA and transportation by laser tweezers; vesicles behaved as barriers, preventing the attack of RNase	Tsumoto et al.[62]
DNA template and the enzyme T7 RNA polymerase microinjected into a selected giant vesicle; nucleotide triphosphates added from the external medium	The permeability of giant vesicles increased in an alternating electric field; mRNA synthesis occurred	Fischer et al.[8]

POPC 1-Palmitoyl-2-oleoyl-phosphatidylcholine.

A more suggestive example was provided shortly thereafter[43] with the use of Qβ replicase, an enzyme that replicates RNA template. Also in this case, replication of a core component was coupled with replication of vesicle shell. With an excess of Qβ replicase/ RNA template, replication of RNA could proceed for a few generations.

This system, as well as the previous one by Walde et al.,[64] is interesting because it represents a case of "core-and-shell replication" in which both the inside of the core and the shell itself undergo duplication. However, limitations of this analogy should be clear; in fact, a real core-and-shell reproduction should be synchronous, which was not the case.

In particular, even if the RNA template and the vesicle shell replicate, the Qβ replicase is not continuously produced in the process; thus, the system undergoes "death by dilution." After a while, new vesicles will

not contain either the enzyme or the template; therefore, the construct cannot reproduce itself completely.

Another complex biochemical reaction implemented in liposomes is polymerase chain reaction (PCR).[43] Liposomes were able to endure the hardships of PCR conditions, with several temperature cycles up to 90°C (liposomes were practically unchanged at the end of the reaction). In addition, nine different chemicals had to be encapsulated in each liposome for the reaction to occur. Depending on liposome formation mechanism and chemical concentration, entrapment efficiency can be different from what is expected on a statistical basis. In particular, it is not obvious that all nine chemicals are simultaneously trapped within one liposome.

Using poly(U) as mRNA, Oberholzer et al.[45] showed the production of poly(Phe), starting from phenylalanine, ribosomes, tRNA^{Phe} and elongation factors entrapped in lecithin vesicles. Compared to the experiment in water without liposomes, the yield was 5%, but the authors argued that the yield was actually surprisingly high, considering that the liposomes occupied only a very small fraction of the total volume and that only a very few of them would contain all ingredients by statistical entrapment.[45]

The table also reports the work of Fischer et al.[8] on mRNA synthesis inside giant vesicles utilizing DNA template and T7 RNA polymerase and that of Tsumoto et al.[62] on DNA transcription. Further considerations on polymerase activity inside vesicles were reported by Monnard.[34]

PROTEIN EXPRESSION IN LIPOSOMES

In "Preliminaries: reactivity in vesicles," we have seen the realization and optimization of rather complex biochemical reactions in liposomes. What could one do to approach the construction of the minimal cell? Theoretically, one should increase the complexity of the core of the liposomes so as to reach the limits on the minimal genome, as outlined previously.

This approach has not been used in the literature until now. Researchers have sought to insert conditions for the expression of a single protein in liposomes. For reasons that are easily understood (mostly for detection), the green fluorescence protein (GFP) has been the target protein.

With how many genes? Well, the answer to this question is also not easy to elicit from current data, as generally a calculation of the genes/enzymes involved has not been performed by the authors. Often, commercial kits are used for protein expression, and these are notoriously black boxes where the number of enzymes is not made known (and occasionally entire E. coli cellular extracts have been utilized). On the other hand, it is fair to say that, for the expression of one single protein, only a minimal part of the E. coli genome is implied.

An overview of the work performed, limited to the expression of proteins in liposomes, is presented in Table 21.6.

Table 21.6. *Protein expression in compartments*

Description of the system	Main goal and results	References
Liposomes from EggPC, cholesterol and DSPE-PEG5000 used to entrap cell-free protein synthesis	Expression of a mutant GFP, determined with flow cytometric analysis	Yu et al.[68]
Small liposomes prepared by the ethanol injection method	Expression of EGFP evidenced by spectrofluorimetry	Oberholzer and Luisi[42]
Gene expression system within cell-sized lipid vesicles	Encapsulation of a gene expression system; high expression yield of GFP inside giant vesicles	Nomura et al.[41]
A water-in-oil compartment system with water bubbles up to 50 μm	Expression of GFP by mixing different compartments that are able to fuse with each other	Pietrini and Luisi[49]
A two-stage genetic network encapsulated in liposomes	A genetic network in which the protein product of the first stage (T7 RNA polymerase) is required to drive the protein synthesis of the second stage (GFP)	Ishikawa et al.[16]
E. coli cell-free expression system encapsulated in a phospholipid vesicle, which was transferred into a feeding solution containing ribonucleotides and amino acids	The expression of the α-hemolysin inside the vesicle solved the energy and material limitations; the reactor could sustain expression for up to 4 days	Noireaux and Libchaber[39]

The common strategy is to entrap all the ingredients for in vitro protein expression (i.e., the gene for the GFP (a plasmid), an RNA polymerase, ribosomes and all the low molecular weight components (amino acids and ATP) needed for protein expression) in the aqueous core of liposomes.

Yu et al.,[68] for example, have reported the expression of a mutant GFP in lecithin liposomes. Large GFP-expressing vesicles, prepared by the film hydration method, were analyzed using flow cytometry as well as confocal laser microscopy.

In the procedure utilized by Oberholzer and Luisi,[42] all ingredients were added to a solution in which the vesicles were formed by the ethanol injection method and enhanced GFP (EGFP) production was then evidenced inside the compartments. In this case, the sample was analysed spectroscopically, monitoring the increase of the fluorescent signal of the EGFP. The disadvantage of this procedure is that entrapping efficiency is generally low due to the small internal volume of liposomes obtained with this method. On the other hand, the observation of EGFP production inside the aqueous core of liposomes confirms that the coentrapment of several different solutes was obtained.

A direct observation of protein expression was accomplished by the procedure utilized by Nomura et al.[41] using giant vesicles. The progress of the reaction is observed by laser scanning microscopy, and it is shown that expression of red shifted GFP (rsGFP) takes place with a very high efficiency (the concentration of rsGFP inside the vesicles was greater, in the first hours, than that in the external environment). The authors also show that vesicles can protect gene products from external proteinase K.

More recently, based on the initial report on the expression of functional protein in liposomes,[68] Ishikawa et al.[16] were able to design and produce experimentally a two-level cascading protein expression. A plasmid containing the T7 RNA polymerase (with SP6 promoter) and a mutant GFP (with T7 promoter) gene was constructed and entrapped in liposomes, together with an in vitro protein expression mixture (of the enzyme SP6 RNA polymerase). In these conditions, SP6 RNA polymerase drives the production of T7 RNA polymerase, which in turn induces the expression of detectable GFP.

Of particular interest is the work by Noireaux and Libchaber.[39] Again, a plasmid encoding for two proteins was used; in particular, the authors introduced EGFP and α-hemolysin genes. In contrast to the cascading network described above, now the second protein (α-hemolysin) does not have a direct role in protein expression, but is involved in a different task. In fact, although α-hemolysin is a water-soluble protein, it is able to self-assemble as a heptamer in the bilayer, generating a pore that is 1.4 nm in diameter (cut-off ~3 kDa). In this way, it was possible to feed the inner aqueous core of the vesicles, realizing a long-lived bioreactor where the expression of the reported EGFP was prolonged up to 4 days. This work certainly represents an important milestone in the road map to the minimal cell because the α-hemolysin pore permitted the uptake of small metabolites from the external medium and thus solved the energy and material limitations typical of impermeable liposomes.

Finally, GFP has also been expressed in another kind of compartment different from vesicles. These are water cavities (aqueous micrometer-sized environments) of water-in-oil emulsion, where it has been shown[49] that a functional protein, representing a tiny volume fraction (~0.5%) of a hydrocarbon sample, can be expressed.

In addition, the desired degree of complexity, intended as the collection of all components required for GFP expression, was obtained by solubilisate exchange and/or fusion between different aqueous compartments, each one carrying a part of the whole biochemical machinery (plasmid; RNA polymerases, ribosomes and cellular extracts; and amino acids).

In summary, in the last few years, a handful of pioneering studies on protein expression within liposomes appeared, and some of these reports evidenced the effect of "compartmentation" (i.e., a higher yield of protein expression in the vesicles compared to the bulk buffer)—a very interesting phenomenon deserving further investigation.

It is also worth mentioning that, to date, only water-soluble proteins have been expressed, and no attention has been devoted to the expression of membrane-soluble proteins.

It is also important to mention some interesting studies, which, although not directly related to the question of the minimal cell, deal with microtubulation. The combination of giant vesicles, minibeads and molecular motors has been studied by a team at the Institut Curie.[52] The authors show that lipid giant unilamellar vesicles, to which kinesin molecules have been attached, give rise to membrane tubes and complex tubular networks that form an original system emulating intracellular transport. Membrane tube formation from giant vesicles through dynamic association of motor proteins has been also studied by Koster et al.,[23] while Glade et al.[14] have shown tubule-mediated collective transport and organization of phospholipid vesicles and other particles.

This kind of work paves the way for the study of intracellular transport and organization at a higher complexity level within semi-artificial cells.

WHAT NEXT?

Keeping in mind the notion of minimal cell, analysis of the data presented in the article reveals what is still needed before we can proceed in this field.

For example, protein expression, as outlined in most salient experiments of Table 21.6, has been carried out without checking the number of enzymes/genes utilized in the work. We believe that it would be appropriate to carry out protein expression by utilizing known concentrations of the single enzymes/genes instead (and forgetting the commercial kits) to know exactly what is in "the pot" and to possibly have a hand in the corresponding chemistry. This operation would correspond to the implementation of the minimal genome inside liposomes and may pave the road for the next steps.

Previous discussions and the data reported in Table 21.6 make clear one other essential element that is still needed before we can reach the ideal case of Fig. 21.1: self-reproduction. In fact, after having produced GFP, none of the systems of Table 21.6 has been found capable of reproducing itself and giving rise to a chain of multiplying GFP-producing systems.

In real biological systems, a cell is capable of duplicating and reproducing itself with the same genetic content. This is due to systems of regulation, and this aspect has not been contemplated yet in the experimental set-up of minimal cells. In this context, besides making reference to prokaryotic cell division, the previously cited work on microtubulin might be quite an interesting insight into the problem.

A very interesting case is achieving vesicle self-reproduction by endogenous synthesis of vesicle lipids. Two strategies can, in principle, be pursued: (1)

incorporating first the enzymes that synthesize the lipids, or (2) starting from the corresponding genes (i.e., expressing those enzymes within the vesicles).

Early attempts have focused on the enzymatic production of lecithin in lecithin liposomes.[54] The metabolic pathway is the so-called salvage pathway, which converts glycerol-3-phosphate to phosphatidic acid, to diacylglycerol and, finally, to phosphatidylcholine. The four enzymes needed to accomplish these reactions were simultaneously inserted into liposomes by the detergent depletion method, and the synthesis of new phosphatidylcholine (10% yield) was followed by radioactive labelling. Liposome transformation, followed by dynamic light scattering, showed that vesicles changed their size distribution during the process.

This was indeed a complex system, and it was realized later that one could theoretically stop at the synthesis of phosphatidic acid, as this compound also formed stable liposomes. Further studies[27] were oriented to characterize the process by means of overexpression in E. coli and reconstitution in liposomes of the first two enzymes of the phospholipid salvage pathway to obtain self-reproducing vesicles with only two enzymes.

Production of the cell boundary (as depicted in Fig. 21.5) from within corresponds to the notion of autopoiesis.[29,30,32,63]

The internal synthesis of lecithin in lecithin liposomes would be a significant step forward. In particular, it will be very interesting to see, given a certain excess of the two enzymes, for how many generations cell self-reproduction could go on. However, it is clear that, after a certain number of generations, the system would undergo "death by dilution."

Finally, to get closer to the real minimal cell, there is the problem of further reduction of the number of genes. In all systems of Table 21.6, we still deal with ribosomal protein biosynthesis, and this implies 100–200 genes. We are still far from our ideal picture of a minimal cell, and we can pose, once more, the question of how to devise actual experiments to reduce this complexity.

As a way of thinking, we must resort to conceptual knock-down experiments (e.g., those outlined in the works by Islas et al.[17] and Luisi et al.[31])—a simplification that also corresponds to movement towards early cells. The simplification of ribosomal machinery and of the enzyme battery devoted to RNA and DNA synthesis has been seen as a necessary step.[2]

Is this experimentally feasible? For example, can simple forms of rigid support for reactions (in particular protein biosynthesis) that are operative in vitro as ribosomes be developed? Think of protein-free ribosomal RNA first. Can one operate, at costs of specificity, with only a very few polymerases? Similarly, it might not be necessary, at first, to have all possible specific tRNA, but a few unspecific ones instead. One might even conceive experiments with a limited number of amino acids. Now, all this must be tested experimentally; there is no other way around it.

CONCLUDING REMARKS

The definition of minimal cell, as given in the beginning of this review, appears simple and is provided with its own elegance. Conversely, experimental implementations of minimal cells may not appear equally satisfactorily and elegant. We have outlined the main difficulties possibly encountered in the construction of an ideal minimal cell, and we have pointed out, for example, that, in the best of hypothesis, death by dilution is one limitation; self-reproduction is one target that has not yet been accomplished.

One problem with the present literature on minimal cells is that the link between the "minimal genome" and the minimal cell is too weak; in other words, there is no direct correlation. It would, of course, be advisable for researchers working on minimal cells to "count" the genes that are active in their conditions and to compare the figures with the figures on the minimal genome given by researchers. Even within these limitations, experimental attempts to build a minimal cell are of great value in—but are not limited to—evaluating the specific simplification of the minimal genome. The use of liposomes as a sophisticated "reaction vessel" is certainly instrumental in the technical realization of the minimal cell, but also has the added value of representing a possible route to the origin of early cells, emphasizing manifold consequences of compartmentation.

Constructs produced in the laboratory still represent poor approximations of a full-fledged biological cell. This distance from fully biologically active cells makes it indeed premature to question possible hazards and bioethical issues in the field of minimal cells.

But there is still another very important topic that has not yet been discussed in due light by authors studying the minimal cell: interaction with the environment. Of course, feeding of the minimal cell is somehow taken into consideration, but only as a passive reservoir of nutrients and/or energy. In fact, we believe that the next generation of studies on the minimal cell should more actively incorporate such interactions with the surroundings, questioning, in particular, in which environmental conditions the minimal cell is able to perform its three basic functions.

Yet, these forms of "limping life," in our opinion, represent a very interesting part of this ongoing research. In fact, these approximations to life are as follows: a cell that produces proteins and does not reproduce itself; or one that does reproduce for a few generations and then dies out of dilution; or a cell that reproduces only parts of itself; and/or one characterized by very poor specificity and metabolic rate.

All these constructs are important because, most probably, similar constructs were intermediates experimented on by nature to arrive at the final goal: a full-fledged biological cell. Thus, the creation of these partially living minimal cells in the laboratory, as well as the historical evolutionary pathway by which this target may have been reached, may be of fundamental importance to understanding the real essence of cellular life. In addition, the construction of semi-synthetic living cells in the laboratory would be a demonstration (if still needed) that life is indeed an emergent property. In fact, in this case, cellular life would be created from non-life, since single genes and/or single enzymes are, *per se*, non-living.

Generally, although the minimal cell can teach us a lot about early cellular life and evolution, it may not necessarily shed light on the origin of life. The reasons for this have been already expressed and lie mostly in the fact that, in our approach to the minimal cell, we start with extant enzymes and genes, where life is already in full expression.

All this is very challenging and, perhaps for this reason, as already mentioned, there has been an abrupt rise of interest in the minimal cell. It appears that one additional reason for this rise of interest lies in a diffused sense of confidence that the minimal cell is indeed an experimentally accessible target.

ACKNOWLEDGMENTS

We thank the "Enrico Fermi" Study Center (Rome) and COST D27 Action for financial support.

NOTE

[a] This chapter originally appeared in *Naturwissenschaften* 93(1) (2006), 1–13.

REFERENCES

1. Bachmann, P. A., Luisi, P. L., & Lang, J. (1992). Autocatalytic self-replicating micelles as models for prebiotic structures. *Nature*, 357, 57–59.
2. Berclaz, N., Mueller, M., Walde, P., & Luisi, P. L. (2001). Growth and transformation of vesicles studied by ferritin labeling and cryotransmission electron microscopy. *Journal of Physical Chemistry B*, 105, 1056–1064.
3. Bloechliger, E., Blocher, M., Walde, P., & Luisi, P. L. (1998). Matrix effect in the size distribution of fatty acid vesicles. *Journal of Physical Chemistry*, 102, 10,383–10,390.
4. Calderone, C. T. & Liu, D. R. (2004). Nucleic-acid-templated synthesis as a model system for ancient translation. *Current Opinion in Chemical Biology*, 8, 645–653.
5. Cello, J., Paul, A. V., & Wimmer, E. (2002). Chemical synthesis of poliovirus cDNA: generation of infectious virus in the absence of natural template. *Science*, 297, 1016–1018.
6. Chakrabarti, A. C, Breaker, R. R., Joyce, G. F., & Deamer, D. W. (1994). Production of RNA by a polymerase protein encapsulated within phospholipid vesicles. *Journal of Molecular Evolution*, 39, 555–559.
7. Dyson, F. J. (1982). A model for the origin of life. *Journal of Molecular Evolution*, 18, 344–350.
8. Fischer, A., Franco, A., & Oberholzer, T. (2002). Giant vesicles as microreactors for enzymatic mRNA synthesis. *ChemBioChem*, 3, 409–417.
9. Fraser, C. M., Gocayne, J. D., White, O., *et al.* (1995). The minimal gene complement of *Mycoplasma genitalium*. *Science*, 270, 397–403.
10. Frick, D. N. & Richardson, C. C. (2001). DNA primases. *Annual Review of Biochemistry*, 70, 39–80.
11. Gavrilova, L. P., Kostiashkina, O. E., Koteliansky, V. E., Rutkevitch, N. M., & Spirin, A. (1976). Factor-free (non-enzymic) and factor-dependent systems of translation of polyuridylic acid by *Escherichia coli* ribosomes. *Journal of Molecular Biology*, 101, 537–552.
12. Gil, R., Sabater-Munoz, B., Latorre, A., Silva, F. J., & Moya, A. (2002). Extreme genome reduction in *Buchnera* spp: Toward the minimal genome needed for symbiotic life. *Proceedings of the National Academy of Sciences*, 99, 4454–4458.
13. Gil, R., Silva, F. J., Peretó, J., & Moya, A. (2004). Determination of the core of a minimal bacteria gene set. *Microbiology and Molecular Biology Reviews*, 68, 518–537.
14. Glade, N., Demongeot, J., & Tabony, J. (2004). Microtubule self-organization by reaction-diffusion processes causes collective transport and organization of cellular particles. *BMC Cell Biology*, 5, 23. DOI 10.1186/1471-2121-5-23.
15. Hutchinson, C. A., Peterson, S. N., Gill, S. R., *et al.* (1999). Global transposon mutagenesis and a minimal *Mycoplasma* genome. *Science*, 286, 2165–2169.
16. Ishikawa, K., Sato, K., Shima, Y., Urabe, I., & Yomo, T. (2004). Expression of a cascading genetic network within liposomes. *FEBS Letters*, 576, 387–390.
17. Islas, S., Becerra, A., Luisi, P. L., & Lazcano, A. (2004). Comparative genomics and the gene complement of a minimal cell. *Origin of Life and Evolution of the Biosphere*, 34, 243–256.
18. Itaya, M. (1995). An estimation of the minimal genome size required for life. *FEBS Letters*, 362, 257–260.
19. Jay, D. & Gilbert, W. (1987). Basic protein enhances the encapsulation of DNA into lipid vesicles: Model for the formation of primordial cells. *Proceedings of the National Academy of Sciences*, 84, 1978–1980.
20. Kolisnychenko, V., Plunkett, G., III, Herring, C. D., *et al.* (2002). Engineering a reduced *Escherichia coli* genome. *Genome Research*, 12, 640–647.
21. Koonin, E. V. (2000). How many genes can make a cell: The minimal-gene-set concept. *Annual Review of Genomics and Human Genetics*, 1, 99–116.
22. Koonin, E. V. (2003). Comparative genomics, minimal gene-sets and the last universal common ancestor. *National Review of Microbiology*, 1, 127–136.
23. Koster, G., van Duijn, M., Hofs, B., & Dogterom, M. (2003). Membrane tube formation from giant vesicles by dynamic association of motor proteins. *Proceedings of the National Academy of Sciences*, 100, 15,583–15,588.
24. Lazcano, A., Guerriero, R., Margulius, L., & Oró, J. (1988). The evolutionary transition from RNA to DNA in early cells. *Journal of Molecular Evolution*, 27, 283–290.
25. Lazcano, A., Valverde, V., Hernandez, G., Gariglio, P., Fox, G. E., & Oró, J. (1992). On the early emergence of

reverse transcription: Theoretical basis and experimental evidence. *Journal of Molecular Evolution*, **35**, 524–536.

26. Lonchin, S., Luisi, P. L., Walde, P., & Robinson, B. H. (1999). A matrix effect in mixed phospholipid/fatty acid vesicle formation. *Journal of Physical Chemistry B*, **103**, 10,910–10,916.

27. Luci, P. (2003). *Gene cloning expression and purification of membrane proteins*. ETH-Z Dissertation No. 15108, Swiss Federal Institute of Technology (ETH), Zurich.

28. Luisi, P. L. (2002). Toward the engineering of minimal living cells. *Anatomical Record*, **268**, 208–214.

29. Luisi, P. L. (2003). Autopoiesis: A review and a reappraisal. *Naturwissenschaften*, **90**, 49–59.

30. Luisi, P. L., & Varela, F. J. (1990). Self-replicating micelles—A chemical version of minimal autopoietic systems. *Origin of Life and Evolution of the Biosphere*, **19**, 633–643.

31. Luisi, P. L., Oberholzer, T., & Lazcano, A. (2002). The notion of a DNA minimal cell: A general discourse and some guidelines for an experimental approach. *Helvetica Chimica Acta*, **85**, 1759–1777.

32. Luisi, P. L., Stano, P., Rasi, S., & Mavelli, F. (2004). A possible route to prebiotic vesicle reproduction. *Artificial Life*, **10**, 297–308.

33. Marchi-Artzner, V., Jullien, L., Belloni, L., Raison, D., Lacombe, L., & Lehn, J. M. (1996). Interaction, lipid exchange, and effect of vesicle size in systems of oppositely charged vesicles. *Journal of Physical Chemistry*, **100**, 13,844–13,856.

34. Monnard, P. A. (2003). Liposome-entrapped polymerases as models for microscale/nanoscale bioreactors. *Journal of Membrane Biology*, **191**, 87–97.

35. Morowitz, H. J. (1967). Biological self-replicating systems. *Progress in Theoretical Biology*, **1**, 35–58.

36. Mushegian, A. (1999). The minimal genome concept. *Current Opinion in Genetics and Development*, **9**, 709–714.

37. Mushegian, A. & Koonin, E. V. (1996). A minimal gene set for cellular life derived by comparison of complete bacterial genomes. *Proceedings of the National Academy of Sciences*, **93**, 10,268–10,273.

38. Nissen, P., Hansen, J., Ban, N., Moore, P. B., & Steitz, T. A. (2000). The structural basis of ribosome activity in peptide bond synthesis. *Science*, **289**, 920–930.

39. Noireaux, V. & Libchaber, A. (2004). A vesicle bioreactor as a step toward an artificial cell assembly. *Proceedings of the National Academy of Sciences*, **101**, 17,669–17,674.

40. Noireaux, V., Bar-Ziv, R., & Libchaber, A. (2003). Principles of cell-free genetic circuit assembly.

Proceedings of the National Academy of Sciences, **100**, 12,672–12,677.

41. Nomura, S. M., Tsumoto, K., Hamada, T., Akiyoshi, K., Nakatani, Y., & Yoshikawa, K. (2003). Gene expression within cell-sized lipid vesicles. *ChemBioChem*, **4**, 1172–1175.

42. Oberholzer, T. & Luisi, P. L. (2002). The use of liposomes for constructing cell models. *Journal of Biological Physics*, **28**, 733–744.

43. Oberholzer, T., Albrizio, M., & Luisi, P. L. (1995). Polymerase chain reaction in liposomes. *Chemistry and Biology*, **2**, 677–682.

44. Oberholzer, T., Wick, R., Luisi, P. L., & Biebricher, C. K. (1995). Enzymatic RNA replication in self-reproducing vesicles: An approach to a minimal cell. *Biochemical and Biophysical Research Communications*, **207**, 250–257.

45. Oberholzer, T., Nierhaus, K. H., & Luisi, P. L. (1999). Protein expression in liposomes. *Biochemical and Biophysical Research Communications*, **261**, 238–241.

46. Ono, N. & Ikegami, T. (2000). Self-maintenance and self-reproduction in an abstract cell model. *Journal of Theoretical Biology*, **206**, 243–253.

47. Pantazatos, D. P. & MacDonald, R. C. (1999). Directly observed membrane fusion between oppositely charged phospholipids bilayers. *Journal of Membrane Biology*, **170**, 27–38.

48. Paul, N. & Joyce, G. F. (2002). A self-replicating ligase ribozyme. *Proceedings of the National Academy of Sciences*, **99**, 12,733–12,740.

49. Pietrini, A. V. & Luisi, P. L. (2004). Cell-free protein synthesis through solubilisate exchange in water/oil emulsion compartments. *ChemBioChem*, **5**, 1055–1062.

50. Pohorille, A. & Deamer, D. (2002). Artificial cells: Prospects for biotechnology. *Trends in Biotechnology*, **20**, 123–128.

51. Rasi, S., Mavelli, F., & Luisi, P. L. (2003). Cooperative micelle binding and matrix effect in oleate vesicle formation. *Journal of Physical Chemistry B*, **107**, 14,068–14,076.

52. Roux, A., Cappello, G., Cartaud, J., Prost, J., Goud, B., & Bassereau, P. (2002). A minimal system allowing tubulation with molecular motors pulling on giant liposomes. *Proceedings of the National Academy of Sciences*, **99**, 5394–5399.

53. Sankararaman, S., Menon, G. I., & Kumar, P. B. (2004). Self-organized pattern formation in motor-microtubule mixtures. Physical Review E, 70, 031905. DOI 10.1103/PhysRevE.70.**031905**.

54. Schmidli, P. K., Schurtenberger, P., & Luisi, P. L. (1991). Liposome-mediated enzymatic synthesis of phosphatidylcholine as an approach to self-replicating liposomes. *Journal of the American Chemical Society*, **113**, 8127–8130.

55. Shimkets, L. J. (1998). Structure and sizes of genomes of the Archaea and Bacteria. In F. J. De Bruijn, J. R. Lupskin, and G. M. Weinstock (Eds.), *Bacterial genomes: physical structure and analysis* (pp 5–11). Boston, MA: Kluwer.

56. Spirin, A. (1986). *Ribosome structure and protein synthesis.* Menlo Park, CA: Benjamin Cummings.

57. Stamatatos, L., Leventis, R., Zuckermann, M. J., & Silvius, J. R. (1988). Interactions of cationic lipid vesicles with negatively charged phospholipid vesicles and biological membranes. *Biochemistry*, **27**, 3917–3925.

58. Suttle, D. P. & Ravel, J. M. (1974). The effects of initiation factor 3 on the formation of 30S initiation complexes with synthetic and natural messengers. *Biochemical and Biophysical Research Communications*, **57**, 386–393.

59. Szathmáry, E. (2005). Life: In search of the simplest cell. *Nature*, **433**, 469–470. DOI 10.1038/433469a.

60. Szostak, J. W., Bartel, D. P., & Luisi, P. L. (2001). Synthesizing life. *Nature*, **409**, 387–390.

61. Thomas, C F. & Luisi, P. L. (2004). Novel properties of DDAB: Matrix effect and interaction with oleate. *Journal of Physical Chemistry B*, **108**, 11,285–11,290.

62. Tsumoto, K., Nomura, S. M., Nakatani, Y., & Yoshikawa, K. (2001). Giant liposome as a biochemical reactor: Transcription of DNA and transportation by laser tweezers. *Langmuir*, **17**, 7225–7228.

63. Varela, F., Maturana, H. R., & Uribe, R. B. (1974). Autopoiesis: The organization of living system, its characterization and a model. *Biosystems*, **5**, 187–196.

64. Walde, P., Goto, A., Monnard, P. A., Wessicken, M., & Luisi, P. L. (1994). Oparin's reactions revisited: Enzymatic synthesis of poly (adenylic acid) in micelles and self-reproducing vesicles. *Journal of the American Chemical Society*, **116**, 7541–7544.

65. Walde, P., Wick, R., Fresta, M., Mangone, A., & Luisi, P. L. (1994). Autopoietic self-reproduction of fatty acid vesicles. *Journal of the American Chemical Society*, **116**, 11,649–11,654.

66. Weiner, A. M. & Maizels, N. (1987). tRNA-like structures tag the 3 ends of genomic RNA molecules for replication: Implications for the origin of protein synthesis. *Proceedings of the National Academy of Sciences*, **84**, 7383–7387.

67. Woese, C. R. (1983). The primary lines of descent and the universal ancestor. In D. S. Bendall (Ed.), *Evolution from molecules to man* (pp. 209–233). Cambridge, UK: Cambridge University Press.

68. Yu, W., Sato, K., Wakabayashi, M., *et al.* (2001). Synthesis of functional protein in liposome. *Journal of Bioscience and Bioengineering*, **92**, 590–593.

69. Zhang, B. & Cech, T. R. (1998). Peptidyl-transferase ribozymes: Trans reactions, structural characterization and ribosomal RNA-like features. *Chemical Biology*, **5**, 539–553.

70. Zimmer, C. (2003). Tinker, tailor: Can Venter stitch together a genome from scratch? *Science*, **299**, 1006–1007.

ENDNOTES

a The international meetings were the Third COST D27 Workshop held in Crete in October 2004 (http://cost.cordis.lu/src/action_detail. cfm?action=D27) and The International School on Complexity held in Erice, Sicily, in December 2004 (http://www.ccsem.infn.it). See also Szathmáry (2005).

b One of the referees, whom we particularly thank for acute comments, suggested that it would be actually useful to define "a hierarchy of 'minimal cells'. Some members of this hierarchy might require extensive resources from the environment, such as high-energy compounds. Others might be able to survive in a nutrient-poor environment, presumably more compatible with the 'primordial soup.' In fact, it would be quite interesting to analyze the differences between different members; they would be quite revealing as far as the nature of life goes." This proposal may indeed be the basis for future developments of this kind of work on the minimal cell, particularly when experimental data become available on these different classes of artificial protocells.

22 · Creating "real life"

EVELYN FOX KELLER

In its modern incarnation, use of the term artificial life was at first confined mainly to the world of computer simulations. But when Langton expressed the hope of building models so lifelike that they would be actual examples of life, he deliberately—and provocatively—left open the possibility of constructing these examples in some other (nonvirtual) medium. Indeed, the very ambition to identify "the essence of life" was from the start—for Langton and his colleagues, just as for their precursors in the early part of the last century—linked to the vision of transcending the gap between the living and the non-living. The hope was to create artificial life, not just in cyberspace but in the real world. Rodney Brooks's contribution to the web-based "World Question Center" makes the link explicit: "What is the mathematical essence that distinguishes living from non-living," he asks, "so that we can engineer a transcendence across the current boundaries?"[1] It is hardly surprising, therefore, that artificial life quickly became the operative term referring indiscriminately to digital organisms and to physically embodied robots inhabiting the same four-dimensional world as biological organisms.

In a recent book entitled *Creation: Life and How to Make It*, Steve Grand writes, "Research into artificial life is inspiring a new engineering discipline whose aim is to put life back into technology. Using A-life as an approach to artificial intelligence, we are beginning to put souls into previously lifeless machines ... The third great age of technology is about to start. This is the Biological Age, in which machine and *synthetic* organism merge."[2] Here, the word synthetic reveals yet another ambiguity, referring simultaneously to artificial structures created in the "mirror-world" of cyberspace, and to those built by engineers, "working with [material] objects and combining them to make

new structures" (p. 83). Yet, for all the realism with which digital organisms may be represented on the screen, and for all the seductiveness of the biological lexicon attached to these simulations, engineers, if they are to succeed with such a task, must still grapple with the difference between cyberspace and real space, and with the formidable difficulties encountered in attempting to bridge that gap. Within the A-Life community, however, where explanatory goals remain more abstract, such difficulties tend to be given only glancing attention. As Howard Pattee noted in the first conference held on the subject, "Very little has been said ... about how we would distinguish computer simulations from realizations of life" (p. 63).

Pattee asserts "a categorical difference between the concept of a realization that is a literal, substantial replacement, and the concept of simulation that is a metaphorical representation." Simulations, he writes, "are in the category of symbolic forms, not material substances" (p. 68). And he reminds his readers of the warning von Neumann himself had issued when he wrote, "By axiomatizing automata in this manner, one has thrown half the problem out the window and it may be the more important half."[3] As Pattee sees it, the problem lies first and foremost in the fundamental relation between symbol and matter, and it shows up with particular urgency for this project in the intrinsic dependence of the "reality" of the organism on the "reality" of its environment.

Yet synthetic life forms, made from material components and assembled in real space, are clearly being built, and in ways that draw directly from work on lifelike simulations in cyberspace. Engineering is a science that specializes in negotiating the gap between symbol and matter, and robotic engineers, like their colleagues in allied disciplines, have well-developed

The Nature of Life, ed. M. A. Bedau and C. E. Cleland. Published by Cambridge University Press. © M. A. Bedau and C. E. Cleland 2010.

techniques for translating from one domain to the other, for realizing the metaphors of simulation in the construction of material objects. In one sense, computer simulations of biological organisms obviously are, as Pattee writes, "metaphorical representations," but they are also models in the time-honored sense of guides or blueprints. In the hands of skillful engineers, they can be, and are, used as guides to construction in altogether different mediums. Here, the simulated organisms of cyberspace are used to guide the synthesis of material objects mimicking the behavior of biological organisms in real space: But can we take such a physically realized synthetic creature to be a "literal, substantial replacement" of the creature it has been designed to mimic? If it walks like a duck and quacks like a duck, is it a duck? For that matter, does it even meet the less demanding criteria that would qualify it as *alive*?

These of course are questions that worry many philosophers, just as they do the rest of us. And while they are not directly germane to the concerns of this book, like most people, I have some views on the matter. Very briefly, I would argue that even though synthetic organisms in physical space-time are no longer computer simulations, they are still simulations, albeit in a different medium. Yet I have no confidence in an ineradicable divide between simulation and realization. For one thing, mediums of construction can change, as they surely will. They might even come to so closely resemble the medium in which, and out of which, biological organisms grow that such a divide would no longer be discernible. For another, convergence between simulation and realization, between metaphoric and literal constructions, can also be approached through the manipulation of existing biological materials. For example, computer scientists might come to give up on the project of the *de novo* synthesis of artificial organisms, just as most of today's biological scientists seem to have done. The engineering of novel forms of life in contemporary biology proceeds along altogether different lines, starting not with the raw materials provided in the inanimate world but with the raw materials provided by existing biological organisms. Techniques of genetic modification, cloning, and "directed evolution" have proven so successful for the engineering of biological novelty from parts given to us by biology that the motivation for attempting the synthesis of life *de novo* has all but

disappeared.[4] The implications of such successes have not gone unnoticed by computer scientists.

In fact, work aimed at bridging the gap between computers and organisms by exploiting techniques of biological engineering is well under way in a number of computer science departments. Some efforts are aimed at harnessing DNA for conventional computational purposes; in others, researchers have begun to use the techniques of recombinant DNA to build specific gene regulation networks, pre-designed to respond to particular stimuli, into real bacteria. One example of the latter is part of a larger and far-reaching project that Tom Knight, Jerry Sussman, and Hal Abelson have recently launched at MIT under the name amorphous computing.[5] The motivation for this project is spelled out in a position paper by Knight and Sussman: "Current progress in biology will soon provide us with an understanding of how the code of existing organisms produces their characteristic structure and behavior. As engineers we can take control of this process by inventing codes (and more importantly, by developing automated means for aiding the understanding, construction, and debugging of such codes) to make novel organisms with particular desired properties."[6]

In efforts such as these, no strictures whatever need obtain against using the biochemical machinery that living organisms have themselves evolved. Nor, for that matter, are there any strictures on what is to count as an organism. In fact, in a subsequent paper on the subject, Abelson and Nancy Forbes write, "The ultimate goal of amorphous computing is to draw from biology to help create an entirely new branch of computer science and engineering concerned with orchestrating the use of masses of bulk computational elements to solve problems." In Freeman Dyson's terms, the novel organisms might be green, or gray, or anything in-between.[7]

The bottom line is that with every passing achievement—in biological computing and computational biology—the gap between computers and organisms becomes both ever narrower and more elusive. Thus, the genetic computer of which Davidson and his colleagues speak need no longer be seen as just a metaphor or even as just a model. In two quite different domains—in the designing of new kinds of computers and in the modification of existing organisms—they have begun to approach at least some sense of literality. The route by which this convergence is occurring, however, bears little resemblance to the story usually told about scientific metaphor.[8] Here, the convergence is simultaneously material and conceptual, and one can find no residually literal sense in which any of the referents remain fixed. Furthermore, the metaphor itself can be seen to play a substantive (one might even say instrumental) role in bringing its referents together. Metaphors do far more than affect our perception of the world. In addition to directing the attention of researchers, metaphors guide their activities and material manipulations. In these ways— in many different kinds of laboratories (biological, computational, and industrial), in efforts directed toward a wide variety of ends (theoretical and practical, academic and commercial)—the metaphoric assimilation of computers and organisms works toward the literal realization of hybrid ends. And conspicuous among these is the production of material objects that resist the very possibility of parsing the categories of computer and organism.

Surely, the new "creatures" coming out of biological computing and computational biology are real, but the more pressing question seems to be whether they are *alive*. Has the gap between the living and the non-living, between organism and machine, already been bridged? And if not yet, must we reconcile ourselves to the inevitable joining of these two realms in the near future? If so, how soon? How closely, and in what ways, must the new kinds of entities resemble the products of biological evolution to qualify for the designation "life"?

Such questions are as troubling as they are compelling, and at least part of what makes them so is the anxiety they generate: How will these new creatures threaten our own status on earth? Are we really in danger of being replaced, outpaced in the evolutionary race of the future by a new kind of species?[9] Where, apart from science fiction, are we to look for answers to such questions? Marc Lange argues that the importance of the distinction between living and non-living things is "an empirical question. The answer is for science to discover."[10] Alternatively, Steven Levy claims it is a question for technology, that "by making life we may finally know what life is."[11]

I suggest, however, that it is a mistake to look either to science or to technology. In fact, there are peculiarities to these questions that might disqualify

them as having any place at all in the realm of science. For example, when people ordinarily ask if something is alive, the object at issue is already assumed to belong to the biological realm. The question is thus a diachronic one: Is the object (now, organism) either still or yet alive? Has its life ended, or has it begun? Here, however, the question is aimed not diachronically but synchronically: How is this object to be taxonomically classified? Is it to be grouped with the living or with the non-living? But the very asking of the question in this form depends on a prior assumption—namely, that a defining, essential property for the category of life objectively exists, or that life is what philosophers call a "natural kind." Is life in fact a natural kind, and not merely a human kind? Is it not the case, as Foucault so provocatively argued, that the demarcation between life and non-life ought better to be viewed as a product of human than of evolutionary history?[12] Did not the very notion that it is possible to find "a true definition of life" begin only two centuries ago, with the advocacy of Jean-Baptiste Lamarck?[13]

François Jacob, following Foucault, is one of many who believe that it did. He claims that, prior to the nineteenth century, "The concept of life did not exist." What does he mean by this? Clearly not that the term "life" had not been used earlier, for he proceeds by opposing his concept to such notions of life as had already been in use: he writes that his claim is "shown by the definition in the *Grande Encyclopédie*, an almost self-evident truth: life 'is the opposite of death.'"[14] Jacob's complaint with earlier definitions is that they are not constitutive; they do not provide us with a positive characterization of "the properties of living organisms"; they do not tell us what life *is*.[15] Lamarck had stated the problem in similar terms: "A study of the phenomena resulting from the existence of life in a body provides no definition of life, and shows nothing more than objects that life itself has produced."[16] But in order for a characterization to tell us what life *is*, it must presuppose a modal (and structural) essence of life, a defining property of living beings that is not in itself alive but nonetheless absent from all non-living things.

Just as Lamarck's did, Jacob's concept of life depends on a particular taxonomy of natural objects—one which singles out the boundary between living and non-living as primary and relegates all other distinctions between different modes of living to insignificance. This of course was the taxonomy which Lamarck had so strenuously urged at the beginning of the nineteenth century and which Foucault credited with marking the beginnings of "biology." It contrasts the living not with the dead but with the "inorganic."[17] It highlights one distinction at the expense of others—submerging not only earlier boundaries (most notably, between plants and animals) but also differences of kind between the various sorts of structures that were subsequently to come into prominence (genes, gametes, cells, tissues, organisms, and perhaps even auto-catalytic systems and cellular automata). As long as they possessed the essential defining characteristic, all these structures could—as it were, equally—qualify as instantiations of life.

But by far the most interesting feature of the quest for the defining essence of life, and surely its greatest peculiarity, is that, even while focusing attention on the boundary between living and non-living, emphasizing both the clarity and importance of that divide, this quest for life's essence simultaneously works toward its dissolution. Rodney Brooks and his colleagues are not the first to link the question of what life is with the ambition to transcend current boundaries—the same duality of impulse can already be seen in the writings of Lamarck, and indeed it might be said to inhere in the very demarcation of biology as a separate science. For Lamarck, biology was to be "an enquiry into the physical causes which give rise to the phenomena of life."[18] "Nature has no need for special laws," he wrote, "those which generally control all bodies are perfectly sufficient for the purpose."

Why then demarcate biology as a separate science in the first place? Where, if not to its causes or laws, should we look for the properties that so critically and decisively distinguish the subject matter of this science from that of the physical sciences, that account for the "immense difference," the "radical hiatus" between inorganic and living bodies? (p. 194). Lamarck's answer was to look to the "organization" of living matter: "it is in the simplest of all organisations that we should open our inquiry as to what life actually consists of, what are the conditions necessary for its existence, and from what source it derives the special force which stimulates the movements called vital" (p. 185). Yet, because of his commitment to the adequacy of physical causes and laws, he believed that organization too must have—and must have had—purely physical origins. Hence his interest in

spontaneous generation and the origin of life. In other words, just as for many of his nineteenth- and twentieth-century counterparts, the very demarcation of life as a separate domain served Lamarck as an impetus for the breaching of that boundary—if not practically, then at least conceptually.[19] Those who are currently most interested in the distinguishing properties of organization, however, tend to focus more on the construction of material bridges. But either way, conceptually or materially, such bridges invite the formation of new groupings—groupings that necessarily violate older taxonomies. Instead of linking together in a single category plants and animals, they might conjoin computers and organisms; thunderstorms, people, and umbrellas; or animals, armies, and vending machines.[20]

Should we call these newly formed categories by the name of "life"? Well, that depends. It depends on our local needs and interests, on our estimates of the costs and benefits of doing so, and also, of course, on our larger cultural and historical location. The notion of doing so would have seemed absurd to people living not so long ago—indeed, it seems absurd to me now. But that does not mean either that we will not, or that we should not. It only means that the question "What is life?" is a historical question, answerable only in terms of the categories by which we as human actors choose to abide, the differences that we as human actors choose to honor, and not in either logical, scientific, or technical terms. It is in this sense that the category of life is a human rather than a natural kind. Not unlike explanation.

NOTES

This chapter originally appeared as part of chapter 9 in Evelyn Fox Keller, *Making sense of life: Explaining biological development with models, metaphors, and machines*, pp. 285–294, Cambridge, MA: Harvard University Press, 2002.

REFERENCES

1. Abelson, H. & Forbes, N. (2000). Amorphous computing. *Complexity*, 5(3), 22–25.
2. Arnold, F. H. (2001). Combinatorial and computational challenges for biocatalyst design. *Nature*, **409**, 253–257.

3. Arnold, F. H. & Volkov, A. A. (1999). Directed evolution of biocatalysts. *Current Opinion in Chemical Biology*, 3(1), 54–59.

4. Bennett, C. H. (1986). On the nature and origin of complexity in discrete, homogeneous, locally-interacting systems. *Foundations of Physics*, 16(6), 585–592.

5. Brooks, R. (1997). *World question center*. Edge Foundation, Inc. Available at http://www.edge.org/documents/archive/edge31.html (accessed August, 2008).

6. Doyle, R. (1997). *On being living*. Stanford: Stanford University Press.

7. Dyson, F. (1985). *Infinite in all directions*. New York: Harper & Row.

8. Foucault, M. (1966). *Les mots et les choses: Une archéologie des sciences humaines*. Paris: Gallimard.

9. Grand, S. (2000). *Creation: Life and how to make it*. London: Weidenfeld and Nicholson.

10. Hesse, M. (1980). The explanatory function of metaphor. In M. Hesse, *Revolutions and reconstructions in the philosophy of science* (pp. 111–124). Bloomington, IN: Indiana University Press.

11. Jacob, F. (1976). *The logic of life*. New York: Pantheon.

12. Joyce, G. F. (1992). Directed molecular evolution. *Scientific American*, 267(6), 48–55.

13. Joyce, G. F. (1997). Evolutionary chemistry: Getting there from here. *Science*, 276, 1658–1659.

14. Kauffman, S. A. (1971). Gene regulation networks: A theory for their global structure and behavior. *Current Topics in Developmental Biology*, 6, 145–182.

15. Keller, E. F. (1995). *Refiguring life: Metaphors of twentieth century biology*. New York: Columbia University Press.

16. Knight, T. F., Jr. & Sussman, G. J. (1998). Cellular gate technology. In C. Calude, J. L. Casti, and M. J. Dinneen (Eds.), *Unconventional models of computation* (pp. 257–272). New York: Springer.

17. Lamarck, J.-B. (1809/1984). *Philosophical zoology: An exposition with regard to the natural history of animals*. Chicago: University of Chicago Press.

18. Lange, M. (1996). Life, 'artificial life,' and scientific explanation. *Philosophy of Science*, 63, 135–144.

19. Levy, S. (1993). *Artificial life: The quest for a new creation*. New York: Pantheon Books.

20. Medawar, P. B. (1977). *The life science: Current ideas of biology*. New York: Harper & Row.

21. Moravec, H. (1988). *Mind children: The future of robot and human intelligence*. Cambridge, MA: Harvard University Press.

22. Pattee, H. H. (1989). Simulations, realizations, and theories of life. In C. G. Langton (Ed.), *Artificial life* (Santa Fe Institute studies in the sciences of complexity, proceedings vol. IV) (pp. 63–77). Redwood City, CA: Addison-Wesley.

23. Pirie, N. W. (1937). The meaninglessness of the terms 'life' and 'living.' In J. Needham and D. E. Green (Eds.), *Perspectives in biochemistry* (pp. 11–22). Cambridge, UK: Cambridge University Press.

24. Schiller, J. (1978). *La notion d'organisation dans l'histoire de la biologie*. Paris: Maloine.

25. Sugita, M. (1963). Functional analysis of chemical systems in vivo using a logical circuit equivalent, II: The idea of a molecular automation. *Journal of Theoretical Biology*, 4(2), 179–192.

26. Thomas, R. (1973). Boolean formalization of genetic control circuits. *Journal of Theoretical Biology*, 42(3), 563–585.

27. von Neumann, J. (1966). *Theory of self reproducing automata*, Ed. A. Burks. Urbana: University of Illinois Press.

28. Weiss, R., Homsy, G., & Knight, T. F. (1999). *Toward in vivo digital circuits. Presented at DIMACS workshop on evolution as computation*, Princeton, NJ. Available at http://www.swiss.ai.mit.edU/projects/amorphous/paperlisting.html#invivo-circuits (accessed August, 2008).

ENDNOTES

1 Rodney Brooks (1997), http://www.edge.org/documents/archive/edge31.html.

2 Grand (2000), pp. 7–8.

3 von Neumann (1966), quoted in Pattee (1989), p. 69.

4 Interestingly, work on "directed evolution" also grew out of discussions originally held at the Santa Fe Institute. In directed evolution, enzymes designed to perform specific tasks are produced either by bacteria that have been brought into existence by sequential selection, under conditions ever more closely approximating the targeted task, or by direct selection of proteins produced by laboratory recombination of homologous genes; see, for example, Joyce (1992), (1997); Arnold (2001); and Arnold and Volkov (1999).

5 Weiss *et al.* (1999). This work is of particular interest because it draws its inspiration directly (and explicitly) from the early efforts of Motoyosi Sugita (1963), Stuart Kauffman (1971), and Rene Thomas (1973) to construct formal models of genetic regulatory networks.

6 Knight and Sussman (1998).

7 Abelson and Forbes (2000), p. 25. Dyson (1985).

8 See Mary Hesse's (1980) discussion of scientific metaphors.

9 See Moravec (1988).

10 Lange (1996), p. 231.

11 Levy (1993), p. 10.

12 In *Les mots et les choses* (1966), Foucault made the claim that, in the eighteenth century, "Life itself did not exist" (p. 139), a claim to which many historians have since objected. Joseph Schiller (1978), for example, argued, "The opposite is nearer the truth: the inanimate did not exist but life there was to excess, penetrating everywhere and animating everything" (p. 79). I suggest, however, that Foucault's claim does make historical sense if read as a claim about "life itself," that is, as a claim about life as a natural kind.

13 Lamarck (1984 [1809]) wrote: "A study of the phenomena resulting from the existence of life in a body provides no definition of life, and shows nothing more than objects that life itself has produced. The line of study which I am about to follow has the advantage of being more exact, more direct and better fitted to illuminate the important subject under consideration; it leads, moreover, to a knowledge of the true definition of life" (p. 201).

14 Jacob (1976), p. 89.

15 It is noteworthy that, after the flurry of essays and books by that title in the early part of the twentieth century, the question "What is life?" faded from view among biologists. It was resurrected by Erwin Schrödinger with the publication of his famous book on the subject in 1943 and has remained, ever since, most commonly associated with Schrödinger's name—only rarely if ever posed by contemporary experimental biologists. To P. B. Medawar (1977), such discussions indicate "a low level in biological conversation" (p. 7). By tacit consent, today's biologists appear to concur with the judgment of Norman Pirie from the 1930s that the question is "meaningless." "Nothing turns," wrote Pirie (1937), "on whether a virus is described as a living organism or not" (p. 22). Where the question of what life is does arise today is mainly in A-Life studies and robotics. And like Pirie, we might ask: What hangs on whether these creatures are described as living or not, for either the scientists, the engineers, the industry, or the consumers of their products?

16 Lamarck (1984 [1809]), p. 201.

17 As Lamarck (1984 [1809]) wrote, "If we wish to arrive at a real knowledge of what constitutes life, what it consists of, what are the causes and laws which control so wonderful a natural phenomenon, and how life itself can originate those numerous and astonishing phenomena exhibited by living bodies, we must above all pay very close attention to the differences existing between inorganic and living bodies; and for this purpose a comparison must be made between the essential characters of these two kinds of bodies" (p. 191).

18 Lamarck (1984), p. 282.

19 My argument here is closely related to that of Richard Doyle (1997). Doyle claims that, instead of constituting the actual object of biology, life is (merely) its "sublime" object.

20 The reference to thunderstorms, people, and umbrellas comes from Charles Bennett (1986): "In the modern world view, dissipation has taken over one of the functions formerly performed by God: It makes matter transcend the clod-like nature it would manifest at equilibrium, and behave instead in dramatic and unforeseen ways, molding itself for example into thunderstorms, people and umbrellas" (p. 586), while the reference to animals, armies, and vending machines is from the definition of a system in a 1950 progress report to the U.S. Air Force; see Keller (1995), pp. 90–91.

Section IV
Defining and explaining life

The chapters in this section focus on the big question: What is life? This question cannot be answered by providing a list of things that are or were once alive. It is concerned with the very nature of life, with what qualifies something as a living thing; life in this sense is not contrasted with death but rather with nonlife. This section contains a diverse collection of views, but the authors all share the conviction that there is something special and fundamental about living systems, and hence that a general, unified explanation of life is in order. Not everyone shares this view. As the chapters in Section III by Sober and Keller illustrate, some researchers are skeptical about the wisdom of formulating general theories of life.

Running through this section is a disagreement over whether the stuff that composes life is essential to it. Some authors simply assume that this is the case, although they might also emphasize the importance of the way in which it is structured (spatially arranged). Others explicate life in terms of abstract organizational or functional properties that are independent of the material out of which it is composed. How one decides this issue has critical consequences for the possibility of strong artificial life (see Section III) as well as the possibilities of truly alien forms of natural life (see Section II). Authors (e.g., Dawkins (Ch. 29) and Bedau (Ch. 31)) who identify life with certain abstract properties (respectively, Darwinian evolution and supple adaptation) are open to the possibility of life made of very different kinds of stuff, including stereotypical robots (made of silicon, metal, and plastic), purely informational (e.g., digital) structures, and extraterrestrial silicon-based creatures such as the Horta of Star Trek fame. Among those who believe that there are compositional limitations on the possibilities for life, there is disagreement over just how restrictive they are. Some authors (Koshland (Ch. 24), Ruiz-Mirazo et al.

(Ch. 25), Sterelny & Griffiths (Ch. 28), and Kauffman (Ch. 30)) require only that life be chemical, which rules out informational forms of life and stereotypical robots. At the other extreme, Pace (Ch. 11) restricts life to carbon-containing molecules, which consigns the poor Horta to the class of nonliving things. Benner and colleagues (Ch. 12) defend an intermediate position, exploring some specific possibilities for non-carbon-based, chemical life while dismissing others. More attention needs to be focused on the question of whether and to what extent the stuff of which life is composed is essential to life. Many writers simply assume an answer without bothering to explore and defend it.

The authors in this section also disagree about the proper method for answering the question "what is life?" As many of the chapters in this anthology illustrate, speculation about the origin, extent, and nature of life is commonly framed in terms of a "definition" of life. In a now-classic essay from the 1970s, reprinted as Chapter 23 in this section, astronomer Carl Sagan canvasses the then-popular definitions of life, namely, "physiological," "metabolic," "biochemical," "genetic" (Darwinian natural selection), and "thermodynamic." Sagan points out that none of these definitions is satisfactory; they all face robust counterexamples, ranging from inappropriately including candle flames to inappropriately excluding mules. Moreover, as he points out, they are all based upon a single example of life, namely, contemporary life on Earth. As Sagan points out, and Cleland and Chyba also emphasize in Chapter 26, one cannot safely generalize from a single example of life to all life. It is difficult if not impossible to tell which features of familiar Earth life are essential to *all* life and which are the result of mere evolutionary contingencies.

Sagan's essay generated a flurry of activity on definitions of life. Attempts were made to salvage

The Nature of Life, ed. M. A. Bedau and C. E. Cleland. Published by Cambridge University Press. © M. A. Bedau and C. E. Cleland 2010.

some of the definitions he discussed by refining them in ways that avoid the counterexamples; Bedau's evolutionary definition of life in terms of supple adaptation (vs. natural selection), which is presented in Chapter 31, provides a good example. Less sanguine about the prospects for salvaging them, others proposed new definitions, e.g., ones grounded in autopoieis (Maturana & Varela, 1980) or cybernetics (Korzeniewski, 2001).

Conceding the failures of the most popular definitions of life, biologist Daniel Koshland opts in Chapter 24 for defining life in terms of a cluster of characteristics or, in his terminology, "principles" or "pillars." Koshland takes seven characteristics to be essential to life. To the extent that they focus on how life "operates" these characteristics are functional. Some of them are very abstract, e.g., the requirement that life have a "program," whereas others place (fairly loose) restrictions upon the composition of life, e.g., the requirement for "seclusion," which, as he explains, presupposes chemical pathways. As a consequence, Koshland's definition is not fully abstract. It requires that life be a chemical system of some sort, thus ruling out "soft" artificial life (consisting of informational simulations and other digital constructions) and "hard" artificial life (mechanical robots). Nevertheless he is open to the possibility that these principles could be implemented on chemical systems quite different from Earth life. As Ruiz-Mirazo and colleagues (Ch. 25) point out, however, definitions like Koshland's are redundant and unsystematic, amounting to gerrymandered lists, and hence lacking in explanatory power; see Bedau (1996) for a similar argument.

Philosopher Kepa Ruiz-Mirazo and colleagues attempt to formulate a more logically sophisticated definition of life in Chapter 25. They begin by distinguishing "essential" definitions from "descriptive" definitions. Both sorts of definition are complete in the logical sense of supplying necessary and sufficient conditions (fully determining the membership of the class of items falling under the term being defined). They contend, however, that only essential definitions are capable of explaining life. Following Oparin (Ch. 5), they also insist that a satisfactory definition of life be "genealogical" in the sense of shedding light on the process by which life originates; as discussed in the introduction to Section II, this is a contentious issue.

After criticizing some popular definitions of life, Ruiz-Mirazo and colleagues focus on two definitions that they believe are more promising, namely, NASA's chemical Darwinian definition (Joyce, 1994) and autopoietic definitions (which identify life with the minimal network of processes required for self-maintenance and self-production). They argue that both definitions are problematic. Darwinian definitions cannot explain metabolism, which they maintain is much richer than sometimes portrayed. They also contend that, although they capture much of the richness of metabolism, autopoietic definitions suffer from being too abstract (detached from the material and energetic requirements for implementing self-producing systems) and cannot account for the diversity and growth of biological complexity. Rather than treating one as fundamental and the other as derivative, Ruiz-Mirazo and colleagues argue that a good definition of life should explain how the "individual-metabolic" aspects of life, captured in autopoietic definitions, are physically realized and linked to the "collective-ecological" aspects of life, captured in Darwinian definitions. To achieve this end, they formulate a new definition combining both of these features: "a living system is any autonomous system with open-ended evolutionary capacities." Because it explicitly requires that life be chemical, their definition rules out "soft" and "hard" forms of artificial life; the chemical possibilities, however, are left wide open. Ruiz-Mirazo and colleagues argue that their definition has significant practical advantages for clarifying how life originated on Earth and searching for extraterrestrial life.

Not everyone is convinced that a definition can provide a scientifically compelling explanation of life. Cleland and Chyba (Ch. 26), Sterelny and Griffiths (Ch. 28), and Bedau (Ch. 31; see also Lange (Ch. 17) in Section III) concur that a satisfactory scientific explanation of life requires a universal theory of life. Sterelny and Griffiths and Bedau also believe that provisional definitions can play important roles in the pursuit of such a theory. Cleland and Chyba disagree. They believe that definitions of life are likely to mislead scientists about the essential properties of life, and run the risk of blinding them to truly "weird" forms of life should they encounter them in space missions.

Philosopher Carol Cleland and planetary scientist Christopher Chyba mount a sustained argument against the use of definitions in explanations of life.

They point out that definitions cannot go beyond our current concepts and tell us about a mind-independent world of nature. As a consequence, definitions do not provide very compelling scientific answers to questions about life: Most scientists do not want to know what we currently *believe* about life; they want to know what life *really is*, a point also emphasized by Bedau (1996). Cleland and Chyba discuss a number of different kinds of definition, including operational, stipulative (scientific or theoretical), and "ideal" (complete in the technical logical sense). They also briefly consider and reject a popular alternative to the definitional approach, namely, Wittgensteinian family-resemblance relations, which substitute "loose" clusters of descriptions for necessary or sufficient conditions (or both). On their view, cluster theories still suffer from the central difficulty with the definitional approach: explicating the *nature* of life in terms of our current *concept* of life.

Cleland's and Chyba's discussion of definitions and family resemblance relations raises the perplexing question of how one can identify essential properties of life independently of our current concept of life. On their view, this requires formulating a "theoretical identity statement" (which is not the same as a theoretical definition) derived from an empirically well-grounded, general theory of life. As they discuss, the problem is that we are currently in no position to formulate such a theory because our experience with life is limited to a single example, familiar life on Earth; despite its striking morphological diversity, all known life on Earth is astonishingly similar at the molecular and biochemical level (see Section II). We need additional examples of life in order to formulate a truly general theory of life.

Cleland and Chyba close with a novel proposal for searching for unfamiliar forms of life in the absence of a definition or theory. The basic idea is to utilize familiar Earth life to formulate a diversity of "tentative criteria" for identifying "anomalies," suspicious physical systems that resemble life as we know it in provocative ways and yet also differ from it in perplexing ways. Unlike definitions, tentative criteria are incapable of settling the issue of whether a "suspicious" physical system is alive; their purpose is to identify physical systems that are worthy of further scientific investigation. As Cleland and Chyba discuss, tentative criteria have significant advantages over definitions. They are revisable in light of new information, and

there is no need to pick out which characteristics (e.g., metabolism or reproduction) of familiar Earth life are most fundamental. Tentative criteria might even include characteristics that are not universal to Earth life, e.g., adaptations present in microbes coping with special environmental conditions.

The chapter by Cleland and Chyba raises a number of philosophical questions about the structure of scientific theories and how they explain natural phenomena. Their argument explicitly assumes that life is a *natural kind*—a category of nature, analogous to water, gold or lightning, that would exist even if there had been no human beings to talk or think about it; for more on natural kinds, see Bird and Tobin (2009). If life is not a natural kind then it is doubtful that scientists will be able to formulate a successful universal theory of life. Cleland and Chyba concede that it is an open question whether life is a natural kind. Keller (Ch. 22) disagrees, contending that life is not a natural kind; significantly, her argument does not rest upon an all-encompassing (anti-realist) skepticism about natural kinds. (Anti-realism is discussed below.) Cleland and Chyba raise additional open questions concerning how theoretical identity statements are inferred from scientific theories and justified in terms of empirical evidence, and whether there is really a principled distinction between a theoretical identity statement and a theoretical definition.

Until fairly recently philosophers of science have looked to physics for paradigmatic examples of scientific theories. Isaac Newton's theory provides a familiar example. It unifies an enormous variety of physical phenomena (the motion of freely falling bodies, colliding billiard balls, swinging pendulums, the rise and fall of the tides, orbits of the planets, etc.) under a logical system consisting of three fundamental laws of motion and the law of universal gravitation and a class of theoretical concepts (force, mass, space, and time). But as philosophers of science have come to recognize, not all scientific theories closely resemble those of physics. Laws of physics are traditionally characterized as exceptionless, (physically vs. logically) necessary, universal generalizations; see Harré (2000) for more detail. But no one has been able to identify any biologically distinctive generalizations of this sort (Beatty, 1995). The same is true for the distinctive generalizations of other "special sciences" such as geology and psychology. While it is commonly thought that the generalizations of the special sciences are "reducible" to those of fundamental physics in principle but not in practice (because they are too complex), recent work in philosophy of science suggests that the traditional analysis of natural laws might be mistaken. Nancy Cartwright (e.g., 1983) has persuasively argued that even the most fundamental generalizations of physics are subject to exceptions. In light of these and other considerations (such as the problem of making sense of what constitutes physical, as opposed to logical or metaphysical, necessity), some philosophers have proposed less demanding conceptions of natural law (see, e.g., Mitchell, 2002). Others argue against a one-size-fits-all conception of scientific theories, contending that while all scientific theories require general principles of *some* sort, they take different forms in different scientific disciplines, sometimes describing robust patterns or causal mechanisms instead of exceptionless regularities (see Sterelny and Griffiths, Ch. 28).

The items related by the general principles of a scientific theory include a class of *theoretical kinds*. Some examples are mass (Newton's theory), gene (genetics), and ego (Freudian psychology). Theoretical kinds cannot be directly "observed" with the unaided human senses, and yet it seems clear that they are crucial to the explanatory power of scientific theories. As a consequence, there is a lively debate among philosophers over their metaphysical status. Should we construe them as natural kinds—categories that are (so to speak) carved out by nature, as opposed to human interests and concerns—or merely as useful tools for systematizing and unifying phenomenal experience? Scientific realists take the latter position, whereas anti-realists (e.g., instrumentalists) take the former position; for more on this debate, see Psillos (1999). In this context it is important to keep in mind that no realist believes that all the theoretical kinds postulated by current scientific theories represent genuine divisions in nature. As history counsels us, some theories will eventually be replaced by better ones appealing to new theoretical kinds.

Much of the explanatory power of a scientific theory derives from the manner in which it subsumes phenomenal properties and categories under basic theoretical principles and categories. Cleland and Chyba contend that the vehicles for such explanations are theoretical identity statements—made famous by philosophers Saul Kripke (1972) and Hilary Putnam (1973, 1975) in their discussions of natural kinds. The well-worn example is "water is H_2O," which ostensibly identifies the water we experience in everyday life in terms of theoretical categories from contemporary chemistry (e.g., hydrogen, oxygen, and covalent bonding).

Theoretical identity statements closely resemble logically complete definitions insofar as they provide necessary and sufficient conditions for membership in the extension of a general term. As a consequence, some philosophers suspect that the distinction between theoretical identity statements and theoretical definitions is not very significant. Because they are stipulative, theoretical definitions no more dissect the concept that we associate with a general term, such as 'life,' than theoretical identity statements; theoretical definitions represent little more than explicit decisions to change the concept that we associate with a term. Moreover, it has been argued that once scientists became convinced that water is H_2O, the claim that water is H_2O took on the status of a definition—it literally became our *scientific* concept of water. Cleland and Chyba reject such arguments, arguing that theoretical identity statements are established through empirical investigation, as opposed to stipulation or conceptual analysis. In other work, Cleland (2006, forthcoming) goes further, arguing that theoretical identity statements are not even true identities, which further undermines the claim that they are theoretical definitions. On the other hand, many scientists who talk about defining life are not familiar with the technical logical notion of definition. It might therefore seem more appropriate to reinterpret their "definitions" as theoretical identity statements. Cleland and Chyba would reject such a move on the grounds that successful theoretical "identities" presuppose (are articulated in terms of) scientific theories, and we currently lack even an empirically compelling framework for a general theory of life. This raises the intriguing question of what distinguishes a complex theoretical definition or theoretical identity statement from a theory. These multifaceted issues are in need of more attention from philosophers of science; see Psillos (1999, Sec. IV) and Cleland (2011) for more discussion.

Chapter 27 is a poetic essay by microbiologist Lynn Margulis and science writer Dorion Sagan, taken from their book about the diverse interlocking web of life on Earth. They emphasize that material and energy are ceaselessly flowing through all organisms, and so life should be viewed as a process. Organisms are viewed as regular, developing, and adapting patterns that emerge from the flux of material and energy through certain kinds of highly complex chemical systems. This perspective considers the autopoietic self-maintenance and self-production achieved by the process of metabolism to be essential to life. But because life is continually expanding and developing in unpredictable directions, Margulis and Sagan conclude that "autopoiesis, reproduction, and evolution only begin to encompass the fullness of life." They doubt that life is the sort of thing that can be captured in a definition or other fixed generalization, because life is continually changing and evolving into new forms that engage in ever more complex interactions.

Margulis has been a stalwart champion of the Gaia hypothesis, first formulated by scientist James Lovelock (1979). This hypothesis proposes that the surface of the Earth, together with the air, water, and life forms covering it, form a complex, self-regulating and self-maintaining system, dubbed "Gaia." Whether the chemical self-maintenance and self-regulation postulated by this hypothesis actually happens is still debated. Going beyond the Gaia hypothesis, Margulis and Sagan also express the more controversial belief that Gaia (the actual global self-maintaining chemical and physical system) should be considered to be an atypical but genuine form of life. Sterelny and Griffiths discuss some empirical and conceptual problems that arise from viewing Gaia as a bona fide living organism in Chapter 28.

From the perspective of Margulis and Sagan it is misleading to think of life primarily in terms of *individual* organisms such as a (single) bacterium, redwood tree, or antelope. On their view, the Earth is a four-dimensional (space-time), interwoven and integrated, whole living community simultaneously existing at many levels of organization. It is thus a mistake to think that the properties traditionally attributed to life, e.g., metabolism, reproduction, and evolution, exhaust the essence of life. All organisms lead multiple lives as individuals and members of different communities: "[H]umans are animals are microbes are chemicals." Indeed, they contend that humanity itself is transforming from a society into a new level of organic being and that Gaia could grow beyond the Earth, into the solar system and beyond.

The holistic Gaian perspective is a radical break from traditional biological theorizing about life, which focuses on compact individual organisms and their

populations, but the hypothesis might yet become a convincing scientific theory with general principles and theoretical kinds. In fact, something very much like this famously happened with an earlier theory championed by Margulis. The endosymbiotic theory that eukaryotic life forms arose from the symbiotic fusion of free-living bacteria (Margulis, 1981) was initially ridiculed when she introduced it, but it is now standard fare in biology textbooks.

Like Cleland and Chyba, philosophers Kim Sterelny and Paul Griffiths (Ch. 28) are less concerned with providing an actual explanation of life than laying out conditions for formulating one. On the assumption that life is a natural kind, they argue that its essence is discoverable through normal empirical science, analogous to the way in which we discovered that water is H_2O. They emphasize that a definition of life is not required for biologists to make scientific progress, any more than ordinary speakers of English need to consult a dictionary in order to use the word 'life' in ordinary conversations. In their view, having a definition will not help settle odd, controversial, borderline cases like viruses, prions, social insect colonies, and Gaia. Instead, pre-theoretic intuitions about these cases will tend to determine how we evaluate candidate definitions. Nevertheless, they believe that definitions are useful for marking distinctions that might otherwise be confused. Most importantly, however, they believe that one can pursue an empirical program of universal biology, and attempt to study the essential characteristics of all actual and possible forms of life, in the absence of a definition of 'life.'

Sterelny and Griffiths point out that the program of universal biology and the search for a universal theory of life face some serious obstacles. First, there is the challenge of discovering biologically distinctive natural laws. As mentioned above, philosophers have traditionally considered natural laws to be exceptionless, necessary, universal generalizations, and biologists have yet to discover any distinctively biological generalizations of this sort. But as Sterelny and Griffiths note, this conception of the principles making up scientific theories has been challenged in recent years, and theorizing in biology seems to fit some of the looser conceptions that have been proposed. Sterelny and Griffiths also stress that life on Earth provides a poor foundation for generalizing to all life. They criticize well-known attempts by Gould (1989), Dawkins (Ch. 29), and Kauffman (1993, 1995; see also Ch. 30) to identify universal principles for life, and one of their central themes is that the program of universal biology needs to be more concerned with how biological complexity is physically realized. They would thus presumably approve of the general biochemical orientation of the discussions of universal features of life by Pace (Ch. 11) and Benner and colleagues (Ch. 12), though they might question some specific conclusions.

Sterelny and Griffiths are skeptical about artificial life simulations, describing the field as "over-hyped." They think that artificial life simulations can supply valuable information about life by drawing out empirically testable consequences from theoretical speculations. However, they reject the strong (soft) artificial life thesis that a computer simulation could literally be alive, and the functionalist perspective on life with which it is associated. They doubt that artificial life computer systems will ever produce a second example of life, because they find strong (soft) artificial life's distinction between form (function, structure, and organization) and matter (material composition) to be naïve and insufficiently hierarchical. It is an open question whether more sophisticated hierarchical forms of functionalism about life, incorporating more specific physical constraints at the lower levels, could escape their criticism and win favor.

The remaining chapters in this section (Dawkins, Kauffman, and Bedau) represent three different contemporary approaches to formulating a general theory of life. These theories attempt to unify a wide range of relevant phenomena (from extraterrestrial life to artificial life) under a single theoretical framework. The reader is urged to apply the considerations about scientific theories and life discussed in earlier chapters of this section to them.

Evolutionary biologist Richard Dawkins argues in Chapter 29 that we need look no further for an adequately general theory of life than Darwin's theory of evolution by natural selection. He defends the widely accepted biological view that natural selection provides the best explanation of the properties that give individual organisms the appearance of design (self-organization, self-reproduction, self-maintenance, adaptation, etc.). Dawkins thinks that adaptive complexity is the best diagnostic of the presence of life itself. Dawkins intends this thesis to be interpreted very broadly, and to cover any form of life that could exist anywhere in the universe, thus contributing to

"universal biology," which he defines as the study of what is uniquely true of all actual and possible forms of life. He expands on this thesis as follows: "If you find something, anywhere in the universe, whose structure is complex and gives the strong appearance of having been designed for a purpose, then that something is either alive, or was once alive, or is an artifact created by something alive." The bulk of this chapter is devoted to Dawkins's argument for a second thesis: that Darwinian evolution (natural selection operating on heritable random variation) is the only mechanism ever suggested that adequately explains organized, adaptive complexity. His argument consists of considering and rejecting each of what he thinks are the viable alternatives to Darwinian natural selection. Considerable space is spent arguing against the possibility of Lamarckian evolution (the inheritance of acquired characteristics); Sterelny and Griffiths (Ch. 28), however, are not convinced by these arguments.

In several influential works, biochemist and theoretical biologist Stuart Kauffman (1993, 1995) has developed a different foundation for a general theory of life; the essay reprinted in this section as Chapter 30 provides a good overview of his theory. Kauffman rejects the widespread view that evolutionary processes of some sort (for Dawkins, natural selection) are the primary source of organized adaptive complexity, which encompasses self-organization, self-reproduction, self-maintenance, and adaptation. For Kauffman, organized adaptive complexity is best explained in terms of physics, namely, the spontaneous emergence of organization at a certain level of complexity in non-equilibrium (i.e., open thermodynamic) chemical reaction systems. As the molecular diversity of the system increases, a critical threshold ("phase transition") is reached at which a collectively autocatalytic, self-reproducing chemical reaction network emerges spontaneously, enabling the system to act on its own behalf (to exhibit the organizational features traditionally attributed to life). Thus, on Kauffman's theory, organized adaptive complexity arises prior to natural selection; natural selection, which he believes plays an important role in the history of life, begins with more order than many suppose. Life, for Kauffman, is a "collectively autocatalytic molecular system."

Although not an advocate of strong (soft) artificial life, Kauffman is one of the founding fathers of artificial life, and his counterarguments to Darwinian views of life rest on a family of sophisticated computer simulations of complex adaptive systems. Kauffman emphasizes that these soft artificial life models generate a large amount of unexpected order, and that this order comes "for free" without any need for a Darwinian explanation. Kauffman's argument applies to open thermodynamic systems, and he considers the expected behavior of the systems in different parameter regimes. In one model of a random Boolean network, if the system is above a certain minimal threshold of complexity, then it will spontaneously tend to exhibit a collective stable dynamic, because there are "small attractors" in the space of the system's possible states. A second model represents a set of reacting chemical species that is autocatalytic in the sense that each chemical in the set is produced by reactions involving, and catalyzed by, only the other chemicals in the set. So, in the aggregate, the set of chemicals catalyzes its own production.

Kauffman thinks that these abstract autocatalytic sets provide theoretically fruitful models of minimal chemical living systems. He argues that such models can explain the emergence of life from nonliving chemicals. Autocatalytic sets also give rise to an interesting non-Darwinian form of adaptive evolution. This evolution consists merely of change in concentration of chemical species. Sometimes old chemical species go extinct and disappear from the set, and new chemical species sometimes arise and grow in concentration. This model generates order "for free" because large enough chemical networks are virtually guaranteed on statistical grounds to be autocatalytic. Since Kauffman's collectively autocatalytic systems can evolve even though devoid of macromolecules, Kauffman disputes Schrödinger's famous claim in Chapter 4 that life requires a macromolecule to serve as genome. (Many of Kauffman's central arguments are explained and evaluated by Burian and Richardson (1991).)

In Chapter 31 philosopher and artificial life scientist Mark Bedau proposes what he calls a "definition" of life, but he is clear that this definition is not concerned with the meaning of the word 'life' or the analysis of anyone's current conception of life. Bedau's definition is proposed as the core of a general theory of life, one taken to unify and explain the central phenomena of life. Bedau contends that a final definition of life should be expected only after we have constructed a fully adequate theory of life. But while some philosophers and scientists (e.g. Cleland and Chyba) think that we lack sufficient evidence to construct useful theories of life, Bedau believes that proposing and testing concrete hypotheses today will accelerate our understanding of life. While we should expect that any theories we formulate today will be judged false in the future, today's proposals can still aid our search for better theories (Wimsatt, 1987).

The theory of life as supple adaptation shares a general evolutionary orientation with Dawkins' Darwinian theory in Chapter 29, but the details in the two theories are quite different. Bedau's central substantive proposal is that life, at least in its primary form, is the open-ended and indefinitely creative evolutionary process of supple adaptation. Supple adaptation involves natural selection, but elsewhere Bedau (1996, 2009) has disputed Dawkins's conclusion that natural selection sufficiently explains life's adaptive complexity; he concludes that supple adaptation involves other creative global processes besides natural selection. Bedau tentatively suggests that we view the process of supple adaptation to be life's *primary* form, on the grounds that this process provides a natural and generally compelling explanation of life's hallmarks and puzzles. Individual organisms are genuine forms of life on this theory, but they count as *secondary* products of the primary process of supple adaptation. The theory of life as supple adaptation explains the phenomena of life by focusing not on individual organisms but rather on the supply adapting interconnected network of organisms. This proposal is holistic, like the Gaian perspective of Margulis and Sagan, and this holism leaves Bedau unperturbed by the traditional problems for evolution-centered accounts of life, such as organisms like mules that cannot reproduce. Since mules are natural secondary components of the larger process of supple adaptation, they qualify as secondary forms of life, alongside reproductively active organisms.

Although many authors in this book talk about "theories" of life, it is sometimes unclear what such theories are supposed to explain. The central methodological thesis in Bedau's chapter is that theories of life should be judged by how well they can account for life's hallmarks and puzzles; Lange also endorses this approach in Section III (Ch. 17). Bedau's chapter illustrates this methodological thesis by showing how the theory of life as supple adaptation explains four specific puzzles about life: (a) how different forms of life at different hierarchical levels are related, (b) whether the distinction between life and non-life is a dichotomy or a matter of degree, (c) whether the nature of life concerns its material composition or its form, and (d) whether life and mind are intrinsically related. Bedau thinks that this approach is the most useful way to evaluate all general theories of life, including those proposed by philosophers (like Ruiz-Mirazo and colleagues) or by scientists (including Dawkins and Kauffman), as well as the compositional explanations of Pace (Ch. 11) and Benner and colleagues (Ch. 12) and the minimal chemical cell proposals of Deamer (Ch. 20), Luisi and colleagues (Ch. 21) and Gánti (Ch. 7). Bedau's theory might have to confront a lack of consensus about which of the features of familiar life are most deeply puzzling and in need of explanation. It is instructive to compare Bedau's list with those of Mayr (Ch. 6), Gánti (Ch. 7), Sagan (Ch. 23), and Koshland (Ch. 24). It should be noted, however, that Bedau's methodological approach is not tied to any particular set of hallmarks. Theories of life can be evaluated by how well they explain the hallmarks of life, as Bedau suggests, whatever those hallmarks turn out to be.

REFERENCES

Beatty, J. (1995). The evolutionary contingency thesis. In G. Wolters & J. G. Lennox (Eds.), *Concepts, theories, and rationality in the biological sciences* (pp. 45–81). Pittsburgh: University of Pittsburgh Press.

Bedau, M. A. (1996). The nature of life. In M. A. Boden (Ed.), *The philosophy of artificial life* (pp. 332–357). New York: Oxford University Press.

Bedau, M. A. (2009). The evolution of complexity. In A. Barberousse, M. Morange, & T. Pradeu (Eds.), *Mapping the future of biology: Evolving concepts and theories* (pp. 111–131). Berlin: Springer.

Bird, A. & Tobin, E. (2009). Natural kinds. Forthcoming in E. N. Zalta (Ed.), *The Stanford encyclopedia of philosophy* (Spring2009 edition). Will be available on the web at http://plato.stanford.edu/archives/spr2009/entries/natural-kinds/.

Burian, R. M. & Richardson, R. C. (1991). Form and order in evolutionary biology. In A. Fine, M. Forbes, & L. Wessels (Eds.), *PSA 1990: Proceedings of the 1990 Biennial Meeting of the Philosophy of Science Association, II*(East Lansing, MI: Philosophy of Science Association) (pp. 267–287). Reprinted (with minor revisions) in M. Boden (Ed.), *The philosophy of artificial life* (pp. 146–172). New York: Oxford University Press, 1996.

Cartwright, N. (1983). *How the laws of physics lie.* Oxford: Oxford University Press.

Cleland, C. E. (2006). Understanding the nature of life: A matter of definition or theory? In J. Seckbach (Ed.), *Life as we know it* (pp. 589–600). Dordrecht: Springer.

Cleland, C. E. (forthcoming). *The quest for a universal theory of life: Searching for life as we don't know it.* Cambridge: Cambridge University Press.

Gould, S. J. (1989). *Wonderful life: The Burgess shale and the nature of history.* New York: Norton & Co.

Harré, R. (2000). Laws of nature. In W. H. Newton-Smith (Ed.), *A companion to the philosophy of science* (pp. 213–223). Oxford: Blackwell Pubs Ltd.

Kauffman, S. A. (1993). *The origins of order: Self-organization and selection in evolution.* New York: Oxford University Press.

Kauffman, S. A. (1995). *At home in the universe: The search for laws of self-organization and complexity.* New York: Oxford University Press.

Korzeniewski, B. (2001). Cybernetic formulation of the definition of life. *Journal of Theoretical Biology,* **209,** 275–286.

Kripke, S. A. (1972). *Naming and necessity.* Cambridge: Harvard University Press.

Lovelock, J. E. (1979). *Gaia: A new look at life on Earth.* Oxford: Oxford University Press.

Margulis, L. (1981). *Symbiosis in cell evolution.* New York: Freeman.

Maturana, H. R. & Varela, F. J. (1980). *Autopoiesis and cognition: The realization of the living.* Dordrecht: D. Reidel Publishing Co.

Mitchell, S. D. (2002). *Ceteris paribus:* An inadequate representation for biological contingency. *Erkenntnis,* **57,** 329–350.

Psillos, S. (1999). *Scientific realism: How science tracks the truth.* London: Routledge.

Putnam, H. (1973). Explanation and reference. In R. J. Pearce & P. Maynard (Eds.), *Conceptual change* (pp. 199–221). Dordrecht: D. Reidel.

Putnam, H. (1975). The meaning of 'meaning.' In K. Gunderson (Ed.), *Language, mind, and knowledge* (pp. 215–271). Minneapolis: University of Minnesota Press.

Wimsatt, W. C. (1987). False models as means to truer theories. In M. Niteckiand & A. Hoffman (Eds.), *Neutral modes in biology* (pp. 23–55). Oxford: Oxford University Press.

23 · Definitions of life

CARL SAGAN

A great deal is known about life. Anatomists and taxonomists have studied the forms and relations of more than a million separate species of plants and animals. Physiologists have investigated the gross functioning of organisms. Biochemists have probed the biological interactions of the organic molecules that make up life on our planet. Molecular biologists have uncovered the very molecules responsible for reproduction and for the passage of hereditary information from generation to generation, a subject that geneticists had previously studied without going to the molecular level. Ecologists have inquired into the relations between organisms and their environments; ethologists, the behavior of animals and plants; embryologists, the development of complex organisms from a single cell; evolutionary biologists, the emergence of organisms from preexisting forms over geological time. Yet despite the enormous fund of information that each of these biological specialties has provided, it is a remarkable fact that no general agreement exists on what it is that is being studied. There is no generally accepted definition of life. In fact, there is a certain clearly discernible tendency for each biological specialty to define life in its own terms. The average person also tends to think of life in his own terms. For example, the man in the street, if asked about life on other planets, will often picture life of a distinctly human sort. Many individuals believe that insects are not animals, because by "animals" they mean "mammals." Man tends to define in terms of the familiar. But the fundamental truths may not be familiar. Of the following definitions, the first two are in terms familiar in everyday life; the next three are based on more abstract concepts and theoretical frameworks.

PHYSIOLOGICAL

For many years a physiological definition of life was popular. Life was defined as any system capable of performing a number of such functions as eating, metabolizing, excreting, breathing, moving, growing, reproducing, and being responsive to external stimuli. But many such properties are either present in machines that nobody is willing to call alive, or absent from organisms that everybody is willing to call alive. An automobile, for example, can be said to eat, metabolize, excrete, breathe, move, and be responsive to external stimuli. And a visitor from another planet, judging from the enormous numbers of automobiles on the Earth and the way in which cities and landscapes have been designed for the special benefit of motorcars, might well believe that automobiles are not only alive, but are the dominant life form on the planet. Man, however, professes to know better. On the other hand, some bacteria do not breathe at all, but instead live out their days by altering the oxidation state of sulfur.

METABOLIC

The metabolic definition is still popular with many biologists. It describes a living system as an object with a definite boundary, continually exchanging some of its materials with its surroundings, but without altering its general properties, at least over some period of time. But again there are exceptions. There are seeds and spores that remain, so far as is known, perfectly dormant and totally without metabolic activity at low temperatures for hundreds, perhaps thousands, of years, but that can revive perfectly well upon being subjected to more clement conditions. A flame, such as

The Nature of Life, ed. M. A. Bedau and C. E. Cleland. Published by Cambridge University Press. © M. A. Bedau and C. E. Cleland 2010.

that of a candle in a closed room, will have a perfectly defined shape with fixed boundary, and will be maintained by the combination of its organic waxes with molecular oxygen, producing carbon dioxide and water. A very similar chemical reaction, incidentally, is fundamental to most animal life on Earth. Flames also have a well-known capacity for growth.

BIOCHEMICAL

A biochemical or molecular biological definition sees living organisms as systems that contain reproducible hereditary information coded in nucleic acid molecules, and that metabolize by controlling the rate of chemical reactions using proteinaceous catalysts known as enzymes. In many respects, this is more satisfying than the physiological or metabolic definitions of life. There are, however, even here, the hints of counter-examples. There seems to be some evidence that a virus-like agent called scrapie contains no nucleic acids at all, although it has been hypothesized that the nucleic acids of the host animal may nevertheless be involved in the reproduction of scrapie. Furthermore, a definition strictly in chemical terms seems peculiarly vulnerable. It implies that were man able to construct a system that had all the functional properties of life, it would still not be alive if it lacked the molecules that earthly biologists are fond of—and made of.

GENETIC

All organisms on Earth, from the simplest cell to man himself, are machines of extraordinary powers, effortlessly performing complex transformations of organic molecules, exhibiting elaborate behavior patterns, and indefinitely constructing from raw materials in the environment more or less identical copies of themselves. How could machines of such staggering complexity and such stunning beauty ever arise? The answer, for which today there is excellent scientific evidence, was first discerned by Charles Darwin in the years before the publication in 1859 of his epoch-making work, the *Origin of Species*. A modern rephrasing of his theory of natural selection goes something like this: Hereditary information is carried by large molecules known as genes, comprised in part of nucleic acids. Different genes are responsible for the expression of different characteristics of the organism.

During the reproduction of the organism the genes also reproduce, or replicate, passing the instructions for various characteristics on to the next generation. Occasionally, there are imperfections, called mutations, in gene replication. A mutation alters the instructions for a particular characteristic or characteristics. It also breeds true, in the sense that its capability for determining a given characteristic of the organism remains unimpaired for generations until the mutated gene is itself mutated. Some mutations, when expressed, will produce characteristics favourable for the organism; organisms with such favourable genes will reproduce preferentially over those without such genes. Most mutations, however, turn out to be deleterious and often lead to some impairment or to death of the organism. To illustrate, it is unlikely that one can improve the functioning of a finely crafted watch by dropping it from a tall building. The watch may run better, but this is highly improbable. Organisms are so much more finely crafted than the finest watch that any random change is even more likely to be deleterious. The accidental beneficial and hereditable change, however, does on occasion occur; it results in an organism better adapted to its environment. In this way organisms slowly evolve toward better adaptation, and, in most cases, toward greater complexity. This evolution occurs, however, only at enormous cost; man exists today, complex and reasonably well adapted, only because of billions of deaths of organisms slightly less adapted and somewhat less complex. In short, Darwin's theory of natural selection states that complex organisms developed, or evolved, through time because of replication, mutation, and replication of mutations. A genetic definition of life therefore would be a system capable of evolution by natural selection.

This definition places great emphasis on the importance of replication. Indeed, in any organism enormous biological effort is directed toward replication, although it confers no obvious benefit on the replicating organism. Some organisms, many hybrids for example, do not replicate at all. But their individual cells do. It is also true that life defined in this way does not rule out synthetic duplication. It should be possible to construct a machine that is capable of producing identical copies of itself from preformed building blocks littering the landscape, but that arranges its descendants in a slightly different manner if there is a random change in its instructions. Such a machine would, of course, replicate its instructions as well. But the fact that such a machine would satisfy the genetic definition of life is not an argument against such a definition; in fact, if the building blocks were simple enough, such a machine would have the capability of evolving into very complex systems that would probably have all the other properties attributed to living systems. The genetic definition has the additional advantage of being expressed purely in functional terms: It does not depend on any particular choice of constituent molecules. The improbability of contemporary organisms—dealt with more fully below—is so great that these organisms could not possibly have arisen by purely random processes and without historical continuity. Fundamental to the genetic definition of life then is the belief that a certain level of complexity cannot be achieved without natural selection.

THERMODYNAMIC

Thermodynamics distinguishes between open and closed systems. A closed system is isolated from the rest of its environment and exchanges neither light, heat, nor matter with its surroundings. An open system is one in which such exchanges do occur. The second law of thermodynamics states that, in a closed system, no processes can occur that increase the net order (or decrease the net entropy) of the system. Thus the universe taken as a whole is steadily moving toward a state of complete randomness, lacking any order, pattern, or beauty. This fate has been known since the nineteenth century as the "heat death" of the universe. Yet living organisms are manifestly ordered and at first sight seem to represent a contradiction to the second law of thermodynamics. Living systems might then be defined as localized regions where there is a continuous increase in order. "Living systems," however, are not really in contradiction to the second law. They increase their order at the expense of a larger decrease in order of the universe outside. Living systems are not closed, but rather open. Most life on Earth, for example, is dependent on the flow of sunlight, which is utilized by plants to construct complex molecules from simpler ones. But the order that results here on Earth is more than compensated by the decrease in order on the Sun, through the thermonuclear processes responsible for the Sun's radiation.

H. Morowitz has argued on grounds of quite general open-system thermodynamics that the order of a system increases as energy flows through it, and moreover that this occurs through the development of cycles. A simple biological cycle on the Earth is the carbon cycle. Carbon from atmospheric carbon dioxide is incorporated by plants and converted into carbohydrates through the process of photosynthesis. These carbohydrates are ultimately oxidized by both plants and animals to extract useful energy locked in their chemical bonds. In the oxidation of carbohydrates, carbon dioxide is returned to the atmosphere, completing the cycle. Morowitz believes that very similar cycles develop spontaneously and in the absence of life by the flow of energy through a chemical system. In this view, biological cycles are merely an exploitation by living systems of those thermodynamic cycles that pre-exist in the absence of life. It is not known whether open-system thermodynamic processes in the absence of replication are capable of leading to the sorts of complexity that characterize biological systems. It is clear, however, that the complexity of life on Earth has arisen through replication, although thermodynamically favored pathways have certainly been used.

SUMMARY

The existence of five such diverse definitions of life surely means that life is something complicated. A fundamental understanding of biological systems has existed since the second half of the nineteenth century. But the number and diversity of definitions suggest something else as well. As detailed below, all the organisms on the Earth are extremely closely related, despite superficial differences. The fundamental ground pattern, both in form and in matter, of all life on Earth is essentially identical. As will emerge below, this identity probably implies that all organisms on Earth are evolved from a single instance of the origin of life. It is difficult to generalize from a single example, and in this respect the biologist is fundamentally handicapped as compared, say, to the chemist, or physicist, or geologist, or meteorologist, who now can study aspects of his discipline beyond the Earth. If there is truly only one sort of life on Earth, then perspective is lacking in the most fundamental way. It is not known what aspects of living systems are necessary in the sense that living systems everywhere must have them; it is not known what aspects of living systems are contingent in the sense that they are the result of evolutionary accident, so that somewhere else a different sequence of events might have led to different characteristics. In this respect the possession of even a single example of extraterrestrial life, no matter how seemingly elementary in form or substance, would represent a fundamental revolution in biology. It is not known whether there is a vast array of biological themes and counterpoints in the universe; whether there are places that have fugues, compared with which our one tune is a bit thin and reedy. Or it may be that our tune is the only tune around. Accordingly the prospects for life on other planets must be considered in any general discussion of life.

NOTES

This chapter originally appeared as the first section in the entry for "Life" in *Encyclopedia Britannica*, pp. 1083–1083A, Chicago: Encyclopædia Britannica Incorporated, 1970.

24 · The seven pillars of life

DANIEL E. KOSHLAND

What is the definition of life? I remember a conference of the scientific elite that sought to answer that question. Is an enzyme alive? Is a virus alive? Is a cell alive? After many hours of launching promising balloons that defined life in a sentence, followed by equally conclusive punctures of these balloons, a solution seemed at hand: "The ability to reproduce—that is the essence characteristic of life" said one statesman of science. Everyone nodded in agreement that the essence of a life was the ability to reproduce, until one small voice was heard. "Then one rabbit is dead. Two rabbits—a male and female—are alive but either one alone is dead." At that point, we all became convinced that although everyone knows what life is there is no simple definition of life.

If I were forced to rush in where angels fear to tread, I would offer "a living organism is an organized unit, which can carry out metabolic reactions, defend itself against injury, respond to stimuli, and has the capacity to be at least a partner in reproduction." But I'm not happy with such a brief definition. When allowed more extensive reflection, however, I think the fundamental pillars on which life as we know it is based can be defined. By "pillars" I mean the essential principles—thermodynamic and kinetic—by which a living system operates. Current interest in discovering life in other galaxies and in recreating life in artificial systems indicates that it would be desirable to elucidate those pillars, their operation, and why they are essential to life. In this chapter, I will refer to the particular mechanisms by which those principles are implemented in life on Earth, while reserving the right to suggest that there may be other mechanisms to implement the principles. If I were in ancient Greece, I would create a goddess of life whom I would call PICERAS, for reasons that will become clear.

The first pillar of life is a Program. By program I mean an organized plan that describes both the ingredients themselves and the kinetics of the interactions among ingredients as the living system persists through time. For the living systems we observe on Earth, this program is implemented by the DNA that encodes the genes of Earth's organisms and that is replicated from generation to generation, with small changes but always with the overall plan intact. The genes in turn encode for chemicals—the proteins, nucleic acids, etc.—that carry out the reactions in living systems. It is in the DNA that the program is summarized and maintained for life on Earth.

The second pillar of life is Improvisation. Because a living system will inevitably be a small fraction of the larger universe in which it lives, it will not be able to control all the changes and vicissitudes of its environment, so it must have some way to change its program. If, for example, a warm period changes to an ice age so that the program is less effective, the system will need to change its program to survive. In our current living systems, such changes can be achieved by a process of mutation plus selection that allows programs to be optimized for new environmental challenges that are to be faced.

The third of the pillars of life is Compartmentalization. All the organisms that we consider living are confined to a limited volume, surrounded by a surface that we call a membrane or skin that keeps the ingredients in a defined volume and keeps deleterious chemicals—toxic or diluting—on the outside. Moreover, as organisms become large, they are divided into smaller compartments, which we call cells (or organs, that is, groups of cells), in order to centralize and specialize certain functions within the larger organism. The reason for compartmentalization is that

The Nature of Life, ed. M. A. Bedau and C. E. Cleland. Published by Cambridge University Press. © M. A. Bedau and C. E. Cleland 2010.

life depends on the reaction kinetics of its ingredients, the substrates and catalysts (enzymes) of the living system. Those kinetics depend on the concentrations of the ingredients. Simple dilution of the contents of a cell kills it because of the decrease in concentration of the contents, even though all the chemicals remain as active as before dilution. So a container is essential to maintain the concentrations and arrangement of the interior of the living organism and to provide protection from the outside.

The fourth pillar of life is Energy. Life as we know it involves movement—of chemicals, of the body, of components of the body—and a system with net movement cannot be in equilibrium. It must be an open and, in this case, metabolizing system. Many chemical reactions are going on inside the cell, and molecules are coming in from the outer environment—O_2, CO_2, metals, etc. The organism's system is parsimonious; many of the chemicals are recycled multiple times in an organism's lifetime (CO_2, for example, is consumed in photosynthesis and then produced by oxidation in the system), but originally they enter the living system from the outside, so thermodynamicists call this an open system. Because of the many reactions and the fact that there is some gain of entropy (the mechanical analogy would be friction), there must be a compensation to keep the system going and that compensation requires a continuous source of energy. The major source of energy in Earth's biosphere is the Sun—although life on Earth gets a little energy from other sources such as the internal heat of the Earth—so the system can continue indefinitely by cleverly recycling chemicals as long as it has the added energy of the Sun to compensate for its entropy changes.

The fifth pillar is Regeneration. Because a metabolizing system composed of catalysts (enzymes) and chemicals (metabolites) in a container is constantly reacting, it will inevitably be associated with some thermodynamic losses. Because those losses will eventually change the kinetics of the program adversely, there must be a plan to compensate for those losses, that is, a regeneration system. One such regeneration system is the diffusion or active transport of chemicals into the living organism. For example, CO_2 and its products replace the losses inevitable in chemical reactions. Another system for regeneration is the constant resynthesis of the constituents of the living system that are subject to wear and tear. For example, the heart

muscle of a normal human beats 60 times a minute—3600 times an hour, 1,314,000 times a year, 91,980,000 times a lifetime. No man-made material has been found that would not fatigue and collapse under such use, which is why artificial hearts have such a short utilization span. The living system, however, continually resynthesizes and replaces its heart muscle proteins as they suffer degradation; the body does the same for other constituents—its lung sacs, kidney proteins, brain synapses, etc.

This is not the only way the living system regenerates. The constant resynthesis of its proteins and body constituents is not quite perfect, so the small loss for each regeneration in the short run becomes a larger loss overall for all the processes in the long run, adding up to what we call aging. So living systems, at least the ones we know, use a clever trick to perfect the regeneration process—that is, they start over. Starting over can be a cell dividing, in the case of *Escherichia coli*, or the birth of an infant for *Homo sapiens*. By beginning a new generation, the infant starts from scratch, and all the chemical ingredients, programs, and other constituents go back to the beginning to correct the inevitable decline of a continuously functioning metabolizing system.

The sixth pillar is Adaptability. Improvisation is a form of adaptability, but is too slow for many of the environmental hazards that a living organism must face. For example, a human that puts a hand into a fire has a painful experience that might be selected against in evolution—but the individual needs to withdraw his hand from the fire immediately to live appropriately thereafter. That behavioral response to pain is essential to survival and is a fundamental response of living systems that we call feedback. Our bodies respond to depletion of nutrients (energy supplies) with hunger, which causes us to seek new food, and our feedback then prevents our eating to an excess of nutrients (that is, beyond satiety) by losing appetite and eating less. Walking long distances on bare feet leads to calluses on one's feet or the acquisition of shoes to protect them. These behavioral manifestations of adaptability are a development of feedback and feedforward responses at the molecular level and are responses of living systems that allow survival in quickly changing environments. Adaptability could arguably include improvisation (pillar number 2), but improvisation is a mechanism to change the fundamental program, whereas adaptability (pillar number 6) is a behavioral response that is part of the program. Just as these two necessities are handled by different mechanisms in our Earth-bound system, I believe they will be different concepts handled by different mechanisms in any newly devised or newly discovered system.

Finally, and far from the least, is the seventh pillar, Seclusion. By seclusion, in this context, I mean something rather like privacy in the social world of our universe. It is essential for a metabolizing system with many reactions going on at the same time, to prevent the chemicals in pathway 1 ($A{\rightarrow}B{\rightarrow}C{\rightarrow}D$ for example) from being metabolized by the catalysts of pathway 2 ($R{\rightarrow}S{\rightarrow}T{\rightarrow}U$). Our living system does this by a crucial property of life—the specificity of enzymes that work only on the molecules for which they were designed and are not confused by collisions with miscellaneous molecules from other pathways. In a sense this property is like insulating an electrically conducting wire so it isn't short-circuited by contact with another wire. The seclusion of the biological system is not absolute. It can be interrupted by feedback and feedforward messages, but only messages that have specifically arranged conduits can be received. There is also specificity in DNA and RNA interactions. It is this seclusion of pathways that allows thousands of reactions to occur with high efficiency in the tiny volumes of a living cell, while simultaneously receiving selective signals that ensure an appropriate response to environmental changes.

These seven pillars of life—P(rogram), I(improvisation), C(ompartmentalization), E(nergy), R(egeneration), A(daptability), S(eclusion), PICERAS, for short—are the fundamental principles on which a living system is based. Further examination makes it clear how life on Earth has implemented these principles. But these mechanisms may not be perfect and might be improved. For example, the regeneration system used by life on Earth is imperfect for any particular individual and hence requires a "starting over." That mechanism in turn requires a device for heredity to maintain continuity in the program for the next generation. Suppose the proteins, hormones, and cells had a better feedback system so that the gradual decay with age was constantly being corrected by feedback. Then, the need to start over would disappear. That would eliminate the death and hereditary needs

of the current system. It would also mean that a single individual could live forever without aging. There would be a problem, however, because the starting over (death and a new birth) provide an opportunity for improvisations (mutations in the DNA), and that pillar would need to be replaced by a new mechanism to achieve the same advantage.

Such dilemmas make us confront another reality. At the present time the way in which mutation and selection (survival of the fittest) has worked over evolutionary time no longer seems to apply to *Homo sapiens*. We have become more compassionate, less demanding. Perhaps a newer approach—longer life and deliberate changes in the program by a supreme council of wise Solomons—could be substituted for the cruder survival-of-the-fittest scenario. I do not necessarily advocate such a drastic change in the current mechanism of improvisation, which has served us well over the centuries, but am only pointing out that there is the possibility to change particular mechanisms as long as we maintain the pillars.

This listing of the seven foundations of life allows us to think differently about the goals and therapeutic approaches of current research. The adaptability concept, for example, is certainly one in which better mechanisms could be devised, probably by adjusting existing mechanisms to allow these to work more effectively in real living systems. For example, the eye can adapt to outside light levels that range over 10 orders of magnitude (10^{10}), whereas the other organs of the human body have much smaller ranges. Perhaps other organs such as the lungs, kidneys, or spleen could be improved so that they would function over larger concentrations of regulators and aging would be less harmful to them.

Thus, the PICERAS principles seem to be necessary for the operation of a living system. Mechanisms to achieve such a system can be varied as long as they satisfy the thermodynamic and kinetic requirements. We have one example, life on Earth, showing how it can be done. It will be interesting to see whether a different, self-consistent set of mechanisms could yield a model with life as an outcome.

NOTES

This chapter originally appeared in *Science* 295(22) (2002), 2215–2216.

25 · A universal definition of life: autonomy and open-ended evolution

KEPA RUIZ-MIRAZO, JULI PERETÓ, AND ALVARO MORENO

INTRODUCTION

Definitions of life are highly controversial. And it is not just a question of lack of consensus among the different proposals formulated so far. Some authors are very skeptical about the actual possibility of grasping "in any scientifically relevant language" such a complex and multifarious phenomenon. Others think that we have to wait until biological theory(ies) become more rigorous, more encompassing and meaningful. And some others consider that it is not worth undertaking the challenge since, even if we could obtain a proper definition of life, it would still be a rather conventional one and would probably have little influence on the development of specific research programs in biology.

The living phenomenology shows, indeed, many different sides (that appear at various levels of organization) and it is not easy to capture all of them in a single conceptual scheme. This is made even more difficult by life's ability to diversify and explore its own limits (always producing border-line cases, exceptions to the rule, ...). Last century's impressive advances in molecular biology have revealed a great underlying biochemical unity of all living forms, but it is not clear to what extent this is the result of contingency or of real necessity: i.e., whether that unity can serve to extract general biological principles or just derives from having a universal common ancestor of all terrestrial life. In addition, since the problem of the origin of life is also far from being solved, it is not at all obvious how those "biological principles" would relate to the general laws of physics and chemistry, i.e., if they would be subject to an eventual reduction to the latter, or should have their own "status" (with their own explanatory power, degree of abstraction, etc.) as scientific laws.

However, despite these and other difficulties that we could think of, there are also good arguments that now is a suitable time to tackle the question (like some authors are, in fact, doing, see, e.g., Páyli *et al.* (2002)). Certainly, no general full-fledged theory of biology is available yet (as Cleland & Chyba (2002) highlight in order to support their skeptical view), but the insightful research carried out during the last decades in areas such as bioenergetics, enzymology or genetics provide us with a body of knowledge which is deep and wide enough to try structuring it around some fundamental "tentatively universal" concepts. This effort, in itself, can be very helpful for the development of a general, and better formalized, biology. The lack of success of previous attempts at defining life (in the sense that they have not led to a well-established consensus) should not discourage us, especially after noticing that some confusion might still exist in the field (see the article on the subject recently published in *Science* by Koshland (2002), as well as our critical remarks below). And neither should the skeptical minds of those who do not regard the task of defining life as a possible or useful one for biology. Their claims, based on the continuity between physico-chemical and biological phenomena, or on the irreducibility/reducibility of biology to physics and chemistry, are not conclusive.

In this chapter we suggest a concrete definition of life (of minimal and universal life), with the aim of opening it to discussion. Our proposal is meant not only to provide some conceptual clarification on this particular issue, but to contribute to guiding other lines of research, like specific programs in the fields of origin of life, astrobiology or artificial life. The first step, anyway, is to determine what kind of definition serves the purpose and, then, justify in what sense the

The Nature of Life, ed. M. A. Bedau and C. E. Cleland. Published by Cambridge University Press. © M. A. Bedau and C. E. Cleland 2010.

proposal here suggested means an advance with regard to previous attempts.

REQUIREMENTS A DEFINITION OF LIFE SHOULD FULFILL

In general terms, definitions can be made with two main different purposes: (i) to demarcate or classify a certain type of phenomenon, and (ii) to make manifest—and, perhaps, even explain—the fundamental nature of that type of phenomenon. The first purpose normally leads to *descriptive* definitions that consist in a set of properties (typically, a "check-out" list) containing all that is required to determine if a phenomenon belongs to a particular kind or not, whereas the second involves a completely different way of formulating the question: *essentialist* definitions characterize a given phenomenon in terms of its most basic dynamic mechanisms and organization.

Several lists of properties have been suggested in the literature to discern "the inert" from "the living," e.g., self-organization, growth, development, functionality, metabolism, adaptability, agency, reproduction, inheritance or susceptibility to death. See, among others, Mayr (1982), Farmer and Belin (1992), or the so-called "seven pillars" of de Duve (1991). However, since these catalogues do not provide a hierarchy or an account through which the chosen properties can be related to each other, it is hard to tell whether any of the proposals includes just the necessary and sufficient ones (if there are no properties deriving from others, i.e., the set would be redundant—or if some additional fundamental property ought to be introduced—i.e., there is something important missing in it). Besides, lists do not offer any hint to clarify the source or process of integration of the system/phenomenon under analysis. This is crucial because a definition of life that is expected to be truly universal must be built from general principles, some of which should stem from physics and chemistry. In other words, the definition has to include primitive concepts that help to bridge the gap between physico-chemical and biological phenomenology. This was already highlighted by Oparin (1961), who claimed that the problem of defining life is tightly intertwined with the problem of its origin.

Therefore, we must look for an essentialist type of definition that, at the same time, is "genealogical,"

in the sense that it explains—or at least throws some light on—the process that leads to the constitution of the phenomenon, starting from well-determined conditions. In this way, it should offer a natural framework for generalization (i.e., for selecting which are the universal features of all possible life). More schematically, the set of requirements that a definition of life should meet in order to be of use for the development of present biology—and other related fields of research—can be condensed in the following points (by and large equivalent to Emmeche's (1998)):

The definition should:

(a) be fully coherent with current knowledge in biology, chemistry, and physics;
(b) avoid redundancies and be self-consistent;
(c) possess conceptual elegance and deep explanatory power (i.e., it must provide a better understanding of the nature of life, guiding our search into its origins and its subsequent maintenance and development);
(d) be universal (in the sense that it must discriminate the necessary from the contingent features of life, selecting just the former);
(e) be minimal but specific enough (i.e., it should include just those elements that are common to all forms of life—not being, in principle, restricted to life on Earth—and, at the same time, it must put forward a clear operational criterion to tell the living from the inert, clarifying border-line cases, contributing to determine biomarkers, etc.).

According to these general points, we can assess the validity or usefulness of different proposals. Lists, for instance, show difficulties to fulfill "c" and typically also "b" (and, thereafter, "d" and "e"). Vitalist approaches neglect "a." So those are already "out of the game." Let us briefly check out a recent example: Koshland's proposal (2002). This author offers a definition that is not intended to be a mere list, but it ends up exhibiting very similar weaknesses. His suggestion of a set of seven "principles" or "pillars" of life (program, improvisation, compartmentalization, energy, regeneration, adaptability and seclusion[1]) is not really satisfactory: it not only lacks elegance and explanatory power, but is clearly redundant (several of Koshland's "pillars" can be included in the concept of "autopoiesis," see below and also contrast with Table I in Margulis (1990)).

Obviously, we can find other options in the literature, which fit better into the group of essentialist definitions. To be highlighted here is Emmeche's *bio-semiotic* definition (1998): "life is the functional interpretation of signs in self-organizing material 'code-systems' that construct their own 'umwelts'." This proposal is certainly self-consistent and conceptually elegant (even more so if one looks into the bio-semiotic theory), but has an important drawback: it assumes "signs" (i.e., information) as a primitive natural kind, when physics and chemistry do not (the latter just accept: time, matter, energy, charge, fields, . . .). Therefore, although it does not directly contradict physico-chemical knowledge, it does not come to terms with it, and requirement "a" is not strictly fulfilled.

It is also worth mentioning here Shapiro and Feinberg's definition (1990), according to which life is the activity of a "biosphere," i.e., the activity of "a highly ordered system of matter and energy characterized by complex cycles that maintain or gradually increase the order of the system through an exchange of energy with its environment." This proposal is formulated in very broad terms and with a well-intended purpose of bringing biology closer to physics. However, the apparent advantage of this formulation (particularly in relation to requirements "a" and "d") turns out to be a shortcoming: where is the specificity of the living phenomenon? In this scheme, where can we look for properties like self-production, reproduction, adaptation, heredity, selective evolution . . . , which are so characteristic of biological systems? The motivation to establish a link between the inert and the living domains should not lead us to forget all that we actually know about biology (or chemistry!). There are many ways of maintaining or increasing the order of a system whose dynamics takes place in far from equilibrium conditions (self-organizing phenomena), but living systems have developed their own, and that is why they are so distinct from anything else. Thus, the definition above is not suitable since it should, at least, give an indication of where to search in this direction (i.e., requirement "e" is not adequately fulfilled).

The issue is quite tricky: we cannot expect that all biologically relevant items are touched or made manifest in a brief statement like a definition. But, nevertheless, a good definition must be well balanced, containing the key conceptual tools to develop a theory or scheme around it that is coherent and provides specific enough hints to establish a natural connection between the physico-chemical and biological realms. In the following section we will review two examples that get closer to this ideal. Even if they are "subject to improvement," we can say that, by and large, they meet all the standards above. And, in any case, the discussion on these two definitions will be very helpful to introduce our own later.

BRIEF REVIEW OF TWO SIGNIFICANT ATTEMPTS

The main message to be gathered from the last section is that a proper definition of life should not be merely descriptive but "essentialist-generational". In particular, we must search for definitions that contain high explanatory power/deepness, i.e., that are supported by or tightly linked to a well-founded conceptual framework about biological phenomenology. On these lines, two major proposals will be selected from the literature.

Standard definition

"Life is a system capable of evolution by natural selection" (Sagan 1970) or, in a more elaborate and precise version: "Life is a self-sustained chemical system capable of undergoing Darwinian evolution" (NASA's "working definition"; Joyce 1994; see also Luisi 1998).

The underlying conception is very close to the standard view that supports the neo-Darwinian paradigm in biology, where the stress is put on the evolutionary dynamics of biological systems. According to this view, the key properties that a system must show in order to evolve through natural selection are "reproduction, variability, and inheritance" (Lewontin 1970; Maynard Smith 1986). It follows, then, that any system with those properties, including a population of replicating molecules, could readily fulfill the definition.

Some authors, like Wicken (1987), Brooks and Wiley (1988), Luisi (1998) or Kauffman (2000), are critical of this perspective, arguing that a proper definition of living being must take into account the characteristic way in which the components of such a system get organized as a coherent whole. The second version of the definition might seem less vulnerable to this criticism, since it introduces the idea of "self-maintenance" (apparently linked to that of metabolism, see below). However, the core idea stays the same. In Joyce's own terms (1994, p. xi) while "the notion of Darwinian evolution subsumes the processes of self-reproduction, material continuity over an historical lineage, genetic variation, and natural selection," "self-maintenance" "refers to the fact that living systems contain all the genetic information necessary for their own constant production (i.e., metabolism)." The geneticist bias is rather obvious: metabolism here seems to be the result of acquiring a complete enough pool of genes so as to achieve the constant production of a system. As Luisi (1998) emphasizes, this definition was created from a conception of life that is fundamentally molecular, in tune with a general research program that looks into the roots of Darwinian evolution in the context of populations of replicating molecules undergoing some selective dynamics (e.g., models of "quasi-species," "hypercycles," etc.; see Eigen and Schuster 1979; Eigen 1992).

Nevertheless, as the next definition will show, the notion of metabolism is much richer than that. It involves a material self-(re-)producing organization, functionally integrated (i.e., not reducible to the properties of its molecular components), which is not really grasped by the term "self-maintenance" (specially if this is understood in a weak sense; see Moreno and Ruiz-Mirazo 1999; Ruiz-Mirazo and Moreno 2000). In summary, the basic problem of this definition is that there is no proper characterization of the type of material organization that would allow the beginning of a process of Darwinian evolution (precisely, some sort of pre-genetic metabolic organization).

Autopoietic definition

"An autopoietic system—the minimal living organization—is defined as a network of processes of production (synthesis and destruction) of components such that these components: (i) continuously regenerate and realize the network that produces them, and (ii) constitute the system as a distinguishable unit in the domain in which they exist" (Varela 1994; originally Maturana & Varela 1973).

This definition, despite its high level of abstraction, is much more explicit than the previous one with regard to capturing the basic organization that constitutes a minimal living being, its metabolism. The main idea put forward in it is that life should be defined

from the perspective of individual organisms. A living being is conceived as a recursive web of component production and transformation, continuously self-generated and self-regenerated, which produces a physical boundary that, in turn, is a necessary condition for the maintenance of the web. All this conveys an organizational "autos" that goes beyond what is usually considered as a phenomenon of "self-organization." This "autos" involves a group of interrelated processes that gets organized according to a global operational logic: the system is, somehow, "closed up in itself," in the sense that it creates its own boundaries and follows its own circular dynamics of self-production.

Yet, the autopoietic definition presents two main problems.

(1) On the one hand, it is excessively abstract. Maturana and Varela's original purpose was to define a minimal biological system in as general terms as possible. However, they go too far in that direction, since they offer a conception of the living which is too detached from the physical—material and energetic—requirements that are crucial for its actual implementation (as Fleischaker (1988) already highlighted). This leads to a characterization of the "minimal organizational logic" of biological systems in which physics and chemistry (thermodynamics in particular) have nothing to say, and in which the interactive-agential relationship between system and environment is disregarded (or, at least, it is regarded as secondary). However, the introduction of material-thermodynamic aspects involves a totally different "organizational logic" of self-production (Moreno and Ruiz-Mirazo 1999; Kauffman 2000), highlighting the relationship with the environment as one of the key ingredients to understand how a minimal self-constructing chemical system operates (Ruiz-Mirazo & Moreno 2000).[2]

(2) On the other hand, even if the autopoietic criteria were modified so as to include some additional ingredients related to the previous point, it would not suffice to characterize "the living" in a complete way. The ability to produce diversity and growth in complexity as a result of a selective evolutionary process, which is another fundamental feature of biological systems, is not at reach for autopoietic systems. In principle, an autopoietic

system is capable of reproduction (through auto-catalytic growth and division), of adapting to external disturbances (organizational homeostasis) and even of modifying its type of organization—its identity, in a broad sense—(through the accumulation of structural changes), but it is not capable of initiating a process of Darwinian evolution because, as such, it does not have the genetic mechanisms required to do so. Thus, the main problem of this definition, the characteristic evolutionary capacity of living beings, is not taken into account.

Quite interestingly, these two definitions (standard and autopoietic) cover the two fundamental aspects of biological phenomenology. As Maynard Smith (1986) already pointed out, life appears both as a collective population of self-reproducing hereditary systems (life as evolution) and as individual self-maintaining dissipative units (life as metabolism). The problem of both definitions is that each focuses on just one of those two dimensions: neither of them works out properly, i.e., in a well-balanced way—the tension between the individual-metabolic and the collective-ecological sides of the phenomenon. Nevertheless, a good definition of life should precisely highlight and contribute to explain the link between the two sides.[3]

In order to do so, it is absolutely necessary to tackle the problem with a methodology that combines a "bottom-up" approach (following the various stages in the process of emergence of living entities from complex physico-chemical phenomena) and a "top-down" approach (specifying which is the final scenario of the journey: i.e., what may be already considered as "minimal life"). This reflects, again, the fact that the definition of life (i.e., the main "top-down" constraint) and its origins (the "bottom-up" sequence of events) are inseparable questions. Accordingly, our proposal is next articulated in two different sections: in the first we introduce and briefly explain our definition (which is meant to be well-rooted in present knowledge of living systems but, at the same time, is also meant to be a projection that captures their most universal properties); and in the second we offer a genealogical account of the cornerstones of this definition, pointing out the major transitions suggested as necessary and sufficient for the unfolding of any biological world. The basic mechanisms underlying each of those transitions will be indicated as we go along, so this second type of analysis will also serve to come out—at the end of that section—with an improved (more specific and operational) version of our definition.

OUR PROPOSAL: AUTONOMY AND OPEN-ENDED EVOLUTION AS THE BASIC INGREDIENTS OF A NEW UNIVERSAL DEFINITION OF LIVING BEING

After reviewing two representative definitions in the field, it is our turn now to offer an alternative that, somehow, introduces advantages with respect to them. Even if these previous proposals roughly fulfill the requirements listed on p. 311, we consider that the following definition does it in a more complete and satisfactory way (particularly concerning requirements "c"—on conceptual elegance and explanatory power— and "e"—on the balance between universality and specificity) and, therefore, means a significant improvement. We will start by defining life from an individual perspective (from the perspective of a single "living being" or "organism") and, later (especially on pp. 319–320) we will discuss the implications of such a definition in a wider collective scale, introducing a conception of "life" as a more encompassing, global phenomenon.

The new proposed definition: *"a living being"* is *any autonomous system with open-ended evolutionary capacities*, where

(i) by *autonomous* we understand a far-from-equilibrium system that constitutes and maintains itself establishing an organizational identity of its own, a functionally integrated (homeostatic and active) unit based on a set of endergonic-exergonic couplings between internal self-constructing processes, as well as with other processes of interaction with its environment, and

(ii) by *open-ended evolutionary capacity* we understand the potential of a system to reproduce its basic functional-constitutive dynamics, bringing about an unlimited variety of equivalent systems, of ways of expressing that dynamics, which are not subject to any predetermined upper bound of organizational complexity (even if they are, indeed, to the energetic-material restrictions imposed by a finite environment and by the universal physico-chemical laws).[4]

At first sight, this definition may seem too generic and not very operational, because it does not specify the type of components or molecular mechanisms that make an autonomous and evolutionary behavior of this kind possible (membrane, catalysts, energy currencies, informational records...). To a great extent, that is the work we have ahead. In the next section we will analyze how this way of *being, doing and changing* in time can progressively be realized and integrated in a system (or group of systems). Through that analysis, we will be able to endow the concepts of "autonomy" and "open-ended evolution" with a more profound and accurate meaning, particularly looking into the connection between the two and discussing why one must come after the other.[5]

In any case, the present version already contains the fundamental (necessary and sufficient) theoretical ingredients to reconstruct the essential steps of the origins of life, from self-organizing physico-chemical phenomena until the constitution of systems with a level of complexity equivalent to that of the last universal common ancestor of all terrestrial forms of life. As we mentioned before, the autopoietic definition, by itself, would not lead us all the way up to that point (since an open-ended type of evolution requires the development of non-trivial mechanisms, e.g., those supporting hereditary reproduction), whereas the standard definition is not satisfactory because it does not provide the key conceptual tools to characterize the basic material organization required to get there ("self-maintenance" is, in that sense, much weaker and limited than "autonomy").[6]

BUILDING A GENEALOGICAL EXPLANATION OF OUR DEFINITION

The starting point of our account is a scenario in which self-organizing phenomena (complex physico-chemical phenomena like the so-called "dissipative structures," Nicolis and Prigogine 1977) take place. By "self-organization" we refer to a phenomenon occurring when a series of non-linear microscopic processes generate a global–macroscopic correlation (a new "pattern of dynamic behavior") in far from equilibrium thermodynamic conditions that are maintained by the continuous action of a set of constraints, one of which—at least—is a result of the actual phenomenon.

It is quite clear that living beings cannot do without self-organizing processes (which are ubiquitous at all levels of biological organization), but the constraints that allow a living organism to stay in a far from equilibrium situation are much more elaborate. In fact, the organism *itself* elaborates the different constraints that control the flow of matter and energy through the system, unlike what happens in the typical examples of dissipative structures (Bénard convection patterns, B-Z chemical reaction oscillations), where the flow that keeps the phenomenon running is externally harnessed. Thus, we can safely say that life is self-organization (we disagree with those who think that the "self should be dropped"; see, for instance, Margulis & Sagan 2002) but it is, indeed, much more than self-organization.

So, what is the main step after self-organization? Self-maintenance. Someone could say that in pure self-organizing phenomena it is already possible to start speaking about "self-maintenance" (insofar as the generation of the macroscopic pattern contributes to its own maintenance by means of its continuous constraining action on the microscopic dynamics). However, in that case, we would be using the term in its weakest sense. A more significant *self*-maintenance cannot take place until a system starts producing some of the constraints that are crucial to control the matter-energy flow through it and, in this way, it begins to develop the capacity to maintain its organization in the face of external perturbations (i.e., a primitive kind of "organizational homeostasis"). A fact to be highlighted here is that this is only possible if the system is chemical, because the variety of constraining mechanisms required to achieve that capacity simply is not at reach for bare physical systems. Dynamics (understood in its classical sense: as the change in time of the position of a body or many-body system) is insufficient for such a task. Besides, as will be more obvious below, it is just through chemistry that a process of open-ended growth in complexity may take place.[7]

Thus, the next scenario is that of self-maintaining chemical networks. Systems made of relatively simple molecular components in which autocatalytic cycles are formed and have the potential to vary and grow in complexity (Kauffman 1993; Wächtershäuser 1988; de Duve 1991) could be good candidates for this stage, provided that they are located in a proper—physically realistic—context: in between a source and a sink of free energy—as Morowitz (1968, 1992) rightly emphasizes through his "cycling theorem." Thermodynamics matters. Although it is not possible to explain the relevant transitions at this stage just by means of thermodynamic tools and theories, it is very important not to forget about the implications of the general thermodynamic framework, which are certainly far-reaching. In particular, in the case of self-maintaining chemical networks it is crucial to consider the energetic requirements (both internal and with regard to the environment) for the reactions involved to actually take place: as we will see next, only those networks that are capable of establishing a set of endergonic-exergonic couplings between their constitutive processes will be candidates for basic autonomy (Ruiz-Mirazo and Moreno 1998, 2000; Kauffman 2000).

The emergence of basic autonomy

The concept of self-maintenance, being important, still does not capture the most characteristic properties of a living organism: in particular, its capacity to build and rebuild continuously all the components and constraints that are responsible for its organization and behavior, together with the capacity to adaptively modify that internal organization (plus the actual relation with the environment) as a response to external changes. Biological systems are component production systems (chemical networks) that manage their material and energetic resources in such a way that they continuously accomplish a global self-construction dynamics in a plastic way. But all this would not be possible if endergonic processes (non-spontaneous processes like the synthesis of polymers, or the transport of a substance against its electrochemical potential gradient) could not profit from the exergonic drive of other spontaneous processes. Thus, the actual generation—and regeneration—of the components that act as functional constraints in the system is based on the establishment of endergonic–exergonic couplings (and hence the deep entanglement between anabolism and catabolism in any metabolism).

We suggest that the transition from a bare self-maintaining chemical system to a full-fledged self-producing system with minimal adaptive capacities requires that the network of reactions gets encapsulated by a semi-permeable boundary of its own making and, at the same time, solves three fundamental problems: (i) an eventual osmotic crisis, (ii) the spatio-temporal coordination of all processes—and their couplings—, and (iii) the achievement of an efficient energy transfer through the components and processes of the system. The minimal set of components necessary to overcome these problems are (Ruiz-Mirazo 2001): a *membrane* (an enclosure through which the system controls concentrations, establishes a clear cut distinction with the environment, and channels the interaction with it), a group of *energy currencies* (at least one soluble in water—like PP_i or ATP—and one directly related to transport processes—like the electrochemical potential gradient of protons or sodium ions across a membrane), and a set of *catalysts* (polymers—or initially just shorter oligomers like de Duve's (1991) "multimers"—responsible for modulating the fates at which reactions take place, for setting up regulation/homeostatic mechanisms, and for carrying out mediated transport processes).

It is only when these three types of components get together in a single system and interrelate adequately that a complete enough set of endergonic-exergonic couplings may be accomplished and, together with it, a minimal autonomous system. It is important to notice that this dynamics involves both a new internal way of operating (functional relations among the components of the system)[8] and a new way of interacting with the environment (agential behavior).[9] Thus, this type of system can, in a basic and primitive sense, create its own "umwelt."

Some authors, like Luisi (1998), may argue that this kind of autonomous system (cellular proto-metabolisms) already contain all that is required for life. Actually, it is difficult to say what we would do if we ever found such systems (in a test tube, on Mars or somewhere else): would we not call them "living," despite the fact that they do not show a complex metabolic behavior based on macromolecules like genes or proteins? With regard to this point, we consider that basic autonomy is a fundamental property of life, but it is not the only one. In fact, the long-term maintenance of basic autonomous systems will not be guaranteed until they achieve more metabolic versatility and robustness, mechanisms for reliable reproduction and heredity, together with an open-ended capacity to adapt and grow in complexity. In other words, the long-term stability of a biological world crucially depends on how living beings change in time and evolve through generations.[10] This is so much so that even what a living organism comes to be and do during its individual lifetime is profoundly influenced by the global evolutionary process in which it is inserted, as we will underline below.

In any case, we must point out that basic autonomy, although insufficient, is necessary for open-ended evolution. Without a component production machinery that solves the basic physical–material–energetic problems involved in the constitution of any (proto-)metabolic system it is impossible to reach the level of molecular and organizational complexity required to start a process of evolution of those characteristics. In addition, the constitution of basic autonomous systems also provides the necessary potential for functional diversification (i.e., a wide enough "phenotypic space"), which is capital for the development of mechanisms of genetic information (Wicken 1987), and which is missing in models of molecular evolution that deal with populations of pure replicators (like typical "RNA-world" models).

Hereditary autonomous systems: The first steps towards open-ended evolution

Basic autonomous systems, as they have been described above, possess a fair degree of organizational homeostasis and adaptability, are capable of reproduction (even if it is just a "statistical" kind of reproduction, by simple growth and division), and show a remarkable potential for diversification (because, in principle, they would be able to create ever-new components). However, they have no way of fixing or recording the novelties that appear in them (e.g., possible innovations in their metabolic dynamics) and, as their level of molecular and organizational complexity rises, their brittleness also increases.

According to some recent models that deal with relatively simple catalytic net-works and study their growth and reproduction potential (see: Segré *et al.* 2000), at this stage there could already be some transfer of "information" from generation to generation (that is the idea behind what they call a "compositional genome"). Of course, the actual autocatalytic dynamics that is responsible for growth (and subsequent division-multiplication) implies that some of the components and features of the "mother system" get produced once and again, and possibly transmitted to the offspring. However, that continuous production of components and their transmission to the "daughter cells" is not reliable.

Sometimes everything will go fine and reproduction will be successful (i.e., mother and daughter cells will be equivalent, or have equivalent levels of complexity); but in some others (a fraction of the whole that cannot be disregarded, especially if we deal with big populations and a big number of generations) the process will lead to unviable systems. And this is a critical question when the complexity of the components and the difficulties to produce them become higher and higher.

Therefore, there is a major complexity bottleneck that can only be overcome if new, more sophisticated mechanisms of autocatalysis are developed by autonomous systems. These mechanisms must guarantee that the complex functional components of the system are well preserved in the on-going metabolic dynamics, as well as reliably transmitted to the next generations through subsequent processes of reproduction. The solution is to generate functional components that perform *"template activity"*: i.e., material structures that can be faithfully *replicated*,[11] regardless of their complexity (i.e., at this stage: the length of the polymer chain) and of the frequency at which the process is carried out. In this way, components become material "records" and it is possible to start speaking about "heredity" in a more rigorous way (Pattee 1967, 1982, 1997).

As a consequence, a new scenario emerges with systems whose metabolism is more complex and robust than before, whose reproduction is more reliable, and whose evolutionary dynamics, apart from being more selective (stronger pressure among synchronous competitors for limited resources), already introduces the possibility of defining—and tracing—"lineages" (i.e., asynchronous relations between systems with non-coherent lifetimes). This stage of "hereditary autonomous systems" (systems with an operational organization based on a single type of functional polymer[12]) is analogous to an "RNA-world" but, of course, provided that such a "world" is embedded in a cellular metabolism, and not simply set up experimentally or theoretically in the context of a flow reactor, as it is commonly done (for a review see Eigen 1992).

The genotype–phenotype decoupling and the origins of open-ended evolution

In any case, hereditary autonomous systems face a serious limitation when it comes to initiating a truly open-ended evolutionary process: the development of mechanisms that allow for a more and more reliable conservation and transmission of components from generation to generation turns out to be structurally incompatible with the development of the functional efficiency and metabolic versatility of those components. This happens because—as we just mentioned above—these are systems based on a single type of complex functional component, which articulates both the *realization* and the *evolution* of their metabolism, and which needs to accommodate simultaneously two very different properties: template activity and catalytic specificity/efficiency (Moreno and Fernández 1990; Benner 1999).

The problem has a rather simple chemical interpretation. Template activity requires a stable, uniform morphology, suitable to be lineally copied (i.e., a monotonous spatial arrangement that favors low reactivity and it is not altered by sequence changes); whereas catalytic diversity requires precisely the opposite: a very wide range of three-dimensional shapes (configuration of catalytic sites), which are highly sensitive to variations in the sequence.

This "trade-off" problem marks out a second major bottleneck in the way to establish the conditions that permit an indefinite growth of complexity. Such a bottleneck can only be overcome by those autonomous systems which start producing *two* different types of polymers: some "new records" specialized in the reliable recording, storage and replication of certain linear structures—or polymer "sequences"—that become crucial to ensure their correct functioning (and that of similar systems in future generations); and some others, the "new catalysts," specialized in carrying out with increasing efficiency the strictly metabolic tasks required for their continuous realization as individual autonomous systems.

The new records would be, somehow, *decoupled* from metabolism, in quasi-inert states, and would have hardly any causal effects on the system's functioning if they did not act as a fundamental *reference* for the new catalysts (Pattee 1982; Moreno and Ruiz-Mirazo 2002). More accurately, the former must *instruct* the synthesis of the latter, i.e., records must become responsible for the correct specification of catalysts. And these, in turn, must control and catalytically regulate all the processes in which records are involved (replication, translation, reparation, etc.), although without taking part directly in the generation and/or

alteration of those records (since this should be determined by the global evolutionary dynamics of the population of autonomous agents).

The key lies, therefore, on the establishment of a certain circularity—a causal connection or correlation—between those two types of components (and the processes associated to each of them). But, given the structural incompatibility between the capacity for an accurate template replication and the development of more proficient catalysts, it is virtually impossible that the connection between the two could be built on their intrinsic properties as polymers. On the contrary, the conditions from which those components stem (i.e., the previous single-polymer world) relations establish through a series of intermediate-components, i.e., in a non-linear and clearly *indirect* way (even if the potential to have causal effects on one another is not lost). Such an indirect, mediated connection is, in fact, a requisite to achieve an effective decoupling between those two operational modes or levels in the system. And this is precisely what is required for the emerging type of metabolic organization (which may be called "instructed metabolism") to combine coherently and fruitfully the two new dimensions of its activity: the (individual) ontogenic-functional one and the (collective) phylogenic-evolutionary one.

The most natural way of understanding this new ordering in the system, this new (partially decoupled but strongly complementary, i.e., *indirectly coupled*) organizational structure is through the idea of "information."[13] Other authors have defended a similar position, holding that what distinguishes life is precisely the fact that there is a code-type relation between template and functional components (Pattee 1977, 1982; Hoffmeyer and Emmeche 1991; Umerez 1995).

It is only once hereditary autonomous systems start producing "informational" components and mechanisms (i.e., once a *translation code*[14] appears between two very different types of functional components in the system) that the "genotype–phenotype" distinction becomes really significant (even if some authors—for example, Eigen 1992—try to anticipate it) and we can start speaking properly about open-ended evolution. This is so because only through the generation of components which can be replicated with very high reliability, which have energetically degenerate states and, most importantly, which are interpreted as instructions by the system (i.e., "informational records"—or "unlimited hereditary replicators" in Maynard Smith and Szathmáry's (1995) terminology) can autonomous systems begin to construct *indefinitely* new forms of molecular and organizational complexity.

Corollary: The definition revisited

In sum, in order to arrive at a fully-fledged biological scenario, it is necessary to go all the way up until some genetically instructed collection of metabolisms appear. If we stopped before (e.g., at the stage of basic or hereditary autonomous systems) we would not have all the key ingredients that support the organization of a minimal living being, and we would have to face the problem of needing two different—and incompatible— conceptions of minimal life. For the selection of those key ingredients must be done taking into account two rules: the "continuity thesis" between the inert and the living worlds, and the fact that they must be mechanisms or features of the organization that are so fundamental that they cannot be erased from then onwards, throughout the subsequent evolutionary history of that type of organization (as has happened throughout the evolutionary history of life on Earth).

Accordingly, we can make our original definition of a living being as "an autonomous system with open-ended evolutionary capacities" more explicit and operational, since it is possible to state now that, in order to accomplish those two properties, any system must have:

> *a semi-permeable active boundary (i.e., a membrane), an energy transduction/conversion apparatus (a set of energy currencies) and, at least, two types of interdependent macromolecular components: some carrying out and coordinating directly self construction processes (catalysts) and some others storing and transmitting information which is relevant to carry out efficiently those processes in the course of subsequent generations (records).*

THEORETICAL IMPLICATIONS: LIFE IS A HISTORICAL-COLLECTIVE PHENOMENON

Probably the most important theoretical conclusion that can be drawn from the last section is that a living being cannot exist but in the context of a global network of similar systems.[15] This is reflected in the fact that informational components (which specify the metabolic components and organization of single biological entities), in order to be functional, have to be shaped through a process that involves a great amount of individual systems and also very many consecutive generations (or reproductive steps). Such a collective process, which has both a synchronic-ecological side and a diachronic-evolutionary one, is actually crucial for the sustainability of the living phenomenon as a whole.

On the one hand, since the material resources of any real physical environment—like a planet—are limited, life must learn how to make the best use of what is available and also of what it continuously produces. In the individual sphere, this means that living agents have to compete for organic compounds or its precursors; so, under selective pressure, depredation strategies are bound to appear and, thereafter, full "food webs," too (initially, of course, just between autotrophic and heterotrophic unicellular systems). In the collective sphere, the ecosystem eventually has to deal with the problem of recycling bioelements at a global scale. Otherwise, there would be a major crisis that would put at risk the persistence of that whole biological world. The solution is to couple with geophysical and geo-chemical processes that take place on the planet and establish global bio-geo-chemical cycles. As life cybernetically controls the environmental conditions of the inhabited planet, such as the composition of the atmosphere (Lovelock and Margulis 1974a,b), this will allow for its long-term maintenance.

On the other hand, an open-ended process of evolution (which does not mean any evolutionary arrow towards higher and higher complexity, but just the possibility that there is a steady increase in it) is precisely what allows the continuous renewal of particular individual organisms and particular types of organism for the sake of keeping the overall process running. This open-ended historical process takes place at a much longer time-scale than the typical lifetimes of individual systems and—as we said in the previous section—strictly speaking, it only begins when those systems are endowed with a genetic machinery that instructs their metabolism and is transmitted reliably through generations. Nevertheless, evolutionary dynamics of the ecosystems does not appear "off the cuff": it must be progressively articulated during stages previous to the origins of genetics (i.e., previous to the origins of life).

In other words, genetic information is the central concept to understand the intricate connection between the functional dynamics of individual organisms and the evolutionary dynamics of the global biological network. However, since that connection cannot be suddenly established, we have to focus the analysis on the *origins* of genetic information (i.e., on which are the most plausible conditions for the emergence of genetic information). And that leads us to study the nature of basic and hereditary autonomous systems, as well as the evolutionary dynamics in which these two types of system can get involved. We dare to say that the fundamental pillars of "ecopoiesis"[16] and, thus, the key—individual and collective—mechanisms required to generate a whole sustainable biosphere, are put together during the development of basic and hereditary autonomy. Therefore, the problems of the origin of life and of the constitution of a biosphere in constant, open-ended evolution would be tightly linked. Further research to support this line of thought is required.

In conclusion, we can say that *"life"*—*is a complex collective network made out of self-reproducing autonomous agents whose basic organization is instructed by material records generated through the evolutionary-historical process of that collective network.*

(i) At the individual level, autonomous agents are a special kind of dissipative self-enclosed chemical organizations that recursively and adaptively maintain themselves thanks to the material records created and transmitted in the course of subsequent generations.

(ii) At the collective level, a lineage of self-reproducing autonomous agents in evolution (diachronic process), competing and cooperating among them (synchronic process), generates new records—genetic information—which, once embedded in the individual organizations, allow for new, open-ended, functional interactions.

There is such a deep entanglement between these two levels that both the collective and the individual organization of life are cause and consequence of each other. Nevertheless, it is important to underline that there is also a basic asymmetry between the individual (metabolic) network and the collective (ecological)

one: both are self-maintaining and self-producing organizations, *but* only individual living beings (organisms) are autonomous *agents* with a self-produced, active physical border, plus a high degree of *functional integration* among components, plus a machinery for *hereditary reproduction*.

PRACTICAL IMPLICATIONS

As a final conclusion to this article, we would like to highlight briefly the main practical implications of our approach and of the general conception that has been offered here for future research. In this sense, four different avenues of research are considered:

(a) *Origins of Life on Earth*. Present research is strongly biased by molecular biology and genomics, and so it focuses on the main structural features, relational properties and possible abiotic synthetic roots of biopolymers (proteins, RNA, DNA) and its precursors (for review see Lazcano 2001). Our proposal, based on the conviction that "energetics" is to be worked out before "genetics" can take over (Morowitz 1981) (and, thus, that a proto-metabolic organization is required before complex biopolymers can arise; Morowitz 1988; de Duve 1991), favors an alternative paradigm: the search into different self-constructing cellular systems, that may later lead to the production of functional macromolecules (always to be regarded in the context of the whole metabolic organization). Our definition allows, in any case, to signalize the general sequence of transitions from the non-living to the living (see Fig. 25.1).

(b) *Astrobiology*. The definition suggested in this article (particularly the extended version) provides the necessary conceptual framework to develop biomarkers, as well as criteria to tell the living from the inert material organizations. (i) At an individual level, a living system must have a semi-permeable active boundary, a set of energy currencies and, at least, two types of interdependent macromolecular components (analogous to proteins and nucleic acids); therefore, a virus would not be living, but a sterile organism would. (ii) At the global level, life would only exist as a long-term collective-historical phenomenon on a planet, which would be made long-term habitable

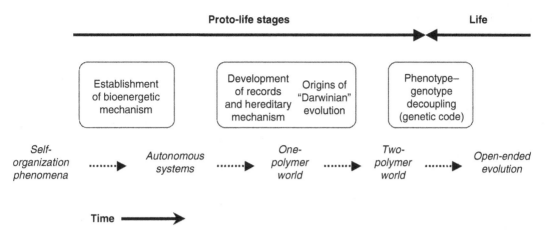

Fig. 25.1. Schematic representation of the sequence of transitions from the non-living to the living proposed in this article. Once bioenergetic mechanisms are established by autonomous systems (that could be regarded as proto-cells in a Shapiro's (2002), "monomer world" or de Duve's (1991), "thioester world") the thermodynamic grounds for the origin of records and the beginning of a cellular "one-polymer world" (like an "RNA world") are set. During the transition to this stage, various recent models can also be relevant, like Lancet's "compositional replication" (Segré *et al.* 2000). This scenario would be the starting point of a process of "Darwinian" evolution (a process of evolution with reproduction, variability and inheritance, but still without genotype–phenotype separation). In any case, according to our definition, the origin of life would strictly take place when a genetic code is established, i.e.: in the transition from the one-polymer world to a "two-polymer world". From this point onwards open-ended evolution capacities take the stage.

as a consequence of the spatial unfolding of life and the operation of its homeorretic mechanisms (Sagan and Margulis 1984; Lovelock 1988). Accordingly, there should be external signs (like the strongly non-equilibrium composition of the atmosphere) that would tell us if there is life or not on the planet (Hitchcock and Lovelock, 1967).

(c) *Artificial Life*. From the three main sub-fields that Artificial Life has developed into (i.e., purely computational models, robotics and *in vitro* experiments), at the moment the most congruent with our perspective is the *in vitro* approach (Deamer 1998; Szostak *et al.* 2001). Computational models tend to disregard the physical-material-energetic aspects of biological organization (which we consider crucial), while robotics—even the so called "autonomous robotics"—pays very little attention to the problem of self-construction (which is, indeed, rather difficult—if not impossible—to deal with outside chemistry). Thus, we consider that the most promising avenue of research in this area is that of *in vitro* experiments, provided that they are primarily focused on the chemical implementation of relatively simple cellular proto-metabolisms (i.e., basic autonomous systems), and specially if they do so without borrowing components or mechanisms from existing biological systems (Bro 1997). This challenge could nowadays be at reach for us (even if bridging the gap from such basic autonomous systems to life—under realistic experimental conditions—will be quite a different issue).

(d) *Biological Theory*. We consider that a profound conceptual debate and clarification is required for the development of a general theory of biology. Attempts to reach consensus on a possible universal definition of life could be very helpful for that task. Our proposal is meant to be a contribution in this direction, putting forward autonomy and open-ended evolution as the key ideas to work on. No general theory of biological organization will be ready until we discover the mechanisms and principles underlying those basic properties and, specially, the way they interweave to produce the living phenomenology.

ACKNOWLEDGMENTS

This article was presented at the Oaxaca ISSOL'02 Meeting, 30 June-5 July, 2002. The authors are grateful to Dr. Jon Umerez for many helpful discussions. K. R.-M. was holding a fellowship from the Instituto Nacional de Técnica Aeroespacial (INTA) of Spain at the Centre of Astrobiology (Madrid) during the preparation and writing of this article. The work was also supported by the Comisión Interministerial de Ciencia y Tecnología (CICYT) of Spain, Grant BMC2000–0764, and by the University of the Basque Country (UPV/EHU) Grant 9/upv 00003.230–13707/2001.

NOTES

This chapter originally appeared in *Origins of Life and Evolution of the Biosphere* 34(3) (2004), 323–346.

REFERENCES

1. Benner, S. A. (1999). How small can a microorganism be? In the Space Studies Board and the National Research Council, *Size limits of very small microorganisms: Proceedings of a workshop* (pp. 126–135). Washington DC: National Academy Press.

2. Bro, P. (1997). Chemical reaction automata. *Complexity*, 2, 38–44.

3. Brooks, D. R. & Wiley, E. O. (1988). *Evolution as entropy: Toward a unified theory of biology*. Chicago: University of Chicago Press.

4. Cleland, C. E. & Chyba, C. F. (2002). Defining 'life.' *Origins of Life and Evolution of the Biosphere*, 32, 387–393.

5. de Duve, C. (1991). *Blueprint for a cell: The nature and origin of life* (p. 9). Burlington NC: Neil Patterson.

6. Deamer, D. W. (1998). Membrane compartments in prebiotic evolution. In A. Brack, (Ed.), *The molecular origins of life: Assembling the pieces of the puzzle* (pp. 189–205). Cambridge, UK: Cambridge University Press.

7. Doolittle, W. F. (1999). Phylogenetic classification and the universal tree. *Science*, 284, 2124–2128.

8. Dyson, F. J. (1985). *Origins of life*. Cambridge, UK: Cambridge University Press.

9. Eigen, M. (1992). *Steps toward life: A perspective on evolution*. New York: Oxford University Press.

10. Eigen, M. & Schuster, P. (1979). *The hypercycle: A principle of natural self-organization*. New York: Springer.

11. Emmeche, C. (1998). *Defining life as a semiotic phenomenon. Cybernet: Human knowing*. New York: Springer.

12. Farmer, J. D. & Belin, A. d' A. (1992). Artificial life: The coming evolution. In C. G. Langton, C. Taylor, J. D. Farmer, and S. Rasmussen (Eds.), *Artificial life II* (Santa Fe Institute studies in the sciences of complexity proceedings, vol. X) (pp. 815–838). Redwood City, CA: Adison-Wesley.

13. Fernández, J., Moreno, A., & Etxeberria, A. (1991). Life as emergence: The roots of a new paradigm in theoretical biology. *World Futures*, 32, 133–149.

14. Fleischaker, G. R. (1988). Autopoiesis: The status of its system logic. *BioSystems*, 22, 37–49.

15. Fleishaker, G. R. (1994). A few precautionary words concerning terminology. In G. R. Fleishaker, S. Colonna, and P. L. Luisi (Eds.), *Self-production of supramolecular structures from synthetic structures to models of minimal living systems* (pp. 33–41). Dordrecht: Kluwer Academic Publishers.

16. Gánti, T. (1987). *The principle of life*. Budapest: OMIKK.

17. Haynes, R. H. (1990). *Ecce eccepoiesis: Playing God on Mars*. In D. MacNiven (Ed.), *Moral expertise: Studies in practical and professional ethics* (pp. 161–183), London: Routledge.

18. Hitchcock, D. R., & Lovelock, J. E. (1967). Life detection by atmospheric analysis. *Icarus*, 7, 149–159.

19. Hoffmeyer, J. & Emmeche, C. (1991). Code-duality and the semiotics of nature. In M. Anderson and F. Merrell (Eds.), *On semiotic modeling* (pp. 117–166). New York: Mouton de Gruyter.

20. Joyce, G. F. (1994). Foreword. In D. W. Deamer and G. R. Fleischaker (Eds.), *Origins of life: The central concepts* (pp. xi–xii). Boston: Jones & Bartlett.

21. Kauffman, S. A. (1993). *The origins of order: Self-organization and selection in evolution*. Oxford: Oxford University Press.

22. Kauffman, S. A. (2000), Autonomous agents. In S. A. Kauffman, *Investigations* (pp. 49–80). Oxford: Oxford University Press.

23. Koshland, D. E., Jr. (2002). The seven pillars of life. *Science*, 295, 2215–2216.

24. Lazcano, A. (2001). Origin of life. In D. E. G. Briggs and P. R. Crowther (Eds.), *Paleobiology II* (pp. 3–8). Oxford: Blackwell Science.

25. Lewontin, R. C. (1970). The units of selection. *Annual Review of Ecology and Systematics*, 1, 1–18.

26. Lovelock, J. E. (1988). *The ages of Gaia*. New York: Norton.

27. Lovelock, J. E. & Margulis, L. (1974a). Atmospheric homeostasis by and for the biosphere: The Gaia hypothesis. *Tellus*, **26**, 2–10.

28. Lovelock, J. E. & Margulis, L. (1974b). Homeostatic tendencies of the Earth's atmosphere. *Origins of Life and Evolution of the Biosphere*, **1**, 12–22.

29. Luisi, P. L. (1994). The chemical implementation of autopoiesis. In G. R. Fleischaker, S. Colonna, and P. L. Luisi (Eds.), *Self-production of supramolecular structures from synthetic structures to models of minimal living systems* (pp. 179–197). Dordrecht: Kluwer Academic Publishers.

30. Luisi, P. L. (1998). About various definitions of life. *Origins of Life and Evolution of the Biosphere*, **28**, 613–622.

31. Margulis, L. (1990). Big trouble in biology: Physiological autopoiesis versus mechanistic neo-Darwinism. In J. Brockman (Ed.), *Doing science: The reality club* **2** (pp. 211–235). New York: Prentice Hall.

32. Margulis, L. & Sagan, D. (2002). Darwin's dilemma. In *Acquiring genomes: A theory of the origin of species* (pp. 25–50). New York: Basic Books.

33. Maturana, H. & Varela, F. J. (1973). *De máquinas y seres vivos—Una teoría sobre la organización biológica*. Santiago de Chile: Editorial Universitaria S. A.

34. Maynard Smith, J. (1986). *The problems of biology*. Oxford: Oxford University Press.

35. Maynard Smith, J. & Szathmáry, E. (1995). *The major transitions in evolution*. Oxford: Freeman and Co.

36. Mayr, E. (1982). *The growth of biological thought*. Cambridge, MA: Harvard University Press.

37. McMullin, B. (2000). John von Neumann and the evolutionary growth of complexity: Looking backward, looking forward... *Artificial Life*, **6**, 347–361.

38. Moreno, A. & Fernández, J. (1990). Structural limits for evolutive capacities in molecular complex systems. *Biology Forum*, **83**, 335–347.

39. Moreno, A. & Ruiz-Mirazo, K. (2002). Key issues regarding the origin, nature and evolution of complexity in nature: Information as a central concept to understand biological organizations. *Emergence*, **4**, 63–76.

40. Moreno, A., Umerez, J., & Fernandez, J. (1994). Definition of life and research program in artificial life. *Ludus Vitalis*, **2**, 15–33.

41. Morowitz, H. J. (1968). *Energy flow in biology*. New York: Academic Press.

42. Morowitz, H. J. (1981). Phase separation, charge separation, and biogenesis. *BioSystems*, **14**, 41–47.

43. Morowitz, H. J. (1992). *Beginning of cellular life*. New Haven: Yale University Press.

44. Morowitz, H. J., Heinz, B., & Deamer, D. W. (1988). The chemical logic of a minimal protocell. *Origins of Life and Evolution of the Biosphere*, **18**, 281–287.

45. Nicolis, G. & Prigogine, Y. (1977). *Self organization in non-equilibrium systems*. New York: Wiley.

46. Oparin, A. I. (1961). The nature of life. In A. I. Oparin, *Life: Its nature, origin and development* (pp. 1–37). New York: Academic Press.

47. Pattee, H. H. (1967). Quantum mechanics, heredity and the origin of life. *Journal of Theoretical Biology*, **17**, 410–420.

48. Pattee, H H. (1977). Dynamic and linguistic modes of complex systems. *International Journal of General Systems*, **3**, 259–266.

49. Pattee, H. H. (1982). Cell psychology: An evolutionary approach to the symbol-matter problem. *Cognition and Brain Theory*, **4**, 325–341.

50. Pattee, H. H. (1997). The physics of symbols and evolution of semiotic controls. In M. Coombs (Ed.), *Workshop on control mechanisms for complex systems: Issues of measurement and semiotic analysis*. Las Cruces, NM: New Mexico State University.

51. Páyli, G., Zucchi, C, & Caglioti, L. (2002). *Fundamentals of life*. Paris: Elsevier.

52. Rosen, R. (1991). *Life itself: A comprehensive inquiry into the nature, origin, and fabrication of life*. New York: Columbia University Press.

53. Ruiz-Mirazo, K. (2001). Condiciones físicas para la aparición de sistemas autónomas con capacidades evolutivas abiertas. Ph.D. dissertation, San Sebastián, Univerisity of the Basque Country.

54. Ruiz-Mirazo, K. & Moreno, A. (1998). Autonomy and emergence: How systems became agents through the generation of functional constraints. In G. L. Farre and T. Oksala (Eds.), *Emergence, complexity, hierarchy, organization* (selected and edited papers from the ECHO III conference) (pp. 273–282). *Acta Polytechnica Scandinavica*, MA91. Espoo-Helsiniki: The Finnish Academy of Technology.

55. Ruiz-Mirazo, K. & Moreno, A. (2000). Searching for the roots of autonomy: The natural and artificial paradigms revisited. *Journal for the Integrated Study of Artificial Intelligence, Cognitive Science, and Applied Epistemology*, **17**, 209–228.

56. Sagan, C, (1970). Life. In *Encyclopedia Britannica*. London: William Benton.

57. Sagan, D. & Margulis, L. (1984). Gaia and philosophy. In L. S. Rouner (Ed.), *On nature* (pp. 60–75). Notre Dame, IN: University of Notre Dame Press.

58. Segré, D., Ben-Eli, D., & Lancet, D. (2000). Compositional genomes: Prebiotic information transfer in mutually catalytic non-covalent assemblies. *Proceedings of the National Academy of Sciences*, 97, 4112–4117.

59. Shapiro, R. (2002). Monomer world. In *Abstracts of the 13th international conference on the origin of life, 10th ISSOL meeting* (p. 60). Oxtaca, Mexico.

60. Shapiro, R. & Feinburg, G. (1990). Possible forms of life in environments very different from the Earth. In J. Leslie (Ed.), *Physical cosmology and philosophy* (pp. 248–255). New York: Macmillan.

61. Szostak, J. W., Bartel, P., & Luisi, P. L. (2001). Synthesizing life. *Nature*, 409, 387–390.

62. Umerez, J. (1995). Semantic closure: A guiding notion to ground artificial life. In F. Moran, A. Moreno, J. J. Merelo, and P. Chaco (Eds.), *Advances in artificial life* (pp. 77–94). Heidelberg: Springer-Verlag.

63. Varela, F. J. (1979). *Principles of biological autonomy*. New York: Elsevier.

64. Varela, F. J. (1994). On defining life. In G. R. Fleischaker, S. Colonna, and P. L. Luisi (Eds.), *Advances in artificial life* (pp. 23–31). Dordrecht: Kluwer Academic Press.

65. von Neumann, J. (1966). *Theory of self reproducing automata*, edited by A. W. Burkes. Urbana, IL: University of Illinois.

66. Wächterchäuser, W. (1988). Before enzymes and templates: Theory of surface metabolism. *Microbiological Reviews*, 52, 452–484.

67. Wicken, J. S. (1987). *Evolution, thermodynamics, and information: Extending the Darwinian program.* Oxford: Oxford University Press.

NOTES

1 Contrast with the "seven pillars" of de Duve (1991): Manufacture of its own components, extract energy from the environment, catalysis, inform cell processes, insulation, regulation and multiplication.

2 Other theoretical models that have addressed the problem of which could be the minimal living organism in general-abstract terms (like Rosen's approach or Gánti's "chemoton theory": see, for instance, Rosen 1991; Gánti 1987) are subject to a similar type of criticism. Any metabolic or proto-metabolic dynamics really takes place in far from equilibrium conditions (and, in order to keep those conditions, all possible minimal organisms must form—and be understood—as thermodynamically open systems). Thus, it is crucial to introduce the energetic–thermodynamic dimension of the problem from the very beginning.

3 This can be achieved even if the definition is formulated from the point of view of individual organisms, i.e., in terms of what a "living being" is and how it organizes and behaves (like authors of the autopoietic school do). The crucial point (which those authors, however, do not consider relevant) is to include in the definition the main feature(s) that will reveal, at the individual-metabolic level of analysis, the existence—and the important implications—of being inserted in a collective-ecological-evolutionary dynamic. Our proposal, in fact, will be elaborated on these lines.

4 According to this criterion, what is really crucial for a system to be considered biological is to have the *potential* to be part of a process of open-ended evolution. In fact, there are living organisms that in natural conditions are not able to reproduce (i.e., sterile organisms, like mules, working bees or plants without seeds) and, thus, cannot continue on the evolutionary process that actually brought them about. Nevertheless, the analysis of the molecular components of any of those organisms would reveal that they have all the mechanisms required for that type of evolution (see below, particularly pp. 317–318). Ontogenic constraints prevent their natural reproduction and their subsequent participation in such a process (even if we could easily think of various ways to overcome those difficulties—through artificial cloning or vegetative reproduction, for instance). Therefore, through the use of the term "capacity" in the second part of our definition we introduce a certain tension between the "actual" and the "in-principle-accessible" properties of living beings, which resembles Gánti's (1987) distinction between *real* (*absolute*) and *potential* criteria for life, and which is necessary—as he showed—to avoid contradictory results.

5 The reader might associate these two fundamental concepts with the names of Varela (1979) and von Neumann (1966), respectively (see also McMullin 2000). However, our formulation is radically different because it is intended to capture the physical-material-energetic requirements behind those concepts (whereas Varela's or von Neumann's approaches are far more abstract).

6 Despite the apparent formal similarity with the "standard definition," our proposal is more restrictive, more demanding and, at the same time, it has a higher explanatory potential. As will be shown with more detail in the next section, autonomy requires much more than the temporary self-maintenance of a chemical network.

7 This issue (i.e., chemistry's relevance for the development of any biological world) would deserve a much longer discussion but, since it is not the central focus of the present article, we just offer a very condensed line of argument. We can add in this brief note that the processes that allow the true *self*-maintenance of a system in far-from-equilibrium conditions must be processes of transformation of matter (i.e., chemical processes), because the components of such a system (including the components that would have to do more directly with the inter-phase or boundary conditions of the system, i.e., with the potential control of the flow of matter-energy through the system) are not stable out of that continuous "turn-over" *chemical* dynamics.

8 Functional relations appear when all the components of a system contribute to or participate in the maintenance of such a system.

9 This "basic agency" has to do with the capacity of the system to control actively some of its boundary conditions—like concentration gradients.

10 Perhaps the exploration of Mars could show traces of proto-metabolisms, of basic autonomous systems, which did not (or have not) overcome the critical transitions to start an open-ended evolutionary dynamics and, therefore, did not (or have not), achieve(d) their long-term maintenance on the planet.

11 Following Dyson (1985), Fleischaker (1994) and Luisi (1994), we distinguish the terms "replication" and "reproduction": Replication is a reliable copying process that takes place at a molecular level, whereas reproduction involves the spatial multiplication of a whole organization.

12 A type of polymer that can already evolve through mutational variations in its space of possible sequences (Benner 1999).

13 By this term we mean a kind of causal connection in a system by which some (quasi) inert material patterns constrain, through a certain mechanism of "translation-interpretation," the metabolic dynamics of the system. In turn, it is only through the participation of the metabolic machinery of the very system that the informational patterns become instructions (see Moreno *et al.* 1994; Moreno and Ruiz-Mirazo 2002).

14 Like the "genetic code" between DNA and proteins present in all living beings.

15 This is fully coherent with recent work that criticizes the traditional idea of "common ancestor" (Doolittle 1999) and suggests that, instead of the common image of *a cell* from which all other living entities on Earth derived (through "vertical evolution"), we should think in terms of a "common ancestral population" of very similar metabolic systems in constant evolution (both "vertical" and "horizontal").

16 Process of constitution of an ecosystem (Haynes 1990).

26 · Does 'life' have a definition?

CAROL CLELAND AND CHRISTOPHER CHYBA

INTRODUCTION

The question "What is life?" is foundational to biology and especially important to astrobiologists who may one day encounter utterly alien life. But how should one approach this question? One widely adopted strategy among scientists is to try to define 'life'.[1] This chapter critically evaluates this strategy. Drawing from insights gained by philosophical investigations into the nature of logic and language, we argue that it is unlikely to succeed. We propose a different strategy, which may prove more fruitful in searches for extraterrestrial life.

We begin by reviewing the history of attempts to define 'life', and their utility in searches for extraterrestrial life. As will become apparent, these definitions typically face serious counterexamples, and may generate as many problems as they solve.

To explain why attempts to define 'life' are fraught with so many difficulties, we must first develop the necessary philosophical background. Therefore, on pp. 328–331 we discuss the general nature of definition and of so-called theoretical identity statements. Pages 332–334 then apply the material developed in these sections to the project of defining 'life'. We argue that the idea that one can answer the question "What is life?" by defining 'life' is mistaken, resting upon confusions about the nature of definition and its capacity to answer fundamental questions about natural categories (Cleland & Chyba 2002).

To answer the question "What is life?" we require not a definition but a general theory of the nature of living systems. In the absence of such a theory, we are in a position analogous to that of a sixteenth-century investigator trying to define 'water' before the advent of molecular theory. The best she or he could do would be to define it in terms of sensible properties, such as its being wet, transparent, odorless, tasteless, thirst quenching, and a good solvent. But no amount of observational or conceptual analysis of these features will reveal that water is H_2O. Yet, as we now know, "H_2O" is the scientifically most informative answer to the question "What is water?" Analogously, in the absence of a general theory of the nature of living systems, analysis of the features that we currently associate with life is unlikely to provide a particularly informative answer to the question "What is life?"

ATTEMPTS TO DEFINE LIFE

The history of attempts to define 'life' is very long, going back at least to Aristotle, who defined 'life' in terms of the capacity to reproduce (Aristotle De *Anima* 415[a22]–415[b2]; Matthews 1977; but see also Shields 1999). To this day, there remains no broadly accepted definition of 'life' (Chyba and McDonald 1995). The scientific literature is filled with suggestions; decades ago Sagan (1970) catalogued physiological, metabolic, biochemical, Darwinian (which he called "genetic"), and thermodynamic definitions, along with their counterexamples. There have been many other attempts[2] (see, e.g., Scluinger 1945 Monod 1971; Feinberg and Shapiro 1980; Dyson 1985; Kamminga 1988; Fleischaker 1990; Joyce 1994, 1995; McKay 1994; Shapiro & Feinberg 1995; Bedau 1996; Rizzotti et al. 1996; Adami 1998; Kauffman 2000; Conrad & Nealson 2001; Harold 2001; Schulze-Makuch et al. 2002) All typically face important problems, in that they include phenomena that most are reluctant to consider alive, or exclude entities that clearly are alive (Chyba & McDonald 1995).

The Nature of Life, ed. M. A. Bedau and C. E. Cleland. Published by Cambridge University Press. © M. A. Bedau and C. E. Cleland 2010.

Consider a few attempted definitions by way of illustration (Sagan 1970). A *metabolic* definition, for example, might be based on the ability to consume and convert energy in order to move, grow, or reproduce. But fire, and perhaps even automobiles, might be said to satisfy some or all of these criteria. A *thermodynamic* definition might describe a living system as one that takes in energy in order to create order locally, but this would seem to include crystals, which like fire would not generally be considered alive. A *biochemical* definition would be based on the presence of certain types of biomolecules, yet one must worry that any such choice could in the future face exceptions in the form of systems that otherwise appear alive but are not made of our particular favored molecules. *Genetic* or *Darwinian* definitions are now more generally favored than any of these other definitions, but these too face drawbacks and will be discussed in detail below.

Another approach has been not so much to "define" life as simply to list its purported characteristics (e.g., Mayr 1982; Koshland 2002). But essentially the same difficulties arise in this approach; for example, Schulze-Makuch (2002) present a list of non-biological parallels to various supposedly distinguishing criteria of life such as metabolism, growth, reproduction, and adaptation to the environment.

Nevertheless, the philosophical question of the definition of 'life' has increasing practical importance, as laboratory experiments approach the synthesis of life (as measured by the criteria of some definitions), and as greater attention is focused on the search for life on Mars (Sullivan & Baross 2007, Ch. 18) and Jupiter's moon Europa (Sullivan & Baross 2007, Ch. 19). In particular, definitions of 'life' are often explicit or implicit in planning remote *in situ* searches for extraterrestrial life. The design of life-detection experiments to be performed on Europa (e.g., Chyba & Philfos 2001, 2002) or Mars (e.g., Nealson & Conrad 1999; Banfield *et al.* 2001; Conrad & Nealson 2001) by spacecraft landers depends on decisions about what life is, and what observations will count as evidence for its detection (Sullivan & Baross 2007, Chs. 22 and 23). This is clearly illustrated by the story of the Viking mission's search for life on Mars.

Viking's search for life on Mars

The Viking mission's search for life on Mars in the mid 1970s remains the only dedicated *in situ* search for extraterrestrial life to date. The basic approach was to conduct experiments with the martian soil to test for the presence of metabolizing organisms, and indeed the results of the labeled release experiment in particular were not unlike what had been expected for the presence of life (Levin & Straat 1979; Levin & Levin 1998). But in the end, The Viking biology team's consensus was for a nonbiological interpretation (Klein 1978, 1979, 1999), strongly influenced by the failure of the Viking gas chromatograph mass spectrometer (GCMS) to find any organic molecules to its limits of detection in the soil with sample heating up to 500 °C (Biemann *et al.* 1977). This instrument had not been intended to conduct a "life-detection" experiment, but de facto did so, implicitly employing a biochemical definition. Moreover, the GCMS would not have detected as many as $\sim 10^6$ bacterial cells per gram of soil (Klein 1978; Glavin *et al.* 2001; Bada 2001), and it now appears that oxidation of meteoritic organics on the martian surface may have produced non-volatile organic compounds that would not have been easily detectable (Benner *et al.* 2000). Correctly interpreted or not, the result was psychologically powerful: no (detected!) organics, no life. Chyba and Phillips (2000, 2002) have presented a list of lessons to be learned from this experience—one lesson is that any *in situ* search for extraterrestrial life should employ more than one definition of life so that results can be intercompared.

Of course, if there were really *one* correct, known definition of 'life', this would be an unnecessary strategy. Currently, it is the Darwinian definition that seems most accepted. We examine this definition below, but shall see that rather than providing us with an unassailable definition, it instead presents fresh dilemmas.

The Darwinian definition

Darwinian (sometimes called *genetic*) definitions of 'life' hold that life is "a system capable of evolution by natural selection" (Sagan 1970). One working version that is popular within the origins-of-life community is the "chemical Darwinian definition" (Chyba & McDonald 1995), according to which "life is a self-sustained chemical system capable of undergoing Darwinian evolution" (Joyce 1994, 1995). Joyce (1994) explains that "the notion of Darwinian evolution subsumes the processes of self-reproduction, material continuity over a historical lineage, genetic variation, and natural selection. The requirement that the system be self-sustained refers to the fact that living systems contain all the genetic information necessary for their own constant production (i.e., metabolism)." The chemical Darwinian definition excludes computer or artificial "life" through its demand that the system under consideration be "chemical", it also excludes biological viruses by virtue of the "self-sustained" requirement.

Some researchers (e.g., Dawkins 1983; Dennett 1995), on the other hand, do not restrict Darwinian evolution to chemical systems, explicitly leaving open the possibility of computer life. This reflects the functionalist view (e.g., Sober 2003) that Darwinian evolution is a more general process that can be abstracted from any particular physical realization. In this view, it is not the computer that is alive but rather the process itself. The artificial vehicle of the computer, produced by human beings, has a status no different from that of the artificial glassware that might be used in a laboratory synthesis of organic life. It is thus not surprising that according to this view, "living" systems or ecosystems can in fact be created in a computer (e.g., Rasmussen 1992; Ray 1992).

Yet this too may seem unsatisfactory: a computer simulation of cellular biochemistry is a simulation of biochemistry, and not biochemistry itself. No computer simulation of photosynthesis, for example, is actually photosynthesis since it does not yield authentic carbohydrates; at best, it yields simulated carbohydrates. So why should a computer simulation of 'life' be called life itself, rather than a *simulation* of life? On the functionalist view, the simulation *is* life, because life is an abstract process independent of any particular physical realization.

There are further problems with Darwinian definitions, in addition to the quandary regarding computer "life." It is possible (though not generally favored among current theories of the origin of life on Earth) that early cellular life on Earth or some other world passed through a period of reproduction without DNA-type replication, during which Darwinian evolution did not yet operate (e.g., Dyson 1985; Rode 1999; New & Pohorille 2000; Pohorille & New 2000). In this hypothesis, protein-based creatures capable of metabolism predated the development of

exact replication based on nucleic acids. If such entities were to be discovered on another world, Darwinian definitons would preclude them from being considered alive.

There is an additional simple objection to the Darwinian definition, namely that individual sexually reproducing organisms in our DNA-protein world do not themselves evolve, so that many living entities in our world are not, by the Darwinian definition, examples of life. The Darwinian definition refers to a *system* that at least in some cases must contain more than one entity; with this reasoning Victor Frankenstein's unique creation (Shelley 1818), for example, is not life even though it is a living entity. But this resolution needs to be explained as more than an ad hoc move to shave from the definition bedeviling entities that we would otherwise call examples of life, but which cause trouble for a particular definition.

Finally, there is a practical drawback to Darwinian definitions. In an *in situ* search for life on other planets, how long would we wait for a system to demonstrate that it is "capable" of Darwinism evolution, and under what conditions (Fleischaker 1990)? This objection, however, is not decisive in itself, since an operational objection is not an objection in principle, and ways (see Chao 2000) might be found to operationalize the definition.

We have focused on Darwinian definitions because they are currently in vogue, especially in light of the great successes of the RNA world model for the origin of life (Gilbert 1986; Sullivan & Baross 2007, Chs. 6 and 8). Nevertheless, as we have discussed, all of the popular versions of the Darwinian definition face similar severe challenges.

DEFINITIONS

To understand why attempts to define 'life' prove so difficult, we now develop the philosophical background for the nature of definition. Definitions are concerned with language and concepts. For example, the definition "'bachelor' means unmarried human male" does not talk about bachelors. Instead, it explains the meaning of a *word*, in this case 'bachelor', by dissecting the *concept* that we associate with it. As this example illustrates, every definition has two parts. The *definiendum* is the expression being defined ('bachelor') and the *definiens* is the expression doing the defining.

Varieties of definition

Many different sorts of things are commonly called "definitions." In this section we will discuss only those that are relevant to understanding the problem of providing a scientifically useful definition of 'life'; for more on definitions, see, for example, Audi (1995).

Lexical definitions report on the standard meanings of terms in a natural language. Dictionary definitions provide a familiar example. Lexical definitions contrast with *stipulative definitions*, which explicitly introduce new, often technical, meanings for terms. The following stipulative definition introduces a new meaning for an old term: 'work' means the product of the magnitude of an acting force and the displacement due to its action. Stipulative definitions are also used to introduce invented terms, e.g., 'electron' (means basic unit of electricity), or 'gene' (means basic unit of heredity). Unlike lexical definitions, stipulative definitions are arbitrary in the sense that rather than reporting on existing meanings of terms, they explicitly introduce new meanings.

Another familiar type of definition is the *ostensive definition*. Ostensive definitions specify the meaning of a term merely by indicating a few (ideally) prototypical examples within its *extension;* the extension of a term is the class of all the things to which it applies. An adult who explains the meaning of the word 'dog' to a child by pointing to a dog and saying "that is a dog" is providing an ostensive definition. Someone who defines 'university' as "an institution such as the University of Colorado, Stanford University, Universidad de Guadalajara, and Cambridge University" is also providing an ostensive definition.

Operational definitions provide an important related form of definition. Like ostensive definitions, operational definitions explain meanings via representative examples. They do not, however, directly indicate examples, but instead specify *procedures* that can be performed on something to determine whether or not it falls into the extension of the definiendum. An example of an operational definition is defining 'acid' as "something that turns litmus paper red." The definiens specifies a procedure that can be used to determine whether an unknown substance is an acid. Operational definitions are particularly important for our discussion since many astrobiologists, e.g., one of these authors (Chyba & McDonald 1995; McKay 1994; Nealson & Conrad 1999; and Conrad & Nealson

2001) have called for the use of operational definitions in searches for extraterrestrial life. The problem with operational definitions is that they do not tell one very much about what the items falling under the definiendum have in common. The fact that litmus paper turns red when placed in a liquid doesn't tell us much about the nature of acidity; it only tells us that a particular liquid is something called 'acid.' In other words, operational definitions differ from ostensive definitions primarily in the manner in which they pick out the representative examples of items falling under the definiendum, namely, indirectly by means of "tests," as opposed to lists or gestures (the "dog" example). We will return to this important point later.

The most informative definitions specify the meanings of terms by analyzing concepts and supplying a noncircular synonym for the term being defined. In philosophy, such definitions are known as *full* or *complete definitions*. But because philosophers sometimes use these expressions to designate more fine-grained distinctions, we shall use the term *ideal definition*.

Ideal definitions

Ideal definitions explain the meanings of terms by relating them to expressions that we already understand. It is thus important that the definiens make use of neither the term being defined nor one of its close cognates; otherwise the definition will be *circular*. Defining 'line' as "a linear path" is an example of an explicitly circular definition, while an implicitly circular definition is defining 'cause' as "something that produces an effect." Someone who does not understand the meaning of 'cause' will also not understand the meaning of 'effect' since 'effect' means something that is caused. Many lexical definitions suffer from the defect of circularity, which is why philosophers dislike dictionary definitions.

The definition of 'bachelor' (as "unmarried human male") with which we began this discussion provides a salient illustration of an ideal definition. It is not circular since the concept of being unmarried, human, and male does not presuppose an understanding of the concept of bachelor. The definiens thus provides an informative analysis of the meaning of 'bachelor.' An ideal definition may thus be viewed as specifying the meaning of a term by reference to a logical conjunction of properties (being unmarried, human, and male), as opposed to representative examples (ostensive definition), or a procedure for recognizing examples (operational definition). The conjunction of descriptions determines the extension of the definiendum by specifying necessary and sufficient conditions for its application. A *necessary condition* for falling into the extension of a term is a condition in whose absence the term does not apply and a *sufficient condition* is a condition in whose presence the term cannot fail to apply.

Most purported ideal definitions face borderline cases in which it is uncertain as to whether something satisfies the conjunction of predicates supplied by the definiens. A good example is the question of whether a 10-year-old boy is a bachelor. Moreover, even if one resolves such cases by adding additional conditions (e.g., adult) to the definiens, there will always be other borderline cases (e.g., the status of 18-year-old males). Language is vague. This is brought forcefully home by the classic example of trying to distinguish a bald man from a man who is not bald in terms of the number of hairs on his head. The fact that we cannot specify a crisp boundary does not show that there is no difference between being bald and not being bald. Ideal definitions that specify both necessary and sufficient conditions are rare. Nevertheless, ignoring the problem of borderline cases, we can often construct fairly satisfactory approximations. If definitions of 'life' faced nothing more serious by way of counterexamples than borderline cases (e.g., viruses), there might not be insurmountable problems. But they have more serious problems. Just as good definitions of 'bachelor' or 'bald man' must deal with, respectively, 40-year-old unmarried men and men sporting thick heads of hair, so good definitions of 'life' must deal with quartz crystals and candle flames, which are (presumably) clearly not alive.

Natural kinds and theoretical identity statements

Ideal definitions specify meanings by providing a "complete" (within the constraints of vagueness) analysis of the concepts associated with terms. They work well for terms such as 'bachelor' or 'fortnight' or 'chair', which designate categories whose existence depends solely on human interests and concerns. Indeed, it is hard to imagine a better answer to the question "What is a bachelor?" than "an unmarried, adult human male."

Ideal definitions do not, however, supply good answers to questions about the identity of *natural kinds*—categories carved out by nature, as opposed to human interests, concerns, and conventions.[3] This issue is particularly important for our purposes since it seems likely (but not certain) that 'life' is a natural kind term—that whether something is living or nonliving represents an objective fact about the natural world. Consider, for example, trying to answer the question "What is water?" by defining the natural kind term 'water'. One could try to define 'water' by reference to its sensible properties, features such as being wet, transparent, odorless, tasteless, thirst quenching, and a good solvent. (This is analogous to some suggested definitions of 'life', e.g., that of Koshland 2002.) Unlike the definition of 'bachelor,' however, this definition of 'water' is not simply a matter of linguistic convention. Nevertheless, reference to a list of sensible properties cannot exclude things that superficially resemble water but are not in fact water. As an example, the alchemists, impressed by water's powers as a solvent, identified nitric acid and mixtures of hydrochloric acid as water, the former being known as *aqua fortis* ("strong water") and the latter as *aqua regia* ("royal water"); *aqua vitae* ("water of life") was a mixture of alcohols (Roberts 1994; Fig. 26.1). Even today we

Fig. 26.1. The distillation of aqua vitae, a form of "water." From *Das Buch zu Distillieren* by Hieronymus Braunschweig (Strassburg, 1519). (From Roberts (1994: p. 100); courtesy British Library.)

commonly classify as "water" various liquids that greatly differ in their sensible properties, e.g., salt water, muddy water, and distilled water. Which of the sensible properties (e.g., transparency or tastelessness) of the various things called "water" are the important ones? Five hundred years ago Leonardo da Vinci (1513) expressed this dilemma well:

> And so it [water] is sometimes sharp and sometimes strong, sometimes acid and sometimes bitter, sometimes sweet and sometimes thick or thin, sometimes it is seen bringing hurt or pestilence, sometimes health-giving, sometimes poisonous. So one would say that it suffers change into as many natures as are the different places through which it passes. And as the mirror changes with the color of its object so it changes with the nature of the place through which it passes: health-giving, noisome, laxative, astringent, sulphurous, salt, incarnadined, mournful, raging, angry, red, yellow, green, black, blue, greasy, fat, thin.

Without an understanding of the intrinsic nature of water, there is no definitive answer to the question "What is water?" Given an understanding of the molecular structure of matter, however, such quandaries disappear. Water *is* H_2O—a molecule made of two atoms of hydrogen and one atom of oxygen. H_2O is what salt water, muddy water, distilled water, and even acidic solutions have in common, despite their obvious sensible differences. The identification of water with H_2O explains why liquids (e.g., nitric acid) that (in some ostensibly important ways) resemble water are not water; their molecular composition is more than H_2O alone. Furthermore, the identification explains the behavior of what we call "water" under a wide variety of chemical and physical circumstances. The identification holds regardless of whether the water is in any of its familiar solid, liquid or vapor phases, and it will hold equally well in less familiar high-pressure solid phases. Indeed, before the advent of modern chemistry, it was not widely recognized that ice, steam, and liquid water are phases of the same kind of stuff. Some ancient Greeks (for example, Anaximenes) believed that steam was a form of "air" (Lloyd 1982: 22). As late as the late seventeenth century, ice and water were thought to be different "species." The Aristotlelian view of water as one of the four basic elements out of which all matter is constructed only began to fall into disfavor in the late eighteenth century with work such as Antoine Lavoisier's paper (1783) entitled "On the nature of water and on experiments which appear to prove that this substance is not strictly speaking an element but that is susceptible of decomposition and recomposition." It took more than analysis of sensible properties to definitively settle questions about the proper classification of such ostensibly different substances as ice and steam.

Notice that the identification of water with H_2O does not have the character of an ideal definition. It cannot be viewed as explicating the concept that has historically been associated with the term 'water' since that concept encompasses stuff varying widely in chemical and physical composition. Moreover, in daily discourse we still use the word 'water' for things that are not pure H_2O. The claim that water is H_2O began as a testable empirical conjecture (situated within Lavoisier's new theoretical framework for chemistry), and it is now considered so well confirmed that most scientists characterize it as a fact. Nevertheless, it remains a scientific hypothesis. It is conceivable (even if extraordinarily unlikely) that we may someday discover that current molecular theory is wrong in some important respect and that water is not H_2O, just as Planck and Einstein showed a century ago that the wave theory of light was incomplete and that light also behaves like a particle. If the claim that water is H_2O represented an ideal definition, we could not admit the possibility that water might not be H_2O any more than we can conceive of a married bachelor or a month-long fortnight.

It is sometimes claimed that *theoretical identity statements* such as "water is H_2O," "temperature is mean kinetic energy," and "sound is a compression wave" represent stipulative definitions. On this view they amount to nothing more than linguistic decisions to take familiar terms from common language and give them wholly new technical meanings within the context of a currently accepted theory (Nagel 1961). The prima facie problem with this account is that it prevents us from making sense of the idea that these statements tell us something new about the stuff designated by the old familiar terms ('water,' 'temperature,' 'sound'). Rather than learning something new, in this view we are merely attaching new concepts (identifying descriptions) to old terms, and hence only changing the way we talk about the world. One might be tempted to say that this is the way language works: if one changes the concept associated with a word radically enough, then one is no longer talking about the same thing. However, such an approach, associated with the philosopher John Locke (and exploited by Thomas Kuhn in his famous arguments for the incommensurability of scientific theories), faces serious logical problems; we discuss these in detail in Appendices 26.1–26.2 at the end of this chapter. For this reason most contemporary philosophers reject the view that theoretical identity statements are stipulative definitions. Some radical changes in the concept of an old word are the result of discovering that we were wrong about the familiar phenomenon that the word designates; for more detail, see Appendix 26.3. Put more concretely, we know something about water that Aristotle and Anaximenes didn't know: water is not a primitive element, but a molecular compound.

WHAT IS LIFE?

Let us return to the definition of 'life'. If (as seems likely, but not certain) life is a natural kind, then attempts to define 'life' are fundamentally misguided. Definitions serve only to explain the concepts that we currently associate with terms. As human mental entities, concepts cannot reveal the objective underlying nature (or lack thereof) of the categories designated by natural kind terms. Yet when we use a natural kind term, it is this underlying nature (not the concepts in our heads) that we are interested in. 'Water' means whatever the stuff in streams, lakes, oceans, and *everything* else that is water has in common. We currently believe that this stuff is H_2O, and our belief is based on a well-confirmed, general scientific theory of matter. We cannot, of course, be absolutely positive that molecular theory is the final word on the nature of matter; *conclusive* proof is just not possible in science. Nevertheless, our current scientific concept of water as H_2O represents a vast improvement over earlier concepts based on superficial sensory experience. If we someday discover that molecular theory is wrong, we will change the concept that we associate with 'water,' but we will still be talking about the same thing.

Analogously to 'water', 'life' means whatever cyanobacteria, hyperthermophilic archaeobacteria, amoebae, mushrooms, palm trees, sea turtles, elephants, humans, and *everything* else that is alive (on

Earth or elsewhere) has in common. No purported definition of 'life' can provide a scientifically satisfying answer to the question "What is life?" because no mere analysis using human concepts can reveal the nature of a world that lies beyond them. The best we can do is to construct and empirically test scientific theories about the general nature of living systems, theories that settle our classificatory dilemmas by explaining puzzling cases—why things that are alive sometimes lack features that we associate with life and why things that are non-living sometimes have features that we associate with life. No scientific theory can be conclusive, but someday we may have a well-confirmed, adequately general theory of life that will allow us to formulate a theoretical identity statement providing a scientifically satisfying answer to the question "What is life?"

Dreams of a general theory of life

In order to formulate a convincing theoretical identity statement for life we need a general theory of living systems. The problem is that we are currently limited to only one sample of life, namely terrestrial life. Although the morphological diversity of terrestrial life is enormous, all known life on Earth is extraordinarily similar in its biochemistry. With the exception of some viruses, the hereditary material of all known life on Earth is DNA of the same right-handed chirality. Furthermore, life on Earth utilizes 20 amino acids to construct proteins, and these amino acids are typically of left-handed chirality. These biochemical similarities lead to the conclusion that life on Earth had a single origin. Darwinian evolution then explains how this common biochemical framework yielded such an amazing diversity of life. But because the biochemical similarities of all life on Earth can be explained in terms of a single origin, it is difficult to decide which features of terrestrial life are common to *all* life, wherever it may be found. Many biochemical features that currently strike us as important (because all terrestrial life shares them) may derive from mere chemical or physical contingencies present at the time life originated on Earth (Sagan 1974). In the absence of a general theory of living systems, how can we discriminate the contingent from the essential? It is a bit like trying to come up with a theory of mammals when one can observe only zebras. What features of zebras should one focus upon—their stripes, common to all, or their mammary glands, characteristic only of the females? In fact, the mammary glands, although present in only some zebras, tell us more about what it means to be a mammal than do the ubiquitous stripes. Without access to living things having a different historical origin, it is difficult and perhaps ultimately impossible to formulate an adequately general theory of the nature of living systems.

This problem is not unique to life. It reflects a simple logical point. One cannot generalize from a single example. What makes the case of life seem different is the amazing diversity of life on Earth today. We risk being tricked into thinking that terrestrial life provides us with a variety of different examples. But biochemical analyses coupled with knowledge of evolution reveals that much of this diversity is a historical accident. Had the history of the Earth been different, life on Earth today would certainly be different. "*How different?*" is a crucial question for astrobiology. In the absence of a general theory of living systems, one simply cannot decide. In essence, the common origin of contemporary terrestrial life blinds us to the possibilities for life in general.

A look at some popular definitions of 'life' illustrates the problem of trying to identify the nature of life in the absence of an adequately general theoretical framework for living things. Many definitions (e.g., Conrad & Nealson 2001; Koshland 2002) cite sensible properties of terrestrial life—features such as metabolism, reproduction, complex hierarchical structure, and self-regulation. But defining 'life' in terms of sensible properties is analogous to defining 'water' as being wet, transparent, tasteless, odorless, thirst quenching, and a good solvent. As we have discussed, reference to sensible properties is unable both to exclude things that are not water (e.g., nitric acid) and to include everything that is water (e.g., ice). Similarly, this approach will be unsuccessful for defining 'life.'[4]

Definitions of 'life' that do not make reference to sensible properties typically suffer from being too general. Definitions of 'life' based on thermodynamics provide good examples. As discussed on pp. 326–328, it is difficult to exclude systems (e.g., crystals) that are clearly non-living without introducing ad hoc devices (Chyba & McDonald 1995). Similarly, the "chemical Darwinian" definition discussed earlier excludes problematic cases (such as artificial or computer life) by simply stipulating that something must be a chemical system in order to qualify as living. If we had an adequate theoretical framework for understanding life, we could avoid the problem of being too general without resorting to ad hoc devices.

New scientific theories change old classifications, for example, by uniting mass and energy under mass–energy, or, less profoundly, by splitting jade into the two minerals jadeite and nephrite. A general theory of living systems might well change our current classifications of living and non-living. These changes in classification will be convincing only if an empirically tested, general theory of living systems can explain, for example, why a system that we once viewed as non-living is really living, or vice versa.[5] But to be in a position to formulate such a theory will require a wider diversity of examples of life. Current laboratory investigations (e.g., research on the hypothesized prebiotic "RNA World" on Earth) and empirical searches for extraterrestrial life are important steps in supplying these examples. Until the formulation of such a theory, we will not know whether such a theoretical identity statement for life exists.

How to search for extraterrestrial life

There remains the problem of how to hunt for extraterrestrial life without either a definition of 'life' or a general theory of living systems. One approach is to treat the features that we currently use to recognize terrestrial life as *tentative criteria* for life (as opposed to *defining criteria*). These features will then necessarily be inconclusive; their absence cannot be taken as sufficient for concluding that something is not alive. Therefore, they cannot be viewed as providing operational definitions of 'life' (in the strict sense of that term). The purpose of using tentative criteria is not to definitively settle the issue of whether something is alive, but rather to focus attention on possible candidates, namely, physical systems whose status as living or non-living is genuinely unclear. Accordingly, the criteria should include a wide diversity of the features of terrestrial life. Indeed, diversity is absolutely crucial (Cleland 2001, 2002) when one is looking for evidence of long past extraterrestrial life, e.g., in the martian meteorite ALH84001 or with instrument packages delivered to ancient martian flood plains or to Europan frozen ice "ponds". Some features for shaping searches for extraterrestrial life (whether extant or extinct) may not even be universal to terrestrial life. For example, features that are common only to life found in certain terrestrial environments may prove more useful for searching for life in analogous extra-terrestrial environments than features that are universal to terrestrial life. Similarly, features that are uncommon or non-existent among non-living terrestrial systems may make good criteria for present or past life, even if they are not universal to living systems, because they stand out against a background of non-living processes. The chains of chemically pure, single-domain magnetite crystals found in ALH84001 provide a potential example. If (as is still quite controversial) it turns out that these chains can only be produced bio-genically (except perhaps under circumstances that are exceedingly unlikely to occur in nature), then they will provide a good biosignature for life, despite the fact that most terrestrial bacteria do not produce them.

The basic idea behind our strategy for searching for extraterrestrial life is to employ empirically well-founded, albeit provisional, criteria that increase the probability of recognizing extraterrestrial life while minimizing the chances of being misled by inadequate definitions. This is similar in spirit (though with greater care given to the limitations of "definition") to suggestions that *in situ* searches for extraterrestrial life should rely when possible on contrasting definitions of life (Chyba & Phillips 2001, 2002). Unlike efforts that focus on a favored definition, our suggestions are perhaps closest to the strategy proposed by Nealson and his colleagues, who (despite their liberal use of the word "definition") emphasize the use of a number of widely diverse biosignatures (atmospheric, hydrospheric, and lithospheric) (Conrad & Nealson 2001; Storrie-Lombardi *et al.* 2001). The important point, however, is that our strategy is deliberately designed to probe the boundaries of our current concept of life. It is only in this way that we can move beyond our Earth-centric ideas and recognize genuinely weird extraterrestrial life, should we be fortunate enough to encounter it. And it is only by keeping the boundaries of our concept of life adaptable and open to unanticipated possibilities that we can accrue the empirical evidence required for formulating a truly general theory of living systems.

APPENDIX 26.1. LOCKE'S THEORY OF MEANING

The idea that theoretical identity statements represent stipulative definitions receives support from a problematic theory of meaning associated with seventeenth-century philosopher John Locke (1689; see Schwartz 1977, for a review). According to this theory, the meaning of *any* term in a language is completely exhausted by the concepts associated with it, and concepts are identified with descriptions. On some versions of the theory, concepts are analyzed as clusters (rather than logical conjunctions) of descriptions. Wittgenstein's oft-cited analysis of the meaning of the word 'game' provides a good illustration (see Wittgenstein 1953; also Schwartz 1977). The items (e.g., chess, solitaire, water polo, charades) that we call "games" are extraordinarily diverse, so diverse that it seems highly improbable that any conjunction of descriptions could distinguish everything that is a game from everything that is not a game. Wittgenstein concludes that there are no necessary and sufficient conditions for being a game. According to Wittgenstein, what distinguishes games from things that are not games is family resemblance: if an item has enough of the pertinent properties, then it is a game. But whether concepts are identified with clusters or conjunctions of descriptions, the question of whether an item falls into the extension of a term is taken to be completely settled by whether it fits the descriptions that we happen to associate with the term. The upshot is that anything that fits our current concept of water qualifies as "water." If our concept of water were completely founded on sensible properties and the sensible properties that we deemed to be most important failed to exclude nitric acid, then not only would we call nitric acid "water" (which, historically speaking, we once did), but on the Lockean view, nitric acid would actually *be* water. On this view, there is no possibility of discovering that we are wrong—that our descriptions are too inclusive or exclusive—since the only thing that qualifies an item as a member of the extension of a term is whether it happens to fit the descriptions that we associate with the term. If we change our concept of water by stipulating, in the context of a new theory, that water is H_2O, then we are no longer talking about the same thing. Thus Aristotle, who held that water is an indivisible element, cannot be interpreted as talking about the same thing that we are talking about when we use the word 'water' because, for us, water is a composite of hydrogen and oxygen atoms.

Locke's theory is unable to distinguish natural kind terms from non-natural kind terms. Locke was fully aware of this; his solution was to bite the bullet, and reject the distinction. In a revealing discussion Locke (1689, Book III, Chapter XI, Section 7) argues that the seventeenth-century debate over whether bats are birds has little scientific merit since the (seventeenth-century) concepts of bat and bird are compatible with either position; for Locke, the debate is merely verbal. Yet in hindsight this seems wrong. The question of whether bats are birds is not merely verbal—a matter of what description we decide to associate with 'bat' and 'bird.' Indeed, we have *discovered* that the things we call "bats" are far more like mammals than birds. It is instructive to compare this situation with an analogous argument over whether bachelors could be married. No one can discover that bachelors are married. Anyone who claims that they have done so either does not understand the meaning of 'bachelor' or, alternatively, is simply stipulating (vs. discovering) a new meaning for 'bachelor'. In other words, unlike the debate over whether bats are birds, the question of whether bachelors can be married *is* purely verbal. An adequate theory of meaning should be able to explain the difference between common nouns like 'bat' and 'bachelor'.

APPENDIX 26.2. JOHN LOCKE AND THOMAS KUHN

The Lockean view of meaning underlies Thomas Kuhn's famous argument for the incommensurability of scientific theories (Kuhn 1962). When the defining descriptions associated with a term drastically change, as happens in scientific revolutions, the Lockean theory says that the meaning of the term also drastically changes. Thus the term 'mass' means something drastically different in Newtonian mechanics (where mass is conserved) than it does in the special theory of relativity (where only mass–energy is conserved). The upshot is that we can't say that the special theory of relativity tells us something new about the thing referred to by the old term 'mass.' Rather than expanding our knowledge of the natural world, on Kuhn's account, new scientific theories only alter our conceptual framework. Yet this conclusion seems wrong. Surely we know more about the natural world than we did a hundred years ago!

The inadequacy of the Lockean framework for meaning cannot, in our view, be overstated. A successful theory of meaning must account for indisputable facts about language and thought; after all, language and thought *are* the subject matter of a theory of meaning. It is undeniable that we speak and think differently about natural kinds than we do about conventional kinds. Because it treats the meaning of *every* term as *just* a matter of convention—as depending only upon the concepts that we happen to associate with it—the Lockean view cannot accommodate this difference; it lacks the resources to explain it.

APPENDIX 26.3. A NEW THEORY OF MEANING

In contemporary philosophy, the Lockean view has been challenged by a new theory of meaning (Schwartz (1997) gives a review). This new theory solves the problems of the old theory by dispensing with the whole project of identifying meanings with concepts, whether construed as conjunctions or as clusters of descriptions.

There are a number of different versions of the new theory of meaning. All of them, however, agree that meaning involves reference, and reference is not determined by concepts. The word 'water' *means* whatever has the same intrinsic nature as the stuff that we typically call "water" regardless of the descriptions that we happen to associate with it. While it is undeniable that we use descriptions (derived from our sensible experiences with paradigmatic examples) to recognize things as water, these descriptions do not (as in the old Lockean view) determine what it is for something *to be* water. Thus something can fit descriptions that we associate with 'water' and yet fail to qualify as water by virtue of having the wrong intrinsic nature.

This point is illustrated by Hilary Putnam (1973, 1975), a founder of the new theory, in a well-known thought experiment. Putnam asks us to suppose that there existed a fantastic planet called "Twin Earth." Twin Earth is like Earth, but the liquid called "water" on Twin Earth is not H_2O but a different liquid whose chemical formula is abbreviated as "XYZ." XYZ and

H_2O have the same sensible properties; XYZ is wet, transparent, odorless, tasteless, and a good solvent. In Putnam's thought experiment, Twin Earthers from the seventeenth century (before molecular theory appeared) and seventeenth-century Earthlings have the same concept of water. A seventeenth-century Earthling might well believe that there is water on Twin Earth. But that conclusion would be wrong. The stuff on Twin Earth that looks like water is not water because it is not H_2O, even though Twin Earthers and Earthlings might not understand this until the end of the eighteenth century.

It is important to understand the point of Putnam's thought experiment. The fact that it makes little scientific sense to speak of Twin Earth being just like Earth *except* for the chemical composition of water is not relevant to his argument. Putnam is making a point about language and concepts. Language is used to describe many kinds of situations, from actual to hypothetical (e.g., what if Al Gore had been the US President in 2003?), to fantastic (e.g., the adventures of the young wizard Harry Potter). An adequate theory of meaning must do justice to hypothetical and fantastic situations as well as factual ones. Putnam's thought experiment about Twin Earth demonstrates that the meaning of a natural kind term is not fully captured by the descriptions that we associate with it. If it were, we would have to conclude that our seventeenth-century Earthling is correct about there being water on Twin Earth.

We now have the tools to evaluate the proposal that theoretical identity statements (e.g., the theoretical identity statement "water is H_2O") are stipulative definitions. On either the old or the new theory of meaning, definitions are concerned only with language and concepts. If statements such as "water is H_2O" are stipulative definitions, then (*à la* Kuhn) they don't tell us anything new about the world of nature. They represent nothing more than linguistic decisions to attach concepts (H_2O) derived from theoretical frameworks (molecular theory) to familiar old terms ('water'). But this interpretation does not do justice to the way we think and speak about theoretical identity statements: we take them to be making defeasible (capable of being invalidated) claims about the old familiar world of experience. On the problematic old theory of meaning, this aspect of our conceptual structure and linguistic behavior could not be explained.

The upshot was that scientific debates over the underlying nature of natural kinds had to be interpreted (e.g., as in Locke's analysis of whether bats are birds) as merely verbal. The new theory of meaning restores the connection between language and the world, and allows us to make good sense of our intuitions about the contingent empirical status of theoretical identities.

Rather than viewing theoretical identity statements as stipulative definitions, it is more accurate to construe them as empirical conjectures, situated within the context of a well-confirmed scientific theory, about a category of items treated in common discourse as a natural kind. Theoretical identities are contingent in the sense that (unlikely as it now seems in some cases) we might someday discover that they are false. Moreover, it is important to keep in mind that there is no guarantee that our best scientific theories will carve up the world in exactly the same way as natural language. As an example, jadeite and nephrite were once included under the common term 'jade', but it is now clear from chemical analysis and microscopic examination that they are different (Bauer 1968; Putnam 1975); the term 'jade' does not designate a (single) natural kind after all. Similarly, in the context of the right theoretical framework, we may discover that what we thought were different natural kinds are actually part of the same natural kind. To cite another example from mineralogy, we now know that rubies and sapphires, despite their striking sensible differences, are members of the natural kind corundum (Al_2O_3) (Bauer 1968). In short, when an old theory is replaced by a new theory, we do not simply change the subject and began talking about something entirely different. We learn something new about an old familiar subject, and this may include discovering that our language and concepts have badly misled us.

NOTES

This chapter originally appeared in Woodruff T. Sullivan III and John A. Baross (Eds.), *Planets and life: The emerging science of astrobiology*, pp. 119–131, Cambridge, UK: Cambridge University Press, 2007.

REFERENCES

1. Adami, C. (1998). *Introduction to artificial life*. New York: Springer-Verlag.

2. Aristotle (1941). De anima, trans. J. A. Smith. In R. McKeon (Ed.), *The basic works of Aristotle* (pp. 535–606). New York: Random House.

3. Audi, R. (Ed.) (1995). *The Cambridge dictionary of philosophy*. Cambridge, UK: Cambridge University Press.

4. Bada, J. (2001). State-of-the-art instruments for detecting extraterrestrial life. *Proceedings of the National Academy of Sciences*, **98**, 797–800.

5. Banfield, J. F., Moreau, J. W, Chan, C. S., Welch, S. A., & Little, B. (2001). Mineralogical biosignatures and the search for life on Mars. *Astrobiology*, **1**(4), 447–465.

6. Bauer, M. (1968). *Precious stones II*. New York: Dover.

7. Bedau, M. (1996). The nature of life. In M. A. Boden (Ed.), *The philosophy of artificial life* (pp. 332–357). Oxford: Oxford University Press.

8. Benner, S., Devine, K., Matveeva, L., & Powell, D. (2000). The missing molecules on Mars. *Proceedings of the National Academy of Sciences*, **97**, 2425–2430.

9. Biemann, K., Oro, J., Toulmin, P., *et al.* (1977). The search for organic substances and inorganic volatile compounds in the surface of Mars. *The Journal of Geophysical Research*, **82**, 4641–4658.

10. Chao, L. (2000). The meaning of life. *Bioscience*, **50**, 245–250.

11. Chyba, C. F. & McDonald, G. D. (1995). The origin of life in the solar system: Current issues. *Annual Review of Earth and Planetary Sciences*, **23**, 215–249.

12. Chyba, C. F. & Phillips, C. B. (2001). Possible ecosystems and the search for life on Europa. *Proceedings of the National Academy of Sciences*, **98**, 801–804.

13. Chyba, C. F. & Phillips, C.B. (2002). Europa as an abode of life. *Origins of Life and Evolution of the Biosphere*, **32**, 47–68.

14. Cleland, C. E. (2001). Historical science, experimental science, and the scientific method. *Geology*, **29**, 978–990.

15. Cleland, C. E. (2002). Methodological and epistemic differences between historical science and experimental science. *Philosophy of Science*, **69**, 474–496.

16. Cleland, C. E. & Chyba, C. F. (2002). Defining 'life.' *Origins of Life and Evolution of the Biosphere*, **32**, 387–393.

17. Conrad, P. G., & Nealson, K. H. (2001). A non-Earth-centric approach to life detection. *Astrobiology*, **1**, 15–24.

18. da Vinci, L. (1513). II codice arundel (no. 263, fol. 57r.). Quoted in Witcombe, C. (Ed.), *Leonardo da Vinci and water* (p. 734), trans. E. MacCurdy. Available at http://witcombe.sbc.edu/water/artleonardo.html (accessed August, 2008).

19. Dawkins, R. (1983). Universal Darwinism. In D. S. Bendall (Ed.), *Evolution from molecules to men* (pp. 403–425). Cambridge, UK: Cambridge University Press.

20. Dennett, D. (1995). *Darwin's dangerous idea: Evolution and the meanings of life*. New York: Simon & Schuster.

21. Dyson, F. (1985). *Origins of life*. Cambridge, UK: Cambridge University Press.

22. Feinberg, G. & Shapiro, R. (1980). *Life beyond Earth: Intelligent Earthlings' guide to the universe*. New York: William Morrow.

23. Fleischaker, G. R. (1990). Origins of life: An operational definition. *Origins of Life and Evolution of the Biosphere*, **20**, 127–137.

24. Gilbert, W. (1986). The RNA world. *Nature*, **319**, 618.

25. Glavin, D., Schubert, M., Botta, O., Kminek, G., & Bada, J. (2001). Detecting pyrolysis products from bacteria on Mars. *Earth and Planetary Science Letters*, **185**, 1–5.

26. Harold, F. M. (2001). Postscript to Schrödinger: So what is life? *American Society for Microbiology News*, **67**, 611–616.

27. Joyce, G. F. (1994). Forward. In D. Deamer and G. Fleischaker (Eds.), *Origins of life: The central concepts* (pp. xi-xii). Boston: Jones & Bartlett.

28. Joyce, G. F. (1995). The RNA world: Life before DNA and protein. In B. Zuckerman and M. Hart (Eds.), *Extraterrestrials: Where are they? II* (pp. 139–151). Cambridge, UK: Cambridge University Press.

29. Kamminga, H. (1988). Historical perspective: The problem of the origin of life in the context of developments in biology. *Origins of Life and Evolution of the Biosphere*, **18**, 1–11.

30. Kauffman, S. (2000). *Investigations*. Oxford: Oxford University Press.

31. Klein, H. P. (1978). The Viking biological experiments on Mars. *Icarus*, **34**, 666–674.

32. Klein, H. P. (1979). Simulation of the Viking biology experiments: An overview. *Journal of Molecular Evolution*, **14**, 161–165.

33. Klein, H. P. (1999). Did Viking discover life on Mars? *Origins of Life and Evolution of the Biosphere*, **29**, 625–631.

34. Koshland, D. E. (2002). The seven pillars of life. *Science*, **295**, 2215–2216.

35. Kuhn, T. S. (1962). *The structure of scientific revolutions*. Chicago: University of Chicago Press.

36. Lahav, N. (1999). *Biogenesis: Theories of life's origins*. New York: Oxford University Press.

37. Lange, M, (1996). Life, 'artificial life,' and scientific explanation. *Philosophy of Science*, **63**, 225–244.

38. Lavoisier, A. L. (1783). On the nature of water and on experiments which appear to prove that this substance is not strictly speaking an element but that it is susceptible of decomposition and recomposition, trans. C. Giunta. *Observations sur la Physique*, **23**, 452–455. Available at http://web.lemoyne.edu/faculty/giunta/laveau.html (accessed August, 2008).

39. Levin, G. V. & Levin, R. L. (1998). Liquid water and life on Mars. *Proceedings of SPIE—The International Society for Optical Engineering*, **3441**, 30–41.

40. Levin, G. V. & Straat, P. A. (1979). Completion of the Viking labeled release experiment on Mars. *Journal of Molecular Evolution*, **14**, 167–183.

41. Lloyd, G. E. R. (1982). *Early Greek science: Thales to Aristotle*. London: Chatto & Windus.

42. Locke, J. (1689). *An essay concerning human understanding*. Oxford: Oxford University Press.

43. MacCurdy, E. (2003). *The notebooks of Leonardo da Vinci, definitive edition in one volume*. Old Saybrook, CT: Konecky & Konecky.

44. Matthews, G. B. (1977). Consciousness and life. *Philosophy*, **52**, 13–26.

45. Mayr, E. (1982). *The growth of biological thought*. Cambridge, MA: Belknap Press.

46. McKay, C. P. (1994). Origins of life. In J. Shirley and R. Fairbridge (Eds.), *Van Nostrand Reinhold encyclopedia of planetary sciences and astrogeology* (pp. 387–391). New York: Van Nostrand.

47. Monod, J. (1971). *Chance and necessity: An essay on the natural philosophy of modern biology*. London: Alfred A. Knopf.

48. Nagel, E. (1961). *The structure of science: Problems in the logic of scientific explanation*. New York: Harcourt, Brace & World.

49. Nealson, K. H., & Conrad, P. G. (1999). Life: Past, present and future. *Philosophical Transactions of the Royal Society of London*, **354**, 1923–1939.

50. New, M. & Pohorille, A. (2000). An inherited efficiencies model of non-genomic evolution. *Simulation Practice and Theory*, **8**, 99–108.

51. Pohorille, A. & New, M. (2000). Models of protocellular structures, functions, and evolution. In G. Palyi, C. Zucchi, and L. Caglioti (Eds.), *Frontiers of life* (pp. 37–42). New York: Elsevier.

52. Putnam, H. (1973). Meaning and reference. *Journal of Philosophy*, **70**, 699–711.

53. Putnam, H. (1975). The meaning of meaning. In H. Putnam (Ed.), *Mind, language and reality: Philosophical papers*, volume II (pp. 215–271). Cambridge, UK: Cambridge University Press.

54. Rasmussen, S. (1992). Aspects of information, life, reality, and physics. *Artificial Life*, **2**, 767–774.

55. Ray, T. S. (1992). An approach to the synthesis of life. *Artificial Life*, **2**, 371–408.

56. Roberts, G. (1994). *The mirror of alchemy: Alchemical ideas in images, manuscripts and books*. Toronto: University of Toronto Press.

57. Rode, B. M. (1999). Peptides and the origin of life. *Peptides*, **20**, 773–786.

58. Rizzotti, M. (Ed.) (1996). *Defining life*. Padova: Padova University Press.

59. Sagan, C. (1970). Life. In *Encyclopedia Britannica* (15th ed.) (pp. 985–1002). Chicago: Encyclopedia Britannica, Inc.

60. Sagan, C. (1974). The origin of life in a cosmic context. *Origins of Life and Evolution of the Biosphere*, **5**, 497–505.

61. Schrödinger, E. (1945). *What is life? The physical aspect of the living cell*. Cambridge, UK: Cambridge University Press.

62. Schulze-Makuch, D., Guan, H., Irwin, L., & Vega, E. (2002). Redefining life: An ecological, thermodynamic, and bioinformatic approach. In G. Palyi, C. Zucchi, and L. Caglioti (Eds.), *Fundamentals of life* (pp. 169–179). New York: Elsevier.

63. Schwartz, S. P. (1977). Introduction. In S. Schwartz (Ed.), *Naming, necessity, and natural kinds* (pp. 13–41). Ithaca: Cornell University Press.

64. Shapiro, R. & Feinberg, G. (1995). Possible forms of life in environments very different from the Earth. In B. Zuckerman and M. Hart (Eds.), *Extraterrestrials: Where are they?* (pp. 165–172), Cambridge, UK: Cambridge University Press.

65. Shelley, M. (1818). *Frankenstein: Or the modern Prometheus*. London: Lackington, Hughes, Harding, Mavor & Jones.

66. Shields, C. (1999). *Order in multiplicity*. Oxford: Oxford University Press.

67. Sober, E. (2003). Learning from functionalism— Prospects for strong artificial life. In C. Langton, C Taylor, J. D. Farmer, and S. Rasmussen (Eds.), *Artificial life II* (Santa Fe Institute studies in the sciences of complexity, proceedings vol. X) (pp. 749–765), Oxford; Boulder, CO: Westview.

68. Storrie-Lombardi, M, Hug, W., McDonald, G., Tsapin, A., & Nealson, K. (2001). Hollow cathode ion laser for

deep ultraviolet Raman spectroscopy and fluorescence imaging. *Review of Scientific Instruments*, **72**, 4452–4459.

69. Sullivan, W. T. & Baross, J. A. (2007). *Planets and Life.* Cambridge: Cambridge University Press.

70. Wittgenstein, L. (1953). *Philosophical investigations*, trans. G. E. M. Anscombe. New York: Macmillan.

ENDNOTES

1 Single quotation marks around a word indicate that it is being mentioned as opposed to being used. Definitions provide one example. As will be discussed on pp. 328–330, definitions are concerned with meaning and language. Another example is the claim that 'life' has four letters; contrast this with the very different claim that life originated on Earth around four billion years ago.

2 Lahav (1999, pp. 117–121) compiles 48 definitions of life (with citations) offered from 1855 to 1997.

3 Some philosophers of science (known as "anti-realists") reject, to greater or lesser degrees, claims that there are knowable, mind-independent facts, entities, or laws. We cannot engage with this literature here; for an introduction see Audi (1995) and references therein.

4 For further discussion of the relation between the concept of life and the features that we use to recognize it, see Lange (1996).

5 Other possibilities include *three* distinct categories of life, or *no* distinct categories, but rather a continuum.

27 · Sentient symphony

LYNN MARGULIS AND DORION SAGAN

Owing to the imperfection of language, the offspring is termed a new animal; but is, in truth, a branch or elongation of the parent, since a part of the embryo animal is or was a part of the parent, and, therefore, in strict language, cannot be said to be entirely new at the time of its production, and, therefore, it may retain some of the habits of the parent system.

Erasmus Darwin

They say that habit is second nature. Who knows but nature is only first habit?

Blaise Pascal

Thinking and being are one and the same.

Parmenides

A DOUBLE LIFE

What is life? Two crucial traits are that life produces (autopoietically self-maintains) and reproduces itself. Then there is inherited change: DNA and chromosome mutation, symbiosis, and sexual fusion of growing life when combined with natural selection means evolutionary change. Nonetheless, autopoiesis, reproduction, and evolution only begin to encompass the fullness of life.

We have glimpsed ways of describing what life is: a material process that sifts and surfs over matter like a strange, slow wave; a planetary exuberance; a solar phenomenon—the astronomically local transmutation of Earth's air, water, and received sunlight into cells. Life can be seen as an intricate pattern of growth and death, dispatch and retrenchment, transformation and decay. Connected through Darwinian time to the first bacteria and through Vernadskian space to all citizens of the biosphere, life is a single, expanding network. Life is matter gone wild, capable of choosing its own

direction in order to forestall indefinitely the inevitable moment of thermodynamic equilibrium—death. Life is also a question the universe poses to itself in the form of a human being.

Life is manifest on Earth as five kingdoms, each revealing from a different angle this mystery of mysteries. In a very real sense, life is bacteria and their progeny. Every available piece of real estate on this planet has become inhabited by subjects of the Kingdom Monera: by the enlightened producer, the tropical transformer, the polar explorer. Life is also the strange new fruit of individuals evolved by symbiosis. Different kinds of bacteria merged to make protoctists. When conspecific protoctists merged the result was meiotic sex. Programmed death evolved. Multicellular assemblages became animal, plant, and fungal individuals. Life is thus not all divergence and discord but also the coming together of disparate entities into new beings. Nor did life stop at complex cells and multicellular beings. It went on, forging societies and communities and the living biosphere itself.

Life is moving, thinking matter, the power of expanding populations. It is the playfulness, precision, and wit of the animal kingdom—which is a marvel of inventions for cooling and warming, moving and holding firm, stalking and evading, wooing and deceiving. It is awareness and responsiveness; it is consciousness and even self-consciousness. Life, historical contingency and wily curiosity, is the flapping fin and soaring wing of animal ingenuity, the avant-garde of the connected biosphere epitomized by members of Kingdom Animalia.

Life is the transmigrator of matter, in which task fungi serve as the closer of loops, making fungal food of plant and animal waste. Life thus seeks out the underworld of soil and rot as much as the sunny vistas enjoyed by photosynthetic beings. Life is a network of

The Nature of Life, ed. M. A. Bedau and C. E. Cleland. Published by Cambridge University Press. © M. A. Bedau and C. E. Cleland 2010.

cross-kingdom alliances, of which Kingdom Mychota is a subtle, seemingly crafty participant. It is an orgy of attractions, from the deceptiveness of counterfeit fungal "flowers" to the delights of truffles and hallucinogens.

Life is the transmutation of energy and matter. Solar fire transmutes into the green fire of photosynthetic beings. Green fire transforms to the red and orange and yellow and purple sexual fire of flowering plants, specialists in cross-kingdom persuasion. Fossilized green fire is hoarded in the human cubicle of the solar economy. Life is incessant heat-dissipating chemistry. And life is memory—memory in action, as the chemical repetition of the past.

These halting descriptions approach but stop short of any final definition of life. We will not proffer any last word, final judgment because life will self-transcend; any definition slips away. In day-to-day adjustment and learning, in long-term action and evolution, in interaction and coevolution, organic beings go beyond themselves in the sense that they become more than what they were. Storing and redistributing the energies of the sun, life displays ever greater levels of activity and complexity. Who can guess what life might make of itself if and when it expands to remake a greater part of the universe into its home?

All organisms lead multiple lives. A bacterium attends to its own needs in the muds of a salt marsh, but it is also shaping the environment, altering the atmosphere. As community member it removes one neighbor's waste and generates another's food. A fungus goes about its business amid forest detritus, as it perforates the leaf of a nearby tree and helps close the loop in the biospheric flow of phosphorus. From one point of view we humans are ordinary mammals; from another, a new planetary force.

Like other animals, we eat, urinate, defecate, copulate; like them, we have descended from merged bacteria and meiotic fertilizing protoctists. Like other mammalian species, *Homo sapiens* should expect to endure for maybe another two million years or so—as the average species life span for mammals in the Cenozoic is less than three million years. All species disappear: they extinguish or diverge to form two or more descendant species. No animal species from Cambrian times is still alive today.

Perhaps *Homo sapiens* will diverge into two offspring species differing as much from us today as we differ from chimps, *Pan troglydites*. Such divergence may even be accelerated by technology. Human descendants, their nervous systems incorporated into durable robotic shells, may observe X-ray emissions of stars with telescope eyes as they cling to interplanetary spacecraft. Perhaps some ex-humans will free their genes of inherited disease or transcend—by genetic manipulation—normal intelligence. Others, dwelling on planets with higher or lower gravities, might undergo dramatic weight gains or losses, with altered bone mass and respiratory systems and rearranged internal organs.

Many scenarios are imaginable. But, whatever we become, our successors will retain traces of the past, which is our present. Consider: even if some new biological weapon were able to instantly vaporize all your animal cells, "you" would not disappear. As Clair Folsome (1932–1988) mused:

> What would remain would be a ghostly image, the skin outlined by a shimmer of bacteria, fungi, round worms, pin worms, and various other microbial inhabitants. The gut would appear as a densely packed tube of anaerobic and aerobic bacteria, yeasts, and other microorganisms. Could one look in more detail, viruses of hundreds of kinds would be apparent throughout all tissues. We are far from unique. Any animal or plant would prove to be a similar seething zoo of microbes.[1]

We share more than 98 percent of our genes with chimpanzees, sweat fluids reminiscent of seawater, and crave sugar that provided ancestors with energy 3000 million years before the first space station had evolved. We carry our past with us.

But now, aggregated into electronically wired cities, humans have begun retooling and transforming life on a planetary scale. Some futurists claim that we have become exempt from mere animal evolution. Are we not more than pretentious apes in fancy clothes? Do we not have music, language, culture, science, computer technology?

Rebuffed by Katharine Hepburn for a gin-drinking binge in the film *The African Queen*, Humphrey Bogart abdicates responsibility for his gruff behavior. "A man takes a drop too much once in a while," he explains, "it's only human nature."

"Nature" replies Hepburn primly, looking up from her Bible, "is what God put us on this Earth to rise *above*."

Self-transcending life never obliterates its past: humans are animals are microbes are chemicals. The view that we are "more" than animals does not contradict the materialistic perspective underpinning science. Life is less mechanistic than we have been taught to believe; yet, since it disobeys no chemical or physical law, it is not vitalistic. While we sense in ourselves a great degree of freedom, all other beings, including bacteria, also make choices with environmental consequences. Stored and transformed in life, the energy of sunlight powers cell growth, sex, and reproduction of highly similar life forms. All living beings may share our own feeling of free will.

Life on Earth is a complex, photosynthetically based, chemical system fractally arranged into individuals at different levels of organization. We cannot rise above nature, for nature itself transcends.

Nature does not end with us, but moves inexorably on, beyond societies of animals. Global markets and Earth-orbiting satellite communication, wireless telephones, magnetic resonance imagery, computer networks, cable television, and other technologies connect us. Indeed, people already form a more-than-human being: an interdependent, technologically interfaced superhumanity. Our activities are leading us toward something as far beyond individual people as each of us is beyond our component cells.

Now, at the end of this century's hot and cold wars, we communicate freely across national boundaries at the speed of light, via telephone and computer. News flashes around the world. But these social changes at the start of the new millennium pale beside the sweeping biological changes. The Phanerozoic eon, which began more than 500 million years ago with widespread predation and defense against that predation in the form of animals with shells, is ending. Evolutionary movements that made eukaryotes out of bacteria and animals from protoctists are now repeating on a planetary scale. Humanity is transforming from a society into a new level of organic being. Our populations are beginning to behave as the brain or neural tissue of a global being. As we become more populous and sedentary, our human and technology-extended intelligence becomes part of planetary life as a whole.

The facts of life, the stories of evolution, have the power to unite all peoples. By integrating the data of thousands of scientists, and by cultivating the doubt and skepticism that is the epitome of scientific inquiry, the cultural invention called science could provide a more compelling, if ever corrigible, description of the world than do parochial myths and divisive, faith-demanding religious traditions. This hardly means that scientists are always correct. Yet the most meaningful story of existence for future humanity is more likely to come from the evolutionary worldview of science than from Hinduism, Buddhism, Judeo-Christianity, or Islam. The dual understanding of scientific inquiry and creation myth could become a single view: a science tale rich both in verifiable fact and personal meaning.

CHOICE

In a truly evolutionary psychology, spirit and mind are not celestial sprinklings but sovereign to living matter. Thought derives from no world but this one; it comes from the activity of cells.

When offered a variety of foodstuffs, swimming bacteria, ciliates, mastigotes, and other mobile microbes make selections—they choose. Squirming forward on retractable pseudopods, *Amoeba proteus* finds *Tetrahymena* delectable but avoids *Copromonas*. *Paramecium* prefers to gobble small ciliates, but if starved for these and other protoctists it reluctantly feeds on aeromonad and other bacteria.

Although "merely" protoctists, foraminifera (one of the most diverse groups of fossil-forming organisms) make an astounding variety of magnificent shells. Foraminifera without their shells resemble amebas with very long, thin pseudopods. The shells are formed from sand, chalk, sponge spicules, even other foram shells. To appropriate their cell-shell homes, some forams agglute whatever particles are available from the environment together with an organic cement. Observations reveal, however, that, when presented with a hodge-podge of different particles, foraminifera make distinct choices based on shape and size—*Spiculosiphon*, for example, passes over much of the motley sediment, selecting only sponge spicules to make its test or shell.[2] Without brains or hands, these determined protoctists choose their building materials.

Smaller still, chemotactic bacteria just two microns long can sense chemical differences. They swim toward sugar and away from acid. A chemotactic bacterium can smell a difference in chemical concentration that is a mere one part in ten thousand more concentrated at one end of its body than at the other.

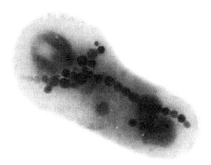

Fig. 27.1. Magnetotactic bacterium remnant showing internal magnetosomes (photo taken with an electron microscope). These cells, able to orient themselves magnetically to the North or South Pole, exemplify the sensitivity of living substance at all levels, scales, and kingdoms. Perception, choice, and sensation apply not just to human beings or animals but, if they apply at all, they apply to all life on Earth.

Biochemist Daniel Koshland explains the spiritual leanings of prokaryotes:

"Choice," "discrimination," "memory," "learning," "instinct," "judgment," and "adaptation" are words we normally identify with higher neural processes. Yet, in a sense, a bacterium can be said to have each of these properties.... it would be unwise to conclude that the analogies are only semantic since there seem to be underlying relationships in molecular mechanism and biological function. For example, learning in higher species involves long-term events and complex interactions, but certainly induced enzyme formation must be considered one of the more likely molecular devices for fixing some neuronal connections and eliminating others. The difference between instinct and learning then becomes a matter of time scale, not of principle.[3]

Microbes sense and avoid heat, move toward or away from light. Some bacteria even detect magnetic fields. They harbor magnets aligned in a row in their tiny, rod-shaped bodies (Fig. 27.1). That bacteria are simply machines, with no sensation or consciousness, seems no more likely than Descartes's claim that dogs suffer no pain. That bacteria sense and act, but with no feeling, is possible—but ultimately solipsistic. (Solipsism is the idea that everything in the world, including other people, is the projection of one's own imagination.) Cells, alive, probably do feel. Indigestible mold spores

and certain bacteria are rejected by protists. Others are greedily gobbled. At even the most primordial level living seems to entail sensation, choosing, mind.

Darwin formally distinguished "natural selection" (referring to interactions between nonhuman life and its environment) from human-generated "artificial selection" (the aesthetic or functional choices of pigeon fanciers, dog breeders, and agriculturalists). But "natural" selection is, in a way, more "artificial," and far less mechanical, than Darwin implied. The environment is not inert. Self-awareness is not confined to the space between human ears. Non-human beings choose, and all beings influence the lives of others.

Humans, we are told, are special. We have upright posture (allowing us to think of ourselves as literally "above" other species). We have opposable thumbs (man the tool user), linguistic abilities (man the story-teller), a superanimalistic soul (Descartes' distinction). We have, at least in Western culture, a tradition of seeing ourselves as being in a position of moral stewardship over the rest of life. Even in the absence of God, we imagine ourselves to possess a unique capacity to destroy the planet (via nuclear weaponry) or to swiftly change atmosphere and climate.

Even such an ardent foe of the idea of progress in evolution as Stephen Jay Gould (and he is not alone) proposes that whereas humans can evolve quickly through "cultural selection," all other forms of life on Earth are shackled to the ancient, plodding system of "natural selection." But the sheer number of traits listed to explain human uniqueness is enough to arouse suspicion. Among the dazzling array of reasons implying our superiority over the rest of life, one scientific argument stands out to us in curious contrast to the rest: humans are the only beings capable of wholesale self-deception.

This claim is based on early humans' presumably delusionary belief in the afterlife. Before written history our ancestors buried their dead with food, weapons, and herbal medicines of little use to corpses. How ironic that we, in seeking examples of our superiority over the rest of life, have finally congratulated ourselves on a trait that threatens to negate all the others! Although members of other species trick one another, humans are the expert self-deceivers: As the best symbol users, the most intelligent species, and the only talkers, we are the only beings accomplished enough to fully fool ourselves.

LITTLE PURPOSES

Freud's understanding of unconsciousness as repression—painful memories are pushed away from the conscious mind—has diverted attention from another way in which actions become unconscious. Not avoidance but extended focus can make an action automatic, second nature. A speech is "learned by heart." A practiced typist no longer glances at the keyboard. Virtually any activity when memorized subsides from conscious attention. The heart pumps, the kidneys filter in autonomic quasi perfection. Over breathing and swallowing, normally automatic, the willful organism can exert some voluntary control and modulation.

Now here is a strange thought: Perhaps we mammals remain unconscious of inborn physiology because, under pressure of survival, our ancestors consciously practiced their skills to unconscious perfection. Although modern science does not yet offer us a mechanism that transmits the learned habits of one generation to the physiology of the next, experience shows that conscious can become unconscious with repeated action. The gulf between us and other organic beings is a matter of degree, not of kind. Taken together, the vast sentience comes from the piling up of little purposes, wants, and goals of uncounted trillions of autopoietic predecessors who exercised choices that influenced their evolution. If we grant our ancestors even a tiny fraction of the free will, consciousness, and culture we humans experience, the increase in complexity on Earth over the last several thousand million years becomes easier to explain: life is the product not only of blind physical forces but also of selection, in the sense that organisms choose. All autopoietic beings have two lives, as the country song goes, the life we are given and the life we make.

In the nineteenth century Samuel Butler (1835–1902), English author, painter, and musician, challenged Charles Darwin's account of evolution. Butler, who had many arguments with his father, left for New Zealand to become a sheep farmer after completing his education at Shrewsbury and St. John's College, Cambridge. Excited by Darwin's *Origin of Species* when he first read the book in New Zealand, Butler gradually grew disenchanted with it. A scholarly rebel who satirized society and explored the origins of religion, Butler accepted evolution but rejected Darwin's presentation. He began to

suspect a dogmatism in the march of science as narrow as, but more insidious than, that of the church. Reading previous evolutionists, including Darwin's grandfather Erasmus, Butler accused the younger Darwin of failing to acknowledge his intellectual debts.

Darwin's schooling at Shrewsbury had been under the famous headmaster Dr. Butler—Samuel Butler's grandfather. The younger Butler claimed Darwin pretended to know little of his predecessors, and in the first editions of *Origin of Species* made it seem, upon returning from his famous Galapagos voyage, that the theory of natural selection had simply "occurred" to him. Butler pilloried Darwin's "Historical Sketch of the Progress of Opinion on the Origin of Species," which Darwin included in the second (1860), third (1861), and fourth (1866) editions of *Origin*. Darwin apologized for this entry, described as "brief but imperfect." By the sixth and final (1872) edition the qualification was simply that it was "brief"—carrying the implication that the historical sketch had perhaps grown more nearly perfect in the interim. Butler disagreed.

What irked Butler most was Darwin's overly mechanical portrayal of the evolutionary process. Darwin, Butler quipped, had "taken the life out of biology." To be palatable to a religious populace in the Victorian era Darwin's evolution needed a credible scientific mechanism. Because the most respected achievement of the time was the physical science of Isaac Newton, Darwin portrayed evolution just as Newton had portrayed gravity: The result of abstract principles and mechanical interactions.

Although best known as author of *Erewhon* and the posthumously published *The Way of All Flesh* (an influential exploration of inter-generational struggle), Butler felt his greatest contribution was to evolution theory. Retreating from Darwin's neo-Newtonian presentation of organic beings as "things" acted on by "forces," Butler presented sentient life as making numberless tiny decisions—and thus responsible in part for its own evolution. Today Butler's view of the sum effect of little purposes escapes reprobation only where humans are concerned. We consider ourselves forward-thinking cultural beings, able to lay the flesh of the future upon the bones of imagination. We even believe we can govern evolution. The rest of life we dismiss as exempt from such Promethean foresight. Other organic beings are portrayed as the result of physicochemical forces or unmediated genetics—too inert, by insinuation, to play a formative role in their own development. Butler begged to differ.

Butler's well-honed arguments and flashy polemics flouted the dry scientific prose of his Victorian day. One Butlerian theme stands out: living matter is mnemic, it remembers and embodies its own past. Life, according to Butler, is endowed with consciousness, memory, direction, goal-setting. In Butler's view all life, not just human life, is teleological; that is, it strives. Butler claimed that Darwinians missed the teleology, the goal-directedness of life acting for itself. In throwing out the bathwater of divine purpose, Darwin discarded the baby of living purposefulness.

No photosynthetic bacterium decided one day to become a willow tree. *Amoeba proteus* does not today set out to make itself into a mouse; it knows only that the swimming *Tetrahymena* it relentlessly pursues is tasty. Ameba-knowledge at this level of sensing and moving generates a million little such willful acts. These are sufficient for evolution to work its wonders.

Life's purposes are grandiose only in the aggregate, and in retrospect. Up close and confined in time they are ordinary. Nonetheless, no organic being is a billiard ball, acted upon only by external forces. All are sentient, possessing the internal teleology of the autopoietic imperative. Each is capable, to varying degrees, of acting on its own.

BUTLER'S BLASPHEMY

In the second of four books published at his own expense, Samuel Butler discussed the evolutionary views of Erasmus Darwin, Jean-Baptiste Lamarck, Georges Louis Leclerc Buffon, and others. *Evolution, Old and New* (published in 1879) was thus a fitting title. In a circuitous series of newspaper letters and essays Butler published criticism of Darwin's work: he accused the great man of evading the credit owed Grandpa Erasmus and of mechanizing life. Butler even questioned Darwin's honesty. Butler, trying to put life back into biology, hoped Darwin would respond to his 1879 book or to his earlier (1877) *Life and habit.*

Butler asked Darwin if his thoughts on evolution were an inspiration without precedent that had appeared to him from on high. Was it just the result of contemplation upon a great number of facts? Butler argued that the aura of poet-evolutionist Erasmus, coupled with far-flung reading of evolutionary ideas,

must have also contributed to Darwin's intellectual development. Whether Darwin was a subtle master of self-promotion or Butler a paranoiac of scholastic patience may never be resolved. But Butler was predisposed, by his own rebellious nature, combined with Darwin's growing status as an intellectual icon, to feel disappointment in the great man. A biographical book about Erasmus Darwin had appeared. It was translated from German, and Darwin approved the translation as accurate. Butler read it and was alarmed to find that the translation into English of the early French evolutionist Lamarck used precisely the same words that Butler had used to translate Lamarck in Butler's own book *Evolution, Old and New.* There was also a comment that those who wished to revive evolutionary thought prior to Charles Darwin showed "a weakness of thought and a mental anachronism which no one can envy." Butler believed he was being attacked obliquely in a fashion that precluded open debate; he believed that Darwin himself might have been responsible for the dismissal of the supposedly antiquated evolutionary ideas. He confronted Darwin first by letter, then in the newspaper. In a letter to Butler, Darwin explained that such alterations in translation as those made in the book about his grandfather were so common that "it never occurred to me to state that the article had been modified."

Darwin's family, and Thomas Huxley, urged Darwin not to respond to any Butler criticism or personal letters. He didn't. Yet biographical records show that he drafted two responses, never sent. In the second Darwin wrote he "could explain distinctly how the accident arose, but the explanation does not seem to me worth giving. This omission, as I have already said, I much regret."[4] Fascinatingly, Darwin vacillates even here, between the two modes of explanation Butler was to neatly summarize in one of his books as *Luck, or Cunning*: on the one hand, something was forgotten, there was a lapse of memory; on the other hand, there was an "accident," something was "accidentally omitted." Even here, close up, we see Darwin hedging, unable to characterize whether an event—and, indeed, one of his own doing—was the result of luck (chance, the "accident") or cunning (choice and design), the very theory Butler accused him of ignoring.

We agree with Butler that life is matter that chooses. Each living being, Samuel Butler argued, responds sentiently to a changing environment and tries during its life to alter itself. But living beings cannot effect changes with great efficiency. No light bulb appeared one day over the mammal that chose to become human. Rather, gradually, in tiny increments, living systems with non-negotiable needs for food, water, and energy transformed themselves in wily and persistent ways.

What theologians called design, and considered otherworldly, was for Butler the result, in part, of Earth-bound thinking matter. The analogy of the writer comes to mind: she is blocked, with only a vague notion of what to write. Nonetheless, by following grammar, spelling, and syntax, by adding word to word, something purposeful emerges. The writerly result is not entirely her own because she complies with the rules of language. Similarly, no life flouts any law of physics, chemistry, or thermodynamics. As the decisions of writers exist in the lexical world, so the choices of living beings exist in the material world. Neither are absolute, and yet the deeper rules of matter, on the one hand, and language, on the other, impose structures that permit overall designs to arise, not perfectly complete but as accumulations of large numbers of minor, individual decisions.

Butler believed minute changes effected by organisms on their environment begin as conscious pursuits, but end as unconscious practice. For Butler, amebas too have their little wants, their little spheres of influence, their little "tool-boxes" with which they materially change their environments, pursue their little goals, and build their little houses. This possibility is not ruled out by modern science. Nobel Prize–winning Danish physicist Niels Bohr (1885–1962), in discussing the striking utilization by organisms of their "past experience for reactions to future stimuli," contended that, despite the success of physics-based mechanistic biology, there was a need for description that includes "purposiveness."

We must realize that the description and comprehension of the closed quantum phenomena exhibit no feature indicating that an organization of atoms is able to adapt itself to the surroundings in the way we witness in the maintenance and evolution of living organisms. Furthermore, it must be stressed that an account, exhaustive in the sense of quantum physics, of all the continually exchanged atoms in the organism not only is

infeasible but would obviously require observational conditions incompatible with the display of life... it is evident that the attitudes termed mechanistic and finalistic [that is, purposeful] do not present contradictory views on biological problems, but rather stress the mutually exclusive character of observational conditions equally indispensable in our search for an ever richer description of life.[5]

For Butler living matter can "memorize" its behavior not only on the ontogenetic level of individual experience but also on the phylogenetic level of species history. The transition between ontogeny (the development of an individual life) and phylogeny (persistence and change of many individuals through time) is relative. The difference between the same individual newly born and at age eighty is greater, Butler argued, than that between a newborn infant of one generation and a newborn infant of the next.

Reptiles shed their skins; insects reshuffle their proteins in the pupa stage. The corpse is replaced, with some overlap, by its grandson. We moderns accept that a caterpillar "metamorphoses" into a butterfly, yet we employ the term "death" for what happens to grandfather's body: Nevertheless, with both the metamorphosing insect and the dying grandparent, the fresh body of youth reappears. Each of us is entitled to think we die, suggests Butler, but the demarcation is highly arbitrary: A parent contributing to the flesh of a child is a prolongation, not an abrupt end, to biological continuity. The "individual" is not so complete, in time, as we have been taught.

Butler believed unconsciousness applies not only to adult human beings but throughout many different levels of living organization. The most important tasks, the ones most often repeated, have become the most unconscious, the most "physiological." The pumping of blood is too ancient and important to be unlearned or modified easily by mere lassitude or an act of will. Divorce from playful consciousness insures that routine but critical activities are competently performed. Treated as work, effected regularly and automatically, important physiology is not exposed to experimentation that might destroy it. The once-conscious process of steering an automobile recedes to the unconscious while the mind attends to other matters.

No concert pianist strives to hit the keys in order. The long periods of former conscious striving ingrain the knowledge into the musician's fingers. Dancers call their talent "muscle memory." Choice and practice become smooth habit.

Cells, with eons of practice, do not consciously decide to respire oxygen or reproduce by mitosis. They, or the bacterial remnants that compose them, may once have so strived. The more recent the addition of a habit to life's physiological repertoire, the more likely it is to be conscious or at least subject to conscious interference. No longer do animals, plants, and fungi directly respond to the metabolic pathways by which their cells obtain energy when oxygen gas reacts with food hydrogen atoms to produce water and carbon dioxide. Such metabolic behavior is submerged in the permanent unconscious of modern organisms. Oxygen-hydrogen reactions performed by mitochondria, which were once free-living bacteria, are a chemical feat that has not ceased for 2,000 million years, ever since oxygen respirers responded to an environment transformed by oxygenic cyanobacteria. By contrast, peristalsis in digestion—also an unconscious phenomenon but one of which mammals tend to be aware—evolved in animal ancestors long after the microbial stage. Swallowing, chewing, and speaking are behaviors that have been learned far more recently, and in that order.

Butler's theory of unconscious memory holds that all beings are capable of forming habits, some of which—upon much repetition—become physiologically ingrained over the course of evolution. We do not remember, writes Butler, when first we grew an eye. Someday, he muses, so many of us will have learned to read and write, so often, that we will be born that way. Perhaps a future William Harvey will be required to uncover the details of how the learning of reading became a physiology, just as Harvey illuminated the circulation of the blood. Here we disagree with Butler. It seems doubtful that children will ever be born reading and writing. However, if one substitutes television for writing one can see that Butler is already on target: children are born, more or less, watching television; and television, put together from technological products made by companies all over the planet, is already so complicated that few humans understand it in detail.

HABITS AND MEMORY

Given free will and the status of living beings as open thermodynamic systems, one should not be too quick to use classical physics to justify an understanding of life as a mechanical phenomenon. A general property distinguishing life from nonliving matter is its historical coherence, including the potential to evolve. By exporting disorder, randomness, and entropy to their surroundings, living systems increase local complexity, intelligence, and beauty, building on the past and planning for the future. Organisms that find new means of extracting energy and matter for the perpetuation of their form will tend to be preserved, leading to increasingly curious and creative beings. The hints and hunches must be replaced by firm detail, yet the extraordinary storage and transmission processes of life for molecular heredity, as well as cultural information, may be robust enough to encompass the phenomenon postulated by Butler: phylogenetic "memorization," the conversion of the conscious strivings of one generation into the activities and, eventually, the physiologies of the next.

Although we fail as yet to see how an organism's or even a species' voluntary habits can become the physiology of a future generation via a material basis of heredity, we are fascinated by Butler's suggestion. We know, for example, that many organic beings acquire new heritable traits by symbiogenesis and that a vast array of others, not only people, are capable of learning. Ecosystems grow increasingly complex and sensitive; processes practiced in them repeatedly by one generation may become easier for the next. More open-minded investigation is needed. Objections may be leveled against Butler's ideas, yet he cannot be accused of the atavistic thinking which clings to humanity's separate status. Covertly considering ourselves divine, under the scientific rubric of "cultural evolution," or by dint of that other desperate euphemism, our "big brains," we are probably now more ecologically impoverished than we would be if, a century ago, we had adopted Butler's notion of all life as a conscious continuum.

Butler did not object to evolution but to the loss of the richness of the earlier, more lively views, in which living beings themselves were involved in natural selection:

> According to Messrs. Darwin and Wallace, we may have evolution, but are on no account to have it as mainly due to intelligent effort, guided by...

sensations, perceptions, ideas. We are to set it down to the shuffling of cards... According to the older men, cards did indeed count for much, but play counted for more. They denied the teleology of the time—that is to say, the teleology that saw all adaptation to surroundings as part of a plan devised long ages since by a quasi-anthropomorphic being who schemed.... This conception they found repugnant alike to intelligence and conscience, but, though they do not seem to have perceived it, they left the door open for a design more true and more demonstrable than that which they excluded... They made the development of man from the amoeba part and parcel of the story that may be read, though on an infinitely smaller scale, in the development of our most powerful marine engines from the common kettle, or of our finest microscopes from the dew drop. The development of the steam-engine and the microscope is due to intelligence and design, which did indeed utilize chance suggestions, but which improved on these, and directed each step of their accumulation, though never foreseeing more than a step or two ahead, and often not so much as this.[6]

EXISTENCE'S CELEBRATION

For nineteenth-century Englishmen of science it was natural and expedient to invoke Newtonian mechanics and conceive of life as Newton's matter: blind bits predictably responding to forces and natural laws. Like some piece of well-made clockwork, the world was donated or its mechanism manufactured by a transcendent god—a god that then stood outside its creation.

This was the new view of evolution: God, if it existed, was Newton's God. Not an active interloper in human affairs, it was the god of the mathematicians, the geometer god, who made the laws and then sat by and watched those laws play out. But an older view left room for a kind of god, too—a more active god. This was the view that Samuel Butler attempted to resuscitate—that life itself was godlike. There was no grand design, but millions of little purposes, each associated with a cell or organism in its habitat.

To the neo-Newtonians, the Darwinians, free will had been all but banished from the universe because the universe was portrayed as a mechanism and

mechanisms do not have consciousness. For Descartes, God continued to have consciousness and people did insofar as they were in touch with God. But when Darwin showed through painstaking work that people too could be explained by the mechanism of natural selection, consciousness suddenly became redundant in the human world as well. Butler brought consciousness back in by claiming that, together, so much free will, so much behavior becoming habit, so much engagement of matter in the processes of life, so many decisions of where, how, and with what or whom to live, had shaped life, over eons producing visible organisms, including the colonies of cells called human. Power and sentience propagate as organisms. Butler's god is imperfect, dispersed.

We find Butler's view—which rejects any single, universal architect—appealing. Life is too shoddy a production, both physically and morally, to have been designed by a flawless Master. And yet life is more impressive and less predictable than any "thing" whose nature can be accounted for solely by "forces" acting deterministically. The godlike qualities of life on Earth include neither omniscience nor omnipotence, although an argument could be made for earthly omnipresence.

Life, in the form of myriad cells, from luminescent bacterium to lily-hopping frog, is virtually everywhere on the third planet. All life is connected through Darwinian time and Vernadskian space. Evolution places us all in the stark but fascinating context of the cosmos. Although something odd may lurk behind and before this cosmos, its existence is impossible to prove. The cosmos, more dazzling than any sect's god, is enough. Life is existence's celebration.

Butler's forgotten theory intrigues us. The mind and the body are not separate but part of the unified process of life. Life, sensitive from the onset, is capable of thinking. The "thoughts," both vague and clear, are physical, in our bodies' cells and those of other animals.

In comprehending these sentences, certain ink squiggles trigger associations, the electrochemical connections of the brain cells. Glucose is chemically altered by reaction of its components with oxygen, and its breakdown products, water and carbon dioxide, enter tiny blood vessels. Sodium and calcium ions, pumped out, traffic across a neuron's membranes. As you remember, nerve cells bolster their connections, new cell adhesion proteins form, and heat dissipates. Thought, like life, is matter and energy in flux; the body is its "other side." Thinking and being are the same thing.

If one accepts this fundamental continuity between body and mind, thought loses any essential difference from other physiology and behavior. Thinking, like excreting and ingesting, results from lively interactions of a being's chemistry. Organism thinking is an emergent property of cell hunger, movement, growth, association, programmed death, and satisfaction. Restrained but healthy former microbes find alliances to make and behaviors to practice. If what is called thinking results from such cell interactions, then perhaps communicating organisms, each of themselves thinking, can lead to a process greater than individual thought. This may have been what Vladimir Vernadsky meant by the noosphere.

Gerald Edelman and William Calvin, both neuroscientists, have each proffered a kind of "neural Darwinism." Our brains, they say, become minds as they develop by rules of natural selection.[7] That idea may provide a physiological basis for Butler's insights. In the developing brain of a mammalian fetus, some 10^{12} neurons each become connected with one another in 10^4 ways. These cell-to-cell adhesions at the surface membranes of nerve cells are called synaptic densities. As brains mature, over 90 percent of the cells die! By programmed death and predictable protein synthesis, connections selectively atrophy or hypertrophy. Neural selection against possibilities, always dynamic, leads to choice and learning, as remaining neuron interactions strengthen. Cell adhesion molecules synthesize and some new synaptic densities form and strengthen as nerve cells selectively adhere and as practice turns to habit. Selection is against most nerve cells and their connections but it is nevertheless for a precious few of them. Of course, more work is needed before the physical basis of thought and imagination can be understood, but selective death in a vast field of proliferating biochemical possibilities may apply to minds as it does to evolution.

The peculiarly curved early embryos of birds, alligators, pigs, and humans are remarkably similar. Developing from a fertile egg, all display a stage with gill slits—whether the hatched or born animal breathes oxygen from air or water. The slits that close behind the ears in the human fetus attest to our common ancestry with fish, whose gill slits function in the adult. Human embryos have tails. Living matter "remembers" and repeats its origins to arrive in the present. In a Butlerian world, the materials of living beings are molded by life, over and over, for millions of generations. Creating a sense of *déjà vu*, the embryo represents a once-unconscious process, now again—at a different level—brought to consciousness.

SUPER HUMANITY

A transhuman being, superhumanity, is appearing, becoming part of the sentient symphony. It is composed not only of people but of material transport systems, energy transport systems, information transport systems, global markets, scientific instruments. Superhumanity ingests not only food but also coal, oil, iron, silicon.

The global network that builds and maintains cities, roadways, and fiber-optic cables grows by leaps and bounds. In Nigeria, for example, the population is expected to reach 216 million by the year 2010, double that of 1988. Unchecked, such growth would bring the number of Nigerians to more than 10,000 million by the year 2110—twice the present global human population. Our stupendous population taps a significant proportion of the solar energy reaching Earth's surface. The raw energy of photosynthesis, past and present, and transformed into edible plants, animal fodder, geological reserves, and human muscle and brain, supports the massive construction of the transcontinental urban ecosystem and even—"biting the hand that feeds it"— the razing of forests capturing and converting solar energies. As the system expands using genetic and atomic technology, its operations become more elegant and cohesive. The potential for disaster also increases.

Superhumanity is neither a simple collection of humans nor something other than aggregated humans and their devices. Plumbing, tunnels, water pipes, electric wires, vents, gas, air conditioning ducts, elevator shafts, telephone wires, fiber-optic cables, and other links enclose humans in a rapidly growing net. The way superhumanity behaves is in part the result of uncountable and unaccountable economic decisions made by people—singly and in groups—within the context of an increasingly planetary capitalism. "The problem with money," says a character in a recent film, "is that it makes us do things we don't want to do."

Whether or not superhumanity's tendencies are conscious beyond us, individual humans should not be surprised if the aggregate of planetary humanity shows unexpected, emergent, seemingly purposeful behaviors. If brainless bacteria merged into fused protists, which cloned and changed themselves over evolutionary time into civilization, what spectacle will emerge from human beings in global aggregation? To deny the existence of superhumanity by insisting it is merely the sum of human actions is like claiming that a person is merely the sum of the microbes and cells that constitute the body.[8]

EXPANDING LIFE

Life today is an autopoietic, photosynthetic phenomenon, planetary in scale. A chemical transmutation of sunlight, it exuberantly tries to spread, to outgrow itself. Yet by reproducing, it maintains itself and its past even as it grows. Life transforms to meet the contingencies of its changing environment and in doing so changes that environment. By degrees the environment becomes absorbed into the processes of life, becomes less a static, inanimate backdrop and more and more like a house, nest, or shell—that is, an involved, constructed part of an organic being. The members of 30 million species interacting at Earth's surface continue to change the world.

Coming to understand life afresh, we find that species of organisms diverge into new kinds, yet earlier patterns never entirely disappear. Old life forms, the bacteria that run the planet's ecology, are supplemented but not replaced. Although every distinct variety of nucleated being—every species of plant, animal, protoctist, and fungus—perishes, similar new taxa evolve from them or from their kind. Meanwhile, the underlying bacteria march symbiogenetically on.

We find that nature is not always "cutthroat," or, as the poet Alfred Tennyson put it, "red in tooth and claw." Living beings are amoral and opportunistic, as befits their needs for water, carbon, hydrogen, and the rest. They are fractally repeating structures of matter, energy, and information, with a very long history. But they are no more inherently bloodthirsty, competitive, and carnivorous than they are peaceful, cooperative, and languid. Lord Tennyson might just as well have cast nature as "green in stem and leaf."

Among the most successful—that is, abundant—living beings on the planet are ones that have teamed up. Moving inside (or perhaps forcibly dragged inside)

another cell, the cyanobacteria that became chloroplasts in protoctist and plant cells weren't lost; they were transformed. So too with the mitochondria—once respiring bacteria—that give your finger muscles the energy to turn this page. Former bacteria, as themselves or parts of larger cells, are still the most abundant forms of life on the planet.

The strength of symbiosis as an evolutionary force undermines the prevalent notion of individuality as something fixed, something secure and sacred. A human being in particular is not single, but a composite. Each of us provides a fine environment for bacteria, fungi, roundworms, mites, and others that live in and on us. Our gut is packed with enteric bacteria and yeasts that manufacture vitamins for us and help metabolize our food. The pushy microbes of our gums resemble department store customers before a holiday. Our mitochondria-laden cells evolved from a merger of fermenting and respiring bacteria. Perhaps spirochetes, symbiotically faded to the edge of detectability, continue to squirm as the undulipodia of our fallopian tubes or sperm tails. Their remnants may move in subtle ways as our microtubule-packed brains grow. "Our" bodies are actually joint property of the descendants of diverse ancestors.

Individuality is not stuck at any one level, be it that of our own species or pondwater *Amoeba proteus*. Most of our dry weight is bacteria, yet as citizens swarming in crowded streets and office buildings, viewing television, traveling in cars, and communicating by cellular and facsimile phone, humans disappear in a global swirl of activity, overwhelmed by emergent structures and abilities that could never be accomplished by individuals or even tribes of human predecessors. No single human can speak to another human, in real time, thousands of miles away. No single human could stand on the moon. These are emergent abilities of superhumanity. Our global activities bring to mind the social insects, except that our "hive" is nearly the entire biosphere.

Inextricably embedded in the biosphere, this superhuman society is not independent. At its greatest extent life on Earth—fauna, flora, and microbiota—is a single, gas-entrenched, ocean-connected planetary system, the largest organic being in the solar system. The upper mantle, crust, hydrosphere, and atmosphere of Earth remain in an organized state very different from that on the surfaces of our neighboring planets. Photosynthesis, respiration, fermentation, bio-mineralization, population expansion, seed germination, stampedes, bird migration, mining, transportation, and industry move and alter matter on a global scale. Life dramatically impacts the environment by producing and storing skeletons and shells of calcium phosphate and carbonate, by caching plant remains as coal and algae residues as oil. Great layers of minerals—sulfides of iron, lead, zinc, silver, and gold—remain in place where they were precipitated by hydrogen gas-producing bacteria.

Minerals not normally associated with life— aragonite, barite, calcite, francolite, fluorite—are produced as crystals and skeletons inside, and exoskeletons or shells outside, living organisms. Plants and microbes induce the formation of "inanimate" substances such as barite, iron oxides, galena (lead sulfide), and pyrite (iron sulfide, or fool's gold). Humanity's cultures are ranked from a stone age through an iron to a bronze age. Some argue that with the advent of computers Earth has entered a "silicon age" But metallurgy preceded us: human metalworking followed the bacterial use of magnetite for internal compasses by 3,000 million years. *Pedomicrobium*, a soil bacterium found fossilized in gold samples, is thought to precipitate gold ions and thus accumulate gold particles in its sheath. Next to hundreds of cubic kilometers of tropical reefs built by coral and entire cliffs of chalk precipitated by foraminifera and coccolithophorids, human technology does not seem uniquely grand.

Our destiny is joined to that of other species. When our lives touch those of different kingdoms— flowering and fruiting plants, recycling and sometimes hallucinogenic fungi, livestock and pet animals, healthful and weather-changing microbes—we most feel what it means to be alive. Survival seems always to require more networking, more interaction with members of other species, which integrates us further into global physiology. Despite the apocalyptic tone of some environmentalists, our species is on its way to becoming better integrated into global functioning. Though technology can poison humans and other organisms and stultify their growth, it also has the capacity to usher in the next major change in biospheric organization.

Teamwork enabled life to spread on Earth: anaerobic microbes joined to make the swimming ancestors of protist colonists, mastigotes ingested but did not digest the mitochondria that allowed them to invade oxygen-rich niches at Earth's surface, fungi and algae combined into lichens that colonized bare rock of dry land. The transport of life to new planetary bodies will also require teamwork. Astronautics, computers, genetics, biospherics, telecommunications, and other forms of human-sired technology will have to combine with the predecessor technology—most significantly photosynthesis—of other planetmates. The ultimate explosion of life onto its next frontier—that of space—will rely on the new technology of life itself. Vivification and terraformation—the coming to life of other planetary bodies—are not only human processes. Someday recycling ecosystems inside spacecraft may feed humans voyaging to other planets. If humans are to reside in space or voyage beyond Earth's orbit, the plants that feed them, the bacteria that digest for them, the fungi and other microbes that recycle their wastes, and the technology that supports them will surely be along for the ride. The extension of our local thermodynamic disequilibrium into space necessarily will involve representatives of all five kingdoms that have forged new ecosystems, able to transfer energy and cycle matter in isolation from the mother planet, the original functioning biosphere of Earth.

The distinction between space colonization by machines alone or by life with machines mirrors a New Zealand newspaper debate Samuel Butler had with himself. Beginning in 1862 Butler contributed an anonymous article to *The Press* of Christchurch, New Zealand. At the time, he was sheep farming in the Upper Rangitata district of Canterbury Province. Entitled "Darwin on the Origin of Species, a Dialogue," the unsigned article generated spirited protest. Butler joined in, criticizing himself as well as others. Signing his divided opinions under different names, Butler ultimately argued for two diametrically opposed interpretations of machines.

In "Darwin among the Machines," a letter signed by one Cellarius and appearing in *The Press* on June 13 1863, Butler held that machines were the latest form of life on Earth, poised to take over and enslave their human masters; the rate of evolution and reproduction of machines was prodigious, and without "war to the death" at once it would be too late to resist their world dominion. Then, in a July 29 1865 article entitled "Lucubratio Ebria," Butler countered Cellarius by saying that human beings were not even human without clothes, tools, and other mechanical accessories/machines were not a threat to human life, but its indissociable natural extension.[9]

If space vehicles do cut free of human influence, voyaging starry skies as they reproduce on their own, then Cellarius and other Luddites will be vindicated. If, however, machines in space flourish not alone but as intelligent enclosures for a wide variety of other life forms, then the author of "Lucubratio Ebria" will be proven correct.

We place our bets on the latter. Machines, we believe, will flourish in a tightly meshed interface with life—not only human life but a rich assortment of starlight-using life forms. People are essential for making the export of life into the night fantastic even possible. But, like the sperm tail which breaks off once the genetic message enters the egg, so human beings are ultimately expendable. Even without us, a hundred million more years of sun-driven planetary exuberance should be enough to get life off the Earth, star-bound. Other technological species might evolve. Besides, not only humans have begun space exploration.

Setting foot on the moon, Neil Armstrong proclaimed, "One small step for man, one giant leap for mankind." True, in a sense; but he overlooked vast numbers of bacteria on his skin and in his intestine that stepped with him. Life has been expansionist from the beginning. Once it gets a firm toehold in space it may kick off its human shoes and run wild.

RHYTHMS AND CYCLES

We and many other animals sleep and wake in cycles that repeat every twenty-four hours. Some ocean protists, dinomastigotes, luminesce when dusk comes, ceasing two hours later. So hooked are they into the cosmic rhythm of Earth that even back in the laboratory, away from the sea, they know the sun has set. Many similar examples abound because living matter is not an island but part of the cosmic matter around it, dancing to the beat of the universe.

Life is a material phenomenon so finely tuned and nuanced to its cosmic domicile that the relatively minor shift of angle and temperature change as the tilted Earth moves in its course around the sun is enough to alter life's mood, to bring on or silence the song of bird, bullfrog, cricket, and cicada. But the steady background beat of Earth turning and orbiting in its cosmic environment provides more than a metronome for daily and seasonal lives. Larger rhythms, more difficult to discern, can also be heard.

Many types of life form encapsulating structures that protect them from temporary dangers of the environment. Propagules of a wide variety of types are miniaturized, viable representatives of mature organic beings. They range from bacterial and fungal spores to protoctist cysts; from plant spores, pollen, seeds, and fruit to the dry eggs of some crustaceans, insects, and reptiles. As such propagules proliferate, natural selection deals with them severely: many die or fail to grow.

Desiccation- and radiation-resistant, most propagules metabolize at an exceedingly slow rate. Spores of bacteria may lay in wait for a hundred years—until rain comes, or phosphorus abounds, or conditions otherwise become less dry and more permissive. Without any dormant seeds or resistant spores, humans survive extraordinary environmental hazards. Houses, clothes, railroads, and automobiles have made possible our expansion from the subtropical home to colder climes. Analogous to spores, cysts, and seeds, these structures protect us from harsh conditions.

Recycling greenhouses are enclosed dwellings that contain representative collections of Earth's life. They detoxify poisons and transform wastes into food and back again. One, designed by Santiago Calatrava, will span the entire roof of the Cathedral of Saint John the Divine, in New York City. Such "artificial biospheres" miniaturize crucial processes of the autopoiesis of the global ecosystem.

The global ecosystem is not an ordinary organic being. The global system, like all living beings, is energetically open: solar radiation comes steadily in, dissipated heat moves steadily out. But unlike other beings, the global system is closed to material exchange. Apart from the occasional incoming meteor or comet, nothing enters. Apart from the occasionally stalled geologic churning here and there of sediments into new crust and cooked gases, nothing leaves. All the matter used by life is recycled matter—reappearing matter that is never consumed.

No living cell or organism feeds on its own waste. Thus artificial ecosystems have biological importance that exceeds architecture or other human concerns. For the first time in evolutionary history, the biosphere has reproduced, or better put, has begun reproducing itself through humankind and technology The generation of new "buds," materially closed living systems, within the "mother biosphere" resembles the structure of a fractal.

From a "green" or "deep ecology" perspective, humans do not dominate but are deeply embedded within nature. Artificial biospheres are the first buds of a planetary organic being, as "man-made" biospheres have the potential to duplicate biospheric, light-dependent self-sufficiency. People in NASA, the European Space Agencies, governments, private industry, academia and elsewhere are pondering these desiccation-resistant structures that sequester, like new Noah's arks, samples of life—not in museums but in a living, self-sustaining form. The largest of the recycling structures is Biosphere 2 in the Sonoran desert at Oracle, Arizona. Ultimately, closed ecosystems are not artificial at all, but part of the natural processes of self-maintenance, reproduction, and evolution in a heat-dissipating universe.

For an organic being to survive in space, food must be replenished and waste disposal systems must work. Photosynthesis has stored solar energy in rocks as reserves of kerogen, oil, gas, iron sulfide, coal, and other substances. The planet's prodigal species now expends these reserves, using the energy to spread its populations. *Homo sapiens* spends the riches of eons; meanwhile, the rhythms of Earth, building up and breaking down for ages, crescendo. Our creative destruction accelerates. But nature has not ended, nor does the planet require saving. The technological dissonance marks no end but a lull, a gathering of forces.

Global life is a system richer than any of its components. We alone among the animals build large telescopes and mine Archean diamonds. But, while our position cannot soon or easily be replaced, we are not in charge. The diamonds are made of carbon, a main element of life since its inception some 4000 million years ago; and telescopes are lenses, parts of the compound eye of a metahuman being that is itself an organ of the biosphere.

The continuous metamorphosis of the planet is the cumulative result of its multifarious beings. Humankind does not conduct the sentient symphony: with or without us, life will go on. But behind the disconcerting tumult of the present movement one can hear, like medieval troubadours climbing a distant hill, a new pastorale. The melody promises a second nature of technology and life, together spreading Earth's multispecies propagules to other planets and the stars beyond. From a green perspective a keen interest in high tech and the altered global environment makes perfect sense. It is high noon for humanity. Earth is going to seed.

NOTES

The epigraphs are from Erasmus Darwin (1794), *Zoonomia; or, the Laws of Organic Life*; Blaise Pascal (1670/1995), *Pensées*, no. 93; and Leonardo Tarán (1965), *Parmenides: A Text with Translation, Commentary, and Critical Essays* (pp. xx–xxi).

This chapter originally appeared as chapter 9 in Lynn Margulis and Dorion Sagan, *What is life?*, pp. 213–243, Berkley: University of California Press, 2000.

REFERENCES

1. Bohr, N. (1958), *Physical science and the problem of life*. New York: Wiley.
2. Butler, S. (1863/1917). Darwin among the machines & Lucubratio ebria. In H. F. Jones (Ed.), The note-books of Samuel Butler, author of 'Erewhon' (pp. 46–53). New York: Dutton.
3. Butler, S. (1967). The deadlock in Darwinism. In R. A. Streatfeild (Ed.), *The humor of Homer and other essays* (pp. 253–254). Freeport, NY: Books for Libraries Press.
4. Calvin, W. (1989). *The cerebral symphony*. New York: Bantam Books.
5. Culver, S. J. (1993). Foraminifera. In J. H. Lipps (Ed.), *Fossil prokaryotes and protists*. Boston: Blackwell Scientific Publications.
6. Darwin, E. (1794). *Zoonomia; or, the laws of organic life* (2 vols.). London: Johnson.
7. Edelman, G. (1987). *Neural Darwinism*. New York: Basic Books.
8. Folsome, C. (1985). Microbes. In T. P. Snyder (Ed.), *The biosphere catalogue* (pp. 51–56). Oracle, AZ: Synergetic Press.
9. Jones, H. F. (Ed.) (1919). *Samuel Butler, author of 'Erewhon' (1835–1902): A memoir* (vol. 2). London: Macmillan.
10. Jones, H. F. & Bartholomew, A. T. (Eds.) (1924). *The Shrewsbury edition of the works of Samuel Butler*. New York: E. P. Dutton.
11. Kelly, K. (1994). *Out of control: The rise of neo-biological civilization*. Reading, MA: Addison-Wesley.
12. Koshland, D. E., Jr. (1992). A response-regulated model in a simple sensory system. *Science*, **196**, 1055–1063.

13. Pascal, B. (1670/1995). *Pensées*, trans. A. J. Krailsheimer. New York: Penguin.

14. Stock, G. (1993). *Metaman: Humans, machines, and the birth of a global superorganism*. London: Bantam.

15. Tarán, L. (1965). *Parmenides: A text with translation, commentary, and critical essays*. Princeton, NJ: Princeton University Press.

ENDNOTES

1 Folsome (1985).

2 Culver (1993), p. 224.

3 Koshland (1992).

4 The most comprehensive source for these little-known events comes from Henry Festing Jones, "The Butler-Darwin Quarrel," pp. 446–447 (in Jones 1919). The "mental anachronism" quote can be found on p. 447; the "it never occurred to me" on p. 448; and "how the accident arose" on p. 425. Butler's fascinating but difficult-to-find "evolution books"—*Life and Habit; Evolution, Old and New;* and *Luck or Cunning*—comprise volumes 4, 5, and 8, respectively, of *The Shrewsbury Edition of the Works of Samuel Butler* (Jones and Bartholomew 1924).

5 Bohr (1958), p. 100.

6 Butler (1967).

7 Calvin (1989). See also Edelman (1987).

8 For more on the emergence of a superhumanity, see Stock (1993). For a relationship between increasingly lifelike machines and increasingly engineered life, between, as the author says, the "born and the made," see Kelly (1994).

9 Butler (1863/1917).

28 · What is life?

KIM STERELNY AND PAUL GRIFFITHS

DEFINING LIFE

Definitions explain the meaning of a term by relating the defined term to other expressions in the language. For example, a definition of *acid* specifies the necessary and sufficient conditions that all, and only, acids share. More generally, definitions relate items in a language to other items in that language. Some of these other terms, in turn, may have their meanings explained through definitions. But at some point the chain of definitions must end. Some concepts must be understood without the help of other verbal formulae. So in semantics and psychology, it is now realized that our capacity to use concepts and refer to kinds need not depend on a grasp, implicit or explicit, on the necessary and sufficient conditions of membership of those categories. Humans have been able to use terms for chemical and physical kinds (*iron, liquid, salt, planet*) long before they understood the nature of those kinds. Though natural kinds may have essences, those essences are discovered not through the construction of definitions at the beginning of inquiry, but, if we are lucky, as the culmination of inquiry.

So biologists do not need a definition of *life* to help them recognize what they are talking about. But definitions are often useful. When categories overlap, or are easily confused with one another, the precision induced by definition is important, for definitions enable us to notice important distinctions that are easily overlooked. Confined as we are to the surface of a near-spherical globe, we can easily overlook the distinction between mass and weight, which is the interaction of mass and a gravitational field. So definitions that made this distinction explicit were important in the development of physics. As we saw in Part II, *gene* has been used to name very different kinds

in biology; making these distinctions explicit avoids confusion. Similarly, different concepts of the organism may be important, and hence it is important that the distinctions among them be explicitly marked. So definition is sometimes an important tool in theoretical advance.

However, we doubt that biology is currently impeded by biologists using *life* for distinct though related kinds. For example, we do not see how a definition of *life* is likely to help us with odd and hard-to-classify cases: prions, viruses, social insect colonies, or the much less plausible idea that the earth itself is a living system. The adequacy of the definition is settled by our view of the case, not vice versa. Consider for a moment the *Gaia hypothesis*, the idea (in one of its forms) that the earth itself, or perhaps just the biosphere, is a living organism. We see no useful role for a definition of life in evaluating this metaphor. In some very important ways, the earth is obviously unlike an organism. It is not the result of evolution through competition within an ancestral population of proto-Gaias. Nor does the biosphere result from a developmental cycle. The biosphere we have now will not produce a world-seed that grows into the biosphere of the earth at some later stage.

If we emphasize the typical histories of living things, then Gaia is not lifelike. But so what? Defenders of the Gaia idea emphasize the interconnections and reciprocal causal influences of living things with one another and the abiotic environment. These reciprocal interactions, they suggest, act like stabilizing or homeostatic mechanisms. There are both conceptual and empirical problems in evaluating this claim. As Kirchner (1991 p. 41) points out, in some respects, clearly life has not been homeostatic. Life, after all, radically altered the composition of the earth's

The Nature of Life, ed. M. A. Bedau and C. E. Cleland. Published by Cambridge University Press. © M. A. Bedau and C. E. Cleland 2010.

atmosphere. So without an exact specification of the particular homeostatic mechanisms under consideration, the idea that the biosphere is a connected set of self-sustaining homeostatic mechanisms is too vague to evaluate. But even if it is made precise, the issue of whether the biosphere is alive is irrelevant. We do not need to detour through that question to evaluate the various Gaia hypotheses about the extent to which living systems and their environment change one another, the extent and ways in which these interactions are stabilized, or the extent to which these mutual changes make the earth more life-friendly.

So defining life is not a prerequisite for determining the scope of biology. The revival of interest in definitions of life has a different source: an interest in *universal biology*. All living systems on earth share many important properties. They are cells or are built from cells. Proteins play an essential role in the metabolism of all living things, and nucleic acids play an essential role in the process through which life gives rise to life. Replication and reproduction results in populations in competition, and natural selection on variation within those populations produces adaptation, sometimes complex adaptation. For all living things live in regimes in which natural selection is at work. But are these and other universal features of life on earth characteristic only of life as we find it here and now? Or are some of these features truly universal: features of life anywhere, any time? Those interested in universal biology seek a characterization not just of life *as it happens to be*, but of life *as it must be*. For them, a definition of life is a specification of life's real essence (Bedau 1996; Langton 1996; Ray 1996).

It is worth pausing for a moment to remind ourselves just how ambitious this project really is. Biologists have always been interested in general principles. We have discussed plenty of candidates from ecology and evolutionary biology. It has been tough enough to find principles that are true of all life here and now. We have argued that adaptive and ecological hypotheses are best seen as hypotheses about particular clades, particular branches in the tree of life, not life as a whole. If that is right, then what price really universal biology: generalizations true not just of our life-world, but of any life-world?

Despite the ambition of the project, a number of biologists have explored the distinction between the specific features of life on earth and those features that

life necessarily has. Gould, Kauffman, Goodwin, and Dawkins have deeply contrasting ideas on evolution, but they share this interest. We considered elsewhere (Sterelny and Griffiths 1999, Ch. 12) both Gould's idea that the array of complex adaptations evolution on earth has produced is contingent, and his idea that the complexity of life tends to drift upward over time as a matter of statistical rather than evolutionary necessity. Gould's main emphasis is on the contingency of life's actual history. In contrast, Dawkins (1986) argues against a "historical accident" view of life's most central mechanisms. The most central features of both developmental biology and genetics are, he claims, features of universal biology. He argues against the possibility of *Lamarckian evolution*, at least if we understand Lamarckian evolution to involve the inheritance of only *adaptive* changes by the next generation. An organism's phenotype can certainly change its germ line genotype. For example, an organism may expose itself to mutagens in the environment, or act in ways that lower the efficiency of its DNA proofreading mechanisms. But that is not yet Lamarckian, for those changes in the stream of influence from parent to offspring do not make the offspring more likely to resemble the parent in this respect. A rat with a taste for nesting in nuclear reactors is unlikely to produce offspring with their DNA altered in such a way as to induce in them the same preference. Dawkins concedes that it is possible, though difficult, to imagine mechanisms in which the acquisition of a novel phenotypic trait changes the replicators responsible for the phenotype of the next generation in ways that make that novel phenotype reappear. In his view, it is much harder to imagine mechanisms that are sensitive to the distinction between adaptive and other novelties, and which make only adaptive changes more likely to reappear in the next generation (Dawkins 1986).

We are skeptical about Gould's ideas on contingency. We are also very wary of plausibility arguments for impossibility claims—"arguments from personal incredulity," as Dawkins himself has called them in a different context. Dawkins, after all, thinks that memes are replicators. Memes—if memes are taken to be the information content of ideas—do change, and sometimes adaptively, during the time they are in a particular interactor. If someone using a stone tool of a standard pattern discovers that grinding its edge on sandstone gives it a sharper cutting surface, that is a change in a specific meme token. It is a mutation, and one likely to be passed on because it is adaptive. So if interaction between phenotype and environment can improve a meme that is carried and transmitted, it is not obvious why Dawkins thinks that no similar mechanism could work with other replicators. Admittedly, if memes are replicators at all, they are late-model replicators. They are replicators that emerge deep into the history of a life-world. So perhaps the mechanisms that permit their evolution to be in this sense "Lamarckian" depend on a rich history of prior evolutionary change. But we do not see why this must be so. After all, the fidelity of genetic replication, and the sequestering of the germ line genes in many species, is itself the product of much evolution.

Despite our skepticism about these particular claims, we agree that there is a very good question lurking behind the idea of a universal biology. We seek not just an account of actual biology in all its diversity, but also an explanation of why that diversity is not greater still. However, we see two problems in asking for an explanation of the limits on life's diversity.

First, we should not conceptualize this question by contrasting chance with necessity. Consider, for example, David Raup's representation of possible and actual shell shapes. He shows that, to a first approximation, shell form can be represented as the outcome of only three different growth parameters. In light of this understanding, actual shells occupy a rather small region of the space of possible shells (for an elegant discussion, see Dawkins 1996, chap. 6). Why? Is this restriction a consequence of function, of subtle constraints on development, or of historical contingency? These are clearly difficult but interesting questions. But it is surely unlikely that most of the unoccupied region is literally impossible to occupy. It is equally unlikely that the occupied region is occupied through nothing but historical chance. Similarly, there are no species with three sexes, and that is no accident. As the literature on the evolution of sex makes clear, sex has a cost, and that cost would increase with the number of sexes. But should we infer that the evolution of three sexes is impossible? That would surely be rash: we can conceive of a developmental biology that might work with three sexes. Nuclear DNA has two parents, so we could have three if mitochondrial DNA came from a third. But an evolutionary trajectory leading to three sexes would be both available to a lineage and favored by selection only in very extraordinary circumstances.

So, as Dennett (1995) has noted, contrasting historical accident with necessity is likely to be the wrong way of posing this problem. It is probably too crude a distinction to get at the questions that really interest us. Instead, we need some notion of a phenomenon's *improbability*. Bats evolved; no marsupial equivalent did. Is there some reason why a flying marsupial is less likely than a flying placental? Difficult though this question is to answer, it is surely a better question than asking whether a flying marsupial is impossible.

A second problem is the difficulty of testing conjectures about universal biology. This problem of testing is one of the fuels of the developing but over-hyped field of *artificial life*. One of the repeated themes of A-life literature is the "$N=1$" problem, the problem of distinguishing between accidental and essential features of life with a sample size of one.

> Ideally, the science of biology should embrace all forms of life. However, in practice, it has been restricted to the study of a single instance of life, life on earth. Because biology is based on a sample size of one, we cannot know what features of life are peculiar to earth, and what features are general, characteristic of all life.   (Ray 1996, p. 111)

One aim of A-life is to increase N, and in doing so, generate a definition of life that tells us which features of life are essential to life in and of itself. Just as "strong AI" claims that some computing systems housed in current or near-current computers are not mere simulations of thought, but instances of it, the defenders of "strong A-life" argue that some computer models of lifelike interactions are not simulations of life, but instances of it. They are alive.

The defenders of strong AI argue that a cognitive system is any system organized in the right way. Whether a system thinks is independent of its physical constitution. The essence of mind is form, organization, or function: some abstract property. Because the essential features of having a mental life are not tied to a specific physical implementation, thinking is *substrate-neutral*. Mental properties are functional properties, not physical ones. Strong A-life models itself on this line of argument. Being alive is substrate-neutral. Life is a feature of form, not matter. A living system is any system with the right organization or structure.

Life is a property of form, not matter, a result of the organization of matter rather than something that inheres in the matter itself. Neither nucleotides nor amino acids nor any other carbon-chain molecule is alive—yet put them together in the right way, and the dynamic behavior that emerges out of their interactions is what we call life. It is effects, not things, upon which life is based—life is a kind of behavior, not a kind of stuff—and as such it is constituted of simpler behaviors, not simpler stuff.

(Langton 1996, p. 53)

and therefore,

it is possible to abstract the logical form of a machine from its physical hardware, it is natural to ask whether it is possible to abstract the logical form of an organism from its biochemical wetware.

(Langton 1996, p. 55)

So in this view, the data structures in, for example, Thomas Ray's famous Tierra program are alive, not merely illustrations of life.

We see no merit at all in these claims. First, the form/matter distinction, the distinction on which the whole idea rests, is an untenable dichotomy. There is no single level of function or organization resting on a single level of matter. Rather, there is a cascade of increasingly or decreasingly abstract descriptions of any one system. In philosophy of psychology, the original home of the function/realization distinction, "two-levelism" has been powerfully criticized by William Lycan (1990). In David Marr's famous description of the structure of psychological theories (1980), there are at least two functional levels alone. The highest level describes the task that the psychological system accomplishes. In the case of vision, Marr claims that the task is to interpret the world in terms of moving, three-dimensional colored objects using patterns of stimulation of the retina as data. An intermediate level might describe how the system processes information in order to accomplish this task. It details the algorithms by which retinal patterns are transformed into representations of the world. The lowest level describes how these computational processes are physically implemented in the brain. Many authors have argued for a number of separate algorithmic or computational levels of description between the superficial level of task description and anything resembling a

direct description of brain structure. Lycan has pointed out that much of what passes for a description of the "physical realization" of the mind is really a description of function. Synapses, the connections between brain cells, come in radically different forms, but for most purposes we can abstract away from this detail and describe them by their function: transferring excitation from one cell to the next.

Exactly the same multilevel picture applies to biological systems. For some purposes, a highly abstract, purely informational description of the genome may be appropriate. For others, we want to know in great physical detail the structure of the DNA molecule; for instance, in explaining its coiling properties. Other needs will call for intermediate degrees of detail. There is a whole language of genetics—of introns and exons, of crossing-over, of gene duplication and gene repair—that is functional in abstracting away from the intricate details of molecular mechanisms (some of which are still, indeed, not known), but which is not wholly abstract. This is certainly not a language of form as opposed to matter. So the substrate neutrality thesis rests on a false dichotomy.

Moreover, we think that the idea that simulations are instances of life is an unnecessary hostage to fortune, for the importance of A-life models does not depend on the claim that they create life. The $N = 1$ problem is indeed a serious obstacle to the testing of conjectures in universal biology. But the $N = 1$ problem has been exaggerated, and in any case, the testing problem is not solved by deeming computer simulations to be alive. Of course exobiology would be great if we could do it; a genuinely independent life-world could scarcely fail to tell us much of importance about what is robust about biological process and what is not. But the problem of universal biology can be attacked here and now by the construction of distinct theories that have different implications for evolutionary, developmental, and ecological possibilities, and which can be tested by their application to the huge and varied experiment we actually have available. We do not have a wonderful array of theories that are well confirmed and empirically equivalent with respect to life on earth, but with different implications about how life might have been. $N = 1$ may begin to bite if and when we have to decide between empirically well-confirmed and locally equivalent theories: theories that make the same predictions about life here—predictions that are confirmed—but which make different predictions about what life might be like elsewhere. But we are yet to be indulged with such choices.

Evolutionary simulations will have an important role to play in constructing these theories of life's robust properties. Such models could test conditions under which particular developmental, genetic, ecological, or evolutionary phenomena would arise. Under what circumstances could a third sex evolve? Under what circumstances could variation be directed rather than random? Well-calibrated models that showed the evolution of exotic phenomena not observed in the natural world would be very suggestive indeed. But they can play that role as representations of biological processes, not manifestations of them.

In running simulations, we are trying to find out what those models predict, when those predictions are inaccessible to analytic techniques. The great virtue of these simulations is that one can play with various parameters and thus get a feel for which outcomes are robust under fine-scale changes in the model, and which are not. Thus, for example, Nilsson's model of the evolution of eyes is impressive because the parameters are chosen conservatively, and yet eyes evolve, by geological standards, with great speed (Nilsson & Pelger 1994; see also Dawkins 1996). Simulations are important, and we will consider their message further in Section 15.3*. But nothing of what these models tell us depends on thinking of them as actually alive. Indeed, we think that, the view that these programs are instances of life rather than representations of it trivializes the real questions that motivate universal biology. Consider, again, three sexes. We would like to know whether there are circumstances that would effectively select for three sexes, It is likely that only evolutionary modeling will advance our grip on this problem. But to do so, such models must be well calibrated. Their assumptions must be realistic. Suppose we were to accept that the data structures manipulated in a Tierra-like program were themselves alive. Suppose, further, that we accept that sex is defined not by the physical exchange of nucleic acids, but abstractly and functionally, as the A-life program urges. Sex, in this abstract conception, is information exchange. So any information exchange between token data structures before they are replicated is sex. There is no doubt that it is possible to develop models with three-way exchange of information between data structures. Hence, by this A-life definition, we could have life with three sexes. But this would be a trivial solution to the problem;

it is too cheap. Unless the model faithfully represented the constraints on physically embodied living things— for example, constraints on development—it would not tell us what we wanted to know about the possibility of three sexes. If it did faithfully represent those constraints, we could learn what we wanted to know. But nothing would be added by insisting that the model manifests as well as represents life.

NOTES

This chapter originally appeared as section 15.1 in Kim Sterelny and Paul Griffiths, *Sex and death: An introduction to philosophy of biology*, pp. 357–364, Chicago: Chicago University Press, 1999.

REFERENCES

1. Bedau, M. A. (1996). The nature of life. In M. A. Boden (Ed.), *The philosophy of artificial life* (pp. 332–360). Oxford: Oxford University Press.
2. Dawkins, R. (1986). *The blind watchmaker*. New York: Norton.
3. Dawkins, R. (1996). *Climbing mount improbable*. New York: W. W. Norton.
4. Dennett, D. (1995). *Darwin's dangerous idea*. New York: Simon & Schuster.
5. Kirchner, J. W. (1991). The Gaia hypotheses: Are they testable? Are they Useful? In S. Schneider and P. J. Boston (Eds.), *Scientists on Gaia* (pp. 38–46). Cambridge, MA: MIT Press.
6. Langton, C. G. (1996). Artificial life. In M. A. Boden (Ed.), *The philosophy of artificial life* (pp. 39–94). Oxford: Oxford University Press.
7. Lycan, W. G. (1990). The continuity of levels of nature. In W. G. Lycan (Ed.), *Mind and cognition* (pp. 77–96). Oxford: Blackwell.
8. Marr, D. (1980). *Vision*. New York: W. H. Freeman.
9. Nilsson, D. -E. & Pelger, S. (1994). A pessimistic estimate of the time required for an eye to evolve. *Proceedings of the Royal Society of London B*, **256**, 53–58.
10. Ray, T. S. (1996). An approach to the synthesis of life. In M. A. Boden (Ed.), *The philosophy of artificial life* (pp. 111–145). Oxford: Oxford University Press.

ENDNOTES

* Ed. note: Section 15.3 is not included in this volume.

29 · Universal Darwinism

RICHARD DAWKINS

It is widely believed on statistical grounds that life has arisen many times all around the universe (Asimov 1979; Billingham 1981). However varied in detail alien forms of life may be, there will probably be certain principles that are fundamental to all life, everywhere. I suggest that prominent among these will be the principles of Darwinism. Darwin's theory of evolution by natural selection is more than a local theory to account for the existence and form of life on Earth. It is probably the only theory that *can* adequately account for the phenomena that we associate with life.

My concern is not with the details of other planets. I shall not speculate about alien biochemistries based on silicon chains, or alien neurophysiologies based on silicon chips. The universal perspective is my way of dramatizing the importance of Darwinism for our own biology here on Earth, and my examples will be mostly taken from Earthly biology. I do, however, also think that "exobiologists" speculating about extraterrestrial life should make more use of evolutionary reasoning. Their writings have been rich in speculation about how extraterrestrial life might work, but poor in discussion about how it might *evolve*. This essay should, therefore, be seen firstly as an argument for the general importance of Darwin's theory of natural selection; secondly as a preliminary contribution to a new discipline of "evolutionary exobiology."

The "growth of biological thought" (Mayr 1982) is largely the story of Darwinism's triumph over alternative explanations of existence. The chief weapon of this triumph is usually portrayed as *evidence*. The thing that is said to be wrong with Lamarck's theory is that its assumptions are factually wrong. In Mayr's words: "Accepting his premises, Lamarck's theory was as legitimate a theory of adaptation as that of Darwin. Unfortunately, these premises turned out to be invalid." But I think I can say something stronger: *even accepting his premises*, Lamarck's theory is *not* as legitimate a theory of adaptation as that of Darwin because, unlike Darwin's, it is *in principle* incapable of doing the job we ask of it—explaining the evolution of organized, adaptive complexity. I believe this is so for all theories that have ever been suggested for the mechanism of evolution except Darwinian natural selection, in which case Darwinism rests on a securer pedestal than that provided by facts alone.

Now, I have made reference to theories of evolution "doing the job we ask of them." Everything turns on the question of what that job is. The answer may be different for different people. Some biologists, for instance, get excited about "the species problem," while I have never mustered much enthusiasm for it as a "mystery of mysteries." For some, the main thing that any theory of evolution has to explain is the diversity of life—cladogenesis. Others may require of their theory an explanation of the observed changes in the molecular constitution of the genome. I would not presume to try to convert any of these people to my point of view. All I can do is to make my point of view clear, so that the rest of my argument is clear.

I agree with Maynard Smith (1969) that "The main task of any theory of evolution is to explain adaptive complexity, i.e., to explain the same set of facts which Paley used as evidence of a Creator." I suppose people like me might be labelled neo-Paleyists, or perhaps "transformed Paleyists." We concur with Paley that adaptive complexity demands a very special kind of explanation: either a Designer as Paley taught, or something such as natural selection that does the job of a designer. Indeed, adaptive complexity is probably the best diagnostic of the presence of life itself.

The Nature of Life, ed. M. A. Bedau and C. E. Cleland. Published by Cambridge University Press. © M. A. Bedau and C. E. Cleland 2010.

ADAPTIVE COMPLEXITY AS A DIAGNOSTIC CHARACTER OF LIFE

If you find something, anywhere in the universe, whose structure is complex and gives the strong appearance of having been designed for a purpose, then that something either is alive, or was once alive, or is an artefact created by something alive. It is fair to include fossils and artefacts since their discovery on any planet would certainly be taken as evidence for life there.

Complexity is a statistical concept (Pringle 1951). A complex thing is a statistically improbable thing, something with a very low *a priori* likelihood of coming into being. The number of possible ways of arranging the 10^{27} atoms of a human body is obviously inconceivably large. Of these possible ways, only very few would be recognized as a human body. But this is not, by itself, the point. Any existing configuration of atoms is, *a posteriori* unique, as "improbable," with hindsight, as any other. The point is that, of all possible ways of arranging those 10^{27} atoms, only a tiny minority would constitute anything remotely resembling a machine that worked to keep itself in being, and to reproduce its kind. Living things are not just statistically improbable in the trivial sense of hindsight: their statistical improbability is limited by the *a priori* constraints of design. They are *adaptively* complex.

The term "adaptationist" has been coined as a pejorative name for one who assumes "without further proof that all aspects of the morphology, physiology and behavior of organisms are adaptive optimal solutions to problems" (Lewontin 1983, and this volume[1]). I have responded to this elsewhere (Dawkins 1982a, Chapter 3). Here, I shall be an adaptationist in the much weaker sense that I shall only be *concerned* with those aspects of the morphology, physiology and behavior of organisms that are undisputedly adaptive solutions to problems. In the same way, a zoologist may specialize on vertebrates without denying the existence of invertebrates. I shall be preoccupied with undisputed adaptations because I have defined them as my working diagnostic characteristic of all life, anywhere in the universe, in the same way as the vertebrate zoologist might be preoccupied with backbones because backbones are the diagnostic character of all vertebrates. From time to time I shall need an example of an undisputed adaptation, and the time-honored eye will serve the

purpose as well as ever (Paley 1828; Darwin 1859; any fundamentalist tract). "As far as the examination of the instrument goes, there is precisely the same proof that the eye was made for vision, as there is that the telescope was made for assisting it. They are made upon the same principles; both being adjusted to the laws by which the transmission and refraction of rays of light are regulated" (Paley 1828, V. 1, p. 17).

If a similar instrument were found upon another planet, some special explanation would be called for. Either there is a God, or, if we are going to explain the universe in terms of blind physical forces, those blind physical forces are going to have to be deployed in a very peculiar way. The same is not true of non-living objects, such as the moon or the solar system (see below). Paley's instincts here were right.

> My opinion of Astronomy has always been, that it is *not* the best medium through which to prove the agency of an intelligent Creator... The very simplicity of [the heavenly bodies'] appearance is against them... Now we deduce design from relation, aptitude, and correspondence of *parts*. Some degree therefore of *complexity* is necessary to render a subject fit for this species of argument. But the heavenly bodies do not, except perhaps in the instance of Saturn's ring, present themselves to our observation as compounded of parts at all
>
> (1828, Vol. 2, pp. 146–147)

A transparent pebble, polished by the sea, might act as a lens, focussing a real image. The fact that it is an efficient optical device is not particularly interesting because, unlike an eye or a telescope, it is too simple. We do not feel the need to invoke anything remotely resembling the concept of design. The eye and the telescope have many parts, all coadapted and working together to achieve the same functional end. The polished pebble has far fewer co-adapted features: the coincidence of transparency, high refractive index and mechanical forces that polish the surface in a curved shape. The odds against such a threefold coincidence are not particularly great. No special explanation is called for.

Compare how a statistician decides what P value to accept as evidence for an effect in an experiment. It is a matter of judgment and dispute, almost of taste, exactly when a coincidence becomes too great to stomach. But, no matter whether you are a cautious statistician or a daring statistician, there are some complex adaptations whose "*P* value," whose coincidence rating, is so impressive that nobody would hesitate to diagnose life (or an artefact designed by a living thing). My definition of living complexity is, in effect, "that complexity which is too great to have come about through a single coincidence." For the purposes of this paper, the problem that any theory of evolution has to solve is how living adaptive complexity comes about.

In the book referred to above, Mayr (1982) helpfully lists what he sees as the six clearly distinct theories of evolution that have ever been proposed in the history of biology. I shall use this list to provide me with my main headings in this paper. For each of the six, instead of asking what the evidence is, for or against, I shall ask whether the theory is *in principle* capable of doing the job of explaining the existence of adaptive complexity, I shall take the six theories in order, and will conclude that only Theory 6, Darwinian selection, matches up to the task.

Theory 1. Built-in capacity for, or drive toward, increasing perfection

To the modern mind this is not really a theory at all, and I shall not bother to discuss it. It is obviously mystical, and does not explain anything that it does not assume to start with.

Theory 2. Use and disuse plus inheritance of acquired characters

It is convenient to discuss this in two parts.

USE AND DISUSE

It is an observed fact that on this planet living bodies sometimes become better adapted as a result of use. Muscles that are exercised tend to grow bigger. Necks that reach eagerly towards the treetops may lengthen in all their parts. Conceivably, if on some planet such acquired improvements could be incorporated into the hereditary information, adaptive evolution could result. This is the theory often associated with Lamarck, although there was more to what Lamarck said. Crick (1982, p. 59) says of the idea: "As far as I know, no one has given *general* theoretical reasons why such a mechanism must be less efficient than natural selection. . ." In this section and the next I shall give two general theoretical objections to Lamarckism of the sort which,

I suspect, Crick was calling for. I have discussed both before (Dawkins 1982b), so will be brief here. First the shortcomings of the principle of use and disuse.

The problem is the crudity and imprecision of the adaptation that the principle of use and disuse is capable of providing. Consider the evolutionary improvements that must have occurred during the evolution of an organ such as an eye, and ask which of them could conceivaby have come about through use and disuse. Does "use" increase the transparency of a lens? No, photons do not wash it clean as they pour through it. The lens and other optical parts must have reduced, over evolutionary time, their spherical and chromatic aberration; could this come about through increased use? Surely not. Exercise might have strengthened the muscles of the iris, but it could not have built up the fine feedback control system which controls those muscles. The mere bombardment of a retina with coloured light cannot call colour-sensitive cones into existence, nor connect up their outputs so as to provide color vision.

Darwinian types of theory, of course, have no trouble in explaining all these improvements. Any improvement in visual accuracy could significantly affect survival. Any tiny reduction in spherical aberration may save a fast flying bird from fatally misjudging the position of an obstacle. Any minute improvement in an eye's resolution of acute colored detail may crucially improve its detection of camouflaged prey. The genetic basis of any improvement, however slight, will come to predominate in the gene pool. The relationship between selection and adaptation is a direct and close-coupled one. The Lamarckian theory, on the other hand, relies on a much cruder coupling: the rule that the more an animal uses a certain bit of itself, the bigger that bit ought to be. The rule occasionally might have some validity but not generally, and, as a sculptor of adaptation it is a blunt hatchet in comparison to the fine chisels of natural selection. This point is universal. It does not depend on detailed facts about life on this particular planet. The same goes for my misgivings about the inheritance of acquired characters.

INHERITANCE OF ACQUIRED CHARACTERS

The problem here is that acquired characters are not always improvements. There is no reason why they should be, and indeed the vast majority of them are injuries. This is not just a fact about life on earth. It has a universal rationale. If you have a complex and reasonably well-adapted system, the number of things you can do to it that will make it perform less well is vastly greater than the number of things you can do to it that will improve it (Fisher 1958). Lamarckian evolution will move in adaptive directions only if some mechanism—selection—exists for distinguishing those acquired characters that are improvements from those that are not. Only the improvements should be imprinted into the germ line.

Although he was not talking about Lamarckism, Lorenz (1966) emphasized a related point for the case of learned behavior, which is perhaps the most important kind of acquired adaptation. An animal learns to be a better animal during its own lifetime. It learns to eat sweet foods, say, thereby increasing its survival chances. But there is nothing inherently nutritious about a sweet taste. Something, presumably natural selection, has to have built into the nervous system the arbitrary rule: "treat sweet taste as reward," and this works because saccharine does not occur in nature whereas sugar does.

Similarly, most animals learn to avoid situations that have, in the past, led to pain. The stimuli that animals treat as painful tend, in nature, to be associated with injury and increased chance of death. But again the connection must ultimately be built into the nervous system by natural selection, for it is not an obvious, necessary connection (M. Dawkins 1980). It is easy to imagine artificially selecting a breed of animals that enjoyed being injured, and felt pain whenever their physiological welfare was being improved. If learning is adaptive *improvement*, there has to be, in Lorenz's phrase, an innate teaching mechanism, or "innate schoolmarm." The principle holds even where the reinforcers are "secondary," learned by association with primary reinforcers (P. P. B. Bateson, this volume[2]).

It holds, too, for morphological characters. Feet that are subjected to wear and tear grow tougher and more thick-skinned. The thickening of the skin is an acquired adaptation, but it is not obvious why the change went in this direction. In man-made machines, parts that are subjected to wear get thinner not thicker, for obvious reasons. Why does the skin on the feet do the opposite? Because, fundamentally, natural selection has worked in the past to ensure an adaptive rather than a maladaptive response to wear and tear.

The relevance of this for would-be Lamarckian evolution is that there has to be a deep Darwinian underpinning even if there is a Lamarckian surface structure: a Darwinian choice of which potentially acquirable characters shall in fact be acquired and inherited. As I have argued before (Dawkins 1982a, pp. 164–177), this is true of a recent, highly publicized immunological theory of Lamarckian adaptation (Steele 1979). Lamarckian mechanisms cannot be fundamentally responsible for adaptive evolution. Even if acquired characters are inherited on some planet, evolution there will still rely on a Darwinian guide for its adaptive direction.

Theory 3. Direct induction by the environment

Adaptation, as we have seen, is a fit between organism and environment. The set of conceivable organisms is wider than the actual set. And there is a set of conceivable environments wider than the actual set. These two subsets match each other to some extent, and the matching is adaptation. We can re-express the point by saying that information from the environment is present in the organism. In a few cases this is vividly literal—a frog carries a picture of its environment around on its back. Such information is usually carried by an animal in the less literal sense that a trained observer, dissecting a new animal, can reconstruct many details of its natural environment.

Now, how could the information get from the environment into the animal? Lorenz (1966) argues that there are two ways, natural selection and reinforcement learning, but that these are both *selective* processes in the broad sense (Pringle 1951). There is, in theory, an alternative method for the environment to imprint its information on the organism, and that is by direct "instruction" (Danchin 1979). Some theories of how the immune system works are "instructive": antibody molecules are thought to be shaped directly by moulding themselves around antigen molecules. The currently favored theory is, by contrast, selective (Burnet 1969). I take "instruction" to be synonymous with the "direct induction by the environment" of Mayr's Theory 3. It is not always clearly distinct from Theory 2.

Instruction is the process whereby information flows directly from its environment into an animal. A case could be made for treating imitation learning, latent learning and imprinting (Thorpe 1963) as instructive, but for clarity it is safer to use a hypothetical example. Think of an animal on some planet, deriving camouflage from its tiger-like stripes. It lives in long dry "grass," and its stripes closely match the typical thickness and spacing of local grass blades. On our own planet such adaptation would come about through the selection of random genetic variation, but on the imaginary planet it comes about through direct instruction. The animals go brown except where their skin is shaded from the "sun" by blades of grass. Their stripes are therefore adapted with great precision, not just to any old habitat, but to the precise habitat in which they have sunbathed, and it is this same habitat in which they are going to have to survive. Local populations are automatically camouflaged against local grasses. Information about the habitat, in this case about the spacing patterns of the grass blades, has flowed into the animals, and is embodied in the spacing pattern of their skin pigment.

Instructive adaptation demands the inheritance of acquired characters if it is to give rise to permanent or progressive evolutionary change. "Instruction" received in one generation must be "remembered" in the genetic (or equivalent) information. This process is in principle cumulative and progressive. However, if the genetic store is not to become overloaded by the accumulations of generations, some mechanism must exist for discarding unwanted "instructions," and retaining desirable ones. I suspect that this must lead us, once again, to the need for some kind of selective process.

Imagine, for instance, a form of mammal-like life in which a stout "umbilical nerve" enabled a mother to "dump" the entire contents of her memory in the brain of her foetus. The technology is available even to our nervous systems: the corpus callosum can shunt large quantities of information from right hemisphere to left. An umbilical nerve could make the experience and wisdom of each generation automatically available to the next, and this might seem very desirable. But without a selective filter, it would take few generations for the load of information to become unmanageably large. Once again we come up against the need for a selective underpinning. I will leave this now, and make one more point about instructive adaptation (which applies equally to all Lamarckian types of theory).

The point is that there is a logical link-up between the two major theories of adaptive evolution—selection and instruction—and the two major theories of embryonic development—epigenesis and preformationism. Instructive evolution can work only if embryology is preformationistic. If embryology is epigenetic, as it is on our planet, instructive evolution cannot work. I have expounded the argument before (Dawkins 1982a. pp. 174–176), so I will abbreviate it here.

If acquired characters are to be inherited, embryonic processes must be reversible: phenotypic change has to be read back into the genes (or equivalent). If embryology is preformationistic—the genes are a true blueprint—then it may indeed be reversible. You can translate a house back into its blueprint. But if embryonic development is epigenetic: if, as on this planet, the genetic information is more like a recipe for a cake (Bateson 1976) than a blueprint for a house, it is irreversible. There is no one-to-one mapping between bits of genome and bits of phenotype, any more than there is mapping between crumbs of cake and words of recipe. The recipe is not a blueprint that can be reconstructed from the cake. The transformation of recipe into cake cannot be put into reverse, and nor can the process of making a body. Therefore acquired adaptations cannot be read back into the "genes," on any planet where embryology is epigenetic.

This is not to say that there could not, on some planet, be a form of life whose embryology was preformationistic. That is a separate question. How likely is it? The form of life would have to be very different from ours, so much so that it is hard to visualize how it might work. As for reversible embryology itself, it is even harder to visualize. Some mechanism would have to scan the detailed form of the adult body, carefully noting down, for instance, the exact location of brown pigment in a sun-striped skin, perhaps turning it into a linear stream of code numbers, as in a television camera. Embryonic development would read the scan out again, like a television receiver. I have an intuitive hunch that there is an objection in principle to this kind of embryology, but I cannot at present formulate it clearly. All I am saying here is that, if planets are divided into those where embryology is preformationistic and those, like Earth, where embryology is epigenetic, Darwinian evolution could be supported on both kinds of planet, but Lamarckian evolution, even if there were no other reasons for doubting its existence, could be supported only on the preformationistic planets—if there are any.

The close theoretical link that I have demonstrated between Lamarckian evolution and preformationistic embryology gives rise to a mildly entertaining irony. Those with ideological reasons for hankering after a neo-Lamarckian view of evolution are often especially militant partisans of epigenetic, "interactionist" ideas of development, possibly—and here is the irony—for the very same ideological reasons (Koestler 1967; Ho & Saunders 1981).

Theory 4. Saltationism

The great virtue of the idea of evolution is that it explains, in terms of blind physical forces, the existence of undisputed adaptations whose statistical improbability is enormous, without recourse to the supernatural or the mystical. Since we *define* an undisputed adaptation as an adaptation that is too complex to have come about by chance, how is it possible for a theory to invoke only blind physical forces in explanation? The answer—Darwin's answer—is astonishingly simple when we consider how self-evident Paley's Divine Watchmaker must have seemed to his contemporaries. The key is that the coadapted parts do not have to be assembled *all at once*. They can be put together in small stages. But they really do have to be *small* stages. Otherwise we are back again with the problem we started with: the creation by chance of complexity that is too great to have been created by chance!

Take the eye again, as an example of an organ that contains a large number of independent coadapted parts, say N. The *a priori* probability of any one of these N features coming into existence by chance is low, but not incredibly low. It is comparable to the chance of a crystal pebble being washed by the sea so that it acts as a lens. Any one adaptation on its own could, plausibly, have come into existence through blind physical forces. If each of the N coadapted features confers some slight advantage on its own, then the whole many-parted organ can be put together over a long period of time. This is particularly plausible for the eye ironically in view of that organ's niche of honor in the creationist pantheon. The eye is, *par excellence*, a case where a fraction of an organ is better than no organ at all; an eye without a lens or even a pupil, for instance, could still detect the looming shadow of a predator.

To repeat, the key to the Darwinian explanation of adaptive complexity is the replacement of instantaneous, coincidental, multi-dimensional luck, by gradual, inch by inch, smeared-out luck. Luck is involved, to be sure. But a theory that bunches the luck up into major steps is more incredible than a theory that spreads the luck out in small stages. This leads to the following general principle of universal biology. Wherever in the universe adaptive complexity shall be found, it will have come into being gradually through a series of small alterations, never through large and sudden increments in adaptive complexity. We must reject Mayr's 4th theory, saltationism, as a candidate for explanation of the evolution of complexity.

It is almost impossible to dispute this rejection. It is implicit in the definition of adaptive complexity that the only alternative to gradualistic evolution is supernatural magic. This is not to say that the argument in favour of gradualism is a worthless tautology, an unfalsifiable dogma of the sort that creationists and philosophers are so fond of jumping about on. It is not *logically* impossible for a full-fashioned eye to spring *de novo* from virgin bare skin. It is just that the possibility is statistically negligible.

Now it has recently been widely and repeatedly publicized that some modern evolutionists reject "gradualism," and espouse what Turner (1982) has called theories of evolution by jerks. Since these are reasonable people without mystical leanings, they must be gradualists in the sense in which I am here using the term: the "gradualism" that they oppose must be defined differently. There are actually two confusions of language here, and I intend to clear them up in turn. The first is the common confusion between "punctuated equilibrium" (Eldredge & Gould 1972) and true saltationism. The second is a confusion between two theoretically distinct kinds of saltation.

Punctuated equilibrium is not macromutation, not saltation at all in the traditional sense of the term. It is, however, necessary to discuss it here, because it is popularly regarded as a theory of saltation, and its partisans quote, with approval, Huxley's criticism of Darwin for upholding the principle of *Natura non facit saltum* (Gould 1980). The punctuationist theory is portrayed as radical and revolutionary and at variance with the "gradualistic" assumptions of both Darwin and the neo-Darwinian synthesis (e.g., Lewin 1980). Punctuated equilibrium, however, was originally conceived as what the orthodox neo-Darwinian synthetic theory should truly predict, on a palaeontological timescale, if we take its embedded ideas of allopatric speciation seriously (Eldredge & Gould 1972). It derives its "jerks" by taking the "stately unfolding" of the neo-Darwinian synthesis, and *inserting* long periods of stasis separating brief bursts of gradual, albeit rapid, evolution.

The plausibility of such "rapid gradualism" is dramatized by a thought experiment of Stebbins (1982). He imagines a species of mouse, evolving larger body size at such an imperceptibly slow rate that the differences between the means of successive generations would be utterly swamped by sampling error. Yet even at this slow rate Stebbins's mouse lineage would attain the body size of a large elephant in about 60,000 years, a time-span so short that it would be regarded as instantaneous by palaeontologists. Evolutionary change too *slow* to be detected by micro-evolutionists can nevertheless be too *fast* to be detected by macroevolutionists. What a palaeontologist sees as a "saltation" can in fact be a smooth and gradual change so slow as to be undetectable to the microevolutionist. This kind of palaeontological "saltation" has nothing to do with the one-generation macromutations that, I suspect, Huxley and Darwin had in mind when they debated *Natura non facit saltum*. Confusion has arisen here, possibly because some individual champions of punctuated equilibrium have also, incidentally, championed macromutation (Gould 1982). Other "punctuationists" have either confused their theory with macromutationism, or have explicitly invoked macromutation as one of the mechanisms of punctuation (e.g., Stanley 1981).

Turning to macromutation, or true saltation itself, the second confusion that I want to clear up is between two kinds of macromutation that we might conceive of. I could name them, unmemorably, saltation (1) and saltation (2), but instead I shall pursue an earlier fancy for airliners as metaphors, and label them "Boeing 747" and "Stretched DC-8" saltation. 747 saltation is the inconceivable kind. It gets its name from Sir Fred Hoyle's much quoted metaphor for his own cosmic misunderstanding of Darwinism (Hoyle & Wickramasinghe 1981). Hoyle compared Darwinian selection to a tornado, blowing through a junkyard and assembling a Boeing 747 (what he overlooked, of course, was the point about luck being "smeared-out" in small steps—see above). Stretched DC-8 saltation is quite different. It is not in principle hard to believe in at all. It

refers to large and sudden changes in *magnitude* of some biological measure, without an accompanying large increase in adaptive information. It is named after an airliner that was made by elongating the fuselage of an existing design, not adding significant new complexity. The change from DC-8 to Stretched DC-8 is a big change in magnitude—a saltation not a gradualistic series of tiny changes. But, unlike the change from junk-heap to 747, it is not a big increase in information content or complexity, and that is the point I am emphasizing by the analogy.

An example of DC-8 saltation would be the following. Suppose the giraffe's neck shot out in one spectacular mutational step. Two parents had necks of standard antelope length. They had a freak child with a neck of modern giraffe length, and all giraffes are descended from this freak. This is unlikely to be true on Earth, but something like it may happen elsewhere in the universe. There is no objection to it in principle, in the sense that there is a profound objection to the (747) idea that a complex organ like an eye could arise from bare skin by a single mutation. The crucial difference is one of complexity.

I am assuming that the change from short antelope's neck to long giraffe's neck is *not* an increase in complexity. To be sure, both necks are exceedingly complex structures. You couldn't go from *no*-neck to either kind of neck in one step: that would be 747 saltation. But once the complex organization of the antelope's neck already exists, the step to giraffe's neck is just an elongation: various things have to grow faster at some stage in embryonic development; existing complexity is preserved. In practice, of course, such a drastic change in magnitude would be highly likely to have deleterious repercussions which would render the macromutant unlikely to survive. The existing antelope heart probably could not pump the blood up to the newly elevated giraffe head. Such practical objections to evolution by DC-8 saltation can only help my case in favour of gradualism, but I still want to make a separate, and more universal, case against 747 saltation.

It may be argued that the distinction between 747 and DC-8 saltation is impossible to draw in practice. After all, DC-8 saltations, such as the proposed macromutational elongation of the giraffe's neck, may appear very complex: myotomes, vertebrae, nerves, blood vessels, all have to elongate together. Why does this not make it a 747 saltation, and therefore rule it out? But although this type of "coadaptation" has indeed often been thought of as a problem for any evolutionary theory, not just macromutational ones (see Ridley (1982) for a history), it is so only if we take an impoverished view of developmental mechanisms. We know that single mutations can orchestrate changes in growth rates of many diverse parts of organs, and, when we think about developmental processes, it is not in the least surprising that this should be so. When a single mutation causes a *Drosophila* to grow a leg where an antenna ought to be, the leg grows in all its formidable complexity. But this is not mysterious or surprising, not a 747 saltation, because the organization of a leg is already present in the body before the mutation. Wherever, as in embryogenesis, we have a hierarchically branching tree of causal relationships, a small alteration at a senior node of the tree can have large and complex ramified effects on the tips of the twigs. But although the change may be large in magnitude, there can be no large and sudden increments in adaptive information. If you think you have found a particular example of a large and sudden increment in adaptively complex information in practice, you can be certain the adaptive information was already there, even if it is an atavistic "throwback" to an earlier ancestor.

There is not, then, any objection in principle to theories of evolution by jerks, even the theory of hopeful monsters (Goldschmidt 1940), provided that it is DC-8 saltation, not 747 saltation, that is meant. Gould (1982) would clearly agree: "I regard forms of macromutation which include the sudden origin of new species with all their multifarious adaptations intact *ab initio*, as illegitimate." No educated biologist actually believes in 747 saltation, but not all have been sufficiently explicit about the distinction between DC-8 and 747 saltation. An unfortunate consequence is that creationists and their journalistic fellow-travellers have been able to exploit saltationist-sounding statements of respected biologists. The biologist's intended meaning may have been what I am calling DC-8 saltation, or even non-saltatory punctuation; but the creationist *assumes* saltation in the sense that I have dubbed 747, and 747 saltation would, indeed, be a blessed miracle.

I also wonder whether an injustice is not being done to Darwin, owing to this same failure to come to grips with the distinction between DC-8 and 747 saltation. It is frequently alleged that Darwin was wedded to gradualism, and therefore that, if some form of evolution by jerks is proved, Darwin will have been shown wrong. This is undoubtedly the reason for the ballyhoo and publicity that has attended the theory of punctuated equilibrium. But was Darwin really opposed to all jerks? Or was he, as I suspect, strongly opposed only to 747 saltation?

As we have already seen, punctuated equilibrium has nothing to do with saltation, but anyway I think it is not at all clear that, as is often alleged, Darwin would have been discomfited by punctuationist interpretations of the fossil record. The following passage, from later editions of the *Origin*, sounds like something from a current issue of *Paleobiology:* "the periods during which species have been undergoing modification, though very long as measured by years, have probably been short in comparison with the periods during which these same species remained without undergoing any change."

Gould (1982) shrugs this off as somehow anomalous and away from the mainstream of Darwin's thought. As he correctly says: "You cannot do history by selective quotation and search for qualifying footnotes. General tenor and historical impact are the proper criteria. Did his contemporaries or descendants ever read Darwin as a saltationist?" Certainly nobody ever accused Darwin of being a saltationist. But to most people saltation means macromutation, and, as Gould himself stresses, "Punctuated equilibrium is not a theory of macromutation." More importantly, I believe we can reach a better understanding of Darwin's general gradualistic bias if we invoke the distinction between 747 and DC-8 saltation.

Perhaps part of the problem is that Darwin himself did not have the distinction. In some anti-saltation passages it seems to be DC-8 saltation that he has in mind. But on those occasions he does not seem to feel very strongly about it: "About sudden jumps," he wrote in a letter in 1860, "I have no objection to them—they would aid me in some cases. All I can say is, that I went into the subject and found no evidence to make me believe in jumps [as a source of new species] and a good deal pointing in the other direction" (quoted in Gillespie (1979). This does not sound like a man fervently opposed, in principle, to sudden jumps. And of course there is no reason why he *should* have been fervently opposed, if he only had DC-8 saltations in mind.

But at other times he really is pretty fervent, and on those occasions, I suggest, he is thinking of

747 saltation: "... it is impossible to imagine so many co-adaptations being formed all by a chance blow" (quoted in Ridley 1982). As the historian Neal Gillespie puts it: "For Darwin, monstrous births, a doctrine favored by Chambers, Owen, Argyll, Mivart, and others, from clear theological as well as scientific motives, as an explanation of how new species, or even higher taxa, had developed, was no better than a miracle: 'it leaves the case of the co-adaptation of organic beings to each other and to their physical conditions of life, untouched and unexplained.' It was 'no explanation' at all, of no more scientific value than creation 'from the dust of the earth'" (Gillespie 1979, p. 118).

As Ridley (1982) says of the "religious tradition of idealist thinkers [who] were committed to the explanation of complex adaptive contrivances by intelligent design," "The greatest concession they could make to Darwin was that the Designer operated by tinkering with the generation of diversity, designing the variation." Darwin's response was: "If I were convinced that I required such additions to the theory of natural selection, I would reject it as rubbish... I would give nothing for the theory of natural selection, if it requires miraculous additions at any one stage of descent."

Darwin's hostility to monstrous saltation, then, makes sense if we assume that he was thinking in terms of 747 saltation—the sudden invention of new adaptive complexity. It is highly likely that that is what he was thinking of, because that is exactly what many of his opponents had in mind. Saltationists such as the Duke of Argyll (though presumably not Huxley!) wanted to believe in 747 saltation, precisely because it *did* demand supernatural intervention. Darwin did not believe in it, for exactly the same reason. To quote Gillespie again (p. 120): "... for Darwin, designed evolution, whether manifested in saltation, monstrous births, or manipulated variations, was but a disguised form of special creation."

I think this approach provides us with the only sensible reading of Darwin's well known remark that "If it could be demonstrated that any complex organ existed, which could not possibly have been formed by numerous, successive, slight modifications, my theory would absolutely break down." That is not a plea for gradualism, as a modern palaeobiologist uses the term. Darwin's theory is falsifiable, but he was much too wise to make his theory *that* easy to falsify! Why on earth *should* Darwin have committed himself to such an arbitrarily restrictive version of evolution, a version that positively invites falsification? I think it is clear that he didn't. His use of the term "complex" seems to me to be clinching. Gould (1982) describes this passage from Darwin as "clearly invalid." So it is invalid if the alternative to slight modifications is seen as DC-8 saltation. But if the alternative is seen as 747 saltation, Darwin's remark is valid and very wise. Notwithstanding those whom Miller (1982) has unkindly called Darwin's more foolish critics, his theory is indeed falsifiable, and in the passage quoted he puts his finger on one way in which it might be falsified.

There are two kinds of imaginable saltation, then, DC-8 saltation and 747 saltation. DC-8 saltation is perfectly possible, undoubtedly happens in the laboratory and the farmyard, and may have made important contributions to evolution. 747 saltation is statistically ruled out unless there is supernatural intervention. In Darwin's own time, proponents and opponents of saltation often had 747 saltation in mind, because they believed in—or were arguing against—divine intervention. Darwin was hostile to (747) saltation, because he correctly saw natural selection as an *alternative* to the miraculous as an explanation for adaptive complexity. Nowadays saltation either means punctuation (which isn't saltation at all) or DC-8 saltation, neither of which Darwin would have had strong objections to in principle; merely doubts about the facts. In the modern context, therefore, I do not think Darwin should be labelled a strong gradualist. In the modern context, I suspect that he would be rather open-minded.

It is in the anti-747 sense that Darwin was a passionate gradualist, and it is in the same sense that we must all be gradualists, not just with respect to life on earth, but with respect to life all over the universe. Gradualism in this sense is essentially synonymous with evolution. The sense in which we may be non-gradualists is a much less radical, although still quite interesting, sense. The theory of evolution by jerks has been hailed on television and elsewhere as radical and revolutionary, a paradigm shift. There is, indeed, an interpretation of it which is revolutionary, but that interpretation (the 747 macromutation version) is certainly wrong, and is apparently not held by its original proponents. The sense in which the theory might be right is not particularly revolutionary. In this field you may choose your jerks so as to be revolutionary, *or* so as to be correct, but not both.

Theory 5. Random evolution

Various members of this family of theories have been in vogue at various times. The "mutationists" of the early part of this century—De Vries, W. Bateson and their colleagues—believed that selection served only to weed out deleterious freaks, and that the real driving force in evolution was mutation pressure. Unless you believe mutations are directed by some mysterious life force, it is sufficiently obvious that you can be a mutationist only if you forget about adaptive complexity—forget, in other words, most of the consequences of evolution that are of any interest! For historians there remains the baffling enigma of how such distinguished biologists as De Vries, W. Bateson and T. H. Morgan could rest satisfied with such a crassly inadequate theory. It is not enough to say that De Vries's view was blinkered by his working only on the evening primrose. He only had to look at the adaptive complexity in his own body to see that "mutationism" was not just a wrong theory: it was an obvious non-starter.

These post-Darwinian mutationists were also saltationists and anti-gradualists, and Mayr treats them under that heading, but the aspect of their view that I am criticizing here is more fundamental. It appears that they actually thought that mutation, on its own without selection, was sufficient to explain evolution. This *could* not be so on any non-mystical view of mutation, whether gradualist or saltationist. If mutation is undirected, it is clearly unable to explain the adaptive directions of evolution. If mutation is directed in adaptive ways, we are entitled to ask how this comes about. At least Lamarck's principle of use and disuse makes a valiant attempt at explaining how variation might be directed. The "mutationists" didn't even seem to see that there was a problem, possibly because they under-rated the importance of adaptation—and they were not the last to do so. The irony with which we must now read W. Bateson's dismissal of Darwin is almost painful: "the transformation of masses of populations by imperceptible steps guided by selection is, as most of us now see, so inapplicable to the fact that we can only marvel . . . at the want of penetration displayed by the advocates of such a proposition. . . " (1913, quoted in Mayr 1982).

Nowadays some population geneticists describe themselves as supporters of "non-Darwinian evolution." They believe that a substantial number of the gene replacements that occur in evolution are non-adaptive substitutions of alleles whose effects are indifferent relative to one another (Kimura 1968). This may well be true, if not in Israel (Nevo, this volume[3]) maybe somewhere in the Universe. But it obviously has nothing whatever to contribute to solving the problem of the evolution of adaptive complexity. Modern advocates of neutralism admit that their theory cannot account for adaptation, but that doesn't seem to stop them regarding the theory as interesting. Different people are interested in different things.

The phrase "random genetic drift" is often associated with the name of Sewall Wright, but Wright's conception of the relationship between random drift and adaptation is altogether subtler than the others I have mentioned (Wright 1980). Wright does not belong in Mayr's fifth category, for he clearly sees selection as the driving force of adaptive evolution. Random drift may make it easier for selection to do its job by assisting the escape from local optima (Dawkins 1982a, p. 40), but it is still selection that is determining the rise of adaptive complexity.

Recently palaeontologists have come up with fascinating results when they perform computer simulations of "random phylogenies" (e.g., Raup 1977). These random walks through evolutionary time produce trends that look uncannily like real ones, and it is disquietingly easy, and tempting, to read into the random phylogenies apparently adaptive trends which, however, are not there. But this does not mean that we can admit random drift as an explanation of real adaptive trends. What it might mean is that some of us have been too facile and gullible in what we think are adaptive trends. That does not alter the fact that there are some trends that really *are* adaptive—even if we don't always identify them correctly in practice—and those real adaptive trends can't be produced by random drift. They must be produced by some non-random force, presumably selection.

So, finally, we arrive at the sixth of Mayr's theories of evolution.

Theory 6. Direction (order) imposed on random variation by natural selection

Darwinism—the non-random selection of randomly varying replicating entities by reason of their "phenotypic" effects—is the only force I know that can, in principle, guide evolution in the direction of adaptive complexity. It works on this planet. It doesn't suffer from any of the drawbacks that beset the other five classes of theory, and there is no reason to doubt its efficacy throughout the universe.

The ingredients in a general recipe for Darwinian evolution are replicating entities of some kind, exerting phenotypic "power" of some kind over their replication success. I have referred to these necessary entities as "active germ-line replicators" or "optimons" (Dawkins 1982a, Chapter 5). It is important to keep their replication conceptually separate from their phenotypic effects, even though, on some planets, there may be a blurring in practice. Phenotypic adaptations can be seen as tools of replicator propagation.

Gould (this volume[4]) disparages the replicator's eye view of evolution as preoccupied with "book-keeping." The metaphor is a superficially happy one: it is easy to see the genetic changes that accompany evolution as book-keeping entries, mere accountant's records of the really interesting phenotypic events going on in the outside world. Deeper consideration, however, shows that the truth is almost the exact opposite. It is central and essential to Darwinian (as opposed to Lamarckian) evolution that there shall be causal arrows flowing from genotype to phenotype, but not in the reverse direction. Changes in gene frequencies are not passive book-keeping records of phenotypic changes: it is precisely because (and to the extent that) they actively *cause* phenotypic changes that evolution of the phenotype can occur. Serious errors flow, both from a failure to understand the importance of this one-way flow (Dawkins 1982a, Chapter 6), and from an over-interpretation of it as inflexible and undeviating "genetic determinism" (Dawkins 1982a, Chapter 2).

The universal perspective leads me to emphasize a distinction between what may be called "one-off selection" and "cumulative selection." Order in the non-living world may result from processes that can be portrayed as a rudimentary kind of selection. The pebbles on a seashore become sorted by the waves, so that larger pebbles come to lie in layers separate from smaller ones. We can regard this as an example of the selection of a stable configuration out of initially more random disorder. The same can be said of the "harmonious" orbital patterns of planets around stars, and electrons around nuclei, of the shapes of crystals, bubbles and droplets, even, perhaps, of the dimensionality of the universe in which we find ourselves

(Atkins 1981). But this is all one-off selection. It does not give rise to progressive evolution because there is no replication, no succession of generations. Complex adaptation requires many generations of cumulative selection, each generation's change building upon what has gone before. In one-off selection, a stable state develops and is then maintained. It does not multiply, does not have offspring.

In life the selection that goes on *in any one generation* is one-off selection, analogous to the sorting of pebbles on a beach. The peculiar feature of life is that successive generations of such selection build up, progressively and cumulatively, structures that are eventually complex enough to foster the strong illusion of design. One-off selection is a commonplace of physics and cannot give rise to adaptive complexity. Cumulative selection is the hallmark of biology and is, I believe, the force underlying all adaptive complexity.

OTHER TOPICS FOR A FUTURE SCIENCE OF UNIVERSAL DARWINISM

Active germ-line replicators together with their phenotypic consequences, then, constitute the general recipe for life, but the form of the system may vary greatly from planet to planet, both with respect to the replicating entities themselves, and with respect to the "phenotypic" means by which they ensure their survival. Indeed the very distinction between "genotype" and "phenotype" may be blurred (L. Orgel, personal communication). The replicating entities do not have to be DNA or RNA. They do not have to be organic molecules at all. Even on this planet it is possible that DNA itself is a late usurper of the role, taking over from some earlier, inorganic crystalline replicator (Cairns-Smith 1982). It is also arguable that today selection operates on several levels, for instance the levels of the gene and the species or lineage, and perhaps some unit of cultural transmission (Lewontin 1970).

A full science of Universal Darwinism might consider aspects of replicators transcending their detailed nature and the time-scale over which they are copied. For instance, the extent to which they are "particulate" as opposed to "blending" probably has a more important bearing on evolution than their detailed molecular or physical nature. Similarly, a universe-wide classification of replicators might make more reference to their dimensionality and coding principles than to their size and structure. DNA is a digitally coded one-dimensional array. A "genetic" code in the form of a two-dimensional matrix is conceivable. Even a three-dimensional code is imaginable, although students of Universal Darwinism will probably worry about how such a code could be "read." (DNA is, of course, a molecule whose 3-dimensional structure determines how it is replicated and transcribed, but that doesn't make it a 3-dimensional code. DNA's meaning depends upon the 1-dimensional sequential arrangement of its symbols, not upon their 3-dimensional position relative to one another in the cell.) There might also be theoretical problems with analog, as opposed to digital codes, similar to the theoretical problems that would be raised by a purely analog nervous system (Rushton 1961).

As for the phenotypic levers of power by which replicators influence their survival, we are so used to their being bound up into discrete organisms or "vehicles" that we forget the possibility of a more diffuse extra-corporeal or "extended" phenotype. Even on this Earth a large amount of interesting adaptation can be interpreted as part of the extended phenotype (Dawkins 1982a, Chapters 11, 12, and 13), There is, however, a general theoretical case that can be made in favour of the discrete organismal body, with its own recurrent life cycle, as a necessity in any process of evolution of advanced adaptive complexity (Dawkins 1982a, Chapter 14), and this topic might have a place in a full account of Universal Darwinism.

Another candidate for full discussion might be what I shall call divergence, and convergence or recombination of replicator lineages. In the case of Earthbound DNA, "convergence" is provided by sex and related processes. Here the DNA "converges" within the species after having very recently "diverged." But suggestions are now being made that a different kind of convergence can occur among lineages that originally diverged an exceedingly long time ago. For instance there is evidence of gene transfer between fish and bacteria (Jacob, this volume[5]). The replicating lineages on other planets may permit very varied kinds of recombination, on very different time-scales. On Earth the rivers of phylogeny are almost entirely divergent: if main tributaries ever recontact each other after branching apart it is only through the tiniest of trickling cross-streamlets, as in the fish/bacteria case. There is, of course, a richly anastomosing delta of divergence and convergence due to sexual recombination *within* the species, but only within the species. There may be planets on which the "genetic" system permits much more cross-talk at all levels of the branching hierarchy, one huge fertile delta.

I have not thought enough about the fantasies of the previous paragraphs to evaluate their plausibility. My general point is that there is one limiting constraint upon all speculations about life in the universe. If a life-form displays adaptive complexity, it must possess an evolutionary mechanism capable of generating adaptive complexity. However diverse evolutionary mechanisms may be, if there is no other generalization that can be made about life all around the Universe, I am betting it will always be recognizable as Darwinian life. The Darwinian Law (Eigen, this volume[6]) may be as universal as the great laws of physics.

As usual I have benefited from discussions with many people, including especially Mark Ridley, who also criticized the manuscript, and Alan Grafen. Dr. F. J. Ayala called attention to an important error in the original spoken version of the paper.

NOTES

This chapter originally appeared in D. S. Bendall (Ed.) *Evolution from molecules to man* pp. 403–425, Cambridge, UK: Cambridge University Press, 1983.

REFERENCES

1. Asimov, I. (1979). *Extraterrestrial civilizations.* London: Pan.

2. Atkins, P. W. (1981). *The creation.* Oxford: W. H. Freeman.

3. Bateson, P. P. G. (1976). Specificity and the origins of behavior. *Advances in the Study of Behavior*, 6, 1–20.

4. Billingham, J. (1981). *Life in the universe.* Cambridge, MA: MIT Press.

5. Burnet, P. M. (1969). *Cellular immunology.* Melbourne: Melbourne University Press.

6. Cairns-Smith, A. G. (1982). *Genetic takeover.* Cambridge, UK: Cambridge University Press.

7. Crick, F. H. C. (1982). *Life itself.* London: MacDonald.

8. Danchin, A. (1979). Themes de la biologie: Theories instructives et theories selectives. *Revue des Questions Scientifiques*, 150, 151–164.

9. Darwin, C. R. (1859/1968). *The origin of species*. London: Penguin.

10. Dawkins, M. (1980). *Animal suffering: The science of animal welfare*. London: Chapman & Hall.

11. Dawkins, R. (1982a). *The extended phenotype*. Oxford: W. H. Freeman.

12. Dawkins, R. (1982b). The necessity of Darwinism. *New Scientist*, **94**, 130–132.

13. Eldredge, N. & Gould, S. J. (1972). Punctuated equilibria: An alternative to phyletic gradualism. In T. J. M. Schopf (Ed.), *Models in paleobiology* (pp. 82–115). San Francisco: Freeman-Cooper.

14. Fisher, R. A. (1958). *The genetical theory of natural selection*. New York: Dover.

15. Gillespie, N. C. (1979). *Charles Darwin and the problem of creation*. Chicago: University of Chicago Press.

16. Goldschmidt, R. (1940). *The material basis of evolution*. New Haven: Yale University Press.

17. Gould, S. J. (1980). *The panda's thumb*. New York: W. W. Norton.

18. Gould, S. J. (1982). The meaning of punctuated equilibrium and its role in validating a hierarchical approach to macroevolution. In R. Milkman (Ed.), *Perspectives on evolution* (pp. 83–102). Sunderland, MA: Sinauer.

19. Ho, M.-W. & Saunders, P. T. (1981). Adaptation and natural selection: Mechanism and teleology. In C. M. Barker, L. Birke, A D. Muir, and S. P. R. Rose (Eds.), *Towards a liberatory biology* (pp. 85–102). London: Allison & Busby.

20. Hoyle, F. & Wickramasinghe, N. C. (1981). *Evolution from space*. London: J. M. Dent.

21. Kimura, M. (1968). Evolutionary rate at the molecular level. *Nature*, **217**, 624–626.

22. Koestler, A. (1967). *The ghost in the machine*. London: Hutchinson.

23. Lewin, R. (1980). Evolutionary theory under fire. *Science*, **210**, 883–887.

24. Lewontin, R. C. (1970). The units of selection. *Annual Review of Ecology and Systematics*, **1**, 1–18.

25. Lewontin, R. C. (1979). Sociobiology as an adaptationist program. *Behavioral Science*, **24**, 5–14.

26. Lorenz, K. (1966). *Evolution and modification of behavior*. London: Methuen.

27. Maynard Smith, J. (1969). The status of neo-Darwinism. In C. H. Waddington (Ed.), *Towards a theoretical biology*. Edinburgh: Edinburgh University Press.

28. Mayr, E. (1982). *The growth of biological thought*. Cambridge, MA: Harvard University Press.

29. Miller, J. (1982). *Darwin for beginners*. London: Writers & Readers.

30. Paley, W. (1828). *Natural theology* (2nd ed.). Oxford: J. Vincent.

31. Pringle, J. W. S. (1951). On the parallel between learning and evolution. *Behaviour*, **3**, 90–110.

32. Raup, D. M. (1977). Stochastic models in evolutionary palaeontology. In A. Hallam (Ed.), *Patterns of evolution* (pp. 59–78). Amsterdam: Elsevier.

33. Ridley, M. (1982). Coadaptation and the inadequacy of natural selection. *British Journal for the History of Science*, **15**, 45–68.

34. Rushton, W. A. H. (1961). Peripheral coding in the nervous system. In W. A. Rosenblith (Ed.), *Sensory communication* (pp. 169–181). Cambridge, MA: MIT Press.

35. Stanley, S. M. (1981). *The new evolutionary timetable*. New York: Basic Books.

36. Stebbins, G. L. (1982). *Darwin to DNA, molecules to humanity*. San Francisco: W. H. Freeman.

37. Steele, E. J. (1979). *Somatic selection and adaptive evolution*. Toronto: Williams & Wallace.

38. Thorpe, W. H. (1963). *Learning and instinct in animals* (2nd ed.). London: Methuen.

39. Turner, J. R. G. (1982). Review of R. J. Berry, *Neo-Darwinism*. *New Scientist*, **94**, 160–162.

40. Wright, S. (1980). Genie and organismic selection. *Evolution*, **34**, 825–843.

ENDNOTES

[1] Ed. note: Lewontin, R. C. (1983). Gene, organism, and the environment. In D. S. Bendall (Ed.), *Evolution from molecules to man* (pp. 273–286). Cambridge, UK: Cambridge University Press.

[2] Ed. note: Bateson, P. P. B. (1983). Rules for changing the rules. In D. S. Bendall (Ed.), *Evolution from molecules to man* (pp. 483–508). Cambridge, UK: Cambridge University Press.

[3] Ed. note: Nevo, E. (1983). Population genetics and ecology: The interface. In D. S. Bendall (Ed.), *Evolution from molecules to man* (pp. 287–322). Cambridge UK: Cambridge University Press.

[4] Ed. note: Gould, S. J. (1983). Irrelevance, submission, and partnership: The changing role of palaeontology in Darwin's three centennials and a modest proposal for macroevolution. In D. S. Bendall (Ed.), *Evolution from molecules to man* (pp. 347–366). Cambridge, UK: Cambridge University Press.

5 Ed. note: Jacob, F. (1983). Molecular tinkering in evolution. In D. S. Bendall (Ed.), *Evolution from molecules to man* (pp. 131–144). Cambridge, UK: Cambridge University Press.

6 Ed. note: Eigen, M. (1983). Self-replication and molecular evolution. In D. S. Bendall (Ed.), *Evolution from molecules to man* (pp. 105–130). Cambridge, UK: Cambridge University Press.

30 · What is life? Was Schrödinger right?

STUART A. KAUFFMAN

In Dublin half a century ago, a major figure in this century's science visited, lectured, and foretold the future of a science which was not his own. The resulting book, *What is Life?*, is credited with having inspired some of the most brilliant minds ever to enter biology to the work which gave birth to molecular biology (Schrödinger 1967). Schrödinger's "little book" is, itself, as brilliant as warranted by its reputation. But, half a century later, and at the occasion of its honouring, perhaps we may dare to ask a new question: is the central thesis of the book right? I mean no dishonour to so superb a mind as Schrödinger's, nor to those properly inspired by him, to suggest that he may have been wrong, or at least incomplete. Rather, of course, like all scientists inspired by his ideas, I too seek to continue the quest.

I am hesitant even to raise the questions I shall raise, for I am also fully aware of how deeply embedded Schrödinger's own answers are in our view of life since Darwin and Weismann, and since the development of the theory of the germ plasma, with the gene as the necessary stable storage form of heritable variation: "Order from order," answered Schrödinger. The large aperiodic solids and the microcode of which Schrödinger spoke have become the DNA and the genetic code of today. Almost all biologists are convinced that such self-replicating molecular structures and such a microcode are essential to life.

I confess I am not entirely convinced. At its heart, the debate centres on the extent to which the sources of order in biology lie predominantly in the stable bond structures of molecules, Schrödinger's main claim, or in the collective dynamics of a system of such molecules. Schrödinger emphasized, correctly, the critical role played by quantum mechanics, molecular stability, and the possibility of a microcode directing ontogeny. Conversely, I suspect that the ultimate sources of self-reproduction and the stability requisite for heritable variation, development and evolution, while requiring the stability of organic molecules, may also require emergent ordered properties in the collective behavior of complex, non-equilibrium chemical reaction systems. Such complex reaction systems, I shall suggest, can spontaneously cross a threshold, or phase transition, beyond which they become capable of collective self-reproduction, evolution, and exquisitely ordered dynamical behavior. The ultimate sources of the order requisite for life's emergence and evolution may rest on new principles of collective emergent behavior in far from equilibrium reaction systems.

In brief, foreshadowing what follows: while Schrödinger's insights were correct about current life, I suspect that, in a deeper sense, he was incomplete. The formation of large aperiodic solids carrying a microcode, order from order, may be neither necessary nor sufficient for the emergence and evolution of life. In contrast, certain kinds of stable collective dynamics may be both necessary and sufficient for life. I would emphasize that I raise these issues for discussion, not as established conclusions.

SCHRÖDINGER'S ARGUMENT

Schrödinger begins his discussion by emphasizing the view of macroscopic order held by most physicists of his day and earlier. Such order, he tells us, consists in averages over enormous ensembles of atoms or molecules. Statistical mechanics is the proper intellectual framework for this analysis. Pressure in a gas confined in a volume is just the average behavior of very large numbers of molecules colliding with and recoiling from the walls. The orderly behavior is an average, and not due to the behavior of individual molecules.

But what accounts for the order in organisms, and, in particular, for rare mutations and heritable variation? Schrödinger then uses current data to estimate the number of atoms which might be involved in a gene, and correctly estimates that the number cannot be more than a few thousand atoms. Order due to statistical averaging cannot help here, he argues, for the numbers of atoms are too small for reliable behavior. In statistical systems the expected sizes of the fluctuations vary inversely with the square root of the number of events. With ten tosses of a fair coin, 80% "heads" is not surprising, with ten thousand tosses, 80% "heads" would be stunning. With a million events, Schrödinger points out, statistical fluctuations would be on the order of 0.001, rather unreliable for the order found in organisms.

Quantum mechanics, argues Schrödinger, comes to the rescue of life. Quantum mechanics ensures that solids have rigidly ordered molecular structures. A crystal is the simplest case. But crystals are structurally dull. The atoms are arranged in a regular lattice in three dimensions. If you know the positions of the atoms in a minimal "unit crystal," you know where all the other atoms are in the entire crystal. This overstates the case, of course, for there can be complex defects in crystals, but the point is clear. Crystals have very regular structures, so the different parts of the crystal, in some sense, all "say" the same thing. In a moment Schrödinger will translate the idea of "saying" into the idea of "encoding." With that leap, a regular crystal cannot encode much information. All the information is contained in the unit cell.

If solids have the order required, but periodic solids such as crystals are too regular, then Schrödinger puts his bet on aperiodic solids. The stuff of the gene, he bets, is some form of aperiodic crystal. The form of the aperiodicity will contain some kind of microscopic code which, somehow, controls the development of the organism. The quantum character of the aperiodic solid will mean that small discrete changes, mutations, will occur. Natural selection, operating on these small discrete changes, will select out favorable mutations as Darwin hoped.

Schrödinger was right. His book deserves its fine reputation. Five decades later we know the structure of DNA. There is, indeed, a code leading from DNA to

RNA and to the primary structure of proteins. This would have been a wonderful success for any scientist, let alone a physicist peering over the wall at biology.

But is Schrödinger's insight either necessary or sufficient? Is the order assembled in the DNA aperiodic crystal either necessary or sufficient for the evolution of life, or for the dynamical order found in current life? Neither, I suspect. The ultimate sources of order may require the discrete order of stable chemical bonds derived from quantum mechanics, but lie elsewhere. The ultimate sources of order and self-reproduction may lie in the emergence of collectively ordered dynamics in complex chemical reaction systems.

The main part of this chapter has two sections. The first examines briefly the possibility that the emergence of life itself is not based on the template replicating properties of DNA or RNA, but on a phase transition to collectively autocatalytic sets of molecules in open thermodynamic systems. The second section examines the emergence of collective dynamical order in complex parallel processing networks of elements. Those elements might be genes whose activities are mutually regulated, or might be the polymer catalysts in an autocatalytic set. Such networks are open thermodynamically, and the core source of the dynamical order they exhibit lies in the way dynamical trajectories converge to small attractors in the phase space of the system.

Since I want to suggest that convergence to small attractors in open thermodynamic systems is a major source of order in living organisms, I want to end this introductory section by laying out the background of Schrödinger's discussion about statistical laws.

The central point is simple: in closed thermodynamic systems, there is no convergence in the appropriate phase space. The character of the resulting statistical laws reflects this lack of convergence. But in some open thermodynamic systems there can be massive convergence of the dynamical flow of the system in its state space. This convergence can engender order rapidly enough to offset the thermal fluctuations which always occur.

The critical distinction between a closed system at equilibrium and an open system displaced from equilibrium is this: in a closed system, no information is thrown away. The behavior of the system is, ultimately, reversible. Because of this, phase volumes are conserved. In open systems, information is discarded into the environment and the behavior of the subsystem of interest is not reversible. Because of this, the phase volume of the subsystem can decrease. I am no physicist, but will try to lay out the issues simply, and I hope, correctly.

Consider a gas confined to a box, closed to exchange of matter and energy. Every possible microscopic arrangement of the gas molecules is as likely as any other. The motions of the molecules are governed by Newton's laws. Hence the motions are microscopically reversible, and the total energy of the system is conserved. When molecules collide, energy is exchanged but not lost. The "ergodic hypothesis," something of a leap of faith that works, asserts that as these molecular collisions occur, the total system visits all possible microstates over time equally often. Thus, the probability that the system is in any macrostate is exactly equal to the fractional number of microstates corresponding to that macrostate.

Liouville's theorem states that volumes in phase space are conserved under the flow of an equilibrium system. For a system with N gas molecules, the current position and momentum of each molecule in three spatial dimensions can be represented by six numbers. Hence in a $6N$-dimensional phase space, the current state of the entire volume of gas can be represented as a single point. Consider a set of nearly identical initial states of the gas in a box. The corresponding points occupy some volume in phase space. Liouville's theorem asserts that, as molecular collisions occur in each copy of the box of gas, the corresponding volume in phase space moves, deforms, and smears out over the phase space. But the total volume in phase space remains constant. There is no convergence in the flow in phase space. Since phase volume is constant, then given the ergodic hypothesis, the probabilities of macrostates are just proportional to the relative numbers of microstates within each macrostate, normalized by the total number of microstates.

Suppose, instead, that the flow of the system in phase space allowed the initial phase volume to progressively contract to a single point, or to a small volume. Then the spontaneous behavior of the system would flow to some unique configuration, or some small number of configurations. Order would emerge! Of course, such convergence cannot happen in a closed, equilibrium thermodynamic system. If it did, entropy would decrease rather than increase in the total system.

Such order can emerge. Clearly, the emergence of this kind of order requires as a necessary condition that the system be thermodynamically open to the exchange of matter and energy. This exchange allows information to be lost from the subsystem of interest into its environment. A physicist would say that "degrees of freedom"—the diverse ways the molecules can move and interact—are lost into the heat bath of the environment.

Thus, the kind of dynamical order we seek can only arise in non-equilibrium thermodynamic systems. Such systems have been termed dissipative structures by Prigogine. Whirlpools, Zhabotinsky reactions, Bénard cells, and other examples are now familiar. However, it is essential to stress that displacement from thermodynamic equilibrium by itself is only a necessary condition, not a sufficient condition, for the emergence of highly ordered dynamics. The fabled butterfly in Rio de Janeiro whose wings beget chaos in the weather can recur in many versions in complex, non-equilibrium chemical systems, begetting chaos which would forbid the emergence and evolution of life. In the third section of this chapter I return to the emergence of collectively ordered dynamical behavior in non-equilibrium open systems.

THE ORIGIN OF LIFE AS A PHASE TRANSITION

The large aperiodic solid about which Schrödinger mused is now known. Since Watson and Crick remarked, with uncertain modesty, that the template complementary structure of the DNA double helix foretold its mode of replication, almost all biologists have fastened upon some version of template complementarity as necessary to the emergence of self-reproducing molecular systems. The current favourite contenders are either RNA molecules or some similar polymer. The hope is that such polymers might act as templates for their own replication, in the absence of any outside catalyst.

So far, efforts to achieve replication of RNA sequences in the absence of enzymes have met with limited success. Leslie Orgel, a speaker at the Trinity College, Dublin conference and an outstanding organic chemist, has worked hard to achieve such molecular replication (Orgel 1987). He, better than I, could summarize the difficulties. But, briefly, the problems are many. Abiotic synthesis of nucleotides is difficult to achieve. Such nucleotides like to bond by $2'$–$5'$ bonds rather than the requisite $3'$–$5'$ bonds. One wants to find an arbitrary sequence of the four normal RNA bases and use these as a template to line up the four complementary bases by hydrogen bond pairing, such that the lined up nucleotides form the proper $3'$–$5'$ bonds and melt from the initial template, and the system cycles again to make an exponential number of copies. It has not worked as yet. The difficulties arise for good chemical reasons. For example, a single strand which is richer in C than G will form the second strand as hoped. But the second, richer in G than C, tends to form G–G bonds which cause the molecule to fold in ways which preclude function as a new template.

With the discovery of ribozymes and the hypothesis of an RNA world, a new, attractive hope is that an RNA molecule might function as a polymerase, able to copy itself and any other RNA molecule. Jack Szostak, at Harvard Medical School, is attempting to evolve such a polymerase *de novo*. If he succeeds, it will be a *tour de force*. But I am not convinced that such a molecule holds the answer to life's emergence. It seems a rare structure to have been by chance as the first "living molecule." And, were it formed, I am not yet persuaded it could evolve. Such a molecule, like any enzyme, would make errors when replicating itself, and hence form mutant copies. These would complete with the "wild type" ribozyme polymerase to replicate that polymerase, and would themselves tend to be more error prone. Hence the mutant RNA polymerases would tend to generate still more badly mutant RNA polymerase sequences. A runaway "error catastrophe," of a type first suggested by Orgel concerning coding and its translation, might occur. I do not know that such an error catastrophe would arise, but believe the problem is worth analysis.

When one contemplates the symmetrical beauty of the DNA double helix, or a similar RNA helix, one is forced to admit the simple beauty of the corresponding hypothesis. Surely such structures were the first living molecules. But is this really true? Or might the roots of life lie deeper? I turn next to explore this possibility.

The simplest free living organisms are the mycoplasma. These derived bacterial forms have on the order of 600 genes encoding proteins via the standard machinery. Mycoplasma possess membranes, but no bacterial cell wall. They live in very rich environments, the lungs of sheep or man, for example, where their requirements for a rather large variety of exogenous small molecules are met.

Why should the simplest free living entities happen to harbour something like 600 kinds of polymers and a metabolism with perhaps 1000 small molecules? And how, after all, does the mycoplasm reproduce? Let us turn to the second question first, for the answer is simple and vital. The mycoplasma cell reproduces itself by a kind of collective autocatalysis. No molecular species within the mycoplasma actually replicates itself. We know this, but tend to ignore it. The DNA of the mycoplasma is replicated thanks to the coordinated activities of a host of cellular enzymes. In turn, the latter are synthesized via standard messenger RNA sequences. As we all know, the code is translated from RNA to proteins only with the help of encoded proteins—namely the amino acid synthetases which charge each transfer RNA properly for later assembly by the ribosome into a protein. The membrane of the cell has its molecules formed by catalysis from metabolic intermediates. We are all familiar with the story. No molecule in the mycoplasm replicates itself. The system as a whole is collectively autocatalytic. Every molecular species has its formation catalysed by some molecular species in the system, or else is supplied exogenously as "food."

If the mycoplasma is collectively autocatalytic, so too are all free living cells. In no cell does a molecule actually replicate itself. Then let us ask why the minimal complexity found in free living cells is on the order of 600 protein polymers and about 1000 small molecules. We have no answer. Let us ask why there should be a minimal complexity under the standard hypothesis that single stranded RNA sequences might serve as templates and be able to replicate without other enzymes. But on this hypothesis there could be no deep answer to why a minimal complexity is observed in all free living cells. Just the simplicity of the "nude" replicating gene is what commends that familiar hypothesis to us. All we might respond is that, 3.45 billion years later, the simplest free living cells happen to have the complexity of mycoplasma. We have no deep account, merely another evolutionary "Just So" story.

I now present, in brief form, a body of work carried out over the past eight years alone and with my colleagues (Kauffman 1971, 1986, 1993; Farmer *et al.* 1986; Bagley 1991; Bagley *et al.* 1992). Similar ideas were advanced independently by Rossler (1971), Eigen (1971), and Cohen (1988). The central concept is that, in sufficiently complex chemical reaction systems, a critical diversity of molecular species is crossed. Beyond that diversity, the chances that a subsystem exists which is collectively autocatalytic goes to 1.0.

The central ideas are simple. Consider a space of polymers including monomers, dimers, trimers, and so forth. In concrete terms, the polymers might be RNA sequences, peptides or other kinds of polymeric forms. Later, the restriction to polymers will be removed and we will consider systems of organic molecules.

Let the maximum length of polymer we will think about be M. And let M increase. Count the number of polymers in the system, from monomers to polymers of length M. It is simple to see that the number of polymers in the system is an exponential function of M. Thus, for 20 kinds of amino acids, the total diversity of polymers up to length M is slightly greater than 20^M. For RNA sequences, the total diversity is slightly more than 4^M.

Now let us consider all cleavage and ligation reactions among the set of polymers up to length M. Clearly, an oriented polymer such as a peptide or RNA sequence of length M can be made from smaller sequences in $M - 1$ ways, since any internal bond in the polymer of length M is a site at which smaller fragments can be ligated. Thus, in the system of polymers up to length M, there are exponentially many polymers, but there are even more cleavage and ligation reactions by which these polymers can interconvert. Indeed, as M increases, the ratio of cleavage and ligation reactions per polymer increases linearly with M.

Define a reaction graph among a set of polymers. In general, we might think of one substrate one product reactions, one substrate two product reactions, two product one substrate reactions, and two substrate two product reactions. Transpeptidation and transesterification reactions are among the two substrate two product reactions that peptides or RNA sequences can undergo. A reaction graph consists of the set of substrates and products, which may be pictured as points, or nodes scattered in a three-dimensional space. In addition, each reaction can be denoted by a small circular "reaction box." Arrows from the substrate(s) leads to the box. Arrows lead from the box to the product(s). Since all reactions are actually at least weakly reversible, the arrow directions are merely useful to indicate which sets of molecules constitute the substrates and which the products in one of the two directions the reaction may take. The reaction graph consists in this entire collection of nodes, boxes, and arrows. It shows all possible reactions among the molecules in the system.

The implication noted above of the combinatorics of chemistry is that, as the diversity of polymers in the system increases, the ratio of reactions to molecules increases. This means that the ratio of arrows and boxes to nodes increases. The reaction graph becomes ever denser, ever more richly interconnected with reaction possibilities, as the diversity of molecules in the system increases.

In such a reaction system it is always true that some reactions happen spontaneously at some velocity. I ask the reader to ignore, for the moment, such spontaneous reactions in order to focus on the following question: *under what conditions will a collectively autocatalytic set of molecules emerge?* I aim to show that, under a wide variety of hypotheses about the system, autocatalytic sets will emerge at a critical diversity.

I begin by drawing attention to well-known phase transitions in random graphs. Throw ten thousand buttons on the floor and begin to connect pairs of buttons at random with red threads. Such a collection of buttons and threads is a random graph. More formally, a random graph is a set of nodes connected at random with a set of edges. Every so often pause to hoist a button and see how many buttons one lifts up with it. Such a connected set of buttons is called a component in a random graph. Erdos and Renyi showed, some decades ago (1960), that such systems undergo a phase transition as the ratio of edges to nodes passes 0.5. When the ratio is lower, when the number of edges is, say 10% of the number of nodes, any node will be directly or indirectly connected to only a few other nodes. But when the ratio of edges to nodes is 0.5, suddenly most of the nodes become connected into a single giant component. Indeed, if the number of nodes were infinite, then as the ratio of edges to nodes passed 0.5 the size of the largest component would jump discontinuously from very small to infinite. The system exhibits a first order phase transition. The point to bear away is simple: when enough nodes are connected, even at random, a giant interconnected component literally crystallizes.

We need only apply this idea to our reaction graph. Looking ahead, we will focus attention on catalysed reactions. We will need a theory about which polymers catalyse which reactions. Given a variety of such theories, we will find a simple consequence: as the diversity of molecules in the system increases, the ratio of reactions to molecules increases. Thus, for almost any model of which polymers catalyse which reactions, at some diversity almost every polymer will catalyse at least one reaction. At that critical diversity a giant component of connected catalysed reactions will crystallize in the system. If the polymers which act as catalysts are themselves the products of the catalysed reactions, the system will become collectively autocatalytic.

But this step is easy. Consider a simple, indeed oversimple, model of which polymer catalyses which reaction. I will relax the idealization I am about to make further below. Let us assume that any polymer has a fixed probability, say one in a billion, to be able to act as a catalyst for any randomly chosen reaction. Now consider our reaction graph at a point when the diversity of molecules in the system is such that there are a billion reactions for every molecule. And let the molecules in question be polymers which are themselves candidates to catalyse the reactions among the polymers. But then about one reaction per polymer will be catalysed. A giant component will crystallize in the system. And, with a little thought, it is clear that the system almost certainly contains collectively autocatalytic subsystems. Self-reproduction has emerged at a critical diversity, owing to a phase transition in a chemical reaction graph.

Figure 30.1 shows such a collectively autocatalytic set. The main point to stress is the near inevitable emergent property of such systems, and a kind of unrepentant holism. At a lesser diversity, the resulting reaction graph has only a few reactions which are catalysed by molecules in the system. No autocatalytic set is present. As the diversity increases, a larger number of reactions are catalysed by molecules in the system. At some point as the diversity increases, a connected web of catalysed reactions springs into existence. The web embraces the catalysts themselves. Catalytic closure is suddenly attained. A "living" system, self-reproducing at least in its silicon realization, swarms into existence.

Further, this crystallization requires a critical diversity. A simpler system simply does not achieve

catalytic closure. We begin to have a candidate for a deep theory for the minimal diversity of free living cells. No "Just So" story here: simpler systems fail to achieve or sustain autocatalytic closure.

The total molecular diversity required to cross the phase transition depends upon two major factors: (1) the ratio of reactions to molecules, and (2) the distribution of probabilities that the molecules in the system catalyse the reactions among the same molecules. The ratio of reactions to molecules depends upon the complexity of the kinds of reactions allowed. For example, if one considers only cleavage and ligation reactions among peptides or RNA sequences, then the ratio of reactions to polymers increases linearly with maximum length, M, of polymer in the system. This is easy to see in outline, since a polymer of length M can be made in $M - 1$ ways. As M increases, the ratio of reactions to polymers increases proportionally. Conversely, one might consider transpeptidation or transesterification reactions among peptides or RNA sequences. In that case the ratio of reactions to polymers increases very much faster than linearly. Consequently, the diversity of molecules requisite for the emergence of autocatalytic sets is much smaller. In concrete terms, if the probability an arbitrary polymer catalysed an arbitrary reaction were one in a billion, then about 18,000 kinds of molecules would suffice for the emergence of collective autocatalytic sets.

The results we are discussing are robust with respect to the over simple idealization that any polymer has a fixed probability of being able to function as a catalyst for any reaction. An alternative model (Kauffman 1993) considers RNA sequences as the potential, simple ribozymes, and supposes that in order to function as a specific ligase, a candidate ribozyme must template match the three terminal 5' nucleotides on one substrate and the three terminal 3' nucleotides on a second substrate. Recently, von Kiederowski (1986) has generated just such specific ligases which actually form small autocatalytic sets! A hexamer ligates two tamers which then constitute the hexamer. More recently, von Kiederowski has created collectively reproducing cross-catalytic systems (personal communication, 1994). In line with von Kiederowski's results, in our model RNA system to capture the fact that other features beyond template matching may be required for the candidate RNA to actually function as a catalyst in the reaction, Bagley and I assumed that any such

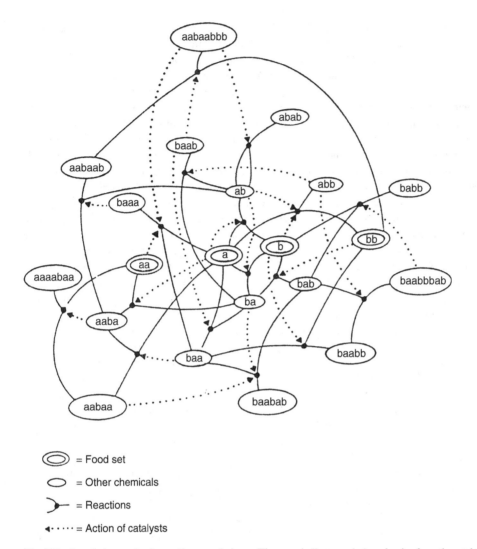

= Food set

= Other chemicals

= Reactions

= Action of catalysts

Fig. 30.1. A typical example of a small autocatalytic set. The reactions are represented by points connecting cleavage products with the corresponding larger ligated polymer. Dotted lines indicate catalysis and point from the catalysts to the reaction being catalysed. Monomers and dimers of A and B constitute the maintained food set (double ellipses).

matching candidate still had only a one in a million chance to be able to function as a specific ligase. Collectively autocatalytic sets still emerge at a critical diversity of model RNA sequences in the system. Presumably, the results are robust and will remain valid for a wide variety of models about the distribution of catalytic capacities among sets of polymers or other organic molecules. I return in a moment to discuss experimental avenues to attempt to create such collectively autocatalytic systems.

If this view is right, then the emergence of life does not depend upon the beautiful templating properties of DNA, RNA, or other similar polymers. Instead, the roots of life lie in catalysis itself and in the combinatorics of chemistry. If this view is right, then the routes to life may be broad boulevards of probability, not back alleys of rare chance.

But can such collectively autocatalytic systems evolve? Can they evolve without a genome in the familiar sense? And if so, what are the implications for our tradition since Darwin, Weismann, and, indeed, since Schrödinger? For, if self-reproducing systems can evolve without a stable large molecular repository of genetic information, then Schrödinger's suggestion about large aperiodic solids is not necessary to the emergence and evolution of life.

At least in computer experiments such collectively autocatalytic systems can evolve without a genome. First, I should stress that my colleagues, Farmer and Packard, and I (1986), have shown, using fairly realistic thermodynamic conditions in model stirred flow reactors, that model autocatalytic systems can, in fact, emerge. Further, Bagley has shown as part of his thesis that such systems can attain and sustain high concentrations of large model polymers in the face of a bias towards cleavage in an aqueous medium. Moreover, such systems can "survive" if the "food" environment is modified in some ways, but are "killed"—i.e., collapse—if other foodstuffs are removed from the flow reactor system. Perhaps the most interesting results, however, show that such systems can evolve to some extent without a genome. Bagley et al. (1992), made use of the reasonable idea that spontaneous reactions which persist in the autocatalytic set will tend to give rise to molecules which are not members of the set itself. Such novel molecules form a kind of penumbra of molecular species around the autocatalytic set, and are present at higher concentrations than they would otherwise be, owing to the presence of the autocatalytic set. The autocatalytic set can evolve by grafting some of these new molecular species into itself. It suffices if one or more of these shadow set molecules fluctuates to a modest concentration, and that these molecules then aid catalysts of their own formation from the autocatalytic set. If so, the set expands to include these new molecular species. Presumably, if some molecules can inhibit reactions catalysed by other molecules, the addition of new kinds of molecules will sometimes lead to the elimination of older kinds of molecules.

In short, at least in silico, autocatalytic sets can evolve without a genome. No stable large molecular structure carries genetic information in any familiar sense. Rather, the set of molecules and the reactions they undergo and catalyse constitutes the "genome" of the system. The stable dynamical behavior of this self-reproducing, coupled system of reactions constitutes the fundamental heritability it exhibits. The capacity

to incorporate novel molecular species, and perhaps eliminate older molecular forms, constitutes the capacity for heritable variation. Darwin then tells us that such systems will evolve by natural selection.

If these considerations are correct, then, I submit, Schrödinger's suggested requirement for a large aperiodic solid as a stable carrier of heritable information is not necessary to the emergence of life or its evolution. Order from order in this sense, in short, may not be necessary.

Finally, I would like to mention briefly some experimental approaches to these questions. The fundamental question is this: if a sufficiently great diversity of polymers together with the small molecules of which they are composed, plus some other sources of chemical energy, were gathered in a sufficiently small volume under the appropriate conditions, would collectively autocatalytic sets emerge? These new experimental approaches rely on new molecular genetic possibilities. It is now feasible to clone essentially random DNA, RNA, and peptide sequences, creating extremely high diversities of these biopolymers (Ballivet & Kauffman 1985; Devlin et al. 1990; Ellington & Szostak 1990). At present, libraries with diversities of up to trillions of sequences are under exploration. Thus, for the first time it becomes possible to consider creating reaction systems with this high molecular diversity confined to small volumes such that rapid interactions can occur. For example, such polymers might be confined not only to continuous flow stirred reactors, but to vesicles such as liposomes, micelles, and other structures, which provide surfaces and a boundary between an internal and external environment. Given von Kiederowski's collectively autocatalytic sets, designed with his chemist's intelligence (personal communication, 1994), we know that such collectively autocatalytic sets of molecules can be constructed de novo. The phase transition theory I have outlined suggests that sufficiently complex systems of catalytic polymers should "crystallize" connected, collectively autocatalytic webs of reactions as an emergent, spontaneous property, without the chemist's intelligent design of the web structure.

The emergence of collective autocatalysis depends upon how easy it is to generate polymers able to function both as substrates and as catalysts. This should not be extremely difficult. The existence of catalytic antibodies suggests that finding an antibody capable of catalyzing an arbitrary reaction might require searching about a million to a billion antibody molecules. The binding site in the V region of an antibody molecule is nearly a set of several random peptides, corresponding to the complement determining regions, held in place by the remaining framework. Thus, libraries of more or less random peptides or polypeptides are reasonable candidates to serve as both substrates and catalysts. Indeed, recent work in collaboration with my graduate student Thomas LaBean and Tauseef Butt has shown that such random polypeptides tend readily to fold into a molten globule state, many of which show cooperative unfolding and refolding in graded denaturing conditions, suggesting modest folding capacities may be common in amino acid sequences (LaBean *et al.* 1992, 1995). The results also suggest that random polypeptides may well exhibit a variety of liganding and catalytic functions. Earlier evidence in support of this rests on the display of random hexapeptides on the outer coat of filamentous phage. The probability of finding a peptide able to bind a monoclonal antibody molecule raised against another peptide is about one in a million (Devlin *et al.* 1990; Scott & Smith 1990; Cwirla *et al.* 1990). Since binding a ligand and binding the transition state of a reaction are similar, these results, coupled with the success in finding catalytic antibodies, suggest that random peptides may rather readily catalyse reactions among peptides or other polymers. Random RNA sequences are interesting candidates as well. Recent results searching random RNA libraries for sequences which bind an arbitrary ligand suggest the probability of success is about one in a billion (Ellington & Szostak 1990). Even more recent results seeking RNA sequences able to catalyse a reaction suggest a probability of about one in a trillion. It may prove easier to find random peptide sequences able to catalyse an arbitrary reaction. These results, coupled with rough estimates of the number of reactions such systems afford, suggest that diversities of perhaps 100,000 to 1,000,000 polymer sequences of length 100 might achieve collective autocatalysis.

THE SOURCES OF DYNAMICAL ORDER

If Schrödinger's suggestion is not necessary to the emergence of life, then might it at least be the case that the aperiodic solid of DNA is either necessary or sufficient to ensure heritable variation? The answer, I shall try to show in more detail than sketched above, is "no." The microcode enabled by the large aperiodic solid is clearly not sufficient to ensure order. The genome specifies a vast parallel processing network of activities. The dynamical behavior of such a network could be catastrophically chaotic, disallowing any selectable heritability to the wildly varying behaviors of the encoded system. Encoding in a stable structure such as DNA cannot, by itself, ensure that the system encoded behaves with sufficient order for selectable heritable variation. Further, I shall suggest that encoding in a large stable aperiodic solid such as DNA is not necessary to achieve the stable dynamical behavior requisite for selectable heritable variation of either primitive collectively autocatalytic sets, or of more advanced organisms. What may be required instead is that the system be a certain kind of open thermodynamic system capable of exhibiting powerful convergence in its state space towards small, stable dynamical attractors. The open system, in another view, must be able to discard information, or degrees of freedom, rapidly enough to offset thermal and other fluctuations.

I now summarize briefly the behavior of random Boolean networks. These networks were first introduced as models of the genomic regulatory systems coordinating the activities of the thousands of genes and their products within each cell of a developing organism (Kauffman 1969). Random Boolean networks are examples of highly disordered, massively parallel processing, non-equilibrium systems, and have become the subject of increased interest among physicists, mathematicians, and others (Kauffman 1984, 1986, 1993; Derrida & Pommeau 1986; Derrida & Weisbuch 1986; Stauffer 1987).

Random Boolean networks are open thermodynamic systems, displaced from equilibrium by an exogenous source of energy. The networks are systems of binary, on–off variables, each of which is governed by a logical switching rule, called a Boolean function. Boolean functions are named in honour of George Boole, a British logician who invented mathematical logic in the last century. Thus, one binary variable might receive inputs from two others, and be active at the next moment only if both input one AND input two were active the moment before. This is the logical or Boolean AND function. Alternatively, a binary

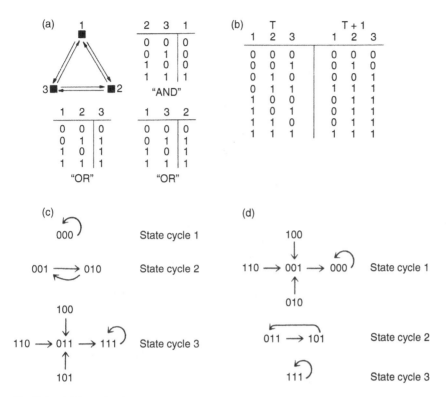

Fig. 30.2. (a) The wiring diagram in a Boolean network with three binary elements, each an input to the other two. (b) The Boolean rules of (a) rewritten to show, for all $2^3 = 8$ states at time T, the activity assumed by each element at the next moment, $T + 1$. Read from left to right, this figure shows the successor state for each state. (c) The state transition graph, or behavior field, of the autonomous Boolean network of (a) and (b) obtained by showing state transitions to successor states connected by arrows. (d) Effects of mutating the rule of element 2 from OR to AND.

variable with two inputs might be active at the next moment if either one OR the other or both of the two inputs is active at the current moment. This is the Boolean OR function.

Fig. 30.2a–c shows a small Boolean network with three variables, each receiving inputs from the other two. One variable is assigned the AND function, the other two are assigned the OR function. In the simplest class of Boolean networks, time is synchronous. At each clocked moment, each element assesses the activities of its inputs, looks up the proper response in its Boolean function, and assumes the specified value. Also, in the simplest case, the network receives no inputs from outside. Its behavior is fully autonomous.

While Fig. 30.2a shows the wiring diagram of interconnections among the three variables, and the Boolean logical rule governing each, Fig. 30.2b shows the same information in a different format. Define a state of the entire network as the current activities of all the binary variables. Thus, if there are N binary variables, then the number of states is just 2^N. In the present case, with 3 variables, there are just 8 states. The set of possible states of the network constitutes its state space. The left column of Fig. 30.2b shows these 8 states. The right column shows the response, at the next moment, of each variable for every possible combination of activities of its inputs. However, another way of reading Fig. 30.2b is to realize that the rows of the right half of the figure correspond to the next activities of all 3 variables. Hence read from left to right, Fig. 30.2b specifies for each state of the entire network what its successor state will be.

Figure 30.2c shows the integrated dynamical behavior of the entire network. This figure is derived from Fig. 30.2b by drawing an arrow from each state to its unique successor state. Since each state has a unique successor state, the system will follow a trajectory of states in its state space. Since there is a finite number of states, the system must eventually re-enter a state previously encountered. But then, since each state has a unique successor state, the system will thereafter cycle repeatedly around a recurrent cycle of states, called a state cycle.

Many important properties of Boolean networks concern the state cycles of the system, and the character of trajectories flowing to such state cycles. Among these properties, the first is the length of a state cycle, which might be a single state which maps to itself forming a steady state, or the state cycle might orbit through all the states of the system. The state cycle length gives information about the recurrence time of patterns of activities in the network. Any Boolean network must have at least one state cycle, but may have more than one. The network in Fig. 30.2c has three state cycles. Each state lies on a trajectory which flows to or is part of exactly one state cycle. Thus, state cycles drain a volume of state space, called a basin of attraction. The state cycle itself is called an attractor. A rough analogy sets state cycles equal to lakes and a basin of attraction equal to the drainage basin flowing into any single lake.

Examination of Fig. 30.2c shows that trajectories converge. Trajectories converge onto one another either before reaching a state cycle, or, of course, converge when they reach the state cycle. This means that these systems throw away information. Once two trajectories have converged, the system no longer has any information to discriminate the pathway by which it arrived at its current state. Consequently, the higher the convergence in state space, the more information the system is discarding. We shall see in a moment that this erasure of the past is essential to the emergence of order in these massive networks.

Another property of interest concerns the stability of state cycles to minimal perturbations, transiently reversing the activity of any single variable. Examination of Fig. 30.2c shows that the first state cycle is unstable to all such perturbations. Any such perturbation leaves the system in the basin of attraction of a different attractor to which the system then flows. In contrast, the third state cycle is stable to any minimal perturbation. Each such perturbation leaves the system in the same basin of attraction, to which the system returns after the perturbation.

THE CHAOTIC, ORDERED, AND COMPLEX REGIMES

After nearly three decades of study, it has become clear that large Boolean networks generically behave in one of three regimes, a chaotic regime, an ordered regime, and a complex regime in the vicinity of the transition between order and chaos. Of the three, perhaps the spontaneous emergence of an ordered regime coordinating the activities of thousands of binary variables is the most stunning for our current purposes. Such spontaneous collective order, I believe, may be one of the deepest sources of order in the biological world.

I shall describe the chaotic regime, then the ordered regime, and end with the complex regime.

Before proceeding, it is important to characterize the kind of questions that are being posed. I am concerned to understand the typical, or generic, properties of large Boolean networks in different classes of networks. In concrete terms, I shall be concerned with networks with a large number of binary variables, N. I shall consider networks classified by the number of inputs per variable, K. And I shall consider networks with specific biases on the set of possible Boolean functions of K inputs. We shall see that, if K is low, or if certain biases are utilized, then even vast Boolean networks, linking the activities of thousands of variables, will lie in the ordered regime. Thus, control of a few simple construction parameters suffice to ensure that typical members of the class, or ensemble, exhibit order. The evolutionary implication is immediate: achieving coordinated behavior of very large numbers of linked variables can be achieved by tuning very simple general parameters of the overall system. Large scale dynamical order is far more readily available than we have supposed.

The aim to study the generic properties of classes, or ensembles, of networks demands that random members of that ensemble be sampled for investigation. Analysis of many such random samples then leads to an understanding of the typical behaviors of members of each ensemble. Thus, we shall be considering randomly constructed Boolean networks. Once constructed, the wiring diagram and logic of the network is fixed.

We consider first the limiting case where $K=N$. Here each binary variable receives inputs from itself and all other binary variables. There is, consequently, only a single possible wiring diagram. However, such systems can be sampled at random from the ensemble of possible $K=N$ networks by assigning to each variable a random Boolean function on its N inputs. Such a random function assigns, at random, a 1 or a 0 response to each input configuration. Since this is true for each of the N variables, a random $K=N$ network assigns a successor state at random from among the 2^N states, to each state. Thus, $K=N$ networks are random mappings of the 2^N integers into themselves.

The following properties obtain in $K=N$ networks. First, the expected median state cycle length is the square root of the number of states. Pause to think of the consequences. A small network with 200 variables would then have state cycles of length 2^{100}. This is approximately 10^{30} states. If it required a mere microsecond for the system to pass from state to state, it would require some billions of times the history of the universe since the big bang 14 billion years ago to orbit the state cycle.

The long state cycles in $K=N$ networks allow me to make a critical point about Schrödinger's argument. Think of the human genome. Each cell in a human body encodes some 100,000 genes. As we all know, genes regulate one another's activities via a web of molecular interactions. Transcription is regulated by sequences of DNA such as *cis* acting promoters, TATA boxes, enhancers, and so forth. The activities of *cis* acting sites, in turn, are controlled by transacting factors, often proteins encoded by other genes, which diffuse in the nucleus or cell, bind to the *cis* acting site, and regulate its behavior. Beyond the genome, translation is regulated by a network of signals, as are the activities of a host of enzymes, whose phosphorylation states govern catalytic and binding activities. In turn, the phosphorylation state is controlled by other enzymes, kinases, and phosphatases which are themselves phosphorylated and dephosphorylated. The genome and its direct and indirect product, in short, constitute an intricate web of molecular interactions. The coordinated behavior of this system controls cell behavior and ontogeny.

Suppose the genome specified regulatory networks which were similar to a $K=N$ network. The time scale to turn a gene on or off is on the order of a minute to perhaps 10 minutes. Let us retain the idealization that genes and other molecular components of the genomic regulatory system are binary variables. A genome with 100,000 genes, harbouring the complexity of the human genome, is capable of a mind boggling diversity of patterns of gene expression: $2^{100,000}$. The expected state cycle attractors of such a system would be a "mere" $2^{50,000}$ or $10^{15,000}$. To sketch the scale, recall that a tiny model genome with only 200 binary variables would require billions of times the age of the universe to traverse its orbit; $10^{15,000}$ is not a number whose meaning we can even roughly fathom. But no organism could be based on state cycles of such unimaginably vast periods.

In short, were the human genome, duly encoded by an aperiodic solid called DNA, to specify a $K=N$ genomic regulatory system, the order enshrined in the aperiodic solid would beget dynamic behavior of no possible biological relevance. Selection due to heritable variations requires a repeated phenotype upon which to operate. A genomic system whose gene activity patterns were a succession of randomly chosen states that only repeated in $10^{15,000}$ steps could not exhibit such a repeated phenotype upon which selection could usefully operate.

$K=N$ networks have state cycles whose expected length scales exponentially with the size of the system. I shall use this scaling to denote one aspect of the chaotic behavior of such networks.

But there is another sense of chaos, closer to the familiar one, which $K=N$ networks exhibit. Such networks show overwhelming sensitivity to initial conditions. Tiny changes in initial conditions lead to massive changes in the subsequent dynamics. The successor state to each state is randomly chosen among the possible states. Consider two initial states which differ in the activity of only one of the N binary variables. The states (000000) and (000001) are an example. The Hamming distance between two binary states is the number of bits which are different. Here, the Hamming distance is 1. If the Hamming distance is divided by the total number of binary variables, 6 in this example, the fraction of sites which are different, here $1/6$, is a normalized Hamming distance. Consider our two initial states differing by a single bit. Their successor states are randomly chosen among the possible states of the network. Hence the expected Hamming distance between the successor states is just half the number of binary variables. The normalized distance jumps from $1/N$ to $1/2$ in a single state transition. In short, $K=N$ networks show the maximum possible sensitivity to initial conditions.

Continuing my disagreement, if such it be, with the thrust of Schrödinger's book, were the human genome a $K = N$ network, not only would its attractor orbits be hyperastronomically long, but the smallest perturbations would lead to catastrophic alterations in the dynamical behavior of the system. Once we have the counterexample of the ordered regime, it will become intuitively obvious that $K = N$ systems, deep in the chaotic regime, cannot be the way the genomic regulatory system is organized. Most importantly, selection operates on heritable variations. In $K = N$ networks, minor alterations in network structure or logic also wreak havoc with all the trajectories and attractors of the system. For example, deletion of a single gene eliminates half the state space, namely those in which that gene is active. This results in a massive reorganization of the flow in state space. Biologists wonder about possible evolutionary pathways via "hopeful monsters." Such pathways are highly improbable. $K = N$ networks could only evolve by impossibly hopeful monsters. In short, $K = N$ networks supply virtually no useful heritable variation upon which selection can act.

A word is needed about the term "chaos." Its definition is clear and established for systems of a few continuous differential equations. Such low dimensional systems fall onto "strange attractors" where local flow is divergent but remains on the attractor. It is not clear, at present, what the relation is between such low dimensional chaos in continuous systems and the high dimensional chaos I describe here. Both behaviors are well established, however. By high dimensional chaos, I shall mean systems with a large number of variables in which the lengths of orbits scale exponentially with the number of variables, and which show sensitivity to initial conditions in the sense defined above.

Order for free: Despite the fact that Boolean networks may harbour thousands of binary variables, unexpected and profound order can emerge spontaneously. I believe this order is so powerful that it may account for much of the dynamical order in organisms. Order emerges if very simple parameters of such networks are constrained in simple ways. The simplest parameter to control is K, the number of inputs per variable. If $K = 2$ or less, typical networks lie in the ordered regime.

Imagine a network with 100,000 binary variables. Each has been assigned at random $K = 2$ inputs. The wiring diagram is a mad scramble of interconnections with no discernible logic, indeed with no logic

whatsoever. Each binary variable is assigned at random one of the 16 possible Boolean functions of two variables, AND, OR, IF, Exclusive OR, etc. The logic of the network itself is, therefore, entirely random. Yet order crystallizes.

The expected length of a state cycle in such networks is not the square root of the number of states, but on the order of the square root of the number of variables. Thus, a system of the complexity of the human genome, with some 100,000 genes and $2^{100,000}$ states, will meekly settle down and cycle among a mere 317 states. And 317 is an infinitesimal subset of the set of $2^{100,000}$ possible states. The relative localization in state space is on the order of $2^{-99,998}$.

Boolean networks are open thermodynamic systems. In the simplest case, they can be constructed of real logic gates, powered by an exogenous electrical source. Yet this class of open thermodynamic systems shows massive convergence in state space. This convergence shows up in two ways. Overall, such systems exhibit a profound lack of sensitivity to initial conditions. The first signature of convergence is that most single bit perturbations leave the system on trajectories which later converge. Such convergence occurs even before the system has reached attractors. Secondly, perturbations from an attractor typically leave the system in a state which flows back to the same attractor. The attractors, in biological terms, spontaneously exhibit homeostasis. Both signatures of convergence are important. The stability of attractors implies repeatable behavior in the presence of noise. But convergence of flow even before attractors are reached implies that systems in the ordered regime can react to similar environments in "the same" way, even if ongoing perturbations by the environmental inputs persistently prevent the system from attaining an attractor. Convergence along trajectories should allow such systems to adapt successfully to a noisy environment.

Such homeostasis, reflecting convergence in state space, stand in sharp contrast to the perfect conservation of phase volume in closed, equilibrium thermodynamic systems. Recall that Liouville's theory ensures such conservation, which, in turn, reflects the reversibility of closed systems and the failure to throw away information into a heat bath. This conservation then underlies the capacity to predict probabilities of macrostates by the fractional number of microstates contributing to each macrostate.

The more important implication of conservation of phase volume in equilibrium systems is the following: Schrödinger correctly drew our attention to the fact that fluctuations in any classical system vary inversely with the square root of the number of events considered. When the system is an equilibrium system, these fluctuations have a given amplitude. However, if we consider an open thermodynamic system with massive convergence in state space, then that convergence tends to offset the fluctuations. The convergence tends to squeeze the system towards attractors, while the fluctuations tend to drive the system randomly in its space of possibilities. But if the convergence is powerful enough, it can confine the noise induced wandering to remain in the infinitesimal vicinity of the attractors of the system. Thus, we arrive at a critical conclusion. The noise induced fluctuations due to small numbers of molecules which concerned Schrödinger can, in principle, be offset by the convergent flow towards attractors if that flow is sufficiently convergent. Homeostasis can overcome thermalization.

But this conclusion is at the heart of the issue I raise with Schrödinger. For I want to suggest the possibility that the use by organisms of an aperiodic solid as the stable carrier of genetic information is not sufficient to ensure order. The encoded system might be chaotic. Nor is the aperiodic solid necessary. Rather, the convergent flow of systems in the ordered regime is both necessary and sufficient for the order required.

LATTICE BOOLEAN NETWORKS AND THE EDGE OF CHAOS

A simple modification of random Boolean networks helps understand the ordered, chaotic, and complex regimes. Instead of thinking of a random wiring diagram, consider instead a square lattice, where each site has inputs from its four neighbors. Endow each binary valued site with a random Boolean function on its four inputs. Start the system in a randomly chosen initial state, and allow the lattice to evolve forward in time. At each time step, any variable may change value from 1 to 0 or 0 to 1. If so, color that variable green. If the variable does not change value, but remains 1 or remains 0, color it red. Green means the variable is "unfrozen" or "moving;" red means the variable has stopped moving and is "frozen."

Random lattice networks with four inputs per variable lie in the chaotic regime. As one watches the lattice, most of the sites remain green; a few become red. More precisely a green unfrozen "sea" spans or percolates across the lattice, leaving behind isolated red frozen islands.

I now introduce a simple bias among all possible Boolean functions. Any such function supplies an output value, 1 or 0, for each combination of values of its K inputs. The set of output values might be nearly half 1 and half 0 values, or might be biased towards all 1 values or all 0 values. Let P measure this bias. P is the fraction of input combination which give rise to the more frequent value, whether it be 1 or 0. For example, for the AND function, three of the four input configurations yield a 0 response. Only if both inputs are 1 is the regulated variable 1 at the next moment. P is therefore 0.75 in this case. Thus, P is a number between 0.5 and 1.0.

Derrida and Weisbuch (1986) showed that a Boolean lattice will lie in the ordered regime if the Boolean functions assigned to its sites are randomly chosen, with the constraint that the P value at each site is closer to 1.0 than a critical value. For a square lattice the critical value, P_c, is 0.72.

Consider a similar "movie" of a network in the ordered regime, in which moving sites are again colored green, and frozen sites are colored red. If P is greater than P_c, then at first most sites are green. Soon, however, an increasing number of sites freeze into their dominant value, 1 or 0, and are colored red. A vast red frozen sea spans or percolates across the lattice, leaving behind isolated green islands of unfrozen variables which continue to twinkle on and off in complex patterns. The percolation of a red frozen sea leaving isolated unfrozen green islands is characteristic of the ordered regime.

A phase transition occurs in such lattice Boolean networks as P is tuned from above to below P_c. As the phase transition is approached from above, the green unfrozen islands grow larger and eventually fuse with one another to form a percolating green unfrozen sea. The phase transition occurs just at this fusion.

With this image in mind, it becomes useful to define "damage." Damage is the propagating disturbances in the network after transiently reversing the activity of a site. To study this, it suffices to make two identical copies of the network, and initiate them in two states which differ in the activity of a single variable. Watch the two copies, and color purple any site in the perturbed copy which is ever in a different activity value than its unperturbed copy. Then a purple stain spreading outward from the perturbed site demarcates the spreading damage from that site.

In the chaotic regime, let a site in the green percolating unfrozen sea be damaged. Then, generically, a purple stain spreads throughout most of the green sea. Indeed, the expected size of the damaged volume scales with the size of the total lattice system (Stauffer 1987). Conversely, damage a site in the ordered regime. If that site lies in the red frozen structure, virtually no damage spreads outward. If the site lies in one of the green unfrozen islands, damage may spread throughout much of that island, but will not invade the red frozen structure. In short, the red frozen structure blocks the propagation of damage, and thus provides much of the homeostatic stability of the system.

At the phase transition, the size of distribution of damage avalanches is expected to be a power law, with many small avalanches and few large ones. The phase transition is the complex regime. In addition to the characteristic size distribution of damage avalanches, the mean convergence along trajectories which are near Hamming neighbors is zero. Thus, in the chaotic regime, initial states which are near Hamming neighbors tend, on average, to diverge from one another as each flows along its own trajectory. This is the "sensitivity to initial conditions" of which I have spoken. In the ordered regime, nearby states tend to converge towards one another, often flowing into the same trajectory before reaching a common attractor. At the edge of chaos phase transition, on average, nearby states neither converge nor diverge.

It is an attractive hypothesis that complex adaptive systems may evolve to the complex regime at the edge of chaos. The properties of the edge of chaos regime have suggested to a number of workers (Langton 1986, 1992; Packard 1988; Kauffman 1993) that the phase transition, or edge of chaos, regime may be well suited to complex computations. On the face of it, the idea is attractive. Suppose one wished such a system to coordinate complex temporal behavior of widely separated sites. Deep in the ordered regime, the green islands which might carry out a sequence of changing activities are isolated from one another. No coordination among them can occur. Deep in the

chaotic regime, coordination will tend to be disrupted by any perturbations which unleash large avalanches of change. Thus, it is very plausible that near the phase transition, perhaps in the ordered regime, the capacity to coordinate complex behaviors might be optimized.

It would be fascinating if this hypothesis were true. We would begin to have a general theory about the internal structure and logic of complex, parallel processing, adaptive systems. According to this theory, selective adaptation for the very capacity to coordinate complex behavior should lead adaptive systems to evolve to the phase transition itself, or to its vicinity.

Tentative evidence is beginning to support the hypothesis that complex systems may often evolve, not precisely to the edge of chaos, but to the ordered regime near the edge of chaos. In order to test this, my colleagues and I at the Santa Fe Institute are allowing Boolean networks to coevolve with one another to "play" a variety of games. In all cases, the games involve sensing the activities of the other networks' elements and mounting an appropriate response to some of a network's own output variables. The coevolution of these networks allows them to alter K, P, and other parameters in order to optimize success at each game by natural selection. In brief summary, such networks do improve at the set of games we have asked them to perform. As always, such an evolutionary search takes place in the presence of mutational random drift processes which tend to disperse an adapting population across the space of possibilities it is exploring. Despite this drift tendency, there is a strong tendency to evolve toward a position within the ordered regime not too far from the transition to chaos. In short, tentative evidence supports the hypothesis that a large variety of parallel processing systems will evolve to the ordered regime near the phase transition in order to coordinate complex tasks.

Future work in this arena will examine the question central to that which I raise with Schrödinger. Two sources of "noise" might occur in such game-playing Boolean networks. The first derives from inputs arriving from other networks. These exogenous inputs drive each system from its current trajectory, and hence perturb its flow towards attractors. The second is thermal noise within any one network. That internal noise will tend to perturb the behavior of the system. In order to compensate and achieve coordination, such systems would be expected to shift deeper into the ordered regime. There the convergence in state space is stronger, and hence provides a more powerful buffer against exogenous noise. Thus, we can ask: how much convergence is required to offset a given amount of internal noise?

The same issue arises in any system whose dynamic behavior is controlled by small numbers of copies of each kind of molecule. This occurs in contemporary cells, where the number of regulatory proteins and other molecules per cell are often in the range of a single copy. The same issue arises in the collectively autocatalytic molecular systems which I suspect may have formed at the dawn of life. How much convergence in state space offsets fluctuations due to the use of small numbers of molecules in a dynamical system, and how does the requisite convergence scale with the decrease in the number of copies of each kind of molecule in the model system? With respect to collectively autocatalytic sets of molecules, presumably some sufficiently high convergence in state space will buffer such systems from fluctuations due to the potentially small numbers of copies of each kind of molecule in the collectively reproducing metabolism. If so, the stable structure of large aperiodic solids is neither necessary nor sufficient to the order required for the emergence of life or those heritable variations upon which selection can successfully act.

ORDER AND ONTOGENY

We have seen that even random Boolean networks can spontaneously exhibit an unexpected and high degree of order. It would simply be foolish to ignore the possibility that such spontaneous order may play a role in the emergence and maintenance of order in ontogeny. While the evidence is still tentative, I believe the hypothesis finds considerable support. I shall briefly describe the evidence that genomic regulatory networks actually lie in the ordered regime, perhaps not too far from the edge of chaos. First, if one examines known regulated genes in viruses, bacteria, and eucaryotes, most are directly controlled by few molecular inputs, typically from 0 to perhaps 8. It is fascinating that, in the on–off Boolean idealization, almost all known regulated genes are governed by a biased subset of the possible Boolean functions which I long ago named "canalizing" functions (Kauffman 1971, 1993; Kauffman and Harris, 1994). Here, at least one molecular

input has one value, 1 or 0, which alone suffices to guarantee that the regulated locus assumes a specific output state, either 1 or 0. Thus, the OR function of four inputs is canalizing, since the first input, if active, guarantees that the regulated element be active regardless of the activities of the other three inputs. Boolean networks with more than $K=2$ inputs per element, but confined largely to canalizing functions, genetically lie in the ordered regime (Kauffman 1993). I have, for some years, interpreted the attractors of a genetic network, the state cycles, as the cell types in the repertoire of the genomic system. Then the lengths of state cycles predict that cell types should be very restricted recurrent patterns of gene expression, and also predict that cells should cycle in hundreds to thousands of minutes. Further, the number of attractors scales as the square root of the number of variables. If an attractor is a cell type, we are led to predict that the number of cell types in an organism should scale as about the square root of the number of its genes. This appears to be qualitatively correct. Humans, with about 100,000 genes, would be predicted to have about 317 cell types. In fact humans are said to have 256 cell types (Alberts *et al.* 1983) and the number of cell types appears to scale according to a relationship that lies between a linear and a square root function of genetic complexity (Kauffman 1993). The model predicts other features such as the homeostatic stability of cell types. The frozen red component predicts, correctly, that about 70 percent of the genes should be in the same fixed states of activity on all cell types in the organism. Further, the sizes of green islands predict reasonably well the differences in gene activity patterns in different cell types of one organism. The size distribution of avalanches seems likely to predict the distribution of cascading alterations in gene activities after perturbing the activities of single randomly chosen genes. Finally, in the ordered regime, perturbations can only drive the system from one attractor to a few others. If attractors are cell types, this property predicts that ontogeny must be organized around branching pathways of differentiation. No cell type should, nor indeed can, differentiate directly to all cell types. Here is a property which presumably has remained true of all multicelled organisms since the Cambrian period or before.

A brief presentation of these ideas is all space allows. However, a fair summary at present is that genomic regulatory systems may well be parallel processing systems lying in the ordered regime. If so, then the characteristic convergence in state space of such systems is a major source of their dynamical order.

But there is a more dramatic implication of the self-organization I discuss here. Since Darwin we have come to believe that selection is the sole source of order in biology. Organisms, we have come to believe, are tinkered together contraptions, *ad hoc* marriages of design principles, chance, and necessity. I think this view is inadequate. Darwin did not know the power of self-organization. Indeed, we hardly glimpse that power ourselves. Such self-organization, from the origin of life to its coherent dynamics, must play an essential role in this history of life, indeed, I would argue, in any history of life. But Darwin was also correct. Natural selection is always acting. Thus, we must rethink evolutionary theory. The natural history of life is some form of marriage between self-organization and selection. We must see life anew, and fathom new laws for its unfolding.

SUMMARY

Schrödinger, writing before he had any right to have guessed so presciently, correctly foresaw that current life is based on the structure of large aperiodic solids. The stability of those solids, he foresaw, would provide the stable carrier material of genetic information. The microcode within such material would specify the organism. Quantum alterations in the material would be discrete, rare, and constitute mutations. He was correct about much of contemporary life.

But at a more fundamental level, was Schrödinger correct about life itself? Is the structural memory of the aperiodic solid necessary for all life? Surely, in the minimum sense that organic molecules with covalent bonds are small "aperiodic solids," Schrödinger's argument has general merit. At least for carbon based life, one needs bonds of sufficient strength to be stable in a given environment But it is the behaviors of collections of those molecules which constitute life on Earth and at least we may presume underlies many potential forms of life anywhere in the universe. Living organisms are, in fact, collectively autocatalytic molecular systems. New evidence and theory, adduced above, suggest that the emergence of self-reproducing molecular systems does not require large aperiodic solids. Limited evolution of such systems does not, in

principle, require large aperiodic solids. Nor is dynamical order and heritable variation assured by an aperiodic solid which encodes the structure and some of the interactions of a large number of other molecules. Rather, heritable variation in self-reproducing chemical systems upon which natural selection can plausibly act requires dynamical stability. This, in turn, can be achieved by open thermodynamic systems which converge sufficiently in their state spaces to offset the fluctuations which derive from the fact that only small numbers of molecules are involved.

It is no criticism of Schrödinger that he did not consider the self-organized behaviors of open thermodynamic systems. Study of such systems had hardly begun fifty years ago and is not much advanced today. Indeed, all we can genuinely say at present is that the kinds of self-organization which we begin to glimpse in such open thermodynamic systems may be changing our view of the origin and evolution of life. It is enough that Schrödinger foresaw so much. We can only wish his wisdom were alive today to help further his and our story.

NOTES

This chapter originally appeared in Michael P. Murphy and Luke A. J. O'Neill (Eds.), *What is life? The next fifty years: Speculations on the future of biology*, pp. 83–114 Cambridge, UK: Cambridge University Press, 1997.

REFERENCES

1. Alberts, B., Bray, D., Lewis, J., Raff, M., Roberts, K., & Watson, J. D. (1983). *Molecular biology of the cell.* New York: Garland.

2. Bagley, R. J. (1991). The functional self-organization of autocatalytic networks in a model of the evolution of biogenesis. Ph.D. thesis, University of California, San Diego.

3. Bagley, R. J., Farmer, J. D., & W. Fontana (1992). Evolution of a metabolism. In C. G. Langton, C. Taylor, J. D. Farmer, and S. Rasmussen (Eds.). *Artificial life II* (Santa Fe Institute studies in the sciences of complexity, proceedings vol, X) (pp. 371–408). Redwood City, CA: Addison-Wesley.

4. Ballivet, M. & Kauffman, S. A. (1985). *Process for obtaining DNA, RNA, peptides, polypeptides or proteins by*

recombinant DNA techniques. International Patent Application, granted in France (1987), United Kingdom (1989), and Germany (1990).

5. Cohen, J. E. (1988). Threshold phenomena in random structures. *Discrete Applied Mathematics*, **19**, 113–118.

6. Cwirla, P., Peters, E. A., Barrett, R. W., & Dower, W. J. (1990). Peptides on phages: A vast library of peptides for identifying ligands. *Proceedings of the National Academy of Sciences*, **87**, 6378–6382.

7. Derrida, B. & Pommeau, Y. (1986). Random networks of automata: A simple annealed approximation. *Europhysics Letters*, **1**, 45–49.

8. Derrida, B. & Weisbuch, G. (1986). Evolution of overlaps between configurations in random Boolean networks. *Journal de Physique*, **47**, 1297–1303.

9. Devlin, J. J., Panganiban, L. C, & Devlin, P. A. (1990). Random peptide libraries: A source of specific protein binding molecules. *Science*, **249**, 404–406.

10. Eigen, M. (1971). Self-organization of matter and the evolution of biological macromolecules. *Naturwissenschaften*, **58**, 465–523.

11. Ellington, A. & Szostak, J. (1990). In vitro selection of RNA molecules that bind specific ligands. *Nature*, **346**, 818–822.

12. Erdos, P. & Renyi, A. (1960). *On the evolution of random graphs (publication no. 5)*. Budapest: Institute of Mathematics, Hungarian Academy of Sciences.

13. Farmer, J. D., Kauffman, S. A., & Packard, N. H. (1986). Autocatalytic replication of polymers. *Physica*, **22D**, 50–67.

14. Kauffman, S. A. (1969). Metabolic stability and epigenesis in randomly connected nets. *Journal of Theoretical Biology*, **22**, 431–467

15. Kauffman, S. A. (1971). Cellular homeostasis, epigenesis and replication in randomly aggregated macromolecular systems. *Journal of Cybernetics*, **1**, 71.

16. Kauffman, S. A. (1984). Emergent properties in random complex automata. *Physica*, **10D**, 145–156.

17. Kauffman, S. A. (1986). Autocatalytic sets of proteins. *Journal of Theoretical Biology*, **119**, 1–24.

18. Kauffman, S, A. (1993). *The origins of order: Self organization and selection in evolution*. New York: Oxford University Press.

19. Kauffman, S. A. & Harris, S. (1994). *Manuscript in preparation*.

20. LaBean, T. H., Kauffman, S. A., & Butt, T. R. (1992). Design, expression and characterization of random sequence polypeptides as fusions with ubiquitin. *FASEB Journal*, **6**, A471.

21. LaBean, T. H., Kauffman, S. A., & Butt, T. R. (1995). Libraries of random-sequence polypeptides produced with high yield as carboxy-terminal fusions with ubiquitin. *Molecular Diversity*, **1**, 29–38.

22. Langton, C. (1986). Studying artificial life with cellular automata. *Physica*, **22D**, 120–149.

23. Langton, C. (1992). *Adaptation to the edge of chaos. Artificial life II (Santa Fe Institute studies in the sciences of complexity, proceedings vol. X) (pp. 11–92)*. Redwood City, CA: Addison-Wesley.

24. Orgel, L. E. (1987). Evolution of the genetic apparatus: A review. *Cold Spring Harbor Symposium on Quantitative Biology*, **52**, 9–15.

25. Packard, N. (1988). Dynamic patterns in complex systems. In J. A. S. Kelso and M. Shlesinger (Eds.), *Complexity in biologic modeling* (pp. 293–301). Singapore: World Scientific.

26. Rossler, O. (1971). A system-theoretic model of biogenesis. *Zeitschrift für Naturforschung A*, B**266**, 741.

27. Schrödinger, E. (1967). *What is life? with mind and matter and autobiographical sketches*. Cambridge, UK: Cambridge University Press.

28. Scott, J. K. & Smith, G. P. (1990). Searching for peptide ligands with an epitope library. *Science*, **249**, 386.

29. Stauffer, D. (1987). Random Boolean networks: Analogy with percolation. *Philosophical Magazine B*, **56**, 901–916.

30. von Kiederowski, G. (1986). A self-replicating hexadesoxynucleotide. *Angewandte Chemie International Edition in English*, **25**, 932–935.

31 · Four puzzles about life

MARK A. BEDAU

WHAT EXPLAINS THE PHENOMENA OF LIFE

Life seems to be one of the most basic kinds of actual natural phenomena. A bewildering variety of forms of life surrounds us, but we usually have no difficulty distinguishing the living from the nonliving. That flower, that mushroom, that worm, that bird are alive; that rock, that mountain, that river, that cloud are not. Just as any attempt to divide nature at its joints must account for mind and matter, so it must account for life.

Yet it is notoriously difficult to say what life is, exactly. Many have noted this (e.g., ref. 1); Farmer and Belin (ref. 2, p. 818) put the point this way:

> There seems to be no single property that characterizes life. Any property that we assign to life is either too broad, so that it characterizes many nonliving systems as well, or too specific, so that we can find counter-examples that we intuitively feel to be alive, but that do not satisfy it.

The fact today is that we know of no set of individually necessary and jointly sufficient conditions for life.

Nevertheless, there is broad agreement that life forms share certain distinctive hallmarks. Various hallmarks are discussed in the literature, and Mayr's list[3] is representative and influential:

1. All levels of living systems have an enormously complex and adaptive organization.
2. Living organisms are composed of a chemically unique set of macromolecules.
3. The important phenomena in living systems are predominantly qualitative, not quantitative.
4. All levels of living systems consist of highly variable groups of unique individuals.
5. All organisms possess historically evolved genetic programs which enable them to engage in teleonomic processes and activities.
6. Classes of living organisms are defined by historical connections of common descent.
7. Organisms are the product of natural selection.
8. Living processes are especially unpredictable.

I agree with Mayr (ref. 3, p. 59) that the coexistence of these properties "make[s] it clear that a living system is something quite different from any inanimate object," so I suspect that there is some unified explanation of vital phenomena. At the same time, it is puzzling that such heterogeneous properties characteristically coexist in nature, especially because each of the hallmarks can be possessed by nonliving things.

Appearances can be deceptive. Vital phenomena might have no unified explanation and life might not be a basic kind of natural phenomena. Skeptics such as Sober[4] think that the question of the nature of life, in general, has no interesting answer. But I suspect otherwise, along with those (e.g., refs. 5,6) searching for extraterrestrial life; they are not searching just for extraterrestrial metabolizers and self-reproducers. Likewise for those searching for the origin of life on Earth (e.g., refs. 7,8). Likewise for those in the field of artificial life attempting to synthesize life in artificial media.[9] This broadly based search for a unified theory of vital phenomena should retreat to skepticism, if at all, only as a last resort.

So, we face a quandary: We expect there is a unified theory of life but we doubt life has necessary and sufficient conditions. We can resolve this quandary if we reconceive our project in two ways. First, we must focus on the *phenomena* of life, not our current concept of life nor the current meaning of our

The Nature of Life, ed. M. A. Bedau and C. E. Cleland. Published by Cambridge University Press. © M. A. Bedau and C. E. Cleland 2010.

word "life." Physicists and chemists want to explain matter itself, not our current concept of matter or the current meaning of the word "matter." I want to explain life itself; such an explanation is what I mean by a *theory* of life. It does not matter whether this theory supports our current preconceptions about life or fits the current meaning of our word "life." Our current concept of life and the current meaning of our word "life" are contingent. They vary across space and time, changing with different human cultures at different places and in different ages. Our theories are connected with our concepts and words, of course, but the connection goes in the other direction, with our concepts and words following the lead of our currently best theories.

The second step in resolving our quandary is to shift our focus from living organisms to the process that *produces* organisms and other living phenomena. The generating process is primary and its products are secondary, for the process provides a unified explanation of the various products. Understanding how organisms and other living entities actually come into existence is the key to understanding what they are.

I believe that the process of supple adaptation is the primary form of life.[36] I defend this proposal here, on the grounds, not that supple adaptation is a necessary and sufficient condition for living organisms (it is not), nor that it matches our current concept of life or the current meaning of the word "life" (it might not), but that it provides the best unified explanation of the phenomena of life. Theories of life should be judged in part by how well they resolve basic puzzles about life. My specific concern here is how supple adaptation resolves four such puzzles. I propose no complete and final theory of life, nor definitive resolutions to the four puzzles. But I show that supple adaptation provides good explanations of the puzzles.

Can any rival theory explain the four puzzles as well? It is easy to dream up rival theories and to imagine that they have good explanations of the puzzles; it is another thing to support such dreams with substantial evidence. The theory of life as supple adaptation does not automatically fend off rival theories. Another theory that provided equally good explanations of the puzzles would be a serious contender. My goal here is not to show that credible contenders are impossible but to establish what standard they must meet to be taken seriously.

THE THEORY OF LIFE AS SUPPLE ADAPTATION

I propose that an automatic and continually creative evolutionary process of adapting to changing environments is the primary form of life. My proposal is broadly in the spirit of genetic definitions of life;[10] various similar proposals occur in the literature (e.g., refs. 7–9). From my perspective, what is distinctive of life is the way in which evolution automatically fashions and refashions appropriate strategies for coping as local contexts change.

The notion of propriety involved in supple adaptation is to be understood teleologically. A response is appropriate only if it promotes and furthers the adapting entity's intrinsic goals and purposes, where those goals and purposes are minimally to survive and, more generally, to flourish. For example, if a clam's shell becomes cracked, then an inappropriate response would be for the clam's soft tissue to ooze out the crack, and an appropriate response would be for the shell to be repaired. By contrast, although water flowing downhill automatically "adapts" to the local landscape's topography, the water has no intrinsic goals or purposes and flowing downhill serves no such goals or purposes. Similarly, a thermostat has no intrinsic goals or purposes, so its "adaptive responses" to local temperature changes can be considered appropriate only relative to the extrinsic goals or purposes that *we* have in using those artifacts. These teleological notions of intrinsic goals and purposes are certainly controversial, and I will not here rehearse my own attempts to resolve these controversies.[11–13] I trust that their connection to the relevant notion of adaptation is clear enough for present purposes.

My proposal is that the thread unifying the diversity of life is the *suppleness* of this process of producing adaptations—its ongoing and indefinitely creative production of significantly new kinds of adaptive responses to significantly new kinds of adaptive challenges and opportunities. A biological arms race[14] is one simple example of supple adaptation. By contrast, a thermostat's response to the ambient temperature is not "supple" in the relevant sense because it is the "same old" kind of response to the "same old" kind of temperature changes. Because the process of supple adaptation involves significantly new kinds of adaptive challenges and opportunities, those challenges and opportunities will be unanticipated by the adapting entities, and they will elicit an open-ended range of appropriate responses. Phrases such as "open-ended evolution" (ref. 15, p. 310; ref. 16, p. 372) or "perpetual novelty" (ref. 17, p. 184) are sometimes used to refer to this process.

Supple adaptation is not to be equated with natural selection. For one thing, natural selection is not necessary for supple adaptation. Other adaptive mechanisms such as Lamarckian selection or Hebbian learning can produce supple adaptation. For another thing, natural selection is not sufficient for supple adaptation. Supple adaptation is the *ongoing* production of significant adaptive novelty. By contrast, the dynamics of natural selection often eventually stabilize in the long run, with the result that significantly new adaptations stop being produced. Even though new mutations continually occur, they yield at best only insignificantly different variants of familiar adaptations. So, natural selection produces supple adaptation only when it is continually creative. Adaptation cannot be continually creative without ongoing environmental change. One way to bring about ongoing environmental change is for the evolving system's own evolution to continually reshape the selection criteria or fitness function,[18] perhaps through some mechanism like this: The organisms in the evolving system interact through their behavior. Each organism's environment consists to a large degree of its interactions with other organisms. So, if one organism evolves an innovative adaptive behavior, this changes the environment of neighboring organisms. This environmental change in turn causes the neighboring organisms to evolve their own new adaptive behaviors, and this finally changes the environment of the original organism. In this way an organism's adaptive evolution ultimately changes the environment of that very organism. The net effect is that the population's adaptive evolution continually drives its own further adaptation.

I should call attention to the difference between a capacity and its exercise, because I hold that life involves the *exercise* of supple adaptation, not just the *capacity* to do so. For me, the key is not supple *adaptability* but actual supple *adaptation*. A system undergoing supple adaptation is not adapting at every moment, of course—the adaptation occurs in fits and starts. But the quiescent periods between adaptive events are transient; every quiescent period is followed by new adaptive events. If a system that *could* undergo supple adaptation never *does*, then by my lights it *could* be alive but never *is*.

Probably the most controversial feature of my theory of life is the claim that supple adaptation does not merely *produce* living entities: The *primary* forms of life are none other than the supply adapting systems themselves. Other living entities are alive by virtue of bearing an appropriate relationship to a supply adapting system; they are *secondary* forms of life. Different kinds of living entities (organisms, organs, cells, etc.) stand in different kinds of relationships to the supply adapting system from which their life ultimately derives. In general, these relationships are ways in which the entity is created and sustained by a supply adapting system. So, the general form of my theory of life can be captured by this definition:

X is living *iff*

1. *X* is a supply adapting system, or
2. *X* is explained in the right way by a supply adapting system.

The effect of this definition is to construe the primary form of life as supply adapting systems.

According to this definition, individual living organisms, organs, cells, and all the other living things count as alive because they are explained in the right way by a supply adapting system. But the definition does not specify which kinds of explanations are the "right" ones. The explanations typically involve the way in which things are created and sustained, but it is not clear whether this is always true. Furthermore, some ways of being generated and sustained are clearly *not* what is intended by the definition, such as the way in which people create and sustain automobiles and garbage dumps, the way in which spiders create and sustain their webs and beavers their dams. I am leaving these details to be settled by whatever in the future best explains living phenomena, so I am not proposing a complete and final theory here. By claiming that the process of supple adaptation is the central explanatory factor underlying and unifying the various phenomena of life, the definition above delineates the central categories to be used in a final definition and proposes boundaries within which to seek that definition. My aim is not to give a particular definition but to set the stage for one to be produced in the future.

One important virtue of the theory of life as supple adaptation is its unified explanation for Mayr's hallmarks of life. The theory implies that we should expect those heterogeneous-seeming properties to coexist in nature. If life consists of supply adapting systems and the entities they generate and sustain, we should expect life to involve the operation of natural selection producing complex adaptive organization in historically connected organisms with evolved genetic programs. Furthermore, the random variation and historical contingency in supple adaptation explains why living phenomena are especially unpredictable and involve unique and variable individuals. Finally, if supple adaptation is produced by a branching process involving birth, reproduction, and death of individuals, such as natural selection, then we can understand why it would give rise to a wealth of qualitative phenomena characterized by frozen accidents like chemically unique macromolecules. The naturalness of all these explanations supports the theory of life as supple adaptation.

Another consideration in favor of the theory is its natural response to potential criticisms. For example, mules, the last living member of an about-to-be extinct species, neutered and spayed animals are all alive, but being infertile, such entities play no role in the supple adaptation of their lineages. However, infertile organisms exist only because of their connections with other fertile organisms that *do* play an active role in a supply adapting biosphere, so they fall within the scope of my theory.

Some might object that an evolving system's supple adaptation has the wrong logical form to be the nature of life. Individual organisms are the paradigmatic living entities and an evolving population of organisms is of a different logical category than an individual organism. So, one might think that life cannot consist in a population-level property like supple adaptation. Now, individual organisms and populations of organisms are of different categories, to be sure, but phenomena from one category can explain phenomena from other categories. The theory of life as supple adaptation denies that individual organisms are the primary forms of life, but it does so consciously and deliberately, out of the conviction that the process of supple adaptation is our current best hope for unifying and explaining the phenomena of life. If the best explanation for life violates some of

our currently dominant paradigms of life, so much the worse for those paradigms.

The possibility of an ecology that has reached a state of stable equilibrium and stopped adapting forever is a more direct challenge to my theory. After all, the organisms in such so-called "climax" ecosystems are certainly alive, yet the ecosystem containing them is not undergoing supple adaptation, so these organisms would seem to fall outside my theory. However, not only do climax ecosystems originate through a process of supple adaptation, but their quiescent periods are transitory. At least, that is the hypothesis behind the theory of life as supple adaptation. If this hypothesis is false and it turns out that climax ecosystems simply do not exhibit supple adaptation, then the theory of life as supple adaptation is also false. It is an empirical question whether the hypothesis is true. My theory implies not that the hypothesis is analytically true (it is not) nor that it is knowable *a priori* (it is not) but only that the nature of life, in fact, is supple adaptation. Being life's nature, it is an essential property of life and so holds necessarily, but it is a necessity that we learn about *a posteriori* through empirical science.

It is easy to *conceive* of circumstances that violate my account of life. Nothing prevents us from entertaining with Boden[19] the scientific fantasy of species that never evolve and adapt. For all I know, this is possible; that is, it is "epistemically" possible, as Kripke[20] might say. So is the possibility that there has been and ever will be only one living organism. So is the possibility that all organisms were created in seven days by an omnipotent, omniscient, and omnibenevolent deity. But these fantasies are just that— fantasies, with no bearing on the true nature of any form of life that we could discover or synthesize. My claim—*a posteriori* to be sure, but still true, I wager—is that all living organisms anywhere in the universe ultimately derive their existence and their characteristic lifelike features from having the right sort of explanatory connection with a system undergoing supple adaptation.

Are there not counterexamples of supply adapting systems devoid of all life? Viruses are adapting against all our best efforts to eradicate them—the AIDS virus evolves remarkably quickly— and viruses are a classic example of entities on some borderline between life and nonlife. Even less lifelike are populations of the tiny clay crystallites that make up mud, yet these seem to have the flexibility to adapt and evolve by natural selection.[7,12] So do autocatalytic networks of chemical species,[22] yet evolving populations of crystals or chemicals are ordinarily thought to involve no life whatsoever. Even more extreme examples are individual human mental activity and collective human intellectual, social, and economic activities; these all look like supple, open-ended capacities to adapt to unpredictably changing circumstances, yet none would ordinarily be called alive. Intellectual and economic activities are generated by living creatures, but the evolving intellectual or economic systems themselves are not thought to be alive. However, I am not offering supple adaptation as an explication of our current concept of life, so unintuitive classifications are no particular concern. These counterintuitive cases do not undermine the fact that supple adaptation is our best explanation of the phenomena of life. If life is supple adaptation, then virus and clay crystallite populations, autocatalytic chemical networks, and human intellectual and economic systems all deserve to be thought of as living if they exhibit supple adaptation. Our ordinary language may well reflect some linguistic pressure from this direction, because we speak of the vitality of such systems (though this might be only a metaphor, of course). If we seek to learn the true nature of the phenomena of life, we must be open to the possibility that life is quite unlike what we now suppose.

FOUR PUZZLES AND PROPOSED SOLUTIONS

We can evaluate a theory of life by how well it resolves persistent puzzles about life. In summary form, this is my present battery of puzzles, along with the resolutions implied by the theory of life as supple adaptation:

Puzzle 1: How are different forms of life at different levels of biological hierarchy related?
Solution: Life must exist at many levels of organization. Different levels involve different but related forms of life.
Puzzle 2: Is the distinction between life and nonlife dichotomous or continuous?

Solution: Various continua and dichotomies separate life and nonlife, but the primary distinction is continuous.
Puzzle 3: Does the essence of life involve matter or form?
Solution: Life is essentially a certain form of process. The suppleness of that form makes the process noncomputational, but a computer simulation of life can create real life.
Puzzle 4: Are life and mind intrinsically related?
Solution: Life and mind are expressions of essentially the same kind of process.

These puzzles are controversial and subtle. A compelling theory must not only resolve the puzzles; it should also explain why they arise in the first place. The theory of supple adaptation does all this.

Levels and dependencies

Living phenomena fall into a complex hierarchy of levels—what I will call the *vital hierarchy*. Even broad brush strokes can distinguish at least eight levels in the vital hierarchy: ecosystems, which consist of communities, which consist of populations, which consist of organisms, which consist of organ systems (immune system, cardiovascular system), which consist of organs (heart, kidney, spleen), which consist of tissues, which consist of cells. Items at one level in the hierarchy constitute items at higher levels. For example, an individual population consists of a lineage of organisms that evolve over time. Individual organisms are born, live for a while, and then die. Taken together over time, these individuals constitute the evolving population. The vital hierarchy raises two basic kinds of questions about the nature of life. First, we may ask whether there is some inherent tendency for living systems to form hierarchies. Why are hierarchies so prevalent in the phenomena of life? The second question (really, set of questions) concerns the relationships among the kinds of life exhibited throughout the vital hierarchy. Are there different forms of life at different levels, and if so then how are these related? How are they similar and different? Which are prior and which posterior? What is the primary form of life? Haldane[21] and Mayr[3] are especially sensitive to these questions, although neither has a ready answer.

The theory of life as supple adaptation involves a two-tier picture with connected but different forms of life. The first tier consists of the primary form of life—the supplely adapting systems themselves. A supplely adapting system is an evolving population of organisms, or a whole evolving ecosystem of many populations, or, in the final analysis, a whole evolving biosphere with many interacting ecosystems. At the second tier, entities that are suitably generated and sustained by such a supplely adapting system branch off as different but connected secondary forms of life. These secondary forms of life include organisms, organs, and cells. So the idea that various forms of life are found at various levels of the biological hierarchy follows from the very structure of the theory of life.

Notice also that the very notion of a supplely adapting system implies simultaneous multiple levels of activity. Adaptive evolution involves the interaction between phenomena at a variety of levels, including at least genes and individual organisms and populations, so the process implies a system with activity at macro, meso, and micro levels. Thus, the theory of life as supple adaptation explains why life involves multiple levels of living phenomena. The agents constituting a supplely adapting population are not in every instance themselves alive. The simplest kind of supplely adapting systems seem to be something like an auto-catalytic network of chemical species, such as those hypothesized to be involved in the origin of life,[22,23] and it is implausible to attribute life to the chemical species that constitute these supplely adapting systems. Nevertheless, the agents in most supplely adapting populations are alive; organisms are the para-digm case of this.

There is another more indirect and much more controversial way in which supple adaptation might explain why there is a vital hierarchy. No one doubts that organisms have parts that function to ensure the organism's survival and reproduction, and no one doubts that in some cases these parts themselves have a complex hierarchical structure (think of the immune system or the brain). The progression of evolution in our biosphere seems to show a remarkable overall increase in complexity, from simple prokaryotic one-celled life to eukaryotic cellular life forms with a nucleus and numerous other cytoplasmic struc-tures, then to life forms composed of a multiplicity of cells, then to large-bodied vertebrate creatures with

sophisticated sensory processing capacities, and ultimately to highly intelligent creatures that use language and develop sophisticated technology. This evidence is consistent with the hypothesis that open-ended evolutionary processes have an inherent, lawlike tendency to create creatures with increasingly complicated functional organization. Just as the arrow of entropy in the second law of thermodynamics asserts that the entropy in all physical systems has a general tendency to increase with time, the hypothesis of the arrow of complexity asserts that the complex functional organization of the most complex products of open-ended evolutionary systems has a general tendency to increase with time. Make no mistake: The arrow of complexity hypothesis is far from settled. Some biologists are sympathetic but plenty are skeptical; see, for example, Gould,[24,25] Maynard Smith and Szathmáry,[9] and McShea,[26] as well as many of the chapters in refs. 27 and 28. I am not trying to resolve this controversy here. In fact, I think we have no compelling evidence either for or against the hypoth-esis right now.[29] My point here is that, *if* the arrow of complexity hypothesis is true, then supplely adapting systems have an inherent, internal tendency to produce entities with a complex, hierarchical structure, and so the theory of life as supple adaptation has a deep explanation of the vital hierarchy.

Whether or not the arrow of complexity hypothesis proves true, the theory of life as supple adaptation resolves the puzzle about the levels of life in a way that provides a natural explanation for why this puzzle arises in the first place.

Continuum or dichotomy

Can things be more or less alive? Serious reflection about life quickly raises the question whether life is a Boolean (black-or-white) property, as it seems at first blush, or whether it is a continuum property, coming in many shades of gray. Common sense leans toward the Boolean view: A rabbit is alive and a rock is not, and there is little apparent sense in the idea of some-thing falling in between these two states, being partly but not fully alive. But the common sense view is put under stress by various borderline cases such as viruses that are unable to replicate without a host and spores or frozen sperm that remain dormant and unchanging indefinitely but then "come back to life" when conditions become suitable. Furthermore, we all agree that the original life forms somehow emerged from a prebiotic chemical soup, and this suggests that there is very little, if any, principled distinction between life and nonlife. Many have concluded this implies that there is an ineliminable continuum of things being more or less alive.[7,22,30,31] But is this right?

If life is viewed as supple adaptation, then the most important life/nonlife distinction involves a continuum because the activity of supple adaptability itself comes in degrees. Different systems can exhibit supple adaptation to different degrees, and a given system's level of supple adaptation can fluctuate over time. A system's level of supple adaptation can smoothly drop to nothing or smoothly rise from nothing. It is obvious enough that evolving systems' level of supple adaptation can rise or fall continuously. In fact, there are methods for quantifying various aspects of an evolving system's level of supple adaptation,[32,33] and this enables the dynamics of supple adaptation in artificial and natural systems to be compared directly.[34,35] Thus, if we view life as supple adaptation, then being alive is a matter of degree. In addition to asking whether something is alive, we can also ask about the extent of its life; indeed, its life might vary along more than one dimension.

It is possible, of course, to define various sharp, Boolean distinctions on top of the continuum of the activity of supple adaptability. One natural distinction is whether a system's level of supple adaptation is positive; this dichotomy marks whether or not a system is alive. But it must be admitted that any such Boolean distinction involves an unmistakable element of arbitrariness; we could just as well focus on whether or not a system's level of supple adaptation exceeds 17 or 3.14159. Furthermore, such dichotomies would be defined in terms of a prior and more fundamental continuum of levels of supple adaptation; a system's level of supple adaptation could be arbitrarily close to our chosen cut-off point. Thus, the continuum is the truth underlying the dichotomies that it can be used to define.

There is a pragmatic dimension of the issue whether life at bottom is Boolean or continuous. If we quantify a system's level of supple adaptation in the way that Norman Packard and I have proposed,[32,33,35,36] then one needs a certain amount of data, and so a certain amount of time to gather the data, to determine (to within a certain level of statistical confidence) whether a system has a given level of supple adaptation. So, a system exhibiting very little supple adaptation will take a long time to generate enough data to distinguish it from the null hypothesis. But on that same time scale the system could exhaust some essential resource and perish. Thus, it might be impossible in practice to detect supple adaptation below a certain level on a certain time scale, and this would create a dichotomy separating detectable life from everything else. Still, this would not lessen the fact that in principle a continuum underlies this dichotomy.

The theory that life is supple adaptation, at least as I construe it, holds that life is the *activity* of supple adaptation, not merely the capacity for it. But the existence of this capacity is more basic than the extent to which it is exercised; the capacity is prior to its exercise. So we might ask whether this capacity is a Boolean property. Even if we do not know exactly what it takes for a system to have this capacity, it might seem that a system either has or lacks that capacity; it might look as if a system either can or cannot undergo open-ended evolution. But the truth seems more complicated. Supple adaptation is the process of producing *significantly new* kinds of adaptive responses to *significantly new* kinds of adaptive challenges and opportunities. Because it is dubious that there is a sharp divide between those challenges and responses that are significantly new and those that are not, the property of having the capacity for life seems to be a matter of degree.

So far we have focused on the supply adapting system itself, as well we should if supple adaptation is the primary form of life, as I have been urging. But other things, such as individual organisms, individual organs, and individual cells, are also alive, if only secondarily, and we should ask whether their life is a matter of degree. Intuitively one would think that a flea or paramecium is no less alive than a cow or human being; likewise, my heart is no less alive than a flea's heart, and a cell in my body is no less alive than a flea's cell. The theory of life as supple adaptation supports these intuitions. The theory attributes different derivative forms of life to entities that have the right connections with a supplely adapting system and in general it is an equally determinate and dichotomous matter for humans and fleas whether such connections obtain. When something definitely does or does not satisfy the conditions of derivative life, it definitely is or is not alive. There still are the familiar borderline cases, though, such as viruses, frozen sperm, and dormant spores. But notice that these are precisely those cases in which connections with the supplely adapting system deviate from the norm. The derivative form of reproduction of viruses makes their participation in the supplely adapting system less autonomous than other organisms. Frozen sperm and dormant spores have become disconnected from the supplely adapting system but when those connections are reestablished they are brought "back to life." In this sort of way the theory of life as supple adaptation offers a natural explanation for why borderline cases *are* borderline cases.

If the theory of life as supple adaptation is right, then both continuous scales and dichotomous divisions separate the living and the nonliving. Given this complexity, it is no wonder that we are puzzled about whether there is a continuum between life and nonlife.

Matter or form

The advent of the field of artificial life has focused attention on a set of questions about the role of matter and form in life.[4,9,31,37,38] On the one hand, certain distinctive carbon-based macromolecules play a crucial role in the vital processes of all known living entities; on the other hand, life seems more like a kind of process than a kind of substance. Furthermore, much of the practice of artificial life research seems to presuppose that life can be realized in a suitably programmed computer (see ref. 39 for a good recent discussion of this). This raises a number of related questions: Does the essence of life concern the material out of which something is composed or the form in which that material is arranged? If the latter, is that form static or is it a process? If the latter, is that process computational? Is the property of being alive a functional property? Is it realizable in an indefinitely long list of different material substrata? Could a computer simulation of a living process ever be a realization of life, that is, literally be alive?

Supple adaptation is a kind of process, not a kind of stuff. Although this process cannot occur unless it is realized in some material, and although it cannot be realized in just *any* kind of material, the range of materials that *can* realize it seems quite open ended. After all, even economic or intellectual systems can exhibit supple adaptation. So, supple adaptation is multiply realizable. What is essential to supple adaptation is the *form* of interactions among the components, not the stuff those components are made from. Thus, what determines whether something is an instance of the process of supple adaptation is whether the right sort of functional structure is present. In other words, the process of supple adaptation has a functional definition.

Of course, the theory of life as supple adaptation leaves room for secondary life forms, which would be delineated by a more specific form of the second clause of the definition on p. 394 above. But it would seem that the clauses in such a definition will also specify structural, causal, or functional conditions and relationships, and these will also be multiply realizable. So the theory of life as supple adaptation construes life entirely as a functional property. So, on this theory, functionalism captures the truth about life. Furthermore, there is no evident reason why the functional structure specified in the theory could not be realized in a suitably structured computational medium. If so, then a computer "simulation" of life could in principle create a real, literally living entity.

A seductive misunderstanding arises at this juncture. In claiming that supple adaptation can be realized in a computational medium I am *not* claiming that the process of supple adaptation corresponds to a fixed algorithm. What blocks this is supple adaptation's *suppleness*—its ability to respond appropriately to an open-ended and unanticipatable range of contingencies. The history of the so-called "frame problem" in artificial intelligence illustrates the problem (see, e.g., the chapters in ref. 40). One could try to embody a supple process in a fixed algorithm, along the lines of traditional artificial intelligence's use of heuristics, expert systems, and the rest. But the empirical fact is that these algorithms do not supplely respond to an open-ended variety of contingencies (see, e.g., refs. 41–43). Their behavior is brittle, lacking the supple sensitivity to context distinctive of intelligence. For the same reason, the suppleness of supple adaptation cannot be captured in a fixed algorithm.

Nevertheless, there is no evident reason why the process of supple adaptation cannot be realized in a computational medium, provided there is a suitably supple mechanism for changing the algorithms involved. This is one of the first important lessons of the field of artificial life. Vital processes typically are supple; think of metabolism or the process of adaptation itself. Successful adaptation depends on the ability to explore an appropriate number of viable evolutionary alternatives; too many or too few can make adaptation difficult or even impossible. In other words, success requires striking a balance between the competing demands for "creativity" (trying new alternatives) and "memory" (retaining what has proved successful). Furthermore, as the context for evolution changes, the appropriate balance between creativity and memory can shift in a way that resists precise and exceptionless formulation. Still, artificial life models can show a supple flexibility in how they balance creativity and novelty[44] because the underlying algorithmic behavior is supplely shaped and reshaped through the process of evolution. The key feature behind the supple vital dynamics produced by genetic algorithms[17] and other supple mechanisms that underlie artificial life models is their "bottom-up" architecture.[9] The supple dynamics is the emergent macro-level effect of a context-dependent competition for influence in a population of micro-level entities in the model. The micro-level models are precise and fixed algorithmic objects, of course, but their emergent macro-level supple dynamics are not. For this reason, supple adaptation can be realized as a nonalgorithmic emergent macro-level effect of an algorithmic micro-level process. Although the multiple realization of supple adaptation implies that life has a functional definition, the suppleness of this functional structure implies that the process of life is not a fixed algorithm. I have elsewhere called this special form of functionalism *emergent functionalism*.[44]

This line of thought identifies three factors that fuel the puzzle about whether life depends on form or matter. One is the inherent subtlety of the relationship functionalism implies between form and matter; what is essential to supple adaptation is a certain form of process, but this form of process cannot exist without being embodied in some matter. No doubt the mechanistic, reductionistic thrust of molecular biology, fueled by the celebrated discovery of DNA's double helix and recently re-energized by the cloning of an adult sheep, also contributes to the puzzle about

whether life is form or matter. The mistaken equation of functionalism and computationalism is a third cause of the puzzle. All of this helps explain why the puzzle about whether life involves form or matter is so animated.

Life and mind

A fourth puzzle is whether there is any intrinsic connection between life and mind. Plants, bacteria, insects, and mammals, for example, have various kinds of sensitivity to the environment, various ways in which this environmental sensitivity affects their behavior, and various forms of interorganism communication. Thus, various kinds of what one could, broadly speaking, call "mental" capacities are present throughout the biosphere. Furthermore, the relative sophistication of these mental capacities seems to correspond to and explain the relative sophistication of those forms of life. So it is natural to ask whether life and mind have some deep connection. The process of evolution establishes a genealogical connection between life and mind, of course, but life and mind might be much more deeply unified. For example, life and mind would be strikingly unified if Beer (ref. 45, p. 11) is right that "it is adaptive behavior, the ... ability to cope with the complex, dynamic, unpredictable world in which we live, that is, in fact, fundamental [to intelligence itself]" (see also[2,46–49]). Because all forms of life must cope in one way or another with a complex, dynamic, and unpredictable world, perhaps this adaptive flexibility inseparably connects life and mind. Resolving how, if at all, life and mind are connected is one of the basic puzzles about life.

If mental capacities are adaptations produced by the process of evolution, then the theory of life as supple adaptation implies that mental capacities are produced by life itself. Some view the evolution of the mind as an entirely unpredictable historical accident[24,25]; or as a plausible adaptation to environmental complexity[50]; or as an almost inevitable consequence of the evolutionary process—what Dennett calls a "forced move."[30] All such views agree that the mind is at most just one adaptation among many. Thus, this line of thought implies that there is a geneological connection between life and mind but it is not unique, so life and mind have no intrinsic unity.

This contrasts with Aristotle's idea that there is a deep unity between life and mind. Code and Moravcsik (51, p. 130) explain Aristotle's position as follows:

> In the case of a living thing, ... its "psychological" activity is the exercise ... of the various capacities and potentialities ... assigned to its soul.... *[F]or a living thing its natural/physical activity just is its psychological activity.* [emphasis added]

An analogously direct connection between life and mind can be grounded on the theory of life as supple adaptation, for one can view the mind as an expression of essentially the same underlying capacity for supple adaptation. It is well known that the emergent dynamical patterns among our mental states are especially difficult to describe and explain. An ineliminable open-ended list of exceptions seems to infect descriptions of all mental patterns, for which reason these patterns are sometimes called "fluid"[42] or "soft".[51] But there are different kinds of fluidity and softness. Fodor[52] and others have emphasized the functionalist point that softness can result from malfunctions in the material and processes implementing mental phenomena. Horgan and Tienson[51,53] have emphasized the softness that results from the indeterminate results of competition among a potentially open-ended range of conflicting desires. But what is most relevant here are specifically those exceptions to the rule that reflect our *ability to act appropriately* in the face of an open-ended range of contextual contingencies. These exceptions occur when we make *appropriate* adjustment to contingencies. Some people conclude that this supple capacity for adaptive behavior is the defining feature of intelligence or mind.[29,44–46,46,48,49,54]

This quasi-Aristotelian view construes the mind as essentially the expression of a form of supple adaptation. Natural selection is not necessarily involved, for Lamarckian selection or some other adaptive process might do the trick. Rather, leaving aside the mechanisms of adaptation, my claim is that the process of having a mind is something quite like the process of being alive. So, the theory of life as supple adaptation is naturally allied with the theory of mind as supple adaptation. Just as the essence of life is the process that generates the phenomena of life, for the same reason the essence of mind is the process that generates intelligent behavior. If life and mind are both produced by basically the same process of supple adaptation, then the mind is not just one adaptation among many. Rather, an essential feature of the mind is involved in the explanation of all other local adaptations, so life and mind could hardly fail to coexist. They exhibit the strong continuity[55] characterized by both exhibiting the abstract pattern of supple adaptation. From the perspective of the theory of life as supple adaptation and the quasi-Aristotelian approach to the mind, it is no wonder that people think that life and mind are deeply connected.

A complete solution to the puzzle about the connection between life and mind should also explain why this connection is largely ignored today, especially among philosophers. The theory of life as supple adaptation combined with the quasi-Aristotelian approach to mind can blame this on Descartes. Contemporary philosophy of mind is a culture deeply influenced by Descartes. Descartes rejected the then orthodox scholastic Aristotelian framework in favor of the view that living substances have no essential connection with mental substances (except for the unmediated causal connection unifying each person's mind and body). Descartes focused on the intrinsic nature of isolated living and mental substances, ignoring the processes that created and sustained them, and concluding that living substances were purely material mechanisms while mental substances are essentially immaterial and spiritual consciousness. Today, even contemporary philosophy of mind that rejects Descartes's dualism of body and mind typically embraces consciousness as the essence of mind and shares Descartes's unconcern about how living and mental substances are produced. One testament to Descartes's persistent influence is the present difficulty of initially motivating the puzzle about how life and mind are connected.

OPEN QUESTIONS AND CONCLUSIONS

I offer no final and complete theory of life and no final and complete solution to the four puzzles about life, but I do defend the general form of the theory of life as supple adaptation. My defense consists of showing the theory's promising and illuminating solutions to four puzzles about life.

This defense highlights three open questions. The first is to determine what, in the end, is the best explanation of the salient phenomena and puzzles concerning life. Even if supple adaptation provides good explanations of these matters, this leaves room for other theories to provide better explanations. Our final understanding of what life really is will turn on which theory in the end provides the best explanations.

When we try to settle exactly how well supple adaptation explains these matters, two more questions arise. For one thing, this theory is no clearer than the notion of supple adaptation itself, and there is still much to learn about supple adaptation. For example, not a single artificial evolutionary model has unambiguously shown the sort of continually growing supple adaptation evident in the biosphere,[34,35] not even those models with unpredictably changing selection criteria and an infinite space of genetic possibilities, such as John Holland's Echo,[17] Kristian Lindgren's evolving strategies for infinite prisoner's dilemmas,[15] and Tom Ray's Tierra.[16] The problem seems to be that no existing model creates a continually unfolding accessible space of new kinds of adaptive innovations. Synthesizing even one demonstrable instance of continually growing supple adaptation would profoundly advance our understanding of this process. The task of producing and certifying such a model falls squarely to the field of artificial life. If life is supple adaptation, finding such a model is one of the field's most pressing current challenges.

Finally, even if our understanding of supple adaptation were complete, we still would need to settle how best to use it to define life. For example, we need to determine the different ways in which different forms of life can be explained by a supplely adapting system. These details will replace the place-holding expression "explained in the right way" in the definition given above. We also need to decide what weight to place on different mechanisms for producing supple adaptation. Natural selection is one such mechanism, but there is an open-ended variety of others (Lamarckian selection, etc.). Once we have delineated all those mechanisms, we will be faced with a choice: Is the primary form of life supplely adapting systems produced by any mechanism? Only by natural selection? The way to settle this question, in the end, is to determine which choice provides the most illuminating understanding of the phenomena and puzzles surrounding life.

I intend the present discussion to establish two main conclusions about the theory of life. The first is

methodological: The search for a theory of life is more productive if it focuses on the best explanation of life, including deep and persistent puzzles about life. This methodology frees us from many traditional worries caused by our current preconceptions about life, including worrying about necessary and sufficient conditions for all and only living organisms. My second conclusion is substantive: The theory of life as supple adaptation deserves our serious consideration. To be sure, the theory generates tension with our present preconceptions of life, but this is no strike against the theory. Rival theories are credible contenders only if they explain living phenomena and resolve the four puzzles at least as well as the theory of supple adaptation.

ACKNOWLEDGMENTS

For helpful discussion of these topics or comments on previous drafts, thanks to Hugo Bedau, Maggie Boden, Andy Clark, Peter Godfrey-Smith, David McFarland, Dan McShea, David Reeve, Francisco Varela, Bill Wimsatt, Marty Zwick, the spring 1997 audience at the conference on the Philosophy of Artificial Life, *PAL97*, Christ Church, Oxford University, the reviewers and summer 1997 audience at the Fourth European Conference on Artificial Life, *ECAL97*, Brighton, England, the winter 1998 audience at my Systems Science colloquium at Portland State University, and the anonymous reviewers for this journal. Thanks to the Santa Fe Institute for support and hospitality during the visits that started and sustained this work. Special thanks to Norman Packard for years of collaboration that have profoundly influenced my thinking about this topic.

NOTES

This chapter originally appeared in *Artificial Life* 4 (2) (1998), 125–140.

REFERENCES

1. Taylor, C. (1992). Fleshing out. In C. G. Langton, C. Taylor, J. D. Farmer, and S. Rasmussen (Eds.). *Artificial life II* (Santa Fe Institute studies in the sciences of complexity, proceedings vol. X) (pp. 371–408). Redwood City, CA: Addison-Wesley.

2. Farmer, D. & Belin, A. (1992). Artificial life: The coming evolution. In C. G. Langton, C. Taylor, J. D. Farmer, and S. Rasmussen (Eds.), *Artificial life II* (Santa Fe Institute studies in the sciences of complexity, proceedings vol. X) (pp. 815–840). Redwood City, CA: Addison-Wesley.

3. Mayr, E. (1982). *The growth of biological thought.* Cambridge, MA: Harvard University Press.

4. Sober, E. (1992). Learning from functionalism—Prospects for strong artificial life. *Artificial Life,* 2, 749–765.

5. Chyba, C. F. & McDonald, G. D. (1995). The origin of life in the solar system: Current issues. *Annual Review of Earth and Planetary Sciences,* 23, 215–249.

6. Pirie, N. W. (1972). On recognizing life. In D. I. Rohlfing and A. I. Oparin (Eds.), *Molecular evolution: Prebiological and biological* (pp. 67–76). New York: Plenum Press.

7. Cairns-Smith, A. G. (1985). *Seven clues to the origin of life.* Cambridge, UK: Cambridge University Press.

8. Eigen, M. (1992). *Steps toward life.* Oxford: Oxford University Press.

9. Maynard Smith, J. (1975). *The theory of evolution* (3rd ed.). New York: Penguin.

10. Langton, C. (1989). Artificial life. In C. Langton (Ed.), *Artificial life* (Santa Fe Institute studies in the sciences of complexity, proceedings vol. IV) (pp. 1–47). Redwood City, CA: Addison-Wesley.

11. Maynard Smith, J. & Szathmary, E. (1995). *The major transitions in evolution.* New York: W. H. Freeman.

12. Sagan, C. (1998). Life. In *The New Encyclopedia Britannica* (15th ed. Macropedia) (pp. 964–981). Chicago: Encyclopedia Britannica, Inc.

13. Bedau, M. A. (1990). Against mentalism in teleology. *American Philosophical Quarterly,* 27, 61–70.

14. Bedau, M. A. (1991). Can biological teleology be naturalized? *The Journal of Philosophy,* 88, 647–655.

15. Bedau, M. A. (1992). Where's the good in teleology? *Philosophy and Phenomenological Research,* 52, 781–805.

16. Dawkins, R. & Krebs, J. R. (1978). Arms races between and within species. *Proceedings of the Royal Society of London B,* 205, 489–511.

17. Lindgren, K. (1992). Evolutionary phenomena in simple dynamics. In C. Langton, C. Taylor, J. D. Farmer, and S. Rasmussen (Eds.), *Artificial life II* (Santa Fe Institute studies in the sciences of complexity, proceedings vol. X) (pp. 295–312). Redwood City, CA: Addison-Wesley.

18. Ray, T. (1992). An approach to the synthesis of life. In C. G. Langton, C. Taylor, J. D. Farmer, and

S. Rasmussen (Eds.), *Artificial life II* (Santa Fe Institute studies in the sciences of complexity, proceedings vol. X) (pp. 371–408). Redwood City, CA: Addison-Wesley.

19. Holland, J. H. (1992). *Adaptation in natural and artificial systems.* Cambridge, MA: MIT Press/Bradford Books.

20. Packard, N. (1989). Intrinsic adaptation in a simple model for evolution. In C. G. Langton (Ed.), *Artificial life* (Santa Fe Institute studies in the sciences of complexity, proceedings vol. IV) (pp. 141–155). Redwood City, CA: Addison-Wesley.

21. Boden, M. A. (1998). *Is metabolism necessary? Cognitive science research paper 482.* University of Sussex, Sussex, UK.

22. Kripke, S. (1980). *Naming and necessity.* Cambridge, MA: Harvard University Press.

23. Haldane, J. B. S. (1937). *Adventures of a biologist.* New York: Macmillan.

24. Bagley, R. & Farmer, J. D. (1992). Spontaneous emergence of a metabolism. In C. G. Langton, C. Taylor, J. D. Farmer, and S. Rasmussen (Eds.), *Artificial life II* (Santa Fe Institute studies in the sciences of complexity, proceedings vol. X) (pp. 93–140). Redwood City, CA: Addison-Wesley.

25. Farmer, J. D., Kauffman, S. A., & Packard, N. H. (1986). Autocatalytic replication of polymers. In J. D. Farmer, A. Lapedes, N. Packard, and B. Wendroff (Eds.), *Evolution, games, and learning: Models for adaptation for machines and nature* (pp. 50–67). Amsterdam: North Holland.

26. Gould, S. J. (1989). *Wonderful life: The Burgess shale and the nature of history.* New York: Norton.

27. Gould, S. J. (1996). *Full house: The spread of excellence from Plato to Darwin.* New York: Harmony Books.

28. McShea, D. W. (1996). Metazoan complexity and evolution: Is there a trend? *Evolution,* 50, 477–492.

29. Nitecki, M. H. (1988). *Evolutionary progress.* Chicago: University of Chicago Press.

30. Barlow, C. (1995). *Evolution extended: Biological debates about the meanings of life.* Cambridge, MA: MIT Press.

31. Bedau, M. A. (1997). Philosophical content and method in artificial life. In T. W. Bynam and J. H. Moor (Eds.), *The digital phoenix: How computers are changing philosophy* (pp. 135–152). New York: Blackwell.

32. Dennett, D. C. (1995). *Darwin's dangerous idea: Evolution and the meanings of life.* New York: Simon & Schuster.

33. Emmeche, C. (1994). *The garden in the machine: The emerging science of artificial life.* Princeton, NJ: Princeton University Press.

34. Bedau, M. A. (1995). Three illustrations of artificial life's working hypothesis. In W. Banshaf and F. Eeckman

(Eds.), *Evolution and biocomputation—Computational models of evolution*. Berlin: Springer-Verlag.

35. Bedau, M. A. & Packard, N. H. (1992). Measurement of evolutionary activity, teleology, and life. In C. G. Langton, C. Taylor, J. D. Farmer, and S. Rasmussen (Eds.), *Artificial life II* (Santa Fe Institute studies in the sciences of complexity, proceedings vol. **X**) (pp. 431–461). Redwood City, CA: Addison-Wesley.

36. Bedau, M. A., Snyder, E., Brown, C. T., & Packard, N. H. (1997). A comparison of evolutionary activity in artificial systems and in the biosphere. In P. Husbands and I. Harvey (Eds.), *Proceedings of the fourth European conference on artificial life ECAL 97* (pp. 125–134). Cambridge, MA: MIT Press/Bradford Books.

37. Bedau, M. A., Snyder, E., & Packard, N. H. (1998). A classification of long-term evolutionary dynamics. In C. Adami, R. Belew, H. Kitano, and C. Taylor (Eds.), *Artificial life VI* (Santa Fe Institute studies in the sciences of complexity, proceedings vol. **IXX**) (pp. 228–237). Cambridge, MA: MIT Press/Bradford Books.

38. Bedau, M. A. (1996). The nature of life. In M. A. Boden (Ed.), *The philosophy of artificial life* (pp. 332–357). New York: Oxford University Press.

39. Emmeche, C (1992). Life as an abstract phenomenon: Is artificial life possible? In F. Varela and P. Bourgine (Eds.), *Towards a practice of autonomous systems* (pp. 466–474). Cambridge, MA: MIT Press/Bradford Books.

40. Pattee, H. H. (1989). Simulations, realization, and theories of life. In C. G. Langton (Ed.), *Artificial life* (Santa Fe Institute studies in the sciences of complexity, proceedings vol. **IV**) (pp. 63–78). Redwood City, CA: Addison-Wesley.

41. Olson, E. T. (1997). The ontological basis of strong artificial life. *Artificial Life*, **3**, 29–39.

42. Pylyshyn, Z. W. (1987). *The robots dilemma: The frame problem in artificial intelligence*. Norwood, NJ: Ablex.

43. Dreyfus, H. (1979). *What computers cannot do* (2nd ed.). New York: Harper & Row.

44. Hofstadter, D. R. (1985). Waking up from the Boolean dream, or, subcognition as computation. In D. R. Hofstadter, *Metamagical themas: Questing for the essence of mind and pattern* (pp. 631–665). New York: Basic Books.

45. Holland, J. H. (1986). Escaping brittleness: The possibilities of general-purpose learning algorithms applied to parallel rule-based systems. In R. S. Michalski, J. G. Carbonell, and T. M. Mitchell (Eds.), *Machine learning II* (pp. 593–623). Los Aires, CA: Morgan Kaufmann.

46. Bedau, M. A. (1997). Emergent models of supple dynamics in life and mind. *Brain and Cognition*, **34**, 5–27.

47. Beer, R. D. (1990). *Intelligence as adaptive behavior: An experiment in computational neuroethology*. Boston: Academic Press.

48. Anderson, J. R. (1990). *The adaptive character of thought*. Hillsdale, NJ: Erlbaum.

49. Clark, A. (1997). *Being there: Putting brain, body, and world together again*, Cambridge, MA: MIT Press.

50. Maturana, H. R. & Varela, F. J. (1987). *The tree of knowledge: The biological roots of human understanding* (revised ed.). Boston: Shambhala.

51. Parisi, D., Nolfi, N., & Cecconi, F. (1992). Learning, behavior, and evolution. In F. Varela and P. Bourgine (Eds.), *Towards a practice of autonomous systems* (pp. 207–216). Cambridge, MA: MIT Press/Bradford Books.

52. Godfrey-Smith, P. (1996). *Complexity and the function of mind in nature*. Cambridge, UK: Cambridge University Press.

53. Code, A. & Moravcsik, J. (1992). Explaining various forms of living. In M. C. Nussbaum and A. O. Rorty (Eds.), *Essays on Aristotle's* De anima (pp. 129–145). Oxford: Clarendon Press.

54. Horgan, T. & Tienson, J. (1990). Soft laws. *Midwest Studies in Philosophy*, **15**, 256–279.

55. Fodor, J. A. (1981). *Representations*. Cambridge, MA: MIT Press/Bradford Books.

56. Horgan, T. & Tienson, J. (1989). Representation without rules. *Philosophical Topics*, **17**, 147–174.

57. Varela, F., Thompson, E., & Rosch, E. (1991). *The embodied mind: Cognitive science and human experience*. Cambridge, MA: MIT Press/Bradford Books.

58. Godfrey-Smith, P. (1994). Spencer and Dewey on life and mind. In R. Brooks and P. Maes (Eds.), *Artificial life IV* (Proceedings of the fourth international workshop on the synthesis and simulation of living systems) (pp. 80–89). Cambridge, MA: MIT Press/Bradford Books.

Supplementary bibliography on life

The literature on the nature of life is extensive. It is sometimes difficult to locate, however, as it is scattered across a wide range of interdisciplinary sources. With this in mind, we have provided a list of further reading to complement and expand on the material in this book. The readings are grouped according to the book's four sections; as with the chapters in the book, many of the readings listed below touch on issues that arise in more than one section. The list is far from exhaustive, but we have tried to include what we consider to be most important, most useful, and most readily accessible.

CLASSICAL DISCUSSIONS OF LIFE

Ablondi, F. (1998). Automata, living and non-living: Descartes' mechanical biology and his criteria for life. *Biology and Philosophy*, **13**, 179–188.

Allen, C., Bekoff, M., & Lauder, G. (Eds.) (1997). *Nature's purposes: Analyses of function and design in biology.* Cambridge: MIT Press.

Bedau, M. A. & Humphreys, P. (Eds.) (2008). *Emergence: Contemporary readings in philosophy and science.* Cambridge: MIT Press.

Boden, M. A. (2006). *Mind as machine: A history of cognitive science.* Oxford: Oxford University Press.

Code, A. & Moravcsik, J. (1992). Explaining various forms of living. In M. C. Nussbaum & A. O. Rorty (Eds.), *Essays on Aristotle's* De anima (pp. 129–145). Oxford: Clarendon Press.

Craver, C. F. & Darden, L. (2005). Introduction: Mechanisms then and now. *Studies in the History and Philosophy of Biological and Biomedical Sciences*, **36**, 233–244.

Des Chene, D. (2000). *Spirits and clocks.* Ithaca, NY: Cornell University Press.

Dick, S. J. (1982). *Plurality of worlds: The origins of the extraterrestrial life debate from Democritus to Kant.* Cambridge, UK: Cambridge University Press.

Gánti, T. (1975). Organization of chemical reactions into dividing and metabolizing units: The chemotons. *Biosystems*, **7**, 15–21.

Gánti, T. (1997). Biogenesis itself. *Journal of Theoretical Biology*, **187**, 583–593.

Gánti, T. (2003). *The principles of life*, with commentary by J. Griesemer and E. Szathmáry. New York: Oxford University Press.

Ginsborg, H. (2001). Kant on understanding organisms as natural purposes. In E. Watkins (Ed.), *Kant and the sciences* (pp. 231–258). Oxford: Oxford University Press.

Gotthelf, A. & Lennox, J. G. (1987). *Philosophical issues in Aristotle's biology.* Cambridge, UK: Cambridge University Press.

Haldane, J. B. S. (1937). What is life? In J. B. S. Haldane, *Adventures of a biologist* (pp. 49–64). New York: Macmillian.

Haldane, J. B. S. (1947). *What is life?* New York: Boni and Gaer.

Haldane, J. B. S. (1929/1967). The origin of life. In J. D. Bernal (Ed.), *The origin of life* (pp. 242–249). London: Weidenfeld & Nicolson.

Hazen, R. (2005). *Genesis: The scientific quest for life's origin.* Washington DC: Joseph Henry Press.

Matthews, G. B. (1996). Aristotle on life. In M. C. Nussbaum & A. O. Rorty (Eds.), *Essays on Aristotle's De anima* (pp. 185–193). Oxford: Clarendon Press

Mayr, E. (1982). *The growth of biological thought.* Cambridge, MA: Belknap Press.

Nussbaum, M. C. & Rortym, A. O. (1992). *Essays on Aristotle's De anima.* Oxford: Clarendon Press.

Oparin, A. I. (1936/1953). *Origin of life*, trans. S. Morgulis. New York: Dover Publications.

Oparin, A. I. (1961). *Life: Its nature, origin, and development*, trans. A. Synge. New York: Academic Press.

Richards, R. J. (2002). *The romantic conception of life: Science and philosophy in the age of Goethe.* Chicago: University of Chicago Press.

Sapp, J. (2003). *Genesis: The evolution of biology*. Oxford: Oxford University Press.

Schrödinger, E. (1944). *What is life? The physical aspect of the living cell*. Cambridge, UK: Cambridge University Press.

Shields, C. (1999). The meaning of life. In C. Shields, *Order in multiplicity* (pp. 176–193). Oxford: Oxford University Press.

Waddington, C. (1961). *The nature of life*. London: George Allen & Unwin Ltd.

Weber, A. & Varela, F. (2002). Life after Kant: Natural purposes & the autopoietic foundations of biological individuality. *Phenomenology and the Cognitive Sciences*, 1, 97–125.

THE ORIGIN AND EXTENT
OF NATURAL LIFE

Alberti, A. (1997). The origin of the genetic code and protein synthesis. *Journal of Molecular Evolution*, 45, 352–358.

Bachmann, P., Luisi, P., & Lang, J. (1992). Autocatalytic self-replicating micelles as models for prebiotic structures. *Nature*, 357, 57–59.

Banfield, J., Moreau, J., Chan, C., Welch, S., & Little, B. (2001). Mineralogical biosignatures and the search for life on Mars. *Astrobiology*, 1 (4), 447–465.

Benner, S. & Hutter, D. (2002). Phosphates, DNA, and the search for nonterran life: A second generation model for genetic molecules. *Bioorganic Chemistry*, 30, 62–80.

Bradley, J., Harvey, R., & McSween, H. (1997). No 'nanofossils' in Martian meteorite. *Nature*, 390, 454–455.

Cairns-Smith, A. G. (1982). *Genetic takeover and the mineral origins of life*. Cambridge, UK: Cambridge University Press.

Cairns-Smith, A. G., Hall, A., & Russell, M. (1992). Mineral theories of the origin of life and an iron sulfide example. *Origins of Life and the Evolution of the Biosphere*, 22, 161–180.

Chyba, C. F. & McDonald, G. (1995). The origin of life in the solar system: Current issues. *Annual Review Earth Planetary Sciences*, 23, 215–249.

Chyba, C. F. & Phillips, C. (2001). Possible ecosystems and the search for life on Europa. *Proceedings of the National Academy of Sciences*, 98, 801–804.

Cleland, C. E. (2007). Epistemological issues in the study of microbial life: Alternative terran biospheres? *Studies in History and Philosophy of Biological & Biomedical Sciences*, 38, 847–861.

Cleland, C. E. & Chyba, C. F. (2002). Defining 'life.'. *Origins of Life and the Evolution of the Biosphere*, 32, 387–393.

Crick, F. (1968). The origin of the genetic code. *Journal of Molecular Biology*, 38, 367–379.

Cronin, J. & Chang, S. (1993). Organic matter in meteorites: Molecular and isotopic analysis of the Murchison meteorite. In J. M. Greenberg, C. X. Mendoza-Gómez, & V. Pironello (Eds.), *The chemistry of life's origin* (pp. 205–258). Dordrecht: Kluwer Academic Publishers.

Darling, D. (2001). *Life everywhere: The maverick science of astrobiology*. New York: Basic Books.

Davies, P. (1995). *Are we alone?* New York: Basic Books.

Davies, P. C. W. & Lineweaver, C. H. (2005). Finding a second sample of life on Earth. *Astrobiology*, 5 (2), 154–163.

Davies, P. C. W., Benner, S. A., Cleland, C. E., Lineweaver, C. H., McKay, C. P., & Wolfe-Simon, F. (2009). Signatures of a shadow biosphere. *Astrobiology*, in press.

Deamer, D. W. (1997). The first living systems: A bioenergetic perspective. *Microbiology and Molecular Biology Review*, 61, 239–261.

de Duve, C. (1995). *Vital dust: Life as a cosmic imperative*. New York: Basic Books.

de Duve, C. (2002). *Life evolving: Molecules, mind, and meaning*. New York: Oxford University Press.

Des Marais, D. & Walter, M. (1999). Astrobiology: Exploring the origins, evolution, and distribution of life. *Annual Review of Ecology and Systematics*, 30, 397–420.

Dick, S. J. (1996). *The biological universe: The twentieth century extraterrestrial life debate and the limits of science*. Cambridge, UK: Cambridge University Press.

DiGiulio, M. (1997). On the origin of the genetic code. *Journal of Theoretical Biology*, 187, 573–581.

DiGregorio, B. E., Levin, G. V., & Straat, P. A. (1997). *Mars, the living planet*. Berkeley: Frog Books.

Donaldson, D. J., Tervahattu, H., Tuck, A. F., & Vaida, V. (2004). Organic aerosols and the origin of life: An hypothesis. *Origins of Life and the Evolution of the Biosphere*, 34, 57–67.

Dyson, F. (1982). A model for the origin of life. *Journal of Molecular Evolution*, 18, 344–350.

Dyson, F. (1985). *Origins of life*. Cambridge, UK: Cambridge University Press.

Eigen, M. (1971). Self-organization of matter and the evolution of biological macromolecules. *Naturwissenschaften*, 58, 465–523.

Eigen, M. (1992). *Steps towards life*. Oxford: Oxford University Press.

Ferris, J. (1987). Prebiotic synthesis: Problems and challenges. *Cold Spring Harbor Symposium on Quantitative Biology*, **55**, 29–35.

Fox, S. (1960). How did life begin? *Science*, **132**, 200–208.

Fry, I. (2000). *The emergence of life on earth: A historical and scientific overview.* London: Rutgers University Press.

Gesteland, R. F., Cech, T. R., & Atkins, J. F. (Eds.) (1999). *The RNA world.* New York: Cold Spring Harbor.

Gilbert, W. (1986). The RNA world. *Nature*, **319**, 618.

Golden, D. C., Ming, D. W., Morris, R. V., *et al.* (2004). Evidence for exclusively inorganic formation of magnetite in Martian meteorite ALH84001. *American Mineralogist*, **89**, 681–695.

Grinspoon, D. (2004). *Lonely planets: The natural philosophy of alien life.* New York: HarperCollins.

Horgan, J. (1991). In the beginning . . . *Scientific American*, *February*, **114**, 125.

Irvine, W. (1998). Extraterrestrial organic matter. *Origins of Life and the Evolution of the Biosphere*, **28**, 365–383.

Kamminga, H. (1982). Life from space—A history of panspermia. *Vistas in Astronomy*, **26**, 67–86.

Kauffman, S. (1986). Autocatalytic sets of proteins. *Journal of Theoretical Biology*, **119**, 1–24.

Kerr, R. (1998). Requiem for life on Mars? Support for microbes fades. *Science*, **282**, 1398–1400.

Klein, H. (1978). The Viking biological experiments on Mars. *Icarus*, **34**, 666–674.

Léger, A. J., Pirre, M., & Marceau, F. (1993). Search for primitive life on a distant planet: Relevance of O_2 and O_3 detections. *Astronomy and Astrophysics*, **277**, 309–313.

Lifson, S. (1997). On the crucial stages in the origin of animated matter. *Journal of Molecular Evolution*, **44**, 1–8.

Luisi, P. (1993). Defining the transition to life: Self-replicating bounded structures and chemical autopoiesis. In W. Stein and F. Varela (Eds.), *Thinking about biology* (pp. 3–23). Redwood City, CA: Addison-Wesley.

Maurette, M. (1998). Carbonaceous micrometeorites and the origin of life. *Origins of Life and the Evolution of the Biosphere*, **28**, 385–412.

Maynard Smith, J. & Szathmáry, E. (1999). *The origins of life: From the birth of life to the origins of language.* New York: Oxford University Press.

McKay, D. S., Gibson, E. K. Jr., Thomas-Keprta, K. L., *et al.* (1996). Search for past life on Mars: Possible relic biogenic activity in Martian meteorite ALH84001. *Science*, **273**, 924–930.

McKay, C. (1997). The search for life on Mars. *Origins of Life and the Evolution of the Biosphere*, **27**, 263–289.

McNichol, J. (2008). Historical review: Primordial soup, fool's gold, and spontaneous generation. *Biochemistry and Molecular Biology Education*, **36**, 255–261.

Miller, S. L. (1953). A production of amino acids under possible primitive Earth conditions. *Science*, **117**, 528–529.

Miller, S. (1992). The prebiotic synthesis of organic components as a step toward the origin of life. In J. Schopf (Ed.), *Major events in the history of life* (pp. 1–28). Boston: Jones and Bartlett Publishers.

Miller, S. & Bada, J. (1988). Submarine hot springs and the origin of life. *Nature*, **334**, 155–176.

Mojzsis, S., Arrhenius, G., McKeegan, K., Harrison, T., Nutman, A., & Friend, C. (1996). Evidence for life on Earth before 3800 million years ago. *Nature*, **384**, 55–59.

Morowitz, H. J. (1992). *Beginnings of cellular life: Metabolism recapitulates biogenesis.* New Haven: Yale University Press.

Morowiz, H. J. (1999). A theory of biochemical organization, metabolic pathways, and evolution. *Complexity*, **4**, 39–53.

National Research Council (2007). *The limits of organic life in planetary systems.* Washington DC: National Academies Press.

Nealson, K. & Conrad, P. (1999). Life: Past, present, and future. *Philosophical Transactions of the Royal Society of London B*, **354**, 1923–1939.

Nealson, K., Tsapin, A., & Storrie-Lombardi, M. (2002). Searching for life in the universe: Unconventional methods for an unconventional problem. *International Microbiology*, **5** (4), 223–230.

O'Malley, M. A. & Dupré, J. (2007). Metagenomics and biological ontology. *Studies in History and Philosophy of Biological and Biomedical Sciences*, **38**, 834–846.

O'Malley, M. A. & Dupré, J. (2007). Size doesn't matter: Towards a more inclusive philosophy of biology. *Biology and Philosophy*, **22**, 155–191.

Orgel, L. E. (1986). RNA catalysis and the origin of life. *Journal of Theoretical Biology*, **123**, 127–144.

Orgel, L. E. (1994). The origin of life on the Earth. *Scientific American*, **271**, 76–83.

Orgel, L. (2004). Prebiotic chemistry and the origin of the RNA world. *Critical Reviews in Biochemistry and Molecular Biology*, **39**, 99–123.

Rode, B. (1999). Peptides and the origin of life. *Peptides*, **20**, 773–786.

Russell, M., Hall, A., Cairns-Smith, A. G., & Braterman, P. (1988). Submarine hot springs and the origin of life. *Nature*, **336**, 117.

Sagan, C. (1974). The origin of life in a cosmic context. *Origins of Life and the Evolution of the Biosphere*, **5**, 497–505.

Sagan, C. (1994). The search for extraterrestrial life. *Scientific American*, **271** (4), 92–99.

Schneider, E. & Kay, J. (1994). Life as a manifestation of the second law of thermodynamics. *Mathematical and Computer Modeling*, **19** (6–8), 25–48.

Schulze-Makuch, D. & Irwin, L. N. (2004). *Life in the universe: Expectations and constraints*. Berlin: Springer-Verlag.

Seckbach, J. (Ed.) (2006). *Life as we know it*. Dordrecht: Springer.

Segré, D. & Lancet, D. (2000). Composing life. *EMBO Reports*, **1**, 217–222.

Segré, D., Ben-Eli, D., Deamer, D. W., & Lancet, D. (2001). The lipid world. *Origins of Life and the Evolution of the Biosphere*, **31**, 119–145.

Shapiro, R. (1986). *Origins: A skeptic's guide to the creation of life on Earth*. New York: Bantam Books.

Shapiro, R. (1999). *Planetary dreams*. New York: Wiley & Sons.

Sullivan, W. T. & Baross, J. A. (2007). *Planets and life: The emerging science of astrobiology*. Cambridge: Cambridge University Press.

Wächtershäuser, G. (1990). The case for the chemoautotrophic origin of life in an iron-sulfide world. *Origins of Life and the Evolution of the Biosphere*, **20** (2), 173–176.

Wächtershäuser, G. (1992). Groundworks for an evolutionary biochemistry: The iron-sulfur world. *Progress in Biophysics and Molecular Biology*, **58**, 85–201.

Ward, P. (2005). *Life as we do not know it: The NASA search for (and synthesis of) alien life*. New York: Viking Penguin.

Ward, P. D. & Brownlee, D. (2000). *Rare Earth: Why complex life is uncommon in the universe*. New York: Springer-Verlag.

Wharton, D. (2002). *Life at the limits: Organisms in extreme environments*. Cambridge, UK: Cambridge University Press.

Whittet, D. (1997). Is extraterrestrial organic matter relevant to the origin of life on Earth? *Origins of Life and the Evolution of the Biosphere*, **27**, 249–262.

Willis, C. & Bada, J. (2000). *The spark of life: Darwin and the primeval soup*. Cambridge, MA: Perseus Publishing.

Woese, C. (1998). The universal ancestor. *Proceedings of the National Academy of Sciences*, **95**, 6854–6859.

Woese, C. (2004). A new biology for a new century. *Microbiology and Molecular Biology Reviews*, **86**, 173–186.

Woese, G. E. & Fox, E. (1977). Phylogenetic structure of the prokaryotic domain: The primary kingdoms. *Proceedings of the National Academy of Sciences*, **74**, 5088–5090.

Yockey, H. (2000). Origin of life on Earth and Shannon's theory of communication. *Computational Chemistry*, **24**, 105–123.

ARTIFICIAL LIFE AND SYNTHETIC BIOLOGY

Adami, C. (1998). *Introduction to artificial life*. New York: Springer-Verlag.

Bagley, R. & Farmer, J. D. (1992). Spontaneous emergence of a metabolism. In C. G. Langton, C. Taylor, J. D. Farmer, & S. Rasmussen (Eds.), *Artificial life II* (Santa Fe Institute studies in the sciences of complexity, proceedings vol. X) (pp. 93–140). Redwood City, CA: Addison Wesley.

Baker, D., Church, G., Collins, J., Endy, D., Jacobson, J., Keasling, J., & Modrich, P. (2006). Engineering life: Building a fab for biology. *Scientific American*, **294**, 44–51.

Bedau, M. A. (2003). Artificial life: Organization, adaptation, and complexity from the bottom up. *Trends in Cognitive Science*, **7** (11), 505–512.

Bedau, M. A. (2007). Artificial life. In M. Matthen & C. Stephens (Eds.), *Handbook of the philosophy of biology* (pp. 585–603). Amsterdam: Elsevier.

Bedau, M., McCaskill, J., Packard, N., *et al.* (2000). Open problems in artificial life. *Artificial Life*, **6**, 363–376.

Bedau, M. A. & Parke, E. (Eds.) (2009). *The prospect of protocells: Social and ethical implications of recreating life*. Cambridge, MA: MIT Press.

Beer, R. D. (1990). *Intelligence as adaptive behavior: An experiment in computational neuroethology*. Boston: Academic Press.

Boden, M. A. (Ed.) (1996). *The philosophy of artificial life*. Oxford: Oxford University Press.

Bonabeau, E., Dorigo, M., & Theraulaz, G. (1999). *Swarm intelligence: From natural to artificial systems*. Oxford: Oxford University Press.

Bongard, J., Zykov, V., & Lipson, H. (2006). Resilient machines through continuous self-modeling. *Science*, **314**, 1118–1121.

Brooks, R. A. (1990). Elephants don't play chess. *Robotics and Autonomous Systems*, **6**, 3–15.

Brooks, R. A. (1991). Intelligence without representation. *Artificial Intelligence Journal*, **47**, 139–160.

Brooks, R. (2001). The relationship between matter and life. *Nature*, **409** (6818), 409–411.

Brooks, R. (2002). *Flesh and machines: How robots will change us.* New York: Pantheon.

Chakrabarti, A., Breaker, R. R., Joyce, G. F., & Deamer, D. (1994). Production of RNA by a polymerase protein encapsulated within phospholipid vesicles. *Journal of Molecular Evolution*, **39**, 555–559.

Cho, M. K., Magnus, D., Caplan, A. L., McGee, D., & The Ethics of Genomics Group (1999). Ethical considerations in synthesizing a minimal genome. *Science*, **286** (5447), 2087–2090.

Clark, A. (1997). *Being there: Putting brain, body, and world together again.* Cambridge, MA: MIT Press.

Deamer, D. (1997). The first living systems: A bioenergetic perspective. *Microbiology and Molecular Biology Review*, **61** (2), 21–38.

Eigen, M. (1971). The molecular quasispecies. *Naturwissenschaften*, **58**, 465–523.

Elman, J. (1998). Connectionism, artificial life, and dynamical systems: New approaches to old questions. In W. Bechtel & G. Graham (Eds.), *A companion to cognitive science* (pp. 488–505). Oxford: Basil Blackwell.

Emmeche, C. (1992). Life as an abstract phenomenon: Is artificial life possible? In F. Varela & P. Bourgine (Eds.), *Towards a practice of autonomous systems* (pp. 466–474). Cambridge, MA: Bradford Books/MIT Press.

Emmeche, C. (1994). *The garden in the machine.* Princeton: Princeton University Press.

Emmeche, C. (1994). Is life a multiverse phenomenon? In C. G. Langton (Ed.), *Artificial life III* (Santa Fe Institute studies in the sciences of complexity, proceedings vol. **XVII**) (pp. 553–568). Redwood City, CA: Addison-Wesley.

Farmer, J. D. & Belin, A. (1992). Artificial life: The coming evolution. In C. G. Langton, C. Taylor, J. D. Farmer, & S. Rasmussen (Eds.) *Artificial Life II* (Santa Fe Institute studies in the sciences of complexity, proceedings vol. X) (pp. 815–840). Redwood City, CA: Addison Wesley.

Farmer, J. D., Lipids, A., Packard, N., & Wendroff, B. (Eds.) (1986). *Evolution, games, and learning: Models for adaptation for machines and nature.* Amsterdam: North Holland.

Fontana, W. & Buss, L. (1994). What would be conserved if the tape were played again? *Proceedings of the National Academy of Sciences*, **91**, 757–761.

Fraser, C. M., Gocayne, J. D., White, O., *et al.* (1995). The minimal gene component of *Mycoplasma genitalium*. *Science*, **270**, 397–403.

Gibson, D. G., Benders, G. A., Andrews-Pfannkoch, C., *et al.* (2008). Complete chemical synthesis, assembly, and cloning of a *Mycoplasma genitalium* genome. *Science*, **319**, 1215–1220.

Gibson, D. G., Glass, J. I., Lartigue, C., *et al.* (2010). Creation of a bacterial cell controlled by a chemically synthetized genome. *Science*, **329**, 52–56.

Hanczyc, M. M., Fujikawa, S. M., & Szostak, J. W. (2003). Experimental models of primitive cellular components: Encapsulation, growth, and division. *Science*, **320**, 618–622.

Holland, J. H. (1992). *Adaptation in natural and artificial systems: An introductory analysis with applications to biology, control, and artificial intelligence* (expanded 2nd ed.). Cambridge, MA: MIT Press.

Holland, J. H. (1995). *Hidden order: How adaptation builds complexity.* New York: Helix Books.

Hutchison, C. A., Peterson, S. N., Gill, S. R., *et al.* (1999). Global transposon mutagenesis and a minimal Mycoplasma genome. *Science*, **286**, 2165–2169.

Kaneko, K. (2006). *Life: An introduction to complex systems biology.* Berlin: Springer.

Kauffman, S. A. (1993). *The origins of order: Self-organization and selection in evolution.* New York: Oxford University Press.

Keeley, B. (1996). Evaluating artificial life and artificial organisms. *Artificial Life*, **5**, 264–271.

Keeley, B. (1998). Artificial life for philosophers. *Philosophical Psychology*, **11** (2), 251–260.

Korzeniewski, B. (2004). Confrontation of the cybernetic definition of a living individual with the real world. *Acta Biotheoretica*, **53** (1), 1–28.

Landecker, H. (2007). *Culturing life: How cells became technologies.* Cambridge, MA: Harvard University Press.

Langton, C. G. (1986). Studying artificial life with cellular automata. *Physica*, **22D**, 120–149.

Langton, C. G. (1989). Artificial life. In C. G. Langton (Ed.), *Artificial life* (Santa Fe Institute studies in the sciences of complexity, proceedings vol. **IV**) (pp. 1–47). Redwood City, CA: Addison-Wesley.

Lartigue, C., Glass, J. I., Alperovich, N., *et al.* (2007). Genome transplantation in bacteria: Changing one species to another. *Science*, **317**, 632–638.

Lee, D. H., Granja, J. R., Severin, K., & Ghadiri, M. R. (1996). A self-replicating peptide. *Nature*, **382**, 525–526.

Lenski, R. E., Ofria, C., Pennock, R. T., & Adami, C. (2003). The evolutionary origin of complex features. *Nature*, **423**, 139–144.

Luisi, P. L. (1996). Self-reproduction of micelles and vesicles: Models for the mechanisms of life from the perspective of compartmented chemistry. *Advances in Chemistry and Physics*, **92**, 425–438.

Maynard Smith, J. (1992). Byte-sized evolution. *Nature*, **355**, 772–773.

Monnard, P. A. & Deamer, D. (2002). Membrane self-assembly processes: Steps toward the first cellular life. *The Anatomical Record*, **268**, 196–207.

Mouritsen, O. G. (2005). *Life—As a matter of fat: The emerging science of lipidomics*. Berlin: Springer-Verlag.

Noireaux, V. & Libchaber, A. (2004). A vesicle bioreactor as a step toward an artificial cell assembly. *Proceedings of the National Academy of Sciences*, **101**, 17,669–17,674.

Nolfi, S. & Floreano, D. (2000). *Evolutionary robotics: The biology, intelligence, and technology of self-organizing machines*. Cambridge, MA: MIT Press.

Nolfi, S. & Floreano, D. (2002). Synthesis of autonomous robots through evolution. *Trends in Cognitive Science*, **6**, 31–37.

Oberholzer, T., Albrizio, M., & Luisi, P. L. (1995). Polymerase chain reaction in liposomes. *Chemistry and Biology*, **2**, 677–682.

Olsen, E. (1997). Ontological basis of strong artificial life. *Artificial Life*, **3**, 29–39.

Patte, H. H. (1996). Simulations, realizations, and theories of life. In M. A. Boden (Ed.), *The philosophy of artificial life* (pp. 379–393). Oxford: Oxford University Press.

Pfeiffer, R., Bongard, J. C., Brooks, R., & Iwasawa, S. (2006). *How the body shapes the way we think: A new view of intelligence*. Cambridge, MA: MIT Press.

Pohorille, A. & Deamer, D. (2002). Artificial cells: Prospects for biotechnology. *Trends in Biotechnology*, **20** (3), 123–128.

Pollack, J. B., Lipson, H., Hornby, G., & Funes, P. (2001). Three generations of automatically designed robots. *Artificial Life*, **7**, 215–223.

Putman, H. (1964). Robots: Machines or artificially created life? *Journal of Philosophy*, **61** (21), 668–691.

Rasmussen, S., Baas, N. A., Mayer, B., Nilsson, M., & Olesen, M. W. (2001). Ansatz for dynamical hierarchies. *Artificial Life*, **7**, 329–353.

Rasmussen, S., Chen, L., Nilsson, M., & Abe, S. (2003). Bridging nonliving and living matter. *Artificial Life*, **9**, 269–316.

Rasmussen, S., Chen, L., Deamer, D., *et al.* (2004). Transitions from nonliving to living matter. *Science*, **303**, 963–965.

Rasmussen, S., Bedau, M. A., Chen, L., *et al.* (2008). *Protocells: Bridging nonliving and living matter*. Cambridge, MA: MIT Press.

Ray, T. S. (1992). An approach to the synthesis of life. In C. G. Langton, C. Taylor, J. D. Farmer, & S. Rasmussen (Eds.), *Artificial life II* (Santa Fe Institute studies in the sciences of complexity, proceedings vol. X) (pp. 371–408). Redwood City, CA: Addison-Wesley.

Ray, T. (1994). An evolutionary approach to synthetic biology: Zen and the art of creating life. *Artificial Life*, **1**, 179–210.

Regis, E. (2008). *What is life? Investigating the nature of life in the age of synthetic biology*. New York: Farrar, Straus, and Giroux.

Sagre, D., Ben-Eli, D., & Lancet, D. (2000). Compositional genomes: Prebiotic information transfer in mutually catalytic noncovalent assemblies. *Proceedings of the National Academy of Sciences*, **97**, 4112–4117.

Sipper, M. (1998). Fifty years of research on self-replication: An overview. *Artificial Life*, **4**, 237–257.

Solé, R. V, Rasmussen, S., & Bedau, M. A. (Eds.) (2007). Towards the artificial cell. *Philosophical Transactions of the Royal Society B*, **362** (1486), 1723–1925.

Szostak, J. W., Bartel, D. P., & Luisi, P. L. (2001). Synthesizing life. *Nature*, **409** (6818), 387–390.

Takakura, K., Toyota, T., & Sugawara, T. (2003). A novel system of self-reproducing giant vesicles. *Journal of the American Chemical Society*, **125**, 8134–8140.

Taylor, T. & Massey, C. (2001). Recent developments in the evolution of morphologies and controllers for physically simulated creatures. *Artificial Life*, **7**, 77–87.

von Kiedrowski, G. (1986). A self-replicating hexadeoxynucleotide. *Angewandte Chemie*, **25**, 932–935.

von Neumann, J. (1966). *Theory of self-reproducing automata*. Urbana-Champagne: University of Illinois Press.

Walde, P., Wick, R., Fresta, M., Mangone, A., & Luisi, P. L. (1994). Autopoietic self-reproduction of fatty acid vesicles. *Journal of the American Chemical Society*, **116**, 11,649–116,454.

Wiener, N. (1948). *Cybernetics, or control and communication in the animal and the machine*. Cambridge, MA: MIT Press.

Wolfram, S. (1994). *Cellular automata and complexity*. Redwood City, CA: Addison-Wesley.

Zimmer, C. (2008). *Microcosmos: E. coli and the new science of life*. New York: Pantheon Books.

DEFINING AND EXPLAINING LIFE

Auffray, C., Imbeaud, S., Roux-Rouquie, M., & Hood, L. (2003). Self-organized living systems: Conjunction of a stable organization with chaotic fluctuations in biological space-time. *Philosophical Transactions of the Royal Society of London A*, **361**, 1125–1139.

Bedau, M. A. (1996). The nature of life. In M. A. Boden (Ed.), *The philosophy of artificial life* (pp. 332–357). New York: Oxford University Press.

Bedau, M. A. (2007). What is life? In S. Sarkar & A. Plutynski (Eds.), *A companion to the philosophy of biology* (pp. 455–471). New York: Blackwell.

Benner, S. & Switzer, C. (1999). Chance and necessity in biomolecular chemistry: Is life as we know it universal? In H. Frauenfelder, J. Deisenhofer, & P. Wolynes (Eds.), *Simplicity and complexity in protein and nucleic acids* (pp. 339–363). Berlin: Dahlem University Press.

Bruggeman, F., Westerhoff, H., & Boogerd, F. (2002). Biocomplexity: A pluralist research strategy is necessary for a mechanistic explanation of the 'live' state. *Philosophical Psychology*, **15**, 411–440.

Clark, B. (2001). Astrobiology's central dilemma: How can we detect life if we cannot even define it? *American Astronomical Society*, **33**, 1152.

Cleland, C. E. (2010). *The quest for a universal theory of life: Searching for life as we don't know it*. Cambridge: Cambridge University Press, forthcoming.

Crick, F. (1981). *Life itself: Its origin and nature*. New York: Simon and Schuster.

Davies, P. (1999). *The fifth miracle*. New York: Simon & Schuster.

Dennett, D. C. (1995). *Darwin's dangerous idea: Evolution and the meanings of life*. New York: Simon and Schuster.

Eigen, M. (1992). *Steps toward life*. Oxford: Oxford University Press.

Farmer, J. D. (2005). Cool is not enough. *Nature*, **436** (7051), 627–628.

Feldman, F. (1992). *Confrontations with the Reaper: A philosophical study of the nature and value of death*. New York: Oxford University Press.

Fleischacker, G. (1989). Autopoiesis: The status of its system logic. *BioSystems*, **22**, 37–49.

Fleischaker, G. (1990). Origins of life: An operational definition. *Origins of Life and the Evolution of the Biosphere*, **20**, 127–137.

Fong, P. (1973). Thermodynamic statistical theory of life: An outline. In A. Locker (Ed.), *Biogenesis, evolution, homeostasis: A symposium by correspondence* (pp. 93–101). Berlin: Springer-Verlag.

Godfrey-Smith, P. (1994). Spencer and Dewey on life and mind. In R. Brooks & P. Maes (Eds.), *Artificial life IV* (proceedings of the 4th international workshop on the synthesis and simulation of living systems) (pp. 80–89). Cambridge, MA: MIT Press/Bradford Books.

Jonas, H. (1966). *The phenomenon of life: Toward a philosophical biology*. New York: Dell.

Kauffman, S. (2000). *Investigations*. New York: Oxford University Press.

Kauffman, S. A. (1995). *At home in the universe: The search for laws of self-organization and complexity*. New York: Oxford University Press.

Keller, E. F. (2002). *Making sense of life: Explaining biological development with models, metaphors, and machines*. Cambridge, MA: Harvard University Press.

Korzeniewski, B. (2001). Cybernetic formulation of the definition of life. *Journal of Theoretical Biology*, **209**, 275–286.

Lovelock, J. (1979). *Gaia: A new look at life on Earth*. New York: Oxford University Press.

Luisi, P. L. (1998). About various definitions of life. *Origins of Life and the Evolution of the Biosphere*, **28**, 613–622.

Luisi, P. L. (2006). *The emergence of life: From chemical origins to synthetic biology*. Cambridge, UK: Cambridge University Press.

Margulis, L. & Sagan, D. (1995). *What is life?* Berkeley: University of California Press.

Maturana, H. & Varela, F. (1980). *Autopoiesis and cognition: The realization of the living*. Boston: D. Reidel.

Maturana, H. R. & Varela, F. J. (1987/1992). *The tree of knowledge: The biological roots of human understanding* (revised ed.). Boston: Shambhala.

Maynard Smith, J. (1995). Life at the edge of chaos? *New York Review of Books*, **2**, 28–30.

Miller, J. G. (1978). *Living systems*. New York: McGraw-Hill.

Morán, F., Moreno, A., Minch, E., & Montero, F. (1997). Further steps towards a realistic description of the essence of life. *Artificial Life*, **5**, 255–263.

Murphy, M. & O'Neill, L. (1995). *What is life? The next fifty years: Speculations on the future of biology.* Cambridge, UK: Cambridge University Press.

Pályi, G., Zucchi, C., & Caglioti, L. (Eds.) (2002). *Fundamentals of life.* New York: Elsevier.

Popa, R. (2004). *Between necessity and probability: Searching for the definition and origin of life.* Berlin: Springer.

Rizzotti, M. (Ed.) (1996). *Defining life.* Padova: Padova University Press.

Rosen, R. (1991). *Life itself: A comprehensive inquiry into the nature, origin, and fabrication of life.* New York: Columbia University Press.

Rosen, R. (2000). *Essays on life itself.* New York: Columbia University Press.

Schneider, E. D. (2004). Gaia: Toward a thermodynamics of life. In S. H. Schneider, J. R. Miller, E. Christ, and P. J. Boston (Eds.), *Scientists debate Gaia* (pp. 45–56). Cambridge, MA: MIT Press.

Schulze-Makuch, D., Guan, H., Irwin, L., & Vega, E. (2002). Redefining life: An ecological, thermodynamic, and bioinformatics approach. In G. Pályi, C. Zucchi, and L. Caglioti (Eds.), *Fundamentals of life* (pp. 169–180). New York: Elsevier.

Solé, R., & Goodwin, B. (2000). *Signs of life: How complexity pervades biology.* New York: Basic Books.

Sterelny, K. (1995). Understanding life: Recent work in philosophy of biology. *British Journal of the Philosophy of Science*, **46**, 115–183.

Sterelny, K. (1997). Universal biology. *British Journal of the Philosophy of Science*, **48**, 587–601.

Thompson, E. (2007). *Mind in life: Phenomenology, and the sciences of the mind.* Cambridge, MA: Harvard University Press.

Thompson, M. (1995). The representation of life. In R. Hursthouse, G. Lawrence, & W. Quinn (Eds.), *Virtues and reasons* (pp. 247–296). Oxford: Clarendon Press.

Varela, F., Maturana, H., & Uribe, R. (1974). Autopoiesis: The organization of living systems, its characterization and a model. *BioSystems*, **5**, 187–196.

Weber, B. (2003). Life. In E. Zalta (Ed.), *The Stanford encyclopedia of philosophy* (Spring 2006 ed.). Available online at http://plato.stanford.edu/archives/spr2006/entries/life/(accessed November 2008).

Index

.

Printed in the United States
By Bookmasters